Design and Applications of Emerging Computer Systems

Weiqiang Liu • Jie Han • Fabrizio Lombardi
Editors

Design and Applications of Emerging Computer Systems

Springer

Editors
Weiqiang Liu
College of Electronic and Information Engineering
Nanjing University of Aeronautics and Astronautics
Nanjing, Jiangsu, China

Jie Han
Department of Electrical and Computer Engineering
University of Alberta
Edmonton, AB, Canada

Fabrizio Lombardi
Department of Electrical and Computer Engineering
Northeastern University
Boston, MA, USA

ISBN 978-3-031-42477-9   ISBN 978-3-031-42478-6 (eBook)
https://doi.org/10.1007/978-3-031-42478-6

This Springer imprint is published by the registered company Springer Nature Switzerland AG
The registered company address is: Gewerbestrasse 11, 6330 Cham, Switzerland

Paper in this product is recyclable.

Preface

Computing has a very different landscape in the era of artificial intelligence (AI). Conventionally, high performance, low power, and uncompromising accuracy are the main pillars of a computing system. With the rise of learning and data-intensive applications, computing is faced with unprecedented and significant challenges that deviate from those that can be coped with by using conventional methodologies.

As Dennard's law is coming to an end, reduction in on-chip power consumption and improvement in throughput due to technology scaling encounter serious challenges; workloads of today's applications (such as for AI, Big Data, and Internet of Things) have also reached extremely high levels of complexity in computation. Power dissipation has become one of the fundamental barrier to scale computing performance across all technology platforms. Therefore it is no surprise that computation at nanoscale requires innovative approaches. Moreover, many so-called emerging computing paradigms have been widely studied (e.g., probabilistic, stochastic, neuromorphic, in-memory), mostly at system level to alleviate these encountered hurdles; however, their successful evolution necessitates implementations that require efficient circuits in a multitude of modules (such as memory, arithmetic, and control). Substantial challenges remain also at architectural and system levels. Although bridging of technology with design has attracted significant attention from academic and industrial communities in the past decade, it still requires considerable efforts to accomplish implementations that are energy-efficient and high-performance for systems in so many diverse applications.

This book addresses the technological contributions and developments at various hardware levels of new computing paradigms that include stochastic, probabilistic, neuromorphic, spintronic, bio-inspired, in-memory computing, and quantum computing. With this book, it is expected to collect state-of-the-art progress in different emerging computing areas; this is achieved by chapters covering the entire spectrum of research activities in such emerging computing paradigms, so bridging the entire system stack, i.e., from devices, circuits, architectures, up to systems. The book covers tutorials, reviews, and surveys of current theoretical/experimental results, design methodologies, and applications over a wide spectrum of scope for an enlarged and specialized readership.

Emerging computing paradigms have shown great potential for many applications; however, they are not yet at a fully mature stage. A comparison of unconventional computing techniques in different dimensions is shown in Fig. 1; hence this book provides a comprehensive reference for current developments and promotes future research on this important research area from which innovative computing schemes can be derived.

The chapters in this book are divided into four parts.

The first part presents in-memory computing for neuromorphic computing and machine learning applications based on emerging magnetic devices [1]. A bottleneck for high performance is the relatively slow communication between memory and processors [2]. As a remedy to this issue, in-memory computing has shown promising prospects and neuromorphic computing has provided a viable means, especially for data-intensive applications such as machine learning [3]. By exploiting the switching properties of magnetic materials and peripheral circuit design, logical and arithmetic operations within the memory array are realized and further applied to neural networks and machine learning, greatly alleviating the

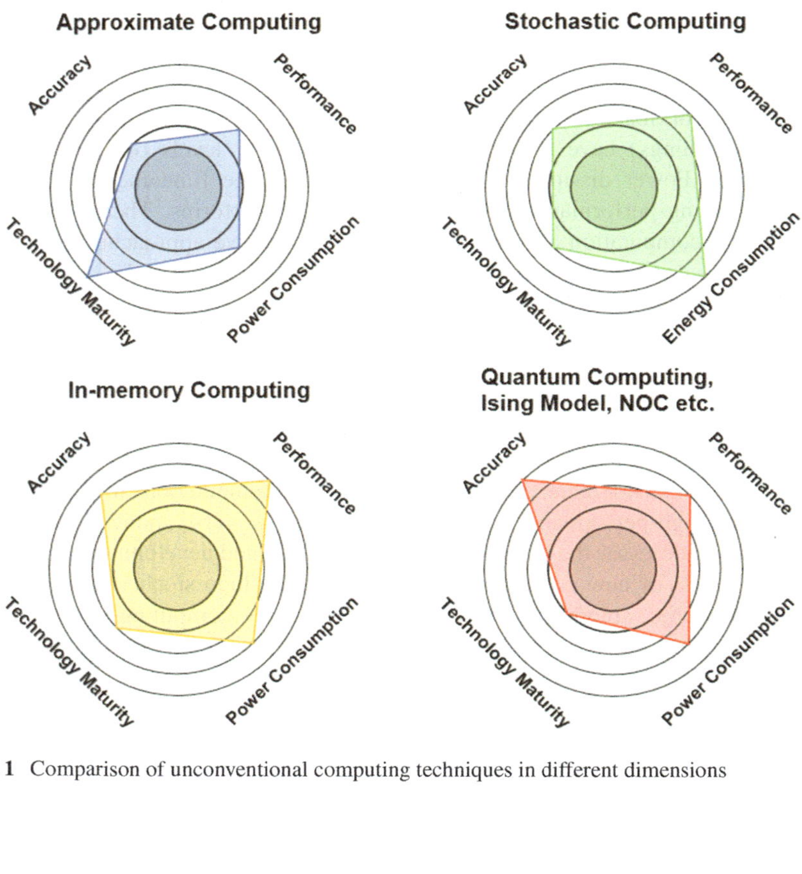

Fig. 1 Comparison of unconventional computing techniques in different dimensions

"Energy Wall" and "Memory Wall" of traditional computing architectures [4]. Several important approaches to achieving in-memory logic and arithmetic computing (e.g., XOR, AND, and Multiply-and-Accumulate) [5], in-memory neuromorphic computing [6], and other new in-memory applications [7] are discussed in Part I of this book.

The second part introduces stochastic computing and its applications. Stochastic computing utilizes random (and sometimes, deterministic) binary bit streams to encode information [8]. Applications of stochastic computing include image and digital signal processing [9]. Recently, it has been shown that it is effective for the inference and training of neural networks [10]. Toward low power, stochastic computing exploits the simplicity in statistical processing using simple logic circuits; however, this benefit comes at the cost of a long latency due to the long stochastic sequence length for achieving a high accuracy [11]. This tradeoff between hardware efficiency and long latency has been addressed using various approaches, including the design and characterization of efficient random number generators, parallel stochastic computing multipliers, and morphological neural networks. These are the topics addressed in Part II of this book.

The third part covers inexact or approximate computing from circuits to AI and communication applications. Emerging computing applications exhibit characteristics of error tolerance or resilience. The requirement of accuracy, therefore, is relieved, especially in the intermediate stages of a computing process, for gains in performance and power dissipation. The paradigm that exploits accuracy as a major design consideration has been referred to as approximate computing [12]. Recently, various approximate arithmetic circuits under different design constraints have been developed [13]. At the same time, substantial research has focused on circuit level techniques for approximating logic and memory design [14]. Approximate computing has found important applications in arithmetic circuits, wireless communication, machine learning accelerators, and neural networks. It is worth mentioning that approximate computing can significantly improve the efficiency of AI applications by exploiting its error-tolerant nature [15]. Part III of this book discusses these important approximate computing approaches.

Finally, the last part includes quantum computing and other emerging computing topics. Quantum computing relies on the superposition and entanglement of physical states at the microscopic scale, so it provides ultimate performance and efficiency by leveraging the quantum mechanical behavior of devices [16]. Various research directions have been pursued by building scalable physical systems for quantum computing and investigating a number of emerging devices [17]. In addition to the challenges in hardware, algorithms are an aspect that is crucial for the scalability and robustness of quantum computers [18]. In contrast, the Ising model provides a mathematical description of the electromagnetic interactions among an array of spins [19]. It has been shown that computers built on the principle of the Ising model, or Ising machines, are efficient solvers of combinatorial optimization problems [20]. Finally, network-on-chips have been developed for multi-core or multi-chip systems for efficient communication and computation [21]. The design and algorithm aspects of these emerging computing paradigms using CMOS implementations and quantum dots are some of the additional subjects of Part IV of this book.

Nanjing, Jiangsu, China Weiqiang Liu
Edmonton, AB, Canada Jie Han
Boston, MA, USA Fabrizio Lombardi

References

1. C. Chang et al., NV-BNN: an accurate deep convolutional neural network based on binary STT-MRAM for adaptive AI edge, in *56th Annual Design Automation Conference*, (2019)
2. M. Horowitz, 1.1 Computing's energy problem (and what we can do about it), in IEEE International Solid-State Circuits Conference (ISSCC), San Francisco, 2014, pp. 10–14
3. S. Jung et al., A crossbar array of magnetoresistive memory devices for in-memory computing. Nature **601**(7892), 211–216 (2022)
4. S. Angizi, Z. He, A. Awad, D. Fan, MRIMA: An MRAM-based in-memory accelerator. IEEE Trans. Comput. Aided Des. Integr. Circuits Syst. **39**(5), 1123–1136 (2020)
5. S. Jain et al., Computing-in-memory with spintronics, in 2018 Design, Automation & Test in Europe Conference & Exhibition (DATE), (2018)
6. L. Chang et al., CORN: In-buffer computing for binary neural network, in *2019 Design, Automation & Test in Europe Conference & Exhibition (DATE)*, (2019)
7. X. Wang et al., TCIM: Triangle counting acceleration with processing-in-MRAM architecture, in *2020 57th ACM/IEEE Design Automation Conference (DAC)*, (2020)
8. B.R. Gaines, Stochastic computing systems, in *Advances in Information Systems Science*, (Springer, New York, 1969), pp. 37–172
9. Alaghi, W. Qian, J.P. Hayes, The promise and challenge of stochastic computing. IEEE Trans. Comput. Aided Des. Integr. Circuits Syst. **37**(8), 1515–1531 (2017)
10. Y. Liu, S. Liu, Y. Wang, F. Lombardi, J. Han, A survey of stochastic computing neural networks for machine learning applications. IEEE Trans. Neural Netw. Learn. Syst. **32**(7), 2809–2824 (2021)
11. S. Liu, W.J. Gross, J. Han, Introduction to dynamic stochastic computing. IEEE Circuits Syst. Mag. **20**(3), 19–33 (2020)
12. W. Liu, F. Lombardi, M. Schulte, Approximate computing: From circuits to applications. Proc. IEEE **108**(12), 2103–2107 (2020)

13. H. Jiang, F.J.H. Santiago, H. Mo, L. Liu, J. Han, Approximate arithmetic circuits: A survey, characterization, and recent applications. Proc. IEEE **108**(12), 2108–2135 (2020)
14. S. Amanollahi, M. Kamal, A. Afzali-Kusha, M. Pedram, Circuit-level techniques for logic and memory blocks in approximate computing systems. Proc. IEEE **108**(12), 2150–2177 (2020)
15. S. Venkataramani et al., Efficient AI system design with cross-layer approximate computing. Proc. IEEE **108**(12), 2232–2250 (2020)
16. H. Mooij, The road to quantum computing. Science **307**(5713), 1210–1211 (2005)
17. J.E. Mooij, T.P. Orlando, L. Levitov, L. Tian, C.H. Van der Wal, S. Lloyd, Josephson persistent-current qubit. Science **285**(5430), 1036–1039 (1999)
18. T.D. Ladd, F. Jelezko, R. Laflamme, Y. Nakamura, C. Monroe, J.L. O'Brien, Quantum computers. Nature **464**(7285), 45–53 (2010)
19. Lucas, Ising formulations of many NP problems. Front. Phys., 5 (2014)
20. K. Yamamoto, K. Kawamura, K. Ando, N. Mertig, T. Takemoto, et al., STATICA: A 512-spin 0.25 m-weight annealing processor with an all-spin-updates-at-once architecture for combinatorial optimization with complete spin–spin interactions. JSSC **56**(1), 165–178 (2020)
21. A.E. Kiasari, Z. Lu, A. Jantsch, An analytical latency model for networks-on-chip. IEEE Trans. Very Large Scale Integr. VLSI Syst. **21**(1), 113–123 (2013)

Contents

Part I In-Memory Computing, Neuromorphic Computing and Machine Learning

Emerging Technologies for Memory-Centric Computing 3
Paul-Antoine Matrangolo, Cédric Marchand, David Navarro, Ian O'Connor, and Alberto Bosio
1 Introduction 3
2 Resistive Random Access Memory (RRAM) 5
  2.1 Device 6
  2.2 Memory Architectures 7
  2.3 IMC Applications 7
3 Spin-Transfer Torque Magnetoresistive Random Access Memory 11
  3.1 Device 11
  3.2 Memory Architectures 12
  3.3 IMC Applications 13
4 Phase-Change Memory 14
  4.1 Device 15
  4.2 Memory Architectures 16
  4.3 IMC Applications 16
5 FeFET 17
  5.1 Device 18
  5.2 Memory Architectures 18
  5.3 IMC Applications 22
6 Comparison and Discussion 24
7 Conclusion 25
References 26

An Overview of Computation-in-Memory (CIM) Architectures 31
Anteneh Gebregiorgis, Hoang Anh Du Nguyen, Mottaqiallah Taouil, Rajendra Bishnoi, Francky Catthoor, and Said Hamdioui
1 Introduction 31
2 Classification of Computer Architectures 32

2.1 Classification Based on Computation Position 33
2.2 Classification Based on Memory Technology 35
2.3 Classification Based on Computation Parallelism 36
3 Computation-in-Memory-Array (CIM-A) 38
3.1 DRISA-3T1C: A DRAM-Based Reconfigurable In Situ Accelerator with 3 Transistors and 1 Capacitor (3T1C) Design 38
3.2 CRS: Complementary Resistive Switch Architecture 41
3.3 CIM: Computation-in-Memory 42
3.4 PLiM: Programmable Logic-in-Memory Computer 43
3.5 MPU: Memristive Memory Processing Unit 45
3.6 ReVAMP: ReRAM-Based VLIW Architecture 46
4 Computation-in-Memory-Peripherals (CIM-P) 48
4.1 S-AP: Cache Automaton 49
4.2 ISAAC: A Convolutional Neural Network Accelerator with In Situ Analog Arithmetic 52
4.3 PRIME: A Processing-in-Memory Architecture for Neural Network Computation in ReRAM-Based Main Memory 53
4.4 STT-CiM: Computing-in-Memory Spin-Transfer Torque Magnetic RAM 55
4.5 DPP: Data Parallel Processor 56
4.6 R-AP: Resistive RAM Automata Processor 58
5 CIM Design-flow 59
5.1 System-Level Design 59
5.2 Circuit-Level Design 60
6 Conclusion 61
References 61

Toward Spintronics Non-volatile Computing-in-Memory Architecture ... 67
Bi Wu, Haonan Zhu, Tianyang Yu, and Weiqiang Liu
1 Introduction 67
2 MRAM 68
3 Implementation of Boolean Logic 69
3.1 Analog Implementation 70
3.2 Read/Write Implementation 71
3.3 Cell Modification 73
3.4 Peripheral Circuit Modification 76
4 MRAM for Neural Networks 78
4.1 CNN and BNN Basics 79
4.2 MRAM for XNOR Operation 81
4.3 MRAM for Bit-Count Operation 81
4.4 MRAM for Max-Pool Operation 83
4.5 MRAM for BNN Training 83
4.6 MRAM for Analog Computing 84
5 MRAM for Other Applications 85
5.1 DNA Read Alignment 85

5.2 Triangle Counting 86
5.3 True Random Count Generators 87
References 88

Is Neuromorphic Computing the Key to Power-Efficient Neural Networks: A Survey 91
Muhammad Hamis Haider, Hao Zhang, S. Deivalaskhmi, G. Lakshmi Narayanan, and Seok-Bum Ko
1 Introduction to Biomorphism 91
2 Understanding the Biological Neuron 92
2.1 The Response to Extra-cellular Stimulus 94
3 Neuromorphic Computing: Learning from the Brain 97
4 Spiking Neural Networks 99
4.1 Spiking Neuron Models 101
4.2 Common Training Algorithms 101
4.3 Applications 103
5 Hardware Accelerators 103
5.1 General Mesh Accelerators 104
5.2 Feedforward Accelerators 106
5.3 Resource Analysis and Discussion 108
6 Limitations and Potential Improvements of SNNs 109
7 Conclusion 111
References 112

Emerging Machine Learning Using Siamese and Triplet Neural Networks 115
Ziheng Wang, Farzad Niknia, Shanshan Liu, Pedro Reviriego, and Fabrizio Lombardi
1 Introduction 115
2 Multi-Branch NNs: Algorithms and Designs 117
2.1 Siamese Networks 117
2.2 Triplet Networks 117
2.3 ASIC-Based Design for a Branch Network 122
3 Error Tolerance of Multi-Branch NNs 126
3.1 Error Tolerance During Inference 126
3.2 Error Tolerance During Training 133
4 Conclusion 138
References 139

An Active Storage System for Intelligent Data Analysis and Management 143
Shengwen Liang, Ying Wang, Lei Dai, Yingying Chen, Renhai Chen, Fan Zhang, Gong Zhang, Huawei Li, and Xiaowei Li
1 Introduction 143
2 Background and Preliminaries 148
2.1 Unstructured Data Analysis System 148

2.2 Near Data Processing and Deep Learning Accelerator 150
2.3 Learned Data Cache and Placement 151
2.4 Active Store 152
3 Motivation 152
3.1 Execution Time Breakdown 153
3.2 ANNS Algorithm Exploration 154
4 Active Storage System 155
4.1 The Active Storage Software: DHS Library 156
4.2 Hardware Architecture: The Active Storage 158
4.3 The Procedure of Data Retrieval in Active Storage 159
5 DHS-x Accelerator 159
5.1 Architecture: Direct Flash Accessing 159
5.2 I/O Path in Active Storage 160
5.3 Hybrid Search Engine 162
5.4 LSTM-Based Data Cache and Placement 167
6 Evaluation 168
6.1 Hardware Implementation 168
6.2 Experimental Setup 168
6.3 Evaluation of DHS Algorithm 169
6.4 Evaluation of DHS-x Accelerator 170
6.5 The Single-Node System Based on Active Storage 175
6.6 Data Cache and Placement 177
7 Conclusion 178
References 178

Error-Tolerant Techniques for Classifiers Beyond Neural Networks for Dependable Machine Learning 185
Shanshan Liu, Pedro Reviriego, Xiaochen Tang, and Fabrizio Lombardi
1 Introduction 185
2 K Nearest Neighbors 187
2.1 Errors in KNNs 187
2.2 Voting-Margin-Based Error-Tolerant KNNs for Binary Classification 189
2.3 $K + 1$ Nearest Neighbors for Multiclass Classification 191
3 Ensemble Classifier 194
3.1 Random Forest 195
3.2 Voting-Margin-Based Error-Tolerant Random Forest 196
4 Support Vector Machines 198
4.1 SVM with Different Kernels 199
4.2 Result-Based Re-Computation Error-Tolerant SVM 200
5 Conclusion 205
References 205

Part II Stochastic Computing

Efficient Random Number Sources Based on D Flip-Flops for Stochastic Computing ... 211
Kuncai Zhong and Weikang Qian
1 Introduction ... 211
2 Guidelines for Applying LFSRs in Stochastic Computing ... 214
2.1 Model and Evaluation Methodology for LFSRs ... 214
2.2 The Selection of the Feedback Polynomials for LFSR Pair ... 215
2.3 The Selection of the Seeds for Identical-Feedback LFSR Pair ... 216
3 Proposed Method for Building Successive RNSs ... 219
3.1 Method for Building Two Successive RNSs ... 219
3.2 Method for Building Multiple Successive RNSs ... 220
4 Method for Building Non-successive RNSs ... 222
5 Efficient Designs of RNSs Based on DFFs ... 224
6 Experimental Results ... 228
6.1 Experimental Setup ... 228
6.2 Accuracy Comparison ... 229
6.3 Area Comparison ... 231
6.4 Area–Accuracy Trade-off Comparison ... 233
7 Conclusion ... 234
References ... 234

Stochastic Multipliers: from Serial to Parallel ... 237
Yongqiang Zhang, Jie Han, and Guangjun Xie
1 Introduction ... 237
2 Background ... 238
2.1 Binary Multiplier ... 238
2.2 Booth Multiplier ... 240
2.3 Stochastic Number ... 242
2.4 SC Correlation and Components ... 242
3 The Design Approaches of Stochastic Multipliers ... 244
3.1 Stochastic Multiplier ... 244
3.2 Exact Stochastic Multipliers ... 253
4 Conclusion ... 263
References ... 263

Applications of Ising Models Based on Stochastic Computing ... 265
Duckgyu Shin, Naoya Onizawa, Warren J. Gross, and Takahiro Hanyu
1 Introduction ... 265
2 Preliminaries ... 266
2.1 SA for the Ising Model ... 266
2.2 P-bit-Based Ising Model ... 267
3 Stochastic Simulated Annealing ... 268
3.1 Spin Operations ... 268
3.2 Annealing Process of SSA ... 270

4 CMOS Invertible Logic 272
4.1 Basics of CIL 272
4.2 Training BNNs Based on CIL 274
5 CIL Training Hardware Design 275
5.1 Architecture of CIL Training Hardware 275
5.2 Performance Evaluation 277
6 Conclusion 278
References 279

Stochastic and Approximate Computing for Deep Learning: A Survey 281
Tina Masoudi, Hao Zhang, Aravindhan Alagarsamy, Jie Han, and Seok-Bum Ko
1 Introduction 281
2 Background 282
2.1 Stochastic Computing 282
2.2 Approximate Computing 285
2.3 Deep Learning Arithmetic Units 286
2.4 Deep Learning Application Accelerators 288
3 Stochastic and Approximate Computing-Based Deep Learning Applications 289
4 Area- and Power-Efficient Hybrid Stochastic-Approximate Designs in Deep Learning Applications 295
4.1 Low Complexity with High-Accuracy Winograd Convolution Based on Stochastic and Approximate Computing 295
4.2 A Hybrid Power- and Area-Efficient Stochastic-Approximate Neuron 297
5 Conclusion 299
6 Future Research Directions 299
References 300

Stochastic Computing Applications to Artificial Neural Networks 303
Josep L. Rosselló, Joan Font-Rosselló, Christiam F. Frasser, Alejandro Morán, Vincent Canals, and Miquel Roca
1 Introduction 303
2 Stochastic Computing Basic Principles 306
2.1 Stochastic Signals and Correlation 307
3 Classical Artificial Neural Networks 307
3.1 Application of Stochastic Computing to Artificial Neural Networks 309
4 Morphological Neural Networks 320
4.1 Application of Stochastic Computing to the Implementation of Morphological Neural Networks 322
5 Conclusions 325
References 326

Characterizing Stochastic Number Generators for Accurate Stochastic Computing 331
Yutao Gong, Heng Shi, and Siting Liu
1 Introduction 331
2 Stochastic Number Generators 333
2.1 Linear Feedback Shift Register (LFSR)-Based SNGs 334
2.2 Low-Discrepancy (LD) Sequence-Based SNGs 338
3 Accuracy Metrics 344
4 Experimental Methods and Results 345
5 Conclusion 348
References 348

Part III Inexact/Approximate Computing

Automated Generation and Evaluation of Application-Oriented Approximate Arithmetic Circuits 353
Ao Liu, Yong Wu, Qin Wang, Zhigang Mao, Leibo Liu, Jie Han, and Honglan Jiang
1 Introduction 353
2 Automated Generation Methodologies for AACs 355
2.1 Statistical Error Metrics 355
2.2 Automated Generation of Generic AACs 357
2.3 Automated Generation of Application-Oriented AACs 361
2.4 Summary 364
3 QoR Evaluation of the AAC-Based Applications 365
3.1 Simulation Acceleration 365
3.2 Prediction Model 367
3.3 Functional Abstraction 368
3.4 Summary 371
4 QoR Recovery 372
4.1 Self-Adjustment 372
4.2 Error-Aware Adjustment 373
4.3 Robustness Enhancement 374
4.4 Summary 375
5 Conclusions and Prospects 376
References 377

Automatic Approximation of Computer Systems Through Multi-objective Optimization 383
Mario Barbareschi, Salvatore Barone, Alberto Bosio, and Marcello Traiola
1 Introduction 383
2 The Approximate Computing Design Paradigm and Its Application 384
2.1 Overview 384
3 Automatic Application-Driven, Multi-Objective Approximate Design Methodology 386

3.1 Identifying Approximable Portions and Suitable Approximation Techniques 386
3.2 Optimization and Design-Space Exploration 388
3.3 Summary 390
4 Automatic Approximation of Combinational Circuits 391
4.1 Approximate Variant Generation 391
4.2 Design-Space Exploration 392
4.3 Experimental Results 393
5 Approximation of Image-Processing Applications 393
5.1 The E-𝕀DE𝔸 Framework 395
5.2 The DCT Case Study 397
6 Automatic Approximation of Artificial Intelligence Applications 403
6.1 Neural Networks 404
6.2 Decision-Tree-Based Multiple Classifier Systems 408
7 Conclusion 414
References 415

Evaluation of the Functional Impact of Approximate Arithmetic Circuits on Two Application Examples 421
Jordi Fornt, Leixin Jin, Josep Altet, Francesc Moll, and Antonio Rubio
1 Introduction 421
2 Description of Approximate Arithmetic Units 423
2.1 Approximate Adders 423
2.2 Approximate Multipliers 424
2.3 Comparison of Approximate Multiplier Approaches 430
3 Application to Digital Filters 431
3.1 Filter Description and Specifications 431
3.2 Effects of Approximate Operators in the Filter Specs 432
4 Application to Deep Neural Networks 435
4.1 Neural Network Basics 435
4.2 YOLO – DCNN for Object Detection 437
4.3 Approximate FP MAD Units 440
4.4 Effects of Approximate FP16 in YOLOv3 441
4.5 Approximate Accelerator Application: Approximate Systolic Array 444
5 Conclusions 449
References 450

A Top-Down Design Methodology for Approximate Fast Fourier Transform (FFT) Design 453
Chenyi Wen, Ying Wu, Xunzhao Yin, and Cheng Zhuo
1 Introduction 453
2 Background 454
2.1 FFT Hardware Implementation 454
2.2 Configurable Floating-Point Approximate Multiplier 455
3 Overview 457

4 Method 458
4.1 Error Modeling 458
4.2 Approximation Optimization 463
4.3 Design Implementation 464
5 Experimental Results 465
5.1 Performance of Error Model 466
5.2 Performance of Optimization 467
5.3 Approximate FFT Design Comparison 468
5.4 System Application 469
6 Conclusions 470
References 471

Approximate Computing in Deep Learning System: Cross-Level Design and Methodology 473
Yu Gong, You Wang, Ke Chen, Bo Liu, Hao Cai, and Weiqiang Liu
1 Introduction 473
2 Related Work 474
2.1 Binary-Weight Neural Networks 475
2.2 Low-Power BWNN System with Approximation 475
2.3 Quality-Driven Approximate Computing System 476
3 Estimation and Evaluation of Low-Power Approximate NN System 477
3.1 Estimation of Approximate Units 477
3.2 Approximation Noise in Datapath 479
3.3 System Evaluation Approach and Mechanism 480
4 Quality Configurable Approximate Computing 481
4.1 Evaluation of Low-Power Approximate Array 483
4.2 Design of Hierarchical Adder Cluster 483
4.3 Pre-analysis for Estimating BWNN Acceleration System 486
5 Reconfigurable Lower-Power NN System for Keyword Spotting Application 489
5.1 Deployment of Approximate BWNN 489
5.2 A 22-nm Low-Power System for Always-On KWS System 491
6 Experimental Results 493
7 Conclusion 498
References 498

Adaptive Approximate Accelerators with Controlled Quality Using Machine Learning 501
Mahmoud Masadeh, Osman Hasan, and Sofiène Tahar
1 Introduction 501
1.1 Approximate Computing Error Metrics 503
1.2 Approximate Accelerators 503
1.3 Quality Control of Approximate Accelerators 504
2 Proposed Methodology 505
2.1 Machine Learning-Based Models 507

3 Software-Based Adaptive Design of Approximate Accelerators 508
3.1 Adaptive Design Methodology 509
3.2 Machine Learning-Based Models 510
3.3 Experimental Results of Image Blending 512
3.4 Summary 516
4 Hardware-Based Adaptive Design of Approximate Accelerators 517
4.1 Dynamic Partial Reconfiguration (DPR) 517
4.2 Machine Learning-Based Models 519
4.3 Adaptive Design Methodology 520
4.4 Experimental Results 523
4.5 Summary 525
5 Conclusions 526
References 526

Design Wireless Communication Circuits and Systems Using Approximate Computing 531
Chenggang Yan, Ke Chen, and Weiqiang Liu
1 Introduction 531
2 Approximate FP Arithmetic Unit in Wireless Communication System 532
2.1 Approximate FP Adder 533
2.2 Approximate FP Multiplier 537
2.3 Application in Wireless Communication System 542
3 Approximate FP FFT Processor 542
3.1 DFT and FFT 544
3.2 Mantissa Bit-Width Adjustment Algorithm 545
3.3 Approximate FP FFT Processor in Channel Estimation 548
4 Approximate Polar Decoder 549
4.1 Approximate SC Polar Code Decoder 550
4.2 Approximate BPF Polar Code Decoder 558
5 Conclusion 561
References 562

Logarithmic Floating-Point Multipliers for Efficient Neural Network Training 567
Tingting Zhang, Zijing Niu, Honglan Jiang, Bruce F. Cockburn, Leibo Liu, and Jie Han
1 Introduction 567
2 Preliminaries 568
2.1 FP Representation 568
2.2 Logarithmic FP Multiplication 569
3 Piecewise Approximation Design Framework 570
3.1 Logarithm Approximation 570
3.2 Anti-logarithm Approximation 573
3.3 Logarithmic FP Multiplication 574
4 Hardware Implementation 575
4.1 The Generic Circuit 575

5 Case Studies of PWLM Designs 580
6 Performance Evaluation and Neural Network Applications 581
6.1 Accuracy Evaluation 581
6.2 Hardware Evaluation 582
6.3 Neural Network Applications 583
7 Conclusions 585
References 586

Part IV Quantum Computing and Other Emerging Computing

Cryogenic CMOS for Quantum Computing 591
Rubaya Absar, Hazem Elgabra, Dylan Ma, Yiju Zhao, and Lan Wei
1 Background 591
1.1 Brief Background on Quantum Computing 591
1.2 Cryo-CMOS in Si QD Quantum Processor 592
2 Transport in Cryogenic CMOS 598
2.1 Cryogenic MOSFET Characteristics 598
2.2 Compact Models 601
2.3 Progress of Cryogenic MOSFET Models 603
3 High-Frequency Noise in Cryogenic CMOS 606
3.1 High-Frequency Noise in MOSFET 606
3.2 Noise in a Mesoscopic View 608
3.3 Experimental Verification 611
3.4 Progress and Challenges on Modeling High-Frequency Noise 612
4 Numerical Simulation for Cryogenic CMOS 613
References 614

Quantum Computing on Memristor Crossbars 623
Iosif-Angelos Fyrigos, Panagiotis Dimitrakis, and Georgios Ch. Sirakoulis
1 Introduction 623
1.1 Quantum Computers' Challenges and the Need of Simulators 624
1.2 Memristor Crossbars as Hardware Accelerators 625
2 Basics of Quantum Computations 626
2.1 Quantum Computations in Simulators 626
2.2 Gates and Qubit Representation 627
3 Performing Quantum Computations on Memristive Crossbars 627
3.1 Simple Example of the Proposed Circuit Operation 628
3.2 Multiple-Crossbar Configuration 631
4 Simulation Results of the Universal Set of Quantum Gates 632
4.1 The Hadamard Gate 632
4.2 The CCNOT Gate 635
5 Simulation Result of Quantum Algorithms' Implementation 635
5.1 Utilized Memristor and Transistor Models 636
5.2 Deutsch Algorithm 636
5.3 Grover Algorithm 639
6 Framework 640

7 Discussion on the Variability and Stochastic Behavior of the Memristor .. 643
8 Conclusions 644
Appendix 644
References 645

A Review of Posit Arithmetic for Energy-Efficient Computation: Methodologies, Applications, and Challenges 649
Hao Zhang, Zhiqiang Wei, Bo Yin, and Seok-Bum Ko
1 Introduction 649
2 Posit Numeric Format 651
 2.1 General Format 651
 2.2 Sign-Magnitude Form Decoding Method 652
 2.3 Two's Complement Form Decoding Method 653
 2.4 Posit to Quire Conversion 653
 2.5 Quire to Posit Conversion 654
3 Posit Applications 654
4 Posit Developing Tools 655
5 Posit-Based Arithmetic Units 656
 5.1 Posit Decoding and Encoding Module 657
 5.2 Posit Arithmetic Units 659
6 Posit-Based Hardware Processors 664
7 Discussion and Perspectives 665
 7.1 Improving the Latency of Posit Arithmetic Operations 665
 7.2 Developing a Practical Tool for Posit Verification 665
 7.3 Designing a Flexible Posit Processor for Applications 666
 7.4 Exploring the Use of Posit in More Fields of Applications 666
8 Conclusion 666
References 667

Designing Fault-Tolerant Digital Circuits in Quantum-Dot Cellular Automata 671
R. Marshal, K. Raja Sekar, Lakshminarayanan Gopalakrishnan, Anantharaj Thalaimalai Vanaraj, and Seok-Bum Ko
1 Introduction 671
2 QCA Operation 672
 2.1 Cell Components and Logic 672
 2.2 Radius of Effect 672
 2.3 Kink Energy 673
 2.4 Wires 673
 2.5 Logic Gates 674
 2.6 Clocking 675
 2.7 Crossover 676
 2.8 Performance Metrics 676
 2.9 Simulation Settings 678
3 Fabrication Defects 679
 3.1 Critical Factor 680

3.2 Tolerance Factor ... 681
3.3 Immunity Percentage ... 681
4 Design Considerations for Fault Tolerance ... 681
4.1 Wires ... 681
4.2 Gates ... 682
4.3 Crossovers ... 682
4.4 Clocking ... 683
4.5 Layout Challenges ... 684
5 Conclusion ... 684
References ... 685

Ising Machines Using Parallel Spin Updating Algorithms for Solving Traveling Salesman Problems ... 687
Tingting Zhang, Qichao Tao, Bailiang Liu, and Jie Han
1 Introduction ... 687
2 Background ... 689
2.1 Problem-Solving via Ising Machines ... 689
2.2 Mapping the Traveling Salesman Problem ... 691
3 Improved Parallel Annealing for TSPs ... 691
3.1 Parallel Annealing ... 692
3.2 Improved Parallel Annealing ... 693
3.3 A Temperature Function ... 693
3.4 A Clustering Approach ... 695
4 Experimental Results ... 696
4.1 Experiment Setup ... 696
4.2 Using Different Incremental Temperatures ... 697
4.3 Comparison ... 697
5 Improved Simulated Bifurcation for TSPs ... 699
5.1 Simulated Bifurcation ... 699
5.2 TSP Solvers Using the Ising Model Without External Fields ... 700
5.3 Improvement Strategies ... 701
5.4 Experimental Results ... 703
5.5 Using Different Dynamic Configurations of the Time Step ... 703
6 Conclusions ... 705
References ... 706

Approximate Communication in Network-on-Chips for Training and Inference of Image Classification Models ... 709
Yuechen Chen, Ahmed Louri, Shanshan Liu, and Fabrizio Lombardi
1 Introduction ... 709
2 Backgrounds ... 711
2.1 Existing Approximate Communication Techniques ... 711
2.2 Existing Sparse Matrix Compression Techniques ... 712
3 Proposed Approximate Communication Technique ... 714
3.1 Approximate Communication for Image Preprocessing (ACT-P) ... 715
3.2 Approximate Communication for Model Inference (ACT-I) ... 719

4 Implementation of the Approximate Communication Technique (ACT) ... 724
4.1 Dual-Matrix Compression Method for Sparse Matrix ... 724
4.2 Software Interface for Approximate Communication ... 727
4.3 Architecture Design of ACT ... 728
5 Evaluation ... 731
5.1 Network Latency ... 732
5.2 Dynamic Power Consumption ... 734
5.3 Accuracy Loss ... 735
5.4 Overall System Performance Evaluation ... 736
5.5 Sensitivity Study ... 737
6 Conclusion ... 737
References ... 738

Index ... 741

Part I
In-Memory Computing, Neuromorphic Computing and Machine Learning

Emerging Technologies for Memory-Centric Computing

Paul-Antoine Matrangolo, Cédric Marchand, David Navarro, Ian O'Connor, and Alberto Bosio

1 Introduction

The advent of the Internet of Things (IoT) has led to a massive production of data to be processed (i.e., data deluge). With the number of sensors and smart devices on the edge of the Internet of Things at 14.4 billion active connections in 2022 [1] and rising, 40 times more data is already generated at the edge than in the cloud data centres, itself around 20 zettabytes per year. In addition, machine learning (ML) applications are nowadays widely used to process data in order to extract information and are indicative of the current trend towards data-intensive applications. Unfortunately, today's computing systems are facing two well-known walls: (1) the memory wall due to the increasing gap between sub-ns processor cycle time and memory latency (itself a function of several factors, including cache tests, processor-memory bandwidth and intrinsic memory access time) which has become a major bottleneck to improving performance as well as energy-efficiency in memory access dominated applications [2] and (2) the power wall as the practical power density limit (around $1\,W/mm^2$) for air-cooling silicon semiconductor chips is reached, meaning processor clock speed is restricted to a few GHz. For these reasons, conventional computing systems cannot provide the required performance for ML applications, especially at the edge, subject to energy consumption constraints.

Current computing paradigms are "Processor-Centric", meaning that the processor cores (general-purpose, application specific, e.g. Graphical Processing Units, GPUs, or dedicated hardware accelerators) are (i) physically distinct and spatially separate from the memory units and (ii) generally the only elements capable of data

P.-A. Matrangolo · C. Marchand · D. Navarro · I. O'Connor · A. Bosio (✉)
Univ Lyon, ECL, INSA Lyon, CNRS, UCBL, CPE Lyon, INL, UMR 5270, Ecully, France
e-mail: paul-antoine.matrangolo@ec-lyon.fr; cedric.marchand@ec-lyon.fr; david.navarro@ec-lyon.fr; ian.o'Connor@ec-lyon.fr; alberto.bosio@ec-lyon.fr

W. Liu et al. (eds.), *Design and Applications of Emerging Computer Systems*,
https://doi.org/10.1007/978-3-031-42478-6_1

processing. A promising alternative is based on the so-called "Memory-Centric" paradigm where some computational tasks are offloaded from the processor core to deployment in enhanced, computation-capable, memory units [3]. The main idea is to decentralize and distribute specific processing hardware from the processor to such memory units. The principle rationale is that the data can thus be processed where they are stored, rather than having to be transferred from/to the memory. These decentralized approach to data processing and significantly reduced use of the core-memory interconnect infrastructure lead to sizeable reductions in energy consumption and latency, thus increasing both overall performance and energy efficiency.

Memory-Centric architectures can be classified according to two criteria: (i) the distance of the computational unit(s) from the memory and (ii) the technology used to fabricate the memory. In Fig. 1a, we depict the two main types of Memory-Centric architectures using the first criteria: (i) *In-Memory Computing (IMC)* where the computational unit(s) is (are) embedded in the memory bank and (ii) *Near-Memory Computing (NMC)* where the computational unit(s) is (are) close to, but outside, the memory bank.

Figure 1b details at finer grain the two types as proposed in [4]:

1. Computation-in-Memory (CiM): Computation is achieved by reading data from the memory through sense amplifiers. The memory access is modified in order to support the computation of Boolean functions. For example in [5], it is possible to access in read mode two or more memory rows. The sensed data is therefore a Boolean function (e.g. NAND, NOR) of the values stored in the accessed rows.

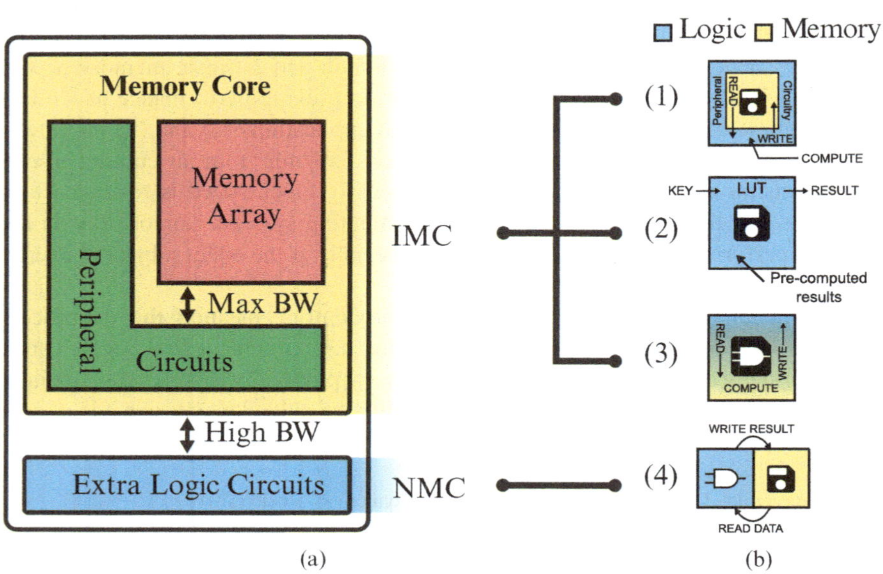

Fig. 1 (**a**) Memory-Centric coarse grain taxonomy, (**b**) Memory-Centric fine grain taxonomy

2. Computation-with-Memory (CwM): the memory unit stores the pre-computed results; the memory addresses correspond to the operands [6].
3. Logic-in-Memory (LiM): memory cells can be accessed in "computation" mode in order to execute a given Boolean function [7].
4. Computation-near-Memory (CnM): the computational unit(s) is(are) outside the memory unit. Operands have to be read from the memory. Performance metrics vary depending on the bandwidth (BW) of the bus connecting the memory and computational unit(s).

Concerning the second classification criteria, the technologies that can be used to implement Memory-Centric applications span from mainstream (i.e. CMOS-based SRAM and DRAM) to emerging. Implementations of CiM based on SRAM or DRAM have been presented in [8, 9]. The LiM implementations have a higher diversity in terms of emerging technologies: we can cite resistive switching technologies (Resistive Random Access Memory (RRAM) or Memristor) [10], magnetic random access memory (MRAM) [11], Phase-Change Memories (PCM) [12], and Ferroelectric (Fe) technologies (FeRAM, Ferroelectric Field Effect Transistor (FeFET) and Ferroelectric Tunnel Junction (FTJ)) [13]. Moreover, such emerging technologies are also non-volatile (NV), thereby enabling aggressive power-down strategies to minimize leakage current concerns [10, 14].

In this chapter, we present Memory-Centric architectures based on the technologies used to implement them. For each technology, we will detail the basic operating principles at device level, then followed by the circuit, architecture and the application levels.

The chapter is organized as follows: from Sects. 2 to 5, we, respectively, present RRAM, Spin-Transfer Torque MRAM (STT-MRAM), PCM and FeFET. Section 6 compares and discusses the reviewed technologies and circuits. The chapter ends with a conclusion in Sect. 7.

2 Resistive Random Access Memory (RRAM)

The memristor (i.e. memory resistor) is a device described as the "missing circuit element" by L.O. Chua in 1971 [15]. It completes the relationship between the four fundamental circuit variables in circuit theory which are the *current*, the *voltage*, the *charge* and the *flux-linkage*. The memristor, as the "missing" relationship between the *flux-linkage* and the *charge*, was demonstrated to behave as a nonlinear resistor with memory. Other terms (RRAM or ReRAM) are also commonly used, in particular when using memristors to build a memory array. Due to the focus of this chapter on memory characteristics, we will use the term RRAM for the rest of the chapter.

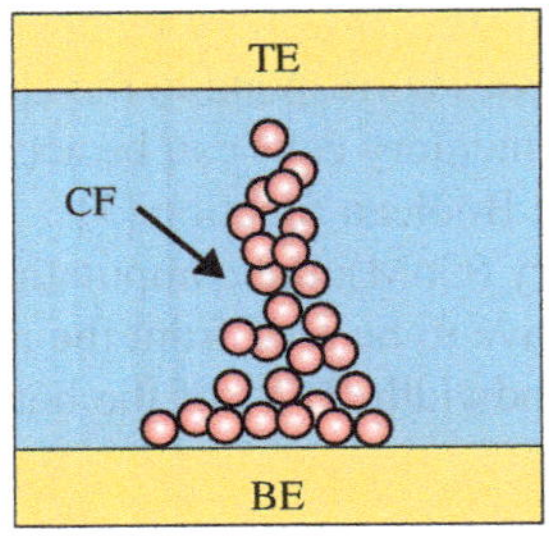

Fig. 2 Schematic of conductive filament switching-based RRAMs. Sandwiched between two metal electrodes (TE and BE), the switching material could be an insulator or a semiconductor. During programming, a conductive filament (CF) appears and changes the resistance of the device; the larger the CF, the lower the resistance. In OxRAM, the CF corresponds to the rupture of oxygen vacancies. In CBRAM, CF is a conductive metallization bridge. Due to this difference, the electrodes of the CBRAM have other names: active electrode as TE and counter electrode as BE

2.1 Device

RRAM implementations are commonly based on the resistance switching effect of certain materials. The RRAM can be described as an insulating layer, sandwiched between a top electrode (TE) and a bottom electrode (BE). Due to the simple structure, it also promises excellent scalability. The resistance of the switching material is varied between non-volatile high-resistance (HRS) and low-resistance (LRS) states in a controlled manner to store non-volatile data. Multiple intermediate resistance states between HRS and LRS can also be achieved in RRAM for multibit storage applications. Based on the basic switching mechanisms, conductive filament (CF)-based RRAM devices can be classified into two types: metal oxide random access memory (OxRAM) and conductive bridge random access memory (CBRAM). These two types of RRAM both exploit filament switching effects: when applying an external stress, the insulating layer changes to increase or decrease resistance [17] by growing or breaking a conductive filament, which forms the current conduction path inside the RRAM.

The appearance of the CF is shown in Fig. 2. In the OxRAM, the major switching mechanism involves the formation and rupture of the oxygen vacancy (V_O)-based CF in the oxide switching layer, while the switching layer in the CBRAM cell can be any solid electrolyte. In this case, the electrolyte switching layer is sandwiched between an electrochemically active electrode as TE and an inert "counter" electrode as BE. For the CE, any conductive material can be used as long as it is inert and does not diffuse easily into the switching layer.

2.2 Memory Architectures

Conventional CMOS-based memory technologies have demonstrated huge progress over the years for all the memory hierarchy. For current data storage needs, they fulfil the requirements to some extent. In the memory hierarchy, the distinction of each level is done by response time, capacity and complexity [18]. Four major storage levels exist. Firstly, there is the central processing unit (CPU) with the CPU caches. They are characterized with a high processing speed but with high cost and small capacities. Secondly, the physical memory (mainly represented with RAM) has good capacities and speed. Thirdly, solid-state memory is fast, but its purpose is to store data "non-volatily" where physical memory is volatile. Finally, virtual memory got a large storage capacity. The aim is to store a high or very high quantity of data (gigabytes to terabytes).

The concept of Storage Class Memory (SCM) [19] gives another hope to RRAM in an intermediate density range target. In order to overcome the memory bottleneck, Freitas et al. propose to combine DRAM and NAND Flash memory.

From early attempts at using RRAM as a digital binary switch element to CiM using RRAM as both storage and computing element, the motivation is always to leverage its simple and high-density crossbar architecture. A cross-point memory array basically consists of a cell stack between two crossing metal lines, called the BL and the WL. Typically, this is the same architecture as the MVM circuit illustrated in Fig. 3b. Each cell stack includes a memory material and a selector material, and, if necessary, additional electrodes may be inserted to work as a diffusion barrier or to enhance the electrical properties of the two active materials. From the technology point of view, encouraging progress has been made from single-bit engineering work with array simulation to kilobit and megabit array-level demonstrations. With challenges ahead to scale up the implementation to even larger bit counts, more efforts are yet to be made.

Among the various emerging memory technologies, CBRAM demonstrates better performances, making it one of the most favourable candidates for future memory applications [20]. Although CBRAM offers better opportunities for future data storage, intensive research has been carried out in recent years to find solutions to challenges and to explore the vast opportunities that CBRAM offers for memory applications.

2.3 IMC Applications

Memristive devices are also used in IMC application because of their non-volatile binary storage. In this section, the main structure common to these technologies will be explore, then the attention will be focus on RRAM applications.

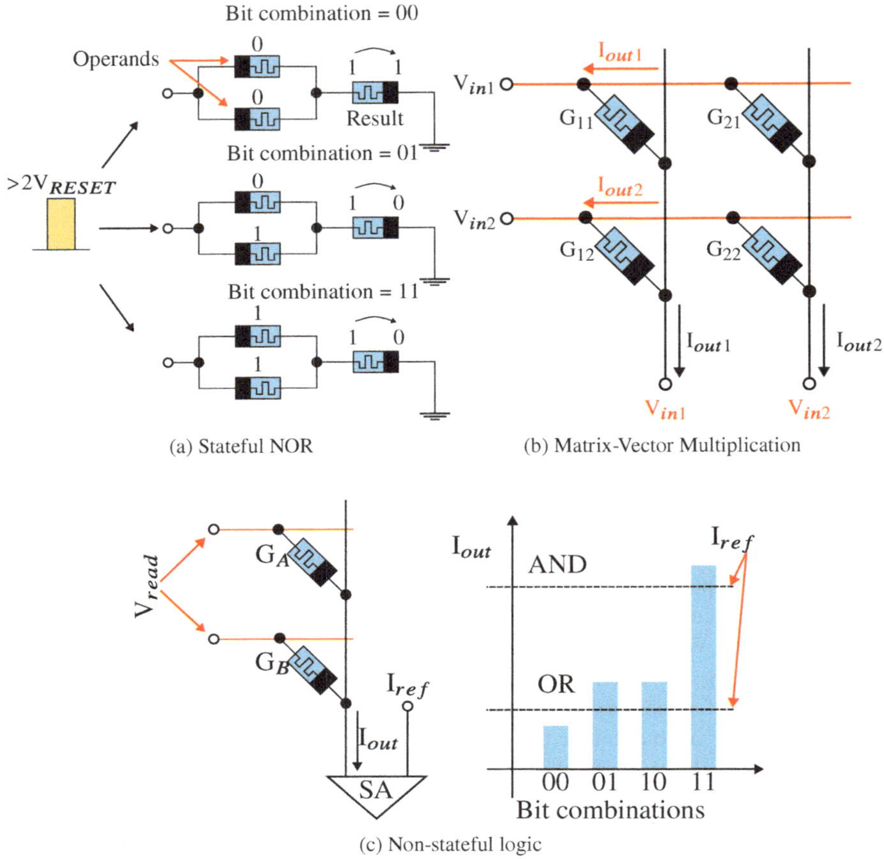

Fig. 3 Memristive device-based circuit designs. (**a**) The stateful NOR circuits perform the NOR operation by using three bipolar memristive devices: two as the operands and one to represent the results as proposed in [16] with MAGIC. With the matrix-vector multiplication (MVM) circuit (**b**), the elements of the constants are mapped linearly to the conductance values of memristive devices organized in a crossbar configuration (i.e. for a matrix A: $A_{11} = G_{11}$, $A_{12} = G_{12}\ldots$ and so on). Then the elements of the multiplied values are mapped linearly to the amplitudes or durations of read voltages and are applied to the crossbar along the rows. (**c**) The non-stateful logic operates with two memristive devices and a variable threshold using a sense amplifier (SA). Depending on the choice of the threshold set by I_{ref}, the circuit performs two operations (AND and OR) between (i) a word mapped on the memristive devices and (ii) another mapped along the row as seen in the MVM

2.3.1 Main Structures of Meristors-based Circuits

IMC circuits with memristive devices could be gathered in three main structures widely studied in literature [21]. It is possible to distinguish three structures:

- **Stateful NOR** (Fig. 3a): In stateful logic, the memristive switches both store logic values and perform logic operations [22]. Recently, it was demonstrated to perform complete IMC [23]. In this circuit the NOR operation is performed using three memristive devices. Two are for the operand and one is for the result. It is a structure proposed by Kvatinsky et al. with MAGIC [7, 16]. This architecture is related to LiM paradigm because the cell can be accessed to perform Boolean functions.
- **Matrix-Vector Multiplication** (Fig. 3b): The interest of this architecture is the possibility to map a constant matrix inside a memory array to perform matrix multiplication. Suppose that $Ax = b$, where A is a matrix and x and b a vector. The elements of A are linearly mapped to the conductance values of memristive devices organized in a crossbar configuration. The elements of x are mapped linearly to the amplitudes or pulse lengths of read voltages and are applied to the crossbar along the rows. It is a good illustration of the CiM paradigm because the mapped element of A must to be read in order to perform a computation.
- **Non-stateful logic** (Fig. 3c): The non-stateful logic operates with two memristive devices and a variable threshold (I_{ref} in the figure). The output value (I_{out}) is compared to the I_{ref} with a SA. Depending on the reference choose, the circuit can perform either the OR operation (in case of a low I_{ref}) or the AND operation (for a high one). It is also a good illustration of the CiM paradigm. This architecture is able to perform logical operations in the memristive devices array.

2.3.2 Vertical Cross-Point Resistive Memory (VRRAM)

Vertical cross-point using RRAM was successfully demonstrated in [24]. It consists of RRAM cells at the crossing of vertical BLs and horizontal WLs. In each plane, WLs are connected in two sets: even and odd WLs, in the aim to reduce the area overhead to connect WLs to the WL drivers on the substrate. Each vertical BL is connected to a column select transistor at the bottom. During operation, only one of the column select line is enabled, and all other BL pillars are floating, reducing the total array leakage. The main difference with 3D cross-point structure is that each cell contains a RRAM element and two-terminal selector to isolate non-address cells so as to suppress leakage current. The main problems with conventional 3D cross-point structure are, in a first hand, it is compromising lateral scaling [25] and, in a second hand, conductive electrodes inside the cell stack would create a short between neighbouring cells. A comparison in [24] showed that VRRAM is more promising than 3D stack. It got more sensitivity to number for layers for read margin, and it reduces the voltage applied to unselected WL, improving the read margin. However, it increases the leakage current of selected WL and the total power consumption.

1Transistor/1RRAM (1T1R) architecture was proposed in Fig. 4 [26] as competitive from crossbar density point of view, while getting rid of any sneak path current, enabling large-scale IMC. Associated with gate-all-around (GAA) stacked nanosheet transistors, OxRAMs are used to create memory arrays in a

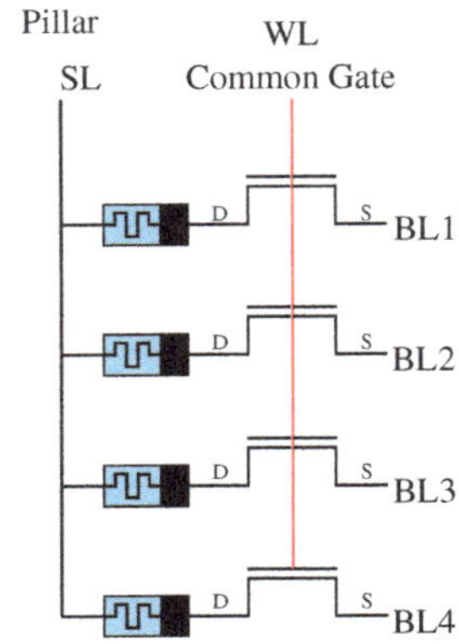

Fig. 4 Schematic of the 1T1R OxRAM memory cell based on GAA nanosheet transistors proposed by Barraud et al. in [26]. Each independent BL controls transistors with a WL common gate. A 3D pillar is used to connect the SL

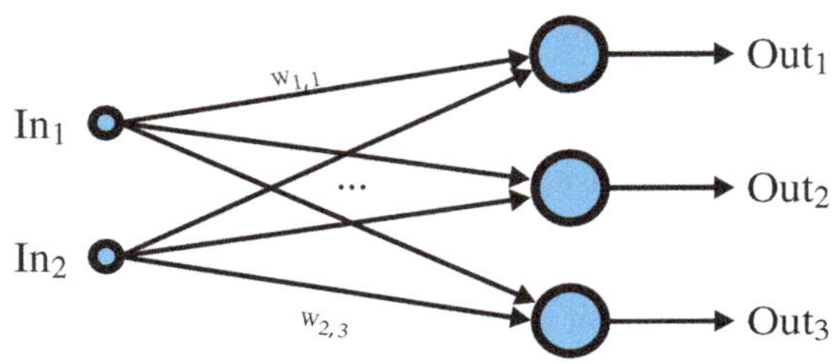

Fig. 5 Schematic of an artificial neural network

3D architecture as a 1T1R memory cube. Each horizontal GAA channel features an independent source connected to a bit line and a drain directly connected to a pillar of RRAM memory cells. This imposes a horizontal WL and vertical SL. To ensure memory endurance, Barraud et al. employed the "SCouting Logic" [27] approach. The main goal is to perform logic operations by modifying the read operation. Circuit architecture is corresponding to the non-stateful-logic (See Fig. 3c). A 4kbits OxRAM array has been validated through SPICE simulations showing high parallelization and flexibility for operands selection. Unlike planar RRAM arrays, SCL can be performed between words of different planes as long as they share the same row levels in each plane.

2.3.3 Neural Network

As stated in the introduction, ML applications are the killer application of memory-centric computing since they require to manipulate a huge amount of data. Neural network (NN), or artificial neural network (ANN), are a subgroup of ML. Their structure is inspired from the human brain with the interconnected neurons sending signals to another. In NN, the neurons (or nodes) connected to another have associated weight and threshold. In Fig. 5, two inputs are connected to the neurons with an associated weight. By performing a sum, the nodes give the output results through the following equations:

$$Out_1 = In_1 W_{1,1} + In_2 W_{1,2} \quad (1)$$

$$Out_2 = In_1 W_{2,1} + In_2 W_{2,2} \quad (2)$$

$$Out_3 = In_1 W_{3,1} + In_2 W_{3,2} \quad (3)$$

Memristor is, like other technologies presented in this review, a good candidate for NN. The architecture corresponds to crossbar section presented in Fig. 3b. Synaptic weights (w) are stored in the memory cells. Inputs data are used as address. The red values on the memory array columns correspond to the analogue product of input and weights. A demonstration of BNN on 16Mb RRAM macro chip was performed and analysed in [28]. The 16Mb RRAM macro chip from Tsinghua corresponds to 16 block of two arrays (512 × 1024) sharing the sense amplifiers with I/O width of 8 bits (architecture of 1T1R). It achieved high learning accuracy (96.5% for MNIST) under the non-perfect bit yield and endurance with a new record of the scale of synaptic array up to 512 × 1024.

3 Spin-Transfer Torque Magnetoresistive Random Access Memory

The Magnetic Tunnel Junction (MTJ) [29] is compound of two ferromagnetic (FM) layers separated by a thin oxide barrier as shown in Fig. 6a. One of the FM layers always has a fixed magnetic direction, which is usually called the reference layer. The other layer, which can be configured using an electrical current, is known as the free layer. Based on the relative directions of the FM layers, an MTJ can have the following two distinct operational modes: parallel (P) mode (when both of the FM layers have the same magnetic direction) and anti-parallel (AP) mode (when the FM layers have an opposite magnetic direction). With electrons flowing from the reference layer to the free layer, the electron spins become polarized in the reference layer and can thus switch the magnetization direction of the free magnetic layer into the "0" state due to the transfer of spin angular momentum to the free layer. Reversing the current direction can switch the free layer into the "1" state.

3.1 Device

In Magnetoresistive Random Access Memory (MRAM), information is stored in the magnetization direction of a free magnetic layer in a MTJ (see Fig. 6b). The device is read by measuring the resistance of the tunnel junction, which depends on if the free layer magnetization is P or AP to the fixed layer magnetization. The resistance changes by a factor of 2–3 between the P and AP to the fixed layer magnetization. The device is written by passing current through the tunnel junction.

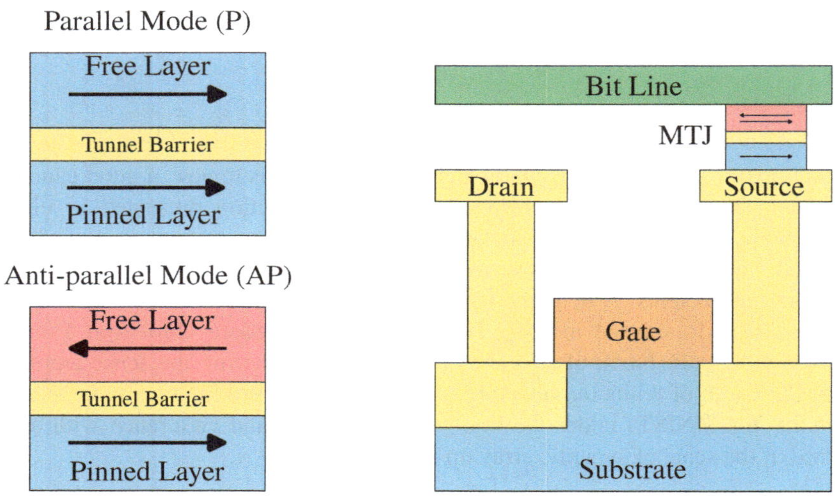

(a) MTJ (b) STT-MRAM.

Fig. 6 (**a**) Illustration of MTJ with its two modes of magnetization directions. A thin oxide barrier (tunnel barrier) is sandwiched between two FM layers. One of the layers is fixed (pinned layer or reference layer) and has a magnetic direction. The other layer is the free layer and can be set using an electrical current. In P mode, the direction of the magnetic field is the same for the two layers. In AP mode, the directions are different. In this schematic, the direction is represented as completely opposed but it is not needed. (**b**) In STT-MRAM, the MTJ is added to the source of MOSFET. It could be to the drain or sometimes both. The device is read by measuring the resistance of the tunnel junction with the BL

A single transistor is used in each memory cell to give access to the magnetic tunnel junction. Since both reading and writing use the same transistor, at low and high voltage, care must be taken to keep the read voltage low, around 50–100 mV. Furthermore, to avoid breaking down the tunnel barrier over the life of the memory, the write voltage must not be too high (around 500 mV) [30].

3.2 Memory Architectures

STT-MRAM could be used as a standalone memory unit. In [30], the device showed an endurance of $1e^{10}$ write and read cycle times in the range 30–100 ns. It is often used to replace battery-backed SRAM and DRAM in order to make product more reliable than possible with either batteries or super capacitors. Seen as the most promising NVM candidate for IMC, the STT-MRAM gives more performance than SRAM when scaled down from 65 to 7 nm.

3.3 IMC Applications

As a promising candidate for IMC, STT-MRAM was enhanced in STT-CiM [31]. According to the paper, the proposed design can perform a range of arithmetic, logic and vector operations. The key idea is to enable multiple WLs simultaneously in an STT-MRAM array, leading to multiple bit-cells being connected to each BL. An enhanced reference generation circuitry enables to natively realize a wider variety of operations.

Highly Flexible In-Memory Computing (HieIM) using STT-MRAM was proposed in [32]. It can perform complete Boolean logic functions between any two bits stored in the same memory array. Here, each memory cell is designed using the 1T1R STT-MRAM bit-cell structure (see Fig. 3b). Each cell is associated with the WLs, BLs and SLs. The WLs and BLs are externally controlled by the row decoder and column decoder, respectively. Additionally, voltage drivers are connected to each BL and SL. Evaluation with In-memory bulk bitwise Boolean vector logic (AND/OR) operation for different vector datasets shows $\sim 8\times$ energy saving and $\sim 5\times$ speed up compared to that using DRAM-based in-memory computing platform. It further has employed in-memory data encryption engine using AES algorithm, which shows 51.5% lower energy consumption compared to CMOS-ASIC according to Parveen et al. IMC paradigm platform using STT-MRAM was proposed in [33] and was demonstrated with the application of image edge extraction. The cells were based on sensing-based in-memory computing scheme (see Fig. 3c). For memory read operation, a single memory cell is addressed and routed in the memory read path to generate a sense voltage which will be compared with a reference voltage. For in-memory Boolean computing, two memory bit-cells are sensed simultaneously. Owing to the different resistance combinations of two selected STT-MRAM bit-cells, three different sense voltages (V_{sense}) could be generated, respectively. When both selected, STT-MRAMs are in AP state ($V_{sense} > V_{ref}$). Thus, this sensing operation with modified reference voltage performs an AND logic operation taken the binary data stored in STT-MRAM as logic inputs. Similarly, when V_{ref} is shifted to $(V_{P,P} > V_{AP,P})/2$, the OR logic operation can be performed as well. Therefore, through tuning the reference voltage for comparison, the sense amplifier can perform reconfigurable in-memory computations.

3.3.1 Binary Neural Network

As shown for the RRAM, STT-MRAMs can be exploited to accelerate the execution of ML applications [34]. It proposes a STT-MRAM in-memory highly scalable architecture for Binary NNs (BNNs), named STT-BNN, with the single-access many-MAC operations, allowing to perform unrestricted accumulation across rows for full utilization of the array and BNN model scalability. Weights and neuron activations are binarized to -1 and $+1$ (i.e. logic "0" and "1"). Each weight is stored in the memory cell that is composed of two transistors and two junctions (2T2J). The

STT-BNN architecture is an array of STT-RAM cells where the BL drivers encode the input features of the neural network being mapped in the form of a signal pair and where the word line is driven by a two-stage buffer to keep the setting time of the WL signal negligible compared to the overall access time. System simulations have shown 80.01% (98.42%) accuracy under the CIFAR-10 (MNIST) dataset under the effect of local and global process variations, corresponding to an 8.59% (0.38%) accuracy loss compared to the original BNN software implementation, while achieving an energy efficiency of 311 TOPS/W. In [35], a 4Mb NMC Macro based on 22 nm STT-MRAM is proposed. The array of STT-MRAM, where the architecture is sensibly the same as Fig. 3b, was demonstrated to reduce latency, energy consumption and memory access. This is mainly due to the use of several things. Firstly, the use of bitwise vertical-weight-mapping Burst-access scheme with a corresponding ReLU-prediction to reduce the frequency of memory access and MAC operations. Secondly, a Bidirectional-BL-access readout scheme to reduce latency. Lastly, a Charge-recycling Voltage-type Small-offset Sense Amplifier to reduce read energy consumption.

Probabilistic programming of STT-MRAM was studied in [36]. It proposed a technique validated by Monte Carlo simulation and fabricated ST-MTJ clusters. This study also shows better programming performance in terms of programming power and delay can be achieved thanks to probabilistic programming in comparison to conventional deterministic programming. The two terminal memory elements are formed by connecting one or more STT-MTJs in various topologies. Due to the probabilistic switching nature of STT-MTJs, a STT-MTJ cluster with N constituent STT-MTJs can reach up to 2^n cluster states when a pulse of programming current is passed through the two terminals of the cluster. Then, STT-MRAM-based Stochastic memristive synapse was proposed in [37]. To do so, STT-MRAM is organized as a crossbar connecting input and output neurons. It consists of an architecture with passive crossbar (1R). The advantage of this structure is the area efficient. When an input neuron spikes, it applies a brief read pulse to the crossbar. This leads to currents that reach the different output neurons simultaneously. The output neurons maintain a constant voltage at their input (limiting sneak path on other STT-MTJs) while reading the current. The current received by an output neuron depends on the state (P or AP). The output neuron provides two features: it can determine if an input was received from a P or AP synapse by using a sense-amplifier; it integrates the information received from their sense-amplifier.

4 Phase-Change Memory

Phase-change materials are materials with the ability to switch between amorphous (disordered) and crystalline (ordered) phases. Those phases got different electrical resistances: the amorphous phase tends to have high electrical resistivity and a low optical reflectivity, while the crystalline phase exhibits a low resistivity and a high optical reflectivity. Figure 7 shows the phase-changing inside a phase-change

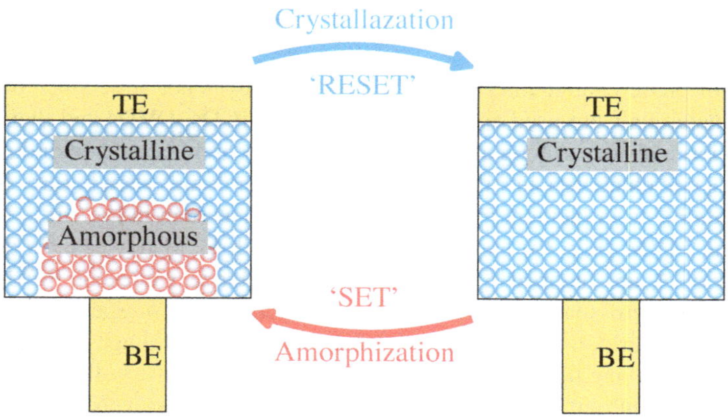

Fig. 7 Schematic of a PCM. A crystalline material is sandwiched between a TE and a BE. To program a PCM, a high temperature during a sufficient amount of time is applied from the BE in the aim to change the phase of the material: crystalline or amorphous

memory (PCM): the change to crystalline phase corresponds to the "crystallization", whereas the change to amorphous phase is the "amorphization". Respectively, the terms as "Potentiation" and "Depression" are also used to describe the changing phenomena [38]. The contrast in optical properties of phase-change materials has been widely employed to enable optical data storage devices such as DVDs and Blu-ray discs.

4.1 Device

A PCM consists of a small active volume of phase-change material sandwiched between two electrodes. In PCM, data is stored by using the electrical resistance contrast between a high-conductive crystalline phase and a low-conductive state, through applying electrical current pulses. See Fig. 7 which represents the two states of the PCM. The stored data can be retrieved by measuring the electrical resistance of the PCM device. An appealing attribute of PCM is that the stored data is retained for a very long time (typically 10 years at room temperature), but is written in only a few nanoseconds. This property enables PCM to be used for non-volatile storage in replacement of Flash and hard disk drives, while operating almost as fast as high-performance volatile memory such as DRAM.

4.2 Memory Architectures

The use of phase-change material as memory devices is not new. Since the phase-change materials were used as optical data storage (CDs, DVDs and then Blu-ray discs), it has a history as long as the RRAM. Memory architectures including PCMs have the same characteristics than the previous technologies seen before. In fact, it is still possible to implement memory units or MVM as seen in Fig. 3b.

Indeed, MVM (Fig. 3b) can be implemented with PCM devices. To do so the elements of A are mapped linearly to the conductance values of PCM devices organized in a crossbar configuration. Let's define $Ax = b$ again. The x values are mapped linearly to the amplitudes or durations of read voltages and are applied to the crossbar along the rows. The result of the computation b will be proportional to the resulting current measured along the columns of the array. Note that it is also possible to perform a matrix-vector multiplication with the transpose of A using the same crossbar configuration. This can be achieved by applying the input voltage to the column lines and measuring the resulting current along the rows. One of the uses of this matrix-vector multiplication is deep learning inference [39] or storing synaptic weights.

4.3 IMC Applications

PCM memories can be used to implement memory-centric architectures by exploiting the same architectures depicted in Sect. 2.

4.3.1 Binary Neural Network

In [38], authors proposed a LIM architecture using PCM for BNNs. The synaptic weight is represented by the combined conductance of N PCM devices. The overall dynamic range and resolution of the synapse are increased by using multiple devices. An input voltage corresponding to the neural activation is applied to all memory cells for the realization of synaptic efficiency. The sum for the individual cell currents forms the synaptic output. For the plasticity, only one out of N cells is selected and programmed at a time (selection with a counter-based arbitration). The PCM devices composing the multi-memristive synapses can be arranged in either a differential (in which two memristive devices are used in a differential configuration showed in Fig. 8 such that one device is used for potentiation and the other for depression) or a non-differential architecture (where each synapse consists of N devices, and one device is selected and potentiated/depressed to achieve synaptic plasticity). In this architecture, by placing the devices that constitute a single synapse along the bit lines of a crossbar as Fig. 3b, it is possible to sum up

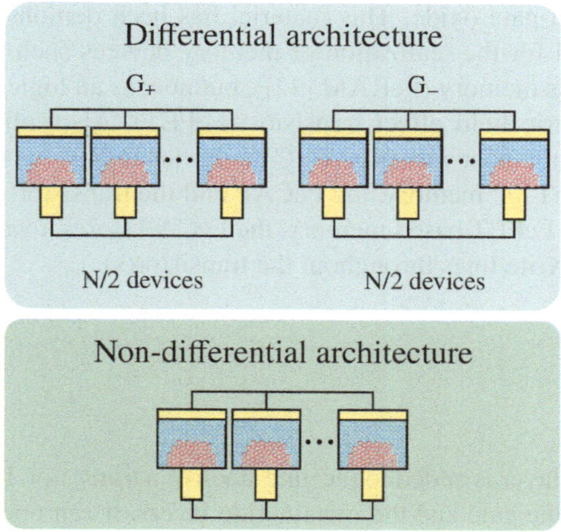

Fig. 8 Schematic of differential and non-differential architecture for multi-memristive synapse using PCM. In the differential architecture, two sets of devices are present, and the synaptic conductance is calculated as $G_+ - G_-$, where G_+ is the total conductance of the set representing the potentiation of the synapse and G_- is the total conductance of the set representing the depression of the synapse. Each set consists of N/2 devices. When the synapse has to be potentiated, one device from the group representing G_+ is selected and potentiated, and when the synapse has to be depressed, one device from the group representing G_- is selected and potentiated

the currents using Kirchhoff's law and obtain the total synaptic current without the need for any additional circuitry.

Introduction of a synaptic unit-cell design for analogue-memory-based BNN training was performed in [40], which combines non-volatile PCM with volatile weight storage using conventional CMOS-based devices. Mixed-hardware-software implementation (hardware PCM-array, software CMOS (processor)) goal was to achieve comparable test accuracy against software baselines that use exactly the same network size but take advantage of software techniques such as unbounded rectified linear unit activation function and cross-entropy training and optimizers such as AdaGrad [41]. Under the assumption that an MNIST example can be processed in 240 ns, the average power consumption for the 2PCM + 3T1C design was calculated to be 54 mW, compared to 22 mW for the PCM design.

5 FeFET

Ferroelectricity is a property of a material where it owns a spontaneous polarization. It is the case of hafnium dioxide or more precisely high-k HfO_2 which is mainly used since it is compatible with silicon technology. It also got a high dielectric permittivity allowing downscaling without the prohibitive leakage currents associated with the traditional gate oxide. This material has been demonstrated to be a very promising material for the realization of memory devices such as the ferroelectric (FE) random access memory (FeRAM [42]), but now as an logic-in-memory device with the ferroelectric field effect transistor (FeFET). Also, other associations of FeCAP and transistor exist with the nT-1C memories in [43] (where n is the number of transistors). In nT-1C memory, the FeCAP and the transistor(s) are independent. In each cell, as the FeFET-based memory, the FeCAP stores a value and is accessed with bit lines and write lines throughout the transistor(s).

5.1 Device

For a FeFET, a FE layer is added to the gate stack of a transistor. Figure 9a shows the FE layer between the gate and the metal/oxide layers. It can operate in at least two different modes: a non-volatile mode, which requires hysteresis-based operation, and a steep switching mode, which can have a hysteresis or non-hysteresis behaviour [44]. As the FE layer thickness increases, and the ratio of ferroelectric capacitance to gate dielectric capacitance is sufficiently low, the polarization of the FE layer can be retained, leading to hysteresis behaviour of the output voltage as a function of the input voltage as shown in Fig. 9b. This is the fundamental characteristic enabling the program/erase operation of the ferroelectric gate layer in the FeFET. To change (program/erase) the stored value (or the polarization direction of the ferroelectric layer), a higher voltage than the coercive voltage must be applied. The aim is to have a sufficiently important potential difference between the transistor pins. Practically, the gates are used to apply a strong positive or negative voltage. It is the easiest way to do, but programming could be performed using the bulk or the source-to-drain voltage.

5.2 Memory Architectures

5.2.1 FeFET-Based Memory

FeFET-based memory structures have been considered since 2016 in [45]. The appeal of FeFETs as memory element relies on their polarization retention capability, high current ratio for the two states and low voltage operation. FeFET is a three-terminal device; thus the memory cell is designed to have separate read and write path, which facilitates simultaneous optimization of the read and write operations. Figure 10 shows the schematic of the 1T-1FeFET memory cell.

The write path has a standard MOSFET used as an access transistor, which is controlled by write select-line enabling selective write operation of the cells in an array. The write bit line is shared among the cells in the same column. The read path consists of the FeFET with the read select connected to the drain and sense line

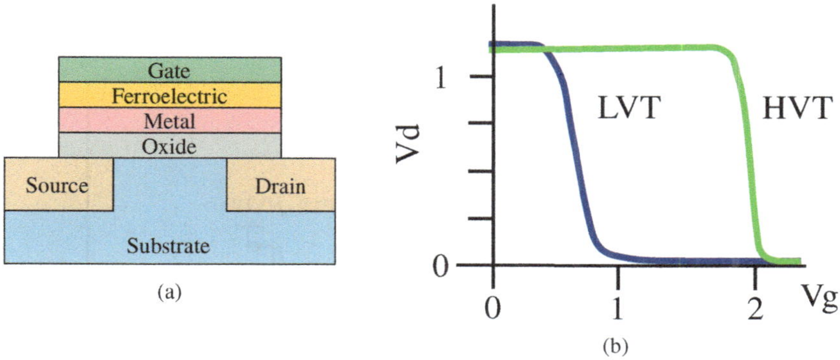

Fig. 9 (**a**) Illustration of the layers of a FeFET. The FE material is sandwiched between the gate and the oxide layer. The FE material has the property to change polarity when applying a strong or sufficiently strong voltage to terminals of the device. The polarization lets appear a hysteresis cycle visible on the characterisation voltage curves in (**b**). This property offers to the FeFET a reprogrammability usable for NV applications

Fig. 10 Schematic of a FeFET-based memory cell when performing reading

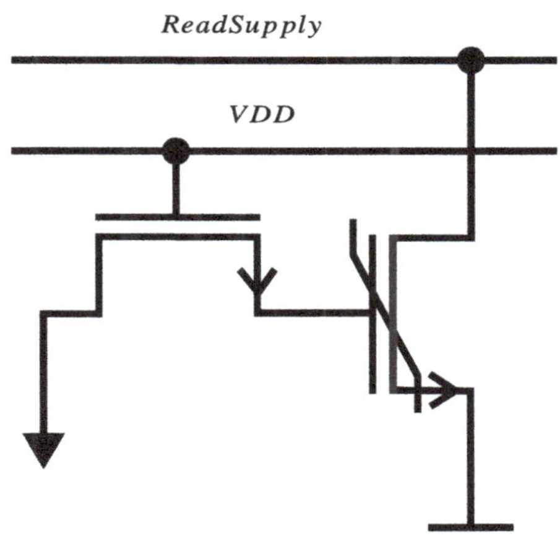

tied to the source. The read select line is shared among the cells in the same row, while the sense line runs along the column; thus, the need for a separate read access transistor is eliminated. Read select also functions as the read supply, reducing metal routing congestion. The FeFET is also more interesting in terms of reading because of the nondestructive read operation where FeRAM is destructive.

5.2.2 Non-volatile Flip-Flop

Proposed in 2016 by Wang et al. in [46], FeFET-based non-volatile flip-flop consists of an n-type pass transistor which connects the D flip-flop (DFF) output Q to the gate of the FeFET during the backup operation. Otherwise, this transistor is turned off

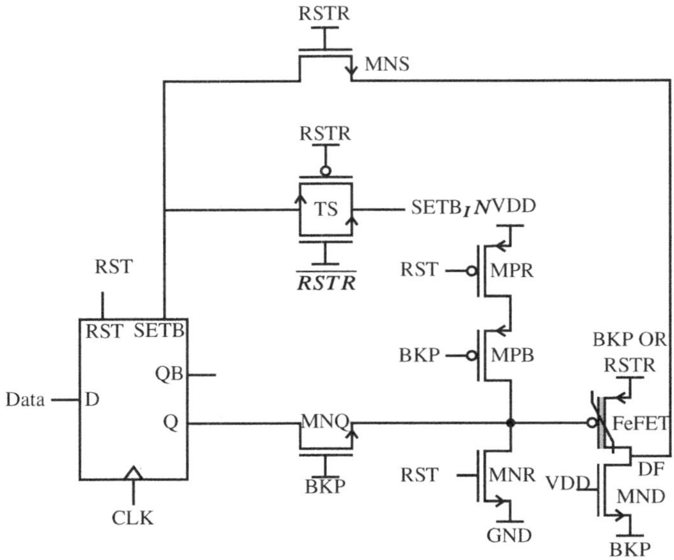

Fig. 11 Schematic of a FeFET-based D Flip-Flop proposed by Wang et al. [46]

to isolate Q from the FeFET gate capacitance. As presented in Fig. 11, the gate of the FeFET is driven to Vdd during the restore and normal operations by a series of transistors controlled by the backup and reset signals. The latter also controls another transistor to pull down the FeFET to "0" when reset is high. The FeFET is connected in series to a transistor, both of which drive the common node DFF to appropriate values depending on the flip-flop operation. The other terminal of the FeFET is driven by the OR operation between the backup and the reset signal. The signal obtained at the output of the FeFET drives the multiplexer controlled by the restore signal. Finally, the multiplexer drives the set input of the flip-flop.

This approach can be employed for supply gating as well as in the design of non-volatile processors integrating non-volatility inside or in the proximity of the flip-flop. Such a technique significantly reduces the overheads associated with the memory-logic data transfer enabling energy-efficient backup/restore. It answers the problem of leakage during stand-by state. Or, it can benefit to systems operating with unreliable power sources such as wearable and implantable device.

FeFETs have been integrated directly into the flip-flop architecture in 2017 by [47] as negative capacitance FET. It proposed a different circuit that incorporates FeFETs inside the flip-flop. It consists of a conventional DFF consisting of a master latch and a slave latch plus accessory circuitry connections for backup and restore operations. Those are associated to the slave latch only. When backup and restore signals are low, the interface transistors between the main body and the auxiliary circuitry are turned OFF by the gate signal, leaving the main body functioning the same as conventional positive-edge triggered DFF. This architecture has shown a

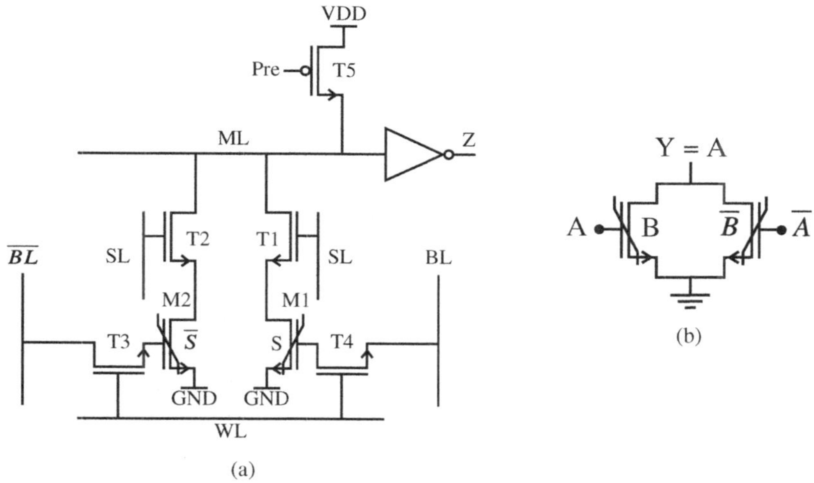

Fig. 12 (**a**) 4T2F FeFET-based TCAM. (**b**) 2F FeFET-based TCAM

reduced area and energy-delay overheads in the normal operation, low energy and low latency in backup and restore operations.

5.2.3 Ternary Content Addressable Memory (TCAM)

Content addressable memory (CAM), also known as Associative Memory, is a type of memory able to perform parallel search of a given data and return the associated information whenever a match occurred. CAMs are well suited for networking hardware in routers and database search [48]. They are also proposed as more energy-efficient, in memory data processing, by reducing the amount of data associated with traditional Von-Neumann processing. A FeFET-based Ternary CAM (TCAM), proposed in [49], consists of two parallel FeFETs that are connected to the matchline via two transistors. Figure 12 shows the circuitry of the TCAM in two ways: a 4 Transistor/2 FeFET (4T2F) and a 2 FeFET (2F). The two FeFETs can both store logic "0" in addition to complementary bits, which facilitates the "don't care" state.

To perform word-wise write operation, the WL is activated for the word to be written, and write voltages are applied to BL per the input data to switch the FE polarization within a FeFET. Negative V_{dd} is applied to WLs of the words that are not to be written to ensure that the gate-source voltages of the access transistors in these words remain less than 0 or at 0 during the write, and no unexpected write occurs to these words. Search-lines are driven to ground during the write to eliminate static current.

The array consists of the TCAM core, the input buffer drivers, the output sense amplifier, the clock signal and the output encoder. The TCAM core contains M words of N bits. The match and word lines are placed horizontally, while search and bit lines are placed vertically within the TCAM cell grid. The search and bit lines are driven by the input buffer, and at the end of each match line, a sense amplifier detects the voltage of the match line and outputs the indicator of match/mismatch to the encoder, which sends a "hit" signal and the corresponding address of the matched entry.

Recently, FeFET-based TCAMs were used to implement the so-called Ternary Content addressable and MEMory (TC-MEM). TC-MEM has two functional modes: TCAM mode, where the match of the data is used to retrieve the word, or the classical memory mode where words are accessed through their addresses. TC-MEM is described in [50]. The 1-bit TC-MEM cell is the 2F TCAM cell in Fig. 12b but with two modes. The first one corresponds to the normal memory mode. This is achieved with a single FeFET transistor. The FeFET, as shown in the Sect. 5.3.1, performs a NAND operation between the value stored and the evaluating input to the gate. The second mode corresponds to the TCAM mode. In the TC-MEM cell, those two modes are accessible thanks to a nMOS transistor between the two FeFETs connected in parallel.

5.3 *IMC Applications*

5.3.1 Reconfigurable Logic Gates

A FeFET-based Boolean gate is implemented by the operation between the polarization and the level of the gate. Described in [51], the concept of a single transistor NAND/NOR gates is based on the idea that the internal polarization state of a FeFET can be used as one input of the Boolean gate. One input (i.e. input A) is permanently saved during the first step. The high or the low V_t corresponds to logic zero or one, respectively. The second input is represented by the gate voltage V_g (i.e. input B) applied during the second step. The reconfigurability of this gate resides in the I_d-V_g curve of the FeFET that can be shifted as shown in Fig. 13. The shift depends on the applied gate input voltage through a source voltage V_s or a suitable back bias V_{bb} at the substrate. This shift is exploited to make the logic FeFET gate reconfigurable. That way, both of NAND and NOR logic can be realized with one device.

5.3.2 FeFET-Based Look-Up Table

As non-volatile storage device, the FeFET was used to implement LUT as proposed in [52]. In this architecture, each row is composed of the n FeFETs one per bit. Their source terminals are connected to a common source line (SL). The gate terminals of the FeFETs in each column are connected to a common Write/Read line (WRL). The

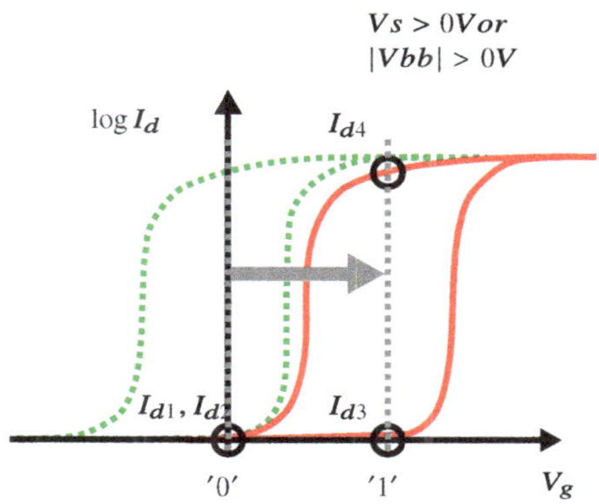

Fig. 13 Schematic of a reconfigurable FeFET-based logic gate

evaluation was performed by simulating the circuit with a 45 nm FeFET predictive model from [53]. According to [52], employing the 1T storage reduces the transistor count by $\sim 63\%$.

5.3.3 Convolution Neural Network

Convolution neural network (CNN) is a feed-forward artificial neural network where the connection between neurons is inspired by the visual cortex of animals. It is exploited in several applications such as image classification, object detection and natural language processing. Even if CNNs showed high accuracy of results, it still needs complex training model and large amount of training data, which in turn require large amounts of memory and energy. Furthermore, using single-precision weight values makes a big neural network, and convolution needs to be calculated with multiplications. To significantly gain space and time, the weight values of CNN can be represented by binary values (BCNN). It makes the neural network significantly smaller, and the convolution is estimated by only addition and subtraction, increasing the processing speed.

FeFET-based architectures can be used to run this application by improving power efficiency of BCNN inference. That was proposed in [54] in crossbar structure using FeFET-based XNOR gate. In the crossbar structure, each cell is a FeFET-based XNOR gate, and it is connected to a horizontal line, its inverse, a vertical line, a word line, a bit line and its inverse. The XNOR cell design, presented in Fig. 14, consists of two FeFETs and two access transistors. The two FeFETs store one weight bit. One bit of the inputs and its complementary bit are applied to the horizontal line and its inverse. The cell performs an XNOR operation between the input bit and the weight bit stored in the two FeFETs. The output bit can be read out from the vertical line by either voltage or current. In the context of crossbars, the current of the vertical lines was sensed and amplified to read the outputs. Two access transistors control read and write operations.

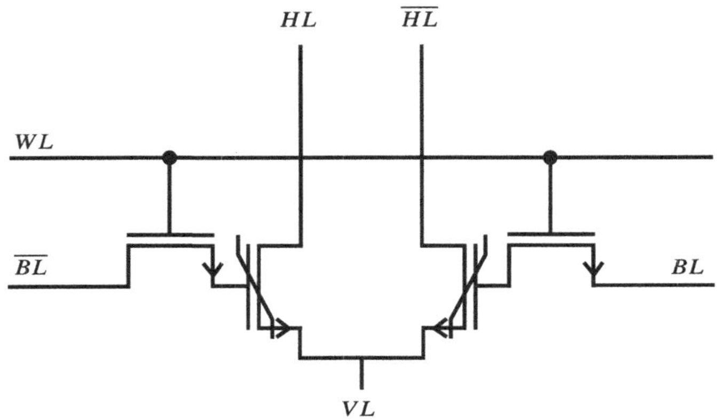

Fig. 14 Schematic of a FeFET-based BNN crossbar cell

Recently, a new cell was proposed for FeFET-based crossbar structure in [55]. It is intended for analogue Compute-in-Memory. This cell consists of 1F1R (1 FeFET, 1 resistance). With a sufficiently high resistance, the output current variability is strongly reduced. Plus, variability originating from the word line is suppressed due to the large operation window of drain to source voltage. By decreasing the current sensing threshold with a novel ADC, it is possible to drastically optimize the power and area efficiency while preserving the accuracy after digitalization.

5.3.4 FeFET-CiM

Reis et al. proposed in [56] a FeFET-based CiM architecture. Again, this architecture is comparable to the memristive device-based crossbar in Fig. 3b. It consists of a matrix of 2T+1FeFET memory cells as a basic memory block. The two transistors are used as the access transistors for write and read, and the FeFET is naturally the storage device. Then a customized sense amplifier, uniquely suited for FeFET-CiM, can be either voltage and/or current-based to perform CiM operation more efficiently. The SPICE model 45 nm predictive model from [53] is used to perform the evaluation.

6 Comparison and Discussion

Table 1 is a comparative table of the devices. The comparison focused on the electrical, writing and reading aspects. The first thing that is possible to conclude from this table is none of the presented devices is perfect. Even if CBRAM is the best in terms of read endurance, the program endurance is very low. This makes

Table 1 Comparative table of the emerging technologies reviewed in this chapter

Technologies	Switching speed (s)	Switching ratio	SET/RESET voltage (V)	Program endurance (cycles)	Read endurance	Retention (s)	Ref
CBRAM	10 n	10^4	$+4/-1.25$	150	10^9	6×10^5	[57]
OxRAM	5×10^{-3}	10^6	$\pm 0, 2$	10^4	10^7	10^4	[58]
STT-MRAM	2 ns [59]	/	$0, 05/0, 5$	$> 10^{12}$	/	$> 10^6$	[30]
PCM	100 n	10^6	$+2, 9/-1, 75$	$> 10^{12}$	/	10^6	[60]
FeFET	100 n	10^6	± 4	10^{10}	10^7	10^9	[61]

Table 2 Comparative table of different circuits seen with their associated emerging technologies

Circuit	Technologies	Type	Endurance	Energy cons.	Flexibility
Stateful NOR	RRAM	LiM	X	Programing	✓
MVM	STT-MRAM	CiM	✓	Read + ADC	X
Non-stateful	PCM	CiM	✓	Read + ADC	–
Reconfigurable logic	FeFET	CiM	X	Programming	✓

CBRAM fitting well for applications with a lot of memory access needed without (a lot) of programming as LUT in CwM application. In terms of retention time, FeFET is the device with the best ability to retain data.

It is also relevant to compare circuit to each other with the associated technologies. Table 2 is showing the difference of the each circuit with different aspects.

Another aspect of the discussion is the comparison with the "volatile" devices as SRAM or DRAM. The aim is to the potential of replacement for same application. In the case of STT-MRAM, a comparison with SRAM in STT-CiM and MAGIC-NOR [7] results in FeFET-CiM achieves $\sim 2\times$ the density of a memory array built on 6T-SRAM cells [31].

7 Conclusion

In this chapter, we presented a review of the emerging technologies used to implement memory-centric architectures. We firstly defined the different flavours of memory-centric applications, IMC and NMC. IMC could be divided in three branches, three ways to make computation inside the memory: CiM, CwM and LiM. NMC, or CnM, is separate of IMC by the fact memory unit is not associated to logic elements but outside the processing unit.

With memristive technologies (RRAM, STT-MRAM and PCM), the implementation of circuits and applications turns around three main structures: the stateful logic (MAGIC), the MVM and the non-stateful logic. These architectures leverage on the resistance (or sometimes said the conductance) of the device to "compute". In each cell, they are associated to an access transistor to control read and write operations. The main interest of those technologies is the high density of implementation that is very useful for MLs applications.

The FeFET is different from other technologies because of the use of the polarization of the FE material. Directly integrated to the stack of a transistor, the circuit could have common structures as memristive devices like crossbar. However, the advantage of FeFET in LiM paradigm is clear. In addition to being compatible with CMOS technologies, the implementation in TCAMs or DFF shows a high ability to build memory with finely integrated logic elements.

References

1. M. Hasan, State of IoT 2022: number of connected IoT devices growing 18% to 14.4 billion globally (2022)
2. W.A. Wulf, S.A. McKee, Hitting the memory wall: implications of the obvious. ACM SIGARCH Comput. Archit. News **23**(1), 20–24 (1995)
3. H.A.D. Nguyen, J. Yu, M.A. Lebdeh, M. Taouil, S. Hamdioui, F. Catthoor, A classification of memory-centric computing. ACM J. Emerg. Technol. Comput. Syst. (JETC) **16**(2), 1–26 (2020)
4. G. Santoro, G. Turvani, M. Graziano, New logic-in-memory paradigms: an architectural and technological perspective. Micromachines **10**(6), 368–392 (2019)
5. M. Kooli, A. Heraud, H.-P. Charles, B. Giraud, R. Gauchi, M. Ezzadeen, K. Mambu, V. Egloff, J.-P. Noel, Towards a truly integrated vector processing unit for memory-bound applications based on a cost-competitive computational SRAM design solution. ACM J. Emerg. Technol. Comput. Syst. (JETC) **18**(2), 1–26 (2022)
6. M. Irfan, A.I. Sanka, Z. Ullah, R.C. Cheung, Reconfigurable content-addressable memory (CAM) on FPGAs: a tutorial and survey. Futur. Gener. Comput. Syst. **128**, 451–465 (2022)
7. N. Talati, S. Gupta, P. Mane, S. Kvatinsky, Logic design within memristive memories using memristor-aided logic (magic). IEEE Trans. Nanotechnol. **15**(4), 635–650 (2016)
8. C.-J. Jhang, C.-X. Xue, J.-M. Hung, F.-C. Chang, M.-F. Chang, Challenges and trends of SRAM-based computing-in-memory for ai edge devices. IEEE Trans. Circuits Syst. I Regul. Pap. **68**(5), 1773–1786 (2021)
9. T. Yoo, H. Kim, Q. Chen, T.T.-H. Kim, B. Kim, A logic compatible 4t dual embedded dram array for in-memory computation of deep neural networks, in *2019 IEEE/ACM International Symposium on Low Power Electronics and Design (ISLPED)* (IEEE, 2019), pp. 1–6
10. D. Ielmini, H.-S.P. Wong, In-memory computing with resistive switching devices. Nat. Electron. **1**(6), 333–343 (2018)
11. W. Kang, L. Zhang, J. Klein, Y. Zhang, D. Ravelosona, W. Zhao, Reconfigurable codesign of STT-MRAM under process variations in deeply scaled technology. IEEE Trans. Electron Devices **62**(6), 1769–1777 (2015)
12. C.D. Wright, P. Hosseini, J.A.V. Diosdado, Beyond von-Neumann computing with nanoscale phase-change memory devices. Adv. Funct. Mater. **23**(18), 2248–2245 (2013)
13. I. O'Connor, M. Cantan, C. Marchand, S.S.B. Vilquin, E.T. Breyer, H. Mulaosmanovic, T. Mikolajick, B. Giraud, J. Noël, A. Ionescu, I. Stolichnov, Prospects for energy-efficient edge computing with integrated HfO2-based ferroelectric devices, in *2018 IFIP/IEEE International Conference on Very Large Scale Integration (VLSI-SoC)* (2018), pp. 180–183
14. N. Talati, R. Ben-Hur, N. Wald, A. Haj-Ali, J. Reuben, S. Kvatinsky, MMPU—a real processing-in-memory architecture to combat the von Neumann bottleneck. Appl. Emerg. Memory Technol. **63**, 191–213 (2020)

15. S. López-Soriano, J. Methapettyparambu Purushothama, A. Vena, CBRAM technology: transition from a memory cell to a programmable and non-volatile impedance for new radio frequency applications. Sci. Rep. **12**, 507–519 (2022)
16. S. Kvatinsky, D. Belousov, S. Liman, G. Satat, N. Wald, E.G. Friedman, A. Kolodny, U.C. Weiser, MAGIC—memristor-aided logic. IEEE Trans. Circuits Syst. II Express Briefs **61**(11), 895–899 (2014)
17. Y. Chen, Reram: History, status, and future. IEEE Transactions on Electron Devices **67**(4), 1420–1433 (2020)
18. B. Alpern, L. Carter, E. Feig, T. Selker, The uniform memory hierarchy model of computation. Algorithmica **12**, 72–109 (1994)
19. R.F. Freitas, W.W. Wilcke, Storage-class memory: the next storage system technology. IBM J. Res. Dev. **52**(4.5), 439–447 (2008)
20. H. Abbas, J. Li, D.S. Ang, Conductive bridge random access memory (CBRAM): Challenges and opportunities for memory and neuromorphic computing applications. Micromachines **13**(5), 725–753 (2022)
21. A. Sebastian, M. Le Gallo, R. Khaddam-Aljameh, E. Eleftheriou, Memory devices and applications for in-memory computing. Nature Nanotechnol. **15**(7), 529–544 (2020)
22. J. Borghetti, G.S. Snider, P.J. Kuekes, J.J. Yang, D.R. Stewart, R.S. Williams, 'memristive' switches enable 'stateful' logic operations via material implication. Nature **464**(7290), 873–876 (2010)
23. Y.S. Kim, M.W. Son, K.M. Kim, Memristive stateful logic for edge boolean computers. Adv. Intell. Syst. **3**(7), 2000278 (2021)
24. L. Zhang, S. Cosemans, D.J. Wouters, B. Govoreanu, G. Groeseneken, M. Jurczak, Analysis of vertical cross-point resistive memory (VRRAM) for 3D RRAM design, in *2013 5th IEEE International Memory Workshop* (2013), pp. 155–158
25. H.S. Yoon, I.-G. Baek, J. Zhao, H. Sim, M.Y. Park, H. Lee, G.-H. Oh, J.C. Shin, I.-S. Yeo, U.-I. Chung, Vertical cross-point resistance change memory for ultra-high density non-volatile memory applications, in *2009 Symposium on VLSI Technology* (2009), pp. 26–27
26. S. Barraud, M. Ezzadeen, D. Bosch, T. Dubreuil, N. Castellani, V. Meli, J. Hartmann, M. Mouhdach, B. Previtali, B. Giraud, J.P. Noël, G. Molas, J. Portal, E. Nowak, F. Andrieu, 3D RRAMS with gate-all-around stacked nanosheet transistors for in-memory-computing, in *2020 IEEE International Electron Devices Meeting (IEDM)*, pp. 29.5.1–29.5.4 (2020)
27. L. Xie, H.A. Du Nguyen, J. Yu, A. Kaichouhi, M. Taouil, M. AlFailakawi, S. Hamdioui, Scouting logic: a novel memristor-based logic design for resistive computing, in *2017 IEEE Computer Society Annual Symposium on VLSI (ISVLSI)* (IEEE, 2017), pp. 176–181
28. S. Yu, Z. Li, P.-Y. Chen, H. Wu, B. Gao, D. Wang, W. Wu, H. Qian, Binary neural network with 16 MB RRAM macro chip for classification and online training, in *2016 IEEE International Electron Devices Meeting (IEDM)* (2016), pp. 16.2.1–16.2.4
29. J.-G.J. Zhu, C. Park, Magnetic tunnel junctions. Mater. Today **9**(11), 36–45 (2006)
30. D.C. Worledge, Spin-transfer-torque MRAM: the next revolution in memory, in *2022 IEEE International Memory Workshop (IMW)* (2022), pp. 1–4
31. S. Jain, A. Ranjan, K. Roy, A. Raghunathan, Computing in memory with spin-transfer torque magnetic ram. IEEE Trans. Very Large Scale Integr. VLSI Syst. **26**(3), 470–483 (2018)
32. F. Parveen, Z. He, S. Angizi, D. Fan, Hielm: Highly flexible in-memory computing using STT MRAM, in *2018 23rd Asia and South Pacific Design Automation Conference (ASP-DAC)* (IEEE, 2018), pp. 361–366
33. Z. He, S. Angizi, D. Fan, Exploring stt-mram based in-memory computing paradigm with application of image edge extraction, in *2017 IEEE International Conference on Computer Design (ICCD)* (2017), pp. 439–446
34. T.-N. Pham, Q.-K. Trinh, I.-J. Chang, M. Alioto, STT-BNN: a novel STT-MRAM in-memory computing macro for binary neural networks. IEEE J. Emerging Sel. Top. Circuits Syst. **12**(2), 569–579 (2022)
35. Y.-C. Chiu, C.-S. Yang, S.-H. Teng, H.-Y. Huang, F.-C. Chang, Y. Wu, Y.-A. Chien, F.-L. Hsieh, C.-Y. Li, G.-Y. Lin et al., A 22 nm 4 mb STT-MRAM data-encrypted near-memory computation

macro with a 192GB/s read-and-decryption bandwidth and 25.1–55.1 TOPS/W 8b MAC for ai operations, in *2022 IEEE International Solid-State Circuits Conference (ISSCC)*, vol. 65 (IEEE, 2022), pp. 178–180
36. W. Wu, X. Zhu, S. Kang, K. Yuen, R. Gilmore, Probabilistically programmed STT-MRAM. IEEE J. Emerging Sel. Top. Circuits Syst. **2**(1), 42–51 (2012)
37. A.F. Vincent, J. Larroque, N. Locatelli, N. Ben Romdhane, O. Bichler, C. Gamrat, W.S. Zhao, J.-O. Klein, S. Galdin-Retailleau, D. Querlioz, Spin-transfer torque magnetic memory as a stochastic memristive synapse for neuromorphic systems. IEEE Trans. Biomed. Circuits Syst. **9**(2), 166–174 (2015)
38. I. Boybat, M. Le Gallo, S. Nandakumar, T. Moraitis, T. Parnell, T. Tuma, B. Rajendran, Y. Leblebici, A. Sebastian, E. Eleftheriou, Neuromorphic computing with multi-memristive synapses. Nat. Commun. **9**(1), 2514 (2018)
39. S.R. Nandakumar, I. Boybat, V. Joshi, C. Piveteau, M. Le Gallo, B. Rajendran, A. Sebastian, E. Eleftheriou, Phase-change memory models for deep learning training and inference, in *2019 26th IEEE International Conference on Electronics, Circuits and Systems (ICECS)* (2019), pp. 727–730
40. S. Ambrogio, P. Narayanan, H. Tsai, R.M. Shelby, I. Boybat, C. Di Nolfo, S. Sidler, M. Giordano, M. Bodini, N.C. Farinha et al., Equivalent-accuracy accelerated neural-network training using analogue memory. Nature **558**(7708), 60–67 (2018)
41. A. Lydia, S. Francis, Adagrad—an optimizer for stochastic gradient descent. Int. J. Inf. Comput. Sci **6**(5), 566–568 (2019)
42. J. Evans, R. Womack, An experimental 512-bit nonvolatile memory with ferroelectric storage cell. IEEE J. Solid State Circuits **23**(5), 1171–1175 (1988)
43. S. Slesazeck, T. Ravsher, V. Havel, E.T. Breyer, H. Mulaosmanovic, T. Mikolajick, A 2TnC ferroelectric memory gain cell suitable for compute-in-memory and neuromorphic application, in *2019 IEEE International Electron Devices Meeting (IEDM)* (IEEE, 2019), pp. 38–6
44. X. Yin, A. Aziz, J. Nahas, S. Datta, S. Gupta, M. Niemier, X.S. Hu, Exploiting ferroelectric fets for low-power non-volatile logic-in-memory circuits, in *2016 IEEE/ACM International Conference on Computer-Aided Design (ICCAD)*, pp. 1–8 (2016)
45. S. George, K. Ma, A. Aziz, X. Li, A. Khan, S. Salahuddin, M.-F. Chang, S. Datta, J. Sampson, S. Gupta, V. Narayanan, Nonvolatile memory design based on ferroelectric FETs (2016)
46. D. Danni Wang, S. George, A. Aziz, S. Datta, V. Narayanan, S.K. Gupta, Ferroelectric transistor based non-volatile flip-flop (2016), pp. 10–15
47. X. Li, J. Sampson, A. Khan, K. Ma, S. George, A. Aziz, S.K. Gupta, S. Salahuddin, M.-F. Chang, S. Datta, V. Narayanan, Enabling energy-efficient nonvolatile computing with negative capacitance FET. IEEE Trans. Electron Devices **64**(8), 3452–3458 (2017)
48. R. Karam, R. Puri, S. Ghosh, S. Bhunia, Emerging trends in design and applications of memory-based computing and content-addressable memories. Proc. IEEE **103**(8), 1311–1330 (2015)
49. X. Yin, M. Niemier, X.S. Hu, Design and benchmarking of ferroelectric FET based TCAM, in *Design, Automation & Test in Europe Conference & Exhibition (DATE), 2017* (2017), pp. 1444–1449
50. C. Marchand, I. O'Connor, M. Cantan, E.T. Breyer, S. Slesazeck, T. Mikolajick, A FEFET-based hybrid memory accessible by content and by address. IEEE J. Explor. Solid-State Comput. Devices Circuits **8**(1), 19–26 (2022)
51. E.T. Breyer, H. Mulaosmanovic, T. Mikolajick, S. Slesazeck, Reconfigurable NAND/NOR logic gates in 28 nm HKMG and 22 nm FD-SOI FEFET technology, in *2017 IEEE International Electron Devices Meeting (IEDM)* (2017), pp. 28.5.1–28.5.4
52. X. Chen, M. Niemier, X.S. Hu, Nonvolatile lookup table design based on ferroelectric field-effect transistors, in *2018 IEEE International Symposium on Circuits and Systems (ISCAS)* (2018), pp. 1–5
53. A. Aziz, S. Ghosh, S. Datta, S.K. Gupta, Physics-based circuit-compatible spice model for ferroelectric transistors. IEEE Electron Device Lett. **37**(6), 805–808 (2016)

54. X. Chen, X. Yin, M. Niemier, X.S. Hu, Design and optimization of FEFET-based crossbars for binary convolution neural networks, in *2018 Design, Automation & Test in Europe Conference & Exhibition (DATE)* (2018), pp. 1205–1210
55. T. Soliman, F. Müller, T. Kirchner, T. Hoffmann, H. Ganem, E. Karimov, T. Ali, M. Lederer, C. Sudarshan, T. Kämpfe, A. Guntoro, N. Wehn, Ultra-low power flexible precision fefet based analog in-memory computing, in *2020 IEEE International Electron Devices Meeting (IEDM)* (2020), pp. 29.2.1–29.2.4
56. D. Reis, M. Niemier, S.X. Hu, Computing in memory with FEFETs. *Proceedings of the International Symposium on Low Power Electronics and Design*, pp. 1–6 (2018)
57. A. Belmonte, U. Celano, A. Redolfi, A. Fantini, R. Muller, W. Vandervorst, M. Houssa, M. Jurczak, L. Goux, Analysis of the excellent memory disturb characteristics of a hourglass-shaped filament in Al2O3/Cu-based CBRAM devices. IEEE Trans. Electron Devices **62**(6), 2007–2013 (2015)
58. J.M. Lopez, L. Hudeley, L. Grenouillet, D.A. Robayo, J. Sandrini, G. Navarro, M. Bernard, C. Carabasse, D. Deleruyelle, N. Castellani et al., Elucidating 1s1r operation to reduce the read voltage margin variability by stack and programming conditions optimization, in *2021 IEEE International Reliability Physics Symposium (IRPS)* (IEEE, 2021), pp. 1–6
59. G. Hu, J. Nowak, M. Gottwald, S. Brown, B. Doris, C. D'Emic, P. Hashemi, D. Houssameddine, Q. He, D. Kim, et al., Spin-transfer torque mram with reliable 2 ns writing for last level cache applications, in *2019 IEEE International Electron Devices Meeting (IEDM)* (IEEE, 2019), pp. 2–6
60. G.W. Burr, M.J. Breitwisch, M. Franceschini, D. Garetto, K. Gopalakrishnan, B. Jackson, B. Kurdi, C. Lam, L.A. Lastras, A. Padilla et al., Phase change memory technology. J. Vac. Sci. Technol. B: Nanotechnol. Microelectron.: Mater. Process. Meas. Phenom. **28**(2), 223–262 (2010)
61. K. Chatterjee, S. Kim, G. Karbasian, A.J. Tan, A.K. Yadav, A.I. Khan, C. Hu, S. Salahuddin, Self-aligned, gate last, FDSOI, ferroelectric gate memory device with 5.5-nm hf0.8zr0.2o2, high endurance and breakdown recovery. IEEE Electron Device Lett. **38**(10), 1379–1382 (2017)

An Overview of Computation-in-Memory (CIM) Architectures

Anteneh Gebregiorgis, Hoang Anh Du Nguyen, Mottaqiallah Taouil, Rajendra Bishnoi, Francky Catthoor, and Said Hamdioui

1 Introduction

For several decades, CMOS down-scaling and architecture improvements have doubled computer performance following Moore law [1]. However, CMOS technology is facing today three main walls, leakage wall, reliability wall, and cost wall [2], while computer architectures also face three walls, memory wall, power wall, and instruction level parallelism (ILP) wall [3]. In order to address these walls and improve the performance, several novel technologies and architectures are being explored actively [4, 5]. As a result, an enormous amount of architectures have been recently proposed.

Since the first Von-Neumann architecture in the early 1950s, computer architectures have been evolved to various complex organizations including pipelined, superscalar, multicore, etc. [6, 7]. Since the energy and performance costs of moving data between the memory subsystem and the CPU have dominated the total costs of computation, architects, and designers are forced to find breakthroughs in computer architecture [8–11], Computation-In-Memory (CIM) paradigms have evolved as a promising solution to circumvent the aforementioned challenges [12]; CIM is a concept of integrating computation and storage of data within the same physical location.The idea of bringing computation closer to the storage location was conceived in 1970 by the so-called Logic-in-Memory computer, which outlines how logic enhanced cache can serve as a high-speed buffer between the CPU

A. Gebregiorgis (✉) · H. A. D. Nguyen · M. Taouil · R. Bishnoi · S. Hamdioui
Delft University of Technology, Delft, The Netherlands
e-mail: A.B.Gebregiorgis@tudelft.nl; H.A.DuNguyen@tudelft.nl; M.Taoui@tudelft.nl; R.K.Bishnoi@tudelft.nl; S.Hamdioui@tudelft.nl

F. Catthoor
IMEC, Leuven, Belgium
e-mail: Francky.Catthoor@imec.be

W. Liu et al. (eds.), *Design and Applications of Emerging Computer Systems*,
https://doi.org/10.1007/978-3-031-42478-6_2

and the main memory [13]. However, the CMOS scaling-driven performance enhancement outshone the hypothesis [14]. Since the emergence of big data and embedded applications, the demand for new computing systems with not only higher performance but also energy efficient increased significantly. In order to fulfill these requirements, several architectures such as multicore processor [7], Graphic Processing Unit (GPU) [15], were explored.

From the year 2008, emerging non-volatile memory technologies (e.g., memristor) have revived the concept of Processing-In-Memory under the new name of CIM [12, 16–19]. The novel CIM architectures utilize new memory technologies such as Spin-Torque Transfer Magnetic Random Access Memory (STT-MRAM), Phase-Change Memory (PCM), and Resistive Random Access Memory (RRAM); they serve as the building blocks and promise a lot of potential in terms of area, performance, and energy efficiency improvements [20–24]. As a result, different CIM-based accelerators have been developed to target different domains, applications, and kernels such as neuromorphic computing [25]. Although these works have demonstrated (at the small scale) the potential of CIM in realizing energy-efficient computing, achieving the maximum attainable potential of CIM is still an open question. For instance, using an appropriate holistic approach to design CIM accelerator for a dedicated application could not only significantly improve the overall energy efficiency but also result in cost-effective and robust hardware implementation [26]. In order to assess the developments in CIM architectures and determine the innovation potentials, it is essential to classify and evaluate the existing CIM architectures. Moreover, selecting the best CIM microarchitecture and applying proper design methodology are essential to harness the full potential of CIM.

The remainder of the chapter is organized as follows. Section 2 gives the metrics used for the classification and an overview of different classes. Section 3 presents a summary of architectures in the CIM-Array (CIM-A) class. Similarly, Sect. 4 presents architectures in CIM-Periphery (CIM-P) class followed by the discussion of the holistic design flow of CIM accelerator architectures in Sect. 5.

2 Classification of Computer Architectures

Evaluation and classification of modern computing systems is a complex process, as several metrics can be used [27]. Among the various metrics, performance, computing power, and resource utilization have been widely used in the literature [28]. However, classification of computing systems based on computation location and computing resource technology, such as memory technology, is another metric that could provide insight information on different classes. This section presents three main metrics used to classify the computer architectures while focusing mainly on CIM architectures [29]. This metrics are as follows:

1. Computation position/location: This metric is used to classify computer architectures based on the location where the computation is performed.
2. Memory technology: This criterion is utilized to classify the architectures based on the memory technology they use.
3. Computation parallelism: This criterion is used to classify the architectures based on the type and level of parallelism they support.

The remainder of this section highlights the main classes of CIM architectures based on the above mentioned three classification metrics.

2.1 Classification Based on Computation Position

Computation position defines where the result of the computation is produced. A computation includes a primitive logic function (e.g., logical operations) or arithmetic operation (e.g., addition, multiplication). Figure 1 shows the classification of computer architectures based on the computation position. The figures use number labels (1, 2, 3, and 4) to indicate the computation positions of the computer architecture classes. If the computation position is within the memory core (labeled as 1 in Fig. 1), the computer architecture is referred to as Computation-In-Memory Array (CIM-A); if the computation position is within the periphery (labeled as 2 in Fig. 1), the architecture is referred to as Computation-In-Memory Periphery (CIM-P); otherwise, the computation position is outside the memory core (labels 3 and 4 in Fig. 1), and the architecture is referred to as Computation-Out-of-Memory (COM).

- **CIM-Array (CIM-A):** In CIM-A, the computation result is produced within the memory array (noted as position 1 in Fig. 1). Note that this is different from a standard write operation. Typical examples of CIM-A architectures use memris-

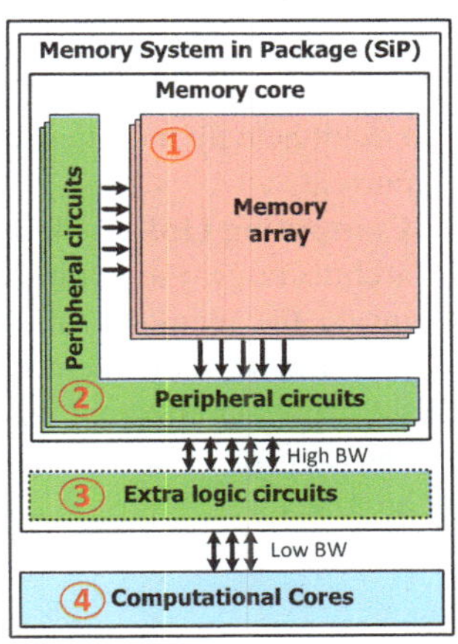

Fig. 1 Computation location of CIM architectures

tive logic designs such as MAGIC [30]. CIM-A architectures mostly require a modification at the cell level to support such logic design, as the conventional memory cell dimensions and their embedding in the bit- and wordline structure do not allow them to be used for logic. In addition, modifications in the periphery are sometimes needed to support the changes in the cell [31]. Therefore, CIM-A architectures can be further subdivided into two groups: (1) Basic CIM-A, where only changes inside the memory array are required. An example of basic CIM-A is an architecture that performs computations using implication logic [32, 33]. (2) Hybrid CIM-A, where, in addition to major changes in the memory array, minimal to medium changes are required in the peripheral circuit. An example of hybrid CIM-A is an architecture that performs computations using MAGIC [30]. In this case, multiple memory rows are written simultaneously; due to the high write currents, modifications are required to the cell, and medium changes in the peripheral circuits are needed to activate the multiple rows.

- **CIM-Periphery (CIM-P):** In CIM Periphery (CIM-P) architectures, the computation result is produced within the peripheral circuitry (noted as position 2 in Fig. 1). Typical examples of CIM-P architectures contain logical operations and vector-matrix multiplications [18, 34]. CIM-P architectures typically contain dedicated peripheral circuits such as DACs and/or ADCs [35–37] and customized sense amplifiers [34]. Even though the computational results are produced in the peripheral circuits for CIM-P, the memory array is a substantial component in the computations. As the peripheral circuits are modified, the currents/voltages applied to the memory array are typically different than in the conventional memory. Hence, similar to the CIM-A sub-class, the CIM-P architectures are also further divided into two groups: (1) Basic CIM-P, where only change in the peripheral is required, which means the current levels should not be affected. An example of basic CIM-P is Pinatubo logic [34]. (2) Hybrid CIM-P, where the majority of the changes take place in the peripheral circuit and minimal to medium changes in the memory array. An example of hybrid CIM-P is ISAAC [35]. ISAAC activates all rows of a memory array at the same time during read operations to perform a matrix vector multiplication using an ADC readout circuit. This architecture accumulates currents in the bitline that impose higher electrical loading in the memory array; hence, not only is the periphery circuit heavily modified, but also the cell requires changes due to the high bitline current.
- **Computing Out-of-Memory (COM):** In Computation Out-of-Memory (COM) architectures, the computation is performed in the extra logic circuit available inside the memory SiP (noted as position 3 in Fig. 1). If the computation is performed by off-memory computational cores (noted as position 4 in Fig. 1), then the architecture is similar to the classical Von-Neumann architecture, and hence, it is not discussed in this paper as the paper focuses on CIM architectures.

2.2 Classification Based on Memory Technology

Memory technologies can be classified as charge-based memories and non-charge-based memories. In charge-based memories such as Dynamic Random Access Memory (DRAM), Static Random Access Memory (SRAM), and Flash, information is stored through the presence of charge [38], whereas the non-charge-based memories include different types of storage elements distinguished by their physical mechanism; these include resistive [39, 40], "magnetic" memories [41, 42]. Other types of memories like molecular memories [43] and mechanical memories [44]) are not discussed in this classification as they are not currently used for CIM.

2.2.1 Charge-Based Memories

The SRAM and DRAM memories are largely adopted by the semiconductor industries. Both of these memories are volatile in nature, which means they require power supply to maintain their state. A six-transistor bit-cell design is commonly used in SRAM, whereas DRAM bit-cell comprises of a capacitor and a transistor. Although SRAM has faster accesses, its bit-cell size is much larger and consumes more leakage than DRAM. Despite the fact that DRAM has a significant advantage in terms of density, it requires periodic refresh in order to retain its data. Due to their volatility, both of these memories are facing serious power dissipation problems. On the contrary, Flash is a non-volatile memory that uses a floating gate transistor which has a charge trapping mechanism. Since Flash uses only a single transistor, its density is significantly higher than that of DRAM. However, it requires high voltage and considerably long duration to write a value. Moreover, it has limited endurance due to gate oxide degradation under strong electric field, meaning it can be only employed for applications where a few write operations are required.

2.2.2 Non-charge-Based Memories

RRAM, MRAM, and PCAM store information in the form of resistance states; these devices are thus also termed as memristors. These devices can be programmed in a high resistance or a low resistance state using reset or set electrical pulses. The RRAM cell consists of a top-electrode and a bottom-electrode that is separated by an oxide layer. Based on the formation/disruption of a Conductive Filament (CF), the resistive switching of RRAM devices takes place. The size of the CF determines the resistance state of the device. When a suitable positive voltage is applied, the breakage of ionic bonds increases the size of the CF, leading to a low resistance state of the device. On the other hand, when a suitable negative voltage is applied, some ions move back into the oxide region, thus reducing the size of the CF, resulting in a high resistance state. RRAMs are capable to perform multi-level bit storage.

Table 1 Design metrics for various memory technologies [45, 46]

Comparison metric	SRAM (6T)	DRAM (1T1C)	Flash (1T)	RRAM (1T1R)	MRAM (1T1R)	PCRAM (1T1R)
Size (F^2)	120-150	10-30	10-30	10-30	10-30	10-30
Volatility	Yes	Yes	No	No	No	No
Write energy	~fJ	~10fJ	~100pJ	~1pJ	~1pJ	~10pJ
Write speed	~1ns	~10ns	0.1-1ms	~10ns	~5ns	~10ns
Read speed	~1ns	~3ns	~100ns	~10ns	~5ns	~10ns
Endurance	10^{16}	10^{16}	10^4-10^6	10^7	10^{15}	10^{12}
Scalability	medium	medium	medium	high	high	high

In MRAM technologies, the value is stored in the Magnetic Tunnel Junction (MTJ) cell that consists of an oxide layer that is sandwiched between two ferromagnetic layers [47]. Out of these two ferromagnetic layers, one is reference layer in which the magnetization is always fixed. The other one is known as free layer, whose magnetic orientation can be freely rotated depending on the direction of current flowing through it. When the magnetic orientation of these two layers is parallel and anti-parallel to each other, the cell exhibits low and high resistance states, respectively. Reading a value from the MTJ cell applies the principle of Tunneling Magneto-Resistance effect.

PCRAM exploits the large resistance contrast between the amorphous and crystalline states of a chalcogenide glass. The change from a high resistive amorphous phase to less resistive crystalline phase is induced by heating the material above its crystallization temperature for a certain duration. The reverse switching can be realized by melting and quenching the material using a reset electrical pulse. Due to these switching mechanisms, PCRAM devices have the capability to achieve a number of intermediate distinct states, enabling the feature of multi-bit storage similar to RRAM.

The unique characteristics associated with each memory technologies are illustrated in Table 1. Besides to the regular memory operations, a range of CIM operations can also be performed using these memory technologies [12].

2.3 Classification Based on Computation Parallelism

Computation parallelism defines the level of parallelism that can be exploited in a computer system, i.e., task-, data-, and/or instruction-level parallelism. In task-level parallelism, a system has multiple independent control units and data memories; examples include multi-threading and multicore systems [7]. In data parallelism, a system with a single control unit is used to apply the same instruction concurrently on a collection of data elements, such as vector and array processor [48]. In instruction-level parallelism, a system with a single control unit is used to execute various instructions concurrently (e.g., VLIW processor) [49].

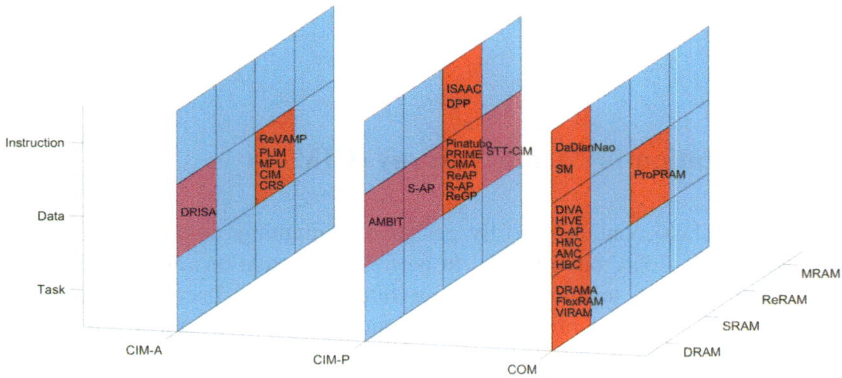

Fig. 2 Computer architecture and proposed classification

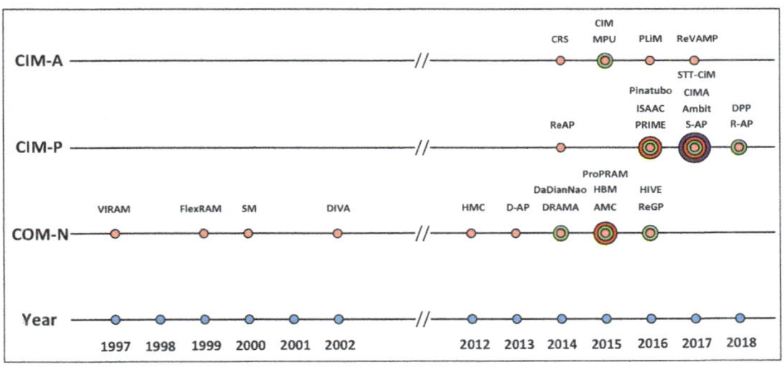

Fig. 3 Evolution timeline of CIM architectures

Based on the above discussed metrics, 36 classes can be differentiated by combining computation location with memory technology as shown in Fig. 2; among those classes, 11 classes are occupied by the existing architectures which are located in red and pink planes (see Fig. 2). The red plane demonstrates that a lot of work has been done for that particular class. The pink plane demonstrates a moderate number of work has been done. The cyan plane demonstrates either unexplored classes due to the lack of attention from the research community or non-existing due to current restrictions of memory technologies. The developments in CIM computing are shown in the timeline of Fig. 3; this shows the trend of computing moving from processor-centric to CIM architectures (CIM-A and CIM-P). In the figure, a larger circle indicates that more work has been proposed in that year. As it can be seen from the figure, the concept of merging computation and memory was introduced back in 1970.

In the following sections, we will discuss the existing architectures classified based on their computation position. Please note that each computation position-based class consists of architectures implemented using different memory technologies.

3 Computation-in-Memory-Array (CIM-A)

The CIM-A class contains mostly resistive computing architectures that use memristive-based logic circuits [50] to perform computations and resistive RAM (RRAM) as memory technology. Few architectures have been proposed in this category. Table 2 shows a brief comparison among the architectures which will be explained in each subsection.

On the one hand, all these architectures have several common advantages:

- Low memory access/bandwidth bottleneck due to computing inside the memory
- High data parallelism due to the possibility of performing concurrent operations inside the crossbars
- Low leakage due to the usage of non-volatile memory technology and small footprint when compared to conventional memory technologies but only in the case of very large memory array.

On the other hand, they all share several limitations:

- High computing latency per access due to the high latency of writing memristors and the need of multiple write steps to perform Boolean functions. Note that despite a high computing latency, the performance can be still high when sufficient parallelism is exploited.
- Higher endurance requirement due to the need of multiple write steps to perform Boolean functions.
- The cell designs are mostly modified in order to make the computing feasible.

The following subsections discuss the details and characteristics of each architecture. In this subsection DRAM-based architecture is presented first, and the remaining RRAM-based architectures are discussed in a chronological order of their date of publication. This ordering technique is also reflected in Table 2.

3.1 DRISA-3T1C: A DRAM-Based Reconfigurable In Situ Accelerator with 3 Transistors and 1 Capacitor (3T1C) Design

DRISA-3T1C was proposed in 2016 by S. Li et al. from University of California [51]. It is a DRAM-based architecture that exploits data parallelism by performing NOR gate inside DRAM cells. The architecture consists of a DRAM memory organized in a hierarchy of banks, subarrays, and mat; each level is

Table 2 Comparison among architectures of CIM-A classes

		Computations			Overheads						Evaluation	
	Hierarchy level	Logic style	Available functions	Memory technology	Periphery	Controller	Sneak path current	Destructive read	Require read-out*	Copy scheme	Simulator	App.
DRISA-3T1C [51]	Accelerator	DRAM	Boolean	DRAM	Modif.	Simple	No	No	No	Both	CACTI-3DD, in-house	CNN
CRS [52]	Accelerator	CRS	Logical, +	RRAM	Conv.	Complex	No	Yes	Yes	Indirect	Hspice	No
CIM [53]	Accelerator	Varied	Logical, +, x	RRAM	Conv.	Varied	Yes	No	Varied	Both	Analytical	Parallel adder and multiplier
PLiM [54]	Main memory	Majority	Majority gates	RRAM	Conv.	Complex	Yes	No	Yes	Both	Analytical	Encryption
MPU [55]	Main memory	MAGIC	Logical, +, x	RRAM	Conv.	Simple	Yes	No	No	Both	Analytical	Image processing
ReVAMP [56]	Main memory	Majority	Majority gates	RRAM	Conv.	Complex	Yes	No	Yes	Both	Analytical	EPFL bench-marks

+ n-bit addition, *Conv.* conventional, (∗) Required read-out during computations, *x* n-bit multiplication, *Modif.* Modified, *App.* Applications and benchmarks

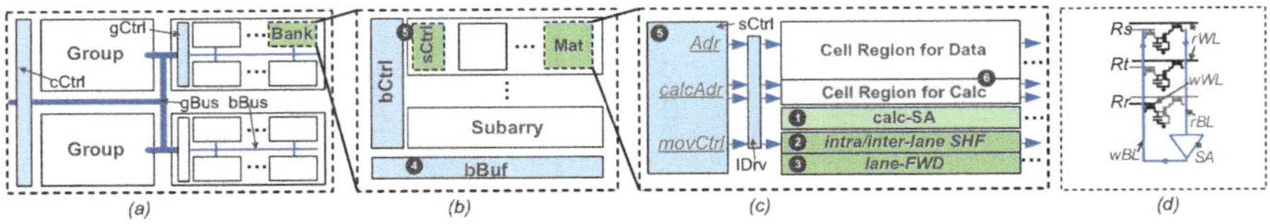

Fig. 4 A DRAM-based Reconfigurable In Situ Accelerator (DRISA) [51]. (**a**) Chip. (**b**) Bank. (**c**) Subarray and mat. (**d**) 3T1C cell structure

controlled by their corresponding controllers as shown in Fig. 4a, b, and c. The banks are connected through global bus (gbus), while communication among subarrays is carried out using bank buffers (bBuf). The mats perform both data storage and computations.

The memory mats consist of cell regions for both data and computations, and peripheral circuits include calc-SA (Sense Amplifier), intra/inter-lance shifter, and lane forwarding unit. The cell regions contain multiple DRAM cells which consist of three transistors connected to form a NOR gate and one capacitor to store the data value (see Fig. 4d). In order to perform computations, two DRAM cells (Rs and Rt) are to be activated simultaneously, and one DRAM cell (Rr) is to store the computation result (as shown in Fig. 4). Read voltages are applied to the sources DRAM cells (Rs and Rt) through the wordline (rWL), while write voltage is applied to the result DRAM cell (Rr). The voltage collected by the sense amplifier (SA) is used to control the transistor in the Rr DRAM cell as shown in Fig. 4d. Due to the NOR organization of these transistors, a NOR operation is realized and produces results in DRAM cells. The SA (also called calc-SA) cooperates with extra logic circuitry such as SHF and FWD to perform complex functions such as addition, copy, and inner product. DRISA-3T1C has the following advantages on top of the general advantages of CIM-A architectures:

- The latency of NOR primitive functions is fixed.
- The data transfer may include both direct and indirect schemes.
- The architecture does not suffer destructive read as in the case of CRS architecture [52]; hence the write energy might be less due to the absence of write-after-read.
- The controller is simpler than for the CRS architecture, as each operation consists of a fixed number of steps while fewer control voltage values are used.
- The architecture uses DRAM technology, which has several benefits such as high maturity and endurance, no sneak path currents, and the accessibility to optimized architectures, technology, and tools.

In spite of those advantages, DRISA-3T1C also has its own set of limitations which impact its effectiveness. The limitations of DRISA-3T1C include the following:

- The latency of complex functions varies depending on the functional complexity as each function needs to be converted into multiple NOR gates.

- The architecture uses DRAM technology which suffers from a low performance, high energy consumption, and large footprint and is difficult to scale down.

The architecture is simulated and evaluated against GPU using four CNN applications [57, 58].

3.2 CRS: Complementary Resistive Switch Architecture

Complementary Resistive Switch (CRS) architecture was proposed in 2014 by A. Siemon et al., from RWTH Aachen University [52]. It is a memristor-based architecture that exploits data-level parallelism using implication logic. The architecture consists of multiple crossbars and a control unit. The crossbar stores and performs logic operations using CRS cells; a CRS cell consists of two resistive switches or resistive RAMs. The control unit distributes signals to the intended addresses (wordlines and bitlines) to perform operations on the crossbars.

The crossbar is controlled by a sequence of operations including write-in (WI), read-out (RO), write-back (WB), and compute (CP). Before the operations can be performed, the crossbar part used for computation is entirely reset to a logic value 0. The WI operation writes a logic value into a memristor. The RO operation reads a logic value from a cell; the logic output value is determined by the sense amplifier. The RO operation is destructive and changes the value of the memristor to logic value 1. The task of the WB operation is to recover the destroyed value. Finally, the CP instructions are used to execute the implication logic gates [52]. The data transfer between CRS cells is carried out through the control unit using a RO and WB operations; in other words, the control unit reads a value of the source CRS cell and writes it into the destination cell. In addition to the general advantages of CIM-A architectures, CRS has the following advantages:

- It is less impacted by the sneak path currents due to the usage of CRS cells. The cell's resistance is always equivalent to high resistance; hence, sneak path currents are eliminated. However, variations in resistances will make such paths practically unavoidable unless a 1T2R cell is used.
- CRS logic requires fewer cells to perform computations than Fast Boolean Logic (FBL) that is require to express the sum-of-product format.
- It is possible to use sufficiently large crossbar arrays, means effective overhead due to control circuits can be minimized.

However, it also has the following limitations:

- The latency of the primitive functions varies and requires extra read-out instructions to determine the voltages that have to be applied.
- The RO operation is destructive; hence, a WB operation is required after each RO operation, which increases the latency and energy of computations.

- The data transfer method is indirect as it is based on the read-out and write-back scheme. As all cells have high resistance, direct copying of cells in the crossbar is not applicable.
- The control unit imposes a high overhead as it is responsible for both controlling the crossbar (requiring a large number of states) and transferring data (which involves the usage of buffers/registers to store temporary values).
- The area of CRS cells is larger than those based on a single memristor cell.
- The architecture requires additional compiling techniques and tools to convert conventional Boolean logic functions to implication logic.

This architecture was only evaluated at circuit level using adders. Therefore, it is hard to make general conclusions on the performance and the applicability of this architecture.

3.3 CIM: Computation-in-Memory

CIM was proposed in 2015 by H.A. Du Nguyen et al., from Delft University of Technology [53]. It is a memristor-based architecture that exploits data-level parallelism by using any memristor logic style; the authors have showed the potential of this architecture using Fast Boolean Logic (FBL) and implication logic. The architecture consists of a memristor crossbar and a control and communication block as shown in Fig. 5 [53]. The memristor crossbar stores data and performs computations. The control and communication block applies appropriate voltages to the memristor crossbar.

The architecture uses state machines stored in the *control and communication block* to compute and transfer data in the crossbar. Once triggered, the state machine applies an appropriate sequence of control voltages to the rows and columns of the memristor crossbar. Depending on the memristor types, the data transfer occurs directly inside the crossbar (for single RRAM cells) or indirectly through the *control and communication block* outside the crossbar (for CRS cells) using the CRS read-out/write-back scheme. In addition to the general advantages of CIM-A architectures, CIM comes with the following set of advantages:

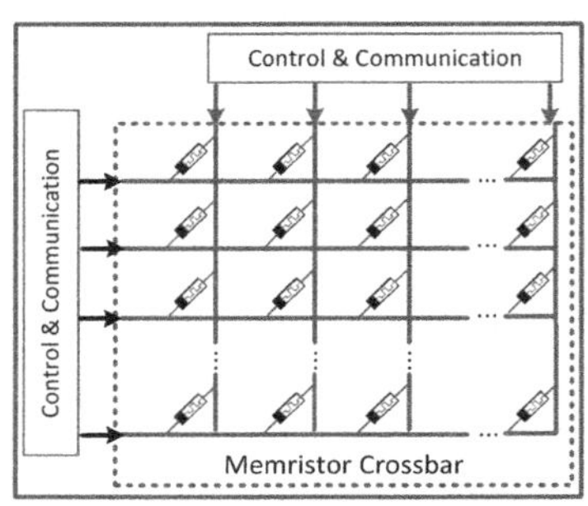

Fig. 5 The Computation-In-Memory (CIM) Architecture [53]

- The architecture can accommodate any type of memristor logic design due to the flexibility of the control and communication block.
- In the case FBL is used, the latency of primitive functions (i.e., addition, multiplication) is a constant number.
- The data transfer using both direct and indirect schemes has been intensively explored in [59].
- The control block of FBL is less complex than the control block of implication logic due to a fixed number of write steps and a simpler control voltage scheme.
- Compared to CRS architecture, CIM architecture has significant area and write energy advantage, as these values are much less at a single respective cell level.

However, it also has the following limitations:

- The architecture has to deal with sneak path currents in case a single RRAM cell (0T1R) is used as multiple rows and columns are activated simultaneously. Possible solutions to alleviate the problem consist of isolating each FBL circuit or the usage of a transistor-memristor (1T1R) structure to actively control each memristor using a transistor [60] or isolated/half select voltages [61].
- In the case FBL is used, typically a lot of cells are required due to LUT-based computing.
- In the case CRS cells are used, the same drawbacks of CRS architecture apply, i.e., in a larger cell area, the control unit imposes a high overhead as the controllers are responsible for both controlling the crossbar and transferring data, and it requires additional compiling techniques and tools to convert conventional Boolean logic functions to implication logic.

The potentials of the architecture are demonstrated using a case study of a binary-tree-based parallel adder and multiplier [53, 62]; the architecture is compared with a conventional multicore architecture. CIM architecture achieves at least one order of magnitude improvements in terms of delay, energy, and area.

3.4 PLiM: Programmable Logic-in-Memory Computer

Similar to the DRAM-based bit-serial addition using majority logic [54], PLiM was proposed in 2016 by P. Gaillardona et al., from EPFL [63]. It is a memristor-based architecture that exploits data parallelism using majority logic to perform elementary Boolean logic such as ***OR*** and ***AND*** operations within the memory array. The architecture consists of a resistive memory organized in banks and a Logic-in-Memory (LiM) controller block as shown in Fig. 6. The memory is a memristive crossbar that stores both instruction and data. The LiM controller is composed of a number of registers and a finite state machine (FSM). The controller functions as a simple processor; it fetches instructions from the memory array and decodes and executes the operation inside the memory.

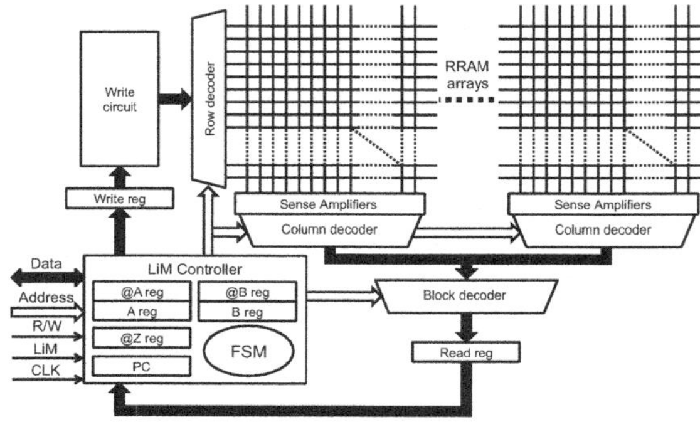

Fig. 6 The Programmable Logic-in-Memory Computer (PLiM) [63]

The LiM controller operates in two modes: conventional memory read/write mode and in-memory instruction mode. In the read/write mode, the FSM is deactivated, and the memory array is read or written in the same manner as a standard memory. In the in-memory instruction mode, FSM is activated, and an instruction is performed using majority logic gates inside the memory. Once the FSM is enabled, the following operations are performed. First, the FSM resets all registers in the LiM controller. Second, an instruction is read from the address in the program counter (PC) and decoded to obtain the addresses of the two operands and output; the addresses of the two operands are stored in registers @A and @B, while the output address is stored in register @Z. Third, the value of the two operands is read using the addresses in registers @A and @B; the obtained values are stored in registers @A and @B, respectively. Fourth, depending on the logic values (0 or 1) of the operands, appropriate voltages are applied to the crossbar at address @Z to perform a majority logic gate. Finally, the PC is incremented by one. On top of the general advantages of CIM-A architectures, PLiM has the following additional advantages:

- The data transfer may include both direct and indirect schemes.
- The write energy and area of a memristor cell are smaller as compared to a CRS cell.

Similarly, it also has its own limitations which are stated as follows:

- The latency of majority primitive functions varies depending on the functional complexity, and some read-outs are required to determine the voltage values to be applied.
- The architecture has to deal with sneak path currents. Possible solutions are mentioned in Sect. 3.3.
- The LiM controller is complex as it has to determine the control voltage values based on the operands' values.
- The architecture requires additional compiling techniques and tools to convert conventional Boolean logic functions to majority logic gates.

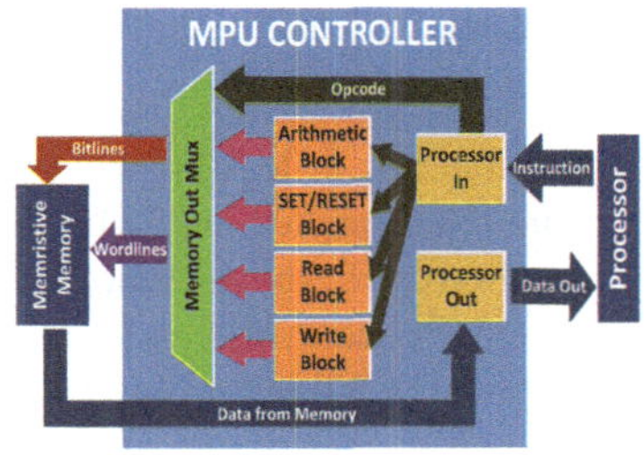

Fig. 7 The Memristor Memory Processing Unit (MPU) [55]

The architecture is evaluated with a PRESENT Block Cipher algorithm [64], which encrypts a 64-bit plain text with an 80- or 128-bit key. The algorithm is compiled into a sequence of majority logic gates and executed on PLiM. Unfortunately, the results show that PLiM's performance is almost a factor of two slower than a 180nm FPGA implementation [64].

3.5 MPU: Memristive Memory Processing Unit

MPU was proposed in 2016 by R. Ben Hur et al., from Technion-Israel Institute of Technology [55]. Similar to the DRAM-based architecture proposed in [65] and SRAM-based solutions in [66, 67], MPU is a memristor-based architecture that exploits data-level parallelism using Memristive-Aid loGIC (MAGIC) [30]. The architecture consists of a conventional processor, MPU controller, and a memristive memory as shown in Fig. 7. The processor contains a control unit, an arithmetic and logic unit, and a memory controller. The MPU controller includes a Processor-In/Out block to interface to the conventional CPU, control blocks to execute specific commands (arithmetic, set/reset, read, and write block), and an output multiplexer to apply appropriate voltages to specific rows/columns of the memristive memory. The conventional processor sends an instruction to the memristive memory using its own memory controller and the MPU controller. The memory controller of the processor recognizes the memristive memory instructions in a similar manner as conventional memory operations, while the MPU controller determines whether to treat the memristive memory as a storage element or a processing element. Based on that, the MPU controller applies read/write signals or a sequence of signals to perform logical or arithmetic operations.

The MPU controller uses the Processor-In unit to divert the instructions to specific blocks (such as arithmetic, read, and write blocks) responsible for the execution of those operations. Each block determines the appropriate voltages that have to be applied to the memristive memory. The set/reset, read, and write blocks have a latency of one cycle, while the arithmetic block requires multiple cycles to execute a vector operations using MAGIC logic [30]. Data movements in the crossbar are performed directly using *copy* (double negation) or single NOT operations. In addition to the general advantages of CIM-A architectures, MPU has the following advantages:

- The latency of MAGIC primitive functions is fixed.
- The data transfer may include direct (based on copying) and indirect (based on read-out/write-back) schemes.
- The MPU controller is simpler than for the CRS architecture, as each operation consists of a fixed number of steps while less controlling mechanisms are used.
- The write energy of a single MAGIC cell is smaller in comparison to those of a CRS cell.
- MAGIC requires in comparison to FBL fewer cells to perform computations.
- MPU can perform many MAGIC gate operations in parallel that makes it very efficient in terms of throughput.
- MPU is compatible with the conventional von-Neumann machines, means it can perform load/store/compute operations.

In spite of the abovementioned benefits, MPU also has its own set of limitations which include the following:

- The architecture has to deal with sneak path currents. Possible solutions are mentioned in Sect. 3.3.
- The control voltages used in MAGIC have to satisfy the constraint $2V_{reset} < V_w < V_{set}$, where V_{set} is the minimum voltage required to switch a memristor from R_H to R_L and V_{eset} is the minimum voltage required to switch a memristor from R_L to R_H; in other words, it requires that the memristors have a higher V_{set} than V_{reset}, leading to an unbalanced hysteresis loop. This limits the types of memristors that can be used for MAGIC.
- The architecture requires additional compiling techniques and tools to convert conventional Boolean logic functions to MAGIC gates.

The potential of the architecture is demonstrated by performing a logical bit-wise OR operation of two 8-bit vectors in 20 steps. The latest research has shown that MAGIC can be used for several arithmetic operations such as addition, multiplication, etc. [68]. In this context, there is an architecture named SIMPLE MAGIC that automatically generates a defined sequence of atomic memristor-aided logic NOR operations [69]. In the same line, a SIMPLER flow is developed that additionally optimizes the mapping executions that reduce complexity and improve throughput [70]. Note a lot of research on cell minimization and cell reuse has been done in the context of MAGIC and other basic cells.

3.6 ReVAMP: ReRAM-Based VLIW Architecture

ReVAMP was proposed in 2017 by D. Bhattacharjee et al. from Nanyang Technological University [56]. It is a memristor-based architecture that exploits data parallelism using majority logic. The architecture shown in Fig. 8 consists of an

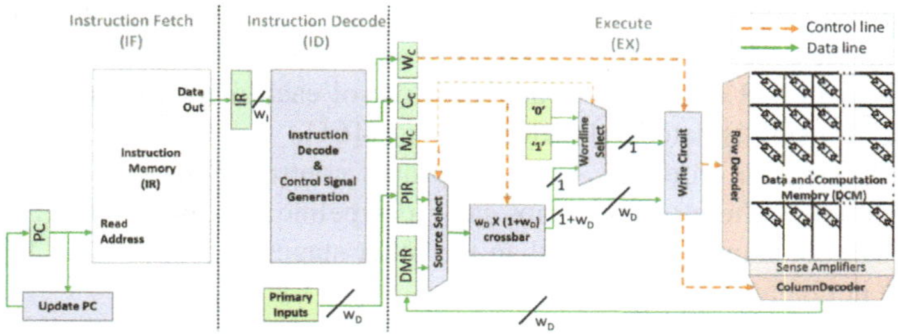

Fig. 8 ReRAM-based VLIW architecture (ReVAMP) [56]

Instruction Fetch (IF), Instruction Decode (ID), and Execute (EX) stage. The IF block fetches instructions from the Instruction Memory using the program counter (PC) as address and stores it in the Instruction Register (IR). The ID block decodes the instruction and generates control signals which are placed in the control registers of the EX block. The EX stage finally executes the instruction.

The IF and ID stages are similar to those of the traditional five-pipelined RISC architectures. The IF stage includes an Instruction Memory (IM) and a Program Counter (PC). The ID stage contains registers (IR and Primary Inputs) and an Instruction Decode and Control Signal Generation. The EX stage consists of several registers (i.e., Data Memory Register (DMR), Primary Input Register (PIR), Mux control (M_c) register, Control (C_c) register, Wordline (W_c) register), as well as a crossbar interconnect, wordline select multiplexer, data Source Select multiplexer, and a Write circuit to control the crossbar that stores data. Once an instruction is fetched and decoded in IF and ID, respectively, the control registers in EX stage are filled with suitable values. These values control the multiplexers that are responsible for applying the right control signals to the crossbar. Depending on the operation, primary inputs from PIR or data retrieved from the crossbar stored in DMR can be used for the next operation. The crossbar interconnect permutes the inputs and control signals (indicated by C_c) to generate the voltages that need to be applied to the memory crossbar. The Write circuit applies these voltages to the targeted wordline address (indicated by W_c). In addition to the general advantages of CIM-A architectures, ReVAMP has the following advantages:

- The data transfer may include direct (within the crossbar based on copying resistance values) and indirect (based on read-out/write-back) schemes.
- The crossbar is based on only one device per cell, resulting in a more compact architecture as compared with other architectures which make use of two devices per cell (i.e., Complementary Resistive Switch (CRS) [52]).

However, it also has the following limitations:

- The latency of majority primitive functions varies depending on the functional complexity; in addition, before any operations are applied to the cells, these cells first have to be read-out in order to determine the appropriate control voltages.

- The architecture has to deal with sneak path currents. Possible solutions to alleviate the problem consist of isolating each tile/crossbar or using a transistor-memristor (1T1R) structure to actively control each memristor using a transistor [60] or using isolated/half select voltages [61].
- The EX stage is complex as it integrates control signals for memory and computations. Therefore, it is not easy to pipeline this architecture, as the EX stage will consume more time than the other stages, i.e., the stages IF, ID, and EX are not balanced.
- The architecture requires additional compiling techniques and tools to convert conventional Boolean logic functions to majority logic gates.

The architecture is simulated and evaluated using EPFL benchmarks [71] and compared against PLiM [63], which is based on a resistive memory with the same logic style.

4 Computation-in-Memory-Peripherals (CIM-P)

The CIM-P class consists of architectures which perform computations during read-out operations (i.e., 2 or more wordlines are activated simultaneously) using special peripheral circuitry. Earlier CIM works explored the potential of shifting data-intensive computations to the memory system with/without reconfigurable logic including Active Pages [72, 73]. These works utilized conventional DRAM and magnetic bubble memories to realize CIM architecture. However, as there are less restrictions on the functionality of the cell, various memory technologies can be used in this category such as DRAM, SRAM, and non-volatile memory technologies. For instance, a medium number of architectures have been proposed in this category. Table 3 shows a brief comparison among the architectures which will be explained in each subsection.

On the one hand, these architectures have several common advantages:

- Low memory access/bandwidth bottleneck as the results are produced in the peripheral circuitry which is connected directly to the memory array.
- High parallelism due to the possibility of performing multiple concurrent operations.
- High performance as computations are performed in a single read step.
- Relatively simple controllers as the operations are constructed in a similar manner as for conventional memory (read/write) operations.
- Higher compatibility with available memory technologies, because redesigning cells would induce a huge cost for the vendors.
- Lower endurance requirement as operations are based on reading instead of writing.

On the other hand, they all share the following limitations:

- Overhead to align data; note that each operation requires the data to be aligned in the memory. Therefore, if the operands are not located in the same crossbar, data transfer operations are required.
- Additional write overhead when the results have to be stored back in to the memory. Note that the outputs are produced as voltages in the peripheral circuit, and therefore, if the results have to be stored back in the memory, extra write operations would be necessary.
- Parallelism is possible but it is achieved at the cost of area and power overhead.

Similar to CIM-A architectures, the following subsections discuss the details and characteristics of each CIM-P architecture. In this section, architectures that utilize charge-based memory technology (DRAM and SRAM) are presented first followed by the discussion of architectures based on non-charge-based memory technology (e.g., resistive, magnetic, etc.). Both charge- and non-charge-based memory technology architectures are discussed in a chronological order of their date of publication. This ordering technique is also reflected in Table 3.

4.1 S-AP: Cache Automaton

S-AP shown in Fig. 9 was proposed in 2017 by A. Subramaniyan et al., from University of Michigan [74]. The architecture targets an automata processor which exploits data-level parallelism by performing computations using state machines. An automata processor contains two main components: the State Transition Elements (STEs) and the routing matrix; the STE stores the accepting states, while the routing matrix stores the state transitions as shown in Fig. 10. The automata processor accepts one input symbol at a time, generates next active states, and decides whether a complete input string is accepted or not.

The architecture consists of STEs and a routing matrix which are implemented using SRAM technology. Each SRAM column corresponds to an STE which stores the accepting states in SRAM cells. The input symbol is fed to all the STEs simultaneously. The sense amplifiers collect a dot-product result of a vector-matrix multiplication. The output of the STE together with the routing matrix is used to determine the next active states; this process is carried on until all input symbols are processed. In case the one or more final active states are part of the acceptance states, it means that the input string has been matched with the corresponding pattern of the acceptance state. Note that data transfer inside the automata processor is carried out using the routing matrix. In addition to the general advantages of CIM-P architectures, S-AP has the following advantages:

- Computations may include logical and arithmetic operations using automata processing.
- Data can be transferred using both direct and indirect schemes.

Table 3 Comparison among architectures of CIM-P classes

		Computations			Overheads						Evaluation	
	Hierarchy level	Logic style	Available functions	Memory technology	Periphery	Controller	Sneak path current	Destructive read	Required read-out*	Copy scheme	Simulator	App.
S-AP [74]	Accelerator	Bool.	Logical,+	SRAM	Modif.	Simple	No	No	No	Both	VASim	ANMLZoo &Regex
ISAAC [35]	Accelerator	NN.	MM.	RRAM	Modif.	Simple	Yes	No	No	Both	Analytical	CNN& DNN
PRIME [18]	Main memory	NN.	MM.	RRAM	Modif.	Simple	Yes	No	No	Both	Analytical	Arithmetic
STT-CiM [75]	Accelerator	Bool.	Logical,+	MRAM	Modif.	Simple	Yes	No	No	Both	STT-CiM Sim.	(1)
DPP [36]	Accelerator	Bool.	Logical,+,x	RRAM	Modif.	Simple	Yes	No	No	Both	TensorFlow, CACTI	PARSEC
R-AP [76]	Accelerator	MM.	Logical,+,x	RRAM	Modif.	Simple	No	No	No	Both	Hspice	No

+ n-bit addition, *CNN* Convolutional Neural Network, (*) Required read-out during computations, *x* n-bit multiplication, *Modif.*: Modified, *App.* Applications and benchmarks, *LUT* Lookup table, *MM.* matrix multiplication, *Analytical* Analytical model, *NN.* Neural network, *Bool.* Boolean, *MRAM*, Magnetic Random Access Memory, *STT-CiM Sim.* STT-CiM device to architecture evaluation framework, (1) string matching, text processing, low-level graphics, data compression, bio-informatic, image processing and cryptography

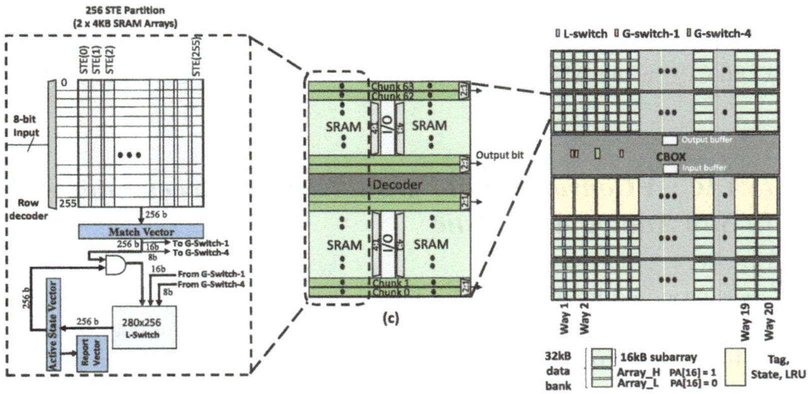

Fig. 9 SRAM Automata Processor (S-AP) [74]

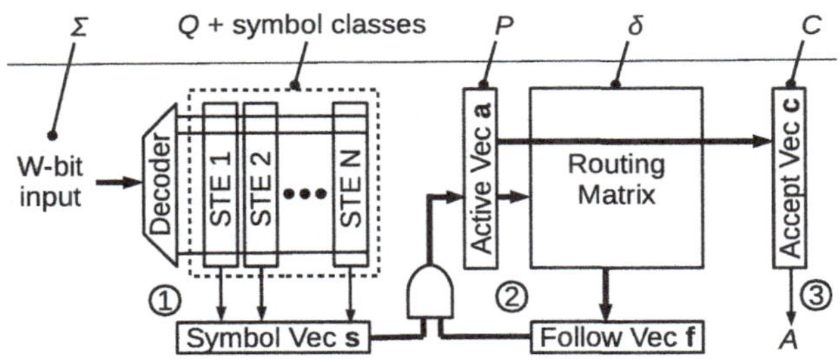

Fig. 10 General architecture for automata processor [76]

- The architecture uses SRAM technology, which has several benefits such as maturity, high endurance, no sneak path currents, and may benefit for the existing optimizing techniques and tools.
- Since the pitch of the SRAM bit-cell is more, it can easily accommodate the modified version of the sense amplifier in a column.
- The automata processing techniques and tooling are quite mature; hence it is feasible to explore many applications using automata processing.

However, it also has the following limitations:

- The architecture uses SRAM technology which suffers from high energy consumption, low scalability, and large footprint.
- The architecture requires additional compiling techniques and tools to perform conventional Boolean logic functions using automata processing.

On the one hand, S-AP has potentials in reducing non-memory components required to implement the automata processor. The D-AP requires most of its resources for routing matrix and other logic, while the S-AP can be implemented on processor which has advantages in realizing logic functions. On the other hand,

S-AP suffers from low frequency, density, and latency due to SRAM intrinsic properties. The S-AP is simulated using VASim [77] and evaluated against DRAM-AP and x86 CPU using ANMLZoo [77] and the Regex [78] benchmark suites.

4.2 ISAAC: A Convolutional Neural Network Accelerator with In Situ Analog Arithmetic

ISAAC was proposed in 2016 by Ali Shafiee et al., from University of Utah [35]. ISAAC is a memristor-based architecture that performs dot-product computations using the memristor crossbar and CMOS peripheral circuitry to exploit instruction-level parallelism. The architecture consists of multiple tiles connected through an on-chip concentrated mesh and an I/O interface, as shown in the left part of Fig. 11. The architecture is only used during the inference phase of machine learning applications, i.e., the phase after training; the inference phase consists of dot-product operations to compute convolutions and shift and add operations and sigmoid operations. ISAAC processes inputs from the I/O interface in multiple tiles. After processing, the outputs are communicated through the I/O interface to the outside world or a different ISAAC chip. Each tile of ISAAC contains multiple In Situ Multiply Accumulate (IMA) units that are connected through a bus, an eDRAM buffer, output register (OR), and computation units (max-pool, sigmoid, and Shift-and-Add (S+A)). Each IMA contains multiple memristor arrays with their DAC and Sample-and-Hold (S+H) units, an Input and Output Register (IR, OR), Shift-and-Add (S+A), and multiple ADC units. Inputs from the I/O interface are delivered to the memristor arrays and are used to perform a dot-product computation with the weights that are already stored in the memristor array. The results thereafter go through the S+H units (to temporarily store data before feeding them to ADCs) and S+A units (to accumulate data) if applicable. Finally, if multiple inputs are fetched, a pipeline is created using IR and OR of the IMAs and tiles. In order to transfer

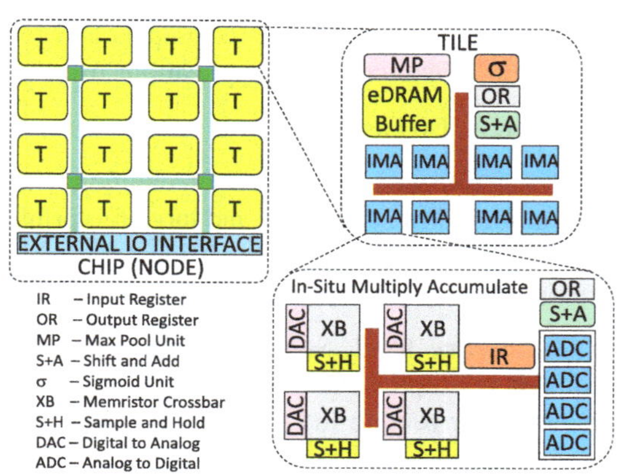

Fig. 11 A convolutional neural network accelerator with In Situ Analog Arithmetic (ISAAC) [35]

data within a single memory array, a controller can be used to apply appropriate voltages to move data directly inside the memory crossbar or use read-out and write-back schemes. The additional merits of ISAAC over the generic CIM-P architecture benefits can be summarized as follows:

- The architecture is used as an accelerator, which has a positive impact on the endurance due to infrequent use [79]. In contrast, some CIM-P architectures are used as main memory and therefore require a much higher endurance.
- The computations for neural networks are quite mature and can benefit from existing neural network techniques and tools.
- The computations for neural networks do not require a high precision; hence, they are more resilient against device variation.
- Data can be transferred in the crossbar using both direct and indirect schemes.
- The architecture uses non-volatile memory; hence it consumes low energy and has a small footprint.

However, ISAAC also has different limitations which need to be addressed properly. Some of the limitations are the following:

- The architecture has to deal with sneak path currents. Possible solutions are mentioned in Sect. 3.3.
- The architecture might suffer from a high overhead due to the need of ADC and DAC converters.
- As the sense amplifiers are complex, a trade-off between area and bandwidth has to be made.
- In case general purpose computing is desired, the architecture requires additional compiling techniques and tools to perform conventional Boolean logic functions using neural network computations.

The architecture is evaluated analytically and compared against DaDianNao architecture (which is an ASIC design with embedded DRAM) using a suite of CNN [57] and SNN workloads [80]. Their analytical result demonstrated that it potentially outperforms DaDianNao in terms of throughput, energy, and computational density. Newton introduced techniques to improve the energy efficiency by adapting sub-block-based ADC precision as well as an algorithm to reduce computations [81].

4.3 *PRIME: A Processing-in-Memory Architecture for Neural Network Computation in ReRAM-Based Main Memory*

PRIME was proposed in 2016 by C. Pinga et al., from the University of California [18]. PRIME is a resistive RAM-based architecture that exploits data-level parallelism to perform computations for neural networks (i.e., weighted vector-matrix multiplication) using high-precision multi-level sense amplifiers and some extra logic circuits. The architecture consists of a CPU and multiple RRAM

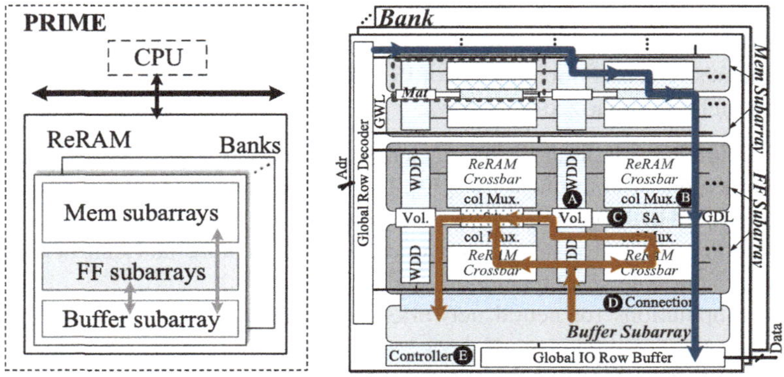

(a) PRIME banks (b) PRIME's Bank architecture

Fig. 12 A Processing-in-Memory architecture for neural network computation (PRIME) [18]: (**a**) PRIME banks organization, (**b**) architecture of PRIME bank

banks; each RRAM bank contains multiple memory crossbars (mem subarrays), full function (FF), and buffer subarrays, as shown in Fig. 12a. The CPU sends instructions to the resistive RAM banks; an instruction is either a memory operation (read/write) or a neural network computation. The memory bank performs the request without blocking the CPU, i.e., the CPU continues executing (different) instructions simultaneously. The results are returned to the CPU for further processing.

In the resistive RAM banks, the memory crossbars store data in multiple mats, while the FF and buffer subarrays serve for computation. Special subarray structures are used to enable both neural network computations and memory operations (blue blocks in Fig. 12b) feasible. The neural network computations are mainly performed in the FF subarray, while the buffer subarray stores temporary data that needs to be processed; this enables a parallel execution between CPU and FF subarrays. Neural network computations are performed using a vector-matrix multiplication between a weighted matrix stored in the FF subarray and a vector stored in the buffer subarray. Additional logic gates such as subtraction and sigmoid units are used to compute negative weights and sigmoid activation functions before the results are sensed by the multi-level sense amplifiers. In order to communicate between the memories, a controller can be used to apply appropriate voltages to the crossbar to move data directly inside it or use read-out and write-back schemes. On top of the generic advantages of CIM-P architectures, PRIME has the following set of unique advantages:

- The computations for neural networks are quite mature and can benefit from existing neural network techniques and tools.
- The computations for neural networks do not require a high precision; hence, they are more resilient against device variations.
- Data can be transferred in the crossbar using both direct and indirect schemes.

- The architecture uses non-volatile memory; hence it consumes a low energy and has a small footprint.

However, the abovementioned merits of PRIME architecture also come with their own challenges and limitations. The limitations of PRIME include the following:

- The architecture uses non-volatile memory as main memory, which may impact the life time due to limited endurance [79].
- The architecture has to deal with sneak path currents. Possible solutions are mentioned in Sect. 3.3.
- As the sense amplifiers are complex, a trade-off between area and bandwidth has to be made.
- In case general purpose computing is desired, the architecture requires additional compiling techniques and tools to perform conventional Boolean logic functions using neural network computations.

The architecture is synthesized using TSMC CMOS library 60nm and modeled using NVSIM, CACTI-3D, and CACTI-IO. It is evaluated using MIBench benchmarks [82] and compared against a CPU-only solution. Their comparison shows that the architecture achieves significant improvements in terms of performance and energy consumption over CPU-only based solution.

4.4 STT-CiM: Computing-in-Memory Spin-Transfer Torque Magnetic RAM

STT-CiM was proposed in 2017 by S. Jain et al., from Purdue University [75]. STT-CiM is a Spin-Transfer Torque Magnetic RAM-based architecture that exploits data-level parallelism by performing computations using both modified sense amplifiers and some additional CMOS gates. The architecture consists of a conventional architecture with a STT-MRAM used as a scratch-pad memory. This scratch-pad memory is equipped with the capability to perform in-memory instructions. These instructions are sent from the main processor.

The STT-CiM contains a CiM decoder, an array of memory cells, enhanced address decoder, and modified sensing circuitry to perform computations, as shown in Fig. 13. Based on the in-memory instruction, the enhanced address decoder activates one (for normal read) or multiple rows (for computations) of the memory array. The CiM decoder determines simultaneously the reference currents of the sense amplifiers. For example, in case an addition is executed, the set of logic gates for addition is enabled. The results are captured by the modified sense amplifiers. Data transfer can be performed by enabling two memory rows for direct copy operations or using the buffers and read-out operations for indirect copy operations. In addition to the general advantages of CIM-P architectures, STT-CiM has the following advantages:

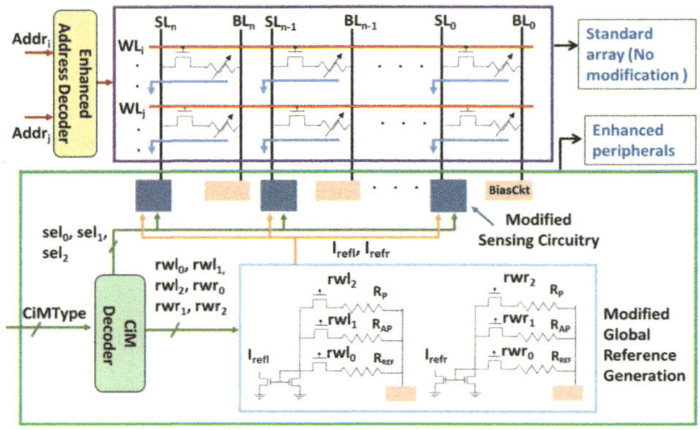

Fig. 13 Spin-Transfer Torque Magnetic RAM-based CIM (STT-CiM) [75]

- The architecture is used as an accelerator (i.e., scratch-pad memory), which has a positive impact on the endurance due to infrequent use [79].
- Computations currently include both logical operations and addition.
- The data transfer may include both direct and indirect schemes.
- The architecture uses non-volatile memory; hence it consumes low energy and has a small footprint.

However, it also has the following limitations:

- The architecture has to deal with sneak path currents. Possible solutions are mentioned in Sect. 3.3.
- As the sense amplifiers are complex, a trade-off between area and bandwidth has to be made.
- Additional software support (i.e., profiling and extracting memory intensive kernels) is required to maximally exploit the accelerator performance.

The architecture is evaluated using the STT-CiM device-to-architecture evaluation framework [75] and a set of benchmarks including string matching, text processing, low-level graphics, data compression, bio-informatic, image processing, and cryptography.

4.5 DPP: Data Parallel Processor

DPP was proposed in 2018 by D. Fujiki, et al. from University of Michigan [36]. DPP is a RRAM-based architecture that exploits instruction and data level parallelism by performing computations using a combination of RRAM-based dot-product operations and LUTs. The architecture consists of multiple RRAM tiles connected as an H-tree; each tile has multiple clusters and some logic units (as

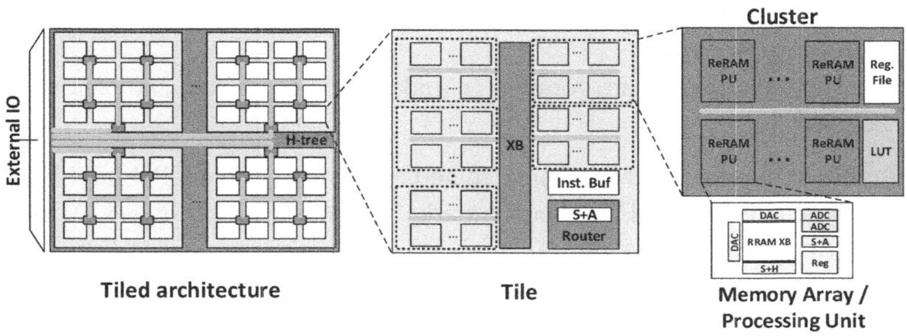

Fig. 14 Data Parallel Processor (DPP) [36]

shown in Fig. 14). Tiles and clusters form a SIMD-like processor that performs the parallel operations. The architecture is considered as a general purpose architecture as it can perform all primitive functions such as logical, arithmetic, shift, and copy operations.

In addition to clusters, each tile has several units to support the computations including instruction buffer, Shift and Add (S+A), and router. Each cluster additionally has one or more computational units; they are Shift and Add (S+A), Sample and Hold (S+H), DAC and ADC, a LUT, and register file (as shown in the right part of Fig. 14). While reading from the high latency RRAM, other units are simultaneously used for processing. Therefore, the S+H is used to read data (in the form of a current) from the RRAM array and temporarily store it. Once that data is needed, it is fed to an ADC to convert the analog value to a digital value. The S+A is used to perform carry propagation in a multiple-bit addition. DAC is used to apply a digital value to the RRAM array with an appropriate control voltage. Some complex functions that cannot be realized with these units are performed using LUTs and register file in each cluster. Data transfer can be performed by enabling two memory rows for direct copy operations or using the buffers and read-out operations for indirect copy operations. In addition to the general advantages of CIM-P architectures, DPP has the following advantages:

- Computations include both logical operations and simple arithmetic operations (e.g., addition and multiplication).
- Data can be transferred using both direct and indirect schemes.
- The architecture uses non-volatile memory; hence it consumes low energy and has a small footprint.
- This architecture is claimed to be general purpose; hence it can exploit the existing instruction set, compiling techniques and tools, as well as applications.

However, it also has the following limitations:

- The architecture uses non-volatile memory as main memory, which may impact the life time due to limited endurance [79].

- The architecture has to deal with sneak path currents. Possible solutions are mentioned in Sect. 3.
- As the sense amplifiers are complex, a trade-off between area and bandwidth has to be made.

The architecture potential was simulated and evaluated against CPU Intel Xeon E5-2697 using a subset of PARSEC benchmarks [83] and against GPU NVIDIA Titan XP using Rodinia benchmarks [84].

4.6 R-AP: Resistive RAM Automata Processor

R-AP was proposed in 2018 by J. Yu et al. from Delft University of Technology [76]. R-AP is an automata processor that exploits data-level parallelism by performing computations similarly as mentioned in Sect. 4.1. The working principle of R-AP is similar to the S-AP. In contrast to S-AP, R-AP uses RRAM-based STEs and routing matrix, as shown in Fig. 15. In addition to the general advantages of CIM-P architectures, R-AP has the following advantages:

- The architecture is used as a read-favored accelerator, which has a positive impact on the endurance due to infrequent use [79].
- Automata processing can be used to perform both logical and arithmetic operations in general.
- Data can be transferred using both direct and indirect schemes.
- The architecture uses non-volatile memory; hence it consumes low energy and has a small footprint.
- The automata processing techniques and tooling are quite mature; hence it is feasible to explore many applications using automata processing.

However, it also has the following limitations:

- The architecture has to deal with sneak path currents. Possible solutions are mentioned in Sect. 3.3.

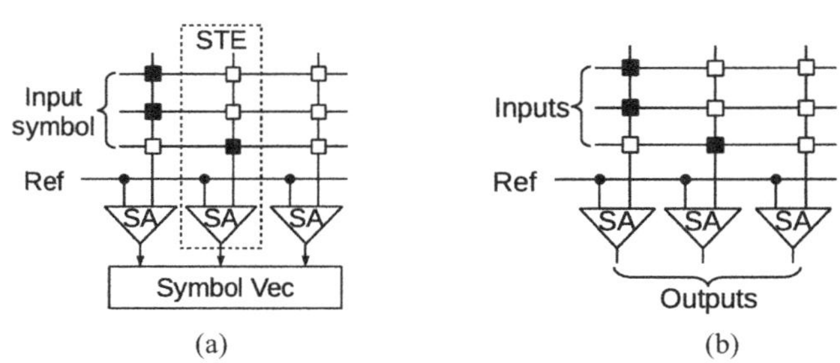

Fig. 15 Resistive RAM Automata Processor (R-AP) [76]. (**a**) Used as STEs. (**b**) Used as routers

- As the sense amplifiers are complex, a trade-off between area and bandwidth has to be made.
- The architecture requires additional compiling techniques and tools to perform conventional Boolean logic functions using automata processing.

The architecture has been validated using circuit-level simulations and evaluated against S-AP.

5 CIM Design-flow

The design of CIM accelerator is strongly application dependent; the kernels/-functions that should be accelerated for one application could be different from those of other applications [85, 86]. Delivering an optimized and cost-effective CIM accelerator implementation requires a holistic approach in which the whole design stack in its entirety should be addressed [87, 88]. Figure 16 illustrates the holistic approach for CIM accelerator design [89]. The remainder of this section discusses CIM design flow from system-level design aspect to circuit-level design of CIM architectures.

5.1 *System-Level Design*

The system-level design aspect of CIM design flow involves two main steps which are crucial for developing efficient CIM architecture. These steps are discussed below.

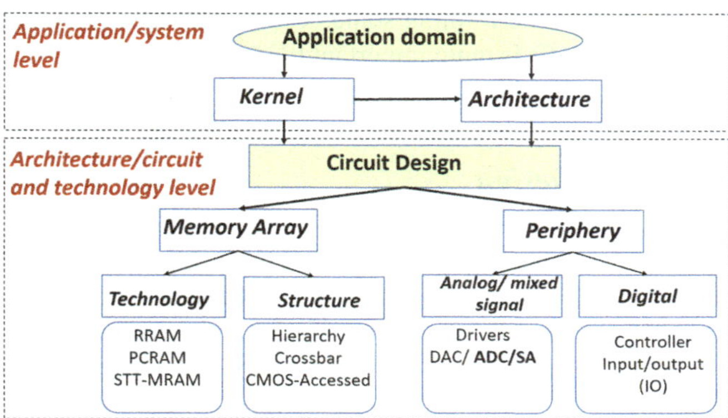

Fig. 16 Illustration of CIM design flow

5.1.1 Application Profiling for Critical Kernel Identification

Application profiling is the process of determining the execution speed and resource utilization of the internal functions of an application. Profiling enables the identification of the critical functions/kernels which have the most significant impact on the performance metrics (e.g., energy, latency) of an application execution [90]. These critical kernels have to be then accelerated by CIM in order to speed up the execution of the application and reduce the overall energy consumption. As already mentioned, different applications may require different kernels to be accelerated while minimizing the data movement. For example, the kernels of a classifier based on neural networks (NN) is vector-matrix multiplication (VMM) [62]; the kernels of data-based application are bit-wise logic operations [91], the kernel of a matching application using an automata processor is the binary vector-matrix multiplication [92], etc. Note that depending on an application, the operands needed by the kernel/function to be accelerated may reside in memory (like it is the case of data-base query), or only one operand resides in the memory, and the other one has to be fed to the CIM core through the external input (as it is the case for VMM used in NNs). Hence, the nature of the kernel contributes to the definition of the CIM (micro) architecture.

5.1.2 Accelerator Configuration Definition

Defining appropriate CIM configuration is critical before starting the circuit design. Different configurations are possible including CIM-A versus CIM-P, on-chip (resides in the same physical chip with the main core) versus off-chip, etc. Each of these has their own pros and cons not only in terms of energy efficiency but also in terms of cost and complexity. Obviously, the size of the problem/application and the kernels that need to be accelerated have a large impact on the selection of the appropriate configuration. Performing some design exploration at this stage while considering some trade-offs is quite important [93].

5.2 Circuit-Level Design

Once the kernels and suitable CIM architecture are identified, the next step is designing the circuit as shown in Fig. 16. Since CIM devices are consisting of memory array and peripheral circuits, the design of both parts has to be considered. The design of the memory array has two aspects: the technology and the structure. The technology refers to the memristive device type to be selected, which could be RRAM, PCRAM, STT-MRAM, etc. The choice of the technology depends on, but not limited to, the number of bits needed per the memory cell (e.g., PCRAM and RRAM can support multi-level storage, while STT-MRAM is binary), the selected CIM architecture (e.g., CIM-A requires higher endurance than CIM-P), etc. The structure refers to the ways both the array (e.g., dual bit line arrays [94], common-source line array [95], crossbar array [96], and the cell (e.g., the one-transistor-one-memristor (1T1R)) are organized. On the other hand, the peripheral CMOS circuits include two parts: analog/mixed signal and digital circuits. The analog/mixed circuits are mainly required for conversions between analog and digital domains; they might be also used for multiplexing signals in analog domain. Examples are digital-to-analog converters (DACs) and analog-to-digital converters (ADCs), customized sense amplifiers (SA), etc. The choice of analog circuits depends on, but not limited to, the kernel and the architecture; e.g., performing analog VMM with CIM-P requires at least the relatively expensive ADCs, while performing OR function with CIM-P architecture requires only customized SAs. The digital CMOS circuits are mainly required for the memory controller and temporary storage (i.e., registers).

6 Conclusion

This chapter presented an overview of Computation-in-Memory (CIM) architectures by classifying them into different classes. The chapter discussed several state-of-the-art architectures from the CIM classes and then presented a cross-layer CIM design flow that needs to be adopted for efficient CIM architecture design. The chapter demonstrated that a potential architecture does not only require to be freed from memory bottleneck but also energy and area efficient. This can be only achieved through the joint effort of both architectural improvement and technology development. Indeed, emerging memory technologies play a vital role in overcoming the energy, performance, and memory bandwidth challenges of Von-Neumann architecture. However, one needs to revisit the architectural and design tool challenges of CIM architectures in order to harness the full potential of CIM.

Acknowledgments This work was supported in part by the EU H2020 grant "DAIS" that has received funding from the ECSEL Joint Undertaking (JU) under grant agreement No. 101007273.

References

1. ITRS, ITRS ERD report (2010)
2. S. Hamdioui et al., Memristor for computing: Myth or reality?, in *DATE* (2017)
3. A. Fuchs, D. Wentzlaff, The accelerator wall: Limits of chip specialization, in *HPCA* (2019)
4. S. Manipatruni, D.E. Nikonov, I.A. Young, Beyond CMOS computing with spin and polarization. Nat. Phys. **14**(4), 338–343 (2018)
5. IRDS, International roadmap for devices and systems, in *IRDS* (2020)
6. J.L. Hennessy, D.A. Patterson, *Computer Architecture: A Quantitative Approach* (2011)
7. S. Gochman et al., Introduction to intel core duo processor architecture. Intel Technol. J. **10**(2) (2006)

8. N.Z. Haron, S. Hamdioui, Why is cmos scaling coming to an end? in *International Design and Test Workshop* (2008)
9. J.A.B. Fortes, Future challenges in vlsi system design, in *Annual Symposium on VLSI* (2003)
10. J. Parkhurst, J. Darringer, B. Grundmann, From single core to multi-core: preparing for a new exponential, in *International Conference on Computer-aided Design* (2006)
11. R.A. Iannucci, Toward a dataflow/von neumann hybrid architecture. ACM SIGARCH Computer Architecture News **16**(2), 131–140 (1988)
12. S. Hamdioui, L. Xie, et al., Memristor based computation-in-memory architecture for data-intensive applications, in *DATE* (2015)
13. H.S. Stone, A logic-in-memory computer. IEEE Trans. Comput. **100**(1), 73–78 (1970)
14. D. Pala et al., Logic-in-memory architecture made real, in *ISCAS* (2015)
15. M. Macedonia, The GPU enters computing's mainstream. Computer **36**(10), 106–108 (2003)
16. M. Di Ventra, Y.V. Pershin, Memcomputing: a computing paradigm to store and process information on the same physical platform. Nat. Phys. 1–2 (2013)
17. A. Yousefzadeh et al., Energy-efficient in-memory address calculation. ACM Trans. Archit. Code Optim. (TACO) **19**(4), 1–16 (2022)
18. P. Chi et al., Prime: a novel processing-in-memory architecture for neural network computation in reram-based main memory, in *Computer Architecture News* (2016)
19. J. Ahn et al., Pim-enabled instructions: a low-overhead, locality-aware processing-in-memory architecture. Computer Architecture News **43**(3S), 336–348 (2015).
20. J. Yue et al., 14.3 a 65 nm computing-in-memory-based CNN processor with 2.9-to-35.8 tops/w system energy efficiency using dynamic-sparsity performance-scaling architecture and energy-efficient inter/intra-macro data reuse, in *ISSCC* (2020)
21. Y.-D. Chih et al., 16.4 an 89tops/w and 16.3 tops/mm 2 all-digital SRAM-based full-precision compute-in memory macro in 22 nm for machine-learning edge applications, in *ISSCC* (2021)
22. S. Rai et al., Perspectives on emerging computation-in-memory paradigms, in *DATE* (2021)
23. Z. Chen, X. Chen, J. Gu, 15.3 a 65 nm 3t dynamic analog ram-based computing-in-memory macro and CNN accelerator with retention enhancement, adaptive analog sparsity and 44top-s/w system energy efficiency, in *ISSCC* (2021)
24. J.-O. Seo et al., Archon: A 332.7 tops/w 5b variation-tolerant analog CNN processor featuring analog neuronal computation unit and analog memory, in *ISSCC* (2022)
25. S. Gupta et al., NNPIM: A processing in-memory architecture for neural network acceleration. IEEE Trans. Comput. **68**(9), 1325–1337 (2019)
26. M.A. Lebdeh et al., Memristive device based circuits for computation-in-memory architectures, in *ISCAS* (2019)
27. A. Shaout, T. Eldos, On the classification of computer architecture. Int. J. Sci. Technol. **14** (2003)
28. K. Hwang, N. Jotwani, *Advanced Computer Architecture, 3e* (McGraw-Hill Education, New York, 2016)
29. A. Gebregiorgis et al., A survey on memory-centric computer architectures. J. Emerging Technol. Comput. Syst. **18**(4), 1–50 (2022)
30. S. Kvatinsky et al., Magic–memristor-aided logic. TTCAS II: Express Briefs **61**(11), 895–899 (2014)
31. A. Singh et al., Cim-based robust logic accelerator using 28 nm stt-mram characterization chip tape-out, in *AICAS* (2022)
32. E. Lehtonen et al., Memristive stateful logic, in *Memristor Networks* (2014)
33. A. Singh et al., Low-power memristor-based computing for edge-ai applications, in *ISCAS* (2021)
34. S. Li et al., Pinatubo: a processing-in-memory architecture for bulk bitwise operations in emerging non-volatile memories, in *DAC* (2016)
35. A. Shafiee et al., ISAAC: a convolutional neural network accelerator with in-situ analog arithmetic in crossbars. Architecture News **44**(3), 14–26 (2016)
36. D.O. Fujiki, In-memory data parallel processor, in *Architectural Support for Programming Languages and Operating Systems* (2018)

37. A. Singh et al., SRIF: Scalable and reliable integrate and fire circuit ADC for memristor-based cim architectures. TCAS I: Regular Papers **68**(5), 1917–1930 (2021)
38. A. Gebregiorgis et al., A comprehensive reliability analysis framework for ntc caches: a system to device approach. IEEE Trans. Comput. Aided Des. Integr. Circuits Syst. **38**(3), 439–452 (2018)
39. G.S. Sandhu, Emerging memories technology landscape, in *NVMTS* (2013)
40. C. Bengel et al., Reliability aspects of binary vector-matrix-multiplications using reram devices. Neuromorph. Comput. Eng. **2**(3), 034001 (2022)
41. S. Bhatti et al., Spintronics based random access memory: a review. Mater. Today **20**(9), 530–548 (2017)
42. A. Gebregiorgis et al., Spintronic normally-off heterogeneous system-on-chip design, in *DATE* (2018)
43. J.E. Green et al., A 160-kilobit molecular electronic memory patterned at 10 11 bits per square centimetre. Nature **445**(7126), 414–417 (2007)
44. R. Cabrera et al., A micro-electro-mechanical memory based on the structural phase transition of vo2. physica status solidi (a) **210**(9), 1704–1711 (2013)
45. S. Salahuddin, K. Ni, S. Datta, The era of hyper-scaling in electronics. Nat. Electron. **1**(8), 442–450 (2018)
46. F. Oboril et al., Evaluation of hybrid memory technologies using sot-mram for on-chip cache hierarchy. IEEE Trans. Comput. Aided Des. Integr. Circuits Syst. **34**(3), 367–380 (2015)
47. A. Gebregiorgis et al., Special session: Stt-mrams: Technology, design and test, in *VTS* (2022)
48. N. Firasta et al., Intel avx: New frontiers in performance improvements and energy efficiency, in *Intel White Paper* (2008)
49. S. Wong, T. Van As, G. Brown, ρ-vex: A reconfigurable and extensible softcore vliw processor, in *FPT* (2008)
50. H.A. Du Nguyen et al., Memristive devices for computing: Beyond cmos and beyond von neumann, in *VLSI-SoC* (2017)
51. S. Li, et al., Drisa: A dram-based reconfigurable in-situ accelerator, in *International Symposium on Microarchitecture* (2017)
52. A. Siemon et al., A complementary resistive switch-based crossbar array adder. IEEE J. Emerging Sel. Top. Circuits Syst. **5**(1), 64–74 (2015)
53. H.A. Du Nguyen et al., On the implementation of computation-in-memory parallel adder. IEEE Trans. Very Large Scale Integr. VLSI Syst. **25**(8), 2206–2219 (2017)
54. M.F. Ali, A. Jaiswal, K. Roy, In-memory low-cost bit-serial addition using commodity dram technology. IEEE Trans. Circuits Syst. I Regul. Pap. **67**(1), 155–165 (2019)
55. R.B. Hur, S. Kvatinsky, Memristive memory processing unit (MPU) controller for in-memory processing, in *ICSEE* (2016)
56. D. Bhattacharjee et al., ReVAMP: ReRAM based VLIW architecture for in-memory computing, in *DATE* (2017)
57. K. Simonyan, A. Zisserman, Very deep convolutional networks for large-scale image recognition. arXiv preprint arXiv:1409.1556 (2014)
58. K. He et al., Deep residual learning for image recognition, in *Computer Vision and Pattern Recognition* (2016)
59. H.A. Du Nguyen et al., Interconnect networks for resistive computing architectures, in *DTIS* (2017)
60. E.J. Merced-Grafals et al., Repeatable, accurate, and high speed multi-level programming of memristor 1t1r arrays for power efficient analog computing applications. Nanotechnology **27**(36), 365202 (2016)
61. L. Xie et al., Boolean logic gate exploration for memristor crossbar, in *DTIS* (2016)
62. A. Haron et al., Parallel matrix multiplication on memristor-based computation-in-memory architecture, in *HPCS* (2016)
63. P.-E. Gaillardon et al., The programmable logic-in-memory (PLiM) computer, in *DATE* (2016)
64. A. Bogdanov et al., Present: an ultra-lightweight block cipher, in *Cryptographic Hardware and Embedded Systems* (2007)

65. F. Gao et al., Computedram: in-memory compute using off-the-shelf drams, in *International Symposium on Microarchitecture* (2019)
66. D. Fujiki et al., Duality cache for data parallel acceleration, in *International Symposium on Computer Architecture* (2019)
67. A.K. Ramanathan et al., Look-up table based energy efficient processing in cache support for neural network acceleration, in *MICRO* (2020)
68. A. Haj-Ali et al., Efficient algorithms for in-memory fixed point multiplication using magic, in *ISCAS* (2018)
69. R.B. Hur et al., Simple magic: synthesis and in-memory mapping of logic execution for memristor-aided logic, in *ICCAD* (2017)
70. R. Ben-Hur et al., SIMPLER MAGIC: synthesis and mapping of in-memory logic executed in a single row to improve throughput. IEEE Trans. Comput. Aided Des. Integr. Circuits Syst. **39**(10), 2434–2447 (2019)
71. L. Amarú et al., The EPFL combinational benchmark suite, in *International Workshop on Logic and Synthesis (IWLS)* (2015)
72. M. Oskin et al., *Active Pages: A Computation Model for Intelligent Memory* (1998)
73. S.Y.W. Su et al., Magnetic bubble memory architectures for supporting associative searching of relational databases. Trans.Comput. **100**(11), 957–970 (1980)
74. A. Subramaniyan et al., Cache automaton, in *International Symposium on Microarchitecture* (2017)
75. S. Jain et al., Computing in memory with spin-transfer torque magnetic RAM. arXiv preprint arXiv:1703.02118 (2017)
76. J. Yu et al., Memristor devices for computation-in-memory, in *DATE* (2018)
77. J. Wadden et al., Anmlzoo: a benchmark suite for exploring bottlenecks in automata processing engines and architectures, in *International Symposium on Workload Characterization (ISWC)* (2016)
78. M. Becchi et al., A workload for evaluating deep packet inspection architectures, in *International Symposium on Workload Characterization (ISWC)* (2008)
79. J. Wang et al., Endurance-aware cache line management for non-volatile caches. ACM Trans. Archit. Code Optim. **11**(1), 1–25 (2014)
80. T. Iakymchuk et al., Simplified spiking neural network architecture and stdp learning algorithm applied to image classification, in *Journal on Image and Video Processing* (2015)
81. A. Nag et al., Newton: Gravitating towards the physical limits of crossbar acceleration. IEEE Micro **38**(5), 41–49 (2018)
82. F. Leisch, E. Dimitriadou, Machine learning benchmark problems, in *R Package, mlbench* (2010)
83. C. Bienia et al., The parsec benchmark suite: Characterization and architectural implications, in *International Conference on Parallel Architectures and Compilation Techniques* (2008)
84. S. Che et al., Rodinia: A benchmark suite for heterogeneous computing, in *International Symposium on Workload Characterization (ISWC)* (2009)
85. M. Zahedi et al., System design for computation-in-memory: from primitive to complex functions, in *VLSI-SoC* (2022)
86. T. Shahroodi et al., KrakenOnMem: a memristor-augmented HW/SW framework for taxonomic profiling, in *Conference on Supercomputing* (2022)
87. A. Gebregiorgis et al., Dealing with non-idealities in memristor based computation-in-memory designs, in *VLSI-SoC* (2022)
88. A.E. Arrassi et al., Energy-efficient SNN implementation using RRAM-based computation in-memory (CIM), in *VLSI-SoC* (2022)
89. A. Gebregiorgis et al., Tutorial on memristor-based computing for smart edge applications. Memories - Mater. Devices Circuits Syst. **4**, 100025 (2023)
90. S. Diware et al., Severity-based hierarchical ECG classification using neural networks. IEEE Trans. Biomed. Circuits Syst. **17**(1), 77–91 (2023)
91. I. Giannopoulos et al., In-memory database query. Adv. Intell. Syst. **2**(12), 2000141 (2020)
92. J. Yu et al., Memristive devices for computation-in-memory, in *DATE* (2018)

93. M. Gomony et al., Convolve: smart and seamless design of smart edge processors. arXiv preprint arXiv:2212.00873 (2022)
94. X. Dong et al., NVSIM: a circuit-level performance, energy, and area model for emerging nonvolatile memory. IEEE Trans. Comput. Aided Des. Integr. Circuits Syst. **31**(7), 994–1007 (2012)
95. Y. Sato et al., Sub-100-μA reset current of nickel oxide resistive memory through control of filamentary conductance by current limit of MOSFET. IEEE Trans. Electron Devices **55**(5), 1185–1191 (2008)
96. L. Zhao et al., Constructing fast and energy efficient 1tnr based reram crossbar memory, in *ISQED* (2017)

Toward Spintronics Non-volatile Computing-in-Memory Architecture

Bi Wu, Haonan Zhu, Tianyang Yu, and Weiqiang Liu

1 Introduction

With the shrinking of integrated circuit technology, traditional complementary metal-oxide-semiconductor (CMOS)-based von Neumann computing architectures are suffering from "power wall" and "memory wall" problems [1]. "Power wall" is mainly static power consumption in the total power consumption overhead accounted for a huge proportion of the progress in technology, as well as the data transfer between the processor and memory power consumption overhead can't be ignored. And "memory wall" is about the ever-increasing gap between processor and memory speed and bandwidth mismatch [1].

The "power wall" and "memory wall" problems are forcing researchers to seek new computing architecture. Computing-in-Memory (CIM) that implements logical and arithmetic operations using inherent memory array and the necessary peripheral circuitry emerges as an effective solution to solve these problems [2]. CIM architecture can take full advantage of the high bandwidth of the memory for more parallel computing, while the data storage and operation occur in the same memory area, greatly reducing the overhead caused by data migration.

Currently, many memory technologies have been employed to design and realize CIM, such as static random access memory (SRAM), dynamic random access memory (DRAM), flash memory, etc. Now there have been some successful cases using traditional memory to achieve computation including neural network inference in memory, such as Samsung's DRAM-based HBM-PIM, Witmem's Flash-based WTM2101, etc. However, some defects of these three kinds of traditional memory limit their further development in computing-in-memory; none of

B. Wu (✉) · H. Zhu · T. Yu · W. Liu
College of Electronic and Information Engineering, Nanjing University of Aeronautics and Astronautics (NUAA), Nanjing, China
e-mail: wubi_sl@nuaa.edu.cn; yutianyang01@nuaa.edu.cn; liuweiqiang@nuaa.edu.cn

W. Liu et al. (eds.), *Design and Applications of Emerging Computer Systems*,
https://doi.org/10.1007/978-3-031-42478-6_3

Table 1 Comparison of existing memory technologies

	Traditional				Emerging		
	SRAM	DRAM	NAND-Flash	NOR-Flash	PCM	STT-MRAM	ReRAM
Cell area(F^2)	>100	6–10	<4	10	4–20	6–50	2
Voltage (V)	<1	<1	<10	<10	<3	<2	<3
Read time	1 ns	10 ns	10 us	50 us	<10 ns	<10 ns	<10 ns
Write time	1 ns	10 ns	100 us–1 ms	10 us–1 ms	50 ns	<10 ns	<10 ns
Write energy (J/bit)	1 fJ	10 fJ	10 fJ	100 pJ	10 pJ	0.1 pJ	0.1 fJ
Retention	N/A	64 ms	>10 year	>10 year	>10 year	>10 year	>10 year
Endurance	$> 10^{16}$	$> 10^{16}$	$> 10^4$	$> 10^5$	10^8–10^{15}	$> 10^{15}$	10^8–10^{12}
Multi-bit	1	1	>4	>4	>2	1	2–7
Non-volatility	No	No	Yes	Yes	Yes	Yes	Yes

these storage technologies is an efficient solution for high static power consumption when considering jointly with the aforementioned "power wall" issue. Table 1 summarizes traditional memory and non-volatile memory (NVM) technologies and their characteristics. Note that the numbers in the table are obtained from most advanced circuit works, which often optimize the design for performance, but the numbers vary considerably depending on the array size, bit per cell, process technology, etc.

As can be seen from Table 1, SRAM's cell area is large, so it is difficult to achieve high density integration due to its 6T structure; DRAM requires frequent refresh, so it consumes a lot of energy; the read/write times of flash are limited. Thus, the non-volatile memory technologies, like Magnetic Random Access memory (MRAM), Resistive Random Access memory (ReRAM), Ferroelectric Random Access memory (FRAM), etc., are becoming an effective way to realize low-power CIM. Among them, MRAM is one of the most promising alternatives due to its near-zero standby power, high integration density, long lifetime, and good compatibility with CMOS [3].

MRAM-based CIM is more mature for Boolean logic and simple arithmetic operations, and in the application-level direction, it mainly focuses on multiplying and accumulating operation, simplified neural network models, such as binary neural networks, triangle count, DNA read alignment, encryption, and true random generator.

2 MRAM

The MRAM storage device appears to be a multi-layer stacked structure from top to bottom. The three layers, named the pinned layer (PL), the oxide layer, and the

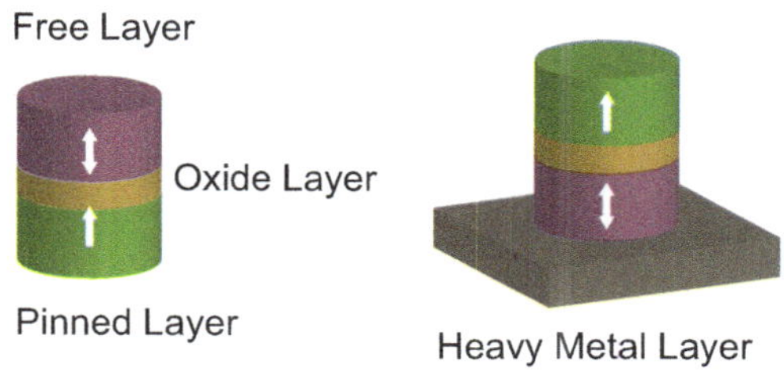

Fig. 1 The structure diagram of the MTJ and SOT-MRAM

free layer (FL), form a magnetic tunnel junction (MTJ) structure. As the name implies, the magnetization direction of the PL is fixed, while that of the FL can be changed through external means. In terms of the magnetization direction of FL and PL, if in parallel, MTJ shows a low resistance state, while anti-parallel means that MTJ resistance is high. The current mature technologies in MRAM are STT-MRAM, which switch the magnetization direction of the free layer using bi-directional currents. The switching principle is that a charge current passes through the fixed layer vertically and could be transferred to a spin-polarized current, whose spin-polarization direction coincides with the magnetization of the FL. The spin-polarized current flowing through the FL produces a spin transfer torque, causing the FL magnetization direction to turn in the direction of spin-polarization direction. Due to the small initial switching angle of the STT-MRAM, its switching speed is typically in the nanosecond range. As a result, the SOT-MRAM with a sub-nanosecond switching delay was proposed. The heavy metal layer (HM) connecting the FL is its characteristic structure as shown in Fig. 1. When the charge current flows horizontally through the HM, a spin orbit torque is produced to assist in switching based on the spin Hall effect [4]. Due to the large initial angle between the horizontal spin orbit torque and the vertical magnetization direction of the FL, the magnetization direction can be rapidly switched to the horizontal state. However, its deterministic switching can only be realized with an assistance, such as the current in a vertical direction, the external magnetic field, and the asymmetry of internal structure [5, 6].

3 Implementation of Boolean Logic

MRAM-based CIM logic design is divided into four categories based on the degree of modification to the array cell and peripheral circuitry: analog implementation, read/write implementation, cell modification, and peripheral circuit modification. Among them, the analog implementation and the read/write implementation are the least modified for the whole memory structure, mainly using the different MRAM characteristics of read/write operations and the high bandwidth of the array to achieve logic operations. For the remaining two categories, as the name implies, the logic is implemented through the modification of the cell and the peripheral circuit of the memory.

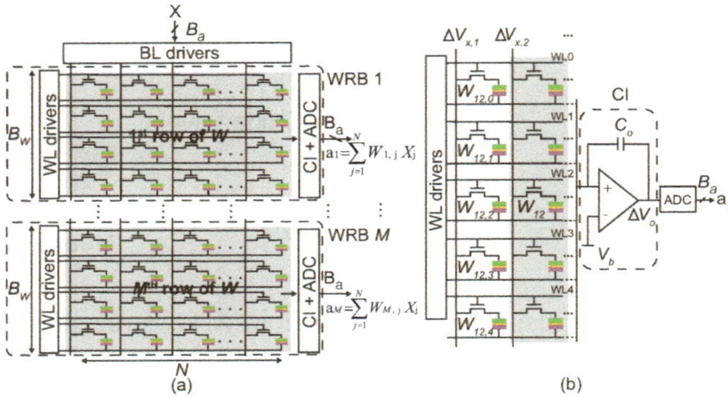

Fig. 2 (**a**) The overall implementation of the analog implementation; (**b**) the diagram of the CI module

3.1 Analog Implementation

Analog implementation generates results via Kirchhoff's law. Generally, through activating multiple lines of word line (WL) at the same time, the voltage on source line (SL) generates corresponding column currents at MTJs of different resistance values, and the calculation is realized in the form of current accumulation.

In Fig. 2, input data are encoded to voltage signal via the digital-to-analog converter (DAC) and applied to multiple bit lines (BLs) simultaneously to generate the corresponding current [7]. And the weight data is stored in MRAM array as the resistant state, which is divided into Bw rows according to the number of columns driven by BL. Each SL generates a multiplication of input data and fixed bit-width weights, and Bw of SLs is summed by a current integrators (CI) module with different bit-width weights, which is finally converted to digital signals by analog-to-digital converter (ADC). In this way, a multiplication-accumulation operation has been implemented.

This method mainly uses the existing structure inside the memory array to implement the function, and there is almost no modification to the memory cell structure. However, for the peripheral circuit structure, CI module as well as high power, high area overhead data conversion module (ADC and DAC) is added. Its greatest advantage is that it takes full advantage of the high bandwidth and high parallelism of the memory array, allowing a large number of operations to be performed simultaneously in one operating cycle. Due to the inherently low on/off ratio nature of MRAM, it severely limits the bandwidth and accuracy of the analogue implementation. Its computation accuracy is highly susceptible to noise and process voltage temperature variations.

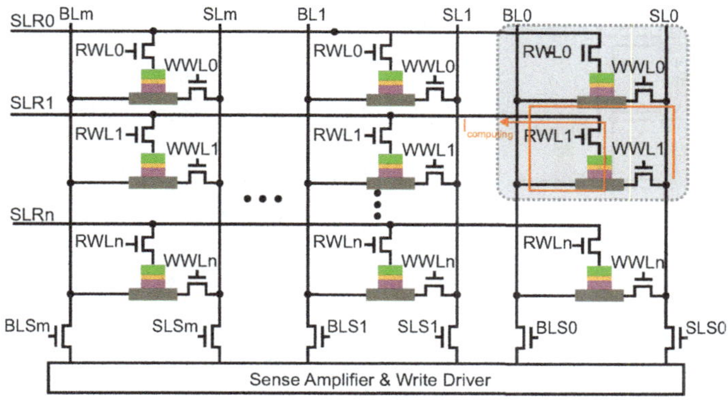

Fig. 3 The current flow of the XOR logic between two MTJs

3.2 Read/Write Implementation

Read/write implementation realizes the logic operations through basic read and write operations within the memory array. The method has a unique implementation for different types of MRAM.

A design adopts emerging unipolar switching spin orbit torque MRAM (US-SOT-MRAM) device, which the magnetization direction of the FL is switched to the opposite state once a SOT current (regardless of the current polarity) exceeds the threshold size flowing the HM [8, 9]. In this way, X(N)OR logic can be implemented in the form of tunnel resistive conduction, which can be treated as a write operation between two MTJs, and the current flow is shown in Fig. 3. More logical operations, such as AND and MAJ, are implemented by sequencing read, write, XNOR without any modifications to cell structure. In addition, it is uncommon to optimize the relaxation time during the MRAM switching by adjusting the sequence of step [8].

Wang et al. proposed using the toggle spin-orbit torque MRAM, which requires not only SOT current assistance for the switching process but also a bi-directional STT for deterministic switching [10]. In this way, the logic operation is completed by a writing between a number of MTJs as shown in Fig. 4. Different logic combinations are used corresponding to the generation of different sizes of SOT currents, which produce differences in the STT switching threshold of the output MTJ when it flows the HM of the output MTJ. For a fixed STT current, different switching results of output MTJ occur, which in combination with the initial state of the output MTJ complete the AND and OR logic. This approach is limited by the low on/off ratio of the MRAM, and the accuracy of the logic operation is compromised.

Alternatively, the another idea of read/write implementation is using logic signal mapping to various control signals to control the write operation, which has been applied to different types of MRAM devices [3, 11, 12].

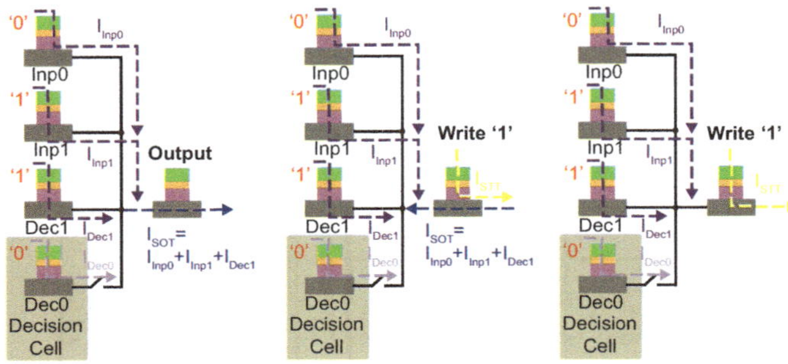

Fig. 4 The current flow of the logic operation of the toggle spin-orbit torque MRAM

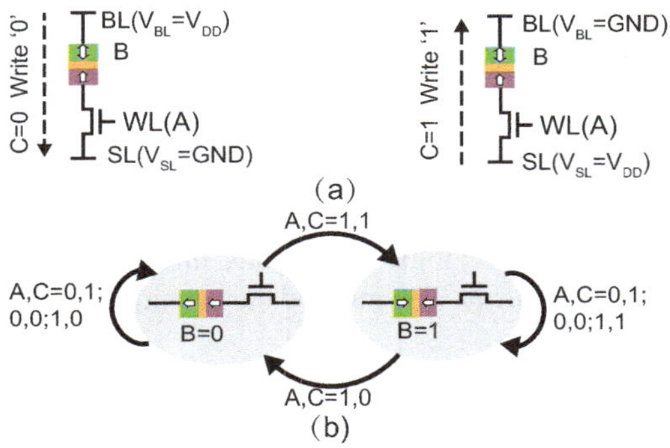

B_i	A	C	B_{i+1}
0	0	0	0
0	1	0	0
1	0	0	1
1	1	0	0

Fig. 5 The diagram of the logic operation of the voltage control and corresponding truth table

A variety of logic operations, including AND, OR, and XOR, are implemented by a single write operation under the conventional array structure of 1T1MTJ using the STT MRAM. The two logic signals are mapped to the WL voltage signal and the resistance state of the MTJ, respectively, plus the voltages across BL and SL as additional control signal C. The logical expressions and truth tables for the implementation process are shown in Fig. 5 [3]. Zhang et al. used the voltage-gated spin Hall effect (VG-SHE)-driven MRAM, where the voltage at both ends of the MTJ controls the threshold current for switching, and the FL is switched to a different direction by a bidirectional current in the HM [11]. The logic operation is completed through the control of three control signals, the voltage across the MTJ, the MTJ resistance state, and the polarity of the current flowing the HM. And Monga et al. similarly implement the logic operation by controlling the BL, SL, SCL, and initial MTJ states of the three-terminal MRAM device with logic signals [12].

Shreya et al. propose a compact structure that can do XNOR operation without complex Write Driver, by using multilevel voltage-controlled SOT memory (MV-SOTM) [13]. The MV-SOTM has two MTJs which share the heavy metal (HM) in one bitcell, and it has two forms as shown in Fig. 6. The initial state of all the output spin-neuron is chosen to be +1 (for anti-parallel, high resistance state), and

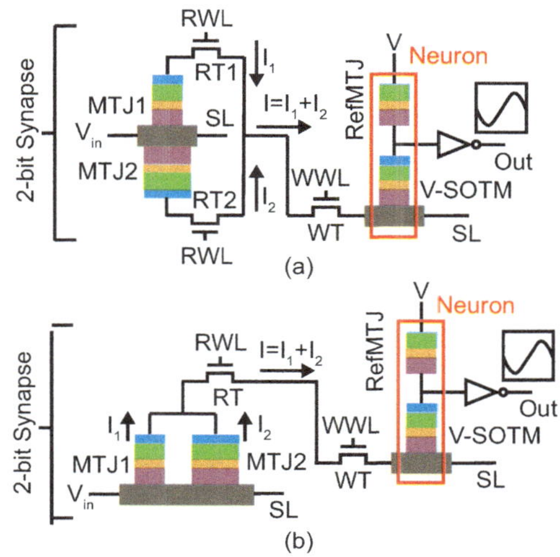

Fig. 6 (**a**) Series MV-SOTM; (**b**) parallel MV-SOTM

the XNOR result is from bitline (BL) current, IBL that is precharged. If both the MTJs in a bitcell are in the same magnetic states, thus, both establish high resistance path. Therefore, IBL cannot discharge and remains high (at the precharged value); thus, spin-neuron is activated (remain +1 state). On the contrary, if both MTJs are in opposite states, then the P state establishes a low resistance path and, thus, starts decreasing the value of IBL. When it crosses the reference value, the spin-neuron switches to −1 state. The spin-neuron's state is just the XNOR result.

3.3 *Cell Modification*

One way to implement logical and arithmetic operations using MRAM is to modify the cell structure of MRAM array.

Zabihi et al. proposed a new cell structure that one transistor is added to each cell to enable data transferring between columns compared to traditional STT MRAM-based 1T1MTJ array architecture, which current can flow through LL in the row as shown in Fig. 7 [14]. In this way, when a voltage is applied to the different BSLs in a row, the current is accumulated on the LL and is controlling the switching of output MTJ in the same line to complete the logic operation ((N)AND, (N)OR, MAJ, INV).

Pham et al. proposed that the binary weights are stored within the bitcells of the STT-MRAM memory array, which are assumed to be of 2T-2J type as in Fig. 8 [15, 16]. The gate of both access transistors M1 and M2 is connected to the WL. The MTJ0 is set either high (anti-parallel magnetization resistance R_{AP}) or low resistance (parallel magnetization resistance R_P) to respectively store $-1/+1$ weight. The other magnetic junction MTJ1 is set to the complementary state. For

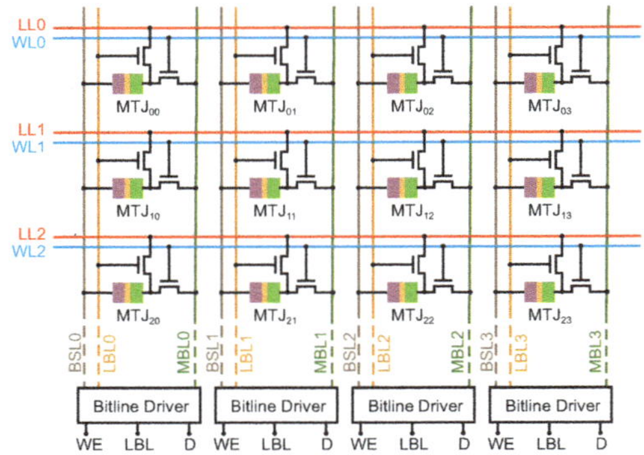

Fig. 7 The structure diagram of the array with 2T1MTJ

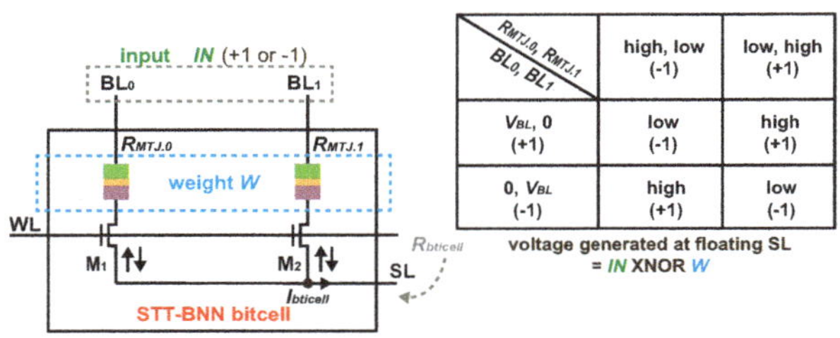

Fig. 8 2T-2J STT-MRAM bitcell for single-bit XNOR of input feature and weight

a single bitcell under a $+1(-1)$ input IN, $V_{BL,0} = V_{BL}$ and $V_{BL,1} = 0$ ($V_{BL,0} = 0$ and $V_{BL,1} = V_{BL}$), and the bitcell short-circuit current $I_{bitcell}$ is expressed as:

$$I_{bitcell}(IN = +1) = \begin{cases} \frac{V_{BL}}{R_P + R_{access,0}}, & W = +1 \\ \frac{V_{BL}}{R_{AP} + R_{access,0}}, & W = -1 \end{cases} \tag{1}$$

$$I_{bitcell}(IN = -1) = \begin{cases} \frac{V_{BL}}{R_P + R_{access,1}}, & W = +1 \\ \frac{V_{BL}}{R_{AP} + R_{access,1}}, & W = -1 \end{cases} \tag{2}$$

where $R_{access,0}$ and $R_{access,1}$ are the resistance of the access transistors M1 and M2, respectively, and there is $R_{access,0} = R_{access,1} = R_{access}$. The advantage of this structure is that the overall resistance $R_{bitcell}$ seen from the SL terminal of the bitcell is data-independent and equal to $(R_{MTJ,0} + R_{access})||(R_{MTJ,1} + R_{access})$. To sum up, the resulting voltage V_{SL} for an isolated active bitcell at position (i, j) within the memory array is:

$$V_{SL} = R_{bitcell} \cdot I_{bitcell,i,j} = \frac{V_{BL}}{X_{ij}} \tag{3}$$

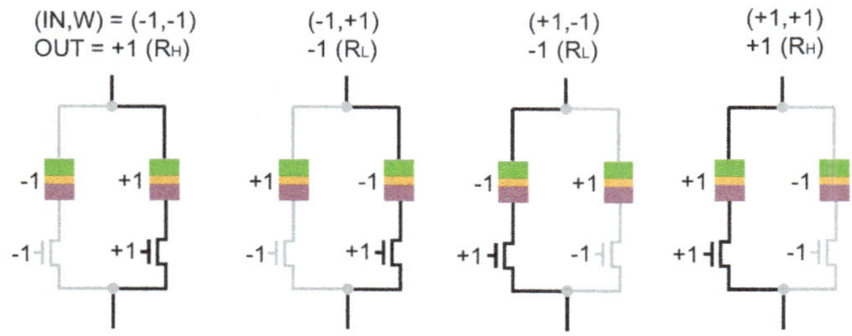

Fig. 9 The XNOR operation of a single cell of the crossbar array

where X_{ij} is defined as follows:

$$X_{ij} = \begin{cases} \frac{R_P + R_{access}}{R_{bitcell}}, & \overline{W_{ij} \oplus IN_j} = +1 \\ \frac{R_{AP} + R_{access}}{R_{bitcell}}, & \overline{W_{ij} \oplus IN_j} = -1 \end{cases} \tag{4}$$

In it, W_{ij} is the weight stored by the bitcell, and IN_j is the input feature map fed to column j. According to the above two formulas, the SL voltage V_{SL} is proportional to the BL voltage V_{BL} and directly depends on the XNOR of the input and the weight stored in the bitcell, as desired to implement multiplication with binary inputs and weights.

Another similar idea for cell structure changes is to implement cell-level operations as shown in Fig. 9 [17]. A 2T2MTJ cell is constructed by adding a MTJ and transistor to the 1T1MTJ cell of the crossbar array. Two columns of MTJs and transistors in the cell are in complementary states, and by controlling of the MTJ resistance and WL according to the input logic, the result of a XNOR operation is reflected in the current of the cell after the voltage is applied.

Cai et al. [18] proposed another way using two MTJs to represent $-1/+1$. As is shown in Fig. 10, one transistor-one MTJ (1T-1J) cell can be implemented with a pseudo 2T-1J structure, where the extra 1T is configured to increase writing current in a bitcell as a collateral access transistor. In this structure, a twin MTJ represents one synaptic weight; +1 is represented by two MTJs as parallel (P) and anti-parallel (AP) states. Its opposite combination represents -1. Two encoded word-lines (WLs) represent the input. The combination of input and weight is able to generate XNOR output within one bitcell. When input is -1, WL0 is within a low level, and WL1 is activated; two access transistors connected to WL1 are turned on. When input is +1, the situation is reversed. The parasitic resistance of source-line (SL as R_{SL}, SL-bar as R_{SLB}) is varied depending on the spin state of MTJ, and it is the output of the XNOR operation.

Double MTJ (DMTJ) is a promising solution since it has an additional reference layer that enhances the total torque acting on the free layer, allowing reduced switching current as compared to the single MTJ. In this case, the DMTJ-based CIM hardware has an advantage over SMTJ-based, and a structure of this kind

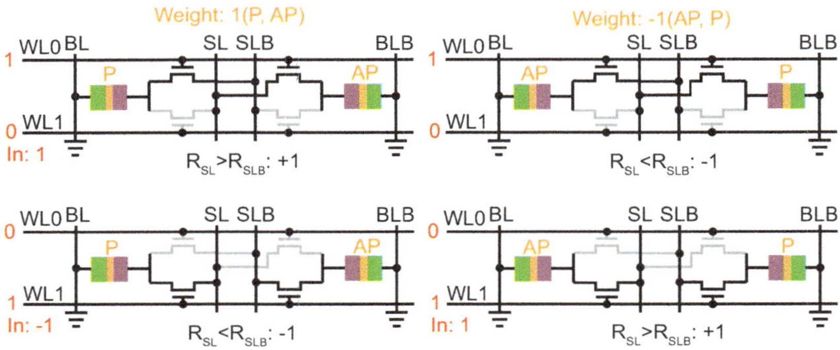

Fig. 10 Bitcell configurations to represent one weight using 4T2J

Fig. 11 XNOR operation exploiting STT-DMTJs

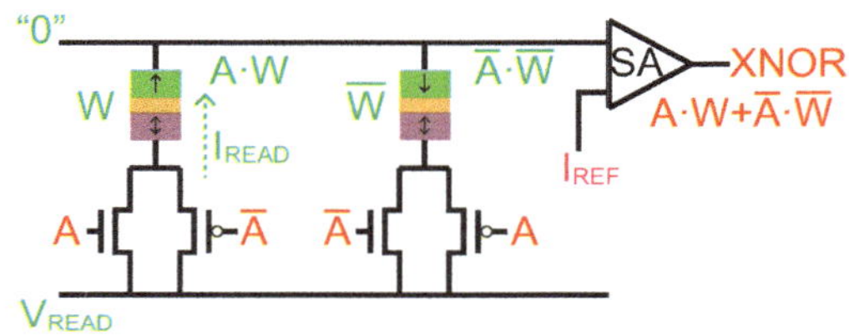

is proposed by Musello et al. as shown in Fig. 11 [19]. Two DMTJs of opposite states store 1-bit weight. By applying 1-bit input to the word-lines and reading both DMTJs simultaneously, the accumulated sensing current in the connected bit-lines represents the XNOR result.

3.4 Peripheral Circuit Modification

Another large category of logical operations for MRAM is by modifying the peripheral circuit structure of the array. And a large part of the modifications to the peripheral circuitry are focused on read circuit. In one category, the judgment criteria are modified, and in the other, the process of reading is controlled.

A method added the transistors between the read circuit and the MTJ to control the read process [20, 21]. As shown in Fig. 12, the MN 2,3,4 are added, and the read process only occurs when A is logic "1." In this way, combined with the resistance status of MTJ and the control signals A, the reading result is the AND result. However, for this method, the final result is generated by the read circuit requiring to write back to the memory. Also, different control transistors are required for different logic operations, resulting in a certain degree of disruption to the consistency of the array.

Zhang et al. proposed a time domain computing and a logic unit to the peripheral circuit [22]. Different logic corresponds to different resistance states, making a

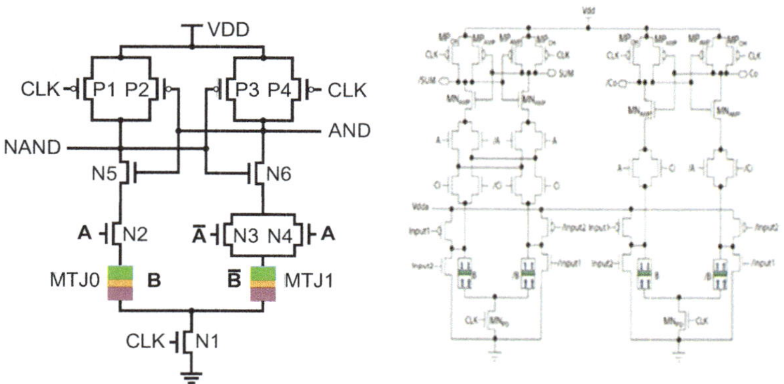

Fig. 12 The structure of the sense amplifier adding the transistor

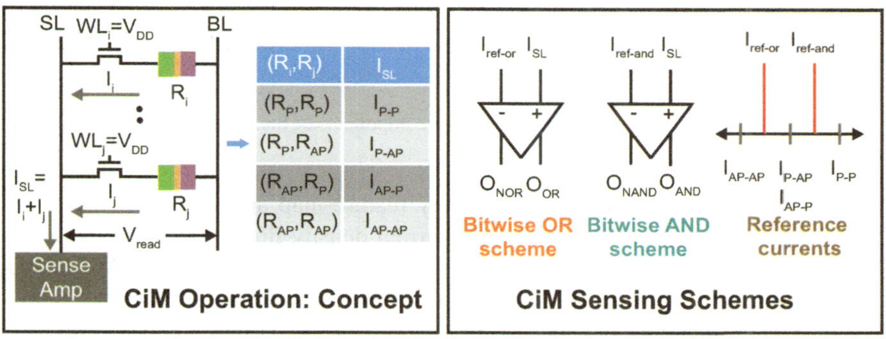

Fig. 13 The schematic diagram of read circuit judgment interval

difference in the discharge rate at the BL voltage, using the time domain computing module to react to the different resistance values as the switching time and then sampling at different times to achieve the corresponding logic.

The method of modifying the reading criteria is generally done by modifying the resistance value in the reference circuit so that the voltage/current during the reading process is within the appropriate range [23–25]. As shown in Fig. 13, when two MTJs are activated simultaneously, the accumulation current caused by the two resistances is divided into three levels, and two types of logic (AND, OR) are implemented using different sizes of judgment currents. For the other logic design, an idea is to use these two logics plus the peripheral CMOS circuitry.

Chang et al. used SOT-MRAM to implement XNOR operation as shown in Fig. 14a, which has three steps. In this work, the AP state of the MTJ is considered as True, while P state as False [26]. In the first step, all the MTJs in the array are reset to AP state. Then, the Write Driver modulates the BL and SL of each bitcell according to the input weight and feature map; then the XNOR result is stored in the MTJ. Finally, the result is read out. The structure of Write Driver is shown in

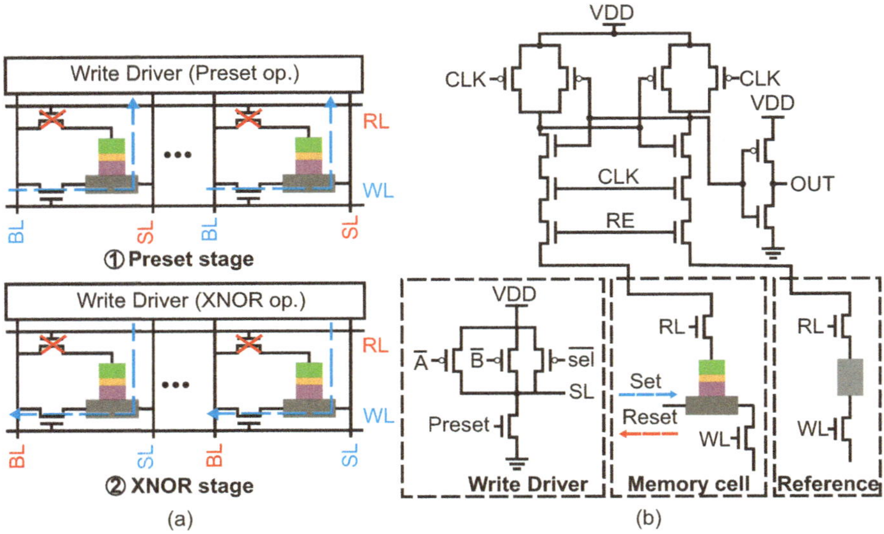

Fig. 14 (**a**) XNOR operation of the array; (**b**) the components of the array

Fig. 14b, in which A and B are the binary feature map and weight, respectively. By changing the states of A and B, the current density flowing through the heavy metal (HM) can be modulated to program the bitcell to different states. When both A and B are False, the current flowing through the HM is so small that the state of bitcell remains unchanged. If both of them are True, the current is excessively large to induce a huge SOT. In this case, the precession dominates the magnetization dynamics since the field-like torque is stronger than the damping-like torque. As a result, the memory cell returns to the initial state. Only one of A and B is True, and an intermediate current could switch the bitcell to the opposite state, as expected by the classic SOT theory. In this approach, the memory array is more for computation rather than storage, because the weights are not stored in the array.

4 MRAM for Neural Networks

In recent years, as a promising intelligent algorithm, neural network has been widely used in basic tasks such as computer vision (CV), natural language processing (NLP), and robot decision-making. Taking computer vision task as an example, convolutional neural network (CNN) has achieved performance beyond humans in sub-tasks such as image classification [27, 28], object detection [29, 30], and semantic segmentation [31, 32] and is playing an important role in emerging applications such as smart cities, autonomous driving, and smart homes. However, the huge amount of computation and parameter of CNN makes it an energy-hungry algorithm, which has high demand for computing power and storage.

In terms of the deployment of neural network, from CPU, GPU to FPGA (or ASIC) accelerator, the speed and energy efficiency of hardware to realize neural network inference is constantly improving. In these hardware based on Von-Neumann architecture, the computing unit is separated from the memory, which results in frequent direct data interaction between them. The research shows that the energy consumption caused by a single data movement is much higher than that caused by a single operation, and the restricted bandwidth between the computing unit and the memory limits the speed of data transmission, thus reducing the efficiency of neural network inference. Although some works achieve the purpose of reducing the data transmission amount and realizing efficient calculation through pruning, quantization, and other model compression methods as well as software-hardware collaborative design [33–36], the data transmission amount is still considerable, and the energy consumption and delay brought by it cannot be ignored.

In this case, computing-in-memory (CIM) techniques are proposed to solve the aforementioned "memory wall" problem. In the CIM-based hardware, the memory array and its external circuits are adjusted to make the modified memory perform computation and store data in the same area, thus realizing computing in the memory and greatly reducing data movement.

In the existing neural network, there is one whose weight and feature map can be represented by 1-bit, which is called binary neural network (BNN). Each 1-bit weight element can be easily stored in each cell of the MRAM, and each 1-bit element of the feature map can be applied to the bit line of the MRAM array. Thus, there is a good match between BNN and MRAM which represents only 1-bit data per cell, and the corresponding MRAM-based digital CIM hardware makes relatively little change to the memory array.

4.1 CNN and BNN Basics

As shown in Fig. 15, the weight parameters of a convolution layer can be divided into C_{out} filters, which correspond to a channel of output feature map. Each filter can be divided into C_{in} convolution kernels corresponding to a channel of input feature map, respectively. The weight parameter W_l shape of each layer can be written as $(C_{out}, C_{in}, K_H, K_W)$, where K_H and K_W are the height and width of each convolution kernel, correspondingly. The shape of the input feature map X for each layer is (C_{in}, H_{in}, W_{in}), while the shape of the output feature map Y for each layer is $(C_{out}, H_{out}, W_{out})$. To sum up, the calculation process is formulated as follows:

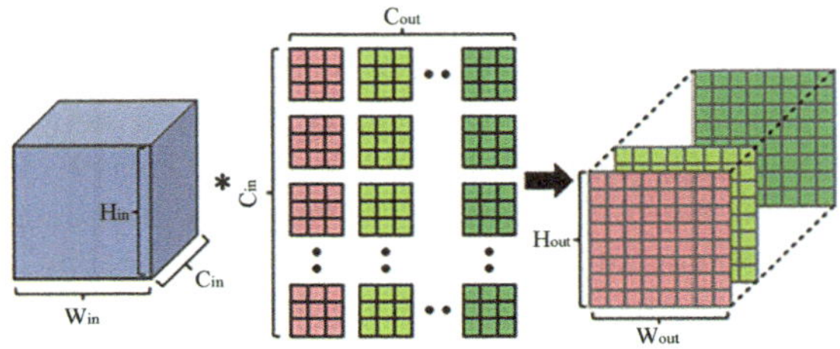

Fig. 15 The calculation flow of convolutional layer

Table 2 Truth table of multiplication and XNOR operation

		Feature map	
Multiplication		+1	−1
Weight	+1	+1	−1
	−1	−1	+1
		Feature map	
XNOR		True	False
Weight	True	True	False
	False	False	True

$$Y_{n,i,j} = \sum_{m=1}^{C_{in}} \sum_{p=1}^{K_H} \sum_{q=1}^{K_W} W_{n,m,p,q} X_{m,i\times s+p,j\times s+q} \tag{5}$$

where n, i, and j are the index in the dimension of channel, height, and width and s is the stride of convolution. The fully connected layer can be viewed as a special case of convolution. The shape of the fully connected layer's weight can be written as $(C_{out}, C_{in}, 1, 1)$, while the shape of the input feature map is $(C_{in}, 1, 1)$. In this case, the calculation formula of the convolutional layer can be multiplexed with the fully connected layer.

In BNN, there are only two possible values for the weight and the input feature map of the convolutional layer, so they can both be represented by 1-bit [37–39]. Usually, the two values are -1 and $+1$. Furthermore, the convolution formula in the form of multiplication-addition above is usually not used in BNN, because the multiplier usually has large area and power consumption and is not conducive to be fused into the memory array. In this case, based on the characteristics of BNN, using bit-wise logic operation to realize convolution becomes a better solution.

As can be seen from Table 2, the multiplication operation is equivalent to the XNOR operation, if 1 is mapped to True and -1 to False. Therefore, multiplication in convolution operations can be replaced by XNOR to reduce hardware overhead. The aforementioned convolution process in BNN is modified as follows:

$$Y_{n,i,j} = bitcount_{m=1,p=1,q=1}^{C_{in},K_H,K_W}(W^b_{n,m,p,q} xnor X_{m,i\times s+p,j\times s+q}) \tag{6}$$

where the result of the bitcount operation is the number of True clipped off the number of False. Usually, this operation requires the digital arithmetic component outside the memory array to complete, which increases the area overhead. Compared with the traditional multiplication-addition convolution operation, the XNOR-bitcount form is significantly smaller in terms of hardware overhead. Then, the binarization layer is attached after the convolutional layer, which is expressed as below:

$$y = \begin{cases} 1, & x > 0 \\ -1, & x \le 0 \end{cases} \quad (7)$$

where x and y are the input and output of the binarization layer.

In the following sections, the implementation of convolutional operations (composed of XNOR operation and bit-count operation), max-pool operations, and online training of neural networks (especially BNN) using MRAM will be introduced. As the implementation of the XNOR logic operation is described in the previous section, it will not be described subsequently. In addition to using MRAM to implement neural network inference in the digital domain, methods for analog computing in the analog domain will also be described.

4.2 *MRAM for XNOR Operation*

When XNOR operation is implemented in memory, the modification of memory array mainly lies in the following: (1) The storage mode of $-1/+1$ determines the organization form of the array. (2) The mechanism by which XNOR is implemented determines the form of the peripheral circuit. The implementation of the XNOR operation is described in detail in Sect. 3.

4.3 *MRAM for Bit-Count Operation*

Bit-count operation is used to sum up the XNOR result, and the following binarization layer is usually fused into it. It is an obvious bottleneck when CIM-based hardware implements BNN. Usually the digital logic component needs to be added in high-density CIM array to do this operation [40], which increases the energy dissipation, area and data movement between the memory array and data buffer.

To deal with this challenge, Chang et al. used low-cost majority-write on SOT-MRAM to replace bit-count, shown in Fig. 16 [41]. The SOT-MRAM is a current-density threshold device, which means the bitcell can be programmed (the state from P to AP or AP to P) only by providing enough current. Firstly, the SOT-MTJ

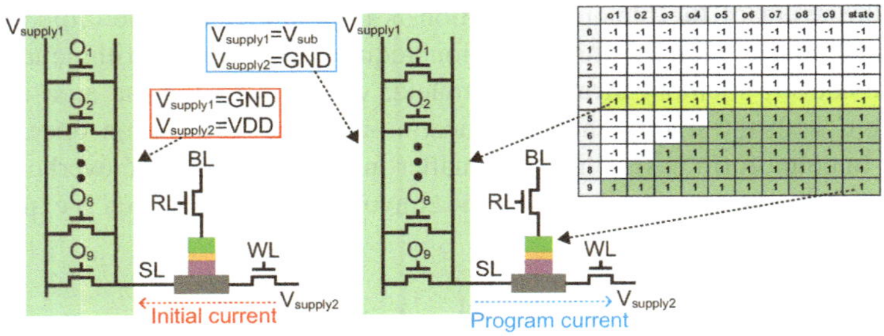

Fig. 16 Majority-write to replace bit-count

is preset to P state (low resistance of SOT-MTJ or binary -1) with an initial current from terminal T3 to T2 of the SOT-MTJ (as shown in the left part of the figure). Then, the $V_{supply1}$ is down to V_{sub}, and several NMOS transistors are activated to provide current for the HM as shown in the right part. The state of the SOT-MTJ can be programmed (from P to AP state) only when more than five NMOS transistors are activated (nine NMOS totally in this example). This writing process matches the truth table described in the figure, and the number of -1 presents deactivation of the NMOS transistor. In this structure, each NMOS transistor is activated by a XNOR result, and the bit-count & binarization result is stored in the SOT-MTJ.

Use the structure shown in Fig. 17 as an example, which shows the memory array which consists of M rows and N columns (i.e., 2N bitlines), with wordlines being used to enable computation in the respective rows and the source lines as outputs. The bitline (BL) drivers encode the binary input feature map elements of the neural network. When considering a single row as depicted in blue in the figure, suppose the number of XNOR outputs equal to $+1(-1)$ across the row be N_1 ($N_0 = N - N_1$). Compared to the single bitcell resistance $R_{bitcell}$ in the previous section, the resistance seen from the SL terminal in a row is reduced to $R_{bitcell}$ /N. The select line voltage $V_{SL,i}$ in the considered row i-th is:

$$V_{SL,I} = \frac{R_{bitcell}}{N} \sum_{j=0}^{N-1} I_{bitcell,ij} = V_{BL} \frac{R_{bitcell}}{N} \sum_{j=0}^{N-1} \frac{1}{X_{ij}}$$

$$= V_{BL} \left(\frac{N_0}{N} \cdot \frac{R_{bitcell}}{R_{AP+R_{access}}} + \frac{N_1}{N} \cdot \frac{R_{bitcell}}{R_{AP+R_{access}}} \right) \tag{8}$$

where X_{ij} is the resistance determining the bitcell current of the bitcell (i, j). Its mid-point value (i.e., 0) is $V_{SL,50\%} = 0.5V_{BL}$. The output binarization is performed by assigning the output OUT_i of the select line in the generic i-th row to +1 if $V_{SL,i} \geq V_{SL,50\%}$ and to -1 if $V_{SL,i} < V_{SL,50\%}$.

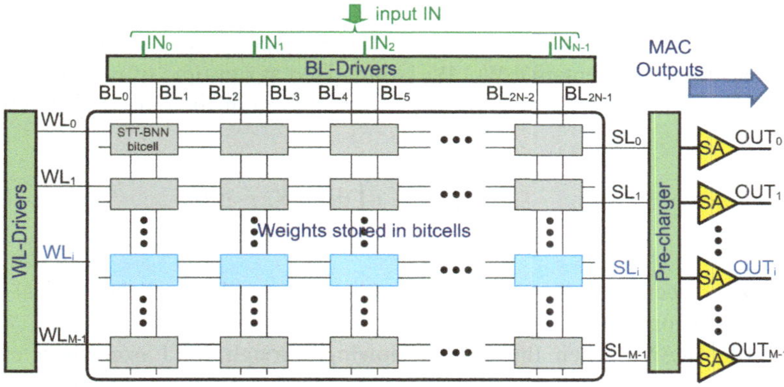

Fig. 17 Memory array using 2T-2J STT-MRAM bitcell

4.4 MRAM for Max-Pool Operation

Max-pool operation is widely used in neural networks to reduce the size of feature maps, and so is BNN. The most common 2x2 max-pool requires only the largest element to be selected from the four adjacent feature map elements, and the overhead of implementing it with digital logic is not obvious. However, it is still necessary to implement this operation efficiently in the form of analog circuits.

Chang et al. also proposed to do max-pool by modifying the structure in Fig. 16 [41]. For instance, a max-pool window consists of x = {x1, x2, x3, x4}, if only all elements of x equal to -1, the max-pool(x) = -1, and max-pool(x) = 1 if x contains +1 element (e.g, max-pool[1, -1, 1, -1] = 1). [o1, o2, o3, o4, o5] is set to [11110], and four bits of the max-pooling window control the other four NMOS transistors. Therefore, the SOT-MTJ can be successfully written only when a "1" exists in the max-pooling window.

4.5 MRAM for BNN Training

Almost all the CIM-based hardware for neural networks inference currently uses offline training, that is, computer completes neural network training, and then the network structure and weight parameters obtained by training are mapped to the hardware. However, offline training may restrict the algorithmic upgrading of intelligent hardware especially when data transmission between Edge and Cloud (upgraded weight from Cloud to Edge, sensor data from Edge to Cloud) is not permitted because of the concerns on energy consumption, latency, and data security. In this case, it is critically important to equip intelligent hardware with not only offline but also online training capability.

Chang et al. made the online training of BNN on STT-MRAM possible [42]. In BNNs, the binary weights W_b are utilized for simplifying convolution calculation. To train a set of representative W_b, W_b is binarized at run-time of training from an float-point weight WFP, for example, by using a deterministic binarization function as:

$$W_b = sign(W_{FP}) = \begin{cases} +1, & W_{FP} \geq 0 \\ -1, & W_{FP} < 0 \end{cases} \tag{9}$$

where W_{FP} accumulates real-valued gradients according to the typical back propagation algorithm during training.

In STT-MRAM, when the current flowing through MTJ exceeds a critical current I_c, sufficient angular momentum transferring from the fixed layer flips the magnetization of the free layer, called spin-transfer torque effect. When the applied current is below I_c, the magnetization flipping in the free layer is assisted by stochastic thermal activation over barrier. By applying a programming current pulse, I with duration Δt to MTJ, the switching probability P_{sw} is expressed as:

$$P_{sw}(\Delta t) = 1 - \exp\left(- \frac{\Delta t}{\tau} \right) \tag{10}$$

Different from the training scheme on computer using W_{FP} to accumulate real-valued gradients, real-valued gradients are accumulated directly on the binary MRAM by using appropriate programming duration t_P and t_D corresponding to the gradients. The cumulative programming duration t_P^i and t_D^i after the i-th training stimulus follows simple monotonic addition rules as:

$$\begin{aligned} t_P^i &= t_P^{i-1} + \Delta t_P^i - 0 \times \Delta t_D^i, W_b^0 = -1 \\ t_D^i &= t_D^{i-1} + \Delta t_D^i - 0 \times \Delta t_P^i, W_b^0 = +1 \end{aligned} \tag{11}$$

where $W_b^0 = -1$ and $+1$ represent the present value of the weight. Only the positive gradient is accumulated when $W_b^0 = -1$, and only the negative gradient is accumulated when $W_b^0 = +1$. When t_P^i and t_D^i accumulate to a certain extent, the status of MRAM is reversed, that is, the weight is updated, and t_P^i and t_D^i are reset to 0.

4.6 MRAM for Analog Computing

Compared with digital CIM hardware, analog CIM hardware has higher on-chip bandwidth and computational area efficiency. ReRAM is the most commonly used device for analog in memory computing. In ReRAM arrays, matrix-vector multiplication can be easily achieved by Kirchhoff's law, since each bitcell can

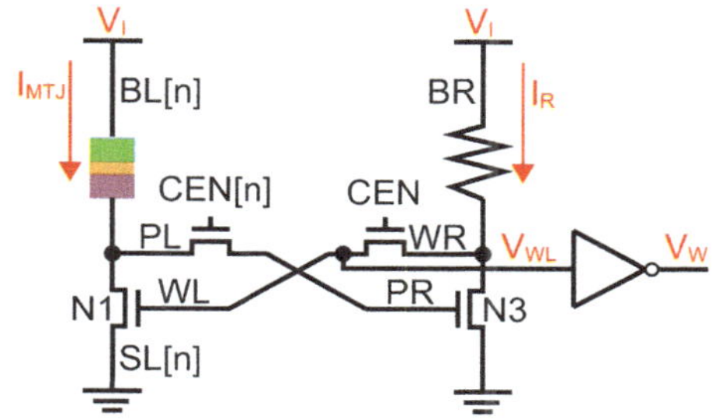

Fig. 18 Bitcell with latch structure

represent multiple bits of data by different conductances. However, ReRAM is limited by its high process/temperature fluctuation, high programming cost, and poor write durability.

MRAM has less process deviation and lower write voltage compared to ReRAM, but MRAM cannot be directly used for analog IMC due to limited tunnel magneto resistance (TMR) ratio. MRAM's 100%–200% TMR ratio is overwhelmingly difficult to fulfill the requirement of high dimensional matrix-vector multiplication. Although TMR has been enhanced to 249% [43], the discrimination of anti-parallel (AP) and parallel (P) resistance is not sufficient, which can be influenced by the process variations. Low TMR could reduce the impact of internally stored data on the calculation results, resulting in a mismatch with external input data.

To solve the aforementioned challenge, Cai et al. proposed to use a latch structure combining with the conventional MRAM bitcell shown in Fig. 18, to realize an enlarged virtual TMR ratio for analog computing [44]. And Patil proposed a new analog computing structure with an operational amplifier integrator circuit, which uses the difference of the current flowing through RP and RAP to complete the calculation [45]. This method alleviates the problem of limited TMR, but it has a high requirement for the performance of operational amplifier.

5 MRAM for Other Applications

MRAM is not only widely used in neural network computing but also in various other scenarios such as triangle counting, true random number generators, and DNA alignment.

5.1 DNA Read Alignment

A DNA sequence is a collection of four nucleotide base pairs—adenine (A), cytosine (C), guanine (G), and thymine (T). The base sequences are identified by the sequencing machine, which encodes the sequence using a string of characters

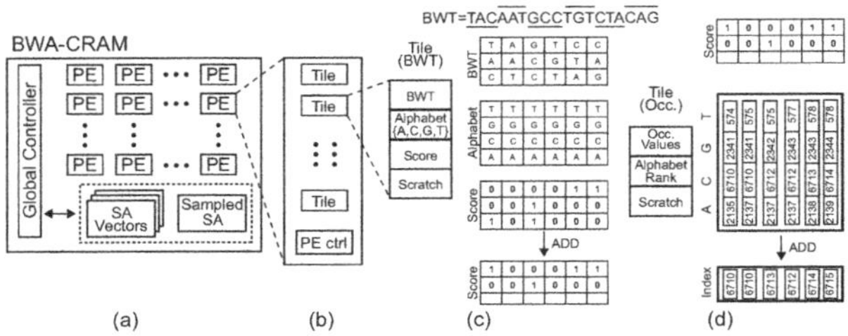

Fig. 19 The high-level architecture of BWA-CRAM. (**a**) Functional blocks. (**b**) Architecture of PE. (**c**) Data layout and BWT search in PE. (**d**) Index computation in PE

A, C, G, T. DNA sequence alignment is the problem of finding out the location of the highest degree of similarity between a very long sequence (a reference) and an order of magnitude shorter sequence (a read). For Burrows–Wheeler transform (BWT) algorithm-based DNA read alignments, the data includes BWT sequences, suffix arrays, incidence tables (OCC), and count tables, which are generated one time for a given reference database on a conventional von Neumann system and stored in the MRAM array.

The key steps in the DNA read alignment are to find the matching position in the BWT based on the letters in the read sequence and to calculate the index value of the matching letters in the BWT [46]. As shown in Fig. 19c, each column in the array has BWT data, Alphabet data, and reserved Score positions. The matching process is carried out by in-memory XOR operation between the BWT data and the Alphabet data, and the result is stored at "Score." Then the number of corresponding letter matches in each column is obtained by in-memory addition. Through the column parallelism of the array, multiple columns can be matched simultaneously within each tile. And with independent control modules between the tiles, the parallelism at the tile level leads to a more efficient matching process. The calculation of the index result is mainly achieved by in-memory addition, by adding the matching value to the corresponding data in the OCC table.

5.2 Triangle Counting

Given a graph, triangle counting (TC) problem seeks to determine the number of triangles. The A is the adjacency matrix representation of a undirected graph G, where $A[i][j] \in \{0, 1\}$ indicates whether there is an edge between nodes i and j. If we compute $A^2 = A * A$, then the value of $A^2[i][j]$ represents the number of distinct paths of length two between nodes i and j. In the case that there is an edge between node i and node j, and i can also reach j through a path of length two, where the intermediate node is k, then the i, j, and k form a triangle. As a result, the number of triangles in G is equal to the number of non-zero elements(nnz) in $A \cap A^2$. After a series of transformations, TC can be completed by only AND and

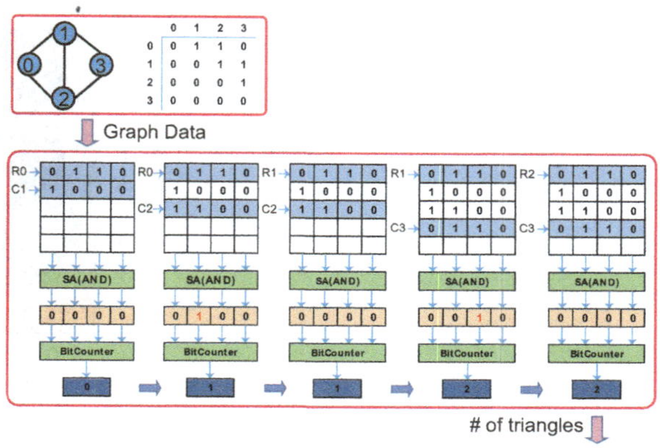

Fig. 20 Demonstrations of triangle counting with AND and BitCount bit-wise operations

BitCount operations as shown in [47]. And the implementation of logical operation is shown in Fig. 20.

$$TC(G) = BitCount(AND(A[i][*], A[*][j]^T)),$$
$$\text{in which } A[i][j] = 1 \tag{12}$$

As shown in Fig. 20, the inter-row AND operation is implemented by the idea of modifying the judgment threshold of read circuit with the peripheral circuit modification raised above. By taking advantage of the parallelism of columns of the array, multi-column simultaneous AND operations are achieved. And the BitCount operation is implemented via the array's peripheral accumulation circuitry.

5.3 True Random Count Generators

True Random Number Generators (TRNG) are specialized circuits used in a wide variety of applications ranging from cryptography and hardware-based security to statistical sampling and advanced simulation techniques. Traditional CMOS-based TRNGs utilize physical noise present in CMOS circuits such as thermal noise, random telegraph noise, and oscillator jitter. However, existing CMOS-based TRNGs require extensive post-processing to ensure a high level of randomness in the output bits, which incurs a significant performance, power, and area overhead. The TRNG based on the random switching probability of Magnetic Tunnel Junctions is proposed [48]. The STT switching phenomenon in an MTJ is subject to random thermal fluctuation noise which gives rise to a switching probability contour map as shown in Fig. 21. By applying an optimal "perturb" pulse whose width and amplitude correspond to the 50% switching probability contour, the final resolved state of the MTJ will depend solely on the random thermal noise, producing an unbiased random output bit.

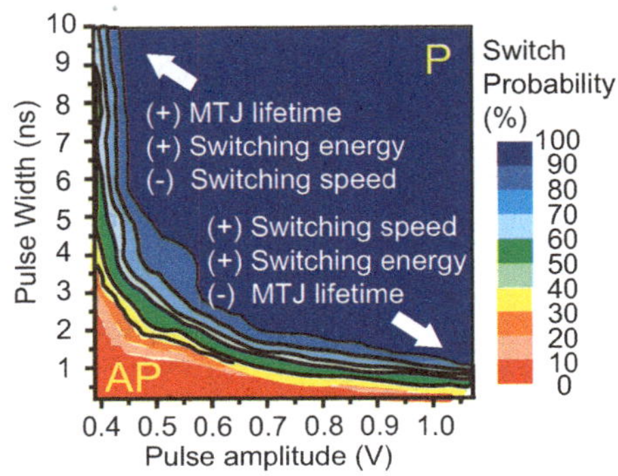

Fig. 21 MTJ switching probability as a function of pulse width and pulse amplitude (AP → P switching direction)

References

1. S. Angizi, Z. He, A. Awad, D. Fan, https://doi.org/10.1109/TCAD.2019.2907886
2. M. Horowitz, https://doi.org/10.1109/ISSCC.2014.6757323
3. H. Zhang, W. Kang, K. Cao, B. Wu, Y. Zhang, W. Zhao, https://doi.org/10.1109/TED.2019.2898391
4. J. Hirsch, https://doi.org/10.1103/PhysRevLett.83.1834
5. Z. Wang, W. Zhao, E. Deng, J. Klein, C. Chappert, https://doi.org/10.1088/0022-3727/48/6/065001
6. A. Brink, S. Cosemans, S. Cornelissen, M. Manfrini, A. Vaysset, W. Roy, T. Min, H. Swagten, B. Koopmans, https://doi.org/10.1063/1.4858465
7. A. Patil, H. Hua, S. Gonugondla, M. Kang, N. Shanbhag, https://doi.org/10.1109/ISCAS.2019.8702206
8. B. Wu, H. Zhu, D. Reis, Z. Wang, Y. Wang, K. Chen, W. Liu, F. Lombardi, X. Hu, https://doi.org/10.1109/TETC.2023.3237541
9. H. Zhu, B. Wu, K. Chen, C. Yan, W. Liu, https://doi.org/10.1109/NANO54668.2022.9928764
10. J. Wang, Y. Bai, H. Wang, Z. Hao, G. Wang, K. Zhang, Y. Zhang, W. Lv, Y. Zhang, https://doi.org/10.1109/TCSI.2022.3192165
11. H. Zhang, W. Kang, B. Wu, P. Ouyang, E. Deng, Y. Zhang, W. Zhao, https://doi.org/10.1109/TNANO.2019.2914009
12. K. Monga, N. Chaturvedi, S. Gurunarayanan, https://doi.org/10.1109/TMAG.2021.3059268
13. S. Shreya, G. Verma, S. Piramanayagam, B. Kaushik, https://doi.org/10.1109/TED.2020.3038140
14. M. Zabihi, Z. Chowdhury, Z. Zhao, U. Karpuzcu, J. Wang, S. Sapatnekar, https://doi.org/10.1109/TC.2018.2858251
15. T. Pham, Q. Trinh, I. Chang, M. Alioto, https://doi.org/10.1109/JETCAS.2022.3169759
16. T. Pham, Q. Trinh, I. Chang, M. Alioto, https://doi.org/10.1109/ISCAS51556.2021.9401695
17. S. Jung, H. Lee, S. Myung, H. Kim, S. Yoon, S. Kwon, Y. Ju, M. Kim, W. Yi, S. Han, B. Kwon, B. Seo, K. Lee, G. Koh, K. Lee, Y. Song, C. Choi, D. Ham, S. Kim, https://doi.org/10.1038/s41586-021-04196-6
18. H. Cai, Z. Bian, Z. Fan, B. Liu, L. Naviner, https://doi.org/10.1109/TED.2021.3134588
19. A. Musello, E. Garzón, M. Lanuzza, L. M. Prócel, R. Taco, https://doi.org/10.1109/TCSII.2023.3241163
20. S. Dikshit, A. Nisar, S. Dhull, N. Bindal, B. Kaushik, https://doi.org/10.1109/OJNANO.2022.3188768
21. E. Deng, Y. Zhang, J. Klein, D. Ravelsona, C. Chappert, W. Zhao, https://doi.org/10.1109/TMAG.2013.2245911

22. Y. Zhang, J. Wang, C. Lian, Y. Bai, G. Wang, Z. Zhang, Z. Zheng, L. Chen, K. Zhang, G. Sirakoulis, Y. Zhang, https://doi.org/10.1109/TCSI.2021.3055830
23. Z. He, S. Angizi, D. Fan, https://doi.org/10.1109/ISVLSI.2018.00103
24. V. Nehra, S. Prajapati, T. Kumar, B. Kaushik, https://doi.org/10.1109/TMAG.2021.3084869
25. S. Jain, S. Sapatnekar, J. Wang, K. Roy, A. Raghunathan, https://doi.org/10.23919/date.2018.8342277
26. L. Chang, X. Ma, Z. Wang, Y. Zhang, Y. Xie, W. Zhao, https://doi.org/10.1109/TVLSI.2019.2926984
27. K. He, X. Zhang, S. Ren, J. Sun, https://doi.org/10.1109/cvpr.2016.90
28. K. Han, Y. Wang, Q. Tian, J. Guo, C. Xu, C. Xu, https://doi.org/10.1109/CVPR42600.2020.00165
29. Q. Chen, Y. Wang, T. Yang, X. Zhang, J. Cheng, J. Sun, https://doi.org/10.1109/CVPR46437.2021.01284
30. A. Bochkovskiy, C. Wang, H. Liao, https://doi.org/10.48550/arXiv.2004.10934
31. M. Fan, S. Lai, J. Huang, X. Wei, Z. Chai, J. Luo, X. Wei, https://doi.org/10.1109/CVPR46437.2021.00959
32. Y. Nirkin, L. Wolf, T. Hassner, https://doi.org/10.1109/CVPR46437.2021.00405
33. T. Yuan, W. Liu, J. Han, F. Lombardi, https://doi.org/10.1109/TCSI.2020.3030663
34. T. Yu, B. Wu, K. Chen, C. Yan, W. Liu, https://doi.org/10.1145/3526241.3530318
35. T. Yu, B. Wu, K. Chen, C. Yan, W. Liu, https://doi.org/10.1109/ICCD56317.2022.00116
36. Y. Song, B. Wu, T. Yuan, W. Liu, https://doi.org/10.1109/ISQED54688.2022.9806216
37. I. Hubara, D. Soudry, R. Yaniv, https://doi.org/10.48550/arXiv.1602.02505
38. M. Rastegari, V. Ordonez, J. Redmon, A. Farhadi, https://doi.org/10.1007/978-3-319-46493-0-32
39. A. Bulat, G. Tzimiropoulos, https://doi.org/10.48550/arXiv.1909.13863
40. T. Na, https://doi.org/10.1109/TCSII.2023.3241446
41. L. Chang, X. Ma, Z. Wang, Y. Zhang, W. Zhao, Y. Xie, https://doi.org/10.23919/DATE.2019.8715265
42. C. Chang, M. Wu, J. Lin, C. Li, V. Parmar, H. Lee, J. Wei, S. Sheu, M. Suri, T. Chang, T. Hou, https://doi.org/10.1145/3316781.3317872
43. M. Wang, W. Cai, K. Cao, J. Zhou, J. Wrona, S. Peng, H. Yang, J. Wei, W. Kang, Y. Zhang, https://doi.org/10.1038/s41467-018-03140-z
44. H. Cai, Y. Guo, B. Liu, M. Zhou, J. Chen, X. Liu, J. Yang, https://doi.org/10.1109/TCSI.2022.3140769
45. A. Patil, H. Hua, S. Gonugondla, M. Kang, N. Shanbhag, https://doi.org/10.1109/ISCAS.2019.8702206
46. Z. Chowdhury, M. Zabihi, S. Khatamifard, Z. Zhao, S. Resch, J. Wang, S. Sapatnekar, U. Karpuzcu, https://doi.org/10.1109/JXCDC.2020.2987527
47. X. Wang, J. Yang, Y. Zhao, Y. Qi, M. Liu, X. Cheng, X. Jia, X. Chen, G. Qu, W. Zhao, https://doi.org/10.1109/DAC18072.2020.9218660
48. W. Choi, Y. Lv, J. Kim, A. Deshpande, G. Kang, J. Wang, C. Kim, https://doi.org/10.1109/IEDM.2014.7047039

Is Neuromorphic Computing the Key to Power-Efficient Neural Networks: A Survey

Muhammad Hamis Haider, Hao Zhang, S. Deivalaskhmi, G. Lakshmi Narayanan, and Seok-Bum Ko

1 Introduction to Biomorphism

Biomorphism is a design philosophy that takes inspiration from nature and its organic forms. It is a concept that has become popular in modern technology to create products that are not only functional but also aesthetically pleasing. One example of biomorphism in technology is the design of airplanes, which was inspired by the flight of birds.

One area where biomorphism has been particularly influential is in the design of transportation vehicles, such as cars, trains, and airplanes. For example, the design of the Japanese Shinkansen bullet train was inspired by the shape of a kingfisher's beak. The streamlined, aerodynamic shape of the beak allows the bird to dive into water without causing a splash, and this same shape has been used to reduce the noise and vibration of the train as it moves through the air at near supersonic speeds [1].

Biomorphism has also been used in the design of medical devices, such as prosthetic limbs and artificial organs. By mimicking the structure and function of natural tissues and organs, designers are able to create devices that are more biocompatible and have better performance. For example, researchers at the Uni-

M. H. Haider · S.-B. Ko (✉)
Department of Electrical and Computer Engineering, University of Saskatchewan, Saskatoon, SK, Canada
e-mail: hamis.haider@usask.ca; seokbum.ko@usask.ca

H. Zhang
Faculty of Information Science and Engineering, Ocean University of China, Qingdao, China
e-mail: hao.zhang@ouc.edu.cn

S. Deivalakshmi · G. Lakshmi Narayanan
National Institute of Technology, Tiruchirapplli, India
e-mail: deiva@nitt.edu; laksh@nitt.edu

W. Liu et al. (eds.), *Design and Applications of Emerging Computer Systems*,
https://doi.org/10.1007/978-3-031-42478-6_4

versity of Michigan have developed an artificial lung that uses micro-channels and hollow fibers to mimic the structure of the human lung. This design allows the lung to exchange gases more efficiently and has the potential to improve the lives of patients with respiratory failure. By taking inspiration from nature, they are able to create products that are more in tune with our natural environment and better serve the needs of society. It is of no surprise that modern artificial intelligence (AI) paradigms are also benefiting from the influence of biomorphism through neuromorphic computing.

Neuromorphic computing is a type of computing that is modelled after the human brain. It is designed to mimic the neural networks of the brain, which are made up of interconnected neurons that communicate through electrical impulses. Neuromorphic computing has the potential to revolutionize the way we process information and solve complex problems. Spiking Neural Networks (SNNs) are a type of neuromorphic computing that is based on the way neurons in the brain communicate through spikes of electrical activity. SNNs are widely regarded as the potential generational successor to DNNs due to their higher computational capabilities and power efficiency. This chapter will highlight different models of SNNs, their features, and their hardware implementation. It will also discuss the higher energy efficiency of SNNs compared to the traditional deep neural networks (DNNs).

2 Understanding the Biological Neuron

Neurons are the primary computational elements packed in a dense network in our brain. The complicated network of neurons in our brains is capable of efficient computations at incredibly low power. The primary biological structure of a neuron consists of three major components: dendrites, cell body (soma), and the axon. The dendrites perform the task of collecting incoming signals and pass it on to the soma. The soma is expected to perform non-linear computations on the signals collected by the dendrites. The output signal is transferred to the connected neurons through the axon.

The long-held scheme for inner working of neurons assumed that they sum all the incoming electrical signals. If the sum exceeds the membrane potential, then the neuron fires a signal through its axon. Existing neural networks perform on the basis of this assumption. However, we've since then found out that this computational scheme is a gross simplification.

The study conducted in the esteemed nature journal [2] concluded that the existing neuron model is misleading. They conducted three experiments to demonstrate that their proposed model for biological neurons was far more plausible than the existing models at that time. However, even in their study, the most fascinating outcome was that it was thoroughly established that even their model fails to completely model the complete intricate nature of the neuron. The authors of this study hypothesize that the biological neuron can be of three possible models, as shown in Fig. 1.

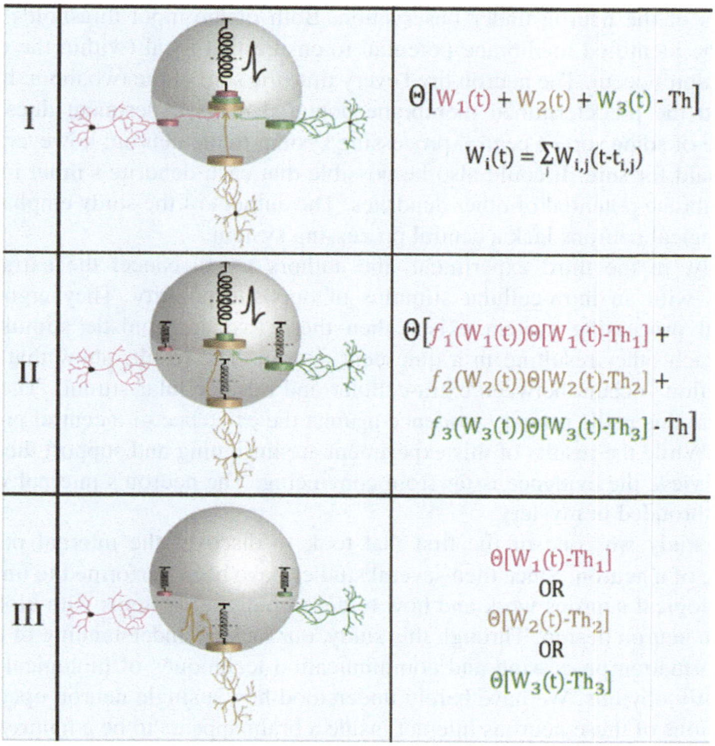

Fig. 1 Three plausible models for a biological neuron hypothesized in [2]. The left-hand side shows a graphical representation of the model, whereas the right-hand side shows an equivalent equation to model the neuron mathematically

The author's first experiment debunked the claim that there exists a unique membrane potential in a neuron that should be dominated to make the neuron fire. The authors applied and increased voltage at one dendrite of the neuron at a time; they increased the voltage until the neuron emits a output waveform. The experiment concluded that the membrane potential would vary depending on which dendrite was receiving the input signals and that the output waveform was dependent on the spatial parameters (where the input was applied) as well. The spatial component of the neuron processing is entirely missing from the traditional artificial neurons used in most deep neural networks (DNN). To a biological neuron, it matters where the input is coming from, and its capability to modify its operation based on this spatial dimension results in a higher-dimension computation than the artificial neurons. Additionally, this experiment hints that there are local computational capabilities in the dendrites of the neurons. How they fit into the overall structure of the neuron is not fully discovered yet.

The second experiment of this study investigates the existence of a central integral threshold in the neuron. They apply separate input waveforms at two dendrites of the neuron under observation. Both of the input thresholds are kept below the identified membrane potential to ensure that local (within the dendrite) triggers don't occur. The neuron fired every time the sum of the two input thresholds exceeded the predetermined membrane potential. This experiment does provide evidence of some sort of central processing system in the neuron; however, nothing can be said for sure. It could also be possible that each dendrite's input influences the membrane potential of other dendrites. The authors of the study emphasize that the biological neurons lack a central processing system.

Finally, in the third experiment, the authors try to cancel the extra-cellular stimulus with an intra-cellular stimulus of opposite polarity. They argue that if a central processing system exists, then these two antagonistic stimuli should cancel each other resulting in a quiescent neuron. The results show that no such "subtraction" occurs between extra-cellular and intra-cellular stimuli. The authors claim that this is the primary evidence against the existence of a central processing system. While the results of this experiment are intriguing and support the author's point of view, the evidence is far from convincing. The neuron's internal workings remain shrouded in mystery.

This study was one of the first that took to discover the internal processing structure of a neuron. Since then several studies have been performed to understand how biological neurons work and how to develop digital systems with biologically plausible neuron design. Through this study, our lack of understanding of the intricate information processing and communication techniques of biological neurons is blatantly obvious. We have barely understood how a single neuron operates and how billions of these neurons interact inside a brain appears to be a fantasy.

2.1 The Response to Extra-cellular Stimulus

Biological neurons have complicated intrinsic properties for information transmission and processing. Every neuron processes the information best suited to its unique type and function. A total of 20 components in a neuron's output signal are discussed in Izhikevich 2004; the most relevant components are tonic behavior, spiking or bursting waveforms, oscillatory frequency response, variable tonic frequency, and accommodation. Through these five components, the neuron masterfully exploits spatial and temporal dimensions concurrently for effective information processing and transmission. This section highlights the intricacy of a biological neuron's internal workings and helps the readers digest the monumental task of simulating this behavior in modern digital circuits. In [3], the author has performed extensive testing over a single biological neuron. The results are shown in Fig. 2; we will discuss the few relevant responses for our discussion with respect to modern spiking neural networks.

Tonic behavior refers to periodic repetition of the neuron's output waveform when it fires (shown in Fig. 2a). Neurons that exhibit this behavior are knows as regular spiking (RS) neurons. An alternative response to tonic spiking is phasic

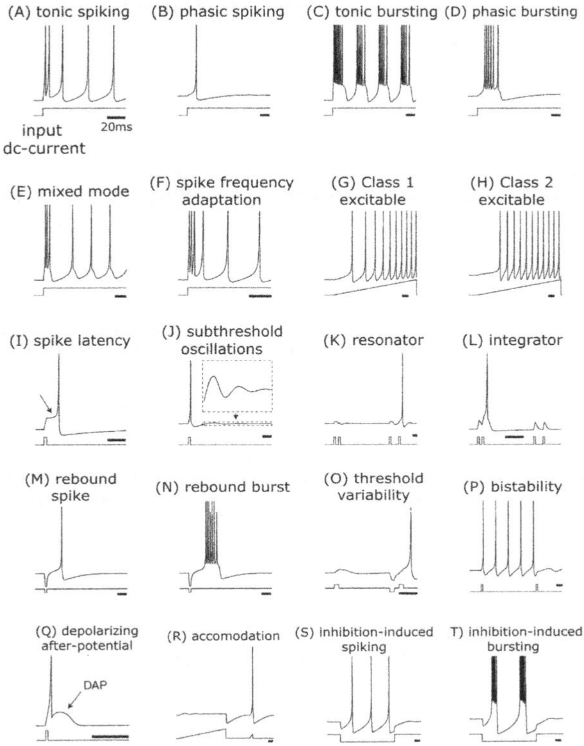

Fig. 2 The multiple responses of a biological neuron mapped by the author in [3]. We refer to the relevant sections of this figure to introduce the readers to the internal workings of an SNN. (**a**) The periodic repetition of the neuron's output waveform when it fires is known as tonic spiking. (**b**) The phasic spiking occurs when the neuron fires a single spike on trigger and then returns to quiescent state. (**c**) The tonic bursting occurs when the neuron fires a quick burst of spikes, and the spiking waveform has a periodic repetition. (**d**) Similar to tonic bursting, the phasic spiking occurs when the neuron fires multiple spikes in quick succession once without periodic repetition. (**e**) In mixed-state behaviour the neurons can exist in multiple states simultaneously, with the states changing rapidly in response to a change in the input signal. (**f**) The neuron firing frequency is relatively high at the onset of stimulation, and then it adapts. (**g**) The frequency of tonic spiking of neocortical RS excitatory neurons depends on the strength of the input, and it may span the range from 2 Hz to 200 Hz, or even greater. (**h**) The neurons that cannot fire low-frequency spike trains. That is, they are either quiescent or fire a train of spikes with a certain relatively large frequency, say 40 Hz. (**i**) Most cortical neurons fire spikes with a delay that depends on the strength of the input signal, also called spike latency, that can be quite large. (**j**) Practically every brain structure has neurons capable of exhibiting oscillatory potentials, as shown. (**k**) Due to the resonance phenomenon, neurons having oscillatory potentials can respond selectively to the inputs having frequency content similar to the frequency of subthreshold oscillations. (**l**) Neurons that prefer high-frequency input; the higher the frequency the more likely they fire. (**m**) When a neuron receives and then is released from an inhibitory input, it may fire a post-inhibitory (rebound) spike. (**n**) Some neurons, including the thalamo-cortical cells, may fire post-inhibitory bursts. (**o**) We apply a subthreshold input (10 mV), followed by an inhibitory pulse. After a delay, a second subthreshold input (10 mV) is applied which results in the neuron firing. (**p**) Some neurons can exhibit two stable modes of operation: resting and tonic spiking (or even bursting). An excitatory or inhibitory pulse can switch between the modes, as shown.

spiking (shown in Fig. 2b); neurons with phasic spiking fire a single spike when triggered by external stimulus and then return to quiescent state. For both the tonic and phasic behaviors, the waveform can be either a spiking waveform (consisting of a spike) or bursting waveform (consisting of multiple spikes in a quick burst fashion), as shown in Fig. 2c and d. These four behavioral classifications are deceptively simple at the first glance; however, they can encode information at a much larger scale than any modern digital communication protocol. These classes are not mutually exclusive, and many neurons are able to switch between them depending on their type and communication method, e.g., they begin the communication with a phasic bursting followed with a tonic spike-train.

The biological neurons introduce intricate variations in the frequencies for each of the four behavior classes aforementioned to smartly encode information. Emergence plays a significant role when multiple neurons are connected together. Theoretically speaking a neuron could encode channel information in the frequency of the bursting waveform and then send the tonic spike-train on that specific frequency channel. Similarly, the constantly decreasing frequency of a tonic spike-train could let the receiver neuron calculate how long the spike-train has been active. Through the combination of these four classes, biological neurons are able to communicate complex information in a short and power-efficient manner.

There are other components to a biological neuron's response which appear not to be involved in the spiking behavior, but they could actually be an integral part of the information transmission. Sub-threshold oscillations refer to the post-spiking oscillations with minimal amplitude in the neuron's response waveform. These oscillations are integral to some neurons capable of frequency modulation; these neurons only fire when input frequency resonates with the sub-threshold oscillations. Quite surprisingly these oscillations are present in all the brain structures that we've discovered to date.

One of the most important aspects to the spiking response of a biological neuron is known as accommodation. An input stimulus that is slowly increased in potential until it reaches membrane potential of the neuron would not be able to trigger a biological neuron. This phenomenon occurs because the neuron is able to accommodate for the change in potential of the input stimulus, and ultimately it doesn't fire. Therefore, a sharp pulse is best suited as an input to a biological neuron; however, it was also discovered that biological neurons have threshold variability. Biological neurons change the membrane potential (threshold) depending on the previous stimulus and its result. For example, a preceding excitatory pulse might

Fig. 2 (continued) (**q**) After firing a spike, the membrane potential of a neuron may exhibit a prolonged after-hyperpolarization (AHP) as shown in (**b**), or a prolonged depolarized after-potential (DAP), as shown. (**r**) Neurons are extremely sensitive to brief coincident inputs, but may not fire in response to a strong but slowly increasing input, as we illustrated. (**s**) A bizarre feature of many thalamo-cortical neurons is that they are quiescent when there is no input, but fire when hyperpolarized by an inhibitory input or an injected current. (**t**) Instead of spiking, a thalamo-cortical neuron can fire tonic bursts of spikes in response to a prolonged hyperpolarization raise the threshold making the neuron less excitable. This property allows the neuron to have the potential of short-term memory processing (think long short-term memory (LSTM) neural networks).

Another important property of neurons is their mixed-state behavior (shown in Fig. 2e). Neurons can exist in multiple states simultaneously, and the specific state of a neuron can change rapidly in response to changes in the input signal. This mixed-state behavior is thought to play an important role in neural processing, allowing neurons to perform complex computations and respond rapidly to changing environmental conditions.

Bistability is another important property of neurons. Bistable neurons can exist in two different stable states, and they can switch between these states in response to changes in the input signal. This property is thought to play an important role in neural processing, allowing neurons to store and retrieve information over long periods of time.

Variable frequency is yet another property of neurons. Neurons can generate spikes at different frequencies, and the frequency of spikes can change rapidly in response to changes in the input signal. This property is thought to play an important role in neural processing, allowing neurons to encode different types of information and respond to different types of stimuli.

3 Neuromorphic Computing: Learning from the Brain

Neural networks designed under neuromorphic computing, called spiking neural networks, are different from traditional artificial neural networks in that they communicate through spikes of electrical activity, rather than continuous signals. This makes them more similar to biological neural networks and allows them to process information more efficiently. SNNs are made up of interconnected neurons that communicate through electrical impulses. These impulses are triggered when a neuron receives enough input from other neurons. The strength of the impulse is determined by the weight of the connection between the neurons.

There are several different models of SNNs (as shown in Fig. 3), each with its own features and hardware implementation. One of the most common types of SNNs is the leaky integrate and fire model. This model is based on the idea that neurons in the brain gradually leak electrical charge over time. When a neuron receives input from other neurons, the charge builds up until it reaches a threshold, at which point the neuron fires an electrical impulse. This model is relatively simple and can be implemented using simple circuitry.

Another type of SNN is the integrate and fire model. This model is similar to the leaky integrate and fire model but does not take into account the gradual leakage of electrical charge. Instead, it simply sums up the input from other neurons and fires an impulse when the sum reaches a threshold. This model is even simpler than the leaky integrate and fire model and can be implemented using even simpler circuitry.

Neuron models	All-or-nothing firing	Refractory period	Excitation Block	Biologically-plausible	Tonic spiking	Phasic spiking	Tonic Bursting	Phasic Bursting	Mixed Mode	Spike frq adaptation	Class 1 excitable	Class 2 excitable	Spike latency	Subthreshold oscillation	Resonator	Integrator	Rebound spike	Rebound burst	Threshold variability	Bistability	DAP	Accomodation	Inhibition-induced spiking	Inhibition-induced bursting	Chaos	# of FLOPS
integrate-and-fire	+	+	-	-	+	-	-	-	-	-	+	-	-	-	-	+	-	-	-	-	-	-	-	-	-	5
integrate-and-fire with adaptation	+	+	-	-	+	-	-	-	-	+	+	-	-	-	-	+	-	-	-	-	+	-	-	-	-	10
integrate-and-fire-or-burst	+	+	-	-	+	+		+	-	+	+	-	-	-	-	+	+	+	-	+	+	-	-	-		13
resonate-and-fire	+	+	+	-	+	+	-	-	-	-	+	+	-	+	+	+	+	-	-	+	+	+	-	-	+	10
quadratic integrate-and-fire	+	+	-	-	+	-	-	-	-	-	+	-	+	-	-	+	-	-	+	+	-	-	-	-	-	7
Izhikevich	+	+	+	-	+	+	+	+	+	+	+	+	+	+	+	+	+	+	+	+	+	+	+	+	+	13
FitzHugh-Nagumo	+	+	+	-	+	+	-		-	-	+	-	+	+	+	-	+	-	+	+	-	+	+	-	-	72
Hindmarsh-Rose	+	+	+	-	+	+	+			+	+	+	+	+	+	+	+	+	+	+	+	+	+		+	120
Morris-Lecar	+	+	+	+	+	+	-		-	-	+	+	+	+	+	+	+		+	+	-	+	+	-	-	600
Wilson	+	+	+	-	+	+	+			+	+	+	+	+	+	+	+	+	+		+	+				180
Hodgkin-Huxley	+	+	+	+	+	+	+			+	+	+	+	+	+	+	+	+	+	+	+	+	+		+	1200
Mihalas-Neibur	+	+		-	+	+	+	+	+	+	+	+	+	-	-	+	+	+	+	+	+	+	+	+		

Fig. 3 Analysis of different models for biological neurons used in SNN hardware accelerators today with respect to their implementational complexity in terms of the number of flops required to simulate their behavior (the lower, the better) [4]

The Izhikevich 2004 [3] model is another popular SNN model. This model is based on the idea that neurons in the brain can exhibit a variety of different firing patterns, including tonic spiking, bursting, phasic spiking, and accommodation. The Izhikevich 2004 model is able to simulate all of these firing patterns and is therefore more complex than the previous two models. The hardware implementation of this model is also more complex, as it requires more circuitry to simulate the different firing patterns. These firing patterns are described in detail in Izhikevich 2004.

The Hodgkin and Huxley model is the most complex SNN model. This model takes into account the different types of ion channels in neurons and how they affect the firing patterns of neurons. The Hodgkin and Huxley model is able to simulate the complex firing patterns of neurons in the brain and is therefore more similar to biological neural networks than the other SNN models. However, the hardware implementation of this model is also the most complex, requiring advanced circuitry to simulate the different types of ion channels.

In terms of hardware implementation, the simpler SNN models such as the leaky integrate and fire and integrate and fire models are relatively easy to implement. They require only a few components and can be implemented using simple circuitry. The more complex models such as the Izhikevich 2004 and Hodgkin and Huxley models require more advanced circuitry and are therefore more difficult to implement.

4 Spiking Neural Networks

It is worth noting that SNNs have several advantages over traditional artificial neural networks (ANNs). One of the key advantages is their ability to process information in parallel. In ANNs, information is processed sequentially, which can be time-consuming and inefficient for certain tasks. SNNs, on the other hand, are designed to process information in parallel, which allows them to process large amounts of data quickly and efficiently. A typical structure of a deep SNN is shown in Fig. 4.

Another advantage of SNNs is their ability to adapt to changing environments. In the brain, neurons are constantly adapting and changing their connections in response to new information. SNNs are designed to mimic this process, which allows them to adapt to new information and learn from experience. This makes them well-suited to tasks such as pattern recognition and classification, where the ability to adapt to new information is critical.

SNNs have several unique features that distinguish them from ANNs. One of these features is tonic spiking, which refers to the regular firing of a neuron at a steady rate. This is in contrast to bursting, which refers to the rapid firing of a neuron followed by a period of quiescence. Another feature is phasic spiking, which refers to the irregular firing of a neuron in response to a stimulus. SNNs are also capable of accommodation, which refers to the gradual reduction in the firing rate of a neuron over time.

Variable frequency is another important feature of SNNs. In the brain, neurons can change their firing rate in response to different stimuli. SNNs are designed to mimic this process, which allows them to adapt to different tasks and stimuli. For example, SNNs can be trained to increase their firing rate in response to a specific stimulus, such as a visual cue or a sound.

The timing of spikes in spiking neural networks (SNNs) is critical for their ability to encode information. In SNNs, information is represented by the timing of spikes, rather than the strength of the connections between neurons as in traditional artificial neural networks (ANNs). This means that the precise timing of spikes at the input

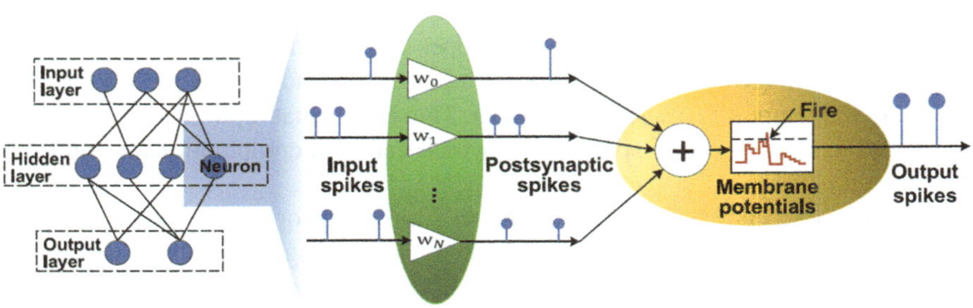

Fig. 4 A typical deep spiking neural network. This figure aims to show the difference in the complexity of the behavior of the hidden layers between SNNs and ANNs. There are infinitely many possibilities due to the involvement of the temporal dimensions (delayed spiking) [5]

and output layers of the network can encode much more information than the firing rates of neurons in ANNs.

For example, in an ANN, a set of inputs may be encoded as a vector of numerical values representing the intensity of each pixel in an image. In contrast, in an SNN, the same image would be encoded as a series of spike trains, where each spike represents the presence of a feature or pattern in the image. The timing of the spikes in the output layer of the network can then be used to identify the object or pattern in the image.

While the timing-based encoding used in SNNs can provide more information than ANNs, it also makes them more difficult to train and calculate errors for [6]. This is because traditional training algorithms such as back-propagation, which rely on the gradient of the error with respect to the network parameters, do not work well with SNNs. This is because the timing of spikes in the network introduces a temporal component that is not present in ANNs, making it more difficult to calculate gradients and propagate errors through the network.

To address these challenges, researchers have developed new training algorithms [7] for SNNs that take into account the timing of spikes. These algorithms use spike-timing-dependent plasticity (STDP) [8], which modifies the strength of connections between neurons based on the precise timing of the spikes. By using STDP, SNNs can learn to recognize temporal patterns in the input, which allows them to process and classify information at a higher dimension than ANNs. Most of the modern research work focuses on developing novel training algorithms for SNNs which are efficient and biologically plausible. We are still unravelling the mysteries of the inner working in our brain, and so the SNNs continue to evolve with each new revelation.

The time dimension is a key advantage of SNNs over ANNs. Because SNNs operate on a temporal scale, they are able to process information in real time, making them well-suited to applications such as speech recognition and natural language processing. SNNs are also able to perform certain computations more efficiently than ANNs. For example, SNNs are able to perform certain types of convolutions more efficiently by using time as an additional dimension, rather than performing a series of matrix multiplications as in ANNs.

However, the time dimension also introduces additional challenges for SNNs. One of the most significant challenges is proper input formatting. SNNs are not able to accept raw images or numerical data as input but instead require spike trains that represent the temporal structure of the data. This means that input data must be preprocessed to convert it into spike trains, which can be a computationally intensive process.

Despite these challenges, SNNs hold great promise for a wide range of applications in healthcare, robotics, and beyond. By mimicking the way that neurons in the brain communicate, SNNs are able to process information in a more efficient and adaptive way than traditional ANNs. As research into SNNs continues, we can expect to see even more innovative applications that take advantage of their unique properties and capabilities.

4.1 Spiking Neuron Models

Another important feature of SNNs is the way that individual neurons fire. In an SNN, neurons can fire in a variety of different ways, depending on their internal dynamics and the input they receive. Some neurons fire tonically, maintaining a steady rate of spiking over time. Others fire phasically, responding only to specific patterns of input. Still others fire in bursts, sending a rapid series of spikes in response to strong input.

The firing behavior of neurons in an SNN can be modelled using a variety of different mathematical equations, each of which captures different aspects of neuronal dynamics. Some of the most commonly used models include the leaky integrate-and-fire (LIF) model, the integrate-and-fire (IF) model, the Izhikevich 2004 model, and the Hodgkin-Huxley model.

The LIF model is one of the simplest and most widely used models for SNNs. In this model, a neuron integrates incoming input over time, and when its membrane potential reaches a threshold, it generates a spike and resets its potential to a lower value. The integrate-and-fire model is similar to the leaky integrate and fire model but does not take into account the gradual leakage of charge. Instead, it simply sums up the input from other neurons and fires an impulse when the sum reaches a threshold.

The Izhikevich 2004 model is a more complex SNN model that is capable of simulating a variety of firing patterns. This model takes into account the fact that neurons in the brain can exhibit different firing patterns, including tonic spiking, bursting, phasic spiking, and accommodation. The Izhikevich 2004 model is able to simulate all of these firing patterns and is therefore more complex than the previous two models.

The Hodgkin and Huxley model is the most complex SNN model and takes into account the different types of ion channels in neurons and how they affect the firing patterns of neurons. This model is able to simulate the complex firing patterns of neurons in the brain and is therefore more similar to biological neural networks than the other SNN models. However, the hardware implementation of this model is also the most complex, requiring advanced circuitry to simulate the different types of ion channels.

4.2 Common Training Algorithms

In [8] the authors have detailed all the training algorithms as they were developed in a chronological order. Many have been abandoned since then in the favor of modern training algorithms. One of the most common training algorithms for SNNs is the spike timing-dependent plasticity (STDP) algorithm. STDP is a form of Hebbian learning that modifies the strength of synaptic connections between neurons based on the timing of their spikes; the overall decay function is shown

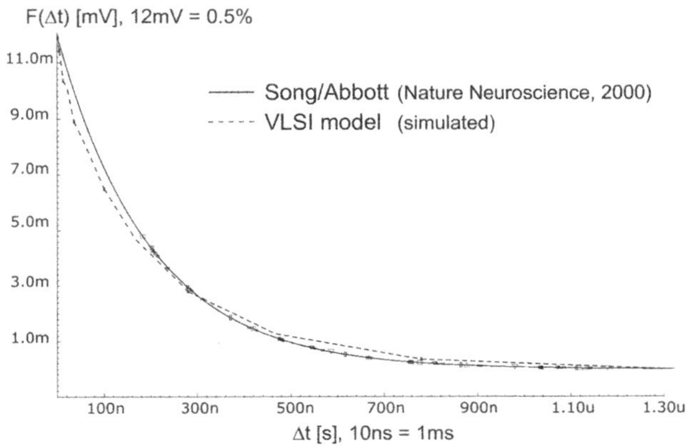

Fig. 5 The biologic spike timing-dependent plasticity (STDP) decay function as originally mentioned in [9] and recreated by simulation in digital systems by the authors of [10]

in Fig. 5. Specifically, if two neurons fire together, the connection between them is strengthened, while if they fire out of sync, the connection is weakened. STDP has been shown to be a powerful learning rule that can enable SNNs to learn a wide range of tasks, including pattern recognition, classification, and prediction.

Another popular training algorithm for SNNs is backpropagation through time (BPTT) [11]. BPTT is an extension of the backpropagation algorithm used in traditional ANNs, but it takes into account the temporal dynamics of SNNs by propagating errors backward through time. BPTT has been shown to be effective for training SNNs for tasks such as speech recognition and image classification, but it can be computationally expensive and difficult to scale to large networks.

Other training algorithms for SNNs include reward-based learning, which uses reinforcement signals to modify the synaptic connections between neurons, and unsupervised learning, which allows SNNs to learn from unlabeled data. Each of these algorithms has its strengths and weaknesses, and the choice of algorithm will depend on the specific task and network architecture.

Despite the success of these training algorithms, there are still several challenges that need to be addressed in order to improve the efficiency and effectiveness of SNN training. One of the biggest challenges is the sparsity of spike data. SNNs typically generate sparse spike trains, with only a small fraction of neurons firing at any given time. This sparsity can make it difficult to train SNNs using traditional optimization algorithms, which are designed for dense data. New algorithms that can handle sparse data, such as the stochastic gradient descent with momentum (SGDM) [1] algorithm, have shown promise for training SNNs more efficiently.

Another challenge with SNN training is the need for specialized hardware. SNNs are computationally intensive and require specialized hardware to implement efficiently. While there has been some progress in developing neuromorphic hardware that can simulate the behavior of SNNs more efficiently than traditional computing systems, more work is needed to make this technology more accessible and affordable.

Finally, there is a need for better evaluation metrics for SNNs. Current metrics, such as accuracy and loss, do not always reflect the performance of SNNs in real-world applications. New metrics that take into account the temporal dynamics of SNNs, such as spike latency and jitter, could provide a more accurate assessment of their performance.

In conclusion, current training algorithms for SNNs have made significant progress in enabling SNNs to learn a wide range of tasks, but there are still several challenges that need to be addressed. These challenges include the sparsity of spike data, the need for specialized hardware, and the need for better evaluation metrics. Future research in these areas could greatly enhance the efficiency and effectiveness of SNN training and make SNNs a powerful tool for modeling and understanding the temporal dynamics of neural processing.

4.3 Applications

SNNs have many potential applications in healthcare, particularly in the field of prosthetics and neural implants. For example, researchers have developed implantable devices that can help people with spinal cord injuries regain control of their limbs. These devices use SNNs to process signals from the brain and send them to electrodes in the limbs, allowing the patient to control their movements. SNNs are also being used in the development of brain-computer interfaces (BCIs).

For example, consider the task of recognizing spoken words in real time. An SNN can process the incoming speech signal as a series of spike trains, where each spike represents a small segment of the signal [12]. By analyzing the timing of the spikes, the SNN can recognize the phonemes of the words and piece them together into a full sentence. In contrast, an ANN would require a much larger amount of data to achieve the same level of accuracy, as it would need to process the entire signal at once rather than in small segments.

SNNs are also quite popular for the detection of irregular heartbeats. They are most suited for applications where the input data is of the form of periodic pulses. Heartbeats detected through ECG can be fed directly into an SNN without any major input pre-processing. In [13], the authors have proposed a unique SNN that incorporates LIF neuron model. However, this work has a narrow focus and fails to provide a generalized solution for this application. Most of the current work for SNNs in healthcare revolves around the optimization of the training algorithms [14].

5 Hardware Accelerators

While deploying SNNs on commercial architectures is a popular approach due to its wide availability and affordability, the computations involved in higher dimensions can be computationally expensive and may result in a significant reduction in throughput. As a result, researchers are exploring alternative methods that are more efficient and scalable.

One of the approaches that have gained attraction in recent years is the design of domain-specific hardware tailored to SNNs. These designs aim to increase efficiency and reduce power consumption in SNNs, making them more viable for real-world applications. For example, Loihi [15] is a neuromorphic chip developed by Intel that emulates biological neurons, while Skydiver is a platform that uses reconfigurable computing to accelerate SNN simulations. In addition to domain-specific hardware, low-power accelerators [16, 17] are also being developed for SNNs. These often employ approximations and lower precision to reduce the complexity of the hardware circuit. Additionally, they optimize the hardware to match the unique computational requirements of SNNs, resulting in increased performance and energy efficiency.

Another approach to improve SNN efficiency is through the use of emerging devices that are better suited for event-based computations. For example, optoelectronics is a technology that leverages light to perform computations, and it has been explored as a potential solution for SNNs due to its high parallelism and low power consumption. Memristors, on the other hand, are electronic components that can store information and perform computation, and they have shown promise in implementing synaptic plasticity in SNNs.

In this section, we will perform a detailed review of the existing domain-specific processors available for SNNs and their preferred topology design (feedforward or general mesh).

5.1 General Mesh Accelerators

General mesh hardware accelerator is a type of hardware accelerator that uses a top-down system to breakdown a problem into simpler tasks which can be run in an efficient and quick manner on its multiple different nodes. Each processing node has its own memory and processing capabilities, and the nodes communicate with each other using a mesh network topology. By distributing the workload in this way, mesh-based computing can achieve higher performance and better scalability than traditional single-processor computing. These accelerators are designed to speed up the execution of computations which benefit from parallelism in the hardware computation unit. These accelerators are particularly useful for tasks that involve large amounts of data and complex computations, such as machine learning, scientific simulations, and data.

SpiNNaker [19] and TrueNorth [20] are two of the older proposed designs that deploy mixed-signal systems to simulate a big portion of the 20 key neuron characteristics highlighted in Fig. 2. However, it is a very complex task as highlighted in the earlier sections. These chips acted as a proof of concept to show that SNN-based hardware accelerators are feasible, albeit, at a lower overall throughput than the traditional accelerators. The designers in TrueNorth acknowledge that their

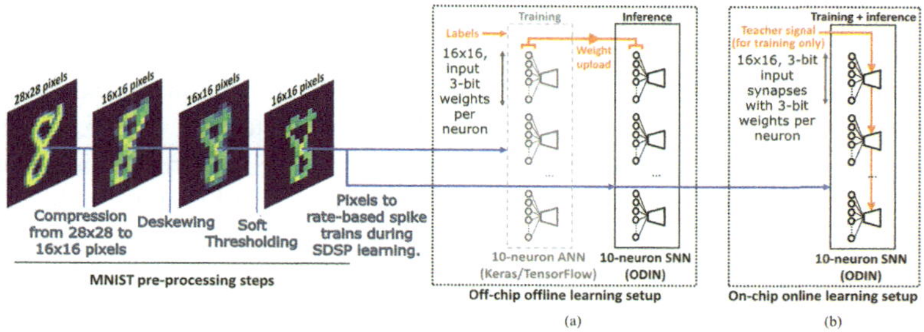

Fig. 6 A common input encoding pipeline to spiking neural networks as implemented in [18]. (**a**) The off-chip learning and loading weights onto the ODIN chip (accuracy, 91.4%) (**b**) The on-chip training of the network (accuracy, 76.4%)

design is a pioneer in the field and still requires major optimizations before it becomes competitive to the "Von Neuman" machines. In TrueNorth, the authors were able to demonstrate that the hardware accelerator can simulate up to 11 of the Izhkevich's models by grouping up three neurons to perform each function. Loihi is Intel's relatively recent attempt at a complete neuromorphic computing chip to help researchers and designers simulate their neuron models. It uses LIF neuron model with threshold adaption, and it can simulate up to 6 of Izhkevich's responses. It is an incredible platform for prototyping, but it also fails fully realize the potential demonstrated by biological neurons.

ODIN Fig. 6 is another hardware chip capable of reflecting all of the 20 characteristics of the neuron characteristics. It uses the Izhkevich's model for the neuron instead of the more common LIF. Referring to Fig. 3, we can see that the Izhkevich's model provides higher biological plausibility at the same hardware complexity. The authors demonstrate the capabilities of the chip using a 10-neuron ANN to SNN conversion with a pre-processed MNIST input encoded signal. The network is able to achieve 91.4% accuracy at its maximum with a maximum energy draw of 12.7 nJ/OP. While neither the energy-efficiency nor the efficiency is groundbreaking, the main highlight of the accelerator is its flexibility.

The most recent attempt for general mesh SNN accelerators was done by the designers of ESSA [21]. ESSA aims to address the challenge of sparsity in neural networks. The accelerator is designed to perform inference tasks efficiently and accurately by exploiting the sparsity in the neural network. It achieves an improvement in throughput of up to 3.2×. It uses an event-driven architecture that is optimized for spiking neural networks, which allows it to handle the unpredictable and sparse firing patterns of the neurons. The accelerator uses a customized algorithm for self-adaptive weight compression and sparse representation of the neural network, which reduces the memory requirements and computational complexity of the inference tasks. The ESSA SNN Accelerator also uses a power management scheme that dynamically adjusts the supply voltage of the accelerator based on the workload, which leads to significant energy savings.

One of the key shortcomings of the available general mesh accelerators is that none of the authors have truly implemented a deep SNN. In ODIN, the authors only implement several single-neuron models. Similarly, Spinnaker and TrueNorth have not been tested on deep SNNs. Loihi is presented as a general-purpose experimentation platform for SNN development. NxTF is a recent attempt at a compiler for optimizing the running on deep SNNs on Loihi, and it achieves one of the lowest error rates on the CIFAR10 dataset. Unfortunately, even this work is only competitive in the limited space of neuromorphic computing and fails to hold its own compared to current efficient DNN networks.

Not all is bad in this review of the current literature. Most of the designs have rightly pointed out that the neuromorphic computing is in its emergent phase right now, and compared to the initial attempts at ANNs, the performance numbers of the SNNs are incredibly promising. Considering these points in the light of the huge potential presented by biological neurons, several other optimizations need to be further developed before the SNN accelerators will be able to compete which is an open context versus the DNN accelerators.

5.2 Feedforward Accelerators

Feedforward accelerators are specialized hardware designed to accelerate the training and inference of artificial neural networks. These accelerators implement the feedforward neural network architecture, which is a type of artificial neural network where the flow of information is unidirectional, from input to output. Feedforward accelerators offer several advantages over traditional CPUs and GPUs, including higher energy efficiency, lower latency, and better throughput.

The architecture of a feedforward accelerator consists of multiple processing elements (PEs) connected in a pipeline. Each PE receives input from the previous PE and produces output for the next PE. The PEs perform the mathematical operations involved in the feedforward neural network, such as matrix multiplication and activation functions.

Skydiver [5] is a modern feedforward accelerator designed specifically for spiking neural networks (SNNs). Spiking neural networks are a type of artificial neural network that simulates the behavior of biological neurons, where information is encoded in the timing and frequency of spikes. SNNs are known for their spatiotemporal sparsity, which means that only a small percentage of neurons fire spikes at any given time. Skydiver was proposed in a research paper titled "Skydiver: a high-throughput and energy-efficient spiking neural network accelerator via workload balancing" by Chang et al. The paper addresses the problem of workload imbalance in SNN accelerators due to the unpredictable sparsity of neuron firing over time.

Spatiotemporal sparsity refers to the fact that only a small percentage of neurons in an SNN fire spike at any given time, and this sparsity is not predictable. In contrast to traditional artificial neural networks, where all neurons are activated in every forward pass, SNNs have a sparse activation pattern. This sparsity is a fundamental property of SNNs and is essential for their energy efficiency and biological plausibility. The sparsity of neuron firing over time in SNNs causes workload imbalance in the feedforward accelerator. Workload imbalance means that some PEs are underutilized while others are overloaded, leading to reduced energy efficiency and throughput. This is because the PEs that receive input from sparsely firing neurons will be underutilized, while the PEs that receive input from densely firing neurons will be overloaded. Workload imbalance ultimately leads to a degradation in energy efficiency. Underutilized PEs still consume energy, and overloaded PEs consume more energy than necessary. This leads to wasted energy and reduced energy efficiency. Energy efficiency is a critical factor in modern computing systems, where energy consumption is a major limiting factor in performance and scalability.

The authors of Skydiver propose two specific algorithms called APRC and CBWS to improve the workload balance in SNN accelerators. APRC stands for Adaptive Potential Range Control, while CBWS stands for Cross-Bar Workload Scheduling. APRC is responsible for adjusting the potential range of each neuron based on its firing rate. The potential range determines the amount of energy required to activate a neuron. By adjusting the potential range based on the firing rate, APRC can balance the workload across the PEs. CBWS is responsible for scheduling the workload across the PEs based on the firing rate of the neurons. CBWS uses a cross-bar architecture to distribute the workload dynamically across the PEs. This ensures that the workload is balanced based on the sparsity of neuron firing over time.

Using the proposed APRC and CBWS algorithms, Skydiver was able to improve the workload balance from 69.19% to 95.69%. This means that the workload is almost evenly distributed across all PEs, which leads to better energy efficiency and throughput. The improvement in workload balance is a significant achievement since it addresses a critical challenge in SNN accelerators. The improvement in workload balance also leads to an increase in the throughput of the accelerator. According to the paper, Skydiver was able to achieve a 1.4× improvement in segmentation and a 1.2× improvement in classification tasks. These results demonstrate the effectiveness of the proposed algorithms in improving the performance of SNN accelerators. The core of the strategy proposed by Skydiver is to establish an approximate proportional ratio between spike firing rate and the membrane potential. The membrane potential is the electric potential difference across the membrane of a neuron. By establishing this ratio, Skydiver can adjust the potential range of each neuron dynamically based on its firing rate, which leads to better workload balance and energy efficiency.

The Lien [22] accelerator is another feedforward accelerator that follows the feedforward topology. It uses compression and sparsity to achieve highly energy-efficient operation. The Lien accelerator was proposed in a research paper titled "A Highly Energy-Efficient Deep Convolutional Neural Network Processor Based on a Compression-Sparsification Unified Framework" by Lien et al.

The Lien accelerator uses a discrete-time approximate LIF (Leaky Integrate-and-Fire) model for spiking neuron simulation. The LIF model is a mathematical model of the behavior of a neuron, where the neuron integrates incoming signals and fires a spike when its membrane potential exceeds a certain threshold. The discrete-time approximation means that the simulation is done in discrete time steps rather than continuous time. The Lien accelerator uses a delta-shaped synaptic kernel, which is a mathematical function that describes the effect of a synapse on the membrane potential of a neuron. The delta-shaped kernel is a simplified version of the more complex functions used in traditional artificial neural networks, which reduces the computational complexity and memory requirements of the accelerator.

The Lien accelerator uses sparsity optimization to reduce the energy consumption of the accelerator. Sparsity optimization means that the accelerator exploits the sparsity of the neural network to reduce the number of computations required. The Lien accelerator uses sparsity optimization techniques when the sparsity ratio is higher than 75%. The Lien accelerator simplifies the model using the proposed gate-to-all product, which uses bit-mas representation to exploit sparsity for low-power computations. The gate-to-all product simplifies the multiplication operation by replacing it with a series of logical operations, which reduces the computational complexity and energy consumption.

The proposed techniques used in the Lien accelerator were able to reduce the dynamic power of the PEs by 46.6% given that the input sparsity is 77.4% and the computing latency is 47.3% due to the 70% sparsity in weights. These results demonstrate the effectiveness of the sparsity optimization techniques used in the Lien accelerator. Feedforward accelerators are specialized hardware designed to accelerate the training and inference of artificial neural networks. Skydiver and Lien accelerators are two modern feedforward accelerators that follow the feedforward topology. Skydiver is designed specifically for spiking neural networks and addresses the challenge of workload imbalance due to spatiotemporal sparsity

5.3 Resource Analysis and Discussion

Compared to deep neural network accelerators, SNN accelerators have lower throughput due to the unpredictable and sparse firing patterns of the neurons in spiking neural networks. This spatiotemporal sparsity makes it challenging to achieve a balanced workload distribution, which can result in degradation of the accelerator's performance and energy efficiency. Additionally, the lack of a widely adopted SNN standard makes it difficult to develop software and optimize hardware for SNN accelerators.

However, SNN accelerators have significant advantages over DNN accelerators in terms of energy efficiency. It can be observed in Table 1 that every SNN design is able to perform at a higher energy-efficiency level compared to the state-of-the-art DNN accelerator designed for low-power applications like IOT. The event-driven architecture and sparsity optimization techniques used in SNN accelerators reduce

Table 1 Resource and performance analysis for state-of-the-art spiking neural networks compared to the state-of-the-art deep neural network accelerators

Work	Skydiver [5]	Lien [22]	ESSA [21]	ODIN [18]	Vesti [23]
Type	SNN	SNN	SNN	SNN	DNN
Clock (MHz)	200	500	140	75	550
On-chip neuron	N/R	N/R	2048	256	N/R
On-chip synapse	N/R	N/R	512K	64K	N/R
Synaptic precision	8-bit	8-bit	8-bit	8-bit	N/R
Max fan-in	N/R	N/R	32768	N/R	N/R
Accuracy (%)	98.5	71.5	96.9	91.4	98.5
Energy-efficiency (GSOP/W)	19.3	18.9 or 35.88	16.5 or 32.1	81.5	9.8
Dataset	MNIST	IVS 3cls	MNIST	MNIST	MNIST

power consumption by exploiting the sparse firing patterns of the neurons. The use of specialized hardware for SNNs also allows for more efficient computation, further reducing energy consumption. SNN accelerators are particularly suitable for low-power and real-time applications, where energy efficiency is a crucial factor. Overall, while SNN accelerators may have lower throughput than DNN accelerators, they excel in energy efficiency and are well-suited for specific applications.

6 Limitations and Potential Improvements of SNNs

SNNs have certain limitations and challenges that need to be addressed in order to improve their performance and practicality. The biggest challenge with SNNs is finding an efficient and effective way to train them. Current training algorithms for SNNs are often slow, computationally intensive, and difficult to scale to large networks. New algorithms that can improve the speed and efficiency of training SNNs could greatly enhance their practicality and applicability [8].

Another way to improve SNNs is to develop more realistic models of biological neurons. Many current SNN models use simple integrate-and-fire or leaky integrate-and-fire models that do not fully capture the complexity and variability of real neurons [3]. More realistic neuron models that incorporate sub-threshold dynamics, spike frequency adaptation, and other biologically relevant features could improve the accuracy and fidelity of SNNs. Additionally, the way that information is encoded and transmitted in SNNs is critical to their performance. Current spike coding schemes often rely on fixed thresholds or pre-defined spike patterns, which can limit the flexibility and adaptability of the network. New spike coding schemes that are more adaptive and can adjust to changing input conditions could greatly enhance the performance of SNNs.

Another way to improve SNNs is to combine them with other types of neural networks, such as convolutional neural networks (CNNs) or recurrent neural networks (RNNs). Hybrid networks that combine the strengths of different types of networks could improve the accuracy and speed of SNNs for specific tasks and applications.

SNNs require specialized hardware to effectively implement the energy-efficient models that are biologically plausible. Designing hardware circuits that can replicate all of these different properties of biological neurons is a significant challenge. While some progress has been made in developing neuromorphic hardware that can replicate some of these properties, such as spike-based communication and variable frequency, replicating the full range of neuronal behaviors is still very difficult [7, 12–14].

One of the main challenges is the sheer complexity of biological neurons. Neurons are highly complex and dynamic systems that are influenced by many different factors, including the structure and properties of the neuron itself, the properties of the input signal, and the interactions between neurons. Replicating all of these interactions and behaviors in a hardware circuit is a monumental task that requires a deep understanding of the underlying mechanisms and principles of neural processing.

Another challenge is the variability and plasticity of biological neurons. Neurons can change their behavior and properties over time, in response to changes in the environment or in the neural network itself. This plasticity allows neurons to adapt to new situations and to learn and store new information. However, replicating this plasticity in a hardware circuit is extremely difficult, as it requires a high degree of flexibility and adaptability that is not typically found in traditional computing systems.

In addition to these challenges, there are also practical limitations to designing hardware circuits that can replicate the properties of biological neurons. One major limitation is the size and complexity of these circuits. Neurons are highly complex and intricate systems, and replicating all of their properties and behaviors requires a large number of components and a high degree of interconnectivity. This can make neuromorphic circuits much larger and more complex than traditional computing systems, which can limit their practicality and scalability.

Another limitation is the energy consumption of neuromorphic circuits. Biological neurons are extremely energy-efficient, using only a small fraction of the energy required by traditional computing systems. Replicating this energy efficiency in hardware circuits is a significant challenge, as it requires the development of new materials, components, and architectures that can operate at low power levels. While some progress has been made in developing low-power neuromorphic circuits, these circuits are still far from the energy efficiency of biological neurons.

Despite these challenges and limitations, there are still many potential benefits to neuromorphic computing. Neuromorphic circuits have the potential to be much faster and more efficient than traditional computing systems, particularly for certain types of applications such as pattern recognition (speech recognition) and sensory processing (ECG). They also have the potential to be more robust and fault-tolerant, as they can operate in a distributed and parallel manner, much like the neural networks in the brain.

To overcome the challenges of designing hardware circuits that can replicate all the different properties of biological neurons, researchers are exploring new approaches to neuromorphic computing. One approach is to use hybrid systems that combine traditional digital computing with neuromorphic hardware. For example, researchers are exploring the use of digital circuits to simulate the complex behavior of neurons, while using neuromorphic circuits to perform the more computationally intensive tasks such as pattern recognition.

Another approach is to use machine learning techniques to optimize the behavior and properties of neuromorphic circuits. By training these circuits on large datasets of neural data, researchers can optimize their performance and energy efficiency and enable them to perform complex computations and tasks that would be difficult or impossible with traditional computing systems [6].

7 Conclusion

In conclusion, spiking neural networks (SNNs) offer a promising approach to modelling the behavior of the brain and developing intelligent systems. SNNs are based on a biologically realistic model of neurons and can handle temporal information more efficiently than traditional artificial neural networks. Furthermore, SNNs have been shown to achieve state-of-the-art performance in various tasks, such as image recognition, speech recognition, and natural language processing. However, SNNs also have some limitations that need to be addressed. One of the main challenges is the difficulty of training SNNs, which require specialized algorithms and hardware. Another limitation is the lack of a comprehensive theory that explains how the brain processes information using spiking neurons. The other challenge is achieving scalability, especially with domain-specific hardware and emerging devices. Additionally, there is a need to develop better algorithms and techniques that can take advantage of the unique characteristics of SNNs to achieve high performance and energy efficiency. Nevertheless, continued research in this field is expected to lead to more efficient and scalable SNN systems that can be applied to a wide range of real-world applications.

At the moment, the SNNs have established themselves as the potential successor to DNN accelerators due to their exceptionally good energy-efficiency. SNNs are able to work at a lower operational power compared to DNNs due to their sparsity and event-based operation. Unlike conventional neural networks, SNNs operate based on the timing of spikes rather than continuous activations. This sparsity arises due to the fact that only a small fraction of neurons are active at any given moment in time. Therefore, SNNs can exploit this sparsity by using event-based processing, which is more energy-efficient than the conventional data-parallel processing used in deep neural networks. In event-based processing, each neuron generates spikes when its membrane potential exceeds a certain threshold. These spikes are communicated to other neurons only when they exceed a certain threshold, leading to reduced communication and computation requirements. SNNs also benefit from the fact that the spike trains generated by neurons are sparse and therefore can be compressed, resulting in lower memory requirements.

SNNs are best suited for applications where the input data is event-based or can be converted to an event-based representation, such as audio, vision, and sensory processing. These applications include speech recognition, object recognition, robot navigation, and other real-time applications that require low-latency and energy-efficient operation. SNNs are also useful for applications that require low power consumption and are constrained by limited resources, such as edge devices and IoT devices.

As research in this field continues to advance, we can expect to see significant progress in the development of intelligent systems that can perform complex tasks with greater efficiency and accuracy. In the future, SNNs could revolutionize fields such as robotics, autonomous vehicles, and healthcare, leading to a more intelligent and interconnected world. If we encourage a better design philosophy, we can avoid ethical dilemmas in the future and instead focus our efforts entirely on improving the healthcare sector which benefits us all.

References

1. J.D. Anderson, Jr., et al., *The Airplane: A History of Its Technology* (AIAA, 2002)
2. S. Sardi, R. Vardi, A. Sheinin, A. Goldental, I. Kanter, New types of experiments reveal that a neuron functions as multiple independent threshold units. Sci. Rep. **7**(1), 18036 (2017)
3. E.M. Izhikevich, Which model to use for cortical spiking neurons? IEEE Trans. Neural Netw. **15**(5), 1063–1070 (2004)
4. J.-Q. Yang, R. Wang, Y. Ren, J.-Y. Mao, Z.-P. Wang, Y. Zhou, S.-T. Han, Neuromorphic engineering: from biological to spike-based hardware nervous systems. Adv. Mater. **32**(52), 2003610 (2020)
5. Q. Chen, C. Gao, X. Fang, H. Luan, Skydiver: a spiking neural network accelerator exploiting spatio-temporal workload balance. IEEE Trans. Comput.-Aided Design Integr. Circuits Syst. **41**(12), 5732–5736 (2022)
6. J.K. Eshraghian, M. Ward, E. Neftci, X. Wang, G. Lenz, G. Dwivedi, M. Bennamoun, D.S. Jeong, W.D. Lu, Training spiking neural networks using lessons from deep learning (2021). arXiv preprint arXiv:2109.12894
7. W. Fang, Z. Yu, Y. Chen, T. Huang, T. Masquelier, Y. Tian, Deep residual learning in spiking neural networks. Adv. Neural Inform. Process. Syst. **34**, 21056–21069 (2021)
8. X. Wang, X. Lin, X. Dang, Supervised learning in spiking neural networks: a review of algorithms and evaluations. Neural Netw. **125**, 258–280 (2020)
9. H. Husi, M.A. Ward, J.S. Choudhary, W.P. Blackstock, S.G.N. Grant, Proteomic analysis of NMDA receptor–adhesion protein signaling complexes. Nat. Neurosci. **3**(7), 661–669 (2000)
10. L.F. Abbott, S.B. Nelson, Synaptic plasticity: taming the beast. Nat. Neurosci. **3**(11), 1178–1183 (2000)
11. T.P. Lillicrap, A. Santoro, Backpropagation through time and the brain. Curr. Opin. Neurobiol. **55**, 82–89 (2019)
12. A. Bittar, P.N. Garner, Surrogate gradient spiking neural networks as encoders for large vocabulary continuous speech recognition (2022). arXiv preprint arXiv:2212.01187
13. H. Chu, Y. Yan, L. Gan, H. Jia, L. Qian, Y. Huan, L. Zheng, Z. Zou, A neuromorphic processing system with spike-driven SNN processor for wearable ecg classification. IEEE Trans. Biomed. Circuits Syst. **16**(4), 511–523 (2022)

14. R. Mao, S. Li, Z. Zhang, Z. Xia, J. Xiao, Z. Zhu, J. Liu, W. Shan, L. Chang, J. Zhou, An ultra-energy-efficient and high accuracy ecg classification processor with SNN inference assisted by on-chip ANN learning. IEEE Trans. Biomed. Circuits Syst. **16**(5), 832–841 (2022)
15. M. Davies, N. Srinivasa, T.-H. Lin, G. Chinya, Y. Cao, S.H. Choday, G. Dimou, P. Joshi, N. Imam, S. Jain, et al., Loihi: a neuromorphic manycore processor with on-chip learning. IEEE Micro **38**(1), 82–99 (2018)
16. Y. Wang, H. Zhang, K.-I. Oh, J.-J. Lee, S.-B. Ko, Energy efficient spiking neural network processing using approximate arithmetic units and variable precision weights. J. Parallel Distrib. Comput. **158**, 164–175 (2021)
17. Y. Wang, K. Shahbazi, H. Zhang, K.-I. Oh, J.-J. Lee, S.-B. Ko, Efficient spiking neural network training and inference with reduced precision memory and computing. IET Comput. Digit. Tech. **13**(5), 397–404 (2019)
18. C. Frenkel, M. Lefebvre, J.-D. Legat, D. Bol, A 0.086-mm^2 12.7-pj/sop 64k-synapse 256-neuron online-learning digital spiking neuromorphic processor in 28-nm CMOS. IEEE Trans. Biomed. Circuits Syst. **13**(1), 145–158 (2019)
19. E. Painkras, L.A. Plana, J. Garside, S. Temple, F. Galluppi, C. Patterson, D.R. Lester, A.D. Brown, S.B. Furber, Spinnaker: A 1-w 18-core system-on-chip for massively-parallel neural network simulation. IEEE J. Solid-State Circuits **48**(8), 1943–1953 (2013)
20. F. Akopyan, J. Sawada, A. Cassidy, R. Alvarez-Icaza, J. Arthur, P. Merolla, N. Imam, Y. Nakamura, P. Datta, G.-J. Nam, et al., Truenorth: design and tool flow of a 65 mw 1 million neuron programmable neurosynaptic chip. IEEE Trans. Comput.-Aided Design Integr. Circuits Syst. **34**(10), 1537–1557 (2015)
21. Y. Kuang, X. Cui, Z. Wang, C. Zou, Y. Zhong, K. Liu, Z. Dai, D. Yu, Y. Wang, R. Huang, ESSA: design of a programmable efficient sparse spiking neural network accelerator. IEEE Trans. Very Large Scale Integr. Syst. **30**(11), 1631–1641 (2022)
22. H.-H. Lien, T.-S. Chang, Sparse compressed spiking neural network accelerator for object detection. IEEE Trans. Circuits Syst. I: Regul. Pap. **69**(5), 2060–2069 (2022)
23. S. Yin, Z. Jiang, M. Kim, T. Gupta, M. Seok, J.-S. Seo, Vesti: energy-efficient in-memory computing accelerator for deep neural networks. IEEE Trans. Very Large Scale Integr. Syst. **28**(1), 48–61 (2019)

Emerging Machine Learning Using Siamese and Triplet Neural Networks

Ziheng Wang, Farzad Niknia, Shanshan Liu, Pedro Reviriego, and Fabrizio Lombardi

1 Introduction

Siamese networks (SNs) and Triplet networks (TNs) are artificial neural networks (ANNs) consisting of multiple branches (two or three) of weight-sharing feedforward networks (also referred to as subnetworks). As multi-branch networks, SNs and TNs have been widely used in various machine learning (ML) tasks (especially similarity measuring) such as face recognition [1], vehicle identification [2], and image retrieval [3]. As an emerging ML technique, multi-branch networks and their varieties are intensively applied in artificial intelligence (AI) systems, such as the two branches of encoders in Contrastive Language-Image Pre-Training (CLIP) networks for image generators [4]. This chapter discusses the basic implementations and advanced techniques; analysis, designs, and evaluations are presented for SNs and TNs with high efficiency and improved performance.

SNs have excellent performance when utilized in tasks that measure the similarity between two data inputs [5]. Different from a traditional ANN, an SN for classification purposes only needs to know whether the two inputs belong to the same category, so it does not require explicit labels; therefore, an SN can be treated as semi-supervised and it usually achieves superior performance in tasks with a large number of classes and one-shot learning [6]. TNs also provide an excellent

Z. Wang · F. Niknia · F. Lombardi
Northeastern University, Boston, MA, USA
e-mail: wang.zihe@northeastern.edu; niknia.f@northeastern.edu; lombardi@ece.neu.edu

S. Liu (✉)
University of Electronic Science and Technology of China, Chengdu, Sichuan, China
e-mail: ssliu@uestc.edu.cn

P. Reviriego
Universidad Politécnica de Madrid, Madrid, Spain
e-mail: pedro.reviriego@upm.es

W. Liu et al. (eds.), *Design and Applications of Emerging Computer Systems*,
https://doi.org/10.1007/978-3-031-42478-6_5

learning performance when there is a paucity of training data available, in which case the traditional ANNs often find difficulty in achieving good performance. A typical TN generates feature embeddings of the original input data; the outputs of different categories are separable in the embedding space and thus, they are available for subsequent classification/recognition. As an emerging ML scheme, the SN/TN achieves better performance than traditional classification/recognition methods with a single network branch [7, 8].

An SN (or TN) presents a unique challenge for error tolerance, because weight matrices are shared between the subnetworks, so memories (storing the weights) are very important for the reliable operation of this type of ANN. Especially in the case when weights are stored in the memory as floating-point (FP) numbers, even a single bit-flip in the exponent part of the FP number can result in a substantial change in weight value, thereby propagating through the SN with catastrophic effects. Therefore, this chapter analyzes bit-flip faults (soft errors) in the inference of both SNs and TNs, and solutions are presented, respectively. Moreover, due to its unique operation, a special case of incorrect convergence has been revealed when stuck-at faults are injected into the negative network; this can lead to a complete system failure. Therefore, the impact of stuck-at faults in the training process should also be evaluated. Traditional error-tolerant schemes such as redundancy [9, 10] and error correction codes (ECCs) [11–13] usually lead to considerable delay and hardware costs [14]. As a large amount of memory is needed to store the parameters in multi-branch networks, its protection has stringent requirements in terms of memory overhead; especially for applications on constrained platforms, such as portable systems.

This chapter introduces the multi-branch NNs (networks have multiple weight-sharing subnetworks), that is, SNs and TNs. The training and inference process of both networks are initially discussed for classification tasks; especially, a separately constrained triple loss (SCTL) is presented for the TNs for better performance. The hardware design of the multi-branch NNs is also provided; considering the similarity between SNs and TNs, the design focuses on the subnetworks.

This chapter then discusses the error tolerance of multi-branch NNs from the aspects of both training and inference. For the inference process, single bit-flip soft errors are evaluated. Initially, the impact of memory errors on the weights of the SN shows that the degradation is mostly caused by outliers in weights. This chapter thus introduces several schemes (using a weight filter and codes) to mitigate the impact of such errors. The training of NNs is usually not sensitive to faults. However, the unique features of the TNs lead to problems in error tolerance during training. An unexpected failure (leading to network convergence to false solutions) can occur when the faults are in the negative subnetwork. This chapter provides an analysis and presents two schemes that can solve such a problem. All the mentioned methods in this chapter are evaluated on popular classification datasets (MNIST [15], Fashion-MNIST [16], CIFAR-10 [17], and SVHN [18]).

2 Multi-Branch NNs: Algorithms and Designs

2.1 *Siamese Networks*

Siamese networks (SNs) are neural networks containing two branches of subnetworks [19]; as shown in Fig. 1, the two subnetworks have identical schemes and share the same weights. The SN feeds two inputs into the two branches and calculates the distance (e.g., the most widely used is the Euclidian distance $\mathrm{dis}\,(p, q) = \sqrt{\sum_{i=1}^{n} (q_i - p_i)^2}$ in the n-space) of the two outputs p and q at the last layer. Predictions are made based on the distances that evaluate the similarity of the inputs; for example, if the calculated distance is smaller than a given threshold, the two inputs are considered as belonging to the same class, not otherwise.

Several flexible schemes can be utilized in the subnetworks, such as the commonly used feedforward networks: the multi-layer perceptron (MLP) [20] or a convolutional neural network (CNN) [21]. Note that these networks usually have a limited number of layers, that is, they are shallow NNs; to the best of the authors' knowledge, the use of deep NNs as subnetworks of an SN has not been investigated in the technical literature.

For classification tasks, contrastive loss [22] is usually used in the training of the SN. As an SN requires a pair of images as inputs for the two branch networks, datasets are preprocessed to balance positive and negative pairs, so to indicate that the two inputs belong to the same or different categories.

2.2 *Triplet Networks*

Network Characteristics

A typical Triplet network (TN) consists of three identical weight-sharing subnetworks for training. Each subnetwork can be any feedforward network (e.g., an MLP, or a CNN); this chapter considers an MLP as an example.

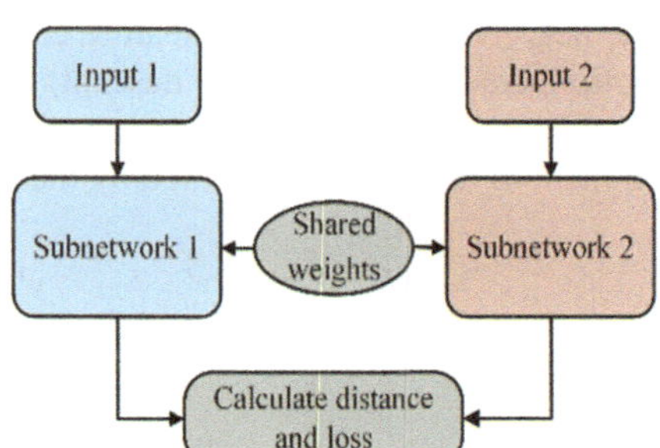

Fig. 1 Block diagram of a conventional Siamese network (SN)

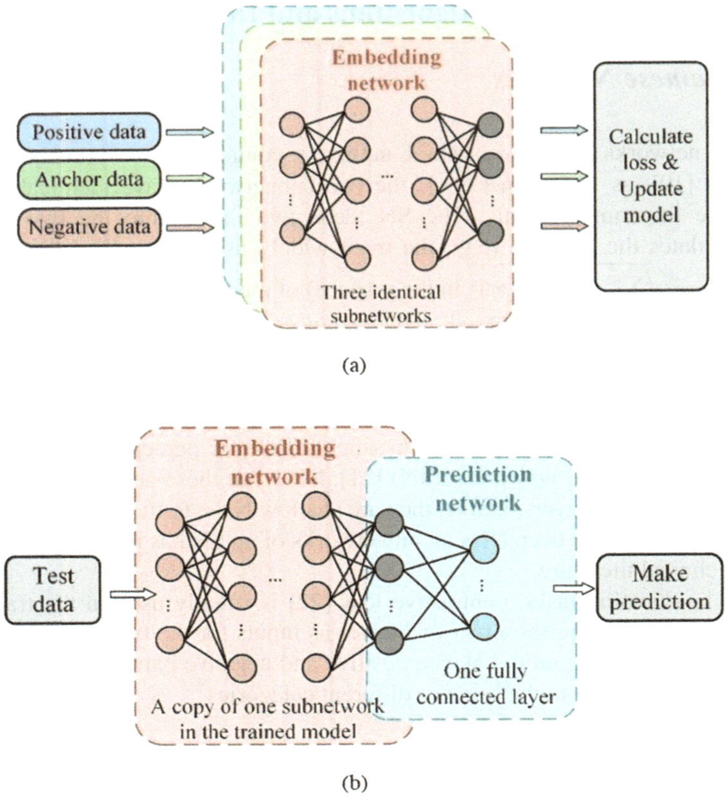

Fig. 2 A triplet network: (**a**) training process; (**b**) inference process

During the training process (Fig. 2a), the three subnetworks of a TN accept the so-called triplets, including anchors, positive samples from the same class, and negative samples from different classes, respectively. The model is trained by mapping the triplets to an embedding space to minimize the distances between the anchor and the positive samples and maximize the distance between the anchor and the negative samples. Such feature embeddings are learned by iteratively applying optimizers (such as gradient descent) to the loss for better performance [7, 8].

Assume that the subnetworks are l-layer MLPs, and the output of the hidden units in the r-th layer can be denoted by the following vectors:

$$\begin{aligned} y_a^r &= \begin{bmatrix} y_{a,1}^r & y_{a,2}^r & \cdots & y_{a,m_r}^r \end{bmatrix}, \quad \text{anchor subnetwork;} \\ y_p^r &= \begin{bmatrix} y_{p,1}^r & y_{p,2}^r & \cdots & y_{p,m_r}^r \end{bmatrix}, \quad \text{positive subnetwork;} \\ y_n^r &= \begin{bmatrix} y_{n,1}^r & y_{n,2}^r & \cdots & y_{n,m_r}^r \end{bmatrix}, \quad \text{negative subnetwork,} \end{aligned} \tag{1}$$

where m_r is the output dimension of the r-th layer. In particular, the outputs of the last layer are represented as y_a^l, y_p^l, and y_n^l. The similarity is evaluated by the Euclidean distance, which is calculated by the positive distance $d_+ = \left\| y_a^l - y_p^l \right\|_2$, and the negative distance $d_- = \| y_a^l - y_n^l \|_2$. Since the objective of training a TN is to minimize the positive distance and maximize the negative distance, the following conditions must be satisfied:

$$d_+ = \sqrt{\sum_{i=1}^{m_r} \left(y_{a,i}^r - y_{p,i}^r \right)^2} < \varepsilon_+; \tag{2}$$

$$d_- = \sqrt{\sum_{i=1}^{m_r} \left(y_{a,i}^r - y_{n,i}^r \right)^2} > \varepsilon_-, \tag{3}$$

where ε_+ and ε_- are constant margins. Moreover, the loss function L usually applies a relaxed constraint that optimizes the difference between the two distances $d_+ - d_-$ [23]. For example, the popular triple loss (TL) [24] can be used and the trained model thus satisfies:

$$L_{\mathrm{TL}} = \max\left(d_+ - d_- + M, 0\right) \leq \varepsilon, \tag{4}$$

where M is the margin and ε is a non-negative constant as the stopping criterion.

For the application of deep metric learning, an additional prediction network is needed to map the embedded data to labels; since the data from the same category has already been clustered in the embedding space by the TN, the prediction can be realized by simpler schemes such as Support Vector Machines (SVM), logistic regression, or K Nearest Neighbors (KNN). This chapter uses a one-hidden-layer MLP as the prediction network.

The trained model of a TN can map the inputs into an embedding space in which samples from different classes can be distinguished; it is then used in the inference process to classify the test dataset. One unknown sample (instead of triplets) is provided as input to the network each time; hence, only one of the three identical subnetworks in the trained model is employed. Let the output of the hidden units in the r-th layer be denoted as $y^r = \left[y_1^r \ y_2^r \ \cdots \ y_{m_r}^r \right]$. The output of the last layer is the feature embedding of the given input; the trained prediction network is applied for prediction. The inference process of TNs is illustrated in Fig. 2b).

Separately Constrained Triple Loss in TNs

The loss function determines the optimization objective of a TN, which greatly affects the overall performance. Define the anchor, positive, and negative outputs (at each subnetwork) as x_a, x_p, and x_n, the positive distance as $d_+ = \|x_a - x_p\|_2$, and negative distance as $d_- = \|x_a - x_n\|_2$, respectively. The traditional triplet loss

(TL) [24] has been reported to have good performance for most tasks, which is given by:

$$L_{\mathrm{TL}} = \max\left(d_{+} - d_{-} + m, 0\right). \tag{5}$$

There are other loss functions found in the technical literature, such as exponential triplet loss (ETL) [7], logarithmic triplet loss (LTL) [25], or ratio triplet loss (RTL) [23]. However, the above loss functions mostly consider a relaxed optimization problem by minimizing the difference term $d_{+} - d_{-}$ (or their ratio). This may lead to false solutions during training; for example, the TL can be zero as long as $d_{-} > d_{+} + m$ independently of a possible very large value of d_{+}. The undesirable intra-class distance can mislead the network; hence, the wrong convergence may cause a fatal failure.

The work presented in [26] also discusses such false solutions and introduces an additional term to d_{+} to address the intra-class difference, so yielding (6), where α is the weight and m_1 (m_2) is the positive (negative) margin; it is noted as the triplet loss addressing intra-class constraints (TLI):

$$L_{\mathrm{TLI}} = \max\left(d_{+} - d_{-} + m_1, 0\right) + \alpha \max\left(d_{+} + m_2, 0\right). \tag{6}$$

However, the "difference" term is still utilized rather arbitrarily in [26] and is not necessarily required. A further issue is represented by the very small value of the applied scalar α (e.g., 0.002); the new term has a small effect compared to the original term as in the standard TL. The feasibility of such a strategy in feature-embedding TNs still needs to be assessed and improved.

The separately constrained triple loss (SCTL) is an improved version of the conventional TL; it is given by:

$$L_{\mathrm{SCTL}} = \max\left(d_{+} - m_1, 0\right) + \beta \max\left(m_2 - d_{-}, 0\right), \tag{7}$$

where β is the parameter determining the weight of the two constraints. SCTL separates the constraint of TL into two terms, that is, the intra-class term $L_{\mathrm{intra}} = \max(d_{+} - m_1, 0)$ and the inter-class term $L_{\mathrm{inter}} = \max(m_2 - d_{-}, 0)$. When using SCTL, TN is trained iteratively until the constraints of $L_{\mathrm{intra}} < \varepsilon_1$ and $L_{\mathrm{inter}} < \varepsilon_2$ are satisfied, where ε_1 and ε_2 are the predefined stopping conditions. These constraints are set for minimizing d_{+} and maximizing d_{-}, which strictly follows the objective of feature-embedding TNs. The parameter β provides flexibility for fine-tuning in different scenarios.

The unique feature of SCTL is separating the intra-class and inter-class constraints. It does not focus on the difference between d_{+} and d_{-} (as TL); instead, it requires minimizing both terms to achieve simultaneously a small intra-class distance and a large inter-class distance. Therefore, the possible undesirable solutions with large intra-class distances can be avoided; moreover, SCTL provides a new perspective to investigate the characteristics of specific tasks. The two independent terms show how the encoded data distributed in the feature-embedding

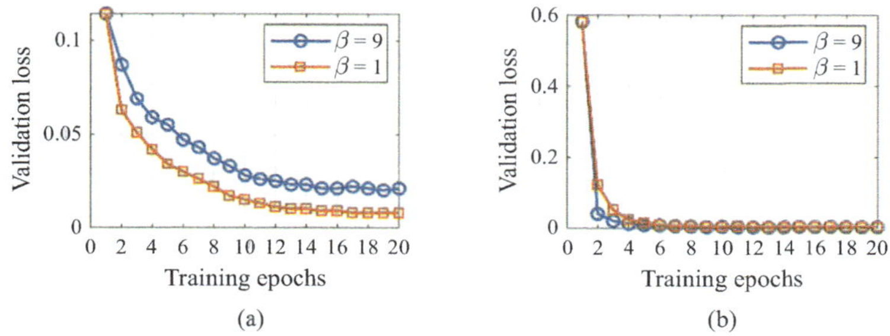

Fig. 3 Performance when using the decaying parameter β for: (**a**) the intra-class term; (**b**) the inter-class term

space can be separately monitored. This reveals the dependency on inter/intra-class similarity; it provides insights for fine-tuning specific tasks and assists in balancing the positive/negative triplets during data preprocessing.

In the SCTL, the parameter β balances the inter-class and the intra-class terms; thus, the selection strategy is initially discussed. According to the experiments, a fixed β has revealed the unbalanced dependency on the two terms L_{intra} and L_{inter}. Therefore, a strategy of a decaying parameter β is applied by the observations of convergence during the training process. Figure 3 shows the loss of the intra-class and inter-class terms during training for the MNIST dataset; large and small β values (9 and 1) are considered for comparison. Those two terms show diverse trends: the inter-class loss decreases very fast in the first several epochs, but it barely changes in the late stages; the intra-class loss decreases at a slower rate at first, but it keeps progressing till the end. This conclusion has also been verified for other datasets with random initialization. Hence, the decaying β with respect to epochs is shown as:

$$\beta = \frac{1}{\beta_0 + \Delta\beta \times n_{\text{epoch}}} - 1, \tag{8}$$

where β_0 is the initial value and $\Delta\beta$ is the step size. It guarantees a linearly changed ratio of the two terms in the SCTL. A large β makes the overall loss converge very fast in the early stages (in which the dependency shows that the inter-class term dominates). However, the inter-class term barely changes after the first few epochs, and a smaller β helps the intra-class term to converge better in the late stages.

The classification performance of TNs using the SCTL with decaying β is then evaluated in terms of inference accuracy. The four classification datasets mentioned previously are also considered in this section. The selection of triplets of the TN follows the strategy of hard negative mining [27, 28], and the embedding network uses 128-dimension outputs to achieve high performance as suggested in [7, 24]. The subnetwork of the embedding network utilizes an MLP that consists of five layers with 1024, 512, 256, 256, and 128 neurons per layer; the additional prediction network utilizes a fully connected layer as shown in Fig. 2b for inference. For all

Table 1 Inference accuracy of TNs with different loss functions

Loss function	MNIST	Fashion-MNIST	CIFAR-10	SVHN
TL	98.99%	90.60%	69.91%	91.35%
ETL	98.81%	89.81%	68.57%	90.64%
LTL	98.96%	90.35%	69.23%	91.52%
RTL	98.09%	85.59%	64.61%	89.29%
TLI	98.99%	90.25%	69.67%	90.85%
SCTL	**99.19%**	**91.45%**	**71.68%**	**92.29%**

loss functions, the same network configuration is applied for 20 training epochs. In the TNs with the SCTL, $m_1 = 0.01$ and $m_2 = 1$ are selected; moreover, different selection strategies of β have been tested and the parameters in each case are set as follows: (i) when using a fixed value, β is set as the value with the best performance; (ii) when using a decaying β as per (8), $\beta_0 = 0.1$ and $\Delta\beta = 0.02$. For existing loss functions, the parameters are selected as per the corresponding reference.

Table 1 reports the inference accuracy of the TNs with different loss functions. Among the existing loss functions, TL generally has excellent performance. SCTL achieves the best performance for all datasets and its benefit is more obvious for datasets with relatively lower accuracy (such as CIFAR-10). The results have confirmed the effectiveness of SCTL in TNs: it accelerates convergence by addressing the inter-/intra-class terms.

2.3 ASIC-Based Design for a Branch Network

In this section, the ASIC-based design is discussed for a branch network (i.e., an MLP) of the SNs and TNs. Single-precision floating-point (FP) data are considered for network computation, and different schemes are investigated for implementation.

Serial Implementation

In a serial implementation of an MLP, all calculations are performed by utilizing a single MAC [29] (as part of the scheme shown in Fig. 4); this design receives two inputs and multiplies them in 4 clock cycles in a pipelined mode. Then, it adds the product with the results from the previous step using the feedback input in the FP adder. Also, in this implementation when the start signal is low, all internal registers of the FP units are reset; otherwise, the multiplier starts the calculation.

Assume that the ith layer has m neurons; for each neuron in the $i+1$th layer, $m+1$ entries (so the number of neurons in the ith layer plus a bias value) is present, and thus $m+1$ pairs of inputs. After calculating the value of each neuron, the MAC is reset, and the calculation for the next neuron can be started. Therefore, a

Fig. 4 Serial MLP implementation (address sizes are for MNIST & FMNIST)

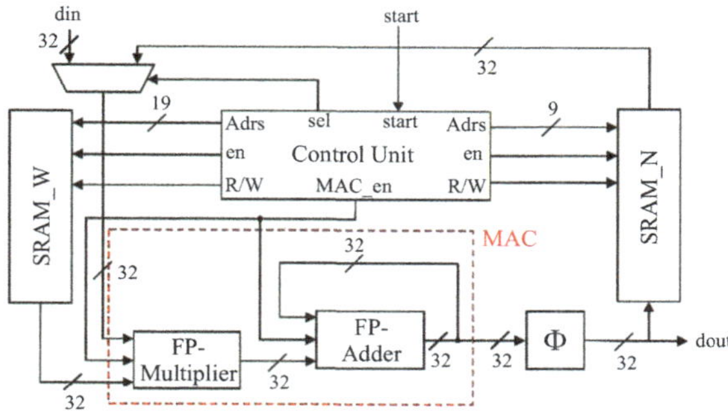

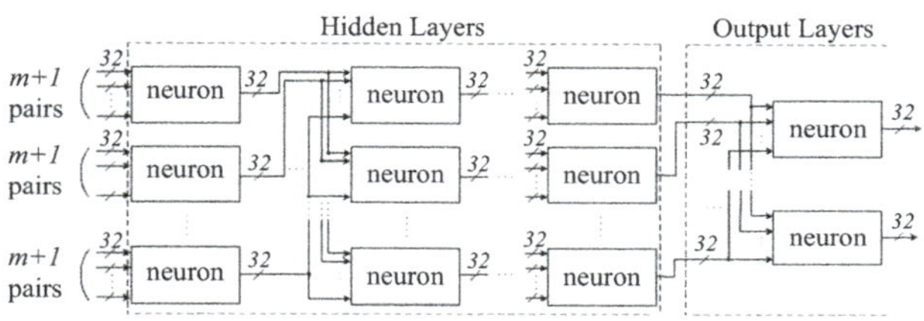

Fig. 5 Parallel MLP implementation

serial implementation computes each neuron value for the hidden and output layers serially; this feature significantly decreases the area, but it increases the latency. Figure 4 shows a schematic diagram of this implementation; an SRAM is required to save the result of each neuron, because this value is needed for the calculations of the neurons in subsequent layers. Also, a portion of the SRAM is used to save the weights related to each neuron; the control unit is responsible for reading/writing from the SRAM and set/reset of the neuron [30].

Hybrid Implementation

A hybrid MLP implementation [45] is based on a combination of parallel and serial implementations; it uses serial neuron hardware (i.e., a MAC). Consider the fully parallel design in Fig. 5 (this design is not presented in detail because its hardware is unfeasible); the hybrid design consists of increasing the level of parallelization by adding more MAC units to the design. Hence, instead of calculating separately, several neurons are calculated at the same time, but still in a serial mode, that is, as each neuron processes only a pair of inputs in each step, the control unit allocates all possible pairs to the neuron during the computational phase. Therefore, the final values of neurons are saved in the SRAM and used for the next layers (because they cannot be processed at the same time due to the limited number of MACs). In this chapter, 16 MAC units are used for the different datasets considered; however, the SRAMs must be implemented such that several banks are provided to either

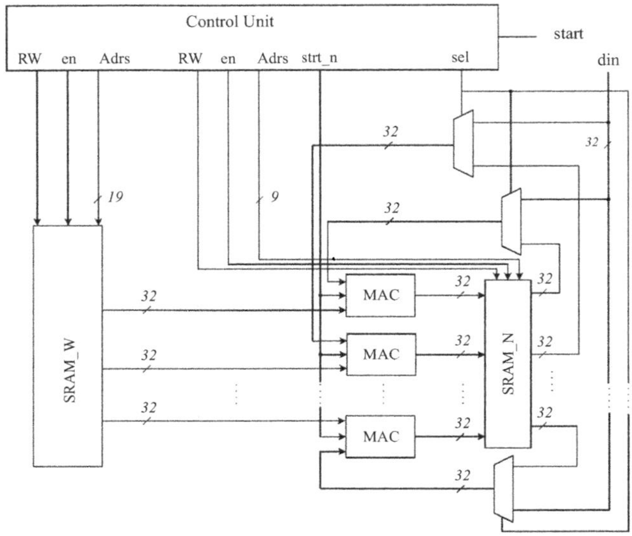

Fig. 6 Hybrid MLP implementation (address sizes are for MNIST & FMNIST)

generate 16 outputs or save 16 inputs at the same time. Therefore, the performance of the SRAM memory plays an important role in the design.

Figure 6 shows the design for the hybrid implementation; in this design, only a single array of MACs is utilized. Therefore, for each layer, the control unit sets the proper inputs for the MACs (weights and input data), and when the computation of that layer is completed, the results are saved in the related SRAM, because they are needed as inputs of the next layer (so different from a fully parallel implementation). For the computation of the next layers, the current array of MACs can be used again. As for the memories, SRAM_W stores the weights and bias values, while SRAM_N saves the values of the neurons.

Evaluation

The same network of an MLP discussed previously is considered for the hardware evaluation. The serial and hybrid designs utilize 1 and 16 MAC units, respectively; since there are 512 neurons in the first hidden layer, the image must be accessed 512 or 32 times for the serial and hybrid designs, respectively, then a small SRAM unit is used to save a pair of images for next time (SRAM_I). All hardware implementations have been designed using Cadence Genus Synthesis Solution and a 32 nm technology library (at 25 °C and TT corner); the optimization effort for area, power, and delay has been set to high for the tool to automatically consider a tradeoff between them and the preset constraints to reach the best result.

Table 2 Synthesis results of an MLP design using different implementation schemes

Dataset	Design	ASIC Power (mW)	Area (mm^2)	Delay (ps)	# Cycles	Memory Size (Mb)	Area (mm^2)
MNIST	Serial	1.91	0.0120	1452	3,725,348	3.547	96.13
	Hybrid	21.69	0.1597	1468	239,262	3.547	96.42
Fashion-MNIST	Serial	1.91	0.0120	1452	3,725,348	3.547	96.13
	Hybrid	21.69	0.1597	1468	239,262	3.547	96.42
CIFAR-10	Serial	2.17	0.0120	1452	4,216,868	4.015	104.17
	Hybrid	24.64	0.1597	1468	271,542	4.015	104.47
SVHN	Serial	2.17	0.0120	1452	4,216,868	4.015	104.17
	Hybrid	24.64	0.1597	1468	271,542	4.015	104.47

Table 3 Comparison of different MLP designs

MLP	Technology	Network	Frequency (MHz)	Power (mW)	Data format
[32]	FPGA-28 nm	7-6-5	100	120	Fixed-Point
[33]	FPGA-28 nm	784-32-10	100	654	Fixed-Point
[34]	FPGA-28 nm	784-600-600-10	490.8	–	Fixed-Point
[35]	FPGA-16 nm	784-12-10	100	568	Floating-Point
[36]	ASIC-28 nm	784-200-100-10	114.7	54	Floating-Point
[45]	ASIC-32 nm	784-512-512-512-2	681.2	21.69	Floating-Point

Table 2 shows the synthesis results for the different datasets for the MLP. The serial design incurs in the least area and power dissipation, but at a higher number of clock cycles. Therefore, the serial implementation is the best design candidate for low-power applications; however, training can be rather time-consuming. The ASIC implementations operate in a pipeline mode at 681.2 MHz frequency, while the total delay during classification is several milliseconds.

Memory is needed for the serial and hybrid designs to save the values of the deeper neurons because these designs do not use all of them at the same time. Although the memory sizes for these two designs are the same, their control units are rather different. In the serial design, the data is read/written serially during each cycle, while for the hybrid design, the data is written in the hybrid mode at different memory banks (16 data banks); during the read cycle, the data is read in the sequence that they have saved, so serially as multiplication is executed between the value of each neuron in the previous layer and its related weight for a neuron in the current layer.

Finally, the hybrid MLP implementation is compared with other existing works (based on FPGA and ASIC [32–35]) found in the technical literature in terms of hardware performance; the MNIST dataset is taken as an example, but the trend of the other datasets is similar, because the benefit of the design comes from a hardware configuration that is mostly independent of the network configuration/datasets. The synthesis results given in Table 3 show that the presented design achieves the highest frequency, with the least power dissipation even though the existing works use more advanced process technology and have an extremely smaller network.

3 Error Tolerance of Multi-Branch NNs

The weight-sharing feature leads to a unique challenge for error tolerance in multi-branch networks (SNs and TNs). This section discusses the error tolerance techniques for the inference and training, respectively; different error/fault models are analyzed and protection schemes are presented. Note that different NNs may be taken as an example in different cases, but the analysis and techniques can be generalized to both SNs and TNs.

3.1 Error Tolerance During Inference

The inference process is critical in the application of emerging machine learning tasks, and it is sensitive to soft errors. This section first introduces the bit-flip fault model and analyzes its behavior in multi-branch networks. Several error tolerance schemes are then presented considering different scenarios, including a weight filter, and optimized error correction codes and parity checks. The performance of those schemes is presented and compared by the error injection experiments.

Bit-Flip Fault Model

For the inference of a multi-branch network, the trained model is usually stored in a memory as weight (and bias) matrices. As introduced previously, memories are prone to suffer different types of errors/faults; one of the most common types is soft errors mainly caused by radiation particles [37]. Such errors can flip the stored bits and then cause data corruption; they have been considered as a significant reliability issue for memories. Since the corrupted data can cause incorrect predictions and possibly, system failure, error protection must be provided for NNs in safety-critical applications [38]. This is especially critical for an SN/TN, because the memory storing the weight matrices is shared by subnetworks and a single error may affect both subnetworks.

Therefore, as the error model during inference, this chapter considers bit-flips in the weight matrices; moreover, it is assumed that the parameters of the SNs/TNs are represented in the IEEE 754 standard single-precision floating-point format [39] (Fig. 7) and stored in a memory with 32-bit words (i.e., each weight is stored in a single memory word). The decimal value of the FP number is calculated as:

$$\text{dec} = (-1)^S \times 2^{E-127} \times \left(1 + M \times 2^{-23}\right), \tag{9}$$

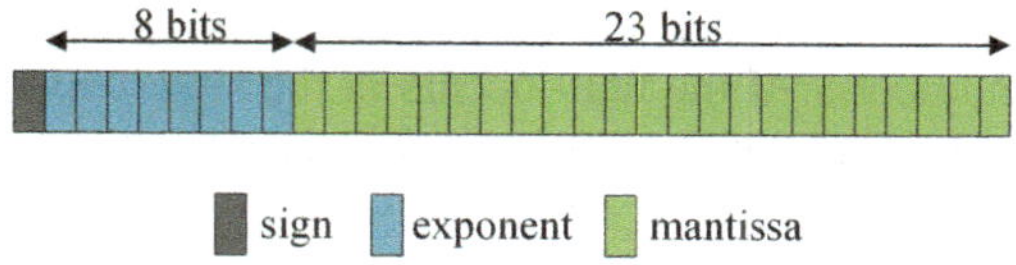

Fig. 7 Data represented in the IEEE standard 754 floating-point single-precision format

Table 4 Different error scenarios

Scenario		# flipped bits	# affected weights
Simple case		1	1
General case	Single-bit errors	1	Up to10
	Multi-bit errors	Up to10	1

where S, E, and M are the decimal number represented by the corresponding sign, exponent, and mantissa bits. A bit-flip on the sign or exponent bits leads to very significant changes in value.

In this chapter, simulation is pursued under several scenarios. The simple case is analyzed under only a one-bit-flip occurrence, while for the general cases, more errors (up to 10) are considered. When more than one error occurs, as multiple data bits in a single memory word, or single data bits in multiple adjacent words [40], two scenarios are defined as general cases: (1) Single-bit errors: multiple errors occur, but each incorrect weight/memory word has only one bit flipped. (2) Multi-bit errors: multiple errors occur but they affect the same weight/memory word. The different error scenarios considered in this chapter are summarized in Table 4.

Impact of Bit-Flip Faults on Inference

During the inference process, the fault model consists of random bit-flips; these faults are transient in nature, and they have been widely accepted as random upset values in memory. Therefore, in the fault model, random bit-flips can occur at any of the bits by reversing the value (0 to 1, or 1 to 0); since predictions are directly determined by the weights stored in memory, bit-flip faults are critical in the inference process. Two scenarios (single-bit error/multi-bit error) are considered as shown in Table 4.

For example, the execution of the inference process in SNs/TNs with bit-flip faults can be represented as patterns. The pattern is represented by $Q_I = (f, r, P)$, where f is the number of faults and r denotes the layer to be injected, $f < m_r$; the set $P = \{p_1, p_2, \cdots, p_f\}$ denotes the position of the corresponding faults, so the location of the bit-flip at the p_i-th bit, $p_i \in \{1, 2, \cdots, 32\}$. The decimal value of a floating number is given by $x = (-1)^S \times 2^{E-127} \times (1 + M \times 2^{-23})$, where S, E, and M are the decimal numbers represented by the sign, exponent, and mantissa bits; hence, faults on the sign or exponent bits ($p_i \leq 9$) lead to more significant changes.

Let x_e denote the erroneous value represented by a single flipped bit of an FP number x. If the fault occurs in the sign, exponent, or mantissa bits, the erroneous value and the absolute error are represented as:

$$x_e = \begin{cases} -x, & \text{sign bit;} \\ 2^{E_i} x, & \text{exponent bits;} \\ \frac{M_i M + 2^{23}}{M + 2^{23}} x, & \text{mantissa bits;} \end{cases} \quad (10)$$

$$|x - x_e| = \begin{cases} |2x|, & \text{sign bit;} \\ \left|\left(2^{E_i} - 1\right) x\right|, & \text{exponent bits;} \\ \left|\frac{(M_i - 1) M}{M + 2^{23}} x\right|, & \text{mantissa bits,} \end{cases} \quad (11)$$

where $E_i = 2^i$ if the i-th bit (counted from the LSB to the most significant bit (MSB)) of the exponent is flipped from 0 to 1; $E_i = 2^{-i}$ if the i-th bit of the exponent is flipped from 1 to 0. Similarly, $M_i = M + 2^i$ when the i-th bit of the mantissa is flipped from 0 to 1; otherwise, $M_i = M - 2^i$. M is the decimal number represented by the mantissa bits; as per (11), the error of an FP number is related to its absolute value $|x|$; therefore, bit-flips in a larger FP number lead to a more significant error.

Weight Filter

It was observed in the error injection experiments that significant changes in the weight value have a larger impact on the performance of the network. Therefore, we focus on reducing the impact of such errors; this is consistent with previous works [41–44]. All these works rely on the same observation that outliers with large absolute values in weights (caused by faults/errors) are the primary causes of performance degradation of NNs.

Consider the characteristics of the weight distribution; the filter can be mathematically represented as a multiplication with a rectangular function that constrains the weight values within a reasonable range. Such function is defined as:

$$\text{weight\_modified} = \text{weight} \times \text{rect}\left(\text{weight}\right), \quad (12)$$

$$\text{rect}(t) = \begin{cases} 1, & \text{if } t \geq \tau_{\min} \text{ or } t \leq \tau_{\max} \\ 0, & \text{otherwise} \end{cases}, \quad (13)$$

where $\tau_{\min}$ and $\tau_{\max}$ determine the lower and upper bounds of the filter, and $\tau_{\min} < \tau_{\max}$. The optimal feasible solution for the filter's upper and lower bounds is given by: (The detailed analysis refers to paper [45])

$$\begin{cases} \tau_{\max} = w_{\max} \\ \tau_{\min} = w_{\min} \end{cases}. \tag{14}$$

This shows that the filter with bounds given by the statistical maximum and minimum of the weight values has the largest probability to maintain the original predictions in the simple case.

The bounds of the filter are recorded when the model is saved in memory. Prior to the inference process, the weight matrices are checked by comparing them with the upper and lower bounds. After the outliers are limited by the filter, the inference process continues with the modified weights.

The above choice of bounds is derived from a simple case that applied the following assumptions:

1. Only one erroneous bit is present in the weight matrices.
2. The output layer of each branch network has only one neuron.
3. The error appears only in the weight of the last layer (i.e., the output layer in the MLP version, or the last fully connected layer in the CNN version).

In the general case, memory errors in a subnetwork may not satisfy the assumptions outlined above. Therefore, it is not possible to prove the optimal bound selection in a generalized case. Instead, this chapter provides a discussion of a heuristic bound selection for the general cases.

In the simple case, the calculation of the Euclidean distance with an error is rather simple; the distance D_0 with no error and the distance D with one bit of error satisfy the following relationship:

$$\begin{aligned} D &= \sqrt{(y_1 - y_2)^2} \\ &= \sum_{j=1}^{n} x_{1,j} w_{1,j} + x_{1,i} e - \sum_{j=1}^{n} x_{2,j} w_{2,j} - x_{2,i} e \\ &= D_0 + \left(x_{1,i} - x_{2,i}\right) e. \end{aligned} \tag{15}$$

If assumptions (1) or (2) are not satisfied, (15) can be changed by substituting the error terms with the sum given by:

$$D = D_0 + \sum_{i=1}^{k} \left(x_{1,i} - x_{2,i}\right) e_i, \tag{16}$$

where k can be the number of soft errors, or the number of output dimensions. In these two cases, the problem is equivalent to minimize $|\sum e_i|$. The optimal solution (14) only holds when all errors have the same sign, which is not necessarily valid in practice; hence, the bounds are not always optimal (depending on the signs of the errors), but it is a feasible solution for all cases.

When assumption (3) is not satisfied, the analysis becomes rather complicated. Even if the filter minimizes the changes in the erroneous layer, the final output cannot be predicted; this also happens to the convolution layers of a CNN. This

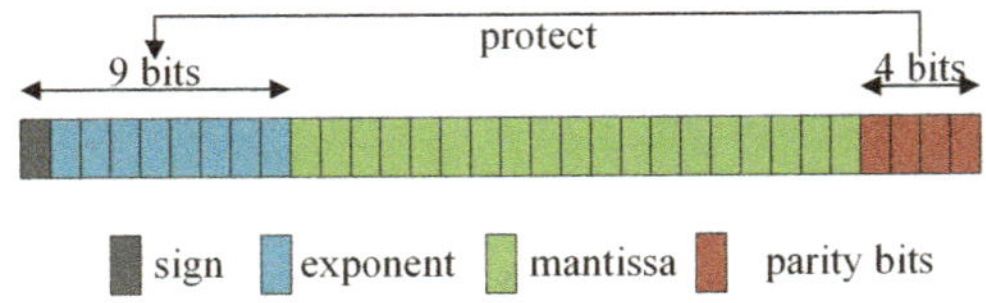

Fig. 8 Data represented in IEEE standard 754 floating-point format protected by the code scheme

occurs because errors cannot be represented as an independent term as in (15) after propagation. In this case, the optimality of the bounds is dependent on the network and datasets; hence, it is difficult to establish a deterministic mathematical model.

Even though optimality does not always hold, the bound selection in (14) can be still used. Considering the diverse datasets, the bounds set by the range of the weights provide good flexibility; moreover, it is generally applicable for all scenarios (optimal in the simple case; feasible and sometimes optimal in the more general cases). As per the observation that outliers are usually extremely large, a small deviation to the optimal bounds brings trivial effects.

Single Error Correction (SEC) Code

A coding scheme is designed to deal with single-bit errors in this section. Since outliers due to errors on the sign and exponent bits of weights cause the most critical changes affecting predictions, the protection of only those bits can greatly reduce the redundancy overhead, while still retaining satisfactory performance. Therefore, in the code scheme, an SEC code is employed to cover only the sign and exponent bits (i.e., 9 bits in total) of each weight; this reduces the required number of parity bits from 6 to 4 compared to a traditional SEC code that covers all 32 floating-point bits [31]. Moreover, the presented scheme stores the parity bits on the 4 least significant bits (LSBs) of the mantissa of each weight (Fig. 8); hence, no memory redundancy/overhead is finally encountered. Even though a very small deviation is introduced by replacing the original 4 LSBs with the parity bits, it leads to negligible changes in values and operations of NNs [46]; this is also applicable to an SN as established in the error injection experiments.

Overall, this scheme is expected to have a significant performance improvement either by itself, or when combined with the filter (when there is more than one erroneous bit on the significant bits, the provided code is not suitable, hence the need for the weight filter).

Parity Code

The parity code can detect the erroneous bits in the weight; it can be applied to the networks for fault tolerance with lower costs. The previous sections showed the effects of random bit-flip faults on the inference process; in particular, large outliers in weights are caused by flips in the significant bits. An efficient way is to

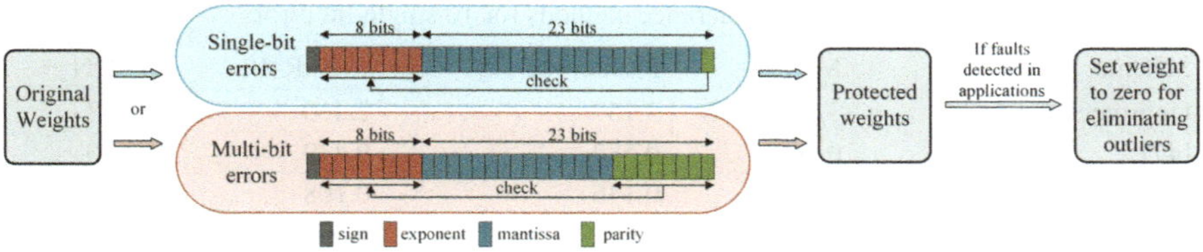

Fig. 9 Parity-based fault tolerance for bit-flips during inference for single-bit or multi-bit errors

detect those errors and directly set the corresponding weight to zero. The presented methods protect the inference process from bit-flips by inserting parity bits for both single-bit and multi-bit errors. The schemes are illustrated in Fig. 9.

Replacing LSB by a Parity Bit for Single-Bit Errors As per the error injection process, the protection of MSBs (exponent bits) is of great importance and can greatly reduce the loss of accuracy. A single parity check is widely used for error detection, and thus, it is applied in this scheme. A parity bit is employed to check the exponent bits (encoded by the XOR operation with each bit); therefore, the parity bit detects a single-bit error when any of the significant bits are changed by faults. Once the bit-flip is detected, the corresponding weight is directly set to zero to prevent outliers; moreover, the parity bit replaces the LSB of the mantissa. This scheme does not require additional memory overhead and shows the applicability of the presented scheme because the data format remains unchanged. Setting the weight to zero and replacing the mantissa bit can lead to a loss in accuracy; this is usually negligible compared to the benefits of protecting the MSBs. With normalization and activation functions (such as *sigmoid* or *tanh*), the weights in NN applications usually have small absolute values (nearly close to zero); the effect of a changed mantissa bit is verified by the analysis and related literature [46]. With large outliers avoided, the degradation caused by bit-flip faults is expected to be greatly mitigated and evaluation of performance is provided by simulation.

Replacing LSBs by Parity Bits for Multi-Bit Errors The case of multiple faults in one weight is rarer than the single case; however, as code-based protection for single-bit errors can fail, this chapter also proposes a parity-based protection method for multi-bit errors. The principle in these methods is similar to detecting single-bit errors by using parity bits (inserted at the LSBs of the mantissa) to detect errors in the exponent bits. However, since more than one fault can occur in these bits, more parity bits are required to check each of them (one parity bit encodes an exponent bit, and 8 bits are required for an exponent in the single-precision FP format). Similarly, if a fault is detected, the weight is directly set to zero to prevent large outliners. Replacing more LSBs in the mantissa decreases the accuracy, but as per the analysis of (11) in Sect. 3.1, such loss is still negligible (<0.001% in the fault-free cases) compared with a possible fatal effect due to these faults. Further evaluation is provided by simulations to verify the feasibility of the presented methods with also a loss in accuracy.

Table 5 Accuracy loss (%) of fault tolerance methods for 10 single-bit errors

Scheme	MNIST	Fashion-MNIST	CIFAR-10	SVHN
Unprotected	3.765	3.157	2.460	2.967
[47]	0.662	0.582	0.449	0.560
[48]	0.178	0.146	0.108	0.117
Error correction code	**0.007**	**0.005**	**0.006**	**0.006**
Parity code (single)	0.118	0.104	0.089	0.086

Table 6 Accuracy loss (%) of fault tolerance methods for a multi-bit error with 10 faulty bits

Scheme	MNIST	Fashion-MNIST	CIFAR-10	SVHN
Unprotected	2.459	2.611	1.932	2.030
[47]	0.388	0.301	0.274	0.319
Filter	0.068	0.080	0.062	0.074
Parity code (multi)	**0.050**	**0.056**	**0.043**	**0.053**

Evaluation

The presented methods and schemes (weight filter, error correction code, and parity codes) and existing related works are compared with bit-flip faults in the inference. The results are evaluated by the accuracy loss compared with the fault-free cases. Multi-layer perceptrons (MLPs) are used as branch networks and each is designed with 5 linear layers (with size 784-512-512-512-2); ReLU is selected as the activation function between layers. The results with no fault tolerance are also provided for comparison.

Two error injection scenarios (single-bit error and multi-bit errors) are applied as shown previously in Table 4. Among the introduced schemes, the error correction code cannot deal with multi-bit errors, while the filter is more suitable for such a scenario. The two schemes of parity code are also designed for each case, respectively.

For the single-bit error evaluation, one bit-flip has been simulated for 10 different random weights; the results are averaged over 10,000 repeated trials and are shown in Table 5. The error correction code achieves the best performance (the lowest accuracy loss), because it performs full correction by SEC; the parity code eliminates on average 96.76% of accuracy loss caused by faults, so better performance than [47] and [48]. Considering that the parity-based method is simpler and replaces fewer bits, it can be thought of as an effective alternative; those schemes can be chosen for different requirements on accuracy or overheads.

For multi-bit errors, 10 bit-flips are applied to one different random weight; the results are averaged over 10,000 repeated trials as shown in Table 6. On average, the parity code reduces accuracy loss by 97.74%, outperforming both [47] and the filter. The replaced bits in the mantissa are thus shown to have a marginal effect compared with the benefits of fault tolerance; therefore, if the application is at risk from multi-bit errors, the parity code provides better protection.

In addition, the weight filter and the error correction code can be applied together as a combined method, which is suitable for more cases. According to the experiments in [45], the combined method can retain the advantages of the filter/code in both cases of single-bit error and multi-bit errors. Therefore, if comprehensive protection is preferred or the error type is unclear, the combined method could be an effective choice.

3.2 *Error Tolerance During Training*

The over-provisioning of neurons in fully connected neural networks provides an inherent fault-tolerant capability, especially for the training process since stochastic techniques are usually applied; a limited number of faults usually lead to a trivial accuracy loss or graceful degradation. However, stuck-at faults in the negative subnetwork of a TN can lead to an unexpected system failure as the network converges to a false solution. This section first introduces the stuck-at fault model and provides a theoretical analysis of such false solutions in TNs. Two error tolerance schemes (regularization to anchor outputs and modified margin) are discussed and evaluated by error injection experiments.

Stuck-at Fault Model

Since the network parameters are changing at each iteration of training, transient faults (such as random bit-flips) barely affect the training process. Therefore, stuck-at faults are considered as fault models during the training process. This model has been originally defined for permanent faults/defects of high/low logic values at physical transistor and circuit levels in Very Large Scale Integration (VLSI) [49]; for ANNs, the model has also been extended as an abstraction as per its impact [50]. As shown in Fig. 10, stuck-at faults in ANNs can be divided into two types: stuck-at 0 faults (missing the neuron) and stuck-at $1/{-1}$ faults (saturating the neuron).

- *Stuck-at 0 faults:* A fault causing a missing neuron means that no signal (and data) is generated irrespective of the input. As it has been abstracted that the output is stuck at an intermediate value [52], the intermediate value after the activation function "*tanh*" (which is used in this chapter) can be treated as "0," because the outputs are with a zero-mean normal distribution. For simplification, the fault causing a missing neuron is denoted as "stuck-at 0 faults" [10].
- *Stuck-at 1/−1 faults:* In a saturated neuron caused by faults, the data or control line is held permanently high or low [51]. Therefore, it can be mapped to the scenario in which outputs of specific neurons are stuck at a maximum or a minimum value. Also, as activation functions (e.g., "*tanh*" with an output range

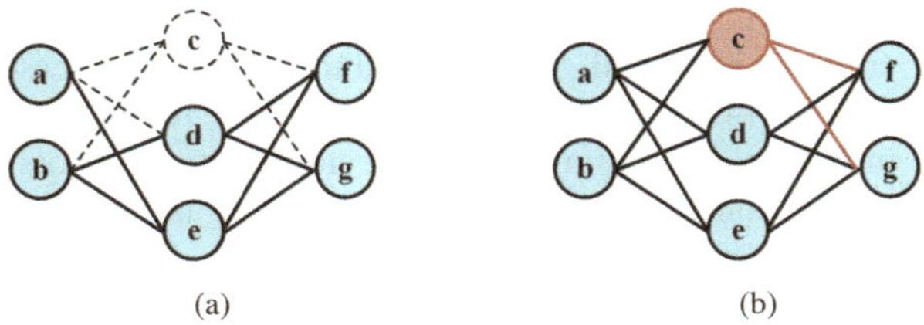

Fig. 10 Examples of the faulty network (with faults in neuron c). (**a**) Stuck-at 0 fault, missing the neuron. (**b**) Stuck-at 1 or -1 fault, saturating the neuron

of $[-1, 1]$) are applied after every neuron, then this model can be simplified as stuck-at 1 or -1.

The execution of the training process in TNs with specific faulty neurons can be represented by patterns [52]. As the exhaustive testing for all possible fault-injection cases is prohibitive, then for simplification purposes, this chapter assumes that all faults are injected into the same layer and the same subnetwork. Hence, the pattern representation is the tuple $Q_T = (f, r, c, v)$, where f is the number of faults and r denotes the layer to be injected (the input layer is not included), $f < m_r$; c is the faulty subnetwork, $c \in$ {anchor, positive, negative}; v denotes the type of the corresponding faults, which means the faulty unit is stuck at v, $v \in \{0, 1, -1\}$.

Impact of Stuck-at Faults on Training

The effect of stuck-at faults in the negative subnetwork of TNs can be analyzed in several different cases:

The incorrect convergence with stuck-at-0 faults in the negative subnetwork is rather intuitive when the faults are in the output layer of MLPs or the last fully connected layer in CNNs (i.e., $r = l$). Due to these faults, a spurious solution can be easily attained in training (so it satisfies the constraints, but it is not a valid solution for the classification problem).

1. *Stuck-at 0 faults in the output layer:* Let I be the set of indices of the stuck-at 0 faults and i denote the index of the faulty neurons, $i \in I$. The false solution only needs to map the anchor and positive subnetworks to similar outputs, while the values of the neurons at the indices of the faulty units ($y^l_{a,i}$) are sufficiently large (the absolute value can be close to 1 after *tanh*). Consider that stuck-at 0 faults are injected, making $y^l_{n,i} = v_i$; as per (3), the constraint of the negative distance in this case is given by:

$$d_- = \sqrt{\sum_{\substack{i=1,\\ i \notin I}}^{m_l} \left(y_{a,i}^l - y_{n,i}^l\right)^2 + \sum_{i \in I} \left(y_{a,i}^l - v_i\right)^2} \geq \sum_{i \in I} \left|y_{a,i}^l - v_i\right| > \varepsilon_-, \tag{17}$$

and it can be satisfied by even only considering the faulty terms $\left|y_{a,i}^l - v_i\right|$, $v_i = 0$. The spurious solution is easy to be achieved; for the anchor and positive subnetworks, the weights address the common part of their inputs, and the negative subnetwork is ignored. Therefore, the training converges at that solution in very few iterations.

2. *Stuck-at 1/−1 faults in the output layer:* Let I be the set of indices of the stuck-at 1/−1 faults. The analysis of false solutions is similar to the case of stuck-at 0 faults. Consider (17) for $v_i = 1$ or -1; the constraint of d_- can also be satisfied by only considering the faulty terms $\left|y_{a,i}^l - v_i\right|$. However, it could be worse because the convergence can get stuck into the zero-loss situation even before training starts. For all loss functions optimizing the relaxed constraint on $d_+ - d_-$, this leads to a false solution from the start. For example, the L_{TL} in (6) can be represented as:

$$\begin{aligned} L_{\mathrm{TL}} &= \max\left(d_+ - d_- + M, 0\right) \\ &= \max\left(d_+ - \sqrt{\sum_{\substack{i=1,\\ i \notin I}}^{m_l} \left(y_{a,i}^l - y_{n,i}^l\right)^2 + \sum_{i \in I} \left(y_{a,i}^l - v_i\right)^2} + M, 0\right) \\ &\leq \max\left(d_+ - \sum_{i \in I} \left|y_{a,i}^l - v_i\right| + M, 0\right), \end{aligned} \tag{18}$$

where $v_i = 1$ or -1. The faulty term leads to a false solution with zero loss if $\sum_{i \in I} \left|y_{a,i}^l - v_i\right| \geq d_+ + M$. This can be achieved under a large number of faults. In practice, the initial outputs v_i are usually close to 0 due to weight normalization, the margin M is usually set to 1 and $|d_+| < 1$; hence, it is likely that the training gets stuck in a zero-loss situation at the beginning.

3. *Stuck-at faults in hidden layers:* The previous false solution (for stuck-at 0/1/−1 faults) can be similarly extended to cases when the faults are in the hidden layers (i.e., $r < l$); however, when considering the complexity of forward propagation, a strict representation of constraints in each layer cannot be established. Next, we provide a heuristic discussion. The propagation between fully connected layers is given by:

$$y_j^{r+1} = \phi\left(\sum_i w_{j,i}^r \bullet y_i^r + b_j^r\right), \tag{19}$$

where $w^r_{j,i}$ and b^r_j are the weight and bias; ϕ is the activation function ("*tanh*" used in this chapter). The propagation in convolutional layers can be represented as:

$$y^{r+1}_{u,v} = \phi\left(\sum_{i=-\infty}^{\infty}\sum_{j=-\infty}^{\infty} y^r_{i+u,j+v} \bullet k^r_{i,j} + b^r_j\right), \tag{20}$$

where $k^r_{i,j}$ is the corresponding entry of the 180° rotated $n \times n$ convolutional kernel K^r (only enabled when $i \geq 0$ and $j \leq n$). The pooling layers do not affect fault propagation; similar to the constraints at the output layer, the faulty terms $\sum_{i\in I}\left(y^r_{a,i} - v_i\right)$ can still dominate at the r-th layer. Such distances in the negative subnetwork can transmit to the following layers (so, no more concentrated on those faulty indices, but separate to all units). However, *tanh*(x) bounds its inputs to the range $[-1, 1]$, and weakens the dominance of large distances caused by faults. This behavior stacks over multiple layers; hence, when the faults occur in the early layers, the negative distance d_- at the spurious solution tends to be smaller. It is possible that $d_- < \varepsilon_-$ and this spurious solution becomes only a local minimum; in this case, the network can sometimes avoid it and continue to converge to the global solution (as the behavior in the fault-free cases). However, with an increasing number of faults, the network is more likely to be stuck at the false solution.

Regularization to Anchor Outputs

From the analysis, the false solution with stuck-at 0 faults occurs when the faulty neurons in the anchor and positive subnetworks have large absolute values; in this case, the constraint of the negative subnetwork can be simply satisfied by the faulty term.

To prevent an incorrect convergence toward large outputs in the anchor or positive terms, an extra penalty term (i.e., the L2-regularization term in the anchor outputs from different layers) must be added to the loss function as:

$$L = \max\left(d_+ - d_- + M, 0\right) + \lambda \sum_{j=1}^{l}\sum_{i=1}^{m_j} {y^j_{a,i}}^2, \tag{21}$$

where m_j is the dimension of layer j and λ is the parameter determining the influence of the penalty term. This chapter assumes that faults occur in any layer; however, if the faulty layer can be specified, only regularization of the corresponding outputs is required. Consider TL as an example only (applicable to other functions too). The regularization term can be added to either the anchor, or the positive subnetworks (their outputs tend to be the same due to the objective function). In this chapter, the regularized anchor outputs affect both positive and negative distances, so they can better accelerate convergence.

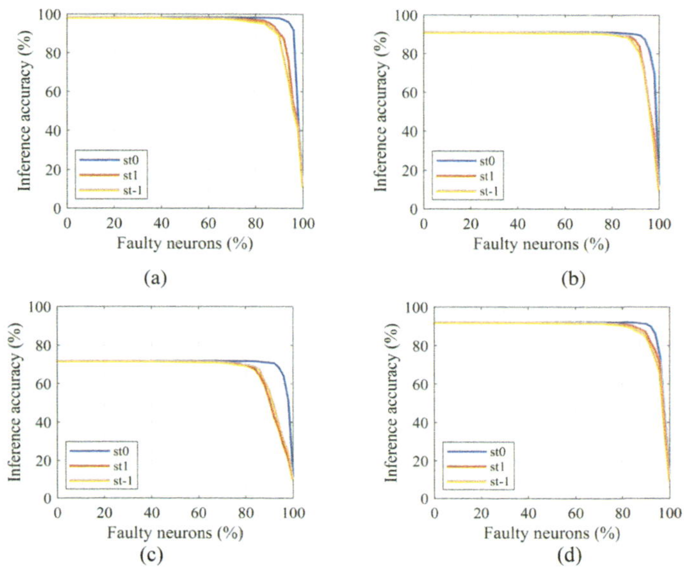

Fig. 11 Inference accuracy of different percentages of faulty neurons with stuck-at 0 faults (regularization on anchor outputs), and stuck-at 1 or −1 faults (modified margin): (**a**) "MNIST"; (**b**) "Fashion-MNIST"; (**c**) "CIFAR-10"; (**d**) "SVHN"

The regularization term on the anchor outputs can restrict the faulty units from taking a large value, thereby preventing the convergence to the false solution. By choosing a proper parameter λ, the additional term is negative to the error-free case; however, it prevents an unexpected system failure to occur when caused by the stuck-at 0 faults in the TNs.

This subsection provides the simulation results for the technique of regularization to anchor outputs (modified margin) to avoid false solutions when the stuck-at 0 (stuck-at 1/−1) faults are in the negative subnetwork. Simulation has been performed for faults that are randomly injected into different layers; hence, it also shows the worst case (when the faults are in the output layer). Consider the MLP implementation as an example, the pattern of the fault-injection simulation is given by $Q_T = (f = m_r \times t, r \in \{5\}, c \in \{\text{negative}\}, v \in \{0, 1, -1\}), t \in [0, 1]$.

The inference accuracy at different percentages of faulty neurons is shown in Fig. 11. The negative subnetwork is no longer sensitive to stuck-at faults (compared with huge degradation with even 1% of faulty neurons shown in paper [53]); hence, the false solutions can be effectively avoided by the proposed approach. More generally, the regularization on the anchor outputs for stuck-at 0 faults operates very well; increasing the margins for stuck-at 1 or −1 faults leads to some accuracy loss, but the performance is still satisfactory. Therefore, such a fault-tolerant method can be utilized in TNs to avoid fatal system failures caused by stuck-at faults. Also, they can be applied simultaneously to protect the network from all other types of stuck-at faults.

Modified Margin

As per the analysis in Sect. 3.2.2, measures should be taken to prevent training to be in a zero-loss situation from the start under stuck-at 1/−1 faults. In this case, consider the objective function TL in (18) as the example; the difference caused by the faulty terms $\sum_{i \in I} \left| y_{a,i}^{l} - v_i \right|$ must be compensated. An intuitive solution is to increase the original margin M; let the number of faulty neurons be denoted by f, so increasing the margin by f can effectively guarantee that the training starts correctly (as the initial outputs $y_{a,i}^{l}$ are usually close to 0).

By the analysis in Sect. 3.2.2, convergence can still lead to a false solution in the training process; in this case, the regularization of the loss function is no longer feasible. A possible reason is that the regularized outputs toward 1 or −1 can cause vanishing gradients with bounded activation functions. Therefore, the method of further increasing the margin is considered; for each faulty term $\left| y_{a,i}^{l} - v_i \right|$, the maximum error is 2 when $v_i = 1$ or $v_i = -1, -1 \leq y_{a,i}^{l} \leq 1$. By (18), if the number of faulty neurons is f, increasing the margin by $2f$ can ensure that the convergence does not achieve a false solution over the entire training.

However, the number of faulty neurons is usually unknown in many applications. An appropriate margin is critical to address the importance of the contributive triplets [23]; setting the margin to an arbitrarily large value slows down the convergence in the error-free case. Therefore, increasing the margin must be thought carefully. We propose to modify the margin at the beginning of the training by:

$$M' = M + 2 \lfloor d_{-} \rfloor , \tag{22}$$

where $\lfloor d_{-} \rfloor$ denotes the round-down to the largest integer smaller than d_{-}. Note that the margin only changes based on the output at the first iteration, and it is fixed in the following training process. This value assumes that the initial negative distance is smaller than 1 in the error-free cases (valid for all applied datasets); thus, it can be an estimate of the number of (stuck-at 1 or −1) faults. Consider the widely used process of weight initialization; this estimate works for most cases, and it can also be modified by the empirical knowledge of different applications. In summary, the modified margin M' is required to correctly start the training process and prevent false solutions when stuck-at 1/−1 faults are present in the negative subnetwork, the evaluation results are also shown in Fig. 11; it significantly improves the tolerance to stuck-at faults compared to the unprotected case [53].

4 Conclusion

Often machine learning (ML) systems encounter difficulties in establishing and executing with a paucity of training data. This chapter discusses advanced machine learning techniques using multi-branch neural networks for performing tasks such as similarity/dissimilarity with limited pre-known information. In particular, the Siamese networks (SNs) that employ two branch networks and Triplet networks (TNs) with three branch networks are considered. Different aspects of these multi-branch networks are comprehensively reviewed from the network features, classification algorithms, hardware design to error tolerance. The excellent classification performance and high-performance hardware design solutions make these multi-branch networks a promising ML technique for similarity-measuring tasks.

References

1. L. Song, D. Gong, Z. Li, et al., Occlusion robust face recognition based on mask learning with pairwise differential siamese network, in *Proceedings of the IEEE/CVF International Conference on Computer Vision*, (2019)
2. Y. Zhang, D. Liu, Z.J. Zha, Improving triplet-wise training of convolutional neural network for vehicle re-identification, in *2017 IEEE International Conference on Multimedia and Expo (ICME)*, (2017)
3. H. Lai, J. Chen, L. Geng, Y. Pan, X. Liang, J. Yin, Improving deep binary embedding networks by order-aware reweighting of triplets. IEEE Trans. Circuits Syst. Video Technol. **30**(4), 1162–1172 (2020)
4. A. Radford et al., Learning transferable visual models from natural language supervision, in *International Conference on Machine Learning*, (PMLR, 2021)
5. D. Chicco, Siamese neural networks: An overview. Artif. Neur. Netw **2190**, 73–94 (2021)
6. G. Koch, R. Zemel, R. Salakhutdinov, et al., Siamese neural networks for one-shot image recognition, in *ICML Deep Learning Workshop*, (Lille, 2015)
7. E. Hoffer, N. Ailon, Deep metric learning using triplet network, in *International Workshop on Similarity-Based Pattern Recognition*, (Springer, Cham, 2015, Nov), pp. 84–92
8. V.B.G. Kumar, G. Carneiro, I. Reid, Learning local image descriptors with deep siamese and triplet convolutional networks by minimising global loss functions, in *Proceedings of the IEEE Conference on Computer Vision and Pattern Recognition*, (2016), pp. 5385–5394
9. D.S. Phatak, I. Koren, Complete and partial fault tolerance of feedforward neural nets. IEEE Trans. Neural Netw. **6**(2), 446–456 (1995)
10. T. Haruhiko, M. Masahiko, K. Hidehiko, H. Terumine, Enhancing both generalization and fault tolerance of multilayer neural networks, in *2007 International Joint Conference on Neural Networks*, (2007, Aug), pp. 1429–1433
11. J.W. Schwartz, J.K. Wolf, A systematic (12,8) code for correcting single errors and detecting adjacent errors. IEEE Trans. Comput. **39**(11), 1403–1404 (1990)
12. S. Pontarelli, P. Reviriego, M. Ottavi, J.A. Maestro, Low delay single symbol error correction codes based on Reed Solomon codes. IEEE Trans. Comput. **64**(5), 1497–1501 (2015)
13. S. Liu, P. Reviriego, F. Lombardi, Detection of limited magnitude errors in emerging multilevel cell memories by one-bit parity (OBP) or two-bit parity (TBP), in *IEEE Transactions on Emerging Topics in Computing*, vol. 9, no. 4, pp. 1792–1802, 1 Oct.-Dec. 2021
14. M. Qin, C. Sun, D. Vucinic, Robustness of neural networks against storage media errors, in *arXiv preprint arXiv:1709.06173*, (2017)
15. Y. LeCun, L. Bottou, Y. Bengio, et al., Gradient-based learning applied to document recognition. Proc. IEEE **86**(11), 2278–2324 (1998)
16. H. Xiao, K. Rasul, R. Vollgraf, Fashion-mnist: A novel image dataset for benchmarking machine learning algorithms, in *arXiv preprint arXiv:1708.07747*, (2017)
17. A. Krizhevsky, G. Hinton, Learning multiple layers of features from tiny images, Citeseer (2009)

18. Y. Netzer, T. Wang, A. Coates, et al., Reading digits in natural images with unsupervised feature learning (2011)
19. J. Bromley, J.W. Bentz, L. Bottou, et al., Signature verification using a "siamese" time delay neural network. Int. J. Pattern Recognit. Artif. Intell. **7**, 669–688 (1993)
20. L. Zheng, S. Duffner, K. Idrissi, C. Garcia, A. Baskurt, Siamese multi-layer perceptrons for dimensionality reduction and face identification. Multimed. Tools Appl. **75**(9), 5055–5073 (2016)
21. D. Yi, Z. Lei, S. Liao, S.Z. Li, Deep metric learning for person re-identification, in *2014 22nd International Conference on Pattern Recognition*, (2014, Aug), pp. 34–39
22. R. Hadsell, S. Chopra, Y. LeCun, Dimensionality reduction by learning an invariant mapping, in *IEEE Computer Society Conference on Computer Vision and Pattern Recognition (CVPR'06)*, (2006)
23. Y. Liu, C. Huang, Scene classification via triplet networks. IEEE J. Selec. Top. Appl. Earth Observ. Remote Sens **11**(1), 220–237 (2018)
24. F. Schroff, D. Kalenichenko, J. Philbin, Facenet: A unified embedding for face recognition and clustering, in *Proceedings of the IEEE Conference on Computer Vision and Pattern Recognition*, (2015), pp. 815–823
25. E. Hoffer, I. Hubara, N. Ailon, Deep unsupervised learning through spatial contrasting. arXiv preprint arXiv:1610.00243 (2016)
26. D. Cheng et al., Person re-identification by multi-channel parts-based cnn with improved triplet loss function, in *Proceedings of the IEEE Conference on Computer Vision and Pattern Recognition*, (2016)
27. O.M. Parkhi, A. Vedaldi, A. Zisserman, Deep face recognition (2015)
28. A. Hermans, L. Beyer, B. Leibe, In defense of the triplet loss for person re-identification. arXiv preprint arXiv:1703.07737 (2017)
29. N. Nedjah, R.M. da Silva, L.M. Mourelle, et al., Dynamic MAC-based architecture of artificial neural networks suitable for hardware implementation on FPGAs. Neurocomputing **72**(10–12), 2171–2179 (2009)
30. S. Liu et al., Stochastic dividers for low latency neural networks, in *IEEE Transactions on Circuits and Systems I: Regular Papers*, vol. 68, no. 10, pp. 4102-4115, Oct. 2021
31. S. Lin, D.J. Costello, *Error Control Coding* (Pretice Hall, Scarborough, 2001)
32. N.B. Gaikwad, T. Varun, K. Avinash, et al., Efficient FPGA implementation of multilayer perceptron for real-time human activity classification. IEEE Access **7**, 26696–26706 (2019)
33. R. Sreehari, D. Vijayasenan, M.R. Arulalan, A hardware accelerator based on quantized weights for deep neural networks, in *Emerging Research in Electronics, Computer Science and Technology*, (Springer, Singapore, 2019), pp. 1079–1091
34. L.D. Medus, T. Iakymchuk, J.V. Frances-Villora, et al., A novel systolic parallel hardware architecture for the FPGA acceleration of feedforward neural networks. IEEE Access **7**, 76084–76103 (2019)
35. W. Isaac, X. Yang, T. Liu, et al., FPGA acceleration on a multi-layer perceptron neural network for digit recognition. J. Supercomput. **77**, 1–18 (2021)
36. Y. Liu, S. Liu, Y. Wang, F. Lombardi, J. Han, A stochastic computational multi-layer perceptron with backward propagation. IEEE Trans. Comput. **67**(9), 1273–1286 (2018)
37. L. Matanaluza et al., Emulating the effects of radiation-induced soft-errors for the reliability assessment of neural networks, in *IEEE Transactions on Emerging Topics in Computing*, (2021) (early access)
38. S. Liu, P. Reviriego, X. Tang, et al., Result-based re-computation for error-tolerant classification by a support vector machine. IEEE Trans. Artif. Intell **1**(1), 62–73 (2020)
39. IS Committee, 754–2008 IEEE standard for floating-point arithmetic. IEEE Computer Society Std (2008)
40. M. Maniatakos, M.L. Michael, Y. Makris, Vulnerability-based interleaving for multi-bit upset (MBU) protection in modern microprocessors, in *IEEE International Test Conference*, (2012), pp. 1–8

41. C.-T. Chin, K. Mehrotra, C.K. Mohan, S. Rankat, Training techniques to obtain fault-tolerant neural networks, in *Proceedings of 24th International Symposium on Fault-Tolerant Computing (FTCS) Dig. Papers*, (1994, Jun), pp. 360–369
42. N. Wei, S. Yang, S. Tong, A modified learning algorithm for improving the fault tolerance of bp networks, in *Proceedings of IEEE International Conference on Neural Network*, vol. 1, (1996, Jun), pp. 247–252
43. N. Kamiura, Y. Taniguchi, T. Isokawa, N. Matsui, An improvement in weight-fault tolerance of feedforward neural networks, in *Proceedings 10th Asian Test Symposium*, (2001, Aug), pp. 359–364
44. T. Haruhiko, K. Hidehiko, H. Terumine, Partially weight minimization approach for fault tolerant multilayer neural networks, in *Proceedings of the 2002 International Joint Conference on Neural Networks. IJCNN'02 (Cat. No.02CH37290)*, vol. 2, (2002), pp. 1092–1096
45. Z. Wang, F. Niknia, S. Liu, P. Reviriego, P. Montuschi, F. Lombardi, Tolerance of siamese networks (SNs) to memory errors: Analysis and design, in *IEEE Transactions on Computers*, vol. 72, no. 4, pp. 1136-1149, 1 April 2023
46. G. Li, S.K.S. Hari, M. Sullivan, et al., Understanding error propagation in deep learning neural network (DNN) accelerators and applications, in *Proceedings of the International Conference for High Performance Computing, Networking, Storage and Analysis*, (2017), pp. 1–12
47. C. Schorn, A. Guntoro, G. Ascheid, An efficient bit-flip resilience optimization method for deep neural networks, in *2019 Design, Automation & Test in Europe Conference & Exhibition (DATE)*, (2019, May), pp. 1507–1512
48. M. Qin, C. Sun, D. Vucinic, Improving robustness of neural networks against bit flipping errors during inference. J. Image Graphics **6**(2), 181 (2018)
49. J.A. Abraham, W.K. Fuchs, Fault and error models for VLSI. Proc. IEEE **74**(5), 639–654 (1986)
50. C.H. Sequin, R.D. Clay, Fault tolerance in artificial neural networks, in *Proc. Int. Joint Conf. Neural Netw. (IJCNN)*, vol. 1, (1990, Jun), pp. 703–708
51. C. Torres-Huitzil, B. Girau, Fault tolerance in neural networks: Neural design and hardware implementation, in *2017 International Conference on ReConFigurable Computing and FPGAs (ReConFig)*, (2017, Dec), p. 1–6
52. B.S. Arad, A. El-Amawy, On fault tolerant training of feedforward neural networks. Neural Netw **10**(3), 539–553 (1997)
53. Z. Wang, F. Niknia, S. Liu, P. Reviriego, A. Louri, F. Lombardi, Fault tolerant triplet networks for training and inference. TechRxiv. Preprint (2022). https://doi.org/10.36227/techrxiv.21251904.v1

An Active Storage System for Intelligent Data Analysis and Management

Shengwen Liang, Ying Wang, Lei Dai, Yingying Chen, Renhai Chen, Fan Zhang, Gong Zhang, Huawei Li, and Xiaowei Li

1 Introduction

Owing to the proliferation of mobile devices and ubiquitous computing, there is a sheer growth of unlabeled data in different categories such as text, audio, image, and video in the edge or cloud computing infrastructures. The massive and diverse unlabeled data is also known as unstructured data. It is reported that unstructured data occupies up to 80% of storage space in commercial data centers [70]. Once being stored and managed in the cloud machines, the massive amount of unstructured data leads to intensive analysis requests issued by the users, which poses a significant challenge to the processing throughput and power consumption of the data center. Moreover, for an unstructured data analysis, users usually pursue to grasp the semantic content lurking behind the data to assist decision-making by adopting deep learning strategies, which are not supported by conventional databases and are much more complicated than database operations as it requires more computational power. Unfortunately, a conventional unstructured data analysis system suffers from high response latency and energy consumption

S. Liang · Y. Wang (✉) · L. Dai · H. Li (✉) · X. Li
State Key Lab of Processors, Institute of Computing Technology, Chinese Academy of Sciences, University of Chinese Academy of Sciences, Beijing, China
e-mail: liangshengwen@ict.ac.cn; wangying2009@ict.ac.cn; dailei@ict.ac.cn; lihuawei@ict.ac.cn; lxw@ict.ac.cn

Y. Chen, F. Zhang, G. Zhang
Theory Lab, 2012 Labs Huawei Technologies Co., Ltd., Hong Kong, China
e-mail: cheng.yingying@huawei.com; zhang.fan2@huawei.com; nicholas.zhang@huawei.com

R. Chen
College of Intelligence and Computing Tianjin University, Tianjin, China

Theory Lab, 2012 Labs Huawei Technologies Co., Ltd., Hong Kong, China
e-mail: renhai.chen@tju.edu.cn

W. Liu et al. (eds.), *Design and Applications of Emerging Computer Systems*,
https://doi.org/10.1007/978-3-031-42478-6_6

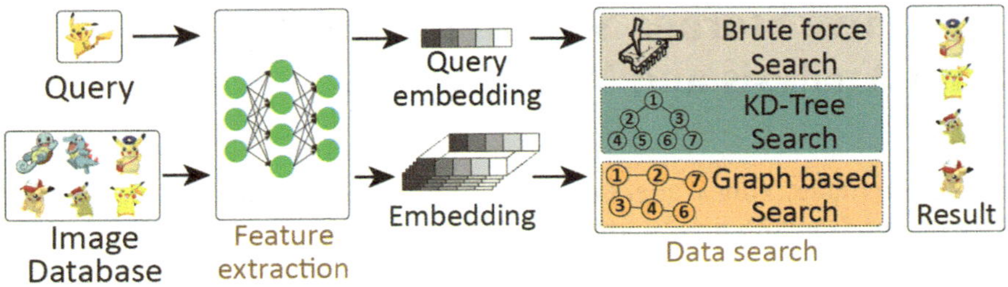

Fig. 1 Content-based unstructured data retrieval system

due to cumbersome data movement path and burdensome memory hierarchies. Consequently, it is critical to have a fast, energy-efficient, and low-cost data analysis system deployed in the cloud service infrastructure to reduce the total cost of ownership (TCO).

A typical unstructured data analysis workload: content-based unstructured data retrieval is given as an example to illustrate the problems faced by the traditional system. As shown in Fig. 1, a typical content-based unstructured data system includes two primary stages: ***feature extraction and data search***. Feature extraction is to capture the semantic feature of input unstructured data and map it to high-dimensional vectors. For example, image and video are usually processed by convolution neural networks and recurrent neural networks, respectively. Data search finds the instance in a given corpus that is similar to input instances quickly and accurately. Unfortunately, conventional content-based multimedia data retrieval systems suffer from the issue of inaccuracy, power inefficiency, and high cost, especially for large-scale unstructured data. Specifically, from an algorithm's perspective, traditional content-based data retrieval systems generally rely on a handcrafted feature such as SIFT and GIST [75] to extract features from data, which fails to capture the semantic feature inside the unstructured data and thus suffers from low retrieval accuracy. With the rapid development of deep learning that can effectively transform unstructured data into learned feature vectors, feature extraction methods gradually transfer to deep learning methods. Despite the deep learning methods providing better feature representation, it still encounters high inference latency and is hard to satisfy the real-time constraint in the edge platform. Besides, in the traditional unstructured data retrieval system, the data search stage often adopts a fixed data retrieval algorithm, overlooking the adaptability between the algorithm and the characteristics of datasets. For example, the brute-force data search is suitable for small-scale datasets with low-dimensional feature vectors, while the graph-based data search prefers a large-scale dataset with high-dimensional feature vectors.

From a system perspective, Fig. 2a briefly depicts a typical content-based data retrieval system composed of a CPU/GPU device and a storage system based on compute-centric architecture [3]. When a data retrieval request arrives from the Internet or the central server, the CPU has to reload massive potential data from disk

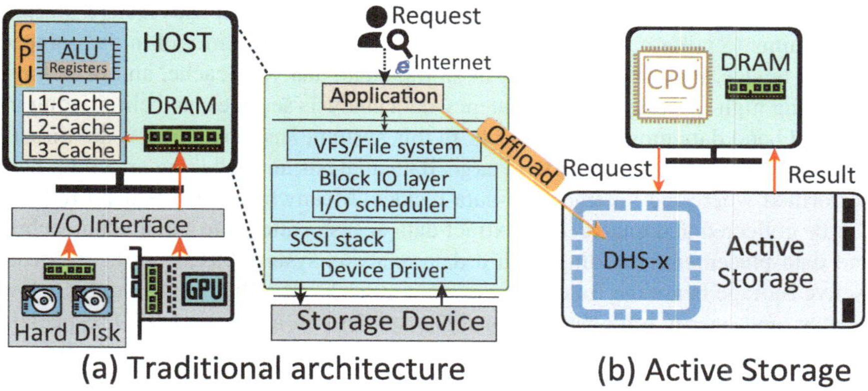

Fig. 2 Traditional architecture (**a**) vs. Active Storage (**b**)

into the temporary DRAM [3] and match the features of the query with the loaded unstructured data to find the relevant targets. This compute-centric architecture is confronted with several critical sources of overhead and inefficiency: (1) the current I/O software stack significantly burdens the data retrieval system when it simply fetches data from the storage devices on retrieval requests [73], as shown in Fig. 2a. The situation is even worse since [60] indicates the performance bottleneck has migrated from hardware (75∼50 us [10]) to software (60.8 us [60]) after traditional HDDs are replaced by non-volatile memory. (2) Massive data movement incurs energy and latency overhead in the conventional memory hierarchy. This issue becomes severe as the scale of data under query increases because the relevant data at the low-level storage must travel across a slow I/O interface (e.g., SATA and SAS), main memory, and multi-level caches before reaching the compute units of CPU or GPUs [13], which is depicted in Fig. 2a.

From a storage perspective, massive unstructured data also poses stringent IO access performance for the storage system. It is unaffordable and unreasonable to use high-performance storage such as (NAND FLASH SSD) to store all unstructured data. Thereby, a multi-tiered storage system that combines various storage devices with distinct performance and cost is a viable strategy because the data can be placed in a suitable tier. In this case, an excellent data caching and data placement strategy are crucial to ensure the performance of multi-tiered storage systems. This is because the former can timely migrate the data to be accessed from the low-level tier with low speed and high capacity to the high-level layer with high speed and limited capacity to reduce the access latency, while the latter can optimize the access latency and storage space occupation according to the frequency of data access. However, conventional data caching and data placement strategies are usually designed for general-purpose scenarios and ignore the access patterns in real-world analysis workload, resulting in low performance when facing complex and changing access patterns.

To address these issues, as shown in Fig. 2b, this work proposes Active Storage, which aims to tailor a unified data storing and analysis system within the compact storage device, realizing intelligent in-storage data analysis, cache, and placement, providing high-accuracy and low-latency data analysis services, and eliminating the major IO and data moving bottleneck. In this system, analysis requests are directly sent to the storage devices, and the target data analysis and indexing are completely performed where the unstructured data resides. Meanwhile, historical IO requests can be collected and analyzed to extract data access patterns to guide data caching and data placement. Building such a data analysis system based on the proposed Active Storage bears the following design goals: (1) providing high-accuracy, low-latency, and energy-efficient data analysis services in a compact SSD, especially selecting the most appropriate retrieval algorithm according to the characteristics of the dataset to provide optimal retrieval latency, (2) exploiting the internal bandwidth of flash devices in a disk for energy-efficient deep learning-based data processing, (3) enabling the developers to customize the data analysis workload for different datasets, and (4) intelligently predicting the hot and cold data block according to historical IO access trace and then achieving intelligent data caching and placement in Active Storage for low access latency. These key design goals are introduced in detail below.

Firstly, instead of relying on the general-purpose CPU or GPU devices in Fig. 2a, we must have a highly computation-efficient yet accurate data analysis architecture in consideration of the SSD form factor and cost. Thereby, in order to support two high-frequency operations including feature extraction and data search in unstructured data analysis workload, we propose a holistic data analysis mechanism by combining the **D**eep neural networks and **H**ybrid data **S**earch algorithms (DHS). The deep neural networks can be attributed to two categories, convolutional neural networks that prefer image data and recurrent neural networks that are in favor of data with time series such as video and audio. Both of them are responsible for extracting the semantic features of unstructured data. Meanwhile, considering many data search algorithms has significant discrepancies in terms of latency and accuracy on different scale datasets with distinct data dimensions. Thereby, we further propose a hybrid data search that unifies the brute-force, KD-tree, and graph search algorithms into one data search paradigm, cooperating with an auto-selection model. The auto-selection model can pick the most appropriate search algorithm according to the inherent characteristics of the dataset to be retrieved.

Secondly, although DHS is a simple and flexible end-to-end data retrieval solution, embedding it into SSDs still takes considerable effort. For instance, it will take about 200~100 ms to perform a convolutional neural network on the Xeon CPU platform [7], which fails to satisfy the requirements of the SLA criterion. This issue is worsened on embedded processors such as ARM Cortex. Thereby, we design a specific hardware accelerator that supports deep neural network and hybrid data search simultaneously, DHS-x, to construct the target Active Storage without using power-unsustainable CPU or GPU solutions. Meanwhile, DHS-x integrates an auto-selection unit to select the best-fit execution data flow on the fly, achieving data search with low latency and high accuracy. However, the limited DRAM inside an

SSD is majorly used to cache the metadata for flash management, leaving no free space for deep learning applications. Fortunately, we have proved that the bandwidth of the internal flash matches the bandwidth demand of DHS-x with proper data layout mapping. By rebuilding the data path in SSD and deliberately optimizing the data layout related to deep learning models and graphs on NAND flash, the DHS-x could fully exploit internal parallelism and directly access data from NAND flash bypassing the on-board DRAM.

Thirdly, as we introduce the deep learning technology into the storage device, we must expose the software abstraction of the Active Storage to the users and developers to process different data structures with different deep learning models. Thus, we abstract the underlying data analysis and data search mechanism as user-visible calls by utilizing the NVMe protocol [63] for command extension. Not only can users' requests trigger the DHS-x accelerator to search the target dataset for query-relevant structures, but also system developers can freely configure the deep neural network architecture with different representation power and overhead for different datasets and performance requirements. In contrast to conventional ad hoc solutions, Active Storage allows the system to be customized to deploy various unstructured data analysis services through the provided APIs.

Finally, unstructured data access workloads usually follow complex IO patterns, especially when multiple users perform independent tasks. More importantly, it also involves huge amounts of data that take up too much storage space to place in the high-level tier. In this case, frequently visiting low-level trier with high access latency can result in performance degradation. Fortunately, although data access requests are unpredictable, the data accessed by multiple requests with similar semantics exists in the locality. Thereby, by learning the IO access pattern using a machine learning algorithm, we can predict the access frequency of data blocks and then cache hot data blocks and place cold blocks to the low-level tier. To this end, we treat the problem of identifying the hot and cold of a data block as a frequency predicate problem and then leverage an LSTM-based neural network to predicate the access frequency of a data block in the next time window. According to the predicted results, we can determine the characteristics of a data block based on predicted access frequency and then place the data in the appropriate storage tier. In summary, we make the following novel contributions:

1. We propose Active Storage, to enable within-SSD deep learning-based unstructured data analysis by integrating a specialized deep learning and hybrid search accelerator (DHS-x). DHS-x directly accesses data from NAND flash without across multiple memory hierarchies for decreasing data movement path and power consumption.
2. We propose an auto-selection strategy to pick the most appropriate search algorithm according to the inherent characteristics of the dataset to be retrieved on the fly.
3. We present an LSTM-based model to predict the access frequency of data blocks in the next window and leverage the predicted result to guide data cache and data placement in Active Storage for optimizing data access latency.

4. We employ Active Storage to build a lightweight data analysis system, which completely abandons the conventional data analysis mechanism in orthodox compute-centric systems. It could independently respond to data analysis requests at real-time speed and low energy cost.
5. We build a prototype of Active Storage on the Cosmos plus OpenSSD platform [32] and use it to deploy a typical data analysis workload: unstructured data search. Our evaluation results show that Active Storage is more energy-efficient than an unstructured data search system implemented on CPU and GPU and reduce search latency by 3.48X compared with a CPU implementation.

2 Background and Preliminaries

2.1 Unstructured Data Analysis System

Take a typical unstructured data analysis system, a content-based unstructured data retrieval system, as an example. It aims to search for certain data entries from the large-scale dataset by analyzing their visual or audio content, and it is replacing the traditional concept or keyword-based data retrieval for its independence of data labels or tags. Figure 3 depicts a typical content-based retrieval procedure that includes two main stages: feature extraction and data search. Feature extraction generates the feature vector for the query data and data search indexes for similar data structures in storage with that feature vector encoded in a semantic space.

Feature Extraction Over the past decades, the researchers have generally relied on handcrafted filters, such as GIST [51] and SIFT [16] to distill the "content representation." Since 2012, the rise of deep learning has transferred the focus of research to the deep neural network (DNN) [43] based features [75] because it provides better mid-level representations [38, 46]. Figure 3 depicts a typical

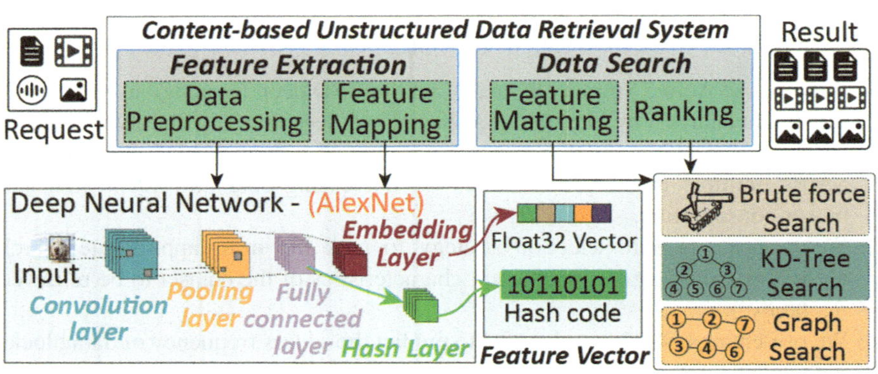

Fig. 3 Content-based unstructured data retrieval system

DNN that contains four key types of network layers: (1) convolution layer, which extracts visual features from input by moving and convolving multidimensional filters across the input data organized into 3D tensors, (2) activation, the nonlinear transformation that we do over the input signal, (3) pooling layers, which down-sample the input channels for scale and other types of invariance, and (4) fully connected (FC) layer, which performs linear operations between the features and the learned weights to predict the categorization or other high-level characteristics of input data. Such a neural network is very flexible and can be designed to have different hyper-parameters, like the number of the convolution and pooling layers stacked together and the dimension and number of the convolution filter. Changing these parameters will impact the generalization ability and also computational overhead of neural networks, which are usually customizable for different datasets or application scenarios [59]. Some prior works directly employ the high-dimensional feature vector generated from an embedding layer that is behind the FC layer to achieve data retrieval. Meanwhile, in order to reduce memory footprint and computation complexity [44], a deep hashing solution is proposed to map a high-dimensional feature vector to hash space for achieving effective yet condensed data representation. Figure 3 exemplifies a convolutional neural network architecture, AlexNet, where an embedding layer and a hash layer follow the last fully connected layer of AlexNet [31] to project the data feature learned from AlexNet into a vector or a hash code. The generated feature vector/hash code could be directly used to index the relevant data structures and get rid of the complex data preprocessing stage.

Data Search Prior deep neural network-based data retrieval works [74] leverage the brute-force method that compares all feature vectors in the database with a query to find similar data structures. However, the brute-force method is subject to the issue of long latency due to the explosive growth of datasets' scale and the curse of dimensionality. Thus, this work adopts approximate nearest neighbor search (ANNS) algorithms to improve search efficiency while relaxing accuracy constraints. ANNS is a task that finds the approximate nearest neighbors among a high-dimensional dataset for a query via a well-designed data structure. Existing ANNS algorithms are mainly divided into the following categories: hashing-based, tree-based, and graph-based [69].

(1) Hashing-based methods generate binary codes for high-dimensional vectors and try to preserve the similarity among the original real vectors. The hash function partitions the original feature space into two parts: the instance in one part is coded as 1, and the instance in the other part is coded as 0. Ideally, if neighbor vectors fall into the same space or the nearby space (measured by the Hamming distance of binary codes), the hashing-based methods can efficiently retrieve the nearest neighbors of an input query. The most popular hashing algorithm is locality-sensitive hashing (LSH) [18]. However, the hashing-based methods suffer from low accuracy when facing large-scale datasets [69].

(2) Tree-based methods continuously partition a dataset into discrete regions at multiple scales, such as KD-tree [15], R-tree [20], and SR-tree [30]. The KD-tree is a binary tree that subdivides the space of the dataset into smaller regions with an approximately equal number of points. These small regions are attached to the leaves of the binary tree, and each represents a local region in space. With a KD-tree, searching a query point involves tree traversal to find the nearest region or bucket and then searching the bucket that is the most likely to contain the nearest neighbors. However, tree-based methods only perform very well when the dimensionality of the data is relatively low. It has been proved to be inefficient when the dimensionality of data is high. Muja and Lowe [50] figured out that when the dimensionality exceeds about 10, existing indexing data structures based on space partitioning are slower than the brute-force method.
(3) Graph-based methods have drawn considerable attention recently, such as NSG [18], HNSW [48], Efanna [17], and FANNG [22]. The basic idea of graph-based methods is that a neighbor's neighbor is also likely to be a neighbor [18]. Specifically, the main idea of the NSG algorithm is mapping the query feature vector into a graph. The vertex of the graph represents an instance, and the edge stands for the similarities between entities. On top of that, the NSG algorithm can iteratively check neighbors' neighbors in the graph to find the true neighbors of the query. In this manner, the NSG algorithm could avoid unnecessary data checking to reduce retrieval latency. However, graph-based methods need to construct a kNN graph whose construction complexity of the kNN graph increases exponentially with the scale of the dataset. Many researchers turn to building an approximated kNN graph, but it is still time-consuming. Meanwhile, graph-based searches yield high memory overhead as they need to maintain both the original vectors and an additional graph structure in memory. In addition, when facing a small-scale dataset, the brute-force methods maybe exhibit better performance than graph-based methods.

In view of different ANNS algorithms prefer distinct dataset scales, this work proposes a hybrid data search mechanism that automatically selects the appropriate ANNS algorithm according to the characteristics of the dataset to be searched. This work further combines the deep neural network and the hybrid data search methods to construct an accurate, computation-efficient, and flexible data retrieval system.

2.2 *Near Data Processing and Deep Learning Accelerator*

Near-data processing (NDP) moves computation from the system's main processors into memory or storage devices to reduce the off-chip data movement [1, 23, 25, 35, 39–42, 53, 54, 56]. While early active disk designs are mostly toward hard drives, which need to address not only the slow IO but also the underlying thrifty disk access bandwidth. Fortunately, the emerging non-volatile storage device provides more promising results and many recent works focus on it [4, 5, 14, 57, 61, 62, 71]. For instance, [65] provides a framework for moving computation to the general-purpose embedded processors on NVMe SSD, and [2] uses low-power processors and flash to handle data processing and focuses on a key-value storage system. Kang et al. [29] introduces the Smart SSD model, which pairs in-device processing with a powerful host system capable of handling data-oriented tasks without modifying the operating system. Wang et al. [67] supports fundamental database operations including sort, scan, and list intersection by utilizing Samsung SmartSSD. Choe et al. [12] investigates by simulation the possibility of employing the embedded ARM processor in SSDs to run SGDs, which is the key component of neural network training. However, all of them cannot handle deep learning processing due to the performance limit of embedded processors. Thereby, [11] presents intelligent solid-state drives (iSSDs) that embed the stream processors into the flash memory controllers to handle linear regression and k-means workloads. Ouyang et al. [52] integrates Programmable Logic into SSDs to achieve energy-efficiency computation for web-scale data analysis. Meanwhile, [27] also uses FPGA to construct BlueDBM that uses flash storage and in-store processing for cost-effective analytics of large datasets, such as graph traversal and string search. GraFBoost [28] focuses on the acceleration of graph algorithms on an in-flash computing platform instead of deep learning algorithms as this work.

2.3 Learned Data Cache and Placement

The rise of machine learning and its remarkable performance in workload characteristic identification drive the researchers to apply machine learning approaches to optimize the module design in the computer architecture domain. As a key part of computer architecture, the storage system that is closely related to the workload has been widely concerned, especially data cache and placement which has a large impact on the performance of storage systems. Popular existing methods apply machine learning to guide data cache and placement strategy in two ways: reinforcement learning based and recurrent neural network based, both of which face challenges to offering desired prediction accuracy, latency, and memory footprint. For example, [66] treat the data cache and replacement problem as a reinforcement learning problem with a regret minimization strategy and construct a general cache replacement framework LeCAR for small cache sizes. However, LeCAR only focuses on types of workloads and hard to accommodate other workloads such as scan and churn. Thereby, [55] presents an enhanced version of LeCAR, CACHEUS. It leverages a reinforcement learning technique that adopts state-of-the-art caching algorithms LFU, LIRA, and ARC as experts to improve the efficiency of eviction decisions. On the other line, [58] leverages a powerful LSTM learning model to build a cache replacement policy and exhibits better accuracy than current hardware predictors. Miura et al. [49] applies LSTM to achieve efficient cache replacement in the virtualized environment.

2.4 Active Store

Prior active disks are integrating either general-purpose processors incapable of handling high-throughput data or specialized accelerators with only the support of simple functions like scanning and sorting. These in-disk computation engines are unable to fulfill the requirement of high-throughput deep neural network (DNN) inference because computation-intensive DNNs generally rely on power-consuming CPU or GPUs in the case of data analysis and query tasks. To enable energy-efficient DNN, prior works propose a variety of energy-efficient deep learning accelerators. Chen et al. [7] maps large DNNs onto a vectorized processing array and employs a data tiling policy to exploit the locality in neural parameters. Eyeriss [8] applies the classic systolic array architecture to the inference of CNN and outperforms CPU and GPU in energy efficiency dramatically. However, these researches focus on optimizing the internal structure of the accelerator and rely on large-capacity SRAM or DRAM instead of external non-volatile memory. In contrast to these works and prior active SSD designs, we propose **Active Store**, which enables the storage device to employ deep learning to conduct the in-storage data analysis. It is designed to replace the conventional data analysis system and contains a flash-accessing accelerator (DHS-x) for deep learning and data search. DHS-x is deliberately reshaped to take advantage of the large flash capacity and high internal bandwidth, and it is also re-architected to enable hybrid data search for target indexing. In addition, Active Store also integrates deep learning-based data cache and placement strategy to improve the effectiveness of the storage system when handling IO workloads.

3 Motivation

To gain insight into the computing characteristics of an unstructured data analysis system, we analyze a representative image retrieval workload to identify the performance bottlenecks. First, we implement a content-based image retrieval system using a classical convolutional neural network AlexNet and graph-based ANNS algorithms named NSG [18]. The dataset used in this experiment is CIFAR-10 and ImageNet dataset. Second, we train the AlexNet model on CIFAR-10 and ImageNet datasets until the accuracy reaches the advertised accuracy. The output of the embedding layer is used as the feature vector. The dimension of the embedding vector on CIFAR-10 and ImageNet is 48. The experiment is built on two platforms: (1) the CPU platform is equipped with Xeon E5-2630, 32 GB, and four 1 TB SSDs and (2) the GPU platform is equipped with NVIDIA 1080Ti GPU, 32 GB host DRAM, and four 1 TB SSDs.

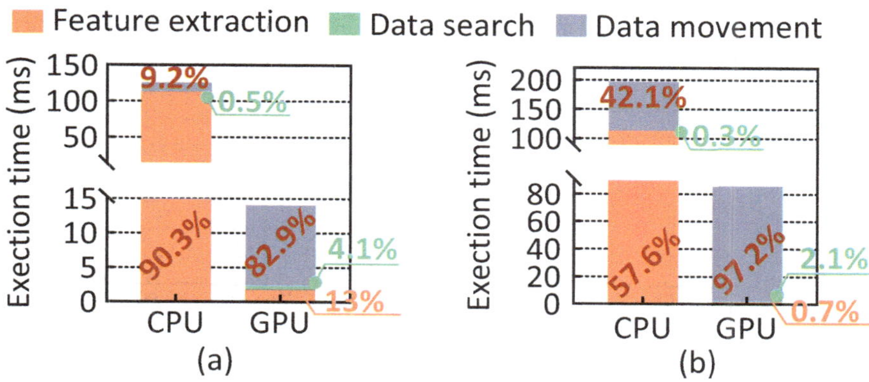

Fig. 4 The time breakdown of unstructured data retrieval system. (**a**) CIFAR-10. (**b**) ImageNet

3.1 Execution Time Breakdown

The query latency is broken down into three parts on the CPU platform: feature extraction time, data search computing time, and data movement time. The profiling results are displayed in Fig. 4a. It can be observed that the data movement only constitutes 9.2% of the total query execution time on the CIFAR-10 dataset. With the dataset scale increasing, the proportion of data movement on the ImageNet dataset rises to 42.1%. Besides, feature extraction is the dominating stage and constitutes over 90.3% and 57.6% of the total time on the CIFAR-10 and ImageNet datasets, respectively. The reason is that the neural network involves complex enormous computation and memory resources, especially with the increase in network depth. What is more, wimpy processors within SSD are not sufficient to perform neural networks [47]. Thereby, it is necessary to accelerate neural networks and alleviate data movement overhead during query execution.

Since the CPU is hard to exploit the parallelism of neural networks and thus suffers from high execution time, we further exploit the execution time breakdown on the GPU platform. As shown in Fig. 4a and b, Since the execution time of the neural network on the GPU platform decreases by 62X compared to the CPU platform, the data movement issue becomes aggravated and constitutes 92.9% of the total query execution time on the CIFAR-10 dataset and climbs to 97.2% on ImageNet dataset.

In summary, to alleviate data movement overhead, we opt to perform unstructured data retrieval directly where the data resides by moving the computation unit into the storage device. In view of the limited performance of wimpy processors within SSD, it is hard to satisfy the requirements of an unstructured data retrieval system. We design and implement a deep neural network and hybrid data search (DHS-x) accelerator to achieve data retrieval with low latency and high energy efficiency.

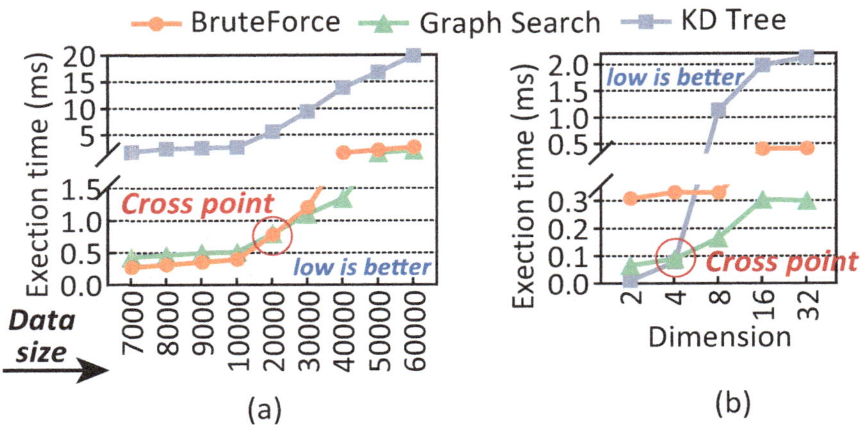

Fig. 5 The execution time of data search methods including brute-force, KD-tree, and graph search. (**a**) Time wrt. data scale. (**b**) Time wrt. dimension

3.2 ANNS Algorithm Exploration

Section 2 mentions the existing approximate nearest neighbor search (ANNS) algorithms can be categorized into three types: hashing-based, tree-based, and graph-based. Different ANNS algorithms exhibit distinct performances when facing various dataset scales. Thereby, in order to analyze the influence of dataset scale on the various ANNS algorithms, we compare the ANNS execution time when the dataset varies in size and vector dimension by generating synthetic datasets. In this experiment, we omit the hashing-based algorithm as it usually suffers from low accuracy and fails to satisfy the requirements of the retrieval system [18]. We add the brute-force search method as it can provide exact results. The typical tree-based methods KD-tree [15] and graph-based methods NSG [18] are used in this experiment. Figure 5a illustrates the execution time of the brute-force, KD-tree, and graph search methods when processing datasets with different scales, where the feature vector dimension is fixed (128). It can be observed that the brute-force is suitable for a small-scale dataset (<20000). With the dataset scale increasing (>20000), the computation complexity ascends dramatically, and the graph-based methods exhibit better performance than brute-force methods. Besides, Fig. 5b depicts the KD-tree works well when searching a dataset (20000 instances) with low feature dimension but quickly loses its effectiveness as the feature vector dimension increases. Based on this observation, we design a hybrid data search strategy to answer the question "What is the fastest approximate nearest-neighbor algorithm for my data?" in this chapter. This strategy can take the feature of the given dataset to automatically determine the best ANNS algorithms.

4 Active Storage System

Target Workload As shown in Fig. 3, this work combines the strength of the deep neural network and hybrid search technique (DHS) to reduce the complexity of unstructured data analysis systems on the premise of high accuracy, which makes it possible to offload data analysis tasks from CPU/GPU into the resource-constraint Active Storage. Based on that, this work builds an end-to-end unstructured data analysis system that supports multimedia data analysis such as audio, video, text, etc. For example, audio can be processed by the recurrent or convolutional neural network models on the Active Storage to generate a feature vector as the index for relevant audio data in the SSD. In this chapter, a typical unstructured data analysis workload, image retrieval is used as a showcase. As shown in Fig. 6, Active Storage is designed to support the major components in the framework of DHS, allowing developers to customize and implement data retrieval solutions in it. Such a near-data retrieval system consists of two major components: the DHS library running in the light server that manages the user requests and the Active Storage that is plugged into the light host server via an interface like PCIe or SATA. As shown in Table 1, the DHS library is established by leveraging the *Vendor Specific Commands* in the I/O command set of *NVMe* protocol. It consists of two parts: *configuration library* and *user library*. The configuration library is provided for the administrator to choose

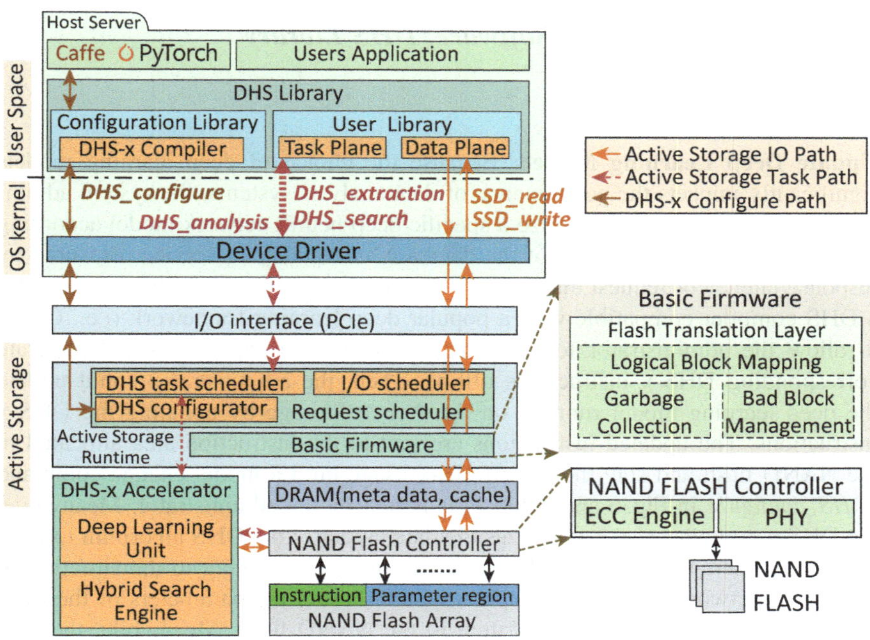

Fig. 6 Overview of Active Storage system

Table 1 DHS Library APIs for Active Storage

	–	API	Description
Configuration Library		*DHS_configure*	Update instruction and model on DHS-x
User Library	Task Plane	*DHS_extraction*	Extract the hashing feature of input data
		DHS_search	Fast data search
		DHS_analysis	User-defined analysis task for input data
	Data Plane	*SSD_read*	Read operation
		SSD_write	Write operation

and deploy different deep learning models on Active Storage quickly according to the demand of the application. After the feature-extracting deep neural network model has been deployed in Active Storage, a data processing request arriving at the host server could send and establish a query session to Active Storage by invoking the APIs provided by the user library. Then, the runtime system on the embedded processor of the Active Storage receives and parses the request to activate the DHS-x module associated with the session created by the user. Next, we elaborate on the design details of the software and hardware design of Active Storage.

4.1 The Active Storage Software: DHS Library

4.1.1 Configuration Library

Update Deep Learning Models Because the choice of deep learning models significantly impacts the performance of data analysis systems, the system administrator must be able to customize a specific deep neural network model according to the complexity and volume of the database and quality of service measured by response latency or request throughput. Thereby, the configuration library provides a DHS compiler compatible with a popular deep learning framework (i.e., Caffe) to allow the administrator to train the new deep learning model and generate corresponding DHS-x instructions offline. Then, the administrator could update the deep learning model running on the Active Storage by updating the DHS-x instructions. The updated instructions are sent to the instruction area allocated in the NAND flash and stay there for DHS-x usage until a model change command (*DHS_configure* in Fig. 6 and Table 1) is issued by the administrator. Meanwhile, the DHS-x compiler also reorganizes the data layout of the DHS algorithm to fully utilize the internal bandwidth of the Active Storage according to the structure of the neural network model and graph structure, before the parameters of the deep learning model and graphs are written to the NAND flash. Meanwhile, the data layout information is also recorded in the instruction of DHS-x. In this manner, the DHS-x obtains the physical address of the data directly at runtime instead of adopting the address translation and look-up operation that could incur extra overhead. More details about data reorganization will be introduced in Sect. 5.

4.1.2 User Library

Data Plane Based on the I/O system call of Linux, the data plane provides *SSD_read* and *SSD_write* API for users to control data transmission between the host server and the Active Storage. The users could invoke these APIs to inject data sent from users to the data cache region or the NAND flash of Active Storage through the file system. Afterward, users could use those addresses to direct the operands in other APIs.

Task Plane To fully improve the scalability of the DHS-x accelerator that supports deep neural network and hybrid data search algorithm, we abstract the function of DHS-x into three APIs in the task plane of user library: *DHS_extraction*, *DHS_search*, and *DHS_analysis*. These APIs are established based on extended commands of the NVMe I/O protocol, respectively. And all of them possess two basic parameters carried by ***DWord*** of NVMe protocol: the *data address* indicating the location of data in the Active Storage and the *data size* measuring the space in bytes taken by the data.

Firstly, the *DHS_extraction* is designed to extract the condensed feature of input data and map it into the semantic space, which is fundamental to the data retrieval system and useful for also other analysis functions like image classification or categorization. This command contains an extended parameter: *feature vector length*, which determines the capacity of the carried information. For example, compared to the database with 500 object types for an instance, the database with 1000 objects needs a longer feature vector length to avoid information loss. Secondly, *DHS_search* is abstracted from the hybrid search function of DHS-x. It includes extended parameters: T, representing the number of search results configured by users based on the applications scenarios. Finally, *DHS_analysis* API allows users to analyze the input data by using the prediction ability of deep neural networks, and it possesses a reserved field for user expansion. These task APIs can be invoked independently or in combination to develop different in-disk data processing functions. For an instance, users could combine the *DHS_extraction* and *DHS_search* to achieve data retrieval from a large-scale database, where the *DHS_extraction* API to map the feature of query data to the feature vector and *DHS_search* could search for top-T similar instances from the large-scale database by using that feature vector.

4.1.3 Active Storage Runtime

The Active Storage runtime deployed on the processor of Active Storage is responsible for managing the incoming extended I/O command via PCIe interface,

Table 2 The I/O Bandwidth of the Active Storage

–	I/O Bandwidth (MB/s) (I/O size = 128 KB)			
–	Write	Random		
Write	Read	Random		
Read				
Method-A	79.79	76.19	72.30	81.13
Method-B	524.86	421.23	654.31	698.98
Peak-Bandwidth	886.58	761.90	903.79	901.41

and it also converts the API-related commands into the instructions of the DHS-x accelerator and handles the basic operations for NAND flash. It includes a request scheduler and the basic firmware. The request scheduler contains three modules: ***DHS task scheduler***, ***I/O scheduler***, and ***DHS configurator***. The *DHS configurator* receives *DHS_configure* command from the host and updates instructions generated by the compiler and parameters of the specified deep learning model for the Active Storage. The *DHS task scheduler* responds to users' requests as supported in the task plane and initiates the corresponding task session in the Active Storage. The I/O scheduler dispatches the I/O request to the basic firmware or the DHS-x. The basic firmware includes the functions of the flash translation layer, garbage collection, and bad block management and also communicates with the NAND flash controller for general I/O requests.

Note that the DHS-x accelerator occupies a noticeable portion of the flash bandwidth once activated, which perhaps degrades the performance of normal I/O requests. To alleviate this problem, instead of letting the task or I/O scheduler wait until the request is completed (denoted as Method A), the DHS task scheduler receives the NVME command sent from the host with a doorbell mechanism and actively polls the completion status of the DHS-x periodically (denoted as Method B) to decide if the next request is dispatchable. We tested the normal read/write bandwidth of the active storage prototype described in Sect. 6.1 with the Flexible IO Tester (Fio) benchmark [34], under the worst-case influence of the DHS-x accelerator operation that occupied all channels of the Active Storage. Experiments (Table 2) demonstrate that adopting Method B only causes a drop of 27%–44% in the normal I/O bandwidth, while using Method A decreases almost 91% of the read/write bandwidth on average when the DHS-x accelerator is busy dealing with the over-committed retrieval tasks.

4.2 Hardware Architecture: The Active Storage

Figure 6 also depicts the hardware architecture of the Active Storage. It is composed of an embedded processor running the Active Storage runtime, a DHS-x accelerator, and NAND flash controllers connected to flash chips. Each NAND flash controller connects to one channel of the NAND flash module and uses an ECC engine for error correction. When the devices in each channel operate in lock-step and are accessed in parallel, Active Storage has a much higher internal bandwidth. More importantly, though SSD often has compact DRAM devices to cache the input data or metadata, the internal DRAM capacity can hardly satisfy the demand of the deep neural network as its numerous neural network parameters. Worse still, the SSD basic firmware like FTL and other components also occupy the majority of the memory resource. Therefore, the NAND flash controller is exposed to the DHS-x accelerator, which enables DHS-x to read and write the related working data directly from the NAND flash instead of the internal DRAM.

4.3 The Procedure of Data Retrieval in Active Storage

Figure 6 also depicts the overall process of Active Storage when users perform unstructured data retrieval tasks. Firstly, assume that the hardware instruction and parameters of the deep neural network model have been generated and written to the corresponding region by leveraging the DHS-x compiler and the DHS API that is shown in Fig. 6. Then, when the DHS library on the host captures a retrieval request, it packages and writes the user input data from the designated host memory space to the Active Storage by invoking *SSD_write* API. Meanwhile, *DHS_extraction* command carrying the address of input data is sent to the Active Storage for feature vector generation. Receiving the command, the request scheduler of the active storage runtime parses it to notify the DHS-x accelerator to start a feature extraction session. Then, DHS-x automatically fetches input query data from the command-specified data address and loads deep learning parameters from NAND flash. When the feature vector of input data begins to be produced, the other command, *DHS_search*, is sent and queued by the task scheduler. After the feature vector is produced, the *DHS_search* will be dispatched to invoke the hybrid search function in DHS-x and uses the feature vector generated by *DHS_extraction* to search the relevant data entries. In this case, DHS-x keeps fetching data to be retrieved from the NAND flash and sends the final retrieval results to the host memory once the task is finished.

5 DHS-x Accelerator

5.1 Architecture: Direct Flash Accessing

In contrast to a traditional hardware accelerator [7], the DHS-x accelerator is designed to directly obtain much of the working-set data from NAND flash. Figure 7 illustrates the high-level diagram of the DHS-x accelerator. The DHS-x accelerator has two InOut buffers and double-banked weight buffers. The intermediate results of

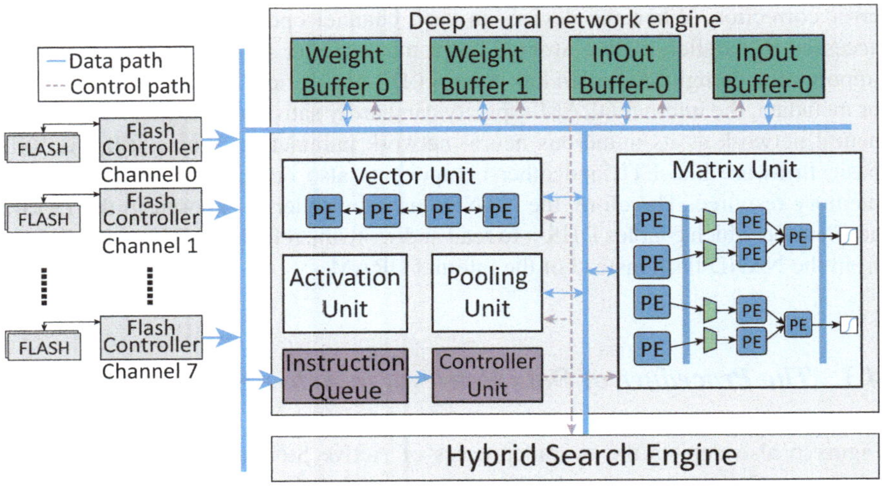

Fig. 7 The Architecture of DHS-x accelerator

each neural network layer are temporarily stored in the InOut buffers or the DRAM memory, while the weight buffers act as a bridge buffer between the deep neural network engine (DNE) and the flash, which stream out a large number of neural parameters to the DNE. The DNE comprises a set of processing engines (PEs), which could perform fixed-point convolutions, pooling, and activation function operations. The Hybrid Search Engine (HSE) is responsible for vector search with the feature vector generated by DNE. Both DNE and HSE are managed by the control unit that fetches instructions from memory. In consideration of the I/O operation granularity of NAND flash, we reorganize the data layout of neural network parameters to exploit the high internal flash bandwidth.

5.2 *I/O Path in Active Storage*

Bandwidth Analysis At first, we must analyze and prove the internal bandwidth of NAND flash could satisfy the demand of deep neural network running on DHS-x accelerator. Assuming that the DHS-x and flash controller operate at the same frequency, and the DNE unit of DHS-x consists of N PEs. If the single-channel bandwidth of NAND flash is BW_{flash}, then the bandwidth of M channels is:

$$BW^{m}_{flash} = M \times BW_{flash} \tag{1}$$

Suppose that a convolution layer convolves an $I_c \times I_h \times I_w$ input feature map (IF) with a $K_c \times K_h \times K_w$ convolution kernel to produce an $O_c \times O_h \times O_w$ output feature map (OF). The subscripts c, h, and w correspond to the channel, height, and width, respectively. The input/weight data uses 8-bit fixed-point representation. It is easy to derive that the computation latency $L_{compute}$ and the data access latency L_{data} from NAND flash to produce one channel of feature map are

$$\begin{aligned} L_{compute} &= \frac{OP_{compute}}{OP_{PE}} \\ &= \frac{2 \times K_c \times K_h \times K_w \times O_h \times O_w}{2 \times N_{PE}} \end{aligned} \tag{2}$$

$$\begin{aligned} L_{data} &= S_{param}/BW^m_{flash} \\ &= (K_c \times K_h \times K_w)/(BW^m_{flash}) \end{aligned} \tag{3}$$

where the $OP_{compute}$ and S_{param} are the operation number and the parameters volume of a convolutional layer. OP_{PE} gauges the performance of DHS-x measured in operations/cycle. To avoid DNE stall, we must have $L_{compute} >= L_{data}$, and O_w is usually equal to O_h, so we have

$$O_w >= \sqrt{N_{PE}/BW^m_{flash}} \tag{4}$$

The above equation indicates that if only the width and height of the output feature map are larger than or equal to the right side of formula 4, which is four in our prototype with $N_{PE} = 256$ and $BW^m_{flash} = 16 bytes/cycle$, the DNE will not stall. For example, in the AlexNet mentioned in Sect. 2.1, the minimum width of the output feature map in convolution layers is 7, which already satisfies in inequality 4. However, in the FC layers, $L_{compute}$ is smaller than L_{data}, so the data transfer time becomes the bottleneck. Thereby, the DHS-x accelerator only uses a column of PEs to deal with an FC layer because our prototype hardware design only supports eight channels, which does not meet inequality 4 with $M = 128$ and consequently causes PE under-utilization. Besides, the parameter-induced flash reads will be minimized if the size of the weight buffer meets the condition: $S_{buffer} >= Max(K_c \times K_h \times K_w)$. The parameters exceeding the size of the weight buffer will be repetitively fetched from the flash. To further improve the performance, we utilize ping-pong weight buffers to overlap the data loading latency with computation.

Data Layout in Flash Devices Owing to the bandwidth analysis on the basics of multi-channel data transmission, we propose flash-aware data layout to fully exploit flash bandwidth with the advanced NAND flash command-*read page cache command* [64]. The read page cache sequential command provided by the NAND flash manufacturer can continuously load the next page within a block into the data register (❷), while the previous page is being read to the buffer of DHS-x or cache region of the SSD from the cache register(❶). Thus, based on the NAND flash architecture with provided page cache command, we choose to split the convolution kernels and store them into flash devices for parallel fetch. As shown in Fig. 8, assuming there are N_{kernel} convolution kernels with the kernel size of S_k and M

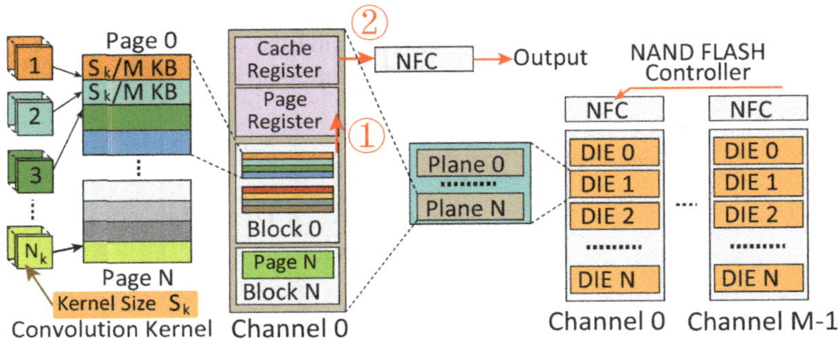

Fig. 8 The Data Layout in NAND flash

NAND flash channels are used by DHS-x accelerator, each convolution kernel is divided into S_K/M sub-blocks and all such sub-blocks are interleaved to the flash channels. The convolution kernels exceeding the size of a page are placed into continuous address space in the NAND flash because the cache command reads the next page automatically without extra address or operation.

Data Flow Taking the AlexNet as an example, when a request arrives at the DHS-x accelerator, the input data and the first kernel of the first convolution layer are transferred in parallel to the InOut buffer-0 and the weight buffer-0. After that, the DHS-x accelerator begins to compute the output feature map and stores them into the InOut buffer-1. When the first kernel is processed, the second kernel is transferred from the NAND flash to weight buffer-1 and then followed by the third and fourth kernels in sequence. Once the feature vector of the incoming query is generated, it is sent to the data search registers of HSE to locate the data structures similar to the query data if the DHS task scheduler decodes and dispatches the following *DHS_search* command.

5.3 Hybrid Search Engine

For fast and accurate feature vector search, the DHS-x accelerator constructs a separated hardware unit, hybrid search engine, to perform data search. In this manner, the feature extraction and data search stage can perform in a pipeline and achieve high throughput. Once the feature vector of the query data has been generated, the DHS-x will use it to index the corresponding feature vector data and search for the closest data entries from the vector corpus.

Hybrid search engine targets three different vector search methods, including brute-force search, KD-tree search, and graph-based search. However, existing accelerator designs only focus on one method. Integrating these accelerators will result in unnecessary areas, power consumption overhead, and low utilization issues.

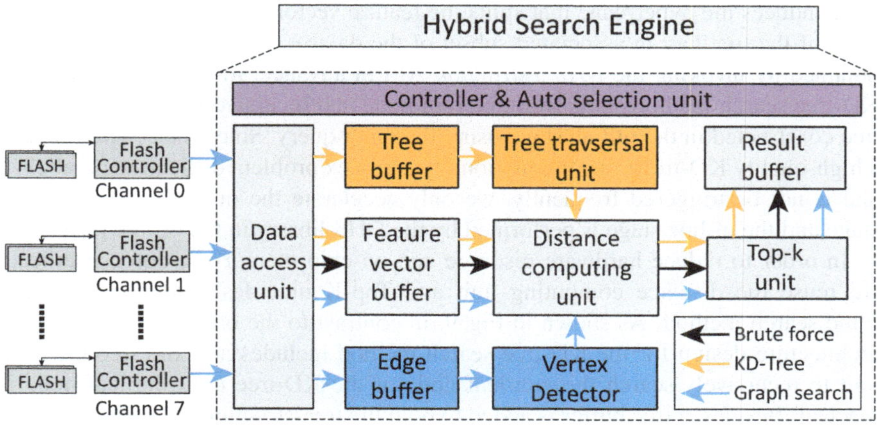

Fig. 9 Hybrid search engine

To address these issues, we unify brute-force search, KD-tree search, and graph-based search into one computing paradigm and design corresponding hardware accelerators. Next, we will introduce the hybrid search engine design process step by step, starting from the brute-force search, showing how to integrate KD-tree and graph search and reuse the computing and on-chip memory resource.

5.3.1 Brute-Force Search

Brute-force search finds the exact k nearest neighbors via scanning through the entire dataset. Figure 9 demonstrates the overview of the architecture design for the brute-force search method. It consists of three primary parts: (1) the distance computing unit is designed to calculate the distances between the input query and the feature vectors of the data corpus, (2) the Top-K unit is designed to find the top k elements according to the calculated distance, and (3) the data access unit is designed to fetch the data from DRAM or NAND flash using the physical address. The physical address is calculated based on the unique ID of each vector instance. The vector buffer is used to store the feature vector. The result buffer is used to store the retrieved result.

5.3.2 KD-Tree Search

Unlike the brute-force search that can directly leverage feature vectors to perform data search, the KD-tree method includes an offline stage that constructs a KD-tree and an online stage that searches elements on KD-tree. In the offline phase, a KD-tree is constructed for feature vectors of the dataset to be retrieved. Each non-leaf node induces the hyperplane that splits the feature vectors into two parts. Each leaf node of the tree has an associated subset of the dataset. The feature vector can be obtained by invoking the *DHS_extraction* API in advance. In the online stage, the KD-tree search method starts from the root node and recursively traverses the KD-tree constructed in the offline stage using the input query. Since the construction of a high-quality KD-tree is a computationally complex problem difficult to parallelize and is not be triggered frequently, we only accelerate the online KD-tree search stage and the offline stage is performed by the DHS library in the offline phase.

In order to reduce hardware resource and on-chip memory resource overhead, we reuse the distance computing unit and Top-K unit designed for the brute-force search method. As shown in Fig. 9, in contrast to the brute-force search, the architecture design for the KD-tree search method includes an extra tree traversal unit to recursively search the non-leaf node of the KD-tree until the leaf node is reached. Besides, a tree buffer is added to hold the tree-structured metadata used by the tree traversal unit.

5.3.3 Graph Search

Similar to the KD-tree search, the graph search method also includes an offline stage and an online stage. A classical graph search method, Navigating Spreading-out Graph [18] (NSG), is used in this chapter. In the offline phase, the NSG method constructs a directed $K_{nbors} - NN$ graph for the storage data structures to be retrieved. In a graph, a vertex represents a data entry by keeping its unique ID and feature vector, where the feature vector can be obtained by invoking the *DHS_extraction* API in advance. The bit-width of ID (W_{id}) and vector item, as well as the length of the feature vector ($W_{feature\_vector}$), are user-configurable parameters in the API. In a graph, a vertex may be connected to many vertices, which have different distances from each other. However, only the top-K_{nbors} closest vertices of a vertex could be defined as its "neighbors," where K_{nbors} is also a reconfigurable parameter and enables users to pursue the trade-off between accuracy and retrieval speed. The DHS-x accelerator only accelerates the online retrieval stage and the database update is offline because the latter is carried out infrequently. The database update consists of the feature vector extraction stage and $K_{nbors} - NN$ graph construction with reorganizing stage, where the former is accelerated by DHS-x accelerator and the latter is completed by DHS library offline. The graph construction takes about 100 seconds to update the $K_{nbors} - NN$ graph on million-scale data offline. The hardware architecture of the graph search engine is presented in Fig. 9.

The graph search function of DHS-x starts from evaluating the distance of random initial vertices in this graph and walks the whole graph from vertex to vertex in the neighborhood to find the closest results. As shown in Fig. 9, to maximize the utilization of on-chip memory, the feature vector buffer and result buffer are reused to store the neighbors of vertices and the search results of the graph search engine, respectively. To support the graph search method, we further add an edge buffer to

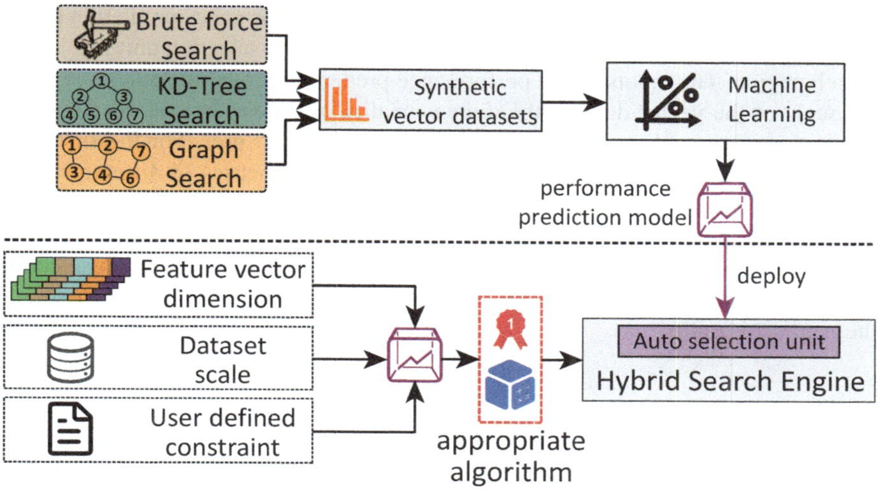

Fig. 10 Auto-selection model

store the edge list of the graph. As shown in Fig. 9, a Vertex Detection Unit (VDU) is inserted to check whether the selected vertex has been evaluated. In VDU, the vertex will be discarded once found already walked before and otherwise sent to the distance computing unit to compute the distance from the query vertex. With the distance provided by the distance computing unit, the Top-K unit can put the vertex into the result buffer. The un-evaluated vertices will be fetched from the result buffer and the neighbors of these vertices are loaded from the neighbor buffer by the control unit. Meanwhile, the control unit will finally return the top-T closest vertices when the number of the vertex in the result buffer reaches the threshold configured by users, where T is also configured by users via *DHS_search* API.

5.3.4 Auto-Selection Model

The hybrid search engine provides three different data search methods. However, there is the fastest approximate nearest neighbor search method for a specific dataset. Therefore, as shown in Fig. 10, we construct an auto-selection model from two perspectives: scene adaptation and data distribution adaptation. The design of the auto-selection model is divided into two phases: the offline model training phase and the online model performance prediction phase. In the offline model training phase, first, we leverage synthetic vector datasets to test the performance of different vector retrieval algorithms at different scales and dimensions by varying the scale of the vector dataset and the dimensionality of the vector features. Second, a performance prediction model is constructed by modeling the collected performance data using machine learning algorithms, and this performance model is validated in the offline phase. For the online model performance prediction phase, the performance prediction model is deployed on the auto-selection unit in a hybrid search engine. The inputs of the performance prediction model include three parts: the scale of the vector dataset, the dimensionality of the vector data features, and the performance and accuracy constraints defined by the user. For the incoming retrieval requests, the execution time and the precision metrics of different vector retrieval algorithms are predicted based on the retrieved data features. Second, the most suitable vector retrieval algorithm is automatically determined based on the performance and accuracy constraints set by the user. Finally, the auto-selection unit automatically controls the execution flow of a hybrid search engine based on the predicted results.

5.3.5 Data Flow

We show an example to brief the overall flow of data retrieval with deep neural networks and hybrid data search. Firstly, when a query comes, the DHS-x accelerator fetches the input data and parameters of the deep learning model from the NAND flash into the InOut buffer and the weight buffer, respectively. Then, the deep neural network engine (DNE) generates the feature vector for the input data and writes it to the data search registers of the HSE. After that, the DHS-x accelerator identifies the data search flow according to the scale of the dataset to be retrieved. When the brute-force method is selected, HSE fetches the feature vector from the NAND flash array to the feature vector buffer and performs a brute-force data search. When the KD-tree or graph search method is selected, HSE transfers the tree metadata or the $K_{nbors} - NN$ graph from the NAND flash array to the tree buffer or the edge buffer, respectively. For the KD-tree method, the tree traversal unit performs tree traversal on the KD-tree according to the metadata in the tree buffer and the feature vector of the incoming query. When the leaf node is reached, the feature vector in this leaf node is fetched from the NAND flash array to the feature vector buffer and then calculated using the feature vector of the incoming query. Then the Top-K results are written to the result buffer. For the graph search method, at the first stage, the initial vertices are calculated and sent into the corresponding areas of the InOut buffer. In the second stage, the DHS-x control unit reads the first un-evaluated vertex from the result buffer in descending order of distance. Then, the hybrid search engine obtains the neighbors of the un-evaluated vertex from the NAND flash and transfers the neighbors to the distance computing unit to generate their distances from the query vertex as well. Next, the Top-K unit writes these vertices to the result buffer according to the calculated distance. Meanwhile, the counter in the graph search engine determines whether the termination signal should be issued by monitoring the total number of vertices stored in the result buffer. Once the termination signal is generated, the Active Storage runtime reads out the vertices from the result buffer and then transfers the ID-directed results stored in NAND flash to the host server via the PCIe interface.

5.4 LSTM-Based Data Cache and Placement

Data caching and data placement optimization are essential to improve the performance of the storage system as the former affects the data access latency of the storage system, while the latter has an impact on the lifespan of the storage device. More importantly, real-world workloads are not random but constrained by the politeness principle. Therefore, by collecting and learning the IO workload behavior imposed by real-world workloads, we can clearly predict the temporal and spatial locality of accessed data, thus ensuring that the data required by the workload is available in a timely manner and thus avoiding massive data migration. Based on these analyses, the goal of our method is to predict hot blocks at the next time moment using the recent data access history and to take them into the internal DRAM with low access latency (~100 ns) from NAND flash with high access latency (~75 us) chip. To this end, borrowing ideas from [9], we treat the problem of identifying hot blocks as a frequency predicate problem. Specifically, instead of directly predicting indexes of hot blocks, our method predicts the frequency of all blocks and then computes the Top-K predicted frequency to identify the hot blocks. After that, the frequency of blocks is counted as historical information to predict new frequency. The original traces including timestamp, access address, access type, and other information are first divided into T time windows according to the fixed window step size. In each time window, the block indexes of all records are counted, and the frequencies of all N blocks are then calculated as data in the current time window. Next, take the continuous window data of fixed size W from the time window dimension in a sliding window manner as the historical input data, and take frequencies in the next time window as their label. With this method, $T - W + 1$ datasets are constructed, each of which contains N data entries of $W * 1$ dimension. However, as the frequency of some blocks is always zero in certain time windows and may harm the training of the network, block entries with a frequency of 0 accounting for more than half and with a label of 0 are excluded to avoid the training data being too sparse.

The network structure is an LSTM augmented with three fully connected (FC) layers, named with LSTM-Reg. The frontier FC transforms the raw input data to the shape of the input of LSTM, the second one transforms the raw input data into a scalar, which is then scattered and added to the outputs of LSTM, and the last FC transforms the output of LSTM into a scalar frequency value. Notably, compared to traditional data cache and placement algorithms such as LRU, the LSTM-based model is computation-intensive, and tedious execution time can affect the data access latency, resulting in performance degradation. Fortunately, as shown in Fig. 11, the customized neural network accelerator integrated into Active Storage can be used to deploy the LSTM-based network, which can reduce the execution time of the neural network model and thus avoid performance degradation issues.

To predict the hot blocks, blocks whose frequencies are the Top-K highest are taken as hot blocks. Those hot blocks will be loaded into the internal DRAM from

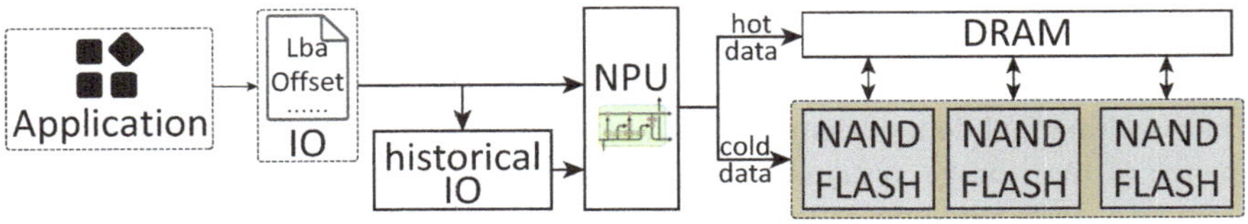

Fig. 11 LSTM-based data cache and placement

Fig. 12 The Active Storage prototype

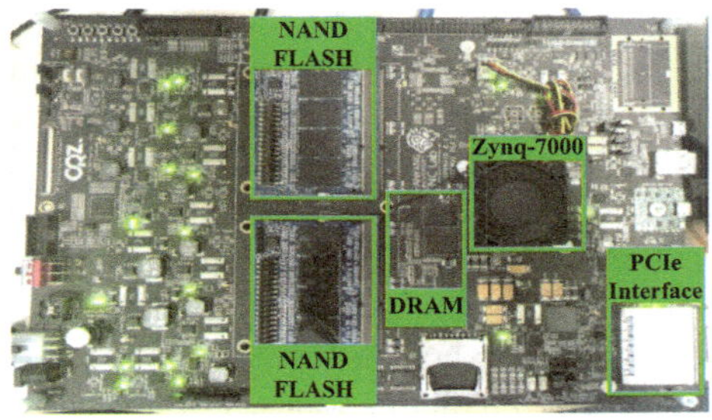

the NAND flash chip. To evaluate the effect, the Recall rate and IO hit rate are calculated between the predicted Top-K hot blocks and real accessed blocks.

6 Evaluation

6.1 Hardware Implementation

To explore the advantages of the Active Storage system, we implemented it on the Cosmos plus OpenSSD platform [32]. The Cosmos plus OpenSSD platform consists of an XC7Z045 FPGA chip, 1 GB DRAM, 8-way NAND flash interface, an Ethernet interface, and a PCIe Gen2 8-lane interface. A DHS-x accelerator and modified NAND flash controllers are implemented on the programmable logic of XC7Z045. The Active Storage runs its firmware on a Dual 1 GHz ARM Cortex-A9 core of XC7Z045. The Active Storage is plugged into the host server via a PCIe link. The host server manages the high-level requests and maintains the DHS library for API calls. Figure 12 shows the Active Storage prototype constructed for this work.

6.2 Experimental Setup

We first selected a content-based image retrieval system (CBIR) based on deep neural networks and hybrid data search (DHS) algorithm as workload and evaluated the performance of the DHS solution compared to other conventional solutions

(Sect. 6.3). Except for the Active Storage prototype, our experimental setup also consists of a baseline server running Ubuntu 14.04 with two Intel Xeon E5-2630 CPU@2.20 GHz, 32 GB DRAM memory, four 1 TB PCIe SSDs, and an NVIDIA GTX 1080ti GPU. Meanwhile, we implemented a CBIR system with C++ language on the baseline server, where the deep neural network is built on top of Caffe [26]. Based on this platform, we constructed four baselines: B-CPU, B-GPU, B-FPGA, and B-DHS-x. For the B-CPU baseline, the DHS algorithm runs on the CPU. For B-GPU, the deep neural network runs on the GPU and hybrid search runs on the CPU. For B-FPGA, we use the ZC706 FPGA board to replace the Active Storage, and the deep neural network runs on the ZC706 FPGA board and hybrid search runs on the CPU. B-DHS-x implements the DHS algorithm on the ZC706 FPGA board without any near-data processing technique compared to Active Storage.

6.3 Evaluation of DHS Algorithm

Experimental Setup We used the precision at top T returned samples (*Precision@T*), measuring the proportionality of corrected retrieved data entries, to verify the performance of our deep neural network method on different models and datasets [37]. The performance is contrasted with traditional feature extraction methods with 512-dimensional GIST feature, including Locality-Sensitive Hashing (LSH) [68] and Iterative Quantization (ITQ) [19]. The used datasets are listed in Table 3.

Evaluation Figure 13a–d shows the *Precision@T* on different datasets with different deep neural network models. We leveraged the output of the embedding layer to extract the feature vector. The dimension of the feature vector is configured to 48 in the experiment. Due to the poor performance of the LSH and ITQ [19] method on ImageNet, we added the AlexNet-CCA-ITQ and AlexNet-ITQ method [72] that uses AlexNet to extract the feature vector for search. Our deep neural network and hybrid data search method perform better than the other approaches on different datasets with different scales regardless of the choice of T value, especially than the conventional deep neural network-based method. It also shows the robustness of DHS solutions when deployed on a real-world system.

To explore the relationship between the feature vector dimensions and the execution time and accuracy, we configured the parameters of the hash layer of the

Table 3 Datasets used in our experiments

Dataset	Total	Train/Validate	Labels
CIFAR-10	60000	50000/10000	10
Caltech256	29780	26790/2990	256
SUN397	108754	98049/10705	397
ImageNet	1331167	1281167/50000	1000

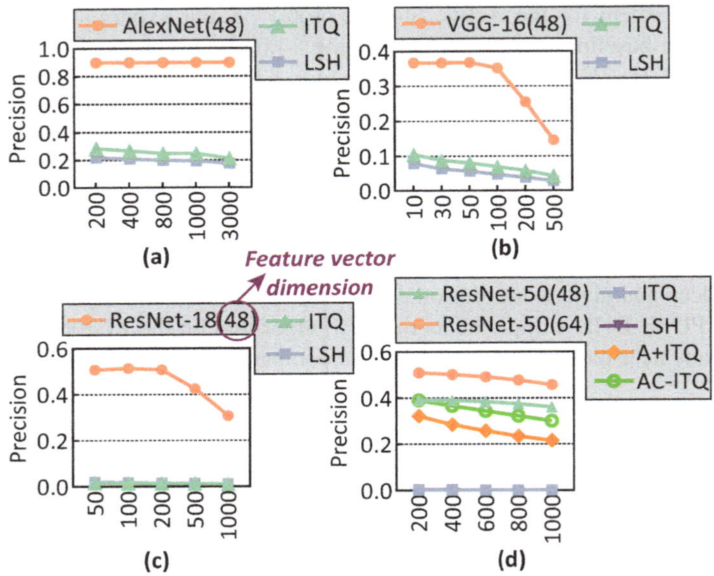

Fig. 13 Precision curves w.r.t top-T. X-axis represents top T returned Images. (**a**) CIFAR-10. (**b**) Caltech256. (**c**) SUN397. (**d**) ImageNet

AlexNet to obtain different feature vector dimensions and test the corresponding execution time and accuracy under the CIFAR-10 dataset. As shown in Fig. 14, as the feature vector dimension rises, the execution time of the AlexNet network slowly rises. The main reason is that the execution time of the feature vector extraction layer is only 0.3% of the total execution time. However, the rising of the feature vectors dimension has a crucial impact on the search accuracy, as shown in Fig. 15. The retrieval accuracy increases from 87.5% to 88.88% as the dimension of the feature vector rises and gradually levels off when the dimension exceeds 48. More importantly, as shown in Fig. 21, the climbing of the feature vector dimension has a critical impact on the computation of the search stage. Thereby, the DHS-x accelerator is configured to support different feature vector lengths to achieve the trade-off between retrieval accuracy and latency.

6.4 *Evaluation of DHS-x Accelerator*

Experimental Setup We implemented the deep neural network and hybrid data search algorithm (DHS) on the DHS-x accelerator of Active Storage and compared the latency and power of the DHS-x accelerator to the solutions based on CPU and GPU, where we ignore the FPGA baseline because its computational units are the same as the DHS-x. Firstly, we only compared the latency and power of the deep

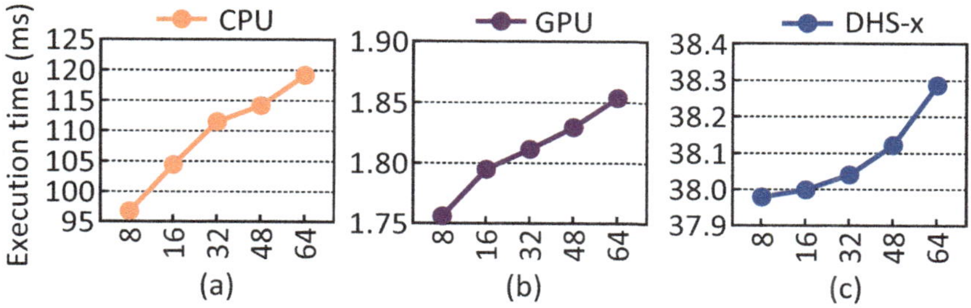

Fig. 14 Execution time w.r.t vector feature dimension on CPU, GPU, and DHS-x

Fig. 15 Accuracy w.r.t vector feature dimension

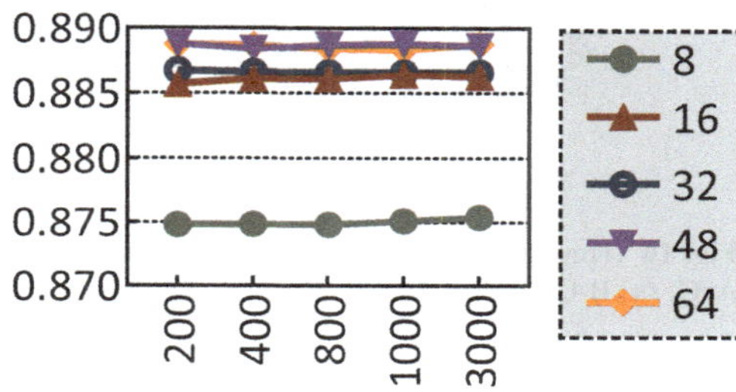

neural network unit on DHS-x running various convolution neural networks to CPU and GPU. The latency of the CPU and GPU is measured by using the command "Caffe time" of the Caffe framework. Secondly, we only evaluated the latency of the hybrid data search unit on DHS-x with respect to different numbers (T) of top retrieved data entries on the CIFAR-10 and ImageNet datasets. Meanwhile, in order to evaluate the efficiency of the hybrid data search function, some synthetic datasets are used to evaluate the performance of the brute-force, KD-tree, and graph search method on the hybrid data search engine. Furthermore, we demonstrate the advantages of the auto-selection model based on synthetic datasets with different scales and feature vector dimensions.

Performance Firstly, Table 4 shows the latency of the DHS-x accelerator on various deep neural network schemes outperforms the solution based on CPU. While the latency of DHS-x is higher than GPU because of the hardware resource and frequency limitation, it consumes less power compared to GPU. It should be noted that the latency of the deep neural network processing decided the capability of the Active Storage system and CPU baseline. Figure 16 shows the feature extraction stage occupies about 98.7% and 98.9% of the total processing time on the DHS-x accelerator and CPU baseline, respectively. Meanwhile, the latency of feature extraction on B-GPU only occupies 61.9% because the processing speed of the deep neural network on the GPU is close to the speed of the graph search algorithm on the CPU.

Secondly, in this experiment, we first utilized the AlexNet model to generate the 48 dimension feature vector database for the construction of the $K_{nbors} - NN$

Table 4 The performance of DHS-x accelerator

Model	–	Latency (ms)	Power (W)
AlexNet	DHS-x	**38**	**9.1**
	CPU	114	186
	GPU	1.83	164
ResNet18	DHS-x	**94**	**9.4**
	CPU	121	185
	GPU	7.13	112

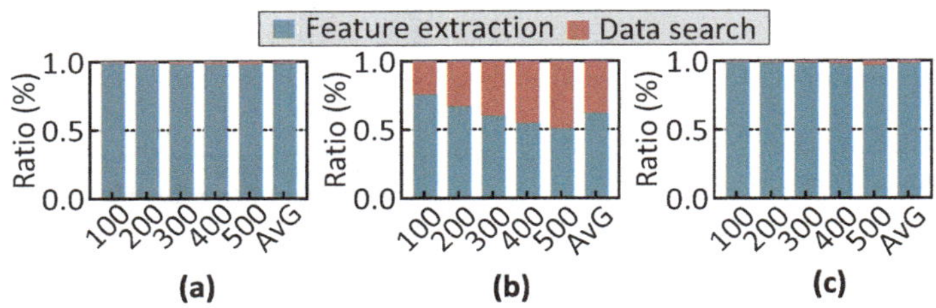

Fig. 16 Time breakdown w.r.t top-T. X-axis represents # of top retrieval. Average is denoted as AvG. (**a**) B-CPU. (**b**) B-GPU. (**c**) Active Storage

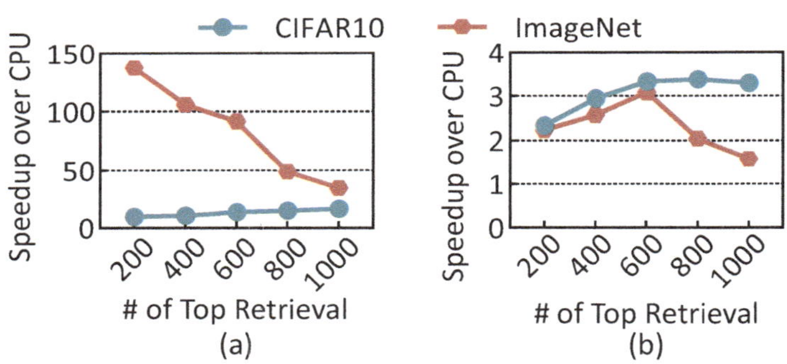

Fig. 17 Data search performance of DHS-x on CIFAR10 and ImageNet dataset. (**a**) Speedup over brute-force on CPU. (**b**) Speedup over graph search on CPU

graph on the CIFAR-10 and ImageNet datasets. We compared the retrieval speed of the DHS-x accelerator with two counterparts: the brute-force search method that evaluates all the feature vectors stored in the database and the CPU-executed hybrid data search algorithm. We deployed Petalinux on OpenSSD and measured the execution time of the DHS-x accelerator. The result is depicted in Fig. 17. For the CIFAR-10 dataset, the DHS-x accelerator is 10.07-16.71x and 2.34-3.30x faster than the brute-force method and the CPU-run graph search algorithm, respectively, while for the ImageNet dataset, it achieves even 137.72-34.93x and 2.23-1.59x speedup over the two baselines, respectively.

To gain insight into the auto-selection model optimizations, we evaluated the DHS-x accelerator under several synthetic datasets with different scales and

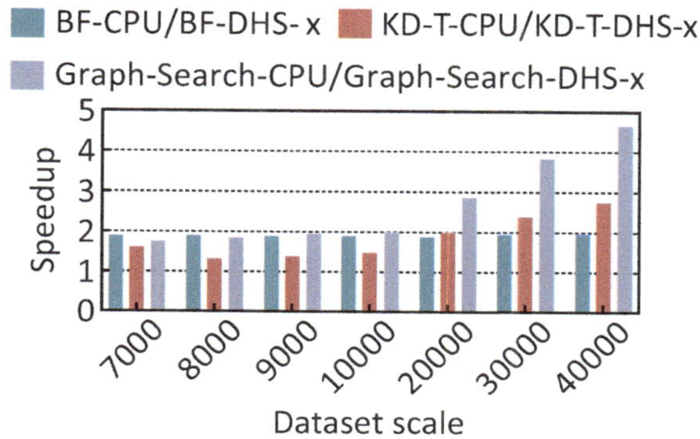

Fig. 18 Data search performance speedup of DHS-x over CPU on different dataset scale

dimensions. The dimension of the dataset varies from 2 to 128, while the number of dataset vertex ranges from 7000 to 40000. First, we evaluated and compared the performance of the brute-force (denoted as BF-CPU), KD-tree (denoted as KD-T-CPU), and graph search method (denoted as Graph-Search-CPU) on the DHS-x accelerator and CPU baseline, respectively. Figure 18 shows that the DHS-x accelerator achieves 1.92X, 1.86X, and 2.71X speedup compared to the CPU baseline when executing the brute-force, KD-tree, and graph search method. The performance improvement is achieved through a parallelized and pipelined datapath, which enhances the efficiency of distance computation. Secondly, in order to illustrate the advantages of auto-selection optimization, we applied this optimization on the CPU baseline (denoted as CPU-Opt) and compared its performance to the CPU baseline without any optimization. Figure 19 figures out that applying auto-selection optimization on CPU baseline can select the appropriate algorithm when facing datasets with different scales. Furthermore, Figs. 20 and 21 show the speedup of DHS-x accelerator with auto-selection optimization over without either optimization when facing datasets with different scales and dimensions. It can be observed that auto-selection optimization exhibits significant performance improvement compared to CPU-Opt. The reason is that the DHS-x accelerator with auto-selection optimization can select the appropriate data search methods according to the feature of a dataset to be retrieved.

Power Consumption We measured and compared the power consumption of the Active Storage system with four baselines: B-CPU, B-GPU, B-FPGA, and B-DHS-x by using a power meter under two different situations: (1) IDLE: No retrieval requests need to respond. (2) ACTIVE: A user continuously accesses the Active Storage. The result is illustrated in Table 5. When the Active Storage is IDLE, the power consumption of the Active Storage is slightly higher than B-CPU and lower than B-GPU because the GPU IDLE power is higher than the Active Storage. For active power, when delivering comparable data retrieval performance, the Active Storage could reduce the total power consumption by up to 28.8% and 58.01% compared to B-CPU and B-GPU. Simply replacing the GPU with an FPGA board could reduce power consumption by 36.79% compared to B-GPU. Furthermore,

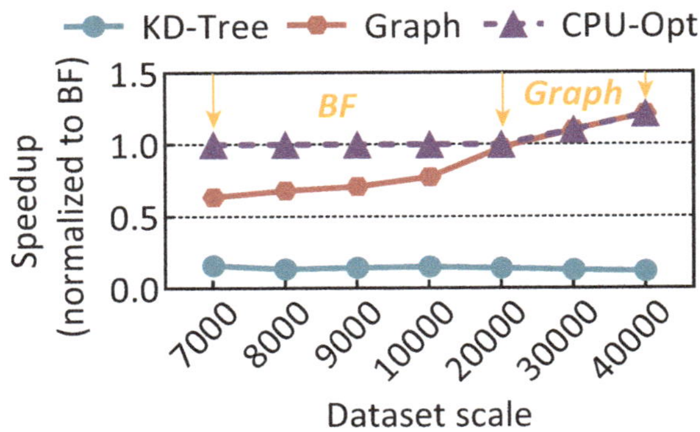

Fig. 19 Data search performance speedup of CPU with optimization

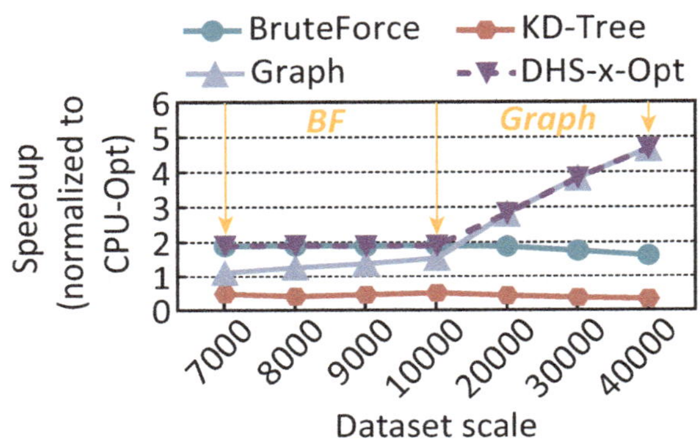

Fig. 20 Data search performance speedup of DHS-x on dataset with different scale

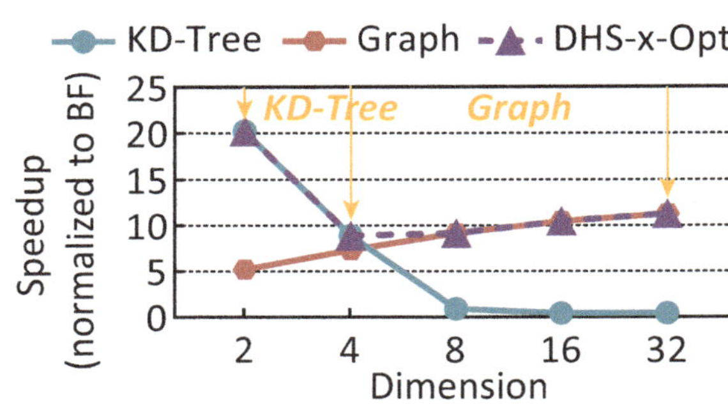

Fig. 21 Data search performance of DHS-x on dataset with different dimension

putting DHS-x on an identical FPGA board without NDP reduces by 38.68% compared to B-GPU, which is attributed to hardware specialization. Placing DHS-x into Active Storage can further improve energy efficiency because of the benefits of near-data processing. In the case of Active Storage plus CPU, the power of the CPU is low because it is only responsible for instruction dispatch without any data transfer between storage and CPU. In other cases, the CPU is not only in charge of

Table 5 Power consumption

Power (Watt)	Active Storage	Active Storage plus CPU	B- DHS-x	B- FPGA	B- CPU	B- GPU
IDLE	18.2	101.5	90	92	78	87
ACTIVE	22	131	191.3	197.2	184	312

Table 6 The hardware utilization of Active Storage

Module	#	LUT	FF	BRAM	DSP
Flash Controller	8	11031	7539	21	0
NVMe Interface	1	8586	11455	28	0
DHS-x Accelerator	1	76344	35342	432	243
`Total`	1	215479	163245	481	243
Percent(%)	–	98.57	37.33	88.26	27

data transfer management but also for instruction dispatch or executing graph search algorithms.

FPGA Resource Utilization The placement and routing were completed with Vivado 2016.2. Table 6 shows the hardware utilization of the Active Storage. It only reports the resources overhead of the flash controller, NVMe controller, and the DHS-x accelerator module. `Total` counts in all FPGA resources spent by the Active Storage.

6.5 The Single-Node System Based on Active Storage

Experimental Setup We implemented the CBIR system by using the ImageNet dataset on the Active Storage with a baseline server, where the baseline server is only responsible for receiving and sending retrieval requests to the Active Storage, and the deep neural network architecture is AlexNet network and the feature vector dimension is 48. We built a web-accessible CBIR system based on web framework CROW [21] to evaluate the latency and query per second (QPS) of the system by simulating the user requests sent to the URL address via the ApacheBench (ab) [45]. The latency measurement indicates the time between issuing a request and the arrival of the result. The QPS is scalability measuring metric characterizing the throughput of the system. The latency and QPS are affected by the software algorithm and the hardware performance of the system. Meanwhile, we leverage the metric of QPS per watt (QPS/W) to evaluate the energy efficiency of the system.

Evaluation We evaluated the performance of a single system on the Active Storage and four baselines under the assumption that data cannot be accommodated in DRAM and must travel across the SSD cache, I/O interface, and HOST DRAM before reaching to compute unit. The performance of the Active Storage system and two baselines are shown in Fig. 22. With the increased number of top images retrieved, the retrieval time spent on the DHS-x accelerator will rise. It leads to

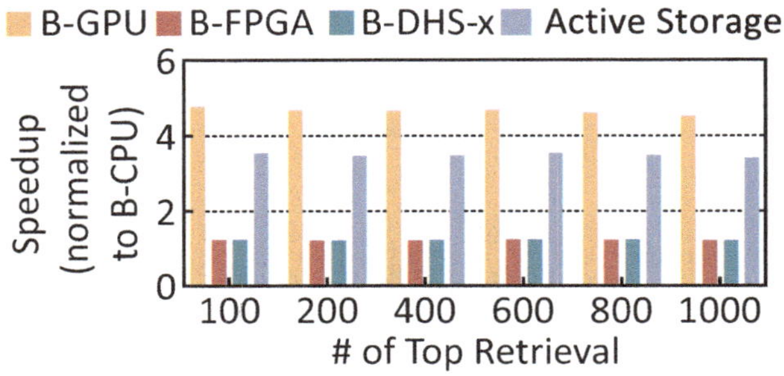

Fig. 22 The speedup of the Active Storage system over B-CPU

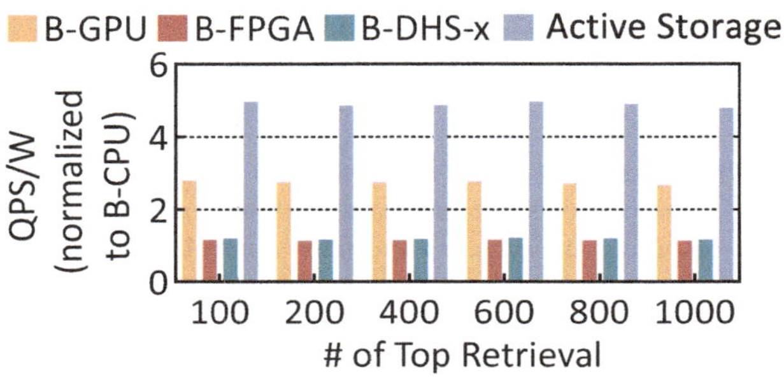

Fig. 23 The energy efficiency of the Active Storage system over B-CPU

increased retrieval latency and decreased QPS of the Active Storage. Meanwhile, we also observe that 95% requests complete in time in experiments when writing operations and garbage collection are inactive on Active Storage. Note that write operation and garbage collection are rare for Active Storage compared to read operations and usually occur offline. The workloads on Active Storage are read-only, which sustains the latency of Active Storage at a steady level with little fluctuation. Besides, in comparison to the B-CPU, the average latency of the Active Storage system is reduced by 3.48x. The performance improvement stems from the high speed of data processing on the DHS-x accelerator compared to B-CPU. Due to the overhead of data movement caused by the bandwidth limitation of the I/O interface and onboard memory, the latency of B-FPGA and B-DHS-x is higher than the B-GPU. Compared to B-FPGA and B-DHS-x baseline, Active Storage could reduce latency by 2.82x and 2.80x, which benefits from the near-data processing. The average retrieval speed of B-GPU is 1.34x faster than the Active Storage system. The reason is that the execution of the deep learning model costs more time on the DHS-x compared to powerful GPUs. However, the Active Storage is more energy-efficient (QPS/w) than a GPU-integrated system by 1.77x, which is shown in Fig. 23. More importantly, the Active Storage is implemented with FPGA and the operating

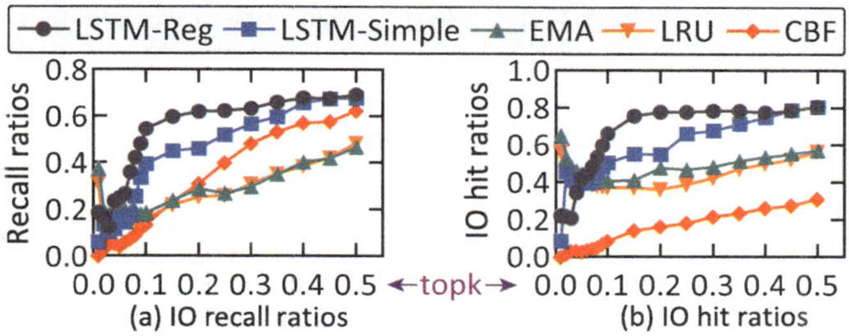

Fig. 24 I/O recall ratios (**a**) and I/O hit ratios (**b**) of LSTM-Reg, LSTM-Simple, EMA, LRU, and CBF

frequency is only 100 MHz. The performance will be better if Active Storage is implemented with ASIC or escalated operating frequency.

6.6 Data Cache and Placement

Experimental Setup We leverage production storage I/O traces Microsoft-1 [33] for the simulation evaluation, which is used to build the training set, testing set, and validating set. The window step size is 100000, the block size is 64MB, and the size of the time window is $W = 20$. The MSE loss and the AdamW optimizer are adopted in our training. We select four baselines to show the effectiveness, including LSTM-Simple, which is an LSTM connected with an FC, and three traditional algorithms including LRU [6], EMA [36], and CBF [24]. The k in Top-K varies from 0.01 to 0.5 to show the effect when different proportions of blocks are selected as hot blocks.

Our performance evaluation employs two metrics, namely, the I/O recall ratio and the I/O hit ratio, to assess the efficacy of our method. The I/O hit ratio gauges the method's ability to accurately predict the Top-K blocks by quantifying the ratio of the intersected blocks between the predicted and actual Top-K sets to the total number of ground-truth Top-K blocks. Conversely, the I/O recall ratio measures the method's ability to correctly identify the Top-K blocks, represented as a proportion of the Top-K set size. These metrics provide a comprehensive means of evaluating the method's performance.

Evaluation Figure 24 shows that when k exceeds 0.03, LSTM-Reg outperforms all other baselines, including LSTM-Simple, EMA, LRU, and CBF, in terms of recall ratio. Similarly, when k exceeds 0.05, LSTM-Reg demonstrates superior I/O hit ratios over all other baselines. These advantages stem from LSTM-Reg's ability to accurately extract load features from historical frequency series and make predictions based on this information. However, for small k values, both LSTM-Reg and LSTM-Simple perform worse than EMA and LRU. The divergent predictive rationales of various methods account for the observed phenomenon.

Specifically, while LRU and EMA adopt a predictive approach that centers on recently accessed blocks to anticipate future access, LSTM-Reg and LSTM-Simple predict the frequency of all blocks and subsequently select the Top-K to scrutinize the most frequently accessed blocks. However, when k is small, the subset of blocks deemed 'hot' is limited, which results in a reduced intersection between predicted and actual hot blocks. As a result, LSTM-Reg and LSTM-Simple may yield inaccurate predictions when k is small. Nevertheless, as k increases, LSTM-Reg offers an advantage in its capacity to extract and predict features with greater accuracy.

7 Conclusion

We have introduced Active Storage, a near-data deep learning device that actively performs low latency, low power, and high accuracy unstructured data analysis. We have designed and implemented Active Storage with direct flash-access deep neural network and hybrid data search accelerator, to combat the complex software stack and inefficient memory hierarchy barriers in conventional multimedia data analysis systems. Our prototype demonstrates that Active Storage provides 3.48x performance speedup over CPU, 4.89X, and 1.77X energy saving against CPU and GPU, respectively. Meanwhile, the LSTM-based data cache and placement policy also exhibits better performance than traditional LRU algorithms.

Acknowledgments We thank professor Jiafeng Guo of the CAS key lab of network data science and technology for his support and suggestions. This chapter is supported in part by the National Key Research and Development Program of China under grant No. 2018YFA0701502 and in part by the National Natural Science Foundation of China (NSFC) under grant Nos. 62090024, U20A20202, 62222411, and 62202453, YESS hip program No. YESS2016qnrc001, and China Post doctoral Science Foundation No.2022M713207.

References

1. A. Acharya, M. Uysal, J. Saltz, Active disks: Programming model, algorithms and evaluation. SIGPLAN Not. **33**(11), 81–91 (1998). http://doi.acm.org/10.1145/291006.291026
2. D.G. Andersen, J. Franklin, M. Kaminsky, A. Phanishayee, L. Tan, V. Vasudevan, Fawn: a fast array of wimpy nodes, in *Proceedings of the ACM SIGOPS 22Nd Symposium on Operating Systems Principles*, SOSP '09 (ACM, New York, 2009), pp. 1–14. http://doi.acm.org/10.1145/1629575.1629577
3. R. Balasubramonian, J. Chang, T. Manning, J.H. Moreno, R. Murphy, R. Nair, S. Swanson, Near-data processing: Insights from a micro-46 workshop. IEEE Micro **34**(4), 36–42 (2014). https://ieeexplore.ieee.org/document/6871738
4. S. Boboila, Y. Kim, S.S. Vazhkudai, P. Desnoyers, G.M. Shipman, Active flash: out-of-core data analytics on flash storage, in *012 IEEE 28th Symposium on Mass Storage Systems and Technologies (MSST)* (2012), pp. 1–12. https://ieeexplore.ieee.org/document/6232366

5. A.M. Caulfield, A. De, J. Coburn, T.I. Mollow, R.K. Gupta, S. Swanson, Moneta: a high-performance storage array architecture for next-generation, non-volatile memories, in *Proceedings of the 2010 43rd Annual IEEE/ACM International Symposium on Microarchitecture*, MICRO '43 (IEEE Computer Society, Washington, 2010), pp. 385–395. https://doi.org/10.1109/MICRO.2010.33
6. L.-P. Chang, T.-W. Kuo, An adaptive striping architecture for flash memory storage systems of embedded systems, in *Proceedings. Eighth IEEE Real-Time and Embedded Technology and Applications Symposium* (IEEE, New York, 2002), pp. 187–196
7. T. Chen, Z. Du, N. Sun, J. Wang, C. Wu, Y. Chen, O. Temam, Diannao: a small-footprint high-throughput accelerator for ubiquitous machine-learning, in *Proceedings of the 19th International Conference on Architectural Support for Programming Languages and Operating Systems*, ASPLOS '14 (ACM, New York, 2014), pp. 269–284. http://doi.acm.org/10.1145/2541940.2541967
8. Y.-H. Chen, J. Emer, V. Sze, Eyeriss: a spatial architecture for energy-efficient dataflow for convolutional neural networks, in *Proceedings of the 43rd International Symposium on Computer Architecture*, ISCA '16 (IEEE Press, Piscataway, 2016), pp. 367–379. https://doi.org/10.1109/ISCA.2016.40
9. Y. Cheng, F. Zhang, G. Hu, Y. Wang, H. Yang, G. Zhang, Z. Cheng, Block popularity prediction for multimedia storage systems using spatial-temporal-sequential neural networks, in *Proceedings of the 29th ACM International Conference on Multimedia*, MM '21 (Association for Computing Machinery, New York, 2021), pp. 3390–3398
10. W. Cheong, C. Yoon, S. Woo, K. Han, D. Kim, C. Lee, Y. Choi, S. Kim, D. Kang, G. Yu, J. Kim, J. Park, K.-W. Song, K.-T. Park, S. Cho, H. Oh, D.D.G. Lee, J.-H. Choi, J. Jeong, A flash memory controller for 15s ultra-low-latency ssd using high-speed 3d nand flash with 3s read time, in *2018 IEEE International Solid—State Circuits Conference—(ISSCC)* (2018), pp. 338–340
11. S. Cho, C. Park, H. Oh, S. Kim, Y. Yi, G.R. Ganger, Active disk meets flash: a case for intelligent SSDS, in *Proceedings of the 27th International ACM Conference on International Conference on Supercomputing*, ICS '13 (ACM, New York, 2013), pp. 91–102. http://doi.acm.org/10.1145/2464996.2465003
12. H. Choe, S. Lee, S. Park, S.J. Kim, E.-Y. Chung, S. Yoon, Near-data processing for machine learning. CoRR, abs/1610.02273 (2016). http://arxiv.org/abs/1610.02273
13. A. De, M. Gokhale, R. Gupta, S. Swanson, Minerva: accelerating data analysis in next-generation SSDS, in *Proceedings of the 2013 IEEE 21st Annual International Symposium on Field-Programmable Custom Computing Machines*, FCCM '13 (IEEE Computer Society, Washington, 2013), pp. 9–16. https://doi.org/10.1109/FCCM.2013.46
14. J. Do, Y.-S. Kee, J.M. Patel, C. Park, K. Park, D.J. DeWitt, Query processing on smart SSDS: opportunities and challenges, in *Proceedings of the 2013 ACM SIGMOD International Conference on Management of Data*, SIGMOD '13 (ACM, New York, 2013), pp. 1221–1230. http://doi.acm.org/10.1145/2463676.2465295
15. J.H. Friedman, F. Baskett, L.J. Shustek, An algorithm for finding nearest neighbors. IEEE Trans. Comput. **C-24**(10), 1000–1006 (1975)
16. C. Fu, D. Cai, EFANNA: an extremely fast approximate nearest neighbor search algorithm based on KNN graph. CoRR, abs/1609.07228 (2016). http://arxiv.org/abs/1609.07228
17. C. Fu, D. Cai, Efanna: An extremely fast approximate nearest neighbor search algorithm based on KNN graph. arXiv preprint arXiv:1609.07228 (2016)
18. C. Fu, C. Wang, D. Cai, Fast approximate nearest neighbor search with navigating spreading-out graphs. *CoRR*, abs/1707.00143 (2017). http://arxiv.org/abs/1707.00143
19. Y. Gong, S. Lazebnik, A. Gordo, F. Perronnin, Iterative quantization: a procrustean approach to learning binary codes for large-scale image retrieval. IEEE Trans. Pattern Anal. Mach. Intell. **35**(12), 2916–2929 (2013). https://doi.org/10.1109/TPAMI.2012.193
20. A. Guttman, R-trees: a dynamic index structure for spatial searching, in *Proceedings of the 1984 ACM SIGMOD International Conference on Management of Data*, SIGMOD '84 (Association for Computing Machinery, New York, 1984), pp. 47–57

21. J. Ha, crow: Crow is very fast and easy to use C++ micro web framework) (2018). https://github.com/ipkn/crow
22. B. Harwood, T. Drummond, Fanng: fast approximate nearest neighbour graphs, in *Proceedings of the IEEE Conference on Computer Vision and Pattern Recognition* (2016), pp. 5713–5722
23. L. He, C. Liu, Y. Wang, S. Liang, H. Li, X. Li, Gcim: A near-data processing accelerator for graph construction, in *2021 58th ACM/IEEE Design Automation Conference (DAC)* (2021), pp. 205–210
24. J.-W. Hsieh, L.-P. Chang, T.-W. Kuo, Efficient on-line identification of hot data for flash-memory management, in *Proceedings of the 2005 ACM symposium on Applied computing* (2005), pp. 838–842
25. A.R. Hurson, L.L. Miller, S.H. Pakzad, M.H. Eich, B. Shirazi, Parallel architectures for database systems, in *Advances in Computers*, vol. 28 (Elsevier, Amsterdam, 1989), pp. 107–151. http://www.sciencedirect.com/science/article/pii/S0065245808600479
26. Y. Jia, E. Shelhamer, J. Donahue, S. Karayev, J. Long, R. Girshick, S. Guadarrama, T. Darrell, Caffe: convolutional architecture for fast feature embedding, in *Proceedings of the 22Nd ACM International Conference on Multimedia*, MM '14 (ACM, New York, 2014), pp. 675–678. http://doi.acm.org/10.1145/2647868.2654889
27. S.-W. Jun, M. Liu, S. Lee, J. Hicks, J. Ankcorn, M. King, S. Xu, Arvind, Bluedbm: an appliance for big data analytics, in *Proceedings of the 42Nd Annual International Symposium on Computer Architecture*, ISCA '15 (ACM, New York, 2015), pp. 1–13. http://doi.acm.org/10.1145/2749469.2750412
28. S.-W. Jun, A. Wright, S. Zhang, S. Xu, Arvind, Grafboost: using accelerated flash storage for external graph analytics, in *Proceedings of the 45th Annual International Symposium on Computer Architecture*, ISCA '18 (IEEE Press, Piscataway, 2018), pp. 411–424. https://doi.org/10.1109/ISCA.2018.00042
29. Y. Kang, Y.-S. Kee, E.L. Miller, C. Park, Enabling cost-effective data processing with smart SSD, in *2013 IEEE 29th Symposium on Mass Storage Systems and Technologies (MSST)* (2013), pp. 1–12. ftp://ftp.cse.ucsc.edu/pub/darrell/kang-msst13.pdf
30. N. Katayama, S. Satoh, The sr-tree: An index structure for high-dimensional nearest neighbor queries, in *Proceedings of the 1997 ACM SIGMOD International Conference on Management of Data*, SIGMOD '97 (Association for Computing Machinery, New York, 1997), pp. 369–380
31. A. Krizhevsky, I. Sutskever, G.E. Hinton, Imagenet classification with deep convolutional neural networks. Commun. ACM **60**(6), 84–90 (2017). http://doi.acm.org/10.1145/3065386
32. J. Kwak, S. Lee, K. Park, J. Jeong, Y.H. Song, Cosmos+ openssd: Rapid prototype for flash storage systems. ACM Trans. Storage **16**(3), 1–35 (2020)
33. M. Kwon, J. Zhang, G. Park, W. Choi, D. Donofrio, J. Shalf, M. Kandemir, M. Jung, Tracetracker: hardware/software co-evaluation for large-scale I/O workload reconstruction, in *2017 IEEE International Symposium on Workload Characterization (IISWC)* (IEEE, New York, 2017), pp. 87–96
34. G. Lee, S. Shin, W. Song, T.J. Ham, J.W. Lee, J. Jeong, Asynchronous I/O stack: a low-latency kernel I/O stack for Ultra-Low latency SSDs, in *2019 USENIX Annual Technical Conference (USENIX ATC 19)* (USENIX Association, Renton, 2019), pp. 603–616
35. H.-O. Leilich, G. Stiege, H.C. Zeidler, A search processor for data base management systems, in *Proceedings of the Fourth International Conference on Very Large Data Bases*, vol. 4, VLDB '78 (VLDB Endowment, New York, 1978), pp. 280–287. http://dl.acm.org/citation.cfm?id=1286643.1286682.
36. J.J. Levandoski, P. Larson, R. Stoica, Identifying hot and cold data in main-memory databases, in *2013 IEEE 29th International Conference on Data Engineering (ICDE)* (IEEE, New York, 2013), pp. 26–37
37. W. Li, Y. Zhang, Y. Sun, W. Wang, W. Zhang, X. Lin, Approximate nearest neighbor search on high dimensional data—experiments, analyses, and improvement (v1.0). *CoRR*, abs/1610.02455 (2016). http://arxiv.org/abs/1610.02455
38. W.-J. Li, S. Wang, W.-C. Kang, Feature learning based deep supervised hashing with pairwise labels, in *Proceedings of the Twenty-Fifth International Joint Conference on Artificial*

Intelligence, IJCAI'16 (AAAI Press, New York, 2016), pp. 1711–1717. http://dl.acm.org/citation.cfm?id=3060832.3060860
39. S. Liang, Y. Wang, H. Li, X. Li, Cognitive SSD+: a deep learning engine for energy-efficient unstructured data retrieval. CCF Trans. High Perform. Comput. **4**(3), 302–320 (2022)
40. S. Liang, Y. Wang, C. Liu, H. Li, X. Li, Ins-dla: An in-ssd deep learning accelerator for near-data processing, in *2019 29th International Conference on Field Programmable Logic and Applications (FPL)* (2019), pp. 173–179
41. S. Liang, Y. Wang, Z. Yuan, C. Liu, H. Li, X. Li, Vstore: In-storage graph based vector search accelerator, in *Proceedings of the 59th ACM/IEEE Design Automation Conference*, DAC '22 (Association for Computing Machinery, New York, 2022), pp. 997–1002
42. C.S. Lin, D.C.P. Smith, J.M. Smith, The design of a rotating associative memory for relational database applications. ACM Trans. Database Syst. **1**(1), 53–65 (1976). http://doi.acm.org/10.1145/320434.320447
43. K. Lin, H. Yang, J. Hsiao, C. Chen, Deep learning of binary hash codes for fast image retrieval, in *2015 IEEE Conference on Computer Vision and Pattern Recognition Workshops (CVPRW)* (2015), pp. 27–35. http://ieeexplore.ieee.org/document/7301269/
44. V.E. Liong, J. Lu, G. Wang, P. Moulin, J. Zhou, Deep hashing for compact binary codes learning, in *2015 IEEE Conference on Computer Vision and Pattern Recognition (CVPR)* (2015), pp. 2475–2483. http://ieeexplore.ieee.org/document/7298862/
45. G. Liu, J. Xu, C. Wang, J. Zhang, A performance comparison of http servers in a 10g/40g network, in *Proceedings of the 2018 International Conference on Big Data and Computing*, ICBDC '18 (Association for Computing Machinery, New York, 2018), pp. 115–118
46. H. Liu, R. Wang, S. Shan, X. Chen, Deep supervised hashing for fast image retrieval, in *2016 IEEE Conference on Computer Vision and Pattern Recognition (CVPR)* (2016), pp. 2064–2072. http://ieeexplore.ieee.org/document/7780596/
47. V.S. Mailthody, Z. Qureshi, W. Liang, Z. Feng, S.G. de Gonzalo, Y. Li, H. Franke, J. Xiong, J. Huang, W.-m. Hwu, Deepstore: In-storage acceleration for intelligent queries, in *Proceedings of the 52nd Annual IEEE/ACM International Symposium on Microarchitecture*, MICRO '52 (Association for Computing Machinery, New York, 2019), pp. 224–238
48. Y.A. Malkov, D.A. Yashunin, Efficient and robust approximate nearest neighbor search using hierarchical navigable small world graphs. IEEE Trans. Pattern Anal. Mach. Intell. **42**(4), 824–836 (2018)
49. T. Miura, K. Kourai, S. Yamaguchi, Cache replacement based on lstm in the second cache in virtualized environment, in *2020 Eighth International Symposium on Computing and Networking Workshops (CANDARW)* (2020), pp. 421–424
50. M. Muja, D.G. Lowe, Fast approximate nearest neighbors with automatic algorithm configuration. VISAPP (1) **2**(331–340), 2 (2009)
51. A. Oliva, A. Torralba, Modeling the shape of the scene: A holistic representation of the spatial envelope. Int. J. Comput. Vision **42**(3), 145–175 (2001). https://doi.org/10.1023/A:1011139631724
52. J. Ouyang, S. Lin, Z. Hou, P. Wang, Y. Wang, G. Sun, Active ssd design for energy-efficiency improvement of web-scale data analysis, in *Proceedings of the 2013 International Symposium on Low Power Electronics and Design*, ISLPED '13 (IEEE Press, Piscataway, 2013), pp. 286–291. http://dl.acm.org/citation.cfm?id=2648668.2648739
53. E. Riedel, C. Faloutsos, G.A. Gibson, D. Nagle, Active disks for large-scale data processing. Computer **34**(6), 68–74 (2001). https://doi.org/10.1109/2.928624
54. E. Riedel, G.A. Gibson, C. Faloutsos, Active storage for large-scale data mining and multimedia, in *Proceedings of the 24rd International Conference on Very Large Data Bases*, VLDB '98 (Morgan Kaufmann Publishers Inc., San Francisco, 1998), pp. 62–73. http://dl.acm.org/citation.cfm?id=645924.671345
55. L.V. Rodriguez, F. Yusuf, S. Lyons, E. Paz, R. Rangaswami, J. Liu, M. Zhao, G. Narasimhan, Learning cache replacement with CACHEUS, in *19th USENIX Conference on File and Storage Technologies (FAST 21)* (USENIX Association, Renton, 2021), pp. 341–354

56. S.A. Schuster, H.B. Nguyen, E.A. Ozkarahan, K.C. Smith, Rap.2 an associative processor for databases and its applications. IEEE Trans. Comput. **28**(6), 446–458 (1979). https://doi.org/10.1109/TC.1979.1675383
57. S. Seshadri, M. Gahagan, S. Bhaskaran, T. Bunker, A. De, Y. Jin, Y. Liu, S. Swanson, Willow: a user-programmable SSD, in *Proceedings of the 11th USENIX Conference on Operating Systems Design and Implementation*, OSDI'14 (USENIX Association, Berkeley, 2014), pp. 67–80. http://dl.acm.org/citation.cfm?id=2685048.2685055
58. Z. Shi, X. Huang, A. Jain, C. Lin, Applying deep learning to the cache replacement problem, in *Proceedings of the 52nd Annual IEEE/ACM International Symposium on Microarchitecture*, MICRO '52 (Association for Computing Machinery, New York, 2019), pp. 413–425
59. K. Simonyan, A. Zisserman, Very deep convolutional networks for large-scale image recognition. *CoRR*, abs/1409.1556 (2014). http://arxiv.org/abs/1409.1556
60. Y. Son, N.Y. Song, H. Han, H. Eom, H.Y. Yeom, A user-level file system for fast storage devices, in *Proceedings of the 2014 International Conference on Cloud and Autonomic Computing*, ICCAC '14 (IEEE Computer Society, Washington, 2014), pp. 258–264. https://doi.org/10.1109/ICCAC.2014.14
61. D. Tiwari, S. Boboila, S.S. Vazhkudai, Y. Kim, X. Ma, P.J. Desnoyers, Y. Solihin, Active flash: towards energy-efficient, in-situ data analytics on extreme-scale machines, in *Proceedings of the 11th USENIX Conference on File and Storage Technologies*, FAST'13 (USENIX Association, Berkeley, 2013), pp. 119–132. http://dl.acm.org/citation.cfm?id=2591272.2591286
62. D. Tiwari, S.S. Vazhkudai, Y. Kim, X. Ma, S. Boboila, P.J. Desnoyers, Reducing data movement costs using energy efficient, active computation on SSD, in *Proceedings of the 2012 USENIX Conference on Power-Aware Computing and Systems*, HotPower'12 (USENIX Association, Berkeley, 2012), pp. 4–4. http://dl.acm.org/citation.cfm?id=2387869.2387873
63. S. Tripathy, D. Sahoo, M. Satpathy, M. Mutyam, Fuzzy fairness controller for NVME SSDS, in *Proceedings of the 34th ACM International Conference on Supercomputing*, ICS '20 (Association for Computing Machinery, New York, 2020)
64. S. Tripathy, D. Sahoo, M. Satpathy, S. Pinisetty, Formal modeling and verification of nand flash memory supporting advanced operations, in *2019 IEEE 37th International Conference on Computer Design (ICCD)* (2019), pp. 313–316
65. H.-W. Tseng, Q. Zhao, Y. Zhou, M. Gahagan, S. Swanson, Morpheus: creating application objects efficiently for heterogeneous computing. SIGARCH Comput. Archit. News **44**(3), 53–65 (2016). http://doi.acm.org/10.1145/3007787.3001143
66. G. Vietri, L.V. Rodriguez, W.A. Martinez, S. Lyons, J. Liu, R. Rangaswami, M. Zhao, G. Narasimhan, Driving cache replacement with ML-based LeCaR, in *10th USENIX Workshop on Hot Topics in Storage and File Systems (HotStorage 18)* (USENIX Association, Boston, 2018)
67. J. Wang, D. Park, Y.-S. Kee, Y. Papakonstantinou, S. Swanson, SSD in-storage computing for list intersection, in *Proceedings of the 12th International Workshop on Data Management on New Hardware*, DaMoN '16 (2016), pp. 4:1–4:7. http://doi.acm.org/10.1145/2933349.2933353
68. J. Wang, H.T. Shen, J. Song, J. Ji, Hashing for Similarity Search: A Survey. *arXiv:1408.2927 [cs]* (2014). http://arxiv.org/abs/1408.2927
69. M. Wang, X. Xu, Q. Yue, Y. Wang, A comprehensive survey and experimental comparison of graph-based approximate nearest neighbor search. Proc. VLDB Endow. **14**(11), 1964–1978 (2021)
70. C. Wei, B. Wu, S. Wang, R. Lou, C. Zhan, F. Li, Y. Cai, Analyticdb-v: A hybrid analytical engine towards query fusion for structured and unstructured data. Proc. VLDB Endow. **13**(12), 3152–3165 (2020)
71. L. Woods, Z. István, G. Alonso, Ibex: An intelligent storage engine with support for advanced sql offloading. Proc. VLDB Endow. **7**(11), 963–974 (2014). https://doi.org/10.14778/2732967.2732972
72. H.-F. Yang, K. Lin, C.-S. Chen, Supervised learning of semantics-preserving hash via deep convolutional neural networks. IEEE Trans. Pattern Anal. Mach. Intell. **40**(2), 437–451 (2018). https://doi.org/10.1109/TPAMI.2017.2666812

73. J. Zhang, M. Kwon, D. Gouk, S. Koh, C. Lee, M. Alian, M. Chun, M.T. Kandemir, N.S. Kim, J. Kim, M. Jung, Flashshare: Punching through server storage stack from kernel to firmware for ultra-low latency SSDS, in *Proceedings of the 12th USENIX Conference on Operating Systems Design and Implementation*, OSDI'18 (USENIX Association, Berkeley, 2018), pp. 477–492. http://dl.acm.org/citation.cfm?id=3291168.3291203
74. F. Zhao, Y. Huang, L. Wang, T. Tan, Deep semantic ranking based hashing for multi-label image retrieval. *CoRR*, abs/1501.06272 (2015). http://arxiv.org/abs/1501.06272
75. L. Zheng, Y. Yang, Q. Tian, SIFT meets CNN: A decade survey of instance retrieval. *CoRR*, abs/1608.01807 (2016). http://arxiv.org/abs/1608.01807

Error-Tolerant Techniques for Classifiers Beyond Neural Networks for Dependable Machine Learning

Shanshan Liu, Pedro Reviriego, Xiaochen Tang, and Fabrizio Lombardi

1 Introduction

In the last decade, Artificial Intelligence (AI) has made major progress in many different applications; it now outperforms humans in many tasks and the number of such tasks increases continuously [1]. Therefore, AI is becoming pervasive and is now used, for example, in applications that range from detecting obstacles and pedestrians in vehicles [2], to manage financial risks [3], or to improve healthcare systems [4]. This AI revolution is now taking a new step and reaching end users. Applications to generate images from text such as DALL-E[1] or MidJourney[2] or to generate text from questions such as ChatGPT[3] have now millions of users and are generating billions of images and text. In the next few years, many new uses and applications of AI will appear making it an integral part of our lives and future society, much like what happened with the Internet in the last decades. This will make the dependability of AI a fundamental issue [5].

[1] https://openai.com/dall-e-2/

[2] https://www.midjourney.com/

[3] https://openai.com/blog/chatgpt/

S. Liu · X. Tang
University of Electronic Science and Technology of China, Chengdu, Sichuan, China
e-mail: ssliu@uestc.edu.cn; xctang@uestc.edu.cn

P. Reviriego (✉)
Universidad Politécnica de Madrid, Madrid, Spain
e-mail: pedro.reviriego@upm.es

F. Lombardi
Northeastern University, Boston, MA, USA
e-mail: lombardi@ece.neu.edu

W. Liu et al. (eds.), *Design and Applications of Emerging Computer Systems*,
https://doi.org/10.1007/978-3-031-42478-6_7

Dependability is a key requirement when a technology is used for critical applications as it now the case for computing [6]. As we depend on that technology, we must ensure that it will not suffer failures or malfunctions that would compromise the operation of a computing system. This is done by first assessing the possible failures, their probabilities, and their effects on the operational conditions of the computing system. When a design is found to have an unacceptable probability of malfunctioning, schemes to detect, correct, or mitigate the effects of errors have to be utilized [7]. This typically implies a cost in terms of circuit area, power dissipation, and speed and thus, a trade-off must be made to achieve the desired level of reliability at a low overhead. There are many techniques to detect errors, for example, error correction codes are widely used to detect and correct errors in memories, while triplication and voting can be used to correct errors on functional units [8]. Error correction codes require adding additional memory bits and encoding/decoding circuitry; triplication has a huge cost, and thus it is only affordable in some systems or elements within a system. Another approach is to exploit the properties of the algorithms to detect or correct errors at a lower overhead [9, 10]. This requires a deep understanding of the effects of faults and errors and the design of ad hoc techniques for each algorithm, but it can reduce the overhead needed to achieve reliable operation and has been successfully used in many signal processing algorithms [11].

Machine learning (ML) is a topical area of AI that develops techniques for computers to learn from data and patterns, and most current AI systems are in fact ML systems [12]. Typically, data is fed to an algorithm based on what is known in the training phase to obtain a model that can then be used to perform the required task. For example, a ML system can be used to classify images in two classes those containing pedestrians and those that do not have them. First, the system is trained with a large set of images of both classes, so it learns to classify. After that phase, a new image can be used as input, and the ML system will output the class of the image. There are many ML algorithms that can be used in such a classifier. From a simple K Nearest Neighbors (KNNs) that just looks for the K closest elements to the new image in the training dataset and outputs the class of the majority of them to a very complex neural network with millions of parameters.

Dependability in ML has been widely studied mostly focusing on neural networks [13–16]. One of the main conclusions in many of the existing works is that ML algorithms tend to exhibit some degree of error tolerance in the sense that many errors do not affect its performance; however, there is typically a fraction of errors that do have a significant impact on the system. Therefore, protection is needed, and it can in many cases be implemented at a lower overhead as only parts of the system must be protected. The properties of neural networks can also be used to reduce the overhead of detecting and correcting errors [14, 17, 18].

In contrast to neural networks, the dependability of other ML algorithms has received significantly less attention from the research community. This is probably due to the popularity that different types of complex neural networks have gained in the recent years, for example, in image-related applications [19]. However, simpler classifiers are still used in many cases, for example, in resource-constraint platforms such as those commonly found in Internet of Things (IoT) applications. In some problems, they achieve an accuracy that is similar to more complex neural networks and can be implemented at a fraction of their cost. Examples of such classifiers are KNNs, Random Forest, or Support Vector Machine (SVM) [20–29].

In this chapter, we review recent studies on error-tolerant techniques for these classifiers showing that they can also be protected at low overhead by using algorithmic-based error tolerance; as with neural networks, these classifiers also have some intrinsic resilience to errors that can be augmented to provide effective protection at low complexity. In the rest of this chapter, error-tolerant KNNs, Random Forest, and SVM are discussed in detail to then present a comparison and summary. The overall goal is to provide a set of tools that readers can use to harden these classifiers when they are used in safety-critical applications.

2 *K* Nearest Neighbors

The *K* Nearest Neighbors (KNNs) is one of the simplest yet powerful ML classification algorithms. Such an algorithm is also referred to as instance-based or lazy learning, because only several samples in the training dataset are selected for predicting the class of a new sample. It does not have a complex prediction model, but only a hyperparameter *K*, which is the number of nearest neighbors around the new sample in the feature space. Depending on the size of the dataset stored and the application constraints, KNNs can be implemented on different platforms, including large computing systems that use MapReduce [30], or specialized processors such as GPUs [31], FPGAs [32], microcontrollers, and ASICs [33]. In this section, the KNNs for both binary and multiclass classifications are introduced and the impact of errors in KNN implementation is discussed. Then, two efficient error-tolerant KNN schemes are presented.

2.1 Errors in KNNs

The KNN classification algorithm is illustrated in Fig. 1 by taking $K = 5$ as an example. They perform classification for a new sample (i.e., the gray sample in Fig. 1), and the distance (typically Euclidean distance) between the new sample and each training sample is first calculated by considering all valid features. Based on the distance results, the *K* nearest neighbors of the new sample are then determined. This can be achieved by exhaustively comparing all samples when the size of the training dataset is not large; else, more sophisticated schemes that use data structures to search the neighbors, are utilized [34]. Finally, the majority voting result among the classes of the *K* nearest neighbors is taken as the predicted result for the new sample [28]. Therefore, in the examples shown in Fig. 1a, b, the classification result for the new sample is B and C, respectively.

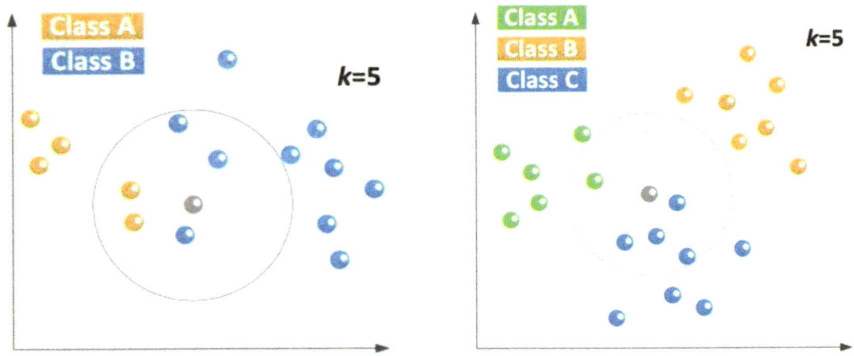

(a) (b)

Fig. 1 Example of the KNNs for (**a**) binary classification; (**b**) multiclass classification

A special case during majority voting occurs when there are equal votes with different classes, leading to a tie. To break a tie in the binary classification scheme, an odd number must be considered when determining the value of K. For the multiclass classification, the class of the nearest or a random neighbor in the tie is generally selected as the final result.

The hardware platforms that are used to implement ML classifiers are prone to suffer a number of errors (e.g., radiation-induced soft errors) [35–37]. They are transient in many cases, but they can corrupt data during computation and finally change the classification results. For example, in the KNN implementation, an error occurring in the arithmetic circuits can result in an incorrect distance and change the set of the K nearest neighbors; otherwise, it can modify the majority voting result. In both cases, the final classification result is changed.

A typical technique to detect computational errors is based on temporal redundancy [8, 38]. In particular, the classification process can be executed again for a comparison with the previous outcome. An error can be detected if the two outcomes are different due to its transient nature. If error correction is required, the classification process is executed for the third time, and a majority voting among the three outcomes is taken as the final result. Despite error tolerance, such a technique incurs in approximately $2\times$ or $3\times$ computational overhead.

An interesting observation is that the KNN classifiers often have some intrinsic redundancy that can be explored for error-tolerant design. This feature is further discussed in the following subsections when presenting error-tolerant KNN schemes, namely, voting-margin-based KNNs and K+1 NNs, respectively. Prior to such discussion, the impact of errors on KNN classification is discussed next. Since distance computation is the main part of the entire KNN classification process, only errors affecting a distance are considered. Furthermore, as soft errors are rare, it is assumed that during the classification of an element, at most a single soft error

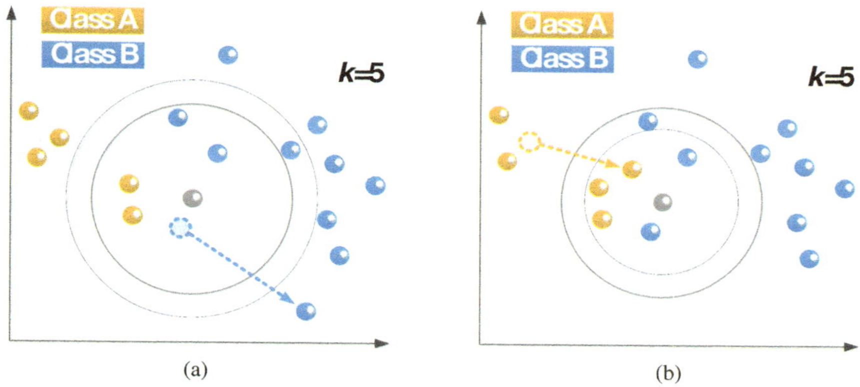

Fig. 2 Examples of a distance computation error: (**a**) an error of type 1, under which the classification result is still Class B; (**b**) an error of type 2, under which the classification result is changed from Class B to Class A

occurs. To handle errors in majority voting, the traditional temporal redundancy method can be employed, because the overhead is low for this computation.

As discussed previously, an error can modify the computed distance between a training sample and the new sample being classified. Therefore, such an error may have a negative impact only when it changes the set of the K nearest neighbors, in the following two scenarios:

- *Type 1*: An error increases the distance of a sample that should be in the set of K nearest neighbors. Therefore, it is pushed out of the set and the $K+1$th nearest neighbor is used in its place (as shown in Fig. 2a).
- *Type 2*: A sample that should not be in the set of K nearest neighbors has a smaller distance as result of an error that puts it into the set. The original K^{th} nearest neighbor is therefore no longer in the set (as shown in Fig. 2b).

Consider the examples shown in Fig. 2; an error does not always change the classification result, even if it has modified the set of K nearest neighbors. This feature will be explored in the error-tolerant KNN schemes.

2.2 *Voting-Margin-Based Error-Tolerant KNNs for Binary Classification*

The voting-margin-based error-tolerant KNN scheme was initially proposed for a binary classification task [39]. The so-called voting margin refers to the case when the difference between the number of neighbors with the majority class and the minority class is greater than one (e.g., 4 vs. 1 when $K = 5$). Since K is always an odd number for binary classification, the voting margin actually has a value of at least 3 for such KNN. In the other cases, we consider that the voting has no margin (e.g., 3 vs. 2 when $K = 5$). Therefore, when a voting margin exists, a single error that changes a distance has no impact on the final classification result. This occurs, because the error can only change the voting between different classes, for example, 4 versus 1 to 3 versus 2, so generating the same predicted class. Hence, errors do not need to be handled when there is a voting margin.

Interestingly, even if the voting margin does not exist, the classification result can also be identified as reliable in many cases. Consider the example of voting between 3 versus 2 for different classes (as shown in Fig. 3).

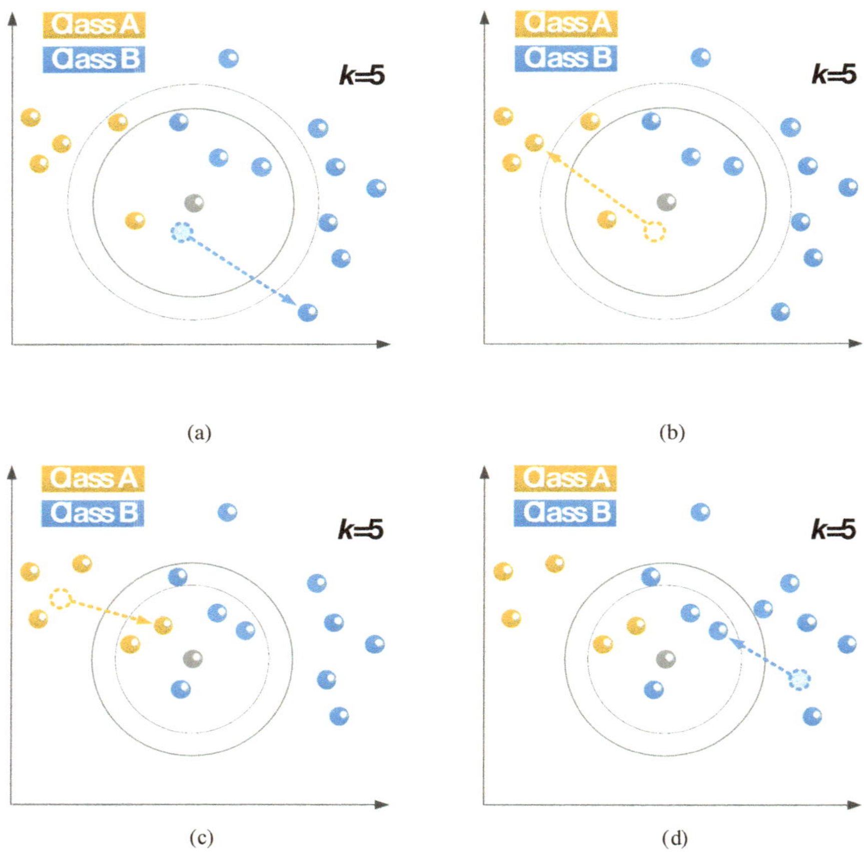

Fig. 3 Examples of reliable classification when there is no voting margin: (**a**) the Kth nearest neighbor belongs to the minority class, under a type 1 error that has affected a majority neighbor; (**b**) the Kth nearest neighbor belongs to the minority class, under a type 1 error that has affected a minority neighbor; (**c**) the $K+1$th nearest neighbor belongs to the majority class, under a type 2 error that has affected a minority neighbor; (**d**) the $K+1$th nearest neighbor belongs to the majority class, under a type 2 error that has affected a majority neighbor

- Suppose an error of type 1 has occurred and pushed a neighbor out of the set. The classification result is reliable when the Kth nearest neighbor of the current set belongs to the minority class. This is valid because no matter whether the original neighbor belongs to the majority class (Fig. 3a) or the minority class (Fig. 3b), the voting result is the same as in the error-free case (i.e., having no impact on the classification result).
- Suppose an error of type 2 has occurred and put another sample into the set. The classification result is reliable when the $K+1$th nearest neighbor of the current set belongs to the majority class, because it should be the original K^{th} neighbor that has been replaced. In this case, regardless of whether the incorrect element belongs to the minority class (Fig. 3c) or the majority class (Fig. 3d), the voting result does not change.

Based on the above discussion, the voting-margin error-tolerant KNN scheme is illustrated in Fig. 4. The first step computes the KNNs as in an unprotected scheme. Then, a voting margin is checked to see if the majority voting result is reliable. If a voting margin exists, the classification result is taken as the final outcome; else, the condition that ensures a reliable vote even if there is no voting margin, is checked. Finally, if the classification cannot be determined to be reliable, the KNN implementation process is computed again or for a third time like in the traditional temporal redundancy scheme for error detection or correction.

The main benefit of the voting-margin scheme is that when a vote can be determined to be reliable, there is no need to re-compute the KNNs. This is achieved in many cases because when the KNNs are designed for high classification accuracy, there is likely a voting margin. Several ML datasets from a public repository [40] are considered to evaluate the benefit of the voting-margin scheme. To inject single errors, a distance of a testing sample is selected randomly, and its value is changed to a random value within the range of all computed distances. This process is repeated 10,000 times for each testing sample of different datasets to obtain an average impact of errors. Table 1 reports the percentage of samples for which re-computation is needed in the voting-margin scheme. For a comprehensive analysis, the value of K in a range from 3 to 19 is considered. As per Table 2, the computational overhead (related to metrics such as latency and power dissipation) to achieve error tolerance can be approximately obtained. Table 2 clearly illustrates the benefits of the voting-margin scheme; it saves at least 60% of the additional computation introduced by the traditional temporal redundancy technique.

2.3 $K+1$ *Nearest Neighbors for Multiclass Classification*

The voting-margin error-tolerant technique is also applicable to KNNs for multiclass classification. To identify reliable classification results in this case, the voting margin must have a value of at least 3 when the two additional conditions (i.e., the $K+1$th nearest neighbor belongs to the majority class and the Kth nearest neighbor

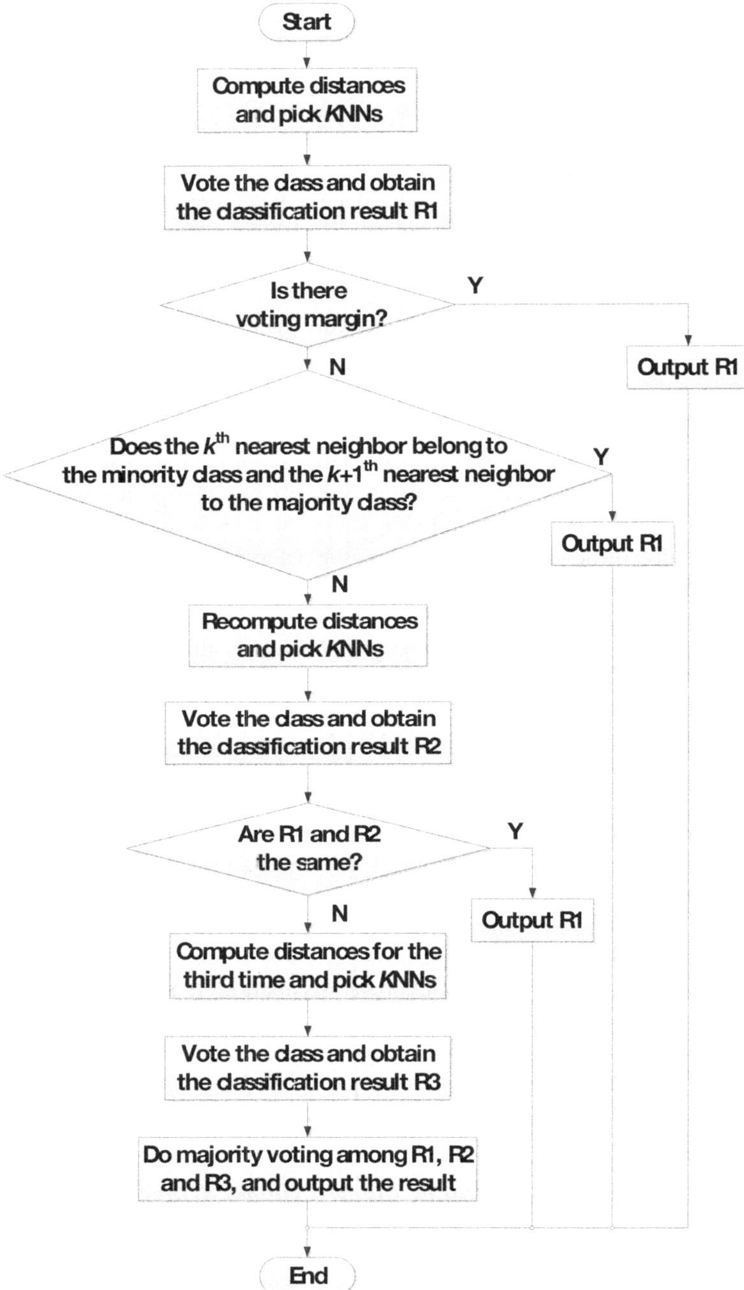

Fig. 4 The voting-margin error-tolerant KNN algorithm

Table 1 Percentage of samples for which re-computation is needed in the voting-margin scheme (results for the optimal K that offers the top classification accuracy are marked in bold)

K	EEG eye state	Sonar	Electrical grid stability simulated data	Banknote authentication	Qualitative bankruptcy	Phishing websites	Climate model simulation crashes
3	**35.11%**	**19.90%**	19.90%	0	**2.69%**	13.49%	17.80%
5	24.48%	12.84%	12.47%	0	2.23%	**9.68%**	**8.37%**
7	19.30%	14.08%	8.95%	**0**	4.33%	7.43%	4.36%
9	16.25%	14.79%	7.38%	0.10%	4.80%	5.21%	2.41%
11	14.16%	14.45%	5.81%	0	5.07%	5.28%	2.46%
13	12.35%	14.68%	5.14%	0	7.24%	4.76%	1.28%
15	11.49%	16.02%	4.94%	0	6.84%	3.63%	0.22%
17	11.31%	12.02%	4.39%	0	9.88%	3.53%	0.69%
19	10.30%	10.02%	**3.50%**	0	11.48%	3.74%	0.04%
Average	17.19%	14.31%	8.05%	0	6.06%	6.31%	4.18%

Table 2 Comparison of different KNN protection techniques for error detection

Scheme	Impact on classification accuracy	Computational overhead
Traditional temporal redundancy	None	2×
Voting-margin	None	<1.4×

belongs to the minority class) are not employed. This is required because a tie may exist in the error-free case when the voting margin is equal to 2 due to an error; if the strategy used to break a tie is to select the class of the nearest neighbor, then the classification result under the error cannot be identified to be the same as in the error-free case. However, this issue can be addressed by using the two additional conditions; the reasons are the same as discussed in Sect. 2.2. Nevertheless, the use of the voting-margin error-tolerant KNN scheme is not very efficient for multiclass classification due to the tight margin, especially for the datasets that do not have a very high classification accuracy (as evaluated in the next section).

An alternative solution for error-tolerant multiclass classification is the $K+1$ NNs scheme [41]. In this scheme, errors are detected by checking twice the $K+1$ nearest distances; it approximately guarantees that the classification result is unchanged by discarding the neighbor with a mismatched distance. Algorithm 1 describes the $K+1$ NNs scheme, including the following steps.

- *Step 1*: When the distance from the new sample being classified to each training sample is computed to select nearest neighbors (as in the standard KNN scheme), the $K+1$th nearest neighbor and its distance are also stored.
- *Step 2*: Compute the distances for the $K+1$ nearest neighbors again.
- *Step 3*: Compare each of the $K+1$ distance pairs. If they are the same (so there is no error affecting these neighbors), a majority voting is taken among the K nearest neighbors to obtain the final classification result. If there is a mismatch on the distance pairs, an error is detected; then, the distance in error is re-

Algorithm 1 $K+1$ NNs scheme

1: Compute distances
2: Select the $K+1$ nearest neighbors and keep their distances **D1**
3: Recompute distances for the $K+1$ nearest neighbors and obtain **D2**
4: **if D1 = D2**
5: Do majority voting among the classes of the K nearest neighbors
6: Output the classification result
7: **else**
8: Recompute distance for the mismatched neighbor
9: Do majority voting among the three computed distances for the mismatched neighbor
10: Obtain the correct distances and the corresponding K nearest neighbors
11: Do majority voting among the classes of the K nearest neighbors
12: Output the classification result
13: **end if**

computed for a third time to obtain the correct value by voting. Once completed, the updated K nearest neighbors are used for classification.

The third time of re-computation for the mismatched distance in step 3 is used to avoid the case in which an error that exchanges two samples with the same classes, but with no impact on the set, is discarded (so the $K+1$th neighbor may change the classification result).

This $K+1$ scheme fully tolerates the errors of type 2 (that decrease a distance), because the sample brought by the error will be discarded, and so keeping the same set of K nearest neighbors. For type 1 errors, the scheme may not be effective in some cases, because an original neighbor of the set in the error-free case may be replaced by the $K+1$th nearest neighbor; in this case, the original set cannot be recovered even if the mismatched distance is re-computed. However, since a type 2 error has a significantly larger probability to occur (i.e., $\frac{N-K}{N}$, where N is the number of training samples and $N \gg K$), the $K+1$ NNs scheme should detect an error in most cases. This is verified by checking several datasets with multiple classes [40]. As shown in Table 3, the percentage of errors that modify the classification result in the $K+1$ NNs scheme is extremely low (i.e., below 0.1%).

3 Ensemble Classifier

The ensemble classifier is another type of voting-based classification technique; it consists of multiple sub-classifiers that act as single voters to make predictions based on different feature importance. Then, the final classification result is obtained by establishing a majority voting among the results of these voters. Compared to single voters (such as KNNs), ensemble classifiers tend to achieve a better classification performance because they perform an additional round of voting. Interestingly,

Table 3 Percentage of errors that modify the classification result in the $K+1$ NNs scheme (results for the optimal K that offers the top classification accuracy are marked in bold)

K	Iris	Forest type mapping [42]	Statlog (Vehicle silhouettes) [43]	Mice protein expression [44]	CNAE-9	Wine quality [45]	Thyroid disease	Nursery
3	0	0.05%	0.06%	**0**	0.07%	0.02%	0	0
5	**0**	0.05%	0.04%	0	0.04%	0.01%	**0.01%**	0
7	0	0.06%	0.06%	0.01%	**0.06%**	0.01%	0	0
9	0	**0.09%**	0.04%	0.01%	0.04%	0.02%	0	0
11	0.02%	0.05%	**0.04%**	0.01%	0.04%	0.02%	0	0
13	0.02%	0.02%	0.01%	0.01%	0.03%	0.02%	0	0
15	0	0.05%	0.06%	0	0.02%	0.02%	0	0
17	0	0.02%	0.06%	0.01%	0.03%	0.02%	0	**0**
19	0	0.02%	0.06%	0.01%	0.04%	**0.02%**	0	0
Average	0	0.05%	0.05%	0.01%	0.04%	0.02%	0	0

the voting-margin-based error-tolerant technique is also applicable to ensemble classifiers due to the nature of their classification procedure. In the following, a widely used ensemble classifier, namely Random Forest, is taken as an example to illustrate the principles of such classifiers. The extension of the voting-margin technique for such classifiers is then discussed.

3.1 *Random Forest*

A Random Forest consists of several decision trees as voters [26]. Once the classification model is trained, labels of training samples are stored on the leaf nodes of the decision trees, and the decision thresholds on the features are stored in the root and the internal (non-leaf) nodes. The decision tree has a flow-chart-like structure, in which each internal node denotes a test on a feature, each branch represents the outcome of a test, and each leaf node holds a class label.

Figure 5 shows an example of the Random Forest with T decision trees for multiclass classification. To predict a class for an incoming sample, each tree makes an initial prediction (e.g., it is predicted by Tree 1 as Class A in Fig. 5). The final result is then obtained by taking the majority voting result among the results of the trees. By establishing the trees based on different importance, a Random Forest comprehensively considers the impact of the features and commonly achieves a higher classification accuracy compared to a single voter-based classifier (e.g., KNNs). This can be found, for example, in Table 4, which reports the top accuracy achieved by using a Random Forest and a KNN classifier; however, a Random Forest usually incurs in a larger execution due to its complexity as each voter (tree) requires significant computation.

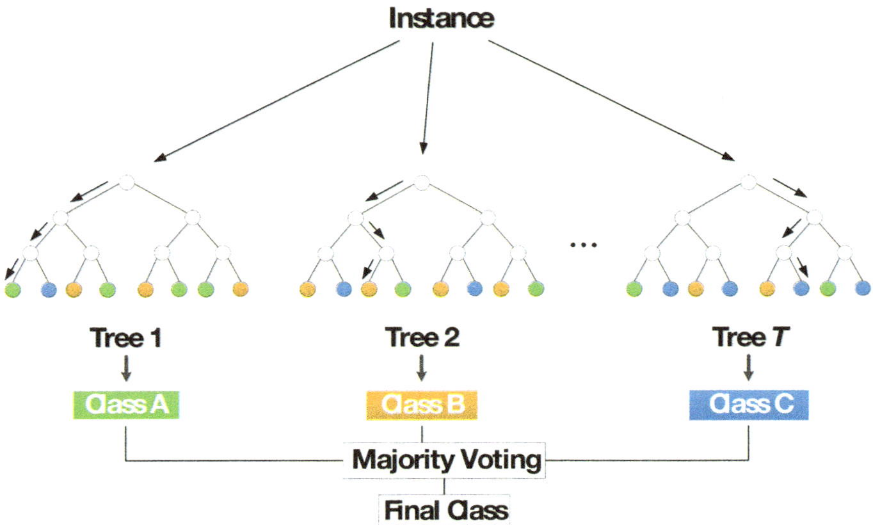

Fig. 5 An example of the Random Forest algorithm that consists of *T* decision trees

3.2 Voting-Margin-Based Error-Tolerant Random Forest

In a Random Forest classifier, single errors on any leaf or root of a decision tree may change the comparison path, leading to a different predicted class. This may affect voting among the trees and finally modify the classification result. However, like the KNN classifier, when the voting has a margin, single errors will not change the classification result; such feature of the Random Forest permits the application of the voting-margin-based error-tolerant technique, as discussed next.

In a KNN classifier, the nearest neighbors considered for a majority voting can be changed due to errors. Differently, the decision trees of a Random Forest classifier keep the same in the presence of errors, albeit their predicted class can be modified. Therefore, when employing the voting-margin scheme in a Random Forest, the error detection process is simpler than for the KNNs [41].

- *Binary classification*: If the voting margin between the number of trees agreeing on a majority class and a minority class is equal to 2 or a larger value, the final classification result is always reliable even if an error has occurred. A special case is that for a margin of 2, a tie may exist in the error-free case. However, since a random class is usually selected to break a tie in a Random Forest, the result with a voting margin of 2 under an error can still be considered as correct. This case can be avoided by utilizing an odd number of trees (i.e., the voting margin is always an odd number); when the voting margin has a value of 1, re-computation is only required for the trees with a majority class to check if

Table 4 Comparison of classification accuracy achieved by using different classifiers

Dataset	Appl.	# Samples	# Features	# Classes	KNNs		Random Forest	
					K	Top accuracy	T	Top accuracy
Iris	Botany	150	4	3	5	97.78%	80	97.78%
Forest typemapping	Ecology	325	27	4	9	77.32%	140	85.57%
Statlog (vehiclesilhouettes)	Vehicle	846	18	4	11	77.17%	90	79.13%
Mice proteinexpression	Biology	1080	80	8	3	99.38%	60	100.00%
CNAE-9	Finance	1080	856	9	7	84.57%	100	92.18%
Wine quality	Life	4898	11	11	19	53.17%	120	66.67%
Thyroiddisease	Medicine	7200	21	3	5	94.34%	110	99.72%
Nursery	Sociology	12,960	8	5	17	96.40%	150	99.38%

Table 5 Percentage of samples for which re-computation is required when employing the voting-margin-based Random Forest for different classifiers

Dataset	KNNs	Random Forest
Iris	11.62%	0.10%
Student academic performance	69.77%	5.74%
Forest type mapping	22.91%	0.45%
Statlog (vehicle silhouettes)	36.66%	4.20%
Mice protein expression	23.13%	0.02%
CNAE-9	28.95%	2.47%
Wine quality	25.07%	3.66%
Thyroid disease	3.74%	0.02%
Nursery	5.46%	0.09%

these trees generate the same class. Then, the error-free majority class can be identified, so a correct classification is obtained.

- *Multiclass classification*: Similar to the above discussion, the classification result should be reliable when the voting margin is equal to 3 or a larger value. Moreover, when the Random Forest has a voting margin of 2, error detection is also not needed, because the majority class should either be the same as in the error-free case, or have a tie with another class (i.e., generating the correct final result). When the voting margin has a value of 1, re-computation is required but again, only for the trees agreeing on the majority class. Then, as per the re-computation results and the number of trees with different classes, the correct classification result can be determined.

Overall, when the voting-margin conditions used to determine a reliable result are not met, re-computation is only required for the partial trees in the Random Forest scheme while it is required for all distances in the KNN scheme. As shown in Table 5, the percentage of samples for which re-computation is needed is significantly smaller in the voting-margin-based Random Forest scheme. This offers another advantage of using a Random Forest classifier compared to a KNN classifier, because it requires a lower computational process to achieve error tolerance.

4 Support Vector Machines

Different from KNNs or Random Forests, a Support Vector Machine (SVM) classification algorithm establishes a decision hyperplane between the training samples of different classes [23, 24]. By maximizing the margin of the hyperplane to the nearest training samples of each class, an SVM avoids the misclassification of data near the decision boundary. Therefore, an SVM tends to achieve a higher classification accuracy compared to KNNs and in some cases even over Random Forests. However, some errors in the calculation of the SVM can cause a sample to jump across the decision boundary, thereby changing the predicted classification result. Recently, an efficient result-based re-computation scheme has been proposed in [41] to achieve error-tolerant classification using an SVM. By investigating the impact of errors, this technique achieves a significant reduction of protection overhead compared to existing schemes by only handling the errors that may cause incorrect classifications. In this section, an SVM with different kernel functions is reviewed. Then, the impact of errors on SVM classification is analyzed and the result-based re-computation technique of [46] for error-tolerant SVM is discussed in detail.

4.1 SVM with Different Kernels

In ML applications, many types of data are not linearly separable in the feature space; in these cases, an SVM tends to provide a better classification performance compared to KNNs or even Random Forests. This is applicable because using different kernel functions, an SVM can map the original feature space to a higher dimensional space, in which the decision hyperplane is obtained to better separate the samples of the different classes.

The margin of the hyperplane is established by some training samples, that is, the so-called Support Vectors (SVs). Based on the parameters associated with each SV that are obtained during the training process, the classification result for a new data sample can be determined by calculating:

$$Y(z) = \operatorname{sign}\left(\sum_{i=1}^{N} w_i K\left(X_i, z\right) + \beta\right) \tag{1}$$

where Y is the classification result of the input sample z based on the sign of the result (i.e., the positive and negative results refer to different classes, respectively), X is the SV, K is the kernel function, and w and β are the modeling parameters.

There are several kernel functions used in the SVM algorithm. Compared to the simplest linear kernel, which typically has no margin and uniform SVs, the non-linear kernels usually have better classification performance. The most widely used non-linear kernels are the Radial Basis Function (RBF) kernel or sigmoid kernel, and these are given by:

$$K_{\mathrm{RBF}}\left(X_i, z\right) = e^{-\gamma \|X_i - z\|^2} \tag{2}$$

$$K_{\mathrm{sigmoid}}\left(X_i, z\right) = \tan h\left(\gamma {X_i}^T z + r\right) \tag{3}$$

where r is the kernel parameter. Figure 6 shows an example of the classification performance of an SVM using a linear kernel function and an RBF kernel function. The RBF kernel is shown to better separate the samples of different classes using its non-linear decision boundary, so offering a better classification performance.

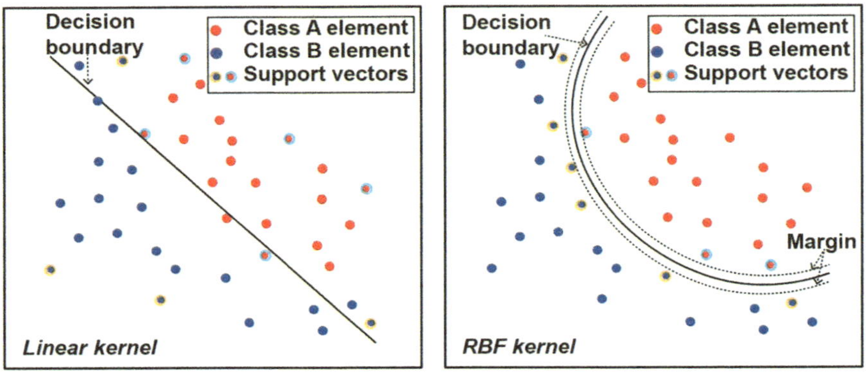

Fig. 6 Comparison of classification performance achieved by an SVM with a linear kernel function and an RBF kernel

4.2 Result-Based Re-Computation Error-Tolerant SVM

An interesting observation is that similar to the KNNs or Random Forest, the SVM computation can also inherently tolerate some errors. Therefore, conventional protection schemes (such as the temporal redundancy technique) introduce over-protection to achieve error-tolerant classification. This feature is explored to design an efficient Result-Based Re-computation (RBR) error-tolerant SVM scheme [46]. Next, an SVM with an RBF kernel function is taken as an example to discuss this technique. However, note that the RBR scheme also applies to some other kernels with similar algorithmic properties, such as the sigmoid kernel.

Let R be the result of Eq. (4) using K_{RBF} in Eq. (2) prior to generating the sign, so given by:

$$R(z) = \sum_{i=1}^{N} w_i e^{-\gamma \|X_i - z\|^2} + \beta. \tag{4}$$

This shows that the summation term is the main component of SVM computation, and the exponent computation (i.e., the RBF kernel) is the most complex part. Therefore, the RBF computation is the most likely part to suffer from computational errors. Figure 7 shows the impact of single bit errors that modify different bits of the RBF result when performing a classification task; various ML datasets from a public repository [40] are considered, and the data is represented with a 16-bit signed integer/binary format as an example. A single bit error is randomly injected on the RBF computation result for a randomly selected SV when inferencing a random sample, and the procedure is repeated 10,000 times to evaluate the average impact of an error. Figure 7 shows that a single error can have a large probability to change the classification result, especially when it affects the most significant

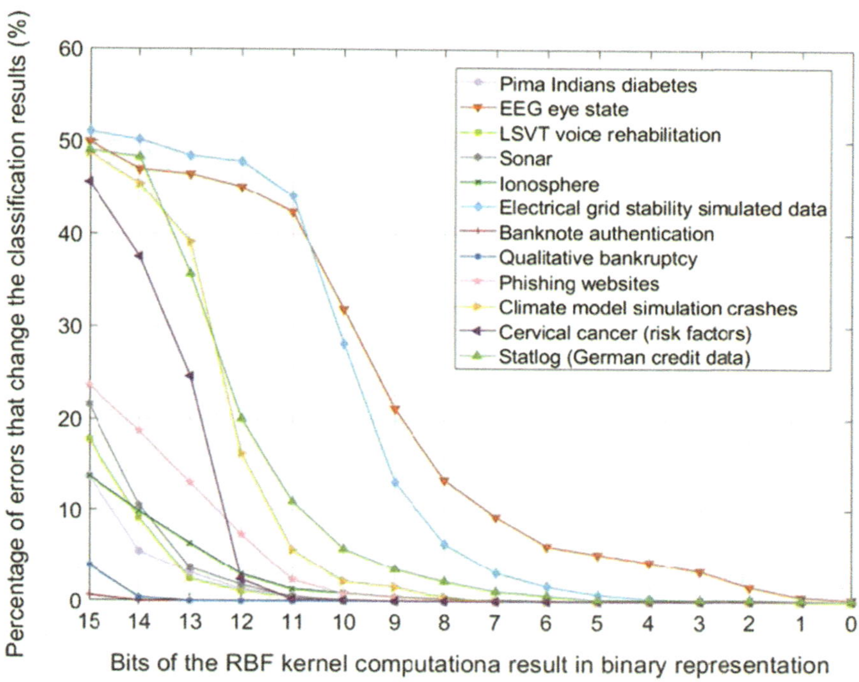

Fig. 7 Percentage of single bit errors that change the classification results

bits. Therefore, error protection is required for an SVM classifier when used in dependable ML applications.

Assume there is an error in the RBF computation for the ith term, changing its value by e_r; then, Eq. (4) can be modified as:

$$R_{\text{err}}(z) = \sum_{i=1}^{N} w_i e^{-\gamma \|X_i - z\|^2} + \beta + w_i e_r \tag{5}$$

where $R_{\text{err}}(z)$ is the value of R affected by an error.

A key observation on RBF is that its result always lies in the interval (0, 1]; hence, the magnitude e_r introduced in the RBF computation by an error should be in the range $(-1, 1)$, causing a difference between R and R_{err} in the range $(-w_i, w_i)$. Based on this observation, the following three scenarios that cover all possible relationships between $|R|$ and $|w_{\max}|$ (where $w_{\max}$ is the weight with the largest absolute value) are considered to discuss the impact of errors on the final classification result. Figure 8 illustrates each case by assuming that the input sample z being classified has only one feature for sake of simplicity.

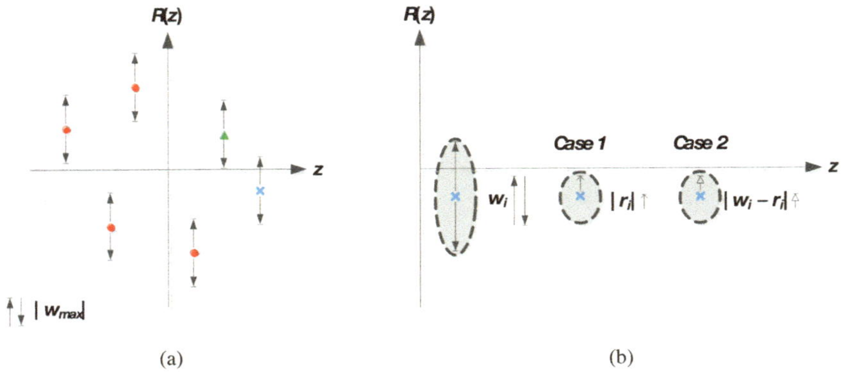

Fig. 8 Impact of computational errors in the RBF kernel function on the value of R: (**a**) general impact for different relationship between $|R|$ and $|w_{\max}|$ (Red: $|R| > |w_{\max}|$, Green:$|R| = |w_{\max}|$, and Blue: $|R| < |w_{\max}|$); (**b**) refined impact when $|R| < |w_{\max}|$ (Case 1: R and r_i have the same sign and $|R| \geq |r_i|$, Case 2: R and r_i have different signs and $|R| \geq |w_i - r_i|$)

- $|R| > |w_{\max}|$: since the result of the RBF is within (0, 1], an error in its computation can bidirectionally change the value of R with a magnitude from 0 to $w_{\max}$ (the red circle markers in Fig. 8a). Therefore, the value of R always keeps the same sign, making the classification result reliable under errors.
- $|R| = |w_{\max}|$: an error makes the value of R to approach 0 in the worst case, while R still keeps the same sign (the green triangle markers in Fig. 8a). Therefore, the classification result is still correct.
- $|R| < |w_{\max}|$: the worst-case error can modify R with a value larger than $|w_{\max}|$, changing the sign of R (i.e., from negative to positive as illustrated by the blue "x" markers in Fig. 8a) and thus, the final classification result. However, a refined impact of errors can be determined in the following two cases (Fig. 8b). *Case 1*: for the ith SV, when R and $r_i = w_i e^{-\gamma \|X_i - z\|^2}$ have the same sign and $|R| \geq |r_i|$, the worst-case error causes r_i to approach 0 without changing the sign of R. *Case 2*: when R and r_i have different signs and $|R| \geq |w_i - r_i|$, the worst-case error changes the value of r_i to w_i without changing the sign of R. Therefore, in these two cases, the computation for the ith support vector can tolerate errors.

Based on the above discussion, the RBR approach using Algorithm 2 is proposed to achieve error tolerance at low computational complexity. In particular, once the SVM classification has been executed, the values of $|R|$ and $|w_{\max}|$ are compared and the two cases when $|R| < |w_{\max}|$ discussed above are further checked if applicable. If the predicted class is identified as reliable under errors, it is used as the final outcome. Otherwise, the terms for some/all SVs in Eq. (1) should be re-computed for error detection; when necessary, the computation is executed for a third time as in the conventional temporal redundancy scheme to correct the error.

Algorithm 2 RBR scheme

Compute Eq. (1) to obtain the classification result C_p;
Keep R and $r_i = w_i e^{-\gamma \|X_i - z\|^2}$, $i = 1, \ldots, N$;
Find the maximum weight $w_{\max}$;
if $|R| \geq |w_{\max}|$
  Output C_p;
  else
    for $i = 1$ to N
      if $|R| \geq |w_i|$
          Output r_i;
       else if $R \bullet r_i > 0$ and $|R| \geq |r_i|$
          Output r_i;
       else if $R \bullet r_i < 0$ and $|R| \geq |w_i - r_i|$
          Output r_i;
       else
          Re-compute $w_i e^{-\gamma \|X_i - z\|^2}$ to obtain r_i';
          if $r_i = r_i$'
            Output r_i;
          else
            Re-compute $w_i e^{-\gamma \|X_i - z\|^2}$ to obtain r_i";
            Do majority voting among r_i, r_i', r_i";
            Obtain the correct r_i;
          end
       end
    end
    Compute Eq. (1) by using the correct r_i;
    Output the correct classification result;
end

Table 6 presents the re-computations for the RBF kernel function saved by employing the RBR scheme; on average the saving in re-computation overhead is significant. For example, for the Banknote authentication dataset, 95.58% of the re-computation can be saved. Even though for a small number of datasets (e.g., the EEG eye state or the Electrical grid stability simulated data), the average percentage of saving is not as large as for others, it is still higher than 12%, so more efficient than the traditional temporal redundancy scheme that has a 0% of saving.

A comparison of the computational overhead in different schemes for error detection is summarized in Table 7. Since both techniques can detect errors, there is no impact of errors on classification accuracy. In terms of computation, the traditional temporal redundancy scheme repeats the entire computation process in all cases, so incurring in a 2× overhead. According to the results presented in Table 6, the RBR scheme only needs to perform re-computation in some cases, incurring in approximately 1.5× overhead on average.

Table 6 Percentage of samples for which RBF re-computations in an SVM can be saved

	$\lvert R\rvert \in [\lvert w_{\max}\rvert, +\infty)$		$\lvert R\rvert \in [\lvert w_{\min}\rvert, \lvert w_{\max}\rvert)$		$\lvert R\rvert \in (-\infty, \lvert w_{\min}\rvert)$		
Dataset	% samples	% SVs	% samples	% SVs	% samples	% SVs	Average
Pima Indians diabetes	40.00%	100.00%	60.00%	51.36%	0%	0%	**70.82%**
EEG eye state	0%	100.00%	93.01%	7.64%	6.99%	2.58%	**7.29%**
LSVT voice rehabilitation	0%	100.00%	100.00%	65.28%	0%	0%	**65.28%**
Sonar	0%	100.00%	100.00%	64.14%	0%	0%	**64.14%**
Ionosphere	0%	100.00%	99.06%	74.48%	0.94%	1.82%	**73.80%**
Electrical grid stability simulated data	0%	100.00%	78.43%	15.77%	21.57%	0.31%	**12.44%**
Banknote authentication	78.40%	100.00%	21.60%	79.56%	0%	0%	**95.58%**
Qualitative bankruptcy	64.00%	100.00%	36.00%	78.23%	0%	0%	**92.16%**
Phishing websites	0%	100.00%	100.00%	55.04%	0%	0%	**55.04%**
Climate model simulation crashes	0%	100.00%	87.04%	24.31%	12.96%	0.28%	**21.20%**
Cervical cancer (risk factors)	0%	100.00%	100.00%	29.85%	0%	0%	**29.85%**
Statlog (German credit data)	0%	100.00%	77.67%	29.69%	22.33%	0.30%	**23.73%**

Table 7 Comparison of different SVM protection techniques for error detection

Scheme	Impact on classification accuracy	Computational overhead
Traditional temporal redundancy	None	2×
Result-Based Re-computation	None	1.5× on average

The savings in computation translate directly in a reduction of power dissipation and thus in energy, as well as computational latency. A reduction in these metrics has benefits in terms of battery life, economic savings, and sustainability. Additionally, lower power dissipation may also reduce the expenses of the components needed to implement the system such as power converters or dissipation and ventilation. Therefore, the RBR scheme is more attractive for systems with these requirements compared to the traditional solution.

5 Conclusion

In this chapter, we have considered dependability in machine learning (ML) focusing on classical algorithms such as *K* Nearest Neighbors (KNNs), Random Forests (RFs), or Support Vector Machines (SVMs). This complements existing works that have concentrated on the analysis of complex neural networks. The analysis of these simple algorithms is of interest as they are used in resource-constrained platforms or in problems in which they achieve sufficient accuracy and more complex models are not needed.

The study, analysis, and simulation results show that ML algorithms considered have some intrinsic degree of error tolerance, but they can still suffer from errors that change the classification results. Therefore, the algorithms must be protected; to reduce the protection overhead, it is of interest to design schemes that exploit their features and intrinsic error tolerance. In this chapter, we have shown that this can be achieved when considering single computational errors. Efficient protection schemes have been proposed for KNNs, RFs, and SVMs to reduce the overhead needed to protect against errors significantly.

The work presented in this chapter can be extended by considering other types of errors or fault models for the classifiers, and also by studying other classical classifiers such as logistic regression or decision trees.

References

1. K. Grace, J. Salvatier, A. Dafoe, B. Zhang, O. Evans, Owain, Viewpoint: When will AI exceed human performance? Evidence from AI experts. J. Artif. Intell. Res. **62**, 729–754 (2018)
2. L. Chen, S. Lin, X. Lu, D. Cao, H. Wu, C. Guo, C. Liu, F. Wang, Deep neural network based vehicle and pedestrian detection for autonomous driving: A survey. IEEE Trans. Intell. Transp. Syst. **22**(6), 3234–3246 (2021)
3. N. Bussmann, P. Giudici, D. Marinelli, J. Papenbrock, Explainable AI in fintech risk management. Front. Artif. Intell **3**, 26 (2020)
4. T. Davenport, R. Kalakota, The potential for artificial intelligence in healthcare. Future Healthcare J **6**(2), 94–98 (2019)
5. S. Liu, P. Reviriego, F. Lombardi, P. Girard, Guest editorial: Special section on "To be safe and dependable in the era of artificial intelligence: Emerging techniques for trusted and reliable machine learning". IEEE Trans. Emerg. Top. Comput. **10**(4), 1668–1670 (2022)
6. J.C. Avizienis, B.R. Laprie, C. Landwehr, Basic concepts and taxonomy of dependable and secure computing. IEEE Trans. Depend. Sec. Comput **1**(1), 11–33 (2004)
7. R. Mariani, An overview of autonomous vehicles safety, in *IEEE International Reliability Physics Symposium (IRPS)*, (Burlingame, CA, USA, 2018, March), pp. 6A–61A
8. M. Nicolaidis, Design for soft error mitigation. IEEE Trans. Device Mater. Reliab. **5**(3), 405–418 (2005)
9. V.S.S. Nair, J.A. Abraham, P. Banerjee, Efficient techniques for the analysis of algorithm-based fault tolerance (ABFT) schemes. IEEE Trans. Comput. **45**(4), 499–503 (1996)
10. G.R. Redinbo, Designing checksums for detecting errors in fast unitary Transforms. IEEE Trans. Comput. **67**(4), 566–572 (2018)

11. Z. Gao, J. Zhu, T.Y. Tyan, A. Ullah, P. Reviriego, Fault tolerant polyphase filters-based decimators for SRAM-based FPGA implementations. IEEE Trans. Emerg. Top. Comput. **10**(2), 591–601 (2022)
12. Simeone, A very brief introduction to machine learning with applications to communication systems. IEEE Trans. Cogn. Commun. Netw **4**(4), 648–664 (2018)
13. D.S. Phatak, I. Koren, Complete and partial fault tolerance of feedforward neural nets. IEEE Trans. Neural Netw. **6**(2), 446–456 (1995)
14. K. Zhao, S. Di, S. Li, X. Liang, Y. Zhai, J. Chen, K. Ouyang, F. Cappello, Z. Chen, FT-CNN: Algorithm-based fault tolerance for convolutional neural networks. IEEE Trans. Parall. Distrib. Syst **32**(7), 1677–1689 (2021)
15. V. Piuri, M. Sami, R. Stefanelli, Fault tolerance in neural networks: Theoretical analysis and simulation results, in *Proceedings, Advanced Computer Technology, Reliable Systems and Applications*, (Bologna, Italy, 1991, May), pp. 429–436
16. M.A. Neggaz, I. Alouani, S. Niar, F. Kurdahi, Are CNNs reliable enough for critical applications? An exploratory study. IEEE Des. Test **37**(2), 76–83 (2020)
17. J. Kosaian, K.V. Rashmi, Arithmetic-intensity-guided fault tolerance for neural network inference on GPUs, in *International Conference for High Performance Computing, Networking, Storage and Analysis*, (St. Louis, MO, USA, 2021), pp. 1–15
18. M. Safarpour, T.Z. Deng, J. Massingham, L. Xun, M. Sabokrou, O. Silvén, Low-voltage energy efficient neural inference by leveraging fault detection techniques, in *IEEE Nordic Circuits and Systems Conference (NorCAS)*, (Oslo, Norway, 2021, Oct), pp. 1–5
19. M.T. McCann, K.H. Jin, M. Unser, Convolutional neural networks for inverse problems in imaging: A review. IEEE Signal Process. Mag. **34**(6), 85–95 (2017)
20. Altaher, Phishing websites classification using hybrid SVM and KNN approach. Int. J. Adv. Comput. Sci. Appl. **8**(6), 90–95 (2017)
21. J.G. Lopez, S. Ventura, A. Cano, Distributed nearest neighbor classification for large-scale multi-label data on spark. Futur. Gener. Comput. Syst. **87**, 66–82 (2018)
22. W. Wang, Y. Li, X. Wang, J. Liu, X. Zhang, Detecting android malicious apps and categorizing benign apps with ensemble of classifiers. Futur. Gener. Comput. Syst. **78**, 987–994 (2018)
23. X. Tang, Z. Ma, Q. Hu, W. Tang, A real-time arrhythmia heartbeats classification algorithm using parallel delta modulations and rotated linear-kernel support vector machines. IEEE Trans. Biomed. Eng. **67**(4), 978–986 (2019)
24. Y. Zheng, L. Sun, S. Wang, J. Zhang, J. Ning, Spatially regularized Structural support vector machine for robust visual tracking. IEEE Trans. Neural Netw. Learn. Syst **30**(10), 3024–3034 (2019)
25. D. Dileep, C.C. Sekhar, GMM-based intermediate matching kernel for classification of varying length patterns of long duration speech using support vector machines. IEEE Trans. Neural Netw. Learn. Syst **25**(8), 1421–1432 (2013)
26. L. Breiman, Random forests. Springer, Mach. Learn **45**(1), 5–32 (2001)
27. G. Gui, F. Liu, J. Sun, J. Yang, Z. Zhou, D. Zhao, Flight delay prediction based on aviation big data and machine learning. IEEE Trans. Veh. Technol. **69**(1), 140–150 (2019)
28. Q. Hu, D. Yu, Z. Xie, Neighborhood classifiers. Expert Syst. Appl **34**, 866–876 (2008)
29. C.J.C. Burges, A tutorial on support vector machines for pattern recognition. Data Min. Knowl. Discov **2**(2), 121–167 (1998)
30. G. Song, J. Rochas, L. El Beze, F. Huet, F. Magoules, K nearest neighbour joins for big data on MapReduce: A theoretical and experimental analysis. IEEE Trans. Knowl. Data Eng. **28**(9), 2376–2392 (2016)
31. V. Garcia, E. Debreuve, M. Barlaud, Fast K nearest neighbor search using GPU, in *Proceedings of the IEEE Computer Society Conference on Computer Vision and Pattern Recognition Workshops*, (2008, June), pp. 1–6
32. E.S. Manolakos, I. Stamoulias, IP-cores design for the KNN classifier, in *Proceedings of the IEEE International Symposium on Circuits and Systems*, (2010, May), pp. 4133–4136
33. N. Attaran, A. Puranik, J. Brooks, T. Mohsenin, Embedded low-power processor for personalized stress detection. IEEE Trans. Circuits Syst. II Express Briefs **65**(12), 2032–2036 (2018)

34. P.N. Yianilos, Data structures and algorithms for nearest neighbor search in general metric spaces, in *SODA*, (1993), pp. 311–321
35. N. Kanekawa, E.H. Ibe, T. Suga, Y. Uematsu, *Dependability in Electronic Systems: Mitigation of Hardware Failures, Soft Errors, and Electro-Magnetic Disturbances* (Springer-Verlag, New York, 2010)
36. M. Nicolaidis, *Soft Errors in Modern Electronic Systems* (Springer Science & Business Media, 2010)
37. T. Karnik, P. Hazucha, Characterization of soft errors caused by single event upsets in CMOS processes. IEEE Trans. Depend. Sec. Comput **1**(2), 128–143 (2004)
38. Oz, S. Arslan, A survey on multithreading alternatives for soft error fault Tolerance. ACM Comput. Surv. **52**, 1 (2019)
39. S. Liu, P. Reviriego, J.A. Hernández, F. Lombardi, Voting margin: A scheme for error-Ttolerant k nearest neighbors classifiers for machine learning. IEEE Trans. Emerg. Topic. Comput **9**(4), 2089–2098 (2021)
40. D. Dua, C. Graff, *UCI Machine Learning Repository* (University of California, School of Information and Computer Science, Irvine, CA, 2019)
41. S. Liu, P. Reviriego, P. Montuschi, F. Lombardi, Error-tolerant computation for voting classifiers with multiple classes. IEEE Trans. Veh. Technol. **69**(11), 13718–13727 (2020)
42. B. Johnson, R. Tateishi, Z. Xie, Using geographically-weighted variables for image classification. Remote Sens. Lett **3**(6), 491–499 (2012)
43. P. Siebert, *Vehicle Recognition Using Rule Based Methods* (Turing Institute Research Memorandum TIRM-87-018, 1987)
44. C. Higuera, K.J. Gardiner, K.J. Cios, Self-organizing feature maps identify proteins critical to learning in a mouse model of down syndrome. PLoS One **10**(6), e0129126 (2015)
45. P. Cortez, A. Cerdeira, F. Almeida, T. Matos, J. Reis, Modeling wine preferences by data mining from physicochemical properties. Elsevier, Decis. Supp. Syst **47**(4), 547–553 (2009)
46. S. Liu, P. Reviriego, X. Tang, W. Tang, F. Lombardi, Result-based re-computation for error-tolerant classification by a support vector machine. IEEE Trans. Artif. Intell **1**(1), 62–73 (2020)

Part II
Stochastic Computing

Efficient Random Number Sources Based on D Flip-Flops for Stochastic Computing

Kuncai Zhong and Weikang Qian

1 Introduction

Stochastic computing (SC) is an unconventional computing paradigm proposed in 1960s [1]. Different from the conventional binary computing, it does computation on stochastic bit streams (SBSs), which encode values by the ratios of ones in the streams. For example, the SBS 10101111 has 6 ones in 8 bits and encodes the value $\frac{6}{8}$. Based on this special encoding format, SC generally benefits from low-computing circuitry and strong fault tolerance. For example, it only needs a single AND gate to implement the multiplication, as shown in Fig. 1a. It has found potential application in several domains, such as decoding of modern error-correcting codes [2], image processing [3–5], and neural networks [6–11].

However, SC is essentially a type of probability computation. The accuracy of SC is highly influenced by the correlation among the SBSs [12]. The SBSs with high correlation generally lead to a considerable error for the SC circuit [12]. For example, as shown in Fig. 1b, if $\frac{6}{8}$ and $\frac{4}{8}$ are encoded as 00111111 and 00001111, respectively, the output SBS of the AND gate will be 00001111, which is a wrong

This work is supported by the National Key R&D Program of China under Grant 2020YFB2205501.

K. Zhong
University of Michigan-Shanghai Jiao Tong University Joint Institute, Shanghai Jiao Tong University, Shanghai, China
e-mail: kczhong@sjtu.edu.cn

W. Qian (✉)
University of Michigan-Shanghai Jiao Tong University Joint Institute, Shanghai, China

MoE Key Laboratory of Artificial Intelligence, Shanghai Jiao Tong University, Shanghai, China
e-mail: qianwk@sjtu.edu.cn

W. Liu et al. (eds.), *Design and Applications of Emerging Computer Systems*,
https://doi.org/10.1007/978-3-031-42478-6_8

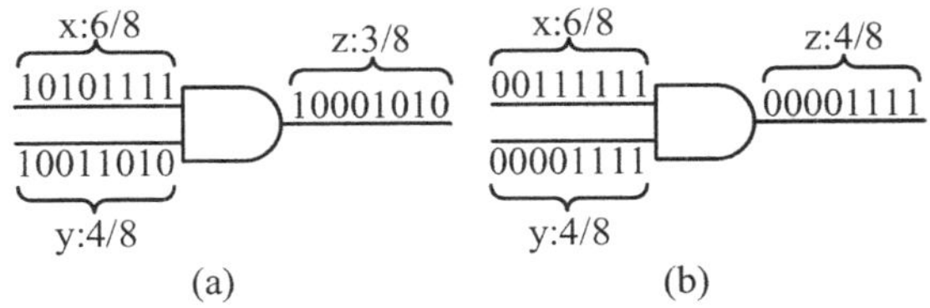

Fig. 1 SC implementation of the multiplication. (**a**) The accurate implementation with 2 uncorrelated SBSs; (**b**) The incorrect implementation with 2 highly correlated SBSs

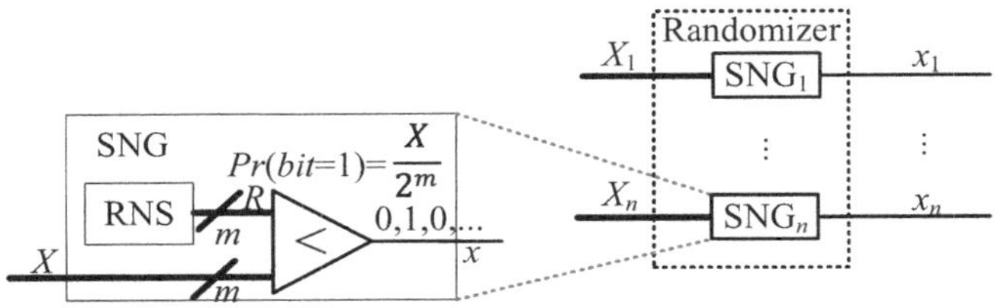

Fig. 2 Illustration of a randomizer

product. Therefore, to achieve a high accuracy for an SC circuit, it is crucial to generate SBSs with low correlation.

SC typically applies a randomizer consisting of n stochastic number generators (SNGs) to generate n SBSs as shown in Fig. 2, where an SNG converts an input binary number into an SBS and consists of a random number source (RNS) and a comparator. The RNS generates an m-bit random binary number R at each clock cycle, and the comparator compares the input binary number X with it to generate a zero if $R \geq X$ and a one otherwise. If the random number R is uniformly distributed in the set $\{0, 1, \ldots, 2^m - 1\}$, then the probability of a bit in the SBS being a one is $\frac{X}{2^m}$. The RNS determines the order and controls the randomness of the bits in the SBS. Hence, to generate SBSs with low correlation, we need to use proper RNSs.

There are several different RNS designs proposed for this purpose, such as low-discrepancy sequence generator [13–15] and pseudo-random sequence generator [16, 17]. A low-discrepancy sequence generator, such as Sobol sequence generator [13, 14], can generate a sequence of uniformly distributed random binary numbers, which can further lead to an SBS with zeros and ones uniformly distributed. In general, the SBSs generated by the low-discrepancy sequence generators have low correlation and lead to a high accuracy for the SC circuit. However, a low-discrepancy sequence generator usually needs complex circuitry to implement, such as least significant zero detectors and storage arrays in Sobol sequence generator [13], and has a large area. A pseudo-random sequence generator can generate random numbers with a small circuit area. The most widely used one is linear feedback shift register (LFSR) as shown in Fig. 3. It consists of m D flip-flops (DFFs) and some 2-input XOR gates to generate the m-bit random binary numbers. We refer to the leftmost DFF as the first DFF and the rightmost one as the m-th DFF. We assume that the output of the i-th ($1 \leq i \leq m$) DFF is the i-th most significant bit in R. The shift direction of an LFSR is from the left to the right. The XOR gates select the outputs of some DFFs to generate a new bit for the first DFF. It is very

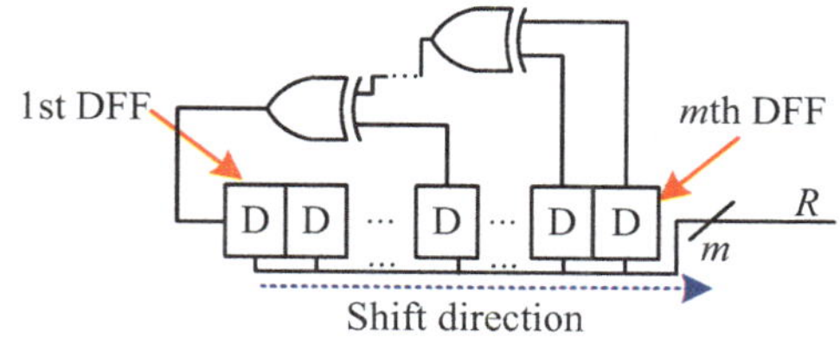

Fig. 3 Illustration of an LFSR

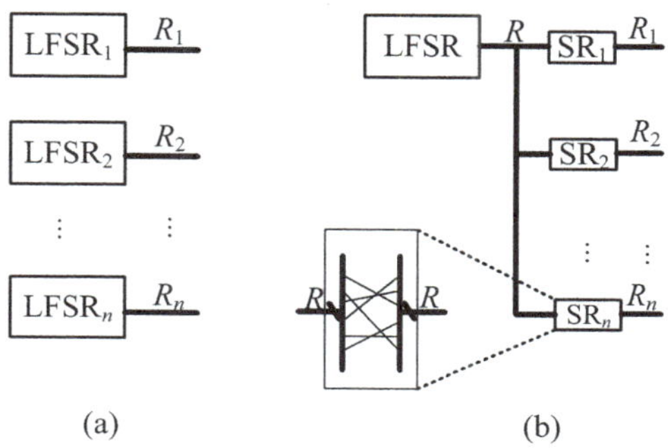

Fig. 4 Illustration of (**a**) the general design of RNSs composed of n LFSRs and (**b**) the min-cost design of RNSs composed of a single LFSR and n scrambling modules (i.e., the SR_i modules in the figure)

promising to achieve a high accuracy and a small area for an SC circuit. In this chapter, we consider exploiting it to design efficient RNSs.

To generate SBSs with low correlation, the randomizer generally applies n LFRSs as RNSs, as shown in Fig. 4a. We call this design of RNSs the *general design of RNSs*. It can lead to a low output error for an SC circuit. However, the total number of DFFs it has is nm, which is very large when n or m is large. To reduce the hardware cost, another design of RNSs based on LFSR is introduced [18], as shown in Fig. 4b. It only uses a single LFSR and n scrambling modules, each of which permutes the output bits of the LFSR to generate a different random binary number. The number of used LFSRs is the minimum. Note that the scrambling module is logic-free. Thus, this design of RNSs has the minimum area. We call it the *min-cost design of RNSs*. Nevertheless, the output error of an SC circuit using the min-cost design is usually large. In summary, the general and the min-cost designs of RNSs are at the two extremes of the RNS design spectrum. The former has a large area and can lead to a low output error for an SC circuit, while the latter has the smallest area and can lead to a high output error. This brings up a question of whether a design of RNSs that has a small area and can lead to a low output error for an SC circuit exists. In what follows, for simplicity, when a design of RNSs can lead to a low output error for SC circuits, we just say that it has a low *SC error*. Furthermore, we call an RNS a *successive RNS*, if it is formed by a successive sequence of DFFs. An example is LFSR. Otherwise, if it is formed by a non-successive sequence of DFFs, we call it a *non-successive RNS*.

In this chapter, we try to address the above question. We first propose several guidelines for applying LFSRs as RNSs for SC circuits. Then, based on these guidelines, we propose a method to build successive RNSs. After that, we propose a method to build non-successive RNSs. Finally, by properly scrambling the RNS outputs [17] and using the negated outputs of DFFs [19], we propose three new designs of RNSs. The experimental results show that our proposed designs can achieve both a high accuracy and a small circuit area. For example, the SNG based on one proposed design of RNSs can reduce 25% area over the SNG based on the general design of RNSs, while achieving a higher accuracy.

The rest of the chapter is organized as follows. Section 2 presents the novel guidelines. Based on these guidelines, Sect. 3 shows the method for building successive RNSs. Then, Sect. 4 presents the method for building non-successive RNSs. Based on these methods, 3 efficient designs of RNSs are proposed in Sect. 5. Section 6 shows the experimental results. Finally, Sect. 7 concludes the chapter.

2 Guidelines for Applying LFSRs in Stochastic Computing

LFSRs are configured by the feedback polynomials and the seeds. The feedback polynomial determines the selection of the XOR gates, and the seed is the initial value of the LFSR. In this section, we will show some guidelines for how to properly configure LFSRs as RNSs in SC circuits. We first show the guidelines on the feedback polynomials of LFSRs. Then, we present the guidelines on the seeds of LFSRs.

2.1 Model and Evaluation Methodology for LFSRs

As shown in Fig. 4a, to generate n independent SBSs, the general design of RNSs needs n LFSRs. The configurations of the n LFSRs can be uniquely identified by their feedback polynomials and their seeds. In the following, we represent an LFSR with the feedback polynomial F and the seed S by a pair (F, S) and a group of n LFSRs by a vector of pairs $[(F_1, S_1), (F_2, S_2), \ldots, (F_n, S_n)]$, where the pair (F_i, S_i) gives the i-th LFSR L_i. We aim to find a group of n LFSRs with low SC error. In this section, we focus on the case where $n = 2$. That is a pair of LFSRs.

Like ref. [12], we measure the SC error of an LFSR pair $[(F_1, S_1), (F_2, S_2)]$ by the *mean absolute error (MAE)* of a 2-input SC multiplier with the pair as the RNSs.

Denote the output of the SC multiplier as $f(a, b)$ when the 2 inputs are a and b. We consider all possible inputs for the SC multiplier in evaluating its MAE. For an m-bit LFSR, it outputs a periodic sequence with the period of $(2^m - 1)$, and the sequence includes values $1, 2, \ldots, 2^m - 1$. Correspondingly, we let the input values only be $1, 2, \ldots, 2^m - 1$. Thus, there are $(2^m - 1)^2$ different input pairs for the SC multiplier. The MAE of the SC multiplier is calculated as

$$MAE = \frac{1}{(2^m - 1)^2} \sum_{a=1}^{2^m-1} \sum_{b=1}^{2^m-1} \left| f(a, b) - \frac{a \cdot b}{(2^m - 1)^2} \right|. \quad (1)$$

Thus, we use Eq. (1) to measure the SC error of an LFSR pair $[(F_1, S_1), (F_2, S_2)]$. Note that SC is not applicable to high-precision computation due to its long latency. Thus, in the following, we focus on the case where the bit width $m \leq 8$.

2.2 *The Selection of the Feedback Polynomials for LFSR Pair*

In this section, we study the selection of the feedback polynomials F_1 and F_2 for a pair of LFSRs $[(F_1, S_1), (F_2, S_2)]$. Note that when $F_1 = F_2$ and $S_1 = S_2$, the pair leads to highly correlated input SBSs and thus should be avoided. Therefore, we will not consider this pair in the following part.

By the relation between F_1 and F_2, all the LFSR pairs can be divided into 2 types: those with $F_1 \neq F_2$ and those with $F_1 = F_2$. We call the former the *distinct-feedback LFSR pairs* and the latter the *identical-feedback LFSR pairs*.

Let the number of different feedback polynomials for an m-bit LFSR be l_m. For $m = 6, 7$, and 8, the values of l_m are 6, 18, and 16, respectively. We further define:

1. $H_{m,t}$: The t-th ($1 \leq t \leq l_m$) feedback polynomial of an m-bit LFSR.
2. $D_{m,t}$: The set of m-bit distinct-feedback LFSR pairs with either $F_1 = H_{m,t}$ or $F_2 = H_{m,t}$.
3. $I_{m,t}$: The set of m-bit identical-feedback LFSR pairs with $F_1 = F_2 = H_{m,t}$.

Then, for each set $D_{m,t}$, we obtain the average SC error and the minimum SC error over all LFSR pairs in the set. We do the same for each set $I_{m,t}$. The average SC errors and the minimum SC errors of $D_{m,t}$ and $I_{m,t}$ for different t's and $m = 6, 7, 8$ are shown in Fig. 5. For a clear comparison, we also show a reference value, which equals the minimum SC error in the set $D_{m,t}$ plus $\frac{1}{2^{m+2}}$. From the figure, we make the following two observations:

Observation 1 For any $6 \leq m \leq 8$ and any $1 \leq t \leq l_m$, the average SC error of the identical-feedback LFSR pairs in set $I_{m,t}$ is less than that of the distinct-feedback LFSR pairs in set $D_{m,t}$. □

Observation 2 For $m = 8$ and any $1 \leq t \leq l_m$, the minimum SC error of the identical-feedback LFSR pairs in set $I_{m,t}$ is less than that of the distinct-feedback LFSR pairs in set $D_{m,t}$. For any $6 \leq m \leq 7$ and any $1 \leq t \leq l_m$, the minimum SC error of the identical-feedback LFSR pairs in set $I_{m,t}$ is less than that of the distinct-feedback LFSR pairs in set $D_{m,t}$ plus $\frac{1}{2^{m+2}}$. □

As the length of the SBS is $(2^m - 1)$, the precision is $\frac{1}{2^m-1}$. Thus, the minimum SC error of the identical-feedback LFSR pairs in set $I_{m,t}$ is very close to that of the

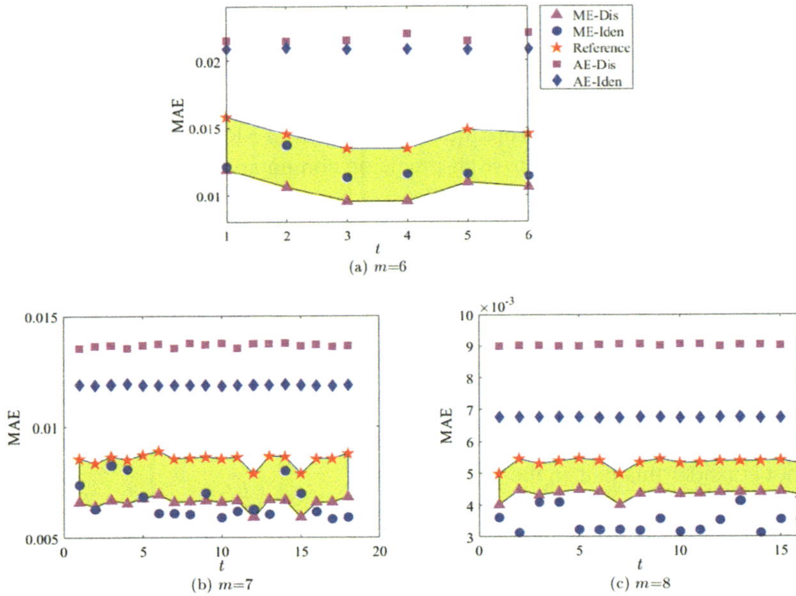

Fig. 5 Comparisons between the distinct-feedback and the identical-feedback LFSR pairs in terms of the average error and the minimum error. In the legend, ME and AE denote the minimum error and the average error, respectively, and Dis and Iden denote the distinct-feedback LFSR pair and the identical-feedback LFSR pair, respectively

distinct-feedback LFSR pairs in set $D_{m,t}$. Therefore, by Observations 1 and 2, we conclude:

Guideline 1 For $6 \leq m \leq 8$, an identical-feedback LFSR pair is a good choice for forming 2 RNSs. □

As will be shown in Sect. 3, an identical-feedback LFSR pair also allows area-efficient implementation. Thus, in the following, we focus on identical-feedback LFSR pairs.

2.3 The Selection of the Seeds for Identical-Feedback LFSR Pair

In this section, we study the selection of the seeds S_1 and S_2 for an identical-feedback LFSR pair $[(F, S_1), (F, S_2)]$, where F denotes the common feedback polynomial.

Suppose that the LFSR $(F, 1)$ outputs N_k at the k-th clock cycle. Then, any LFSR pair $[(F, S_1), (F, S_2)]$ can be represented with the reference of LFSR $(F, 1)$

as $[(F, N_i), (F, N_j)]$. It is easy to see that the two LFSR pairs $[(F, N_i), (F, N_j)]$ and $[(F, N_j), (F, N_i)]$ produce the same MAE for an SC multiplier. Therefore, we only need to focus on the LFSR pair $[(F, N_i), (F, N_j)]$ with $i - j > 0$. We define $c = i - j$. Then, we have $c > 0$, and c gives the difference between the clock cycles when the LFSR $(F, 1)$ produces N_i and N_j. In the following, we call c the *clock cycle difference* of the LFSR pair.

Next, consider two LFSR pairs $C_1 = [(F, N_{1+c}), (F, N_1)]$ and $C_2 = [(F, N_{j+c}), (F, N_j)]$, which have the same clock cycle difference. It is easy to see that for a given input binary number, the SBS produced by the LFSR (F, N_{1+c}) (*resp.*, (F, N_1)) equals that produced by the LFSR (F, N_{j+c}) (*resp.*, (F, N_j)) under a left cyclic shift of $(j - 1)$ cycles. Thus, the two LFSR pairs $C_1 = [(F, N_{1+c}), (F, N_1)]$ and $C_2 = [(F, N_{j+c}), (F, N_j)]$ give the same SC error, and we conclude the following:

Theorem 1 *For the identical-feedback LFSR pair* $[(F, N_i), (F, N_j)]$ *with* $c = i - j > 0$*, the SC error only depends on the choice of the feedback polynomial* F *and the clock cycle difference* c*.* □

Next, we study the proper choice for the clock cycle difference c. We use $T_{m,c}$ $(1 \le c \le 2^m - 2)$ to denote the set of m-bit identical-feedback LFSR pairs with clock cycle difference as c. The LFSR pairs in $T_{m,c}$ have different feedback polynomials. Similar to the analysis in Sect. 2.2, for each set $T_{m,c}$, we obtain the average SC error over all LFSR pairs in the set. In Table 1, for each $m \in \{6, 7, 8\}$, we list the minimum average SC error over all choices of c together with the corresponding c. For each m, we also show a choice of c of interest, i.e., $c = \lfloor \frac{m}{2} \rfloor$, together with the average SC error for it. Finally, for each m, we also list a reference case, which equals $\frac{1}{2^m}$, the precision of a bit stream of length 2^m.

As shown in the table, the average SC error of the clock cycle difference $c = \lfloor \frac{m}{2} \rfloor$ is very close to the minimum value and is no larger than $\frac{1}{2^m}$. Therefore, we can conclude the following:

Guideline 2 For $6 \le m \le 8$, a good choice of the clock cycle difference of an identical-feedback LFSR pair is $c = \lfloor \frac{m}{2} \rfloor$. □

Indeed, why $c = \lfloor \frac{m}{2} \rfloor$ is a good choice can be explained. For two m-bit LFSRs L_1 and L_2, we can plot their output pairs from the clock cycles 1 to $(2^m - 1)$ in a plane with x and y coordinates as the outputs of L_1 and L_2, respectively. For

Table 1 Comparison between the best clock cycle difference and the clock cycle difference $c = \lfloor \frac{m}{2} \rfloor$ in terms of the average SC error

	$m = 6$		$m = 7$		$m = 8$	
Case	Average error	c	Average error	c	Average error	c
Minimum	0.012	3	0.0071	4	0.0037	5
$c = \lfloor \frac{m}{2} \rfloor$	0.012	3	0.0076	3	0.0039	4
Reference $\frac{1}{2^m}$	0.016	–	0.0078	–	0.0039	–

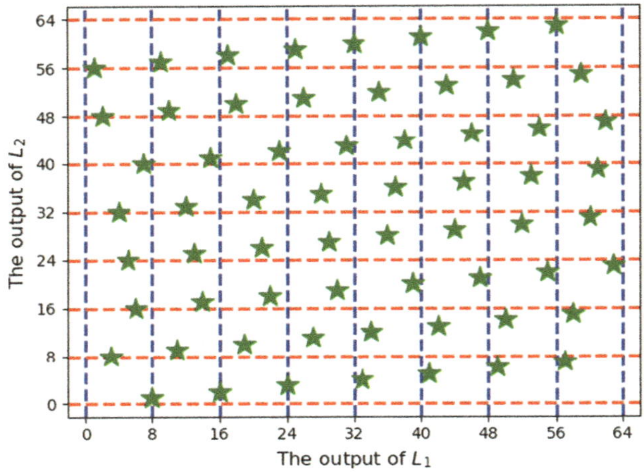

Fig. 6 Distribution of the output pairs from the clock cycles 1 to $(2^6 - 1)$ of a 6-bit identical-feedback LFSR pair with the clock cycle difference $c = \lfloor \frac{6}{2} \rfloor$. Each point corresponds to an output pair

example, Fig. 6 shows the distribution of the output pairs from the clock cycles 1 to $(2^6 - 1)$ of a 6-bit identical-feedback LFSR pair with the clock cycle difference $c = \lfloor \frac{6}{2} \rfloor$. Each point corresponds to an output pair.

We further split the plane into 2^m uniform grids by $(2^{\lceil \frac{m}{2} \rceil} + 1)$ horizontal lines at positions $0, 2^{\lfloor \frac{m}{2} \rfloor}, 2 \cdot 2^{\lfloor \frac{m}{2} \rfloor}, \ldots, 2^m$ and $(2^{\lfloor \frac{m}{2} \rfloor} + 1)$ vertical lines at positions $0, 2^{\lceil \frac{m}{2} \rceil}, 2 \cdot 2^{\lceil \frac{m}{2} \rceil}, \ldots, 2^m$. The total number of output pairs from the clock cycles 1 to $(2^m - 1)$ is $(2^m - 1)$. We say that the output pairs are *uniformly distributed* when $(2^m - 1)$ grids in the plane have exactly one output pair in each of them, and one grid has no output pair in it, where an output pair (o_1, o_2) is in the grid at the i-th column and the j-th row ($1 \le i \le \lfloor \frac{m}{2} \rfloor$ and $1 \le j \le \lceil \frac{m}{2} \rceil$) if $2^{\lceil \frac{m}{2} \rceil + i - 1} \le o_1 < 2^{\lceil \frac{m}{2} \rceil + i}$ and $2^{\lfloor \frac{m}{2} \rfloor + j - 1} \le o_2 < 2^{\lfloor \frac{m}{2} \rfloor + j}$. For the 6-bit identical-feedback LFSR pair with $c = \lfloor \frac{6}{2} \rfloor$, as shown in Fig. 6, except for the grid at the first row and the first column, each grid contains exactly one output pair. Thus, for this particular LFSR pair, its output pairs from the clock cycles 1 to $(2^6 - 1)$ are uniformly distributed. In fact, this is true regardless of m. We have the following theorem.

Theorem 2 *The output pairs from the clock cycles* 1 *to* $(2^m - 1)$ *of an* m*-bit identical-feedback LFSR pair with* $c = \lfloor \frac{m}{2} \rfloor$ *are uniformly distributed.* □

Proof Consider an m-bit identical-feedback LFSR pair $[(F, N_{\lfloor \frac{m}{2} \rfloor + 1}), (F, N_1)]$. Its clock cycle difference $c = \lfloor \frac{m}{2} \rfloor$. Denote the first and the second LFSRs in the pair as L_1 and L_2, respectively. Since the bits of an LFSR are shifted from the first DFF to the m-th DFF as shown in Fig. 3, the value of the i-th ($1 \le i \le \lceil \frac{m}{2} \rceil$) DFF will be shifted into the $(i + \lfloor \frac{m}{2} \rfloor)$-th DFF after $\lfloor \frac{m}{2} \rfloor$ clock cycles. Thus, for L_2, the values of the first to the $\lceil \frac{m}{2} \rceil$-th DFFs at the j-th clock cycle (i.e., the most $\lceil \frac{m}{2} \rceil$

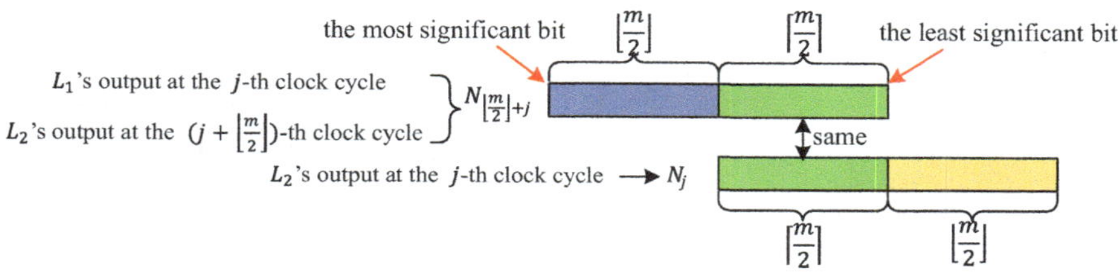

Fig. 7 Illustration of the relation between $N_{\lfloor\frac{m}{2}\rfloor+j}$ and N_j

significant bits of N_j) are the same as the values of the $(1+\lfloor\frac{m}{2}\rfloor)$-th to the m-th DFFs at the $(\lfloor\frac{m}{2}\rfloor+j)$-th clock cycle (i.e., the least $\lceil\frac{m}{2}\rceil$ significant bits of $N_{\lfloor\frac{m}{2}\rfloor+j}$) as illustrated in Fig. 7. Thus, the most $\lfloor\frac{m}{2}\rfloor$ significant bits of $N_{\lfloor\frac{m}{2}\rfloor+j}$ and the most $\lceil\frac{m}{2}\rceil$ significant bits of N_j constitute $N_{\lfloor\frac{m}{2}\rfloor+j}$ as shown in Fig. 7. Since $N_{\lfloor\frac{m}{2}\rfloor+j}$ and N_j are the outputs of L_1 and L_2 at clock cycle j $(1\le j\le 2^m-1)$, respectively, we further have that the most $\lfloor\frac{m}{2}\rfloor$ significant bits of the output of L_1 and the most $\lceil\frac{m}{2}\rceil$ significant bits of the output of L_2 at clock cycle j $(1\le j\le 2^m-1)$ constitute $N_{\lfloor\frac{m}{2}\rfloor+j}$. Thus, in the period of 2^m-1, the set of m-bit binary numbers constituted by the most $\lfloor\frac{m}{2}\rfloor$ significant bits of the output of L_1 and the most $\lceil\frac{m}{2}\rceil$ significant bits of the output of L_2 equals the set $\{1,\ldots,2^m-1\}$. In this case, according to [20], the output pairs of the two RNSs L_1 and L_2 are uniformly distributed. □

If the distribution of the output pairs of two RNSs is more uniform, the output of an SC circuit using the RNSs is generally more accurate [15]. Then, by Theorem 2, we can conclude a guideline as follows.

Guideline 3 An identical-feedback LFSR pair with the clock cycle difference $c=\lfloor\frac{m}{2}\rfloor$ is a good choice for forming 2 RNSs. □

3 Proposed Method for Building Successive RNSs

In this section, based on the guidelines in Sect. 2, we propose a new method for building successive RNSs.

3.1 Method for Building Two Successive RNSs

By Guideline 3, an identical-feedback LFSR pair with the clock cycle difference $c=\lfloor\frac{m}{2}\rfloor$ is a good choice for forming 2 RNSs. In this section, we follow the guideline to build two successive RNSs.

Indeed, we apply a method proposed in [21] to implement two successive RNSs. The implementation is shown in Fig. 8. It consists of an m-bit LFSR together with a

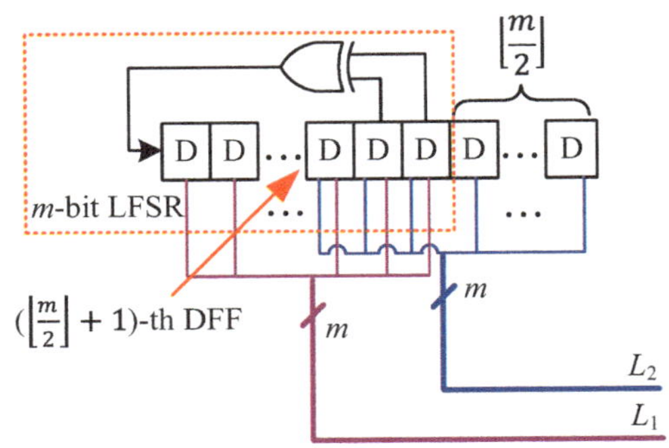

Fig. 8 Building two successive RNSs by $(m + \lfloor \frac{m}{2} \rfloor)$ DFFs

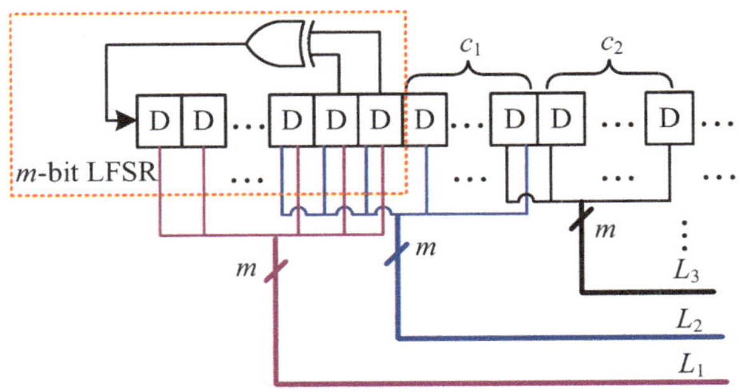

Fig. 9 The design of multiple successive RNSs

sequence of $\lfloor \frac{m}{2} \rfloor$ DFFs inserted after the m-th DFF of the LFSR. The first RNS L_1 is just the m-bit LFSR, while the second RNS L_2 is formed by the $(\lfloor \frac{m}{2} \rfloor + 1)$-th, $(\lfloor \frac{m}{2} \rfloor + 2)$-th, ..., m-th DFFs in the m-bit LFSR together with the newly inserted $\lfloor \frac{m}{2} \rfloor$ DFFs. As shown in the figure, the value given by L_2 at any clock cycle j equals that given by L_1 at the clock cycle $(j - \lfloor \frac{m}{2} \rfloor)$. Thus, L_1 and L_2 are 2 LFSRs with an identical-feedback polynomial, and the clock cycle difference between them is $\lfloor \frac{m}{2} \rfloor$. Furthermore, the two RNSs are successive RNSs by definition. Therefore, the design shown in Fig. 8 implements two successive RNSs, and it is an identical-feedback LFSR pair with the clock cycle difference $c = \lfloor \frac{m}{2} \rfloor$. Note that the design only needs $(m + \lfloor \frac{m}{2} \rfloor)$ DFFs, which is fewer than those needed by two separate LFSRs (i.e., $2m$ DFFs).

3.2 *Method for Building Multiple Successive RNSs*

This section shows how we build n successive RNSs, where $n \geq 2$. We apply the same idea used in Sect. 3.1, i.e., inserting DFFs after an LFSR. The proposed design is shown in Fig. 9. As shown in the figure, the first RNS L_1 is an m-bit LFSR. For any $i > 1$, the i-th RNS L_i is obtained by first inserting c_{i-1} DFFs after the $(i-1)$-th RNS L_{i-1} and then using the last m DFFs of the first $(m + \sum_{j=1}^{i-1} c_j)$ DFFs. In other words, it is formed by the $(\sum_{j=1}^{i-1} c_j+1)$-th, $(\sum_{j=1}^{i-1} c_j+2)$-th, ..., $(\sum_{j=1}^{i-1} c_j+m)$-th DFFs. It can be easily shown that the RNSs $L_1, L_2, \dots$ are LFSRs of an identical-feedback polynomial. If RNS L_1 is LFSR (F, N_s), then RNS L_i $(i > 1)$ is LFSR $(F, N_{s-(\sum_{j=1}^{i-1} c_j)})$. Thus, by a single successive sequence of $(m + \sum_{j=1}^{n-1} c_j)$ DFFs, we can build n different successive RNSs, which are essentially LFSRs. However, different c_j's will lead to different seeds for these LFSRs, which will further lead to different accuracies for SC circuits. Therefore, to obtain a high accuracy, we still need to choose proper values for $c_1, c_2, \dots, c_{n-1}$.

As shown in Sect. 2.3, for the case where $n = 2$, $c_1 = \lfloor \frac{m}{2} \rfloor$ is a good choice. However, for the case where $n \geq 3$, a good choice is still unknown. In the following, we try to find it empirically. For simplicity, we consider the cases where $c_1 = c_2 = \cdots = c_{n-1} = c$. We focus on designs of RNSs with the number of RNSs n satisfying $3 \leq n \leq 6$ and the bit width m satisfying $6 \leq m \leq 8$. The number of used DFFs is $(m + (n-1)c)$.

Similar to Sect. 2.1, we apply an n-input SC multiplier to evaluate the performance of the n RNSs shown in Fig. 9. Denote the set of the SC multipliers, which use n m-bit RNSs formed by inserting $(n-1)c$ DFFs after L_1, as $E_{n,m,c}$, where $1 \leq c \leq 2^m - 2$. Each set $E_{n,m,c}$ consists of SC multipliers using different L_1's configured by different feedback polynomials. For each SC multiplier, we randomly generate 1000 input groups to derive its SC error. Then, for each combination (n, m, c), we obtain the average SC error over all the SC multipliers in set $E_{n,m,c}$. Then, for each pair of n and m, we consider the average SC errors for two special choices of c. The first is the value c giving the minimum average SC error over all choices of c, and the second is $c = \lfloor \frac{m}{2} \rfloor$. Their corresponding average SC errors for each pair of n and m are shown in Fig. 10. Besides, the figure also shows 2 reference cases. The first equals the average SC error of the minimum case plus $\frac{1}{2^{m+2}}$, and the second equals $\frac{1}{2^m}$. We denote the first reference case as *Ref*-1 and the second reference case as *Ref*-2.

As shown in Fig. 10, the average SC error of the case where $c = \lfloor \frac{m}{2} \rfloor$ is always smaller than that of the minimum case plus $\frac{1}{2^{m+2}}$ and is very close to $\frac{1}{2^m}$. Thus, the accuracy of the case where $c = \lfloor \frac{m}{2} \rfloor$ is very close to the best situation we consider. Therefore, the input SBSs generated by the case where $c = \lfloor \frac{m}{2} \rfloor$ can lead to a high accuracy. In summary, we can conclude:

Guideline 4 For $6 \leq m \leq 8$ and $3 \leq n \leq 6$, $c = \lfloor \frac{m}{2} \rfloor$ is a good choice for inserting DFFs to form n successive RNSs. □

The above guideline is consistent with Guideline 3. It means that we can insert $(n-1)\lfloor \frac{m}{2} \rfloor$ DFFs after L_1 to build $L_2, L_3, \dots, L_n$. In the following, for simplicity, we say that $L_2, L_3, \dots, L_n$ are derived by shifting L_1 for $\lfloor \frac{m}{2} \rfloor, 2\lfloor \frac{m}{2} \rfloor, \dots, (n-1)\lfloor \frac{m}{2} \rfloor$ DFFs, respectively. By this construction method, we can build n successive RNSs, which have low correlation and can lead to a high accuracy. The number of used DFFs is $(m + (n-1)\lfloor \frac{m}{2} \rfloor)$, which is approximately $\frac{nm}{2}$. Thus, compared to the general design of RNSs shown in Fig. 4a, the hardware cost is significantly reduced.

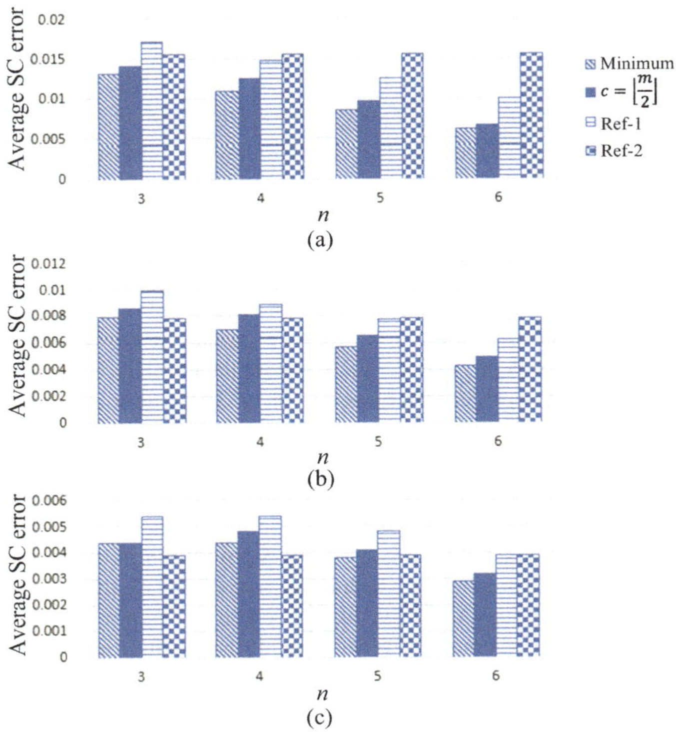

Fig. 10 Comparison of average SC errors for different cases: (**a**) $m = 6$; (**b**) $m = 7$; (**c**) $m = 8$

4 Method for Building Non-successive RNSs

The previous section proposes a method to build successive RNSs. It is natural to further explore how to build non-successive RNSs. In this section, we propose a method to achieve this purpose. We focus on the basic method for building a single non-successive RNS.

Our proposed method builds an m-bit non-successive RNS based on an m-bit LFSR. To minimize the hardware overhead, we add only one extra DFF after the LFSR. This gives us $(m + 1)$ successive DFFs. Our method to build an m-bit non-successive RNS is to select a non-successive sequence of m DFFs from the $(m + 1)$ successive DFFs.

For convenience, we first give some definitions as follows:

1. We call a DFF in the LFSR a *basic DFF*.
2. We call a basic DFF connected to the input of an XOR gate of the LFSR an *XOR DFF*.
3. Since the RNS to be built is non-successive, it must include the extra DFF added and exclude one basic DFF. We call the basic DFF not used by the non-successive RNS the *skipping DFF*.

Fig. 11 Implementation of 2 RNSs based on 7 DFFs

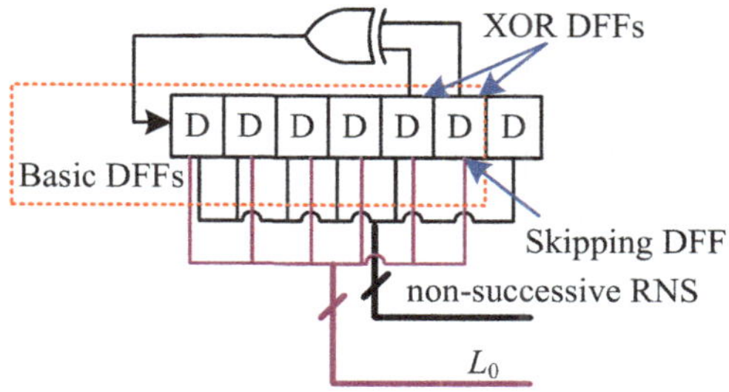

For example, as shown in Fig. 11, L_0 is a 6-bit LFSR, and an extra DFF is added after it. The first to the sixth DFFs are the basic DFFs. The fifth and the sixth DFFs are XOR DFFs. Assume that all of these DFFs except the sixth DFF form a non-successive RNS. Thus, the sixth DFF is the skipping DFF.

Our proposed design method is based on a requirement on the output number sequence of an m-bit non-successive RNS built from $(m + 1)$ successive DFFs. We first introduce this requirement.

Since the underlying m-bit LFSR outputs a random number sequence with a period of $(2^m - 1)$, the m-bit non-successive RNS also outputs a random number sequence with a period of $(2^m - 1)$, with each random number in the range $[0, 2^m - 1]$. For the random numbers in a period of $(2^m - 1)$, it is possible for some of them to repeat. However, to reduce the error of an SC circuit, it is desirable that *the random numbers in a period of* $(2^m - 1)$ *do not repeat*. To understand this, we first consider an RNS outputting a random number sequence of period 2^m, with each number in the range $[0, 2^m - 1]$. If there is a value $k \in [0, 2^m - 1]$ repeating in each period, then there will be another value $l \in [0, 2^m - 1]$ missing in each period. For easy illustration, we consider an example with $m = 3$, where the value 2 repeats twice in each period and the value 4 misses. Then, all the 8 random numbers in a period in the ascending order are 0, 1, 2, 2, 3, 5, 6, 7. Recall that in SC, to generate the SBS for an input binary number X, each random number R in a period of 2^m is compared to X, and a 1 is produced if and only if $R < X$. For the above case, if the input binary number is $X \in \{3, 4\}$, then the number of random numbers in a period less than X is $(X + 1)$. Thus, the number of ones in the SBS is $(X + 1)$, which causes a wrong encoding for X and will further lead to a wrong final output. However, if no value repeats in a period, then all the 8 random numbers in a period in the ascending order are 0, 1, 2, 3, 4, 5, 6, 7. Thus, for each input binary number $X \in [0, 7]$, the number of random numbers in a period less than X is exactly X. Thus, the number of ones in the SBS is X, which gives a correct stochastic encoding for X. For the RNS we consider here, the period of its output random numbers is $(2^m - 1)$ instead of 2^m. By a similar reasoning, it can be proved that if the random numbers in a period of $(2^m - 1)$ do not repeat, then the stochastic encoding is more accurate than that given by the random numbers with some repeat in a period of $(2^m - 1)$.

In summary, to build a non-successive RNS from $(m + 1)$ successive DFFs, we require that its output random numbers in a period of $(2^m - 1)$ do not repeat. We call this requirement *non-repeating requirement* for short. By the definition of skipping DFF, designing a non-successive RNS is equivalent to determining the skipping DFF. The following theorem shows how to select the skipping DFF to satisfy the non-repeating requirement.

Theorem 3 *Suppose that the s-th DFF of the $(m+1)$ successive DFFs is an XOR DFF, and $1 \leq s < m$. If we select the $(s+1)$-th DFF as the skipping DFF, then the remaining DFFs form a non-successive RNS satisfying the non-repeating requirement.* □

Proof Since $1 \leq s < m$, we have $2 \leq (s+1) \leq m$. Since we select the $(s+1)$-th DFF as the skipping DFF, the remaining DFFs are at positions $1, 2, \ldots, s, s+2, \ldots, m+1$, and they form a non-successive RNS.

Next, we prove that the non-successive RNS satisfies the non-repeating requirement. We prove this by contraposition. Assume that the outputs of the non-successive RNS are the same at the j_1-th and the j_2-th clock cycles, where $1 \leq j_1 < j_2 \leq 2^m - 1$. Since the $(s+1)$-th DFF is the skipping DFF, the values of i-th ($1 \leq i \leq m+1$ and $i \neq s+1$) DFF are the same at these two clock cycles. Note that the underlying LFSR satisfies the non-repeating requirement. Thus, the values of the $(s+1)$-th DFF must be different at these two clock cycles. Furthermore, since for each of the last m DFFs, its value at a given clock cycle equals the value of its left DFF at the previous clock cycle, the values of i-th ($1 \leq i \leq m$ and $i \neq s$) DFF are the same at the clock cycles $(j_1 - 1)$ and $(j_2 - 1)$, and the values of s-th DFF are different at the clock cycles $(j_1 - 1)$ and $(j_2 - 1)$. Since the s-th DFF is an XOR DFF, there is only one XOR DFF with different values at the $(j_1 - 1)$-th and the $(j_2 - 1)$-th clock cycles. Thus, the first DFF should have different values at the j_1-th and the j_2-th clock cycles. This is contradictory to the fact that the values of the first DFF are the same at these two clock cycles. Thus, there are no two clock cycles at which the non-successive RNS outputs the same binary number, which means that it satisfies the non-repeating requirement. □

Theorem 3 provides a method to build a non-successive RNS satisfying the non-repeating requirement. For example, for the 7 successive DFFs shown in Fig. 11, the fifth DFF is an XOR DFF, and $1 \leq 5 < 6$. Then, if we select the sixth DFF as the skipping DFF, the remaining DFFs form a non-successive RNS satisfying the non-repeating requirement, which has been verified by our experiment.

5 Efficient Designs of RNSs Based on DFFs

In this section, based on the proposed methods in Sects. 3 and 4, we propose 3 efficient designs of RNSs based on DFFs.

We consider building n RNSs with DFFs. For these n RNSs, we assume that $(n - w)$ of them are implemented by shifting the other w for some DFFs, where $1 \leq w \leq \frac{n}{2}$. In the following, we call the w RNSs the *RNS core* for these n RNSs.

For example, as shown in Fig. 8, 2 RNSs are implemented based on 9 DFFs. For these 2 RNSs, L_2 is implemented by shifting L_1 for 3 DFFs, and L_1 is the RNS core. In general, different RNS cores and different numbers of shifted DFFs will lead to different designs of RNSs, which will eventually cause different accuracies. To obtain a high accuracy, we need to find some proper choices for them.

As revealed in Sect. 3, the case where $c = \lfloor \frac{m}{2} \rfloor$ leads to SBSs with low correlations. Thus, in this chapter, we directly apply $x\lfloor \frac{m}{2} \rfloor$ as the number of shifted DFFs, where $x \geq 1$ is an integer. Then, we only need to focus on the design of the RNS core. In this chapter, we propose 3 designs of RNS core as follows.

RNS Core Design 1 The RNS core is constituted by an m-bit LFSR. □

RNS Core Design 2 The RNS core consists of $(m + 1)$ successive DFFs implementing two RNSs, of which one is an m-bit LFSR and the other is an m-bit non-successive RNS. □

RNS Core Design 3 The RNS core contains 2 RNSs, which are implemented by using the output of an m-bit LFSR twice. □

By shifting these 3 RNS cores for $x\lfloor \frac{m}{2} \rfloor$ DFFs, we can obtain 3 different designs of RNSs. Specifically, for the RNSs based on the RNS core 1, to implement the j-th RNS $(2 \leq j \leq n)$, we shift the RNS in the RNS core for $(j - 1)\lfloor \frac{m}{2} \rfloor$ DFFs. For the RNSs based on the RNS cores 2 and 3, to implement the j-th RNS $(3 \leq j \leq n)$, when j is odd, we shift the first RNS in the RNS core for $\frac{j-1}{2}\lfloor \frac{m}{2} \rfloor$ DFFs, and when j is even, we shift the second RNS in the RNS core for $\frac{j-2}{2}\lfloor \frac{m}{2} \rfloor$ DFFs. In the following, we refer to the designs of RNSs based on the RNS core designs 1, 2, and 3 as *successive design (S-design)*, *non-successive design (NS-design)*, and *double successive design (DS-design)*, respectively.

Furthermore, as the scrambling method, which permutes the outputs of an RNS, can improve the accuracy [17], to obtain a higher accuracy for these designs of RNSs, we consider applying the scrambling method to each RNS [17]. Moreover, note that a DFF simultaneously outputs a signal and its negation. Using some negated signals will lead to a different sequence of random binary numbers, which may also improve the accuracy [19]. Thus, we also try to use the negated signals of DFFs. For simplicity, we assume that we use the negated signals of the first h DFFs for each RNS, where $0 \leq h \leq m$. An example of using the negated signals of the first h DFFs of an RNS followed by scrambling to improve its accuracy is shown in Fig. 12.

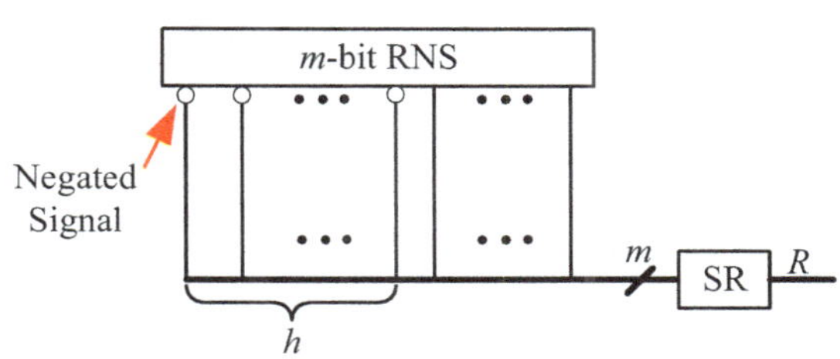

Fig. 12 An example of using the negated signals of the first h DFFs of an RNS followed by scrambling to improve its accuracy. In the figure, SR denotes the scrambling module

To find a proper h, we test the RNSs, which are based on the proposed RNSs and further use the negated signals of the first h DFFs of each RNS followed by scrambling, on n-input SC multipliers. Denote the set of the SC multipliers, which use n m-bit RNSs based on the RNS core design i and using h negated signals of each RNS, as $B_{i,h,n,m}$. Each set $B_{i,h,n,m}$ consists of SC multipliers using different feedback polynomials in their RNSs. For each SC multiplier, we randomly scramble the outputs of n RNSs by 100 times, and for each scrambling, we randomly generate 1000 input groups to obtain the MAE. Among the 100 MAEs corresponding to the 100 scramblings, we record the minimum MAE as the *final* MAE of the SC multiplier. For each combination (i, h, n, m), we obtain the average SC error by averaging the final MAEs of the SC multipliers in $B_{i,h,n,m}$. For each combination of i, n, and m, we consider two special choices of h. The first is the value h giving the minimum average SC error over all choices of h. The second is $h = m - 2$. Besides, we also consider a reference value that equals the minimum average SC error plus $\frac{1}{2^m}$. We do experiments for $m = 6, 7, 8$, $2 \leq n \leq 6$, and three proposed designs of RNSs, i.e., *S-design*, *NS-design*, and *DS-design*. Fig. 13 shows the average SC errors of the two choices of h and the reference value for $m = 8$, $2 \leq n \leq 6$, and

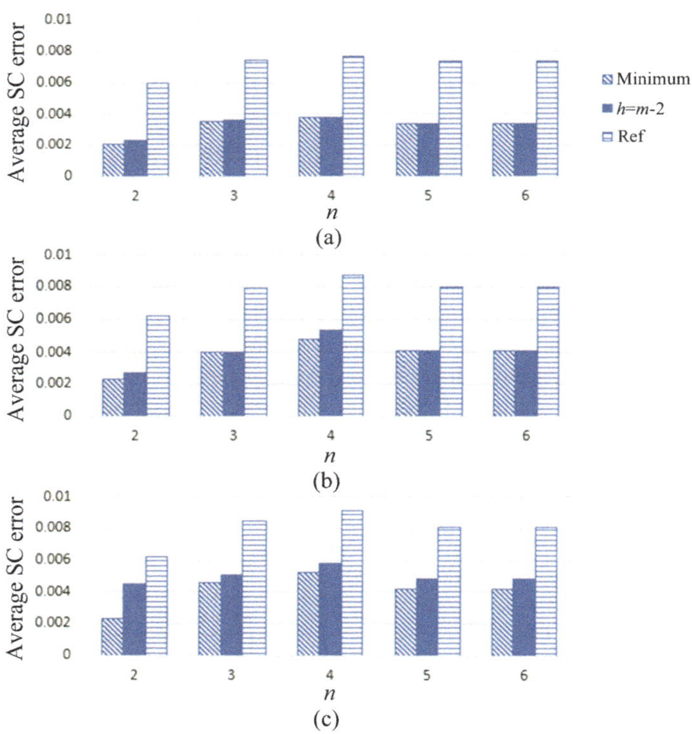

Fig. 13 Comparison of the average SC error of the best choice of h, the average SC error of $h = m - 2$, and a reference value for $m = 8$: (**a**) *S-design*; (**b**) *NS-design*; (**c**) *DS-design*

three proposed designs of RNSs. The comparisons for $m = 6, 7$ are omitted here, which are very similar to Fig. 13.

As shown in the figure, the average SC error of the case where $h = m - 2$ is very close to that of the best choice of h for *S-design* and *NS-design*. Furthermore, the average SC error of the case where $h = m - 2$ is always smaller than the reference value, which equals the average SC error of the minimum case plus $\frac{1}{2^m}$. Thus, we conclude a guideline as follows.

Guideline 5 For *S-design*, *NS-design*, and *DS-design*, if we use the negated signals of the first h DFFs of each component RNS followed by scrambling, then $h = m - 2$ is a good choice. □

Ultimately, by using the negated signals of the first $h = m - 2$ DFFs of each RNS followed by scrambling, we derive the final designs of RNSs. Figure 14 illustrates the proposed designs of RNSs for $m = 6$. Figure 14a shows an example of *S-design*. The RNS core is constituted by a 6-bit LFSR. RNS P_1 is implemented using the negated signals of the first 4 DFFs of this LFSR. RNSs P_2 and P_3 are derived by shifting P_1 for 3 and 6 DFFs, respectively. The final RNSs H_1, H_2, and H_3 are obtained by scrambling the outputs of RNSs P_1, P_2, and P_3, respectively. Figure 14b shows an example of *NS-design*. The RNS core is constituted by a 6-bit LFSR and a 6-bit non-successive RNS. RNSs P_1 and P_2 are implemented using the negated

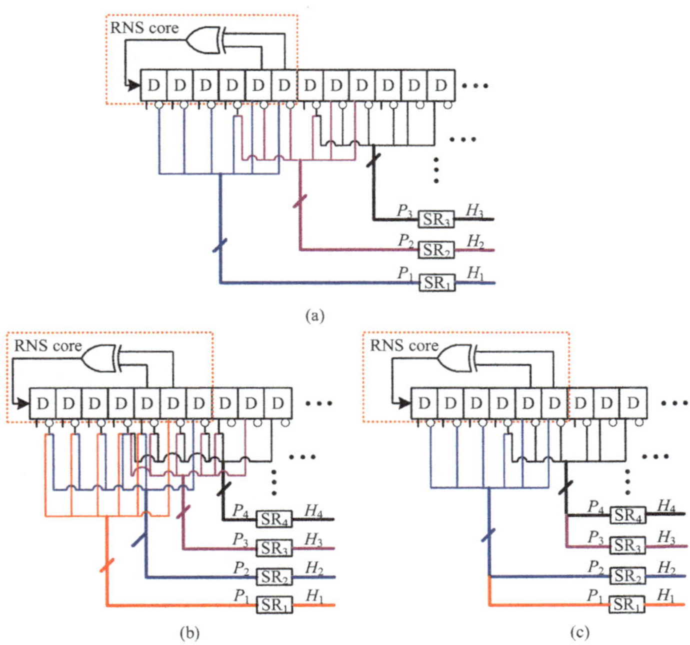

Fig. 14 Illustration of the proposed designs of RNSs with $m = 6$: (**a**) *S-design*, (**b**) *NS-design*, and (**c**) *DS-design*. In the figure, SR denotes the scrambling module

signals of the first 4 DFFs of the LFSR and the non-successive RNS, respectively. RNSs P_3 and P_4 are derived by shifting P_1 and P_2, respectively, for 3 DFFs. The final RNSs H_1, H_2, H_3, and H_4 are obtained by scrambling the outputs of RNSs P_1, P_2, P_3, and P_4, respectively. Figure 14c shows an example of *DS-design*. It is similar to the example of *NS-design* except that the RNS core contains 2 RNSs, which are implemented by using the output of a 6-bit LFSR twice. Consequently, RNSs P_1 and P_2 are both implemented by using the negated signals of the first 4 DFFs of this LFSR.

6 Experimental Results

In this section, we present the experimental results on the performance of 3 proposed designs of RNSs.

6.1 Experimental Setup

We test proposed designs of RNSs on 8 SC circuits. The first four are the circuits implementing $\sin(x)$, $\cos(x)$, $\tanh(x)$, and $\log(1 + x)$, respectively. They are synthesized by the method proposed in [22] with the degree and the precision as 4 and 6, respectively. The following 3 circuits are 3 FIR filters with orders as 8, 16, and 32, respectively. The last circuit is an SC multiplier implementing the function x^4. For simplicity, in the following, we refer to these 8 circuits as $\sin(x)$, $\cos(x)$, $\tanh(x)$, $\log(1 + x)$, f8, f16, f32, and x^4, respectively.

To show the performance of the proposed designs, we compare them to five other designs of RNSs. The first is the general design shown in Fig. 4a, which uses n LFSRs with different feedback polynomials to implement n RNSs. The second is the min-cost design shown in Fig. 4b, which only uses a single m-bit LFSR. The third uses a single $(m + 2)$-bit LFSR to generate multiple m-bit random binary numbers. It randomly chooses m bits from the $(m + 2)$-bit LFSR for n times to implement n m-bit RNSs. Note that the output binary numbers of its RNS may repeat in a period. The fourth and fifth are high-accuracy designs proposed in [13, 15], which use different Sobol sequence generators and Halton sequence generators as the RNSs, respectively. For simplicity, in the following, we refer to these 5 designs as *General*, *Mincost*, *Large*, *Sobol*, and *Halton*, respectively. Among these designs, *Sobol* generally achieves the highest accuracy for SC circuits. Since scrambling can improve the accuracy of the SC circuits, we also scramble the outputs of these designs to achieve a higher accuracy.

Then, recall that for the proposed designs, the numbers of shifted DFFs and whether to use the negated signal of a DFF are deterministic. To more clearly show the advantages of the proposed designs, we also introduce 3 random designs as references. They have the same RNS cores as our proposed designs. However, the numbers of shifted DFFs and whether to use the negated signal of a DFF are randomly determined. Furthermore, note that for these designs, when they shift more than m DFFs to derive a new RNS, the cost of the shifted DFFs will be higher than that of an LFSR. Hence, in this case, we directly apply an independent LFSR to implement the new RNS. In the following, we refer to these random designs based on the RNS core designs 1, 2, and 3 as *SR-design*, *NSR-design*, and *DSR-design*, respectively.

In summary, except for *Sobol* and *Halton*, all of these designs including the proposed ones are based on LFSRs.

6.2 Accuracy Comparison

We first compare the accuracy of the above designs of RNSs. Each design has some configuration knobs. For the designs based on LFSRs, their configuration knobs include the feedback polynomial and the seed for the underlying LFSR and the scrambling ways. For *SR-design*, *NSR-design*, and *DSR-design*, they have extra configuration knobs, i.e., the numbers of shifted DFFs and the DFFs to use negated signals. For *Sobol* and *Halton*, their configuration knobs are the scrambling ways and the corresponding sequence generators.

For each SC circuit with a design of RNSs, we randomly choose the configurations for the RNSs for 1000 times and obtain the corresponding MAEs, where the MAE of a given configuration is measured using 1000 random input groups. We choose the minimum MAE over all the 1000 configurations as the final MAE of a design of RNSs used in an SC circuit. We consider five choices of m: 4, 5, 6, 7, 8. The comparison of the MAEs of different designs of RNSs for $m = 8$ is shown in Fig. 15. As shown in the figure, the proposed designs and the random designs can achieve a very high accuracy, especially for the circuits $\sin(x)$, $\cos(x)$, $\tanh(x)$,

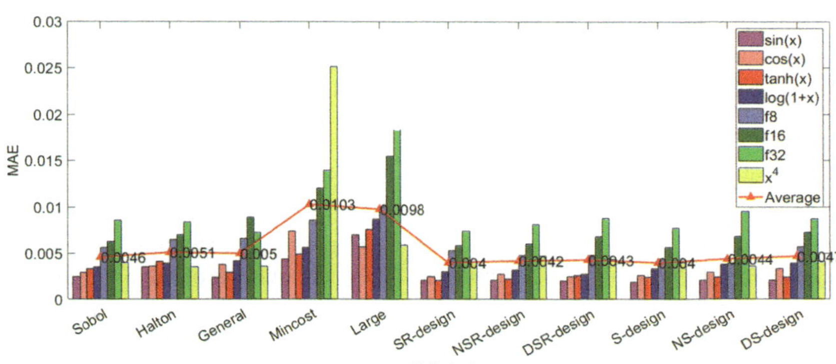

Fig. 15 MAE comparison of different designs of RNSs for $m = 8$

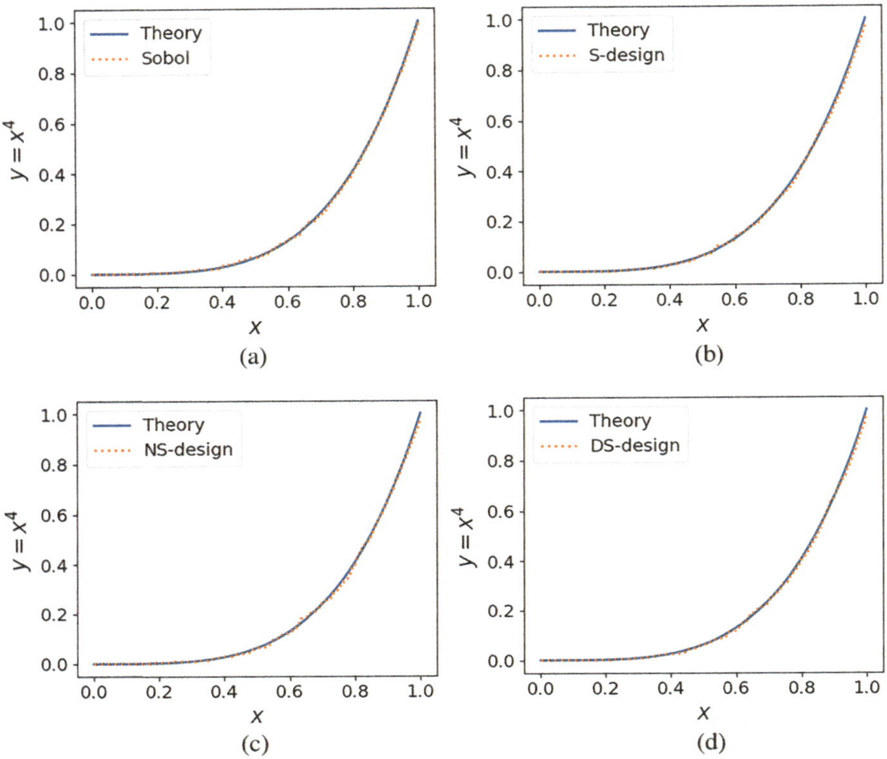

Fig. 16 The outputs of the SC multiplier implementing $y = x^4$ based on (**a**) *Sobol*, (**b**) *S-design*, (**c**) *NS-design*, and (**d**) *DS-design*

and $\log(1 + x)$. *S-design* achieves the highest accuracy on average. *Mincost* has a very low accuracy, especially for f8, f16, f32, and x^4. *Large* also has a low accuracy. This indicates that an RNS design based on scrambling the outputs of a single LFSR multiple times is not suitable for general SC circuits. The comparisons of different designs of RNSs for $m = 4, 5, 6, 7$ are omitted here, which are very similar to Fig. 15.

To illustrate the performance of our proposed designs, we also plot the output curve of the SC multiplier implementing x^4 based on our designs and *Sobol* in Fig. 16. The bit width m is set as 8, and the RNSs of the SC multipliers are configured using the best choice out of 1000 random configurations for these designs. As shown in the figure, similar to *Sobol*, the outputs of the SC multipliers based on the proposed designs are very close to the theoretical value.

For each design of RNSs, by averaging the MAEs over all the benchmarks based on the design, we further derive the average mean absolute error (AMAE) for the design. The AMAE comparison of different designs for $m = 4, 5, \ldots, 8$ is shown in Fig. 17. As shown in this figure, *Mincost* and *Large* have lower accuracies than the

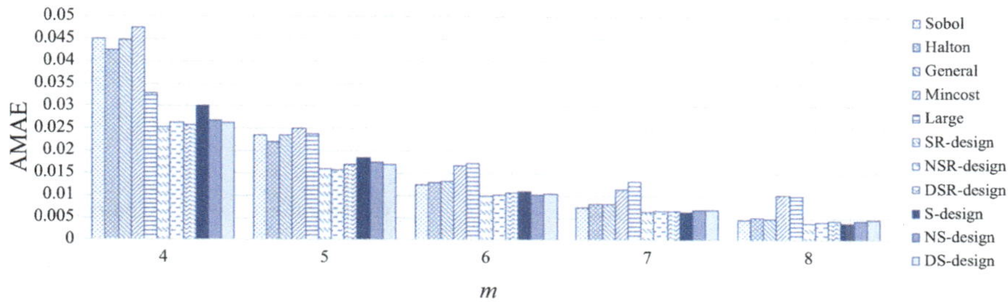

Fig. 17 AMAE comparison for different designs of RNSs

other designs. *General* can achieve a very close accuracy to *Sobol* and *Halton*. The proposed designs, i.e., *S-design*, *NS-design*, and *DS-design*, have a higher accuracy than the other designs except for the random designs. Among the three proposed designs, *S-design* has the lowest accuracy when $m = 4, 5, 6$ and the highest accuracy when $m = 7, 8$. When m is small, the proposed designs and the random designs are even far better than *Sobol* in terms of accuracy, which is the state-of-the-art design with a very high accuracy. This is because using the negated signals can significantly help reduce the correlations among the SBSs, which are generally very high when m is small. A random design generally has a higher accuracy than the corresponding proposed design, since it explores a larger configuration space including the numbers of the shifted DFFs and the DFFs to use negated signals. Yet, the proposed designs can still achieve very close accuracies to their random counterparts.

Therefore, the proposed and the random designs can achieve a high accuracy. By applying the scrambling method and using some negated signals, the proposed designs of RNSs, which are based on LFSRs, can even achieve a higher accuracy than the low-discrepancy sequence generators.

6.3 Area Comparison

In this section, we compare the area of different designs. Since SNGs generally cost most area of an SC circuit [23], in this chapter, we only compare the area of SNGs. Furthermore, the areas of *Halton* and *Sobol* are usually very large [24]. For example, the areas of an 8-bit Sobol sequence generator and an 8-bit base-2 Halton

Table 2 Average number of used XOR gates in an m-bit LFSR

m	4	5	6	7	8	9	10
Number	1	2.33	2.33	3	3.5	4.25	4.53

Table 3 Area of components

Component	DFF	XOR	Comparator	Comparator
area (μm^2)	4.77	1.60	($m = 4$)14.36	($m = 5$)17.55
Component	Comparator ($m = 6$)	Comparator ($m = 7$)	Comparator ($m = 8$)	
area (μm^2)	20.21	23.41	26.87	

sequence generator are 181.68 μm^2 and 65.97 μm^2, respectively.[1] They are much larger than that of an 8-bit LFSR, which is 43.76 μm^2. Therefore, in the chapter, we only compare the areas of the designs of RNSs based on LFSRs.

An SNG is composed of an RNS and a comparator, and a design of RNSs based on LFSR consists of DFFs and XOR gates. The average number of used XOR gates in an m-bit LFSR is listed in Table 2 for $m = 4, 5, \ldots, 10$, where the 9-bit and the 10-bit LFSRs are used for *Large*. For the DFF, the XOR gate, and the comparator, we directly obtain their areas by Synopsys Design Compiler [25] based on the Nangate 45nm library [26]. Their areas are listed in Table 3. Then, based on this table, for each design, we estimate the area of the SNGs used in a benchmark by summing the areas of their components. For example, for an SC circuit with $n = 6$ and $m = 8$, the SNGs based on *General* need 6 8-bit LFSRs and 6 8-bit comparators. The area is $6 \times (8 \times 4.77 + 3.5 \times 1.60) + 6 \times 26.87 = 423.78\ \mu m^2$. After estimating the area, for each design of RNSs and each m, we further derive the average area of the SNGs based on the design by averaging the areas of SNGs over all benchmarks as shown in Fig. 18.

As shown in the figure, the average areas of the SNGs based on the proposed designs are far smaller than that based on *General*. For example, when $m = 8$, the average areas of the SNGs based on *S-design*, *NS-design*, and *DS-design* are smaller than that based on *General* by 18%, 25%, and 26%, respectively. Furthermore, the proposed designs also have smaller areas than the corresponding random designs. For example, the average area of the SNGs based on *S-design* is smaller than that based on *SR-design*. This is because for the random designs, the number of shifted DFFs is generally larger than m, and they need to apply an independent LFSR to implement a new RNS in this case. Moreover, *Mincost* and *Large* have smaller areas than the other designs, while *General* always has the largest area. This is because

[1] Note that the area of a Sobol sequence generator is larger than that of a Halton sequence generator here. This is because a Sobol sequence generator includes a least significant zero detector and the very hardware-demanding storage arrays [13], while a Halton sequence generator only needs several b-bit counters to implement, where b is the base [15]. Thus, when the base is small, the area of a Halton sequence generator is smaller than that of a Sobol sequence generator.

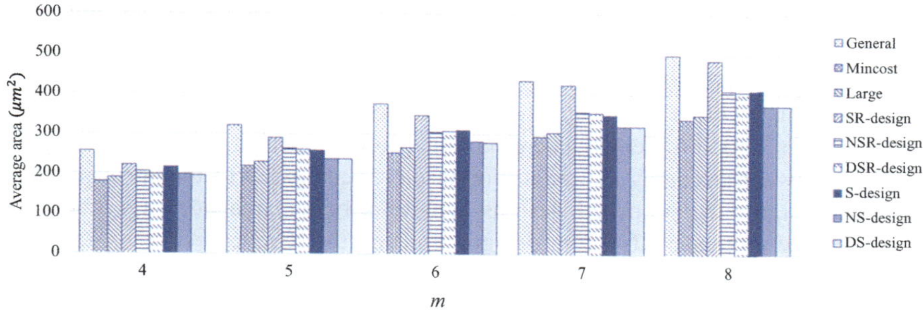

Fig. 18 Comparison of the average areas (μm^2) of the SNGs based on different designs of RNSs

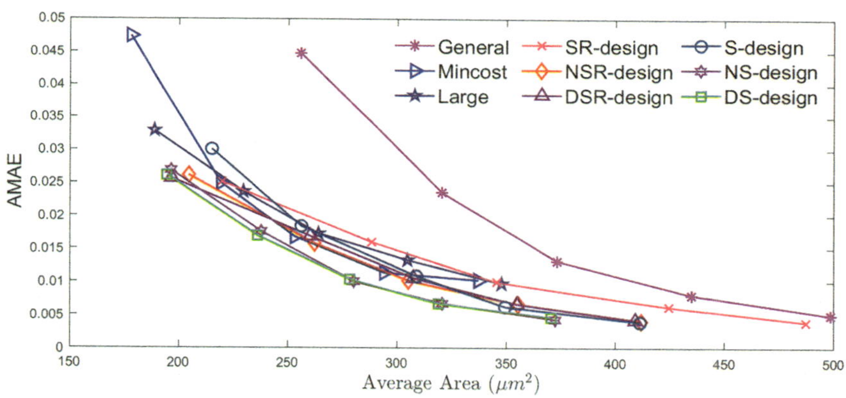

Fig. 19 Comparison of the area–accuracy trade-offs for different designs of RNS, where the accuracy values are the AMAEs from Fig. 17 and the area values are from Fig. 18

Mincost and *Large* only need a single LFSR, and *General* needs n separate LFSRs, to implement.

Therefore, compared to the other designs except for *Mincost* and *Large*, the proposed designs have smaller areas.

6.4 Area–Accuracy Trade-off Comparison

As shown in Sects. 6.2 and 6.3, compared to the proposed designs, *Mincost* and *Large* have a lower accuracy but also a smaller area, while the random designs have a higher accuracy but also a larger area. To understand the performance of the proposed designs of RNSs better, we plot the area–accuracy trade-off curves for different designs of RNSs in Fig. 19, where the accuracy values are the AMAEs from Fig. 17 and the area values are from Fig. 18. Each curve in Fig. 19 corresponds to a design of RNSs, and the five points on each curve correspond to five choices of $m \in \{4, 5, \ldots, 8\}$.

As shown in the figure, *NS-design* and *DS-design* achieve better area–accuracy trade-off than the other designs, and *DS-design* achieves the best trade-off. *S-design* also achieves a good trade-off. Therefore, the proposed designs can achieve a good balance between accuracy and area. Furthermore, compared to the corresponding random designs, the proposed designs achieve a better area–accuracy trade-off. Thus, for the proposed RNS core designs, shifting the RNS cores for $x\lfloor \frac{m}{2} \rfloor$ DFFs and using the negated signals of $(m-2)$ DFFs for an RNS are better than randomly shifting for some DFFs and using the negated signals of some DFFs.

In summary, the proposed RNSs can achieve a high accuracy, a small area, and a good balance between accuracy and area.

7 Conclusion

In this chapter, focusing on designing efficient RNSs based on DFFs, we first present several novel guidelines on using LFSRs as RNSs. Then, based on them, we propose a method to design RNSs based on a successive sequence of DFFs. After that, we also propose a method to design RNSs based on a non-successive sequence of DFFs. It has the property that there is no repeated random binary number in its output period. Eventually, based on these new methods, by applying scrambling and using negated signals, we propose 3 new designs of RNSs based on DFFs. The experimental results show that the proposed designs can achieve a high accuracy, a small area, and a good balance between accuracy and area.

References

1. B.R. Gaines, Stochastic computing systems, in *Advances in Information Systems Science* (Springer, Berlin, 1969), pp. 37–172
2. X.-R. Lee, C.-H. Chen, H.-C. Chang, C.-Y. Lee, A 7.92 Gb/s 437.2 mW stochastic LDPC decoder chip for IEEE 802.15.3c applications. IEEE Trans. Circuits Syst. I: Regul. Pap. **62**(2), 507–516 (2015)
3. P. Li, D.J. Lilja, Using stochastic computing to implement digital image processing algorithms, in *International Conference on Computer Design* (2011), pp. 154–161
4. A. Alaghi, C. Li, J.P. Hayes, Stochastic circuits for real-time image-processing applications, in *Design Automation Conference* (2013), pp. 136:1–136:6
5. P. Li, D.J. Lilja, W. Qian, K. Bazargan, M.D. Riedel, Computation on stochastic bit streams: Digital image processing case studies. IEEE Trans. Very Large Scale Integr. Syst. **22**(3), 449–462 (2014)
6. Z. Li, A. Ren, J. Li, Q. Qiu, Y. Wang et al., DSCNN: hardware-oriented optimization for stochastic computing based deep convolutional neural networks, in *International Conference on Computer Design* (2016), pp. 678–681
7. H. Sim, J. Lee, A new stochastic computing multiplier with application to deep convolutional neural networks, in *Design Automation Conference* (2017), pp. 29:1–29:6

8. A. Ren, Z. Li, C. Ding, Q. Qiu, Y. Wang et al., SC-DCNN: highly-scalable deep convolutional neural network using stochastic computing, in *International Conference on Architectural Support for Programming Languages and Operating Systems* (2017), pp. 405–418
9. K. Kim, J. Kim, J. Yu, J. Seo, J. Lee et al., Dynamic energy-accuracy trade-off using stochastic computing in deep neural networks, in *Design Automation Conference* (2016), pp. 124:1–124:6
10. B. Li, Y. Qin, B. Yuan, D. J. Lilja, Neural network classifiers using stochastic computing with a hardware-oriented approximate activation function, in *International Conference on Computer Design* (2017), pp. 97–104
11. Y. Liu, S. Liu, Y. Wang, F. Lombardi, J. Han, A survey of stochastic computing neural networks for machine learning applications. IEEE Trans. Neural Netw. Learn. Syst. **32**(7), 2809–2824 (2021)
12. A. Alaghi, J.P. Hayes, Exploiting correlation in stochastic circuit design, in *International Conference on Computer Design* (2013), pp. 39–46
13. S. Liu, J. Han, Energy efficient stochastic computing with Sobol sequences, in *Design, Automation & Test in Europe Conference & Exhibition* (2017), pp. 650–653
14. M.H. Najafi, D.J. Lilja, M. Riedel, Deterministic methods for stochastic computing using low-discrepancy sequences, in *International Conference on Computer-Aided Design* (2018), pp. 1–8
15. A. Alaghi, J. Hayes, Fast and accurate computation using stochastic circuits, in *Design, Automation & Test in Europe Conference & Exhibition* (2014), pp. 1–4
16. M.V. Daalen, P. Jeavons, J. Shawe-Taylor, D.A. Cohen, A device for generating binary sequences for stochastic computing. Electron. Lett. **29**, 80–81 (1993)
17. J.H. Anderson, Y. Hara-Azumi, S. Yamashita, Effect of LFSR seeding, scrambling and feedback polynomial on stochastic computing accuracy, in *Design, Automation & Test in Europe Conference & Exhibition* (2016), pp. 1550–1555
18. S.A. Salehi, Low-cost stochastic number generators for stochastic computing. IEEE Trans. Very Large Scale Integr. Syst. **28**(4), 992–1001 (2020)
19. K. Zhong, Z. Li, W. Qian, Towards low-cost high-accuracy stochastic computing architecture for univariate functions: design and design space exploration, in *Design, Automation & Test in Europe Conference & Exhibition* (2022), pp. 346–351
20. K. Zhong, Z. Li, H. Jin, W. Qian, Exploiting uniform spatial distribution to design efficient random number source for stochastic computing, in *International Conference on Computer-Aided Design* (2022), pp. 1–9
21. Z. Li, Z. Chen, Y. Zhang, Z. Huang, W. Qian, Simultaneous area and latency optimization for stochastic circuits by D flip-flop insertion. IEEE Trans. Comput.-Aided Design Integr. Circuits Syst. **38**(7), 1251–1264 (2019)
22. X. Peng, W. Qian, Stochastic circuit synthesis by cube assignment. IEEE Trans. Comput.-Aided Design Integr. Circuits Syst. **37**(12), 3109–3122 (2018)
23. A. Alaghi, W. Qian, J.P. Hayes, The promise and challenge of stochastic computing. IEEE Trans. Comput.-Aided Design Integr. Circuits Syst. **37**(8), 1515–1531 (2018)
24. M.H. Najafi, D. Jenson, D.J. Lilja, M.D. Riedel, Performing stochastic computation deterministically. IEEE Trans. Very Large Scale Integr. Syst. **27**(12), 2925–2938 (2019)
25. Synopsys Inc. http://www.synopsys.com
26. Nangate Inc. http://www.nangate.com

Stochastic Multipliers: from Serial to Parallel

Yongqiang Zhang, Jie Han, and Guangjun Xie

1 Introduction

The importance of multipliers in data-processing-related applications, such as digital signal processing (DSP) and neural networks (NNs), is self-evident. The most commonly studied multipliers are binary and Booth multipliers. However, these multipliers encounter high hardware costs in these complex applications. Thanks to emerging fault-tolerant applications, exact computing results can be relaxed to some extent according to the user's desired computing accuracy. In this way, the hardware costs can then be alleviated accordingly.

In the early 1950s, John von Neumann has found the traces of stochastic computing (SC) and delivered a series of lectures [1]. The basic principle of SC is using redundant error-prone components to generate acceptable results in space or in time. However, this has not drawn much attention until 1967. This year, several individuals have formally and independently put forward the concrete concepts of SC and basic computing elements performing simple arithmetic operations [2–4]. After about 30 years of quiescence later, SC has reemerged and applied to many domains since 2000 [5].

As a common arithmetic component, stochastic multipliers have been widely studied for decades. SC has several number representation formats, including unipolar, bipolar, likelihood ratio, and log-likelihood ratio formats [6]. Without other specifications, all numbers are represented in a unipolar format in this chapter, which is also mostly studied. The designs of stochastic multipliers are divided

Y. Zhang (✉) · G. Xie
Hefei University of Technology, Hefei, China
e-mail: ahzhangyq@hfut.edu.cn; gjxie8005@hfut.edu.cn

J. Han
University of Alberta, Edmonton, AB, Canada
e-mail: jhan8@ualberta.ca

W. Liu et al. (eds.), *Design and Applications of Emerging Computer Systems*,
https://doi.org/10.1007/978-3-031-42478-6_9

into two types, including serial and parallel multipliers. As the suggestion made by Gaines, Poppelbaum, and Ribeiro, an AND gate with two independent input bit sequences can serve as a serial stochastic multiplier, which could provide sufficient computing accuracy as the independent sequences are long enough [2–4]. It is the simplest design by employing only an AND gate to save hardware costs by sacrificing computational throughput. After that, stochastic multipliers have attracted much attention to saving hardware costs and speeding up operations at the same time, while providing the requirements for computing accuracy. However, stochastic multipliers always perform in an inaccurate manner. Thus, deterministic approaches have been proposed to compute exactly, by imitating convolution in mathematics.

Many prior designs mainly focus on the designs of stochastic multipliers in hardware, so this chapter will review them in detail. The remainder of this chapter is organized as follows. Section 2 introduces the principle of the designs of existing binary and Booth multipliers, and the basics of SC. Section 3 details the design approaches of stochastic multipliers. Finally, this chapter is concluded in Sect. 4.

2 Background

Generally, a multiplier consists of three stages, including partial product generation, partial product accumulation, and final addition. Let $x = x_{n-1}x_{n-2} \ldots x_1x_0$ and $y = y_{n-1}y_{n-2} \ldots y_1y_0$, respectively, be multiplicator and multiplicand to be multiplied through an n-bit multiplier. The partial products can be generated through AND gates or Booth encoders, where the former is called a binary multiplier and the latter is a Booth multiplier.

2.1 Binary Multiplier

Figure 1 shows the schematic and working principle of a 4-bit unsigned binary multiplier, where the multiplicator and multiplicand are $x_3x_2x_1x_0$ and $y_3y_2y_1y_0$, resulting in a result $z = z_7z_6z_5z_4z_3z_2z_1z_0 = x_3x_2x_1x_0 \times y_3y_2y_1y_0$. Each pp_{ij} is generated by an AND gate, i.e., $pp_{ij} = y_ix_j$ ($i, j = 0, 1, 2,$ and 3). Generally, the partial product accumulation is mainly based on the Wallace tree or the Dadda tree [7]. Here, full adders, half adders, and 4-2 compressors are used to compress the products based on the Dadda tree. The red text is the generated signals through the full adders, half adders, and compressors in each stage, and serves as inputs for the next components. For example, the ANDed results pp_{02} and pp_{11} generate sum s_0 and carry c_0, where s_0 and c_0, respectively, provide inputs of full adders in the next column and stage. The final addition is accomplished through a ripple carry adder, generating the final multiplied 8-bit result $z_7z_6z_5z_4z_3z_2z_1z_0 = c_9s_9s_8s_7s_6s_5s_4pp_{00}$.

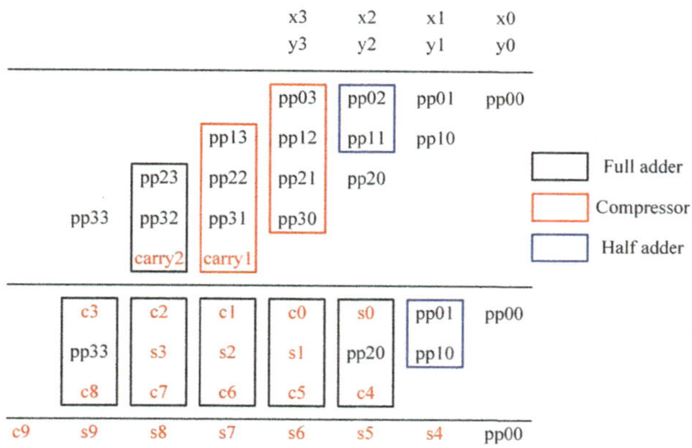

Fig. 1 The schematic of a 4-bit binary multiplier, including partial product generation through AND gates, partial product accumulation through the full adders, half adders, and 4-2 compressors, and a final addition through a ripple carry adder

The logical expression for a half adder is

$$\begin{aligned} c_o &= ab \\ s &= a \oplus b, \end{aligned} \tag{1}$$

where a and b are two inputs, c_o and s, respectively, are outputs carry and sum, and $\oplus$ denotes the XOR operation.

The logic expression for a full adder with inputs a, b, and c_i is

$$\begin{aligned} c_o &= ab + ac_i + bc_i \\ s &= a \oplus b \oplus c_i. \end{aligned} \tag{2}$$

To lower the latency of partial product accumulation, the 4-2 compressor is used to compress products. A 4-2 compressor has five inputs a, b, c, d, and e, and three outputs c_o, s, and carry. The four inputs a, b, c, d, and the output sum s have the same weight, while the outputs c_o and carry are weighted one bit higher. According to the truth table in Table 1, its logical expression is

$$\begin{aligned} c_o &= (d \oplus c)\, b + \overline{(d \oplus c)}a \\ s &= a \oplus b \oplus c \oplus d \oplus c_i \\ \text{Carry} &= (a \oplus b \oplus c \oplus d)\, c_i + \overline{(a \oplus b \oplus c \oplus d)}a. \end{aligned} \tag{3}$$

Obviously, a 4-2 compressor can be designed by cascading two full adders.

Table 1 The truth table of a 4-2 compressor

c_i	a	b	c	d	c_o	carry	s
0	0	0	0	0	0	0	0
0	0	0	0	1	0	0	1
0	0	0	1	0	0	0	1
0	0	0	1	1	1	0	0
0	0	1	0	0	0	0	1
0	0	1	0	1	1	0	0
0	0	1	1	0	1	0	0
0	0	1	1	1	1	0	1
0	1	0	0	0	0	0	1
0	1	0	0	1	0	1	0
0	1	0	1	0	0	1	0
0	1	0	1	1	1	0	1
0	1	1	0	0	0	1	0
0	1	1	0	1	1	0	1
0	1	1	1	0	1	0	1
0	1	1	1	1	1	1	0
1	0	0	0	0	0	0	1
1	0	0	0	1	0	1	0
1	0	0	1	0	0	1	0
1	0	0	1	1	1	0	1
1	0	1	0	0	0	1	0
1	0	1	0	1	1	0	1
1	0	1	1	0	1	0	1
1	0	1	1	1	1	1	0
1	1	0	0	0	0	1	0
1	1	0	0	1	0	1	1
1	1	0	1	0	0	1	1
1	1	0	1	1	1	1	0
1	1	1	0	0	0	1	1
1	1	1	0	1	1	1	0
1	1	1	1	0	1	1	0
1	1	1	1	1	1	1	1

2.2 *Booth Multiplier*

Different from a binary multiplier, a Booth multiplier encodes the multiplicand by grouping the multiplicand according to certain rules before operations to tackle the multiplication of two's complement. Figure 2 shows two various encoding methods

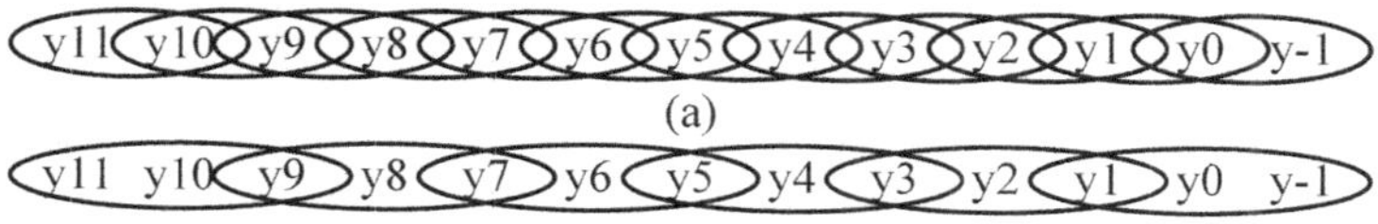

Fig. 2 Booth encoder. (**a**) Radix-2. (**b**) Radix-4

Table 2 Radix-2 booth encoders

y_k	$y_{k\text{-}1}$	d_k	c_k	pp_k
0	0	+0	0	+0
0	1	+1	0	+x
1	0	−1	1	−x
1	1	−0	1	−0

Table 3 Radix-4 booth encoders

y_{2k+1}	y_{2k}	$y_{2k\text{-}1}$	d_k	c_k	pp_k
0	0	0	+0	0	+0
0	0	1	+1	0	+x
0	1	0	+1	0	+x
0	1	1	+2	0	+2x
1	0	0	−2	1	−2x
1	0	1	−1	1	−x
1	1	0	−1	1	−x
1	1	1	−0	1	−0

for a 12-bit multiplicand, respectively, called radix-2 and radix-4 Booth encoders, where a redundant bit $y_{-1} = 0$ is added to the least significant bit. The former in Fig. 2a groups and encodes two bits, while the latter in Fig. 2b groups and encodes three bits. Each grouped bits overlap one bit for two encoding methods.

The encoding rules for radix-2 and radix-4 Booth encoders are, respectively, listed in Tables 2 and 3, where d_k is an encoded value, pp_k is a partial product, and c_k is a sign correction bit. Negation operation in each partial product pp_k is achieved by complementing each bit of x and adding 1 to the least significant bit.

Figure 3 illustrates the partial product accumulation of a 4-bit unsigned radix-4 Booth multiplier, through the full adders, half adders, and 4-2 compressors. Again, the red text is the generated signals through the full adders, half adders, and compressors in each stage, and serves as inputs for the next components. The final addition is also accomplished through a ripple carry adder, generating the final multiplied 8-bit result $z = z_7z_6z_5z_4z_3z_2z_1z_0 = s_{15}s_{14}s_{13}s_{12}s_{11}s_{10}s_9s_8$. Compared to an n-bit binary multiplier, the number of partial products of an n-bit radix-4 Booth multiplier is halved to $n/2$, which will accelerate the partial product accumulation and reduce the hardware costs and latency significantly. With this purpose and feature, high-radix Booth multipliers, i.e., radix-256 Booth multiplier, attract much attention in recent years [8].

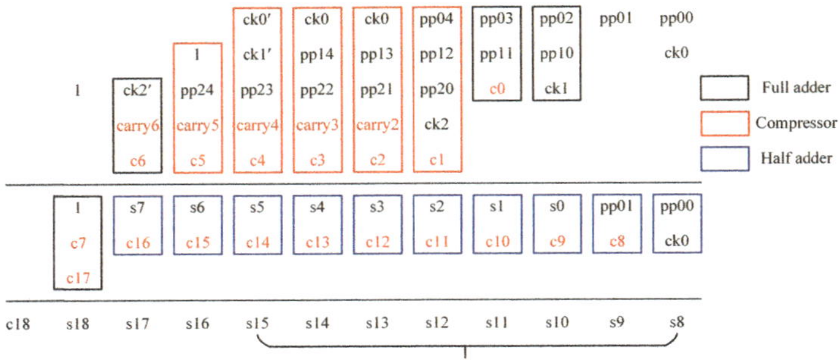

Fig. 3 The partial product accumulation of a 4-bit unsigned radix-4 Booth multiplier, through the full adders, half adders, and 4-2 compressors, and a final addition through a ripple carry adder

2.3 Stochastic Number

SC has the ability to realize arithmetic functions through very simple logic gates, which benefits from its number representation formats. Given a probability p, $0 \leq p \leq 1$, the corresponding stochastic number x in unipolar format is a bit sequence or bitstream of randomly distributed binary numbers $x_0x_1x_2 \ldots x_{n\text{-}1}$, where each x_i ($0 \leq i \leq n$-1) is a 0 or 1, with probability p. The mean value of the sequence is expected to converge to the probability p if the sequence is long enough. Thus, the precision of stochastic numbers improves with their length, which is called progressive precision. Considering the proportion of 1 s, for instance, bit sequences $x = 11{,}101{,}000$ and $y = 1010$ denote $p(x) = p(y) = 1/2$ because they both contain 50% of the bits as 1 s. Neither their length nor the positions of 1 s and 0 s are fixed. Generally, the longer the bit sequence is, the higher the accuracy becomes. However, a longer bit sequence leads to the drawback of a longer processing time. A stochastic number x in bipolar format is represented as $x = 2p\text{-}1$, where p is the probability of 1 s in a bit sequence. If a stochastic number x is in the range of $[0,\infty)$, likelihood ratio and log-likelihood ratio formats are, respectively, represented as $x = p/(1\text{-}p)$ and $x = \log(p/(1\text{-}p))$.

2.4 SC Correlation and Components

Given two stochastic numbers a and b, a simple AND gate functions as a stochastic multiplier in a unipolar format. Two bitstreams $a = 010101101001$ and $b = 110{,}101{,}110{,}110$ in Fig. 4a, respectively, encode values $p(a) = 6/12$, $p(b) = 8/12$. The generated output bitstream is obtained as $c = 01010110000$ encoding the value $p(c) = 4/12$, satisfying $c = a \times b$.

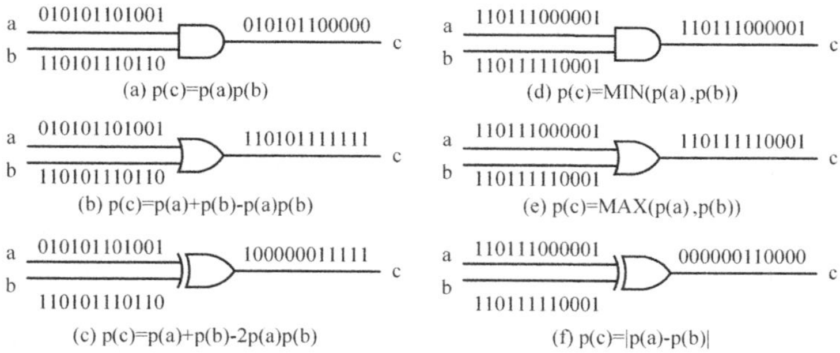

Fig. 4 (**a**), (**b**), and (**c**) Stochastic operators with uncorrelated input bitstreams. (**d**), (**e**), and (**f**) Corresponding operators with correlated input bitstreams

The computing accuracy of SC generally relies on the independence of input bitstreams. Therefore, most designs avoid correlation, by using D flip-flops (DFFs) as a common element to decrease the correlation. The bitstream correlation is called an SC correlation (SCC) and is defined as (4) [9].

$$\mathrm{SCC}\left(S_x, S_y\right) = \begin{cases} \frac{\delta(S_x,S_y)}{\min(P(S_x),P(S_y))-P(S_x)P(S_y)}, & \left(\delta\left(S_x, S_y\right) > 0\right) \\ 0, & \left(\delta\left(S_x, S_y\right) = 0\right) \\ \frac{\delta(S_x,S_y)}{P(S_x)P(S_y)-\max(P(S_x)+P(S_y)-1,0)}, & \left(\delta\left(S_x, S_y\right) < 0\right) \end{cases}, \quad (4)$$

where $\delta(S_x,S_y) = P(S_x \wedge S_y)\text{-}P(S_x)P(S_y)$. The value of SCC is between -1 and 1. When SCC $= 0$, it indicates that the bitstreams are ideally independent. When SCC is 1 or $-$ 1, it indicates that the correlation between bitstreams reaches the maximum. By sharing an RNG, bitstreams with maximal correlation can be obtained [10]. Figure 4d shows that the AND gate can realize the function of MIN(A,B) when its input bitstreams have a maximal correlation. The remaining logic gates in Fig. 4 implement the functions when the input bitstreams of OR and XOR gates are independent and maximally correlated, respectively.

SC uses a stochastic number generator (SNG) to convert a given binary number x within [0, N] to a stochastic number. An SNG generates a random integer value $R \in [0, N]$ by a random number generator (RNG). The value R is compared with the binary number to generate an N-bit sequence with a probability of x/N. To convert a stochastic bitstream back to its binary encoded format, a counter is used, to sum up each bit in the bitstream. The conversion circuits can consume up to 80% of the total area in a stochastic circuit [11].

3 The Design Approaches of Stochastic Multipliers

3.1 Stochastic Multiplier

3.1.1 Shared Stochastic Number Generator-Based Multiplier

The basic serial stochastic multiplier is an AND gate with two independent inputs. The latency of 2^n clock cycles for the serial multiplier exponentially grows as the bitwidth n of binary inputs increases, which leads to a noticeable decrease in energy efficiency. In addition, two SNGs to generate independent inputs consume too large area and energy. The outputs of a linear feedback shift register (LFSR) are reversed and shared between two SNGs to halve the consumed area of a multiplier [12], as shown in Fig. 5. The reversed order of the outputs of the LFSR guarantees that a minimum SCC occurs between input bitstream1 and bitstream2 of the AND gate, resulting a high computing accuracy. This design saves hardware costs while improving computing accuracy, compared to generating bitstreams independently. However, energy efficiency remains at a low level because of the noticeable latency.

3.1.2 Advanced Termination-Based Multiplier

To reduce the latency of the multiplier using an AND gate, a multiplexer (MUX)-based multiplier can terminate the operation in advance by detecting each bit in the multiplicator bitstream [13], as shown in Fig. 6. It uses a finite state machine (FSM) as the multiplexer's select input to generate a relatively uniformly distributed low discrepancy sequence as the bitstream of input multiplicator. The multiplicand bitstream enters a down counter. The multiplicand bitstream is reduced by 1 every clock cycle if the multiplicator bitstream is detected to be 1. If the down counter reaches 0, the multiplier finishes the operations. It shows a high computing accuracy and no longer requires 2^n clock cycles. However, if the multiplicand in the down

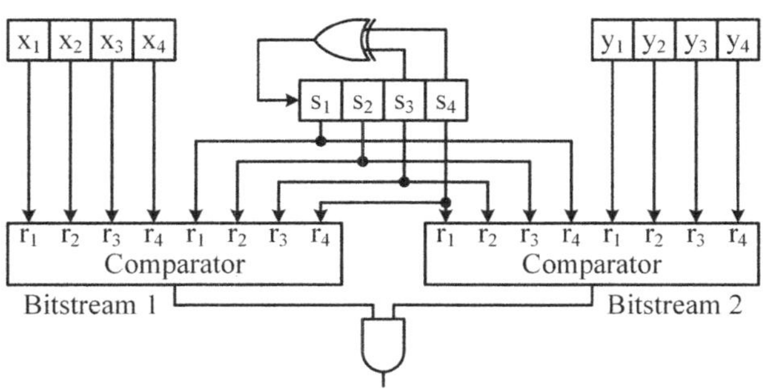

Fig. 5 A serial stochastic multiplier using a shared LFSR

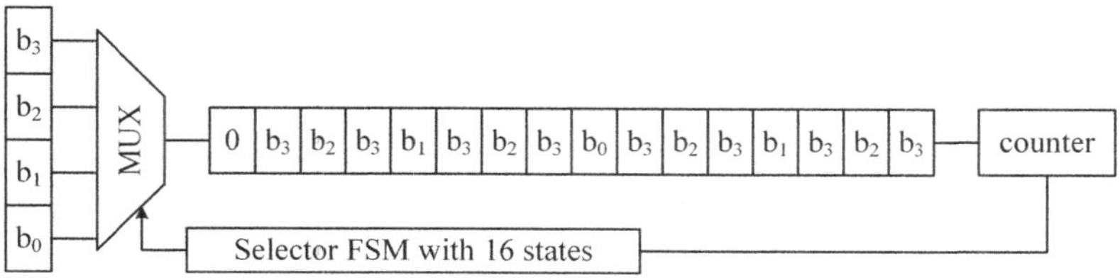

Fig. 6 An advanced termination-based multiplier using a multiplexer and a finite state machine

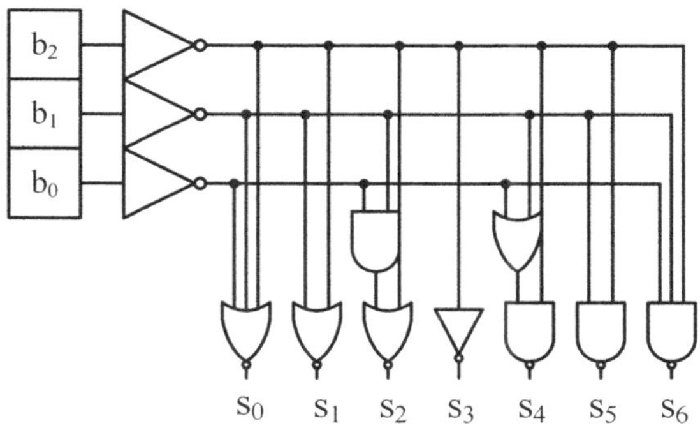

Fig. 7 A parallel thermometer code-based SNG

counter is large or even close to 1, it requires the same number of clock cycles as the serial stochastic multiplier.

Assuming that the multiplicand follows a uniform distribution in the interval $[0,2^n\text{-}1]$, the expectation of the required number of clock cycles is $(1 + 1 + 2 + 3 \ldots + 2^n-1)/2^n = 2^{n\text{-}1}-1/2$, where each element in the parenthesis gives the number of clock cycles for computing each number. This shows that the multiplier approximately reduces the number of clock cycles by half on average, compared to the serial one using a shared stochastic number generator.

3.1.3 Thermometer Code-Based Multiplier

Using parallel SNGs, the latency of a stochastic multiplier can significantly be reduced because the parallel bitstreams are generated in only one clock cycle. An example is the thermometer code-based SNG [14], as shown in Fig. 7. Its truth table is listed in Table 4. It only requires several logical gates to generate parallel bitstreams. However, few studies on parallel designs have been carried out because of the large hardware cost of massively parallel computing units. For example, it only requires one AND gate to perform the multiplication of two n-bit numbers for a serial multiplier, while this number increases to 2^n for a parallel design.

Table 4 The truth table of thermometer code

Decimal	Binary			Thermometer code						
	b_0	b_1	b_2	s_0	s_1	s_2	s_3	s_4	s_5	s_6
0	0	0	0	0	0	0	0	0	0	0
1	0	0	1	0	0	0	0	0	0	1
2	0	1	0	0	0	0	0	0	1	1
3	0	1	1	0	0	0	0	1	1	1
4	1	0	0	0	0	0	1	1	1	1
5	1	0	1	0	0	1	1	1	1	1
6	1	1	0	0	1	1	1	1	1	1
7	1	1	1	1	1	1	1	1	1	1

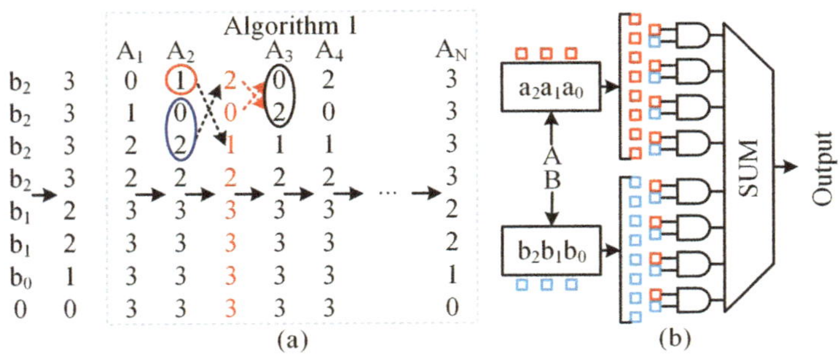

Fig. 8 (**a**) An example of Algorithm 1. (**b**) A 3-bit parallel multiplier

3.1.4 Optimal Multiplicative Bitstream-Based Multiplier

The optimal multiplicative bitstream (OpMulbs)-based SNG provides a significant improvement in the performance of a stochastic multiplier [15]. Algorithm 1 describes the principle of finding OpMulbs for an n-bit binary number from an initial bitstream. Figure 8a illustrates the permutation process of Algorithm 1 for an initial vector. The bitstreams with the smallest mean squared errors (MSEs) are saved to be OpMulbs.

With the OpMulbs, parallel stochastic multipliers are designed to reduce latency from 2^n clock cycles to 1 clock cycle. Figure 8b shows a 3-bit multiplier with the binary inputs $A = a_2a_1a_0$ and $B = b_2b_1b_0$. The multiplier uses hard-wired connections to generate parallel bitstreams, followed by $2^n = 2^3 = 8$ AND gates and a SUM unit to obtain 3-bit binary results in one clock cycle. The use of hard-wired connections exempts the generation of stochastic bitstreams from dedicated SNGs. These simple connections benefit from OpMulbs, which makes the multipliers more hardware efficient.

Algorithm 1: Algorithm for finding optimal multiplicative bitstreams generated by an n-bit binary number

---

Input: n (bitwidth of a binary number)
Output: A (matrix to store all permutations without duplications)
Initialization: $N = C_2^1 \times C_4^2 \times \cdots \times C_{2^n}^{2^{n-1}}, A = \text{zeros}(N, 2^n)$
for $i = 1$ to n do
  $A(1, 2^{i-1} + 1{:}2^i) = i$; % denoting an initial bitstream or initial vector
for $p = 2$ to N do
  $t = A(p\text{-}1,:)$;
  for $p_1 = 2$ to 2^n do
    if $t(p_1-1) < t(p_1)$ then break;
  for $p_2 = 1$ to 2^n do
    if $t(p_2) < t(p_1)$ then break;
  exchange($t(p_1)$,$t(p_2)$);
  for $k = 1$ to $p_1/2$ do
    exchange($t(k)$,$t(p_1\text{-}k)$);
  $A(p,:) = t$;

3.1.5 Evaluation

3.1.5.1 Computing Accuracy

The MSE for an n-bit multiplier is computed by

$$\text{MSE} = \sum_{x=0}^{2^n-1} \sum_{y=0}^{2^n-1} \frac{\left(p_{xy} - p_x \times p_y\right)^2}{2^n \times 2^n}, \tag{5}$$

where p_x and p_y, respectively, represent the binary values of a permutation and the initial bitstream, and p_{xy} denotes their ANDed result. The mean absolute error (MAE) is also employed to evaluate the computing accuracy. The MAE is obtained by one million Monte Carlo trials for each input within the range of [0,1] randomly generated by using the rand function in MATLAB.

Tables 5 and 6 list the smallest MSEs and MAEs of 3-bit to 8-bit multipliers, designed using the shared LFSR (Shared), advanced termination using a

Table 5 The MSEs of 3-bit to 8-bit multipliers

n-bit \ MSE	3 (10^{-3})	4 (10^{-4})	5 (10^{-4})	6 (10^{-5})	7 (10^{-5})	8 (10^{-6})
Shared	2.37	9.20	3.37	11.78	3.97	12.94
MUX	2.26	6.75	1.93	5.38	1.48	4.02
Thermo	11.5	112.2	111.4	1111.8	1111.3	11111.5
OpMulbs	2.26	6.75	1.93	5.38	1.48	4.02

Table 6 The MAEs of 3-bit to 8-bit multipliers

n-bit \ MAE	3 (10^{-3})	4 (10^{-4})	5 (10^{-4})	6 (10^{-5})	7 (10^{-5})	8 (10^{-6})
Shared	3.59	2.23	1.44	8.82	5.26	3.07
MUX	3.52	2.02	1.10	5.89	3.10	1.62
Thermo	8.20	8.30	8.33	83.31	83.33	83.33
OpMulbs	3.52	2.02	1.10	5.89	3.10	1.62

MUX (MUX), thermometer code (Thermo), and OpMulbs. As mentioned before, "Shared" means the outputs of an LFSR are reversed and shared between two SNGs to realize a high computing accuracy [12]. The multipliers using OpMulbs provide the same MSEs and MAEs as the MUX-based designs in [13], while the shared LFSR-based multipliers reach higher values of these metrics. The designs using thermometer code occur the largest MSEs and MAEs [16], since all 1 s in the thermometer code proceed to 0 s and the 0 s involved in multiplication do not affect the number of 1 s in results.

3.1.5.2 Hardware Costs

All circuits are synthesized by the Synopsys Design Compiler with a TSMC 65 nm library at the typical corner and with a nominal supply voltage of 1.2 V. The command "compile_ultra" is used to ungroup all components and automatically synthesize the circuits according to timing constraints. The power is measured by PrimePower using a vector-free power analysis model.

Table 7 shows the hardware costs of 3-bit multipliers. The MUX-based design terminates operations in advance to reduce latency according to the inputs. Its latency is computed by the critical path delay (CPD) multiplied by the expected number of required clock cycles, i.e., $0.46 \times (2^{n-1}\text{-}1/2) = 0.46 \times (2^{3-1}\text{-}1/2)$. In the same way, the shared LFSR-based multiplier is also a serial design, of which the latency is computed by the CPD multiplied by the number of clock cycles, i.e., $0.11 \times 2^{\mathrm{n}} = 0.11 \times 2^3$.

The latencies of the Thermo and OpMulbs multipliers are, respectively, equal to their CPD since they are designed to operate in parallel. Energy efficiency (EE) is denoted as the ratio of the energy of previous multipliers to that of the OpMulbs. With these data, the OpMulbs multiplier is more energy efficient than its stochastic

Table 7 The hardware costs of 3-bit multipliers

Multiplier	Area (μm^2)	Power (mW)	CPD (ns)	Latency (ns)	Energy (pJ)	EE
Shared	112.7	0.012	0.11	0.88	0.010	3.5
MUX	134.3	0.012	0.46	1.61	0.019	6.5
Thermo	85.7	0.005	0.71	0.71	0.004	1.3
OpMulbs	74.2	0.005	0.57	0.57	0.003	1.0

Table 8 The hardware costs of 8-bit multipliers

Multiplier	Area (μm^2)	Power (mW)	CPD (ns)	Latency (ns)	Energy (pJ)	EE
Shared	274.7	0.021	0.17	43.52	0.93	24.1
MUX	379.1	0.033	0.71	90.52	2.95	76.5
Thermo	3543.5	0.113	6.25	6.25	0.70	18.3
OpMulbs	405.7	0.028	1.37	1.37	0.04	1.0

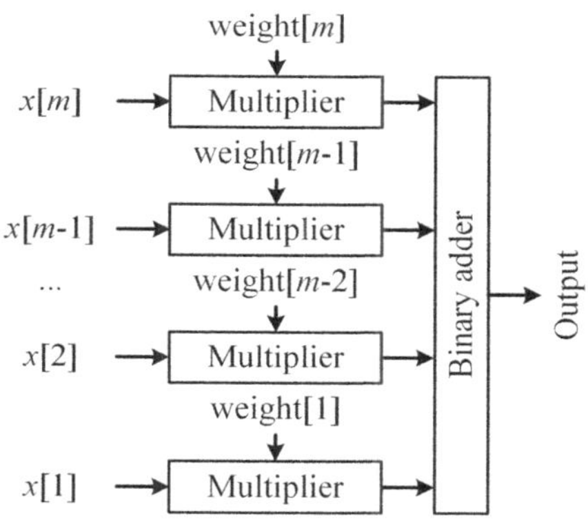

Fig. 9 The architecture for an m-input MAC using the OpMulbs multipliers

counterparts. This benefits from the parallel design and the hard-wired connections dominated by OpMulbs to avoid complex SNGs. In addition, the hardware costs for 8-bit multipliers are listed in Table 8, which indicate the energy efficiency of the OpMulbs multipliers.

3.1.6 Multiply-Accumulate Unit

The 6-bit OpMulbs multiplier shows an MSE of 5.38×10^{-5}, which is low enough for many fault-tolerant applications [12]. Hence, 3-bit to 6-bit multipliers are investigated in various applications to verify the efficacy of the stochastic multipliers.

An m-input MAC using the OpMulbs multipliers requires only cascading m multipliers in parallel, as shown in Fig. 9, where $x[j]$ and weight[j] are n-bit binary inputs of each multiplier $(1 \leq j \leq m)$. All products are summed by using a binary adder, so the results do not need to be scaled.

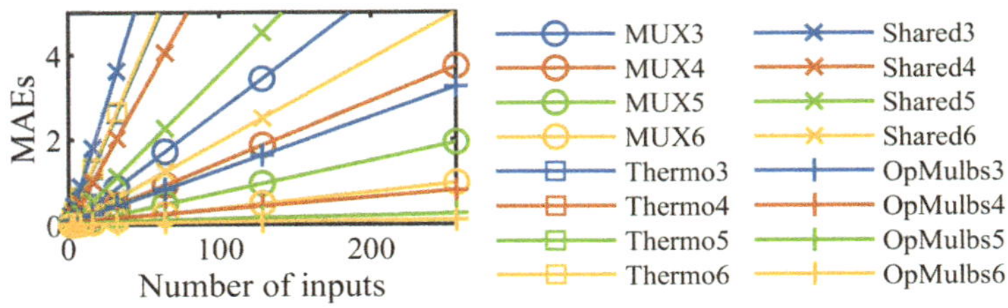

Fig. 10 The MAEs for MACs using different stochastic multipliers

3.1.6.1 Computing Accuracy

Figure 10 shows the MAEs of MACs using different stochastic multipliers versus the number of inputs m ($2 \leq m \leq 2^8$), where the terms "MUX3," "Thermo3," "Shared3," and "OpMulbs3" means a MAC composed of respective 3-bit stochastic multipliers.

The 6-bit OpMulbs multiplier-based MACs show almost the same MAEs as those denoted as MUX6, whereas Thermo6 and Shared6 present higher MAEs. The MACs using the lower-bit OpMulbs multipliers show lower MAEs than other designs, because of the utilization of OpMulbs. The data also show that the MAEs linearly increase as the number of the inputs of MACs increases. The increasing rates of the designed MACs are smaller than others.

3.1.6.2 Hardware Costs

Table 9 lists the hardware measurements of m-input MACs using various 6-bit multipliers ($2 \leq m \leq 2^8$), where the 256-input and 9-input MACs are, respectively, used for image multiplication and image smoothing later. The energy of the OpMulbs multiplier-based MACs is, respectively, reduced by 97.68%, 33.14%, and 98.73% on average compared with the MUX, Thermo, and Shared designs.

3.1.7 Image Processing

Five images including the airplane, cameraman, clock, Lena, and moon from the USC-SIPI Image Database are multiplied in pairs and smoothened to assess the practicability of multipliers [17]. The peak signal to noise ratio (PSNR) and mean structural similarity index (MSSIM) are used to evaluate the processed image quality. The PSNR is given by (6) [18].

$$PSNR = 10\log_{10}\left(w \times r \times \text{MAX}^2 / \sum_{i=0}^{w-1}\sum_{j=0}^{r-1}[S'(i,j) - S(i,j)]^2\right), \tag{6}$$

Table 9 The hardware measurements of m-input MACs using 6-bit multipliers

m		2	4	8	9	16	32	64	128	256
MUX	Area	0.07	0.13	0.27	0.30	0.54	1.07	2.14	4.27	8.53
	Power	0.06	0.11	0.22	0.24	0.43	0.85	1.70	3.40	6.83
	Latency	22.68	21.42	22.68	23.31	25.83	30.87	64.89	75.60	87.26
	Energy	1.36	2.36	5.00	5.61	11.14	26.31	110.59	257.69	597.49
Thermo	Area	0.19	0.38	0.75	0.85	1.50	2.99	5.98	11.96	23.92
	Power	0.07	0.14	0.27	0.30	0.53	1.04	2.07	4.16	8.46
	Latency	1.95	1.94	1.89	2.07	2.05	1.99	2.00	2.10	2.09
	Energy	0.14	0.27	0.50	0.62	1.09	2.06	4.14	8.75	17.67
Shared	Area	0.07	0.14	0.29	0.32	0.58	1.15	2.29	4.56	9.25
	Power	0.05	0.09	0.18	0.20	0.35	0.68	1.35	2.85	5.98
	Latency	14.08	14.72	69.76	81.28	90.24	108.80	132.48	154.88	183.68
	Energy	0.70	1.32	12.56	16.26	31.58	73.98	178.85	441.41	1098.41
OpMulbs	Area	0.06	0.12	0.24	0.27	0.48	0.96	1.92	3.83	7.66
	Power	0.04	0.08	0.16	0.18	0.32	0.62	1.26	2.58	5.34
	Latency	0.97	1.02	1.01	1.00	1.01	1.00	1.13	2.37	2.74
	Energy	0.04	0.08	0.16	0.18	0.32	0.62	1.42	6.11	14.63

Area (μm^2*10^4), Power (mW), Latency (ns), and Energy (pJ)

where w and r are the length and width of images; MAX is the maximum value of pixels; and $S'(i,j)$ and $S(i,j)$ are the exact and stochastic results for each pixel.

The MSSIM is defined as

$$\text{MSSIM}(X, Y) = \frac{1}{k}\sum_{i=1}^{k} \frac{\left(2\mu_x\mu_y + C_1\right)\left(2\sigma_{xy} + C_2\right)}{\left(\mu_x^2 + \mu_y^2 + C_1\right)\left(\sigma_x^2 + \sigma_y^2 + C_2\right)}, \tag{7}$$

where the descriptions for these parameters can be found in [18].

Multiplication: An algorithm for multiplying two images pixel by pixel by using the stochastic multipliers is developed in MATLAB.

Smoothing: The smoothing algorithm is given by [19].

$$Y(i, j) = \sum_{m=-1}^{1}\sum_{n=-1}^{1} X(i + m, j + n)\, K(m + 2, n + 2), \tag{8}$$

where X and Y are, respectively, an input image to be smoothened and an output image, K is the smoothing kernel given by

$$K = \frac{1}{8}\begin{bmatrix} 0 & 1 & 0 \\ 1 & 4 & 1 \\ 0 & 1 & 0 \end{bmatrix}. \tag{9}$$

Table 10 The average PSNRs of images using various stochastic multipliers (dB)

Multiplier	3-bit	4-bit	5-bit	6-bit
MUX	23.34/14.69	29.56/24.01	35.31/28.83	41.17/36.48
Thermo	17.66/9.06	17.97/9.04	18.04/9.06	18.05/9.08
Shared	21.25/9.71	25.92/17.29	31.27/23.70	36.81/29.83
OpMulbs	24.77/15.29	29.68/24.03	35.74/28.83	41.21/37.24

Table 11 The average MSSIMs of images using various stochastic multipliers

Multiplier	3-bit	4-bit	5-bit	6-bit
MUX	0.687/0.532	0.816/0.819	0.930/0.909	0.979/0.959
Thermo	0.683/0.642	0.789/0.671	0.838/0.682	0.857/0.686
Shared	0.692/0.442	0.809/0.735	0.928/0.872	0.976/0.944
OpMulbs	0.725/0.697	0.861/0.846	0.944/0.909	0.982/0.962

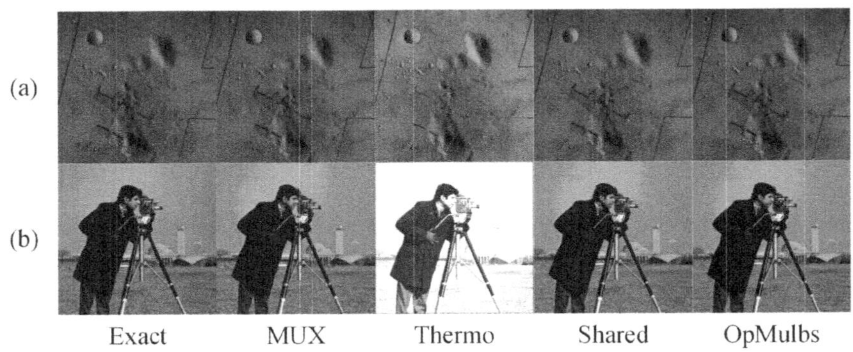

Fig. 11 The processed images using 6-bit multipliers. (**a**) Multiplication. (**b**) Smoothing

3.1.7.1 Computing Accuracy

Tables 10 and 11 list the average PSNRs and MSSIMs for the processed images, where the data before/after the slashes, respectively, are the values for the image multiplication and smoothing. The PSNRs of images multiplied using the OpMulbs multipliers are, respectively, improved by 1.56%, 83.21%, and 14.00% on average, while the MSSIMs are improved by 5.60%, 10.89%, and 5.84% compared with those in [12, 13, 16]. The PSNRs and MSSIMs of the smoothened images are improved by 76.03% and 18.48% on average compared with the others. These results show that the 6-bit OpMulbs multiplier is accurate enough in image processing because a PSNR larger than 30 dB is considered sufficient [20]. Figure 11 shows examples of the multiplied airplane and moon images and the smoothened cameraman images, generated by the considered 6-bit multipliers.

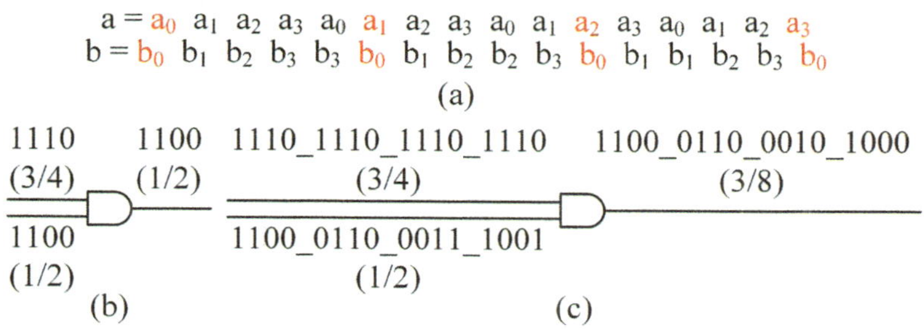

Fig. 12 (**a**) An illustration of the deterministic approach. (**b**) A traditional approach for multiplying two numbers. (**c**) The deterministic approach for multiplying two numbers

3.1.7.2 Hardware Costs

The circuits for multiplying and smoothing images are similar to those of 256-input and 9-input MACs, respectively, as listed in Table 9. Thus, the OpMulbs multipliers achieve a high computing accuracy, while significantly lowering the hardware costs in fault-tolerant applications, compared with the MUX, Thermo, and Shared-based designs [12, 13, 16].

3.2 Exact Stochastic Multipliers

3.2.1 Deterministic Approaches

Typically, SC suffers from inherent random fluctuations, which make it inaccurate as the designed stochastic multipliers above. Deterministic approaches have been developed to alleviate this issue [21]. Similar to convolution in mathematics, each bit in one bitstream interacts with all bits in the other one so that the generated outputs are accurate. For example, when multiplying two bitstreams a and b, b_0 interacts with a_0, a_1, a_2, and a_3, as illustrated in Fig. 12a. Figure 12b, c illustrates the comparison between the traditional and deterministic approaches when multiplying two 2-bit binary numbers in SC. In Fig. 12c, 1110 (encoding 3/4) is repeated 4 times and 1100 (encoding 1/2) is rotated 4 times. It is shown that the result generated using the deterministic approach is completely accurate, whereas the traditional one produces inaccurate results. However, the bitstream length (BSL) in the deterministic approach exponentially increases as the precision increases, resulting in a much longer delay.

Figure 17a–c illustrates three different deterministic approaches, including relatively prime stream length (RPSL), rotation (Rot), and clock division (CloDiv). For the RPSL approach, the lengths of two bitstreams are prime to each other. Two bitstreams are repeated several times to reach the same length. For the Rot approach, one bitstream is repeated while the other is rotated and repeated for the same number

$a_0\ a_1\ a_2\ a_3\ a_0\ a_1\ a_2\ a_3\ a_0\ a_1\ a_2\ a_3$

$b_0\ b_1\ b_2\ b_0\ b_1\ b_2\ b_0\ b_1\ b_2\ b_0\ b_1\ b_2$

(a)

$a_0\ a_1\ a_2\ a_3\ a_0\ a_1\ a_2\ a_3\ a_0\ a_1\ a_2\ a_3\ a_0\ a_1\ a_2\ a_3$

$b_0\ b_1\ b_2\ b_3\ b_3\ b_0\ b_1\ b_2\ b_2\ b_3\ b_0\ b_1\ b_1\ b_2\ b_3\ b_0$

(b)

$a_0\ a_1\ a_2\ a_3\ a_0\ a_1\ a_2\ a_3\ a_0\ a_1\ a_2\ a_3\ a_0\ a_1\ a_2\ a_3$

$b_0\ b_0\ b_0\ b_0\ b_1\ b_1\ b_1\ b_1\ b_2\ b_2\ b_2\ b_2\ b_3\ b_3\ b_3\ b_3$

(c)

Fig. 13 The deterministic approaches using (**a**) relatively prime stream length, (**b**) rotation, and (**c**) clock division

of times. For the CloDiv approach, two bitstreams are both repeated, while only one bit in one of them appears each time. In addition, different types of RNSs have been developed for generating bitstreams with better progressive precision.

It seems that the RPSL approach produces the lowest accuracy since it truncates one of the bitstreams (b_3 is truncated, as shown in Fig. 13a), which results in a slight loss of computing accuracy. However, recent work has shown that it facilitates the design of time-based computation. This method translates analog signals into time-based signals represented by the duty ratio to perform computation in the SC domain. The RPSL approach is utilized to adjust the duty cycle to improve accuracy [22]. The computing latency can then be reduced to $1/2^n$ of that of traditional SC. However, this method has several limitations. For example, it is hardly applicable to sequential circuits. For the Rot approach, in most cases, the outputs are more accurate because the input bitstreams show better progressive precision. Hence, it has a better performance in terms of accuracy than the RPSL and CloDiv approaches [23].

3.2.2 Counter-Based Multiplier

In [21], a binary counter has been exploited as an RNS, as shown in Fig. 14. The generated bitstreams are sequences of consecutive 1 s followed by sequences of 0 s, thus leading to a low computing accuracy if a computation is terminated early. In addition, a drawback of the design in Fig. 14a is that the inputs have relatively prime lengths.

3.2.3 Linear Feedback Shifter Register-Based Multiplier

The RNS has been replaced by an LFSR in [24], where 1 s and 0 s are randomly distributed, in contrast to counter-based SNGs, as shown in Fig. 15. The latency can

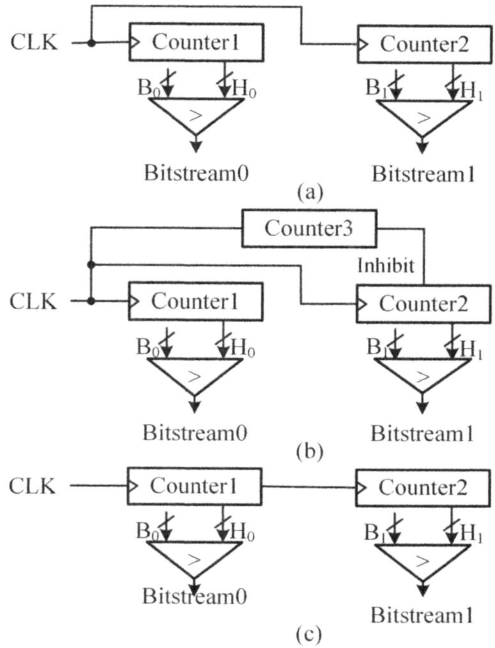

Fig. 14 The architectures of counter-based deterministic SNGs using (**a**) relatively prime stream length, (**b**) rotation, and (**c**) clock division

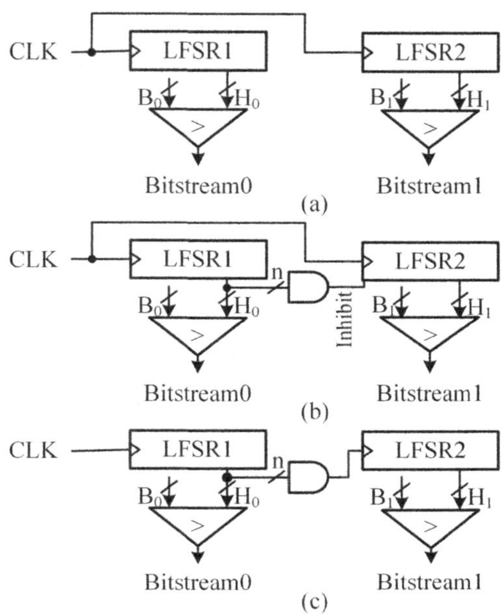

Fig. 15 The architectures of LFSR-based deterministic SNGs using (**a**) relatively prime stream length, (**b**) rotation, and (**c**) clock division

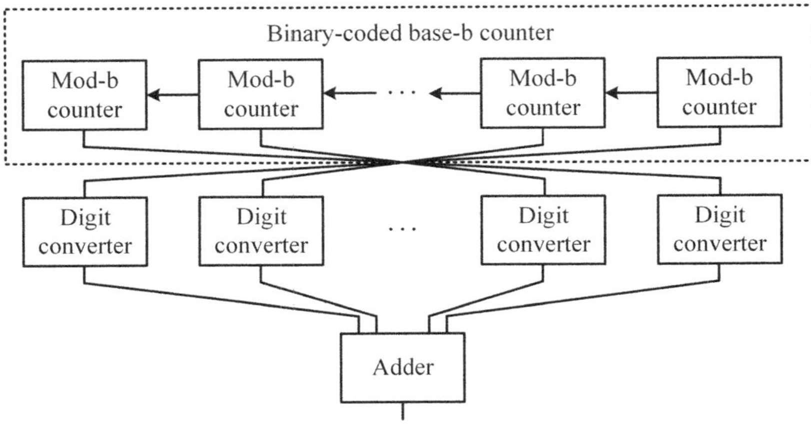

Fig. 16 A Halton sequence generator

be significantly reduced by terminating a computation in time if a slight inaccuracy is acceptable.

3.2.4 Halton Sequence-Based Multiplier

Low-discrepancy (LD) sequences such as Halton and Sobol sequences have been utilized to improve accuracy because they enable better progressive precision properties in generated bitstreams [25]. Sobol sequence generators have been exploited as RNSs to obtain a high accuracy [26]. In a Halton sequence-based generator, 1 s and 0 s in the bitstreams are uniformly distributed, thereby producing better progressive precision [27]. Figure 16 shows the Halton sequence-based RNS; its output is connected to the comparator in an SNG. Different mod-b counters are required for generating bitstreams with better uniform distributions of 1 s and 0 s.

In [23], Halton sequence-based deterministic SNGs have been proposed for image processing. The latency is reduced by up to 128× compared with existing accelerators. Figure 17 shows the architectures of Halton-based SNGs for generating three different deterministic bitstreams. If a slight inaccuracy is acceptable, the latency can be greatly reduced by terminating a computation in advance. Note that the two LD sequence-based solutions provide similar accuracy improvements, while the hardware cost of the Halton-based SNG is lower than that of a Sobol-based one [25].

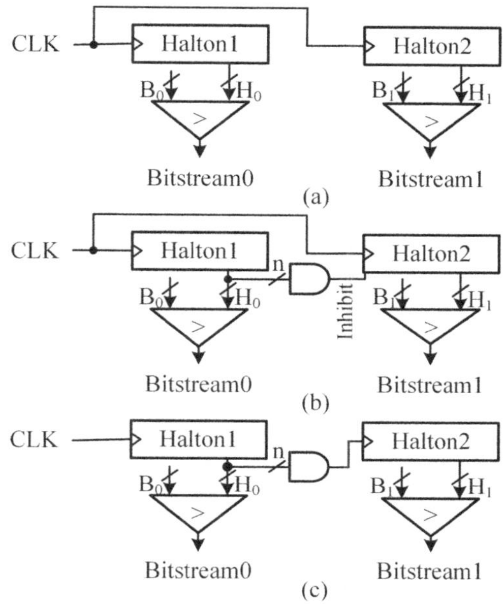

Fig. 17 The architectures of Halton-based deterministic SNGs using (**a**) relatively prime stream length, (**b**) rotation, and (**c**) clock division

Table 12 The bitstreams generated by SNGs with different RNSs

SNG	Bitstream
Counter based	11,110,000 (1/2)
LFSR based	10,101,100 (1/2)
Halton based	10,101,010 (1/2)

3.2.5 Evaluation

Table 12 lists the 3-bit precision bitstreams with the probability of 1/2 generated by SNGs with different RNSs, including counter, LFSR, and Halton [21, 23, 24]. The sub-bitstreams of lengths 2 and 4 generated by a Halton-based SNG are completely accurate, while the counter and LFSR-based ones are not. The sub-bitstreams generated by Halton-based SNGs are more accurate than those generated by the LFSR and counter-based ones because they encode values closer to the expected ones. This also indicates that the bitstreams generated by the Halton-based SNG have the advantage of a fast convergence, so the computation can be terminated early to reduce latency.

3.2.5.1 Hardware Costs

The hardware costs of these deterministic SNGs are synthesized using the Synopsys Design Compiler, with the 45 nm NanGate library [28]. Synthesized results are shown in Table 13. All SNGs are of an 8-bit precision. Columns 3–8 show the area, power, delay, area power product (ADP), power delay product (PDP), and energy

Table 13 The hardware area, power, delay, ADP, PDP, and EDP for SNDs

Approach	SNG	Area (μm^2)	Power (uW)	Delay (ns)	ADP ($\mu m^2 ns$)	PDP (10^{-3}pJ)	EDP (10^{-24}Js)
RPSL	Counter	156.67	1.89	2.26	354.08	4.26	9.63
	LFSR	193.65	2.36	2.43	470.56	5.73	13.92
	Halton	156.14	1.90	2.24	349.76	4.25	9.52
Rot	Counter	151.89	1.89	2.54	385.79	4.78	12.13
	LFSR	192.58	2.35	2.43	467.98	5.71	13.88
	Halton	154.28	1.91	2.24	345.59	4.29	9.61
CloDiv	Counter	154.01	1.88	2.54	391.20	4.77	12.10
	LFSR	192.58	2.35	2.43	467.98	5.71	13.88
	Halton	153.75	1.90	2.24	344.40	4.25	9.52

Table 14 The MAE (%) for multiplication using deterministic approaches versus BSL

Approach	SNG	2^{16}	2^{15}	2^{14}	2^{13}	2^{12}	2^{11}	2^{10}	2^{9}	2^{8}
RPSL	Counter	0	3.1500	4.9200	6.2500	7.1900	7.7800	8.1200	8.3000	8.4000
	LFSR	0.1000	0.1700	0.2200	0.2800	0.3300	0.3900	0.4300	0.4800	0.5500
	Halton	0	0.0004	0.0019	0.0074	0.0297	0.1200	0.4800	1.9900	8.4000
Rot	Counter	0	3.1300	4.8800	6.2000	7.1400	7.7200	8.0600	8.2400	8.3330
	LFSR	0	0.1100	0.1700	0.2300	0.2800	0.3400	0.3900	0.4500	0.5100
	Halton	0	0.0004	0.0018	0.0073	0.0294	0.1200	0.4800	1.9800	8.3300
CloDiv	Counter	0	12.450	18.68	21.790	23.35	24.120	24.510	24.710	24.810
	LFSR	0	1.5100	2.450	3.660	5.360	7.240	9.890	14.410	24.810
	Halton	0	0.0973	0.290	0.680	1.460	3.020	6.130	12.350	24.810

delay product (EDP) of the three designs with various deterministic approaches. It is evident that, for the three deterministic approaches, the hardware cost of the Halton-based designs is the lowest when the ADP, PDP, and EDP are considered. The counter-based designs are a little bit higher than that of the Halton-based designs, and the LFSR-based designs incur the highest hardware overhead. Thus, the Halton-based designs have superior performance in terms of hardware overhead.

3.2.5.2 Computing Accuracy

A half-length bitstream is used each time to verify that the bitstreams generated by the Halton-based SNGs have better truncation error properties. The MAEs of multiplying two 8-bit precision bitstreams generated by SNGs with various deterministic approaches are compared. Actually, the BSL in the RPSL approach is $2^8 \times (2^8-1), 2^7 \times (2^8-1), \cdots 2^0 \times (2^8-1)$ by each truncation. For simplicity, record them as $2^8 \times 2^8 = 2^{16}, 2^7 \times 2^8 = 2^{15}, \cdots, 2^0 \times 2^8 = 2^8$, as illustrated in Table 14.

The MAE becomes 0 when the BSL is 2^{16} except for the LFSR-based SNG with the RPSL approach. It is more efficient to use the Rot approach, especially for the Halton-based SNG. As BSL decreases, the MAE of the Halton-based designs

Table 15 The MAE (%) for edge detection versus BSL

Approach	SNG	2^{16}	2^{15}	2^{14}	2^{13}	2^{12}	2^{11}	2^{10}	2^{9}	2^{8}
RPSL	Counter	0.0969	1.0600	1.5700	1.7800	1.8500	1.8800	1.8900	1.9000	1.9000
	LFSR	0.0966	0.1000	0.1100	0.1300	0.1500	0.1800	0.2300	0.3000	0.4000
	Halton	0.0969	0.0969	0.0969	0.0970	0.0973	0.0980	0.1000	0.1000	1.9000
Rot	Counter	0.0957	1.0600	1.6400	1.8200	1.8700	1.8800	1.8900	1.8900	1.8900
	LFSR	0.0957	0.1000	0.1100	0.1300	0.1500	0.1800	0.2300	0.3000	0.4000
	Halton	0.0957	0.0957	0.0957	0.0957	0.0957	0.0957	0.0957	0.0957	1.8900
CloDiv	Counter	0.0957	2.2100	2.2100	2.2100	2.2100	2.2100	2.2100	2.2100	2.2100
	LFSR	0.0957	0.1700	0.2500	0.3200	0.4600	0.6300	0.8700	1.1800	2.2100
	Halton	0.0957	0.0957	0.0957	0.0957	0.0957	0.0957	0.0957	0.0957	2.2100
Classical	10-bit LFSR	—	—	—	—	—	—	0.1300	0.5300	0.8800

grows more slowly when the BSL is greater than 2^{10}. For example, in the RPSL approach, the BSL of the Halton-based design decreases to 2^{11} while it is 2^{16} for the prior designs if a 0.1% error is considered. For the Rot approach, it is 2^{11} in the Halton-based design, while it is, respectively, 2^{16} and 2^{15} for the counter and LFSR-based SNGs. For the CloDiv approach, the Halton-based design also realizes acceptable results by using shorter BSL. Thus, the SNGs can dramatically reduce the processing time and further save energy.

3.2.6 Robert's Cross Edge Detection

Robert's cross edge detector [29] is a well-known digital image processing algorithm, which implements

$$Z_{i,j} = 0.5 \left(|X_{i,j} - X_{i+1,j+1}| + |X_{i+1,j} - X_{i,j+1}| \right), \tag{10}$$

where $X_{i,j} \sim X_{i+1,j+1}$ are the current pixels in the input photo to be processed, $Z_{i,j}$ is the new pixel output generated by this algorithm.

A photo of 128 × 128 pixels is used to evaluate the performance of various designs. The MAEs are listed in Table 15. All the deterministic SNGs are of an 8-bit precision. If a 0.1% error is considered, for the RPSL and Rot approaches, the BSL is 2^9 for the Halton-based design, which is only 1/64 and 1/128 of the BSL in the LFSR-based and counter-based design, respectively. For the CloDiv approach, this ratio is 1/128. Compared with the classical 10-bit LFSR-based design without using deterministic approaches, the Halton-based designs are also more accurate for the same BSL. Hence, the Halton-based SNGs are much more accurate and efficient.

It turns out that the computing accuracy of the Halton-based SNGs will not decrease when the BSL is not smaller than 2^9 except for the one using the RPSL approach. Hence, one concludes that for a MUX (or scaled addition), the output of two n-bit input MUX using the Halton-based SNG is completely accurate when

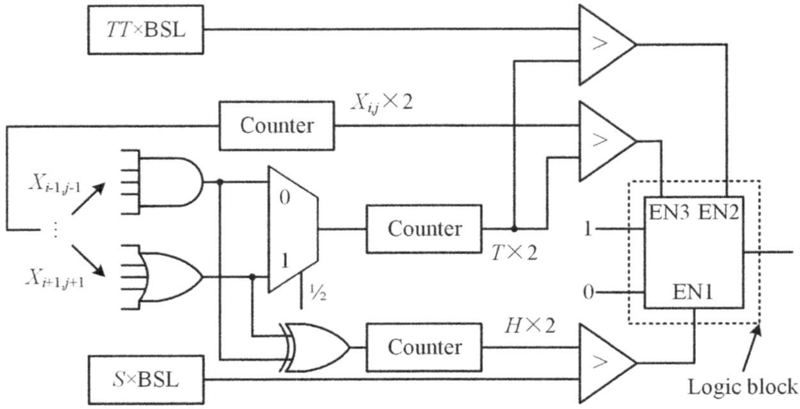

Fig. 18 The stochastic circuit for the Bernsen binarization algorithm

the BSL is not smaller than 2^{n+1}. Because the SNs generated by the Halton-based SNG are uniformly spaced, the input bitstreams of the MUX will be "completely selected" with deterministic approaches if the BSL is not smaller than 2^{n+1}; thus, a scaled addition is correctly executed.

3.2.7 Bernsen Binarization Algorithm

Another digital image processing algorithm called the Bernsen binarization algorithm [30] in SC is used to mitigate the problem of uneven lighting in photos.

The hardware design is shown in Fig. 18 [23]. Set the BSL to 2^{n+1} for accelerating computation as the MUX gate is a key component in this circuit. *TT* and *S* are the user-defined thresholds, $X_{i\text{-}1,j\text{-}1} \sim X_{i+1,j+1}$ are the pixel values in a $k \times k$ window centered on $X_{i,j}$ (for the trade-off between accuracy and processing time, set $k = 3$). Notice that the AND gate and OR gate obtain maximum and minimum, respectively, if their inputs are correlated by simply sharing an RNS. The three parameters *T*, *H*, and $X_{i,j}$ are doubled because the BSL is 2^{n+1}. The counters are used as stochastic-to-binary converters and connected in the comparators. Then the output will be 0 or 1 via a logical processing structure (that contains three enable signals in this figure) to implement

$$\text{OUT} = (EN1 \cdot EN3) + \left(\overline{EN1} \cdot EN2\right). \tag{11}$$

A photo with uneven lighting is used as the input for this circuit. Figure 19 shows the original unevenly lit photo and the image processed by the conventional binary method. Figure 20 shows the MAEs (%) of three SNGs with different deterministic approaches versus the bitwidth.

Fig. 19 Original photo and one processed by the conventional binary method

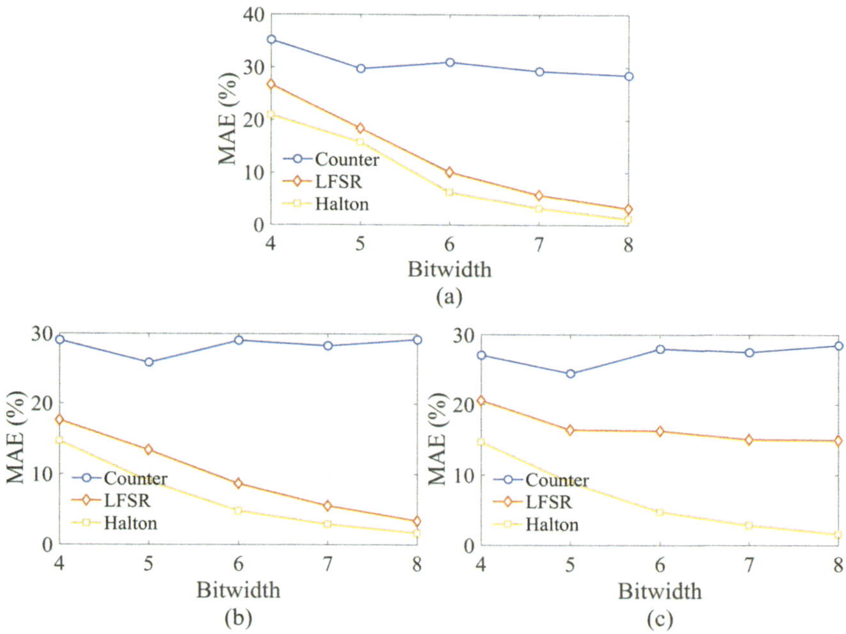

Fig. 20 The MAEs (%) for the Bernsen binarization algorithm using (**a**) relatively prime stream length, (**b**) rotation, and (**c**) clock division

As can be seen from Fig. 20, the larger the bitwidth, the more accurate the results are for all designs except for the counter-based ones. It can be seen that the Halton-based design works very well for this algorithm. For example, when the bitwidth is 8, the error is around 1% by the Halton-based design for various deterministic approaches. For the RPSL approach, the Halton-based design improves by nearly 22× and 2× compared to the counter and LFSR-based designs, respectively, in terms of computing accuracy. It reaches 17× and 2× accuracy in the Rot approach compared to those two approaches. For the CloDiv approach, the Halton-based design achieves nearly 17× and 9× improvement.

This conclusion can also be concluded from Fig. 21. The details in these photos processed by the Halton-based designs are clearer in the texture. Therefore, the Halton-based designs are more effective for this application.

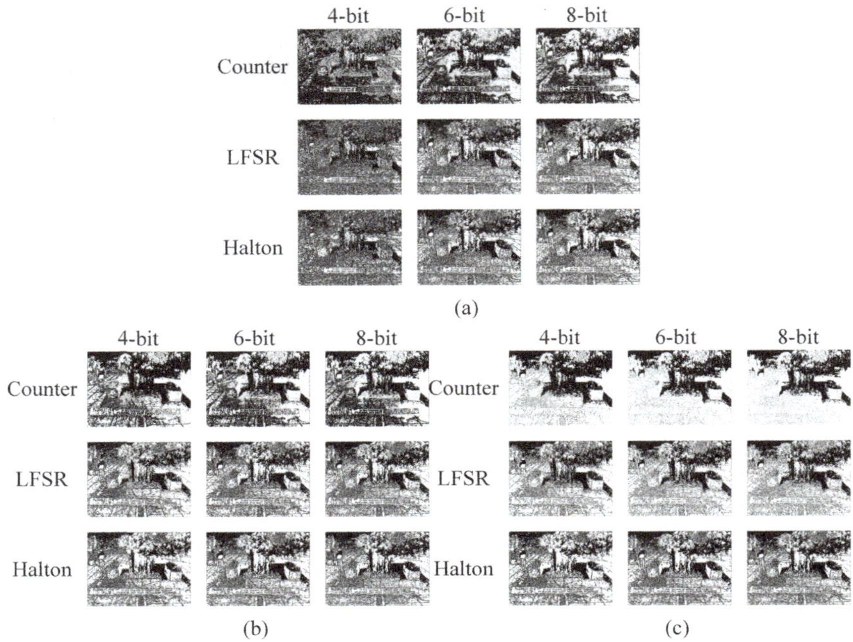

Fig. 21 Part of the processed photos using (**a**) relatively prime stream length, (**b**) rotation, and (**c**) clock division

Fig. 22 The performance of different schemes with different noise rates

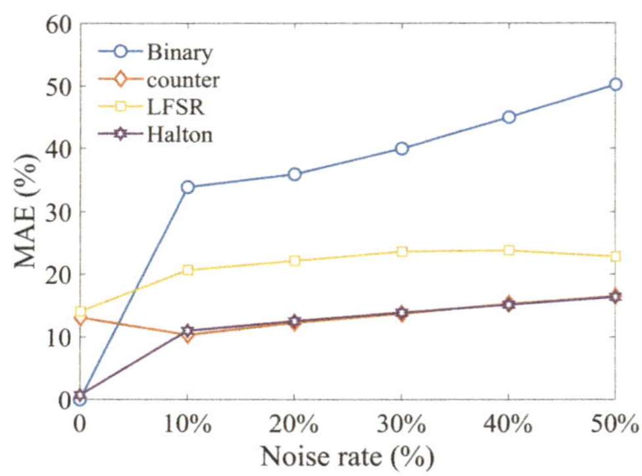

The fault tolerance is also evaluated by injecting different levels of noise between 0% and 50% into the computation. The counter-, LFSR-, and Halton-based SNGs using the Rot approach, and conventional binary method are evaluated. The BSL is 2^{8+1} in all SC implementations. As shown in Fig. 22, the Halton-based design achieves a lower MAE, so it is more robust than other methods.

4 Conclusion

In this chapter, designs of stochastic multipliers and exact stochastic multipliers are reviewed. Their computing accuracy, hardware costs, and applications in image processing are evaluated and synthesized. For stochastic multipliers, the designs using OpMulbs reach the same computing accuracy while incurring much smaller hardware costs, compared with previous designs. This benefits from the parallel design and the hard-wired connections dominated by OpMulbs to avoid complex SNGs. For the exact stochastic multipliers, the designs using Halton sequences can reduce the processing time, and provide more accurate, more energy efficient, and more robust results. Nevertheless, these designs still require a long processing time compared with their binary counterparts.

References

1. J.V. Neumann, Probabilistic logics and the synthesis of reliable organisms from unreliable components, in *Automata Studies*, ed. by C.E. Shannon, J. McCarthy, (Princeton University Press, Princeton, 1956), pp. 43–98
2. B. Gaines, "Stochastic Computing," presented at the *Proceedings of the Spring Joint Computer Conference on – AFIPS'67*, Atlantic City, New Jersey, 18–20 Apr 1967
3. W. Poppelbaum, C. Afuso, J. Esch, "Stochastic computing elements and systems," presented at the *Proceedings of the Fall Joint Computer Conference on – AFIPS'67*, Anaheim, California, 14–16 Nov., 1967
4. S. Ribeiro, Random-pulse machines. IEEE Trans. Electron. Comput. **EC-16**(3), 261–276 (Jun 1967)
5. B. Brown, H. Card, Stochastic neural computation I: Computational elements. IEEE Trans. Comput. **50**(9), 891–905 (2001)
6. W. Gross, V. Gaudet, *Stochastic Computing: Techniques and Applications* (Springer, 2019)
7. P. Meher, T. Stouraitis, *Arithmetic circuits for DSP applications* (Wiley, Hoboken, 2017)
8. F. Zhu, S. Zhen, X. Yi, H. Pei, B. Hou, Y. He, Design of approximate radix-256 booth encoding for error-tolerant computing. IEEE Trans. Circuits Syst. II-Exp. Brief. **69**(4), 2286–2290 (Apr 2022)
9. A. Alaghi, J. Hayes, "Exploiting correlation in stochastic circuit design," presented at the *2013 IEEE 31st International Conference on Computer Design (ICCD)*, Asheville, 6–9 Oct 2013
10. R. Budhwani, R. Ragavan, O. Sentieys, "Taking advantage of correlation in stochastic computing," presented at the *2017 IEEE International Symposium on Circuits and Systems (ISCAS)*, Baltimore, 28–31 May 2017
11. W. Qian, X. Li, M. Riedel, K. Bazargan, D. Lilja, An architecture for fault-tolerant computation with stochastic logic. IEEE Trans. Comput. **60**(1), 93–105 (Jan 2011)
12. S. Salehi, Low-cost stochastic number generators for stochastic computing. IEEE Trans. Very Large Scale Integ. (Vlsi) Syst. **28**(4), 992–1001 (Apr 2020)
13. H. Sim, J. Lee, "Cost-effective stochastic MAC circuits for deep neural networks," *Neural Networks* vol. 117, pp. 152–162, Sep. 2019
14. Y. Zhang et al., "A parallel Bitstream generator for stochastic computing," presented at the *2019 Silicon Nanoelectronics Workshop*, Kyoto, 9–10 June 2019
15. Y. Zhang, L. Xie, J. Han, X. Cheng, G. Xie, Highly accurate and energy efficient binary-stochastic multipliers for fault-tolerant applications. IEEE Trans. Circuits Syst. II: Exp. Brief. **70**(2), 771–775 (2023)

16. Y. Zhang, R. Wang, X. Zhang, Y. Wang, R. Huang, "Parallel hybrid stochastic-binary-based neural network accelerators," IEEE Trans. Circuits Syst. II: Exp. Brief, vol. 67, no. 12, pp. 3387–3391, Dec. 2020
17. *The USC-SIPI Image Database*. Available: https://sipi.usc.edu/database/
18. Z. Wang, A. Bovik, H. Sheikh, E. Simoncelli, Image quality assessment: From error visibility to structural similarity. IEEE Trans. Image Process. **13**(4), 600–612 (Apr 2004)
19. C. Solomon, T. Breckon, *Fundamentals of digital image processing: a practical approach with examples in matlab* (Wiley, The Atrium, Southern Gate, Chichester, 2011)
20. M. Ansari, H. Jiang, B. Cockburn, J. Han, Low-power approximate multipliers using encoded partial products and approximate compressors. IEEE J. Emerg. Select. Topic. Circuit. Syst. **8**(3), 404–416 (2018)
21. D. Jenson, M. Riedel, "A deterministic approach to stochastic computation," presented at the *2016 IEEE/ACM International Conference on Computer-Aided Design*, Austin, 7–10 Nov 2016
22. M. Najafi, S. Jamali-Zavareh, D. Lilja, M. Riedel, K. Bazargan, R. Harjani, An overview of time-based computing with stochastic constructs. IEEE Micro **37**(6), 62–71 (Nov–Dec 2017)
23. Z. Lin, G. Xie, W. Xu, J. Han, Y. Zhang, Accelerating stochastic computing using deterministic Halton sequences. IEEE Trans. Circuit. Syst. II-Exp. Brief. **68**(10), 3351–3355 (Oct 2021)
24. H. Najafi, D. Lilja, High quality down-sampling for deterministic approaches to stochastic computing. IEEE Trans. Emerg. Top. Comput. **9**(1), 7–14 (Mar 2018)
25. S. Liu, J. Han, "Energy efficient stochastic computing with Sobol sequences," presented at the *Design, Automation & Test in Europe Conference & Exhibition (DATE), 2017*, Lausanne, 27–31 Mar 2017
26. M. Najafi, D. Lilja, M. Riedel, "Deterministic methods for stochastic computing using low-discrepancy sequences," presented at the *2018 IEEE/ACM International Conference on Computer-Aided Design (ICCAD)*, San Diego, 5–8 Nov 2018
27. A. Alaghi, J. Hayes, "Fast and accurate computation using stochastic circuits," presented at the *2014 Design, Automation & Test in Europe Conference & Exhibition (DATE)*, Dresden, 24–28 Mar 2014
28. *Nangate open cell library*. Available: https://projects.si2.org. (2014, Jun 18)
29. A. Alaghi, L. Cheng, J. Hayes, "Stochastic circuits for real-time image-processing applications," presented at the *2013 50th ACM/EDAC/IEEE Design Automation Conference (DAC)*, Austin, 29 May – 7 Jun 2013
30. J. Bersen, "Dynamic thresholding of grey-level images," presented at the *Eighth International Conference on Pattern Recognition*, Paris, 27–31 Oct 1986

Applications of Ising Models Based on Stochastic Computing

Duckgyu Shin, Naoya Onizawa, Warren J. Gross, and Takahiro Hanyu

1 Introduction

Combinatorial optimization problems (COPs) are related to several practical real-world applications such as robot path planning [1] and very-large-scale-integrated (VLSI) circuit design [2]. Simulated annealing (SA), one of the probabilistic algorithms, is a possible candidate for solving these COPs [3]. However, the computation time of SA exponentially increases as the size of the COP grows, where some COPs belong to NP-hard problems [4]. Recently, nonconventional annealing methods for Ising models have been studied such as quantum annealing [5] and probabilistic bits (p-bits) [6] for solving COPs with the short computation time. The p-bit can be implemented using a probabilistic magnetic tunnel junction [7, 8]. On the other hand, the p-bit device model can be approximated using stochastic computing [9] and the approximated model can be implemented using standard CMOS digital circuits.

In this chapter, we introduce stochastic simulated annealing (SSA) for Ising model [10] and CMOS invertible logic [9], which are typical applications of stochastic computing for Ising model. SSA approximates the p-bit model using stochastic computing, which is a probabilistic computing method using stochastic random bit streams [11, 12]. In the p-bit model, tanh function is calculated for

D. Shin (✉)
Graduate School of Engineering, Tohoku University, Sendai, Japan
e-mail: duckgyu.shin.p4@dc.tohoku.ac.jp

N. Onizawa · T. Hanyu
Research Institute of Electrical Communication, Tohoku University, Sendai, Japan
e-mail: naoya.onizawa.a7@tohoku.ac.jp; takahiro.hanyu.c4@tohoku.ac.jp

W. J. Gross
Department of Electrical and Computing Engineering, McGill University, Montreal, QC, Canada
e-mail: warren.gross@mcgill.ca

W. Liu et al. (eds.), *Design and Applications of Emerging Computer Systems*,
https://doi.org/10.1007/978-3-031-42478-6_10

transitions of spin state. The tanh function can be approximated to a finite state machine using stochastic computing. Because of the approximation, SSA can be implemented using CMOS digital circuit with a hardware-friendly manner [13]. CMOS invertible logic (CIL) is a practical application of SSA that provides a probabilistic bi-directional operation between the inputs and the outputs. Typical CMOS digital circuits only generate outputs according to given inputs. In contrast, the circuit based on CIL can generate the output when the inputs are given but also generates the inputs when the outputs are given. Using this unique feature of CIL, training neural networks (NNs) can be processed with the same precision of inference [14]. As a design example, the training hardware, based on CIL, for simple 2-layer binarized NN (BNN) [15] is implemented on a field-programmable gate array. The performance of the training hardware is compared with a backpropagation algorithm [16] in terms of computation time and training accuracy.

The rest of this chapter is as follows. In Sect. 2, SA for the Ising model is briefed, and p-bit-based SA is reviewed. In Sect. 3, the algorithm and the architecture of SSA are explained. In Sect. 4, the basic concept of CIL is provided, and the training BNNs based on CIL are represented. The training hardware based on CIL is introduced including performance comparisons in Sect. 5. Section 6 concludes this chapter.

2 Preliminaries

2.1 *SA for the Ising Model*

SA is the probabilistic algorithm for optimization problems [3]. It decreases a cost function of a given problem to the optimal solutions. Let us denote by $S = \{x_0, x_1, \ldots, x_n\}$ the state and $E(S)$ the cost of the given state. The cost difference ΔE is given by $\Delta E = E(S_{new}) - E(S)$, where S_{new} is a randomly chosen neighborhood state. In SA, an acceptance probability of the randomly chosen neighborhood state is given by

$$P(\Delta E) = \exp\left(-\frac{\Delta E}{k \cdot T}\right), \tag{1}$$

where k is a Boltzmann constant and T is pseudo-temperature. When the new state has smaller cost than that of the previous state ($\Delta E < 0$), so that the acceptance probability is bigger than 1, therefore, the new state is always accepted. In contrast, when $\Delta E >= 0$, $P(\Delta E)$ is compared with a random value. The new state is accepted, when $P(\Delta E)$ is bigger than the random value. The acceptance probability is controlled by the pseudo-temperature, T. With high T, the $P(\Delta E)$ becomes larger; thus, the new states with the high cost are accepted frequently. In contrast, the state transitions are settle down when the T is low. With this probabilistic

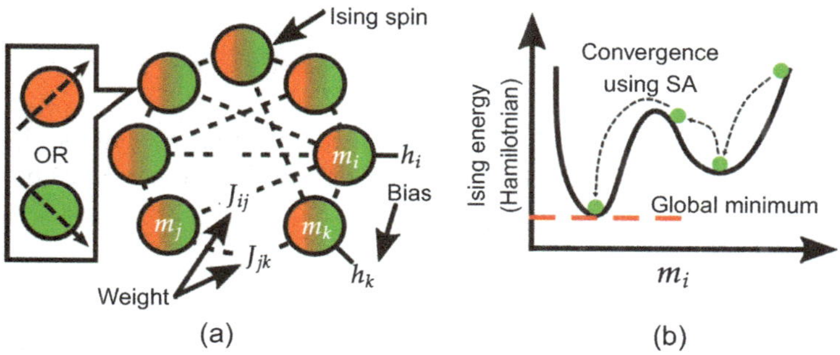

Fig. 1 Ising model and its energy landscape: (**a**) The configuration of Ising model. (**b**) Convergence of the Ising energy

state transitions, SA escapes local minimums and keeps converging to the optimal minimum.

The COPs can be converted to the Ising model. The COP is converted to a quadratic unconstrained binary optimization (QUBO) model [17]. The QUBO model is equivalent to the Ising model; thus, the Ising model is obtained from the COP. The Ising model is a network of the Ising spins and their interconnections [18]. Figure 1a shows the configuration of the Ising model. The Ising spin has a binary state, $m_i = [-1, +1]$, and it has a bias, h_i. Each spin is connected to each other with interconnection weight, J_{ij}. Hamiltonian is the energy of the Ising model, and it is given by

$$H = -\sum_{i} m_i h_i - \frac{1}{2} \sum_{i,j} m_i m_j J_{ij}. \tag{2}$$

The optimal solution of the given COP is embedded to a global minimum of the Hamiltonian; thus, SA can find the optimal solution by converging the Hamiltonian to the global minimum [19]. Figure 1b shows the convergence of the Ising energy using SA. Because SA accepts the transitions that increase the Ising energy, the Ising energy can escape the local minimum.

2.2 P-bit-Based Ising Model

Typical SA calculates the Ising energy to converge to the global minimum. On the other hand, the Ising model based on the p-bits [6] calculates the probability of the spin state rather than the Ising energy. The p-bit-based Ising model is based on a Boltzmann machine [20]. In the Boltzmann machine, the probability of a spin state of "1" is given by

$$P_{i=1} = \frac{1}{1 + \exp\left(-\frac{\Delta H_i}{T}\right)}, \tag{3}$$

where ΔH is the difference of the Ising energy from the transition of the single spin state. Let us denote $I_0 = \frac{1}{T}$ and $I_i = I_0 \Delta H$, then Eq. (3) is represented by:

$$P_{i=1} = \frac{1}{1 + \exp(-I_i)}. \tag{4}$$

Equation (4) is a sigmoid function; thus, it can be represented by a hyperbolic tangent. Using ΔH_i and the hyperbolic tangent, Eq. (4) is represented by:

$$I_i = I_0 \left(h_i + \sum_j J_{ij} m_j \right), \tag{5}$$

$$m_i = \mathrm{sgn}(\tanh(I_i)), \tag{6}$$

where sgn is a sign function. The spin state is directly calculated using Eqs. (5) and (6) without the Ising energy calculations.

3 Stochastic Simulated Annealing

3.1 Spin Operations

In [8], the p-bit is implemented using a nanomagnetic device model which is implemented using the magnetic tunnel junction [7]. The p-bit model can be approximated by integral stochastic computing [9], and the approximated model is implemented using typical CMOS digital circuits. Stochastic computing is a probabilistic computing method that represents values using the frequencies of “1” in the stochastic bit streams [11, 12, 21]. It can be categorized into binary stochastic computing and integral stochastic computing [22] by the range of the represented values. In binary stochastic computing, a real value $X_{binary} \in [-1 : 1]$ is defined by $2 \cdot P_x - 1$, where P_x is the frequency of “1” in the stochastic bit stream $x \in \{0, 1\}$. On the other hand, the real value $X_{integral} \in [-r : r]$, where $r \in \{1, 2, ...\}$, is represented by one or several stochastic bit streams.

By using stochastic computing, some complicated functions such as multiplication or hyperbolic tangent are implemented in the hardware in an area-efficient manner [23]. Based on integral stochastic computing, Eqs. (5) and (6) are approximated and redefined in the proposed SSA method as follows:

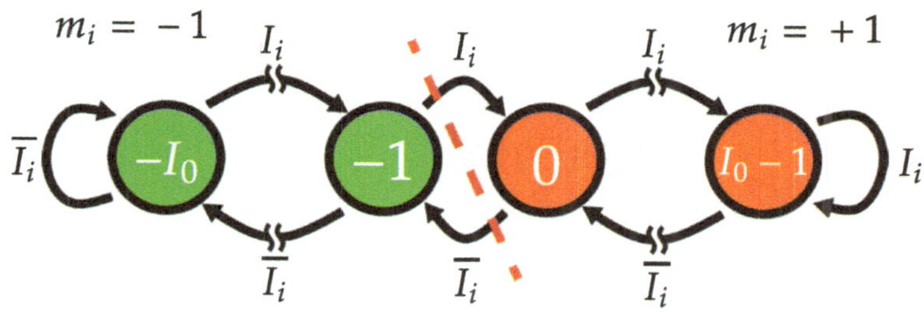

Fig. 2 Finite state machine of approximated Itanh function

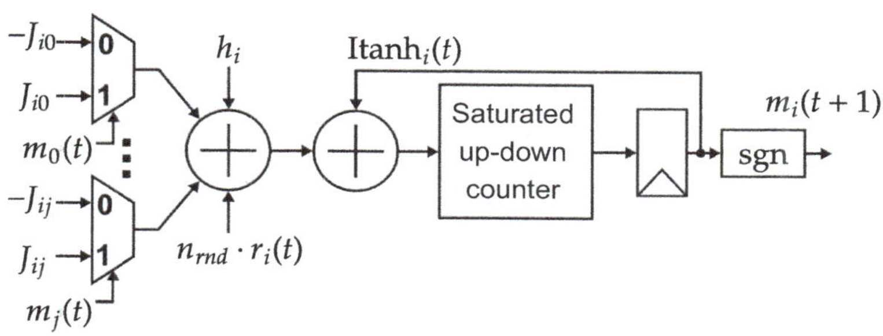

Fig. 3 The architecture of the spin gate

$$I_i(t+1) = h_i + \sum_j J_{ij} m_j + n_{rnd} \cdot r_i(t) + \text{Itanh}_i(t), \tag{7}$$

$$\text{Itanh}_i(t+1) = \begin{cases} I_0(t) - 1, & \text{if } I_i(t+1) >= I_0(t), \\ -I_0(t), & \text{else if } I_i(t+1) < -I_0(t), \\ I_i(t+1), & \text{otherwise,} \end{cases} \tag{8}$$

$$m_i(t+1) = \text{sgn}(\text{Itanh}_i(t+1)) = \begin{cases} +1 & \text{if Itanh}_i(t+1) >= 0, \\ -1 & \text{otherwise,} \end{cases} \tag{9}$$

where t is the number of annealing steps, n_{rnd} is a magnitude of a noise signals, and $r_i(t) \in \{-1, +1\}$ are the random noise signals. Equations (8) and (9) are implemented in a finite state machine (FSM) shown in Fig. 2. Figure 3 shows a spin-gate circuit implemented using Eqs. (7), (8), and (9) based on integral stochastic computing. The output of the spin gate is logical "0" or "1"; thus, logical "0" and "1" correspond to "−1" and "+1," respectively. The multiplications of $m_j(t+1)$ and J_{ij} in Eq. (7) are performed by a two-input multiplexer, and the accumulation is performed by a binary adder. The random signals $r_i(t)$ are generated by pseudo-random generators. The FSM of Itanh_i shown in Fig. 2 is designed using a saturated up-down counter in the spin-gate circuit. The saturated up-down counter truncates

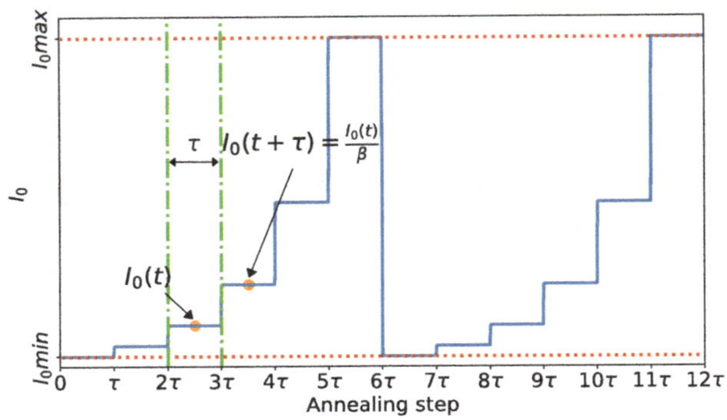

Fig. 4 Pseudoinverse temperature transition during the annealing process of SSA

$I_i(t+1)$ with a range of $2 \cdot I_0(t)$ states and generates $\mathrm{Itanh}_i(t+1)$. The output of the saturated up-down counter determines the output of the spin gate, $m_i(t+1)$.

3.2 Annealing Process of SSA

The SSA method converges to the Ising energy by calculating Eqs. (7), (8), and (9) to obtain the optimal solution of the given Ising model. To converge the Ising energy to the global minimum, a pseudoinverse temperature $I_0(t)$ is controlled by several hyperparameters shown in Fig. 4. I_{0min} is the initial value of I_0 and I_{0max} is the maximum value. τ is the number in which I_0 maintains a stable value, and β is the increased ratio of I_0. The pseudoinverse temperature gradually increases from I_{0min} to I_{0max} according to the following equation:

$$I_0(t+\tau) = \frac{I_0(t)}{\beta}. \tag{10}$$

I_0 is the range of the FSM for the approximated $\mathrm{Itanh}_i(t+1)$, so that the spin state transition occurs frequently when $I_0(t)$ is small. In contrast, the spin state converges to the optimal state when $I_0(t)$ is large. In conventional SA, the pseudo-temperature, T, gradually decreases from its initial value to the minimum value during the annealing process. However, the pseudoinverse temperature, $I_0(t)$, increases to its maximum within a relatively short period, and this operation is repeated during the annealing process. The iteration of $I_0(t)$ is controlled using the hyperparameter m_{shot}.

As a demonstration of SSA, the maximum cut problem (MAX-CUT) is solved using SSA and conventional SA. One of the COPs, the MAX-CUT, is a graph partitioning problem that maximizes the sum of the edge's weight between two partitioned graphs [24, 25]. Figure 5 shows an example of the MAX-CUT which

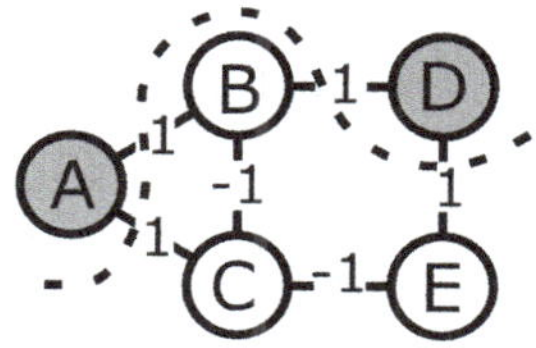

Fig. 5 An example of the maximum cut problem which contains five vertices

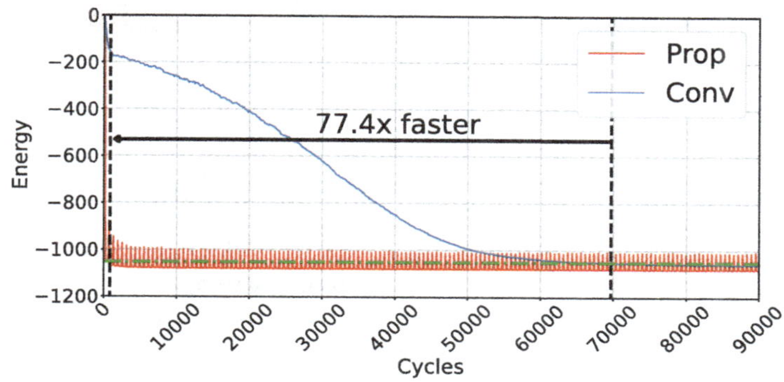

Fig. 6 Comparison of the Ising energy convergence between SSA and conventional SA

contains five vertices. The sum of the edge's weight is maximized when the example shown in Fig. 5 is partitioned into group 1 (A and D) and group 2 (B, C, and E). Therefore, the optimal solution of the given example can be represented in Fig. 5.

SSA and conventional SA solve G11 MAX-CUT from a dataset of MAX-CUT called G-set [26]. G11 contains 800 vertices and each vertex is connected to 4-neighborhood vertices. The number of edges is 1600, the value of the edge's weights is -1 or $+1$, and the maximum cut value of G11 (the optimal solution) is 564. The number of annealing steps is 90,000, and the annealing process is repeated 100 times because SSA and SA are probabilistic algorithms. Figure 6 shows a comparison of the average of Ising energy during the annealing process of SSA and SA. The average of the Ising energy obtained by conventional SA reaches 90% of the optimal solution at 69,700 cycles. In contrast, SSA reaches 90% of the optimal solution at 900 cycles, which is 77.4 times faster than conventional SA.

Figure 7 shows histograms of the cut values obtained by SA and SSA during 100 trials, respectively, as also the average- and the best-cut value. The average-cut value of SSA is 557, which is 11 higher than that of conventional SA. During 100 trials, SSA achieves the optimal solution of G11 (564); however, the best-cut value of conventional SA is 556.

The SSA algorithm can solve a large MAX-CUT, which contains 2000 vertices [27]. K2000 is a fully connected MAX-CUT that contains 2000 vertices, and the edge's weights are -1 or $+1$. SSA achieves the near-optimal solution 1000 times faster annealing step than that of conventional SA. Furthermore, SSA obtains the best-known solution (33,3337) of K2000.

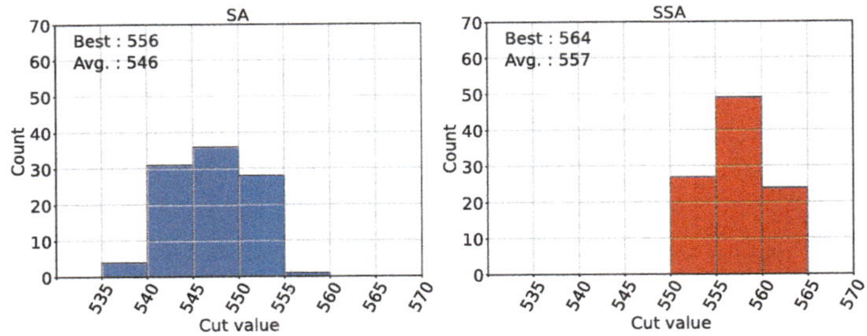

Fig. 7 Histograms of the cut values obtained by SA and SSA, respectively

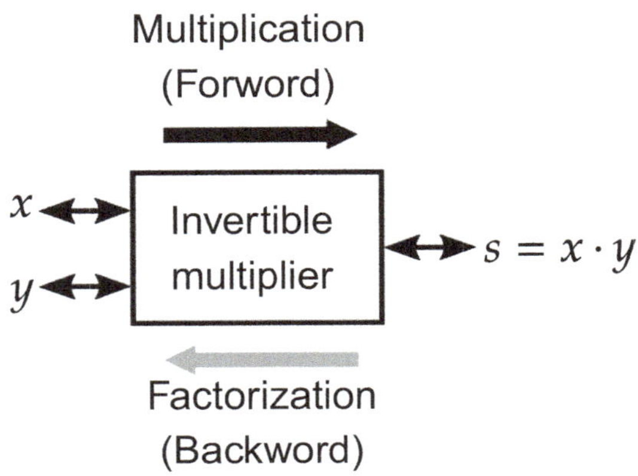

Fig. 8 The bi-directional operation of an invertible multiplier

4 CMOS Invertible Logic

4.1 Basics of CIL

Figure 8 shows a concept of the probabilistic bi-directional operations of CIL. The multiplier based on typical digital circuit can calculate s using the given inputs, x and y. However, it cannot calculate a correct combination of the inputs, (x, y), even if the output s is provided. The CIL circuit can operate in a forward mode or a backward mode [9, 28]. In the forward mode, the CIL circuit calculates the outputs same as the typical digital circuit. On the other hand, the CIL circuit can determine the correct combination of the inputs stochastically, when it operates in the backward mode. For instance, the invertible multiplier shown in Fig. 8 multiplies x and y in the forward mode, and it can factorize the given outputs s in the backward mode. Using the bi-directional-computing capability, CIL is applied to the invertible multiplier (the factorizer) [9], training a perceptron [29], and training the neural networks [14, 30].

The bi-directional operation is derived from the SSA and ground-state spin logic [31, 32]. In SSA, COPs are converted to the Ising model, and the spin state converges

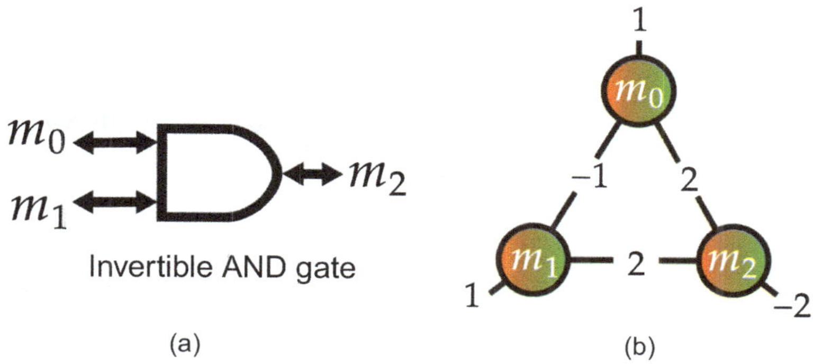

Fig. 9 The Ising model of CIL: (**a**) The invertible AND gate. (**b**) The Ising model of the invertible AND gate

Table 1 The truth table of AND gate and the Ising energy that correspond to the spin state

Truth table			Spin state				
m_0	m_1	m_2	m_0	m_1	m_2	H	Validity
0	0	0	−1	−1	−1	−3	Valid
1	0	0	1	−1	−1	−3	Valid
0	1	0	−1	1	−1	−3	Valid
1	1	0	1	1	−1	1	Invalid
0	0	1	−1	−1	1	9	Invalid
1	0	1	1	−1	1	1	Invalid
0	1	1	−1	1	1	1	Invalid
1	1	1	1	1	1	−3	Valid

to the optimum when the Ising energy converges to the global minimum. Same as the Ising model, basic logic gates, such as AND gate, are converted to the Ising model by ground-state spin logic. Figure 9a shows an invertible AND gate, whose inputs are m_0 and m_1 and output is m_2, and Table 1 shows a truth table of the AND gate. The Ising model (h_i and J_{ij}) is determined so that the Ising energy is the global minimum in the valid cases for the AND gate. For instance, when $[m_0, m_1, m_2] = [-1, -1, -1]$, the Ising energy, H, is -3 because the state is valid for the AND gate. The determined Ising model for the invertible AND gate is shown in Fig. 9b. It contains three spins, and its h_i and J_{ij} are given by

$$h_i = \begin{bmatrix} 1 & 1 & -2 \end{bmatrix},$$

$$J_{ij} = \begin{bmatrix} 0 & -1 & 2 \\ -1 & 0 & 2 \\ 2 & 2 & 0 \end{bmatrix}.$$

In SSA, all Ising spins fluctuate using their interconnections and random noise signals. However, some Ising spins of CIL are fixed in their state according to the

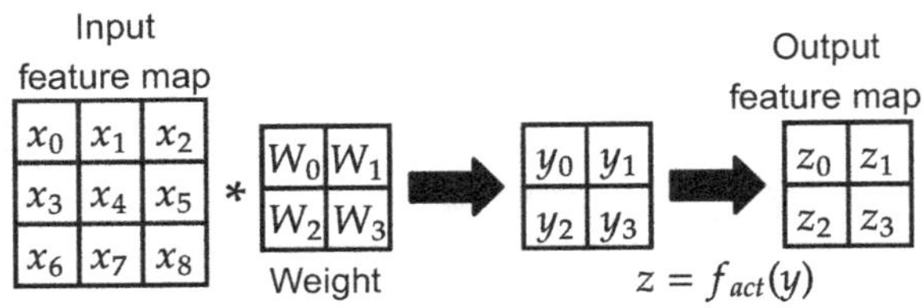

Fig. 10 Computations of the convolution layer in NNs

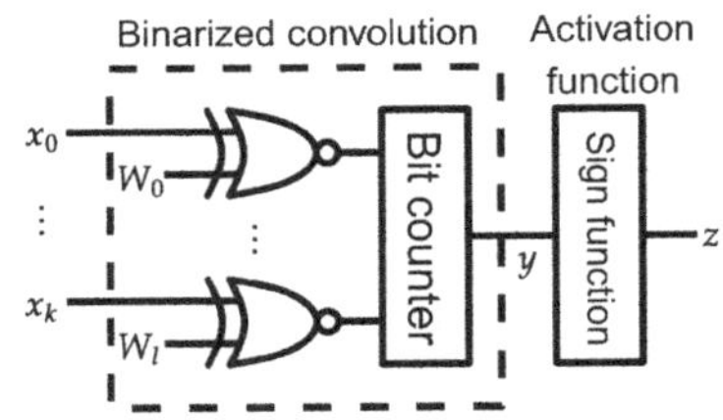

Fig. 11 A computation block of the binarized convolutional layer

operation mode to obtain the outputs of the CIL circuit. In the forward mode, the input spins of the invertible gate are fixed according to the given input. For example, the input spins of the invertible AND gate, m_0 and m_1, are fixed in the forward mode. In contrast, the output spin, m_2, is fixed regarding the given output of the AND gate. The state of the unfixed spins is determined using the connections of the fixed spins and the noise signals.

4.2 Training BNNs Based on CIL

Figure 10 shows the computations of the convolutional layer, which are the fundamental components of NNs [33]. In the convolutional layer, the weight map is multiplied with the input feature map while shifting over the input and the results of multiplications are accumulated. An activation function, f_{act}, is applied to the output of the convolution, y, and generates the output of the convolutional layer, z. Binarized neural networks (BNNs) have been studied to reduce memory usage and power consumption of NNs with low-precision data representation [15]. The weights and the inputs of the convolutional function are represented using binary precision (-1 and $+1$) during inference. The inputs are the outputs from a previous convolutional layer; thus, a sign function is used for the activation function. The weights are binarized during training using the sign function; however, gradients for the update are not binary precision [34]. The convolution of the binarized inputs and weights can be performed by XNOR gates and a bit counter [35]. Therefore, forward propagation of the binarized convolutional layer can be performed using a computation block shown in Fig. 11.

The computation block for inference of BNNs consists of simple logic circuits, such as the XNOR gate and the bit counter. The Ising models of each component are obtained by ground-state spin logic. Furthermore, the Ising model for a complex

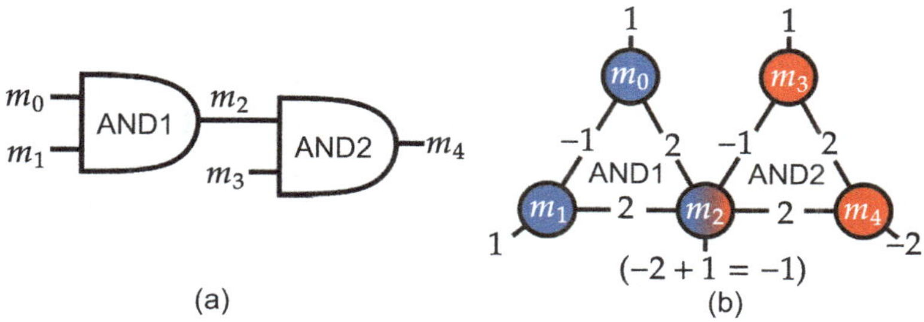

Fig. 12 Combining of the Ising model for CIL: (**a**) Circuit diagram of two AND gates (AND1 and AND2). (**b**) The configuration of the combined Ising model from two AND gates

logic circuit can be composed of a combination of the small Ising models [9]. Figure 12a shows a logic circuit composed of two AND gates, AND1 and AND2. The output of AND1, m_2, is connected to the input of the AND2. In this case, the Ising model of the entire logic circuit is shown in Fig. 12b. The spin m_2 is the output spin for AND1; also it is the input spin for AND2. Therefore, the bias of m_2 is $-2 + 1 = -1$. Using this combining process, the Ising model of the computation block for inference of BNNs can be obtained.

After obtaining the Ising model, the computation of the inference block can be performed in the bi-directional operations using CIL. Therefore, the weights, W, can be calculated by the bi-directional operation when the inputs, x, and the outputs, z, are fixed. However, the calculated W is valid only for the given input data and not for the entire training data. The trained weights W' for the entire training data have to be determined from each W obtained from the bi-directional calculations. In [14], the final weights, W', are determined by averaging all temporal W from each training datum.

5 CIL Training Hardware Design

5.1 Architecture of CIL Training Hardware

The training hardware is designed based on the Ising model which is converted from a BNN model. For evaluations, a 2-layer BNN model shown in Fig. 13 is converted to the Ising model. The 2-layer model contains two binarized convolutional layers (C1 and C2) without biases. The input feature map of the first C1 layer is a 9×9-pixel binarized image, and it is multiplied with 5×5 binarized filter. The output of the C1 layer is 5×5-pixel size, and it is used for the input of the C2 layer. The C2 layer has 3 channels 5×5 weights map, and its output is 1×3 pixels. Because the size of the output from the C2 is only 3 pixels, the output is used as classification labels.

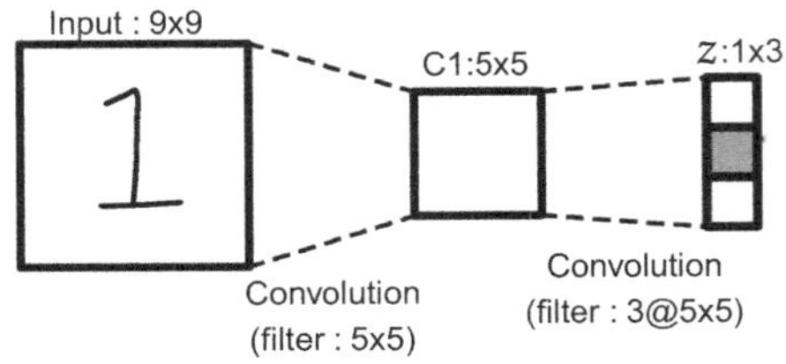

Fig. 13 Description of the 2-layer BNN model which is trained by the CIL training hardware

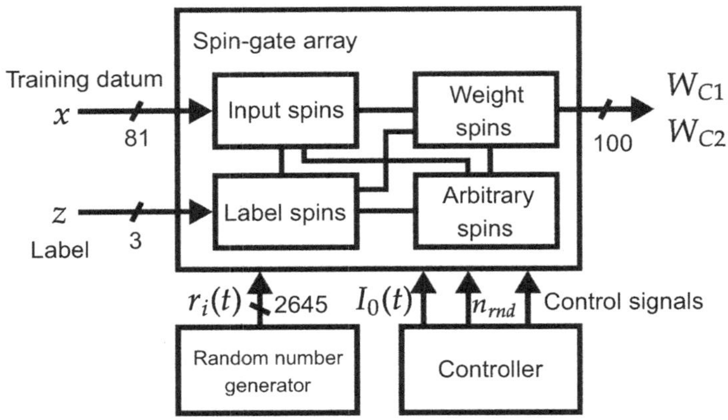

Fig. 14 The architecture of the CIL training hardware for the 2-layer BNN model

The Ising model which is converted from the 2-layer BNN model consists of 2645 spins, and the spins are categorized into input spins, weight spins, label spins, and arbitrary spins. Each spin corresponds to 1 pixel; thus, the number of the input spins is 81 and the number of the label spins is 3. The number of the weight spins for the C1 and C2 layers is 25 and 75, respectively, and the remaining spins are the arbitrary spins. In SSA, the Ising spin is implemented using the spin-gate circuit shown in Fig. 3. However, the spin in CIL generates different outputs depending on its operating mode. When the spin operates in an unfixed mode, the spin state is fluctuated by the interconnections with other spins and the noise signal. In contrast, the spin state is fixed depending on the given value in a fixed mode. The CIL training hardware receives the training datum and the classification label as the inputs and calculates the weights. Therefore, the input spins and the label spins operate in the fixed mode, so that their state is fixed according to the given datum.

Figure 14 shows an architecture of the CIL training hardware for the 2-layer BNN model. The spin-gate array contains 2645 CIL spin gates, and it receives the training datum and the label. The weights of each layer, W_{C1} and W_{C2}, are generated from the spin-gate array using the bi-directional operation. The training hardware also contains a controller and a random number generator (RNG). In Eq. (8), the noise signals, $r_i(t)$, are applied to the Ising spin for the probabilistic operation. The RNG is an XOR-shift [36], and it generates the 2645-bit random signals because the number of the spin is 2645. The controller controls the training process of

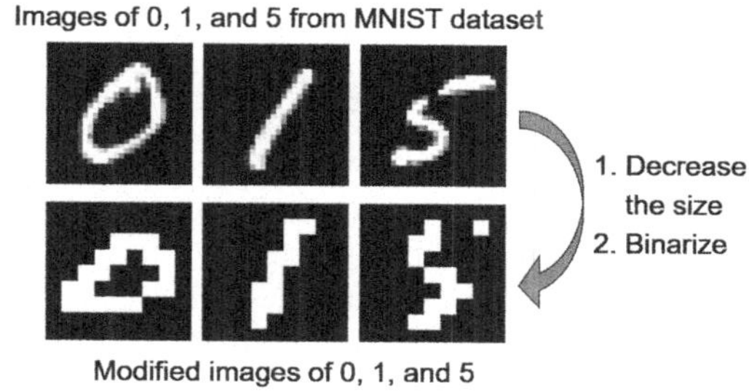

Fig. 15 The modified MNIST dataset [37] for the CIL training hardware

the hardware. Additionally, it generates the pseudoinverse temperature, $I_0(t)$, and controls the magnitude of the noise signals.

5.2 *Performance Evaluation*

The 2-layer BNN model can classify 9×9-pixel size binarized images with 3 labels. The training and test dataset are modified from the Modified National Institute of Standards and Technology (MNIST) dataset [37]. Figure 15 shows the original images from the MNIST data and the modified images for the proposed CIL training hardware. The MNIST dataset contains images of numbers from 0 to 9. In these images, the images of 0, 1, and 5 are selected to compose the new dataset. The size of the MNIST datum is 28×28-pixel; thus, the images are shrunk to the 9×9-pixel images. Finally, the images are binarized using a Gaussian filter. The modified dataset includes 10,800 training data and 3600 test data.

The CIL training hardware is implemented using a field-programmable gate array. The FPGA board is a Digilent Genesys 2 which is powered by a Xilinx Kintex-7. The hardware is designed using SystemVerilog, and it is synthesized and implemented by Xilinx Vivado 2018.3. The usage of lookup tables and flip-flops is 170,317 and 42,951, respectively. The clock frequency of the implemented hardware is 12.5 MHz, and the power dissipation is 0.913 W. The training datum and the label are transmitted from the PC via a universal asynchronous receiver/transmitter (UART) interface. After training the received datum, the training hardware returns the obtained temporal weights to the PC. In addition, the PC controls the training hardware using Python 3.6. The PC which is used for communication with the training hardware is also used for conventional training based on backpropagation [38]. The CPU of the PC is an Intel Core i7 7800X @ 4.4 GHz, and its power dissipation is 122 W.

Table 2 summarizes the performance comparisons between the CIL training hardware and conventional training using backpropagation on the CPU. The trained model is BNN; hence, the precision of inference is binary. However, the conventional training algorithm is backpropagation; thus, the precision is floating point. In contrast, the precision of the proposed hardware is the same as the precision

Table 2 Performance comparisons between the CIL training hardware and conventional training based on backpropagation

	This work	Conventional training	
Platform	FPGA	CPU	
Core	Xilinx Kintex-7	Intel Core I7 7800X	
Clock frequency [MHz]	12.5	4400	
Accuracy [%]	87.97	87.64	90.89
Training time [s]	66.85×10^{-3}	1.30	2.68
Power dissipation [W]	0.913	122	
Energy consumption [J]	0.061	158.60	326.96
# of trained data	100	10,800 (5 epochs)	10,800 (10 epochs)
Precision of training	Binary	Floating point	
Precision of inference	Binary	Binary	
Algorithm	CMOS invertible logic	Backpropagation	

of inference, because it is based on CIL. The maximum cognition accuracy of conventional training based on backpropagation is 90.89% with a training time of 2.68 s. Conventional training trains the entire training dataset (10,800 images) and it repeats the training process 10 times. The training hardware achieves a cognition accuracy of 87.97%, with a training time of 66.85 ms which is approximately 40 times faster than that of conventional training. Even if the training time for the same accuracy of 87& is compared, the CIL training hardware is still 19 times faster than conventional training. Additionally, the CIL training hardware achieves its maximum accuracy when it trains only 100 training data without iterated training. The implemented training hardware consumes 0.913 W, which is 134 times higher energy efficiency than the CPU.

6 Conclusion

In this chapter, we have reported simulated annealing based on stochastic computing for the Ising model. The basics of SSA are presented, and the annealing process of SSA is demonstrated using the MAX-CUT problem. Additionally, CIL, which provides a capability of the probabilistic bi-directional operations, is introduced as one of the application of SSA. The training hardware based on CIL is implemented in the FPGA, and its performance is compared with that of conventional training based on backpropagation.

The implementation of the larger SSA or CIL hardware, which includes more than 2000 spins, and applying SSA to real-world applications are our future prospects.

Acknowledgments This work was supported by a JSPS Grant-in-Aid for Scientific Research (B) (grant number JP21H03404), JST CREST (grant number JPMJCR19K3), and the WISE Program for AI Electronics, Tohoku University.

References

1. H. Martıinez-Alfaro, S. Gómez-Garcıia, Mobile robot path planning and tracking using simulated annealing and fuzzy logic control. Expert Syst. Appl. **15**(3), 421–429 (1998)
2. F. Barahona, M. Grötschel, M. Jünger, G. Reinelt, An application of combinatorial optimization to statistical physics and circuit layout design. Oper. Res. **36**(3), 493–513, 1988.
3. S. Kirkpatrick, C.D. Gelatt, M. P. Vecchi, Optimization by simulated annealing. Science **220**(4598), 671–680 (1983)
4. E.E. Reiter, C.M. Johnson, *Limits of Computation: An Introduction to the Undecidable and the Intractable* (CRC Press, Boca Raton, 2012)
5. T. Kadowaki, H. Nishimori, Quantum annealing in the transverse Ising model. Phys. Rev. E **58**, 5355–5363 (1998)
6. K.Y. Camsari, B.M. Sutton, S. Datta, p-bits for probabilistic spin logic. Appl. Phys. Rev. **6**(1), 011305 (2019)
7. R. Faria, K.Y. Camsari, S. Datta, Low-barrier nanomagnets as p-bits for spin logic. IEEE Magn. Lett. **8**, 1–5 (2017)
8. W.A. Borders, A.Z. Pervaiz, S. Fukami, K.Y. Camsari, H. Ohno, S. Datta, Integer factorization using stochastic magnetic tunnel junctions. Nature **573**(7774), 390–393 (2019)
9. S.C. Smithson, N. Onizawa, B.H. Meyer, W.J. Gross, T. Hanyu, Efficient CMOS invertible logic using stochastic computing. IEEE Trans. Circuits Syst. I: Regul. Pap. **66**(6), 2263–2274 (2019)
10. N. Onizawa, K. Katsuki, D. Shin, W.J. Gross, T. Hanyu, Fast-converging simulated annealing for Ising models based on integral stochastic computing. IEEE Trans. Neural Netw. Learn. Syst. 1–7 (2022)
11. B.R. Gaines, *Stochastic Computing Systems* (Springer US, Boston, 1969), pp. 37–172
12. B.R. Gaines, Stochastic computing, in *Proceedings of the April 18–20, 1967, Spring Joint Computer Conference*, AFIPS '67 (Spring) (ACM, New York, 1967), pp. 149–156
13. D. Shin, N. Onizawa, W.J. Gross, T. Hanyu, Memory-efficient FPGA implementation of stochastic simulated annealing. IEEE J. Emer. Sel. Topics Circuits Syst. **13**(1), 108–118 (2023)
14. D. Shin, N. Onizawa, W.J. Gross, T. Hanyu, Training hardware for binarized convolutional neural network based on CMOS invertible logic. IEEE Access **8**, 188004–188014 (2020)
15. I. Hubara, M. Courbariaux, D. Soudry, R. El-Yaniv, Y. Bengio, Binarized neural networks, in *Advances in Neural Information Processing Systems 29*, ed. by D.D. Lee, M. Sugiyama, U.V. Luxburg, I. Guyon, R. Garnett (Curran Associates, 2016), pp. 4107–4115
16. Y. LeCun, Y. Bengio, G. Hinton, Deep learning. Nature **521**(7553), 436–444 (2015)
17. F. Glover, G. Kochenberger, Y. Du, A tutorial on formulating and using QUBO models (2018)
18. S.G. Brush, History of the lenz-ising model. Rev. Mod. Phys. **39**, 883–893 (1967)
19. S.V. Isakov, I.N. Zintchenko, T.F. Rønnow, M. Troyer, Optimised simulated annealing for ising spin glasses. Comput. Phys. Commun. **192**, 265–271 (2015)
20. D.H. Ackley, G.E. Hinton, T.J. Sejnowski, A learning algorithm for Boltzmann machines. Cogn. Sci. **9**(1), 147–169 (1985)
21. V.C. Gaudet, W.J. Gross, *Stochastic Computing: Techniques and Applications*. (Springer, Cham, 2019)
22. A. Ardakani, F. Leduc-Primeau, N. Onizawa, T. Hanyu, W.J. Gross, VLSI implementation of deep neural network using integral stochastic computing. IEEE Trans. Very Large Scale Integr. Syst. **25**(10), 2688–2699 (2017)

23. P. Li, D.J. Lilja, W. Qian, K. Bazargan, M.D. Riedel, Computation on stochastic bit streams digital image processing case studies. IEEE Trans. Very Large Scale Integr. Syst. **22**(3), 449–462 (2014)
24. C.W. Commander, *Maximum Cut Problem, MAX-CUT* (Springer, Boston, 2009), pp. 1991–1999
25. H.L. Bodlaender, K. Jansen, On the complexity of the maximum cut problem. Nordic J. Comput. **7**(1), 14–31 (2000)
26. Y. Ye, Computational optimization laboratory (1999)
27. K. Katsuki, D. Shin, N. Onizawa, T. Hanyu, Fast solving complete 2000-node optimization using stochastic-computing simulated annealing, in *2022 29th IEEE International Conference on Electronics, Circuits and Systems (ICECS)* (2022), pp. 1–4
28. N. Onizawa, A. Tamakoshi, T. Hanyu, Scalable hardware architecture for invertible logic with sparse hamiltonian matrices, in *2021 IEEE Workshop on Signal Processing Systems (SiPS)* (2021), pp. 1–6
29. N. Onizawa, D. Shin, T. Hanyu, Fast hardware-based learning algorithm for binarized perceptrons using CMOS invertible logic. J. Appl. Logics **6**(7), 41–58 (2020)
30. N. Onizawa, S.C. Smithson, B.H. Meyer, W.J. Gross, T. Hanyu, In-hardware training chip based on CMOS invertible logic for machine learning. IEEE Trans. Circuits Syst. I: Regul. Pap. **67**(5), 1541–1550 (2020)
31. J.D. Biamonte, Nonperturbative k-body to two-body commuting conversion Hamiltonians and embedding problem instances into Ising spins. Phys. Rev. A **77**, 052331 (2008)
32. J.D. Whitfield, M. Faccin, J.D. Biamonte, Ground-state spin logic. Europhys. Lett. **99**(5), 57004 (2012)
33. S. Albawi, T. A. Mohammed, S. Al-Zawi, Understanding of a convolutional neural network, in *2017 International Conference on Engineering and Technology (ICET)* (2017), pp. 1–6
34. M. Courbariaux, Y. Bengio, Binarynet: training deep neural networks with weights and activations constrained to $+1$ or -1 (2016). CoRR, abs/1602.02830
35. M. Rastegari, V. Ordonez, J. Redmon, A. Farhadi, Xnor-net: imagenet classification using binary convolutional neural networks, in *Computer Vision—ECCV 2016*, ed. by B. Leibe, J. Matas, N. Sebe, M. Welling (Springer, Cham, 2016), pp. 525–542
36. S. Vigna, Further scramblings of Marsaglia's xorshift generators. J. Comput. Appl. Math. **315**, 175–181 (2017)
37. Y. Lecun, C. Cortes, The MNIST database of handwritten digits. http://yann.lecun.com/exdb/mnistl/
38. Y. LeCun, L. Bottou, Y. Bengio, P. Haffner, Gradient-based learning applied to document recognition. Proc. IEEE **86**(11), 2278–2324 (1998)

Stochastic and Approximate Computing for Deep Learning: A Survey

Tina Masoudi, Hao Zhang, Aravindhan Alagarsamy, Jie Han, and Seok-Bum Ko

1 Introduction

Deep learning applications are being used in image processing [1], audio and speech recognition [2], and several other areas. These applications have large and complex computations. Hence, they usually consume a large number of hardware resources and they have large areas. According to the fault-tolerant nature of the deep learning applications, the use of inexact computing methodologies like stochastic and approximate computing has gained a vast popularity to reduce the hardware complexity and power consumption [3–6].

Stochastic computing methodology is an inexact methodology in which the computations are done on stochastic bitstreams. In this methodology, binary numbers are converted to the stochastic bitstreams and then the arithmetic operations are done on these bitstreams by simple arithmetic circuits. Approximate computing

T. Masoudi · S.-B. Ko (✉)
Department of Electrical and Computer Engineering, The University of Saskatchewan, Saskatoon, SK, Canada
e-mail: Uka071@usask.ca; seokbum.ko@usask.ca

H. Zhang
Department of Electronic Engineering, Ocean University of China, Qingdao, China
e-mail: hao.zhang@ouc.edu.cn

A. Alagarsamy
Faculty of Electronics and Communication Engineering, Koneru Lakshmaiah Education Foundation, Guntur, India
e-mail: drarvindhan@kluniversity.in

J. Han
Department of Electrical and Computer Engineering, University of Alberta, Edmonton, AB, Canada
e-mail: jhan8@ualberta.ca

W. Liu et al. (eds.), *Design and Applications of Emerging Computer Systems*,
https://doi.org/10.1007/978-3-031-42478-6_11

methodology, mostly, tries to simplify the circuits so that they would consume less area and power. This simplification is mostly done by reducing or eliminating the least important part of the design so that the accuracy loss would not be much. Although both of these computing methodologies are highly power- and area-saving, they both have some disadvantages against each other. For instance, the stochastic multiplier is just a simple AND or XNOR gate that consumes less power and area than all other approximate multipliers. On the other hand, the computational speed in the stochastic computing is so slow and it consumes a large number of registers. In contrast, approximate computing is much faster and it consumes much less registers than stochastic computing methodology. However, stochastic and approximate computing methodologies might not be appropriate for being used in all of the deep learning applications and the use of them is completely related to the accuracy that these applications need and they should be used after considering several criteria.

In this chapter, first, we give a brief explanation about the deep learning computing elements, the advantages and disadvantages of the approximate and stochastic computing methodologies, and deep learning computations. Next, we review the current status of approximate and stochastic computing methodologies in the literature. Then, we propose a methodology for a hybrid stochastic-approximate area- and power-efficient neural network. In the next section we present the future research directions, and finally, this chapter would be concluded in the last section.

2 Background

Stochastic and approximate computing methodologies are imprecise computing methodologies which reduce the area of the design at the cost of losing a small amount of accuracy [7–9].

2.1 *Stochastic Computing*

Stochastic bitstreams are made of 0s and 1s. Stochastic bitstreams whose values range between 0 and 1 are called unipolar numbers [7], while the stochastic numbers between -1 and 1 are called bipolar numbers. The values of the unipolar stochastic numbers are defined by the number of the ones in the stochastic bitstream, and the values of the bipolar numbers are calculated by Eq. (1), in which p is the unipolar stochastic number and $p*$ is the bipolar number [7]:

$$p* = 2p - 1 \tag{1}$$

A $\frac{x}{y}$ number can have $\binom{y}{x}$ different unipolar stochastic presentations [10]. For example, number $\frac{2}{3}$ can be presented in three ways: {110 011 101}.

Fig. 1 Stochastic number generator (SNG)

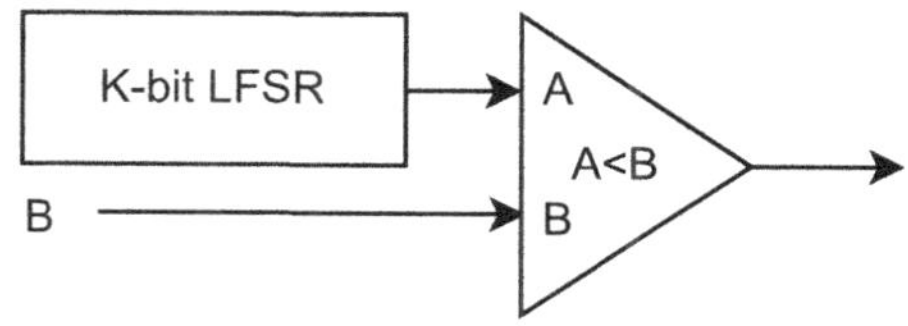

Fig. 2 (**a**) Stochastic computing multiplier for the unipolar representation. (**b**) Stochastic computing multiplier for the bipolar representation [7]

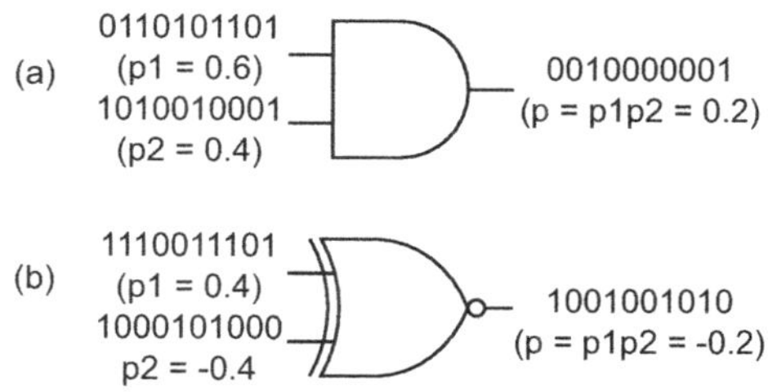

Fig. 3 Multiplexer-based stochastic adder [7]

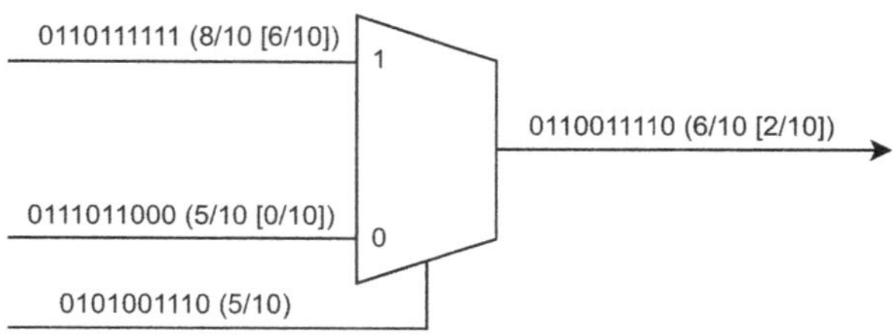

There are several stochastic arithmetic elements. For example, for using stochastic computing methodology, first, we need to convert the binary numbers to stochastic bitstreams by stochastic number generator (SNG) which is made of one k-bit linear feedback shift register (LFSR) and one comparator as shown in Fig. 1 [7]. The LFSR module produces a k-bit random number in each clock cycle, and the comparator compares this number with the k-bit binary input. If the input number is bigger than the number produced by the LFSR, the comparator generates "1," and it generates "0" otherwise. Finally, all these "0" and "1" make the stochastic bitstream [7].

Stochastic number decoder is used for converting binary numbers to stochastic bitstreams which is a simple binary counter. This counter counts the number of the ones in the stochastic bitstream and produces its equal binary number [7].

Stochastic multiplier is a simple AND gate and a simple XNOR gate for unipolar and bipolar numbers, respectively [7], as shown in Fig. 2.

There are three kinds of stochastic adders: multiplexer based [10], the simplest stochastic adder, T flip-flop based [5], and accumulative parallel counter (APC) [5].

The structure of the multiplexer-based adder is shown in Fig. 3. In this adder, the probability of the select signal is equal to 0.5 and is generated by a random number generator (RNG). The output of this adder is not affected by the correlation between the inputs. Thus, different LFSRs are not needed for the inputs. However, the LFSR generates the select signal and should be different from the inputs' LFSR.

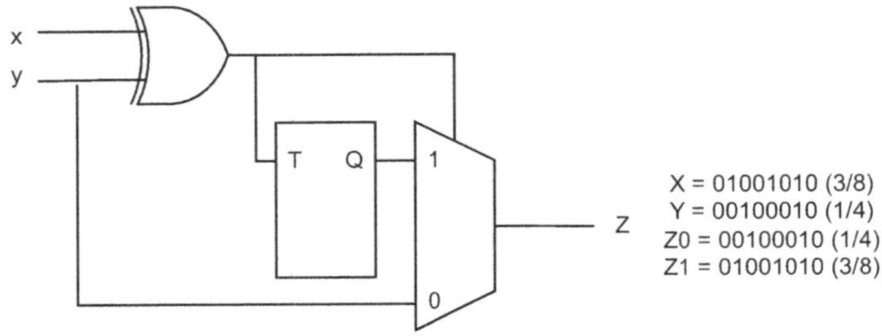

Fig. 4 T flip-flop-based stochastic adder [5]

APC's output is in binary format. Therefore, a separate counter would not be needed for converting the stochastic bitstream into the binary format. In addition, the output of this adder is not affected by the correlation between its inputs. Hence, different LFSRs would not be needed for this adder's inputs.

The T flip-flop-based adder is shown in Fig. 4. The output bitstreams of the T flip-flop adder are always uncorrelated with each other. Therefore, there is no need for different LFSRs for each of its inputs. Moreover, another advantage of this adder is that it does not need an additional LFSR for the select signal like the multiplexer-based adder.

The advantages of stochastic computing methodology are not negligible. This methodology has become widely popular in highly power-consuming and fault-tolerant applications with large areas. Sacrificing a small amount of accuracy is tolerable in fault-tolerant applications and what has the highest significance in these applications is reducing the area as well as the power consumption, in which stochastic computing methodology has one of the best performances [7]. One of the reasons leading to the popularity of the stochastic computing methodology is its multipliers, since they are just a simple AND gate for unipolar numbers and a simple XNOR gate for bipolar numbers. Compared to the exact multipliers which have complex designs, these simple gates reduce the area of the design significantly [7].

Besides the advantages, stochastic computing has some disadvantages. One of the challenges in stochastic computing methodology is its inaccuracy, which can be caused by several reasons. For example, when fluctuations happen in the representation of the stochastic numbers, correlation among the different stochastic bitstreams and sometimes physical errors change the numbers [10]. There are $\binom{y}{x}$ different ways representing a $\frac{x}{y}$ number with stochastic bitstreams and for that reason just a few numbers can have an exact stochastic representation [10]. Correlation is another factor which causes inaccuracy in stochastic bitstreams [10]. The correlation usually happens when the same LFSR is used for generating the random numbers in that case, the bitstreams would be similar to each other, and doing the arithmetic operations on them such as multiplication which is ANDing the bitstreams with each other would contribute to a completely wrong answer [10].

2.2 Approximate Computing

Similar to stochastic computing methodologies, approximate computing is highly popular in the hardware implementation of neural networks because, despite its inaccuracy, it is efficient in terms of area and power consumption [11]. Assorted kinds of approximate adders and multipliers are proposed [12–15]. Approximate computing mostly eliminates the least significant elements and computations in a design [14]. In that case, the area would be reduced with a marginal loss in accuracy. Thus, this computing methodology is popular in fault-tolerant designs such as deep learning applications [16].

Several approximate computational elements have been proposed recently [17, 18]. Most of these computational elements are made by eliminating some significant parts of the exact computational elements or just modifying the exact ones so that they consume less hardware resources by sacrificing a marginal amount of accuracy [9]. For instance, an approximate multiplier is introduced in [19], in which the number of partial products' stages has been reduced to one from two and it still produces the exact output and the approximate multiplier is made out of this exact one by eliminating the computational elements which are for the production of the least significant bits. The partial products of this multiplier for 4×2 multiplication are shown in Eqs. (2) to (7). For a 4×4 multiplication 2 4×2 multipliers can be used. A 4×4 multiplier is produced by using two 4×2 multipliers. As shown in Fig. 5, this multiplier is specifically proposed for FPGA (field programmable gate array) which reduces the number of LUTs (Look Up Tables) compared to the exact multipliers. In addition to approximate multipliers, some approximate adders have been proposed. For example, in [20], some approximate adders with different rates of accuracy have been proposed. The adders are called APAD1 to APAD4, and each of them are made by making 1 to 4 errors in the truth tables of the exact adders. In these adders, as the level of accuracy decreases, the circuit becomes simpler.

$$P_0 = A_0 + B_0 \tag{2}$$

$$P_1 = B_1{}'B_0A_1 + B_1B0'A_0 + B_1A_1{}'A_0 + B_0A_1A_0{}' \tag{3}$$

$$\begin{aligned} P_2 = {} & B_1{}'B_0A_2 + B_1B0'A_1 \\ & +B_0A_2A_1{}' + B_1A_2{}'A_1A_0{}' + B_1A_2A_1A_0 \end{aligned} \tag{4}$$

$$\begin{aligned} P_3 = {} & B_1{}'B_0A_3 + B_1B0'A_2 + B_1A_3{}'A_2A_1 \\ & {}' + B_0A_3A_2{}'A_1{}' + B_1B_0A_3{}'A_2{}'A_1A_0 \\ & +B_0A_3A_2A_1 + B_0A_3A_1A_0{}' \end{aligned} \tag{5}$$

$$\begin{aligned} P_4 = {} & B_1B_0{}'A_3 + B_1A_3A_2{}'A_1{}' \\ & +B_1A_3A_2{}'A_0{}' + B_1B_0A_3{}'A_2A_1 \end{aligned} \tag{6}$$

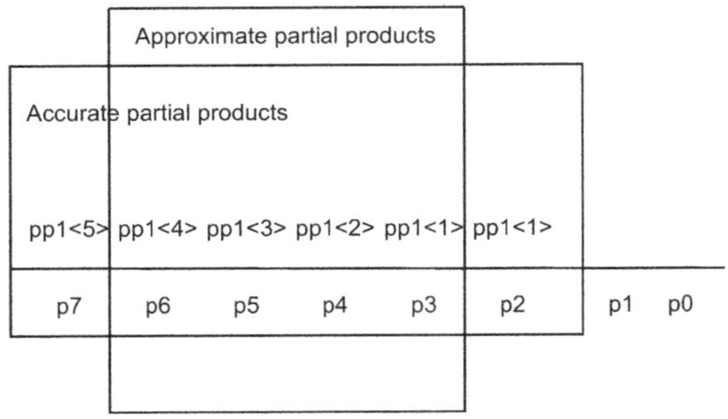

Fig. 5 A 4×4 accurate and approximate multipliers [19]

$$P_5 = B_1 B_0 A_3 A_2 + B_1 B_0 A_3 A_1 A_0 \tag{7}$$

Just like stochastic computing, approximate computing has gained a wide popularity among fault-tolerant applications with large areas and high power consumption. Different methodologies can be used in approximate computing according to the performance of the design. In most of the approximate computing methodologies, the least significant elements would be eliminated. In that case, area and power would be reduced while the accuracy would not change significantly [8]. One of the advantages of this method is its flexibility, so that one can change the design according to its performance. For example, if the accuracy is very important in a design, the number of elements that are going to be eliminated would be less compared to a design in which the accuracy has less importance [8].

Besides the advantages, one of the challenges in approximate computing is that, for every kinds of deep learning applications, different kinds of approximate arithmetic circuits are needed, which means that the arithmetic circuit should be designed first and then be used in the application to test if it suits well or not [8]. Which would be highly difficult, costly, and time-consuming. In addition, approximate arithmetic circuits are more susceptible to hardware Trojan attacks compared with exact arithmetic units [8]. This happens because approximate circuits are usually controlled with an additional signal which does not belong to the original circuit and that can be the good target for the hardware Trojan attacks. Moreover, some of the approximate circuits might need error correction units which can be another target for these kinds of hardware attacks [8].

2.3 Deep Learning Arithmetic Units

There are several deep learning arithmetic units such as MAC (multiply accumulator). In neural networks, several weights dedicated to each output and all of these weights should be multiplied by the inputs and then all of the outputs of the

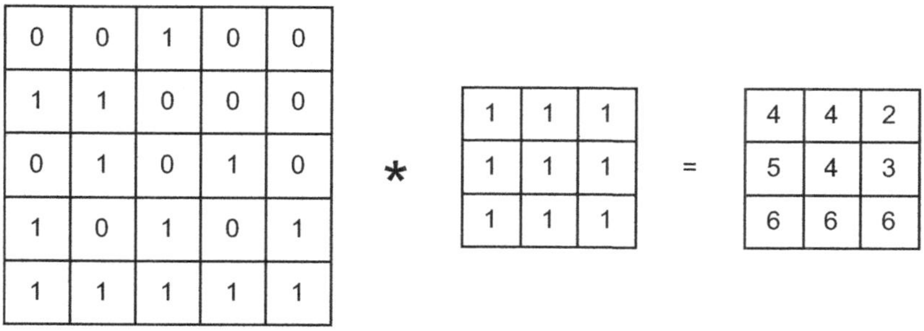

Fig. 6 Convolutional operation in the convolutional layer of the convolutional neural network

multiplication should be summed up together in order to produce the output. These complex operations contribute to the large area and power consumption. Thus, using inexact computing methodologies in this unit could reduce the area as well as power consumption [21]. MAC arithmetic operations are shown in Eq. (8).

$$\sum_{i=1}^{n} w_i x_i \tag{8}$$

Another arithmetic unit in neural networks which is used in CNNs (convolutional neural networks) is called convolutional operations which is done in the convolutional layer of the CNNs. The outputs of these operations are the dot productions of the filters with inputs as shown in Fig. 6. These arithmetic operations are highly area- and power-consuming since they contain several multiplications [22] and for that reason using the inexact arithmetic circuits is highly popular in these operations. Another way which has gained much popularity during the recent years is using Winograd algorithm for the convolutional operations. Winograd algorithm was first proposed in [23]. This algorithm is used in the convolutional layer of the CNNs in order to reduce the number of multiplications by increasing the number of additions [23]. Since multiplication is highly complex and consumes a large area and addition is simple and does not need much hardware resources, this algorithm would reduce the area of the design significantly. The number of multiplications for a Winograd unit with m outputs and r-tap filters is shown in Eq. (9) [24]. The multidimensional algorithm can also be made of one-dimensional algorithm by nesting two one-dimensional algorithms with each other [24]. For example, for F(m,r) and F(n,s), for computing m×n outputs and r×s filters, the number of multiplications would be like Eq. (10). The Winograd algorithm for F(2,3) is shown in Eqs. (11) and (12), which has 6 multiplications. Winograd algorithm can also be written in matrix form shown in Eqs. (13) and (14) [24].

$$\mu F((m, r)) = m + r - 1 \tag{9}$$

$$\mu(F(m \times n, r \times s)) = \mu(F(m, r))\mu(F(n, s)) = (m + r - 1)(n + s - 1) \tag{10}$$

$$F(2,3) = \begin{bmatrix} d_0 \; d_1 \; d_2 \\ d_1 \; d_2 \; d_3 \end{bmatrix} \begin{bmatrix} g_0 \\ g_1 \\ g_2 \end{bmatrix} = \begin{bmatrix} m_1 + m_2 + m_3 \\ m_2 - m_3 - m_4 \end{bmatrix} \tag{11}$$

$$\begin{aligned} m_1 &= (d_0 - d_2) g_0 \\ m_2 &= (d_1 + d_2) \frac{g_0 + g_1 + g_2}{2} \\ m_3 &= (d_2 - d_1) \frac{g_0 - g_1 + g_2}{2} \\ m_4 &= (d_1 - d_3) g_2 \end{aligned} \tag{12}$$

$$Y = A^T [(G_g) \odot (B^T \odot d)] \tag{13}$$

$$B^T = \begin{bmatrix} 1 & 0 & -1 & 0 \\ 0 & 1 & 1 & 0 \\ 0 & -1 & 1 & 0 \\ 0 & 1 & 0 & -1 \end{bmatrix} \quad G = \begin{bmatrix} 1 & 0 & 0 \\ \frac{1}{2} & \frac{1}{2} & \frac{1}{2} \\ \frac{1}{2} & -\frac{1}{2} & \frac{1}{2} \\ 0 & 0 & 1 \end{bmatrix} \quad A^T = \begin{bmatrix} 1 & 1 & 1 & 0 \\ 0 & 1 & -1 & -1 \end{bmatrix} \tag{14}$$
$$g = \begin{bmatrix} g_0 \; g_1 \; g_2 \end{bmatrix}^T \quad d = \begin{bmatrix} d_0 \; d_1 \; d_2 \; d_3 \end{bmatrix}^T$$

Some of the activation functions are other costly arithmetic operations in NNs. For example, sigmoid and tanh activation functions are highly area- and power-consuming since they have complex architectures and hard to be implemented, and even if possible, the area as well as the power consumption would be significantly high. Hence, inexact computing methodologies are highly popular for being used in these activation functions because although they lead to a marginal loss in accuracy, they reduce the hardware and area consumption significantly. There are several stochastic and approximate activation functions proposed in the literature.

2.4 Deep Learning Application Accelerators

One of the basic hardware accelerators is General Purpose (GP) CPU. Since (GP) CPUs do not have enough bandwidth and computational complexity, they cannot be used for the state-of-the-art deep learning applications [25]. In addition, according to the Von Neumann bottleneck, these accelerators consume much more energy than other accelerators [25]. Moreover, most of the resources in the GP CPUs would remain underutilized which would increase the expenses and would not be efficient [25].

GPUs are better accelerators for deep learning applications since they are highly beneficial for parallel computations. In addition, they use larger memory bandwidth which leads to the higher throughput [25], but since GPUs are not designed specifically for DL applications, they are not efficient in terms of area, power, and energy [25]. Moreover, tasks such as pruning and weight decomposition cannot be accelerated ideally in GPUs. Generally both GPUs and CPUs are not much ideal for DL applications, but compared to CPUs, GPUs are better for this task.

Another hardware accelerator is tensor processing unit (TPU) which was specifically designed by Google company for machine learning (ML) applications. TPUs have large cache memory, high memory bandwidth, and hardware acceleration for complex activation functions. Therefore, TPUs are much better accelerator for DL applications compared to CPUs and GPUs [25].

Among all CPU, GPU, and TPU, field programmable gate arrays (FPGAs) and application-specific integrated circuits (ASIC) showed the best performance and efficiency [25]. Several architectures with stochastic and approximate computing have been proposed on FPGA and ASIC which would be reviewed in the next sections of this chapter.

3 Stochastic and Approximate Computing-Based Deep Learning Applications

Stochastic and approximate computing methodologies are widely used in different computational units of deep learning applications. For example, a new stochastic adder was proposed in [5]. This T flip-flop-based adder is shown in Fig. 4. In this adder, if the value of the two inputs are the same, they are going to be propagated to the output. Otherwise, the T flip-flop would be the output and the state of the T flip-flop would be toggled. Although the overall design [5] has achieved many benefits over the previous works, the stochastic adder proposed in this work does not seem the best. LFSR+TFF-based adder is better than the proposed adder in terms of mean squared error (MSE).

Another stochastic adder was also proposed in [26]. In this work, the APC adder is optimized to increase its power and energy consumption and reduce its area. In [26], instead of using conventional full adders in the adder tree, inverse mirror full adders proposed in [27] were used. APC adder was also optimized in [28]; this APC uses 4-1 multiplexers instead of AND/OR gates, which significantly reduces the area and is better than the conventional APC in terms of area, power, and delay. The proposed T flip-flop-based adder in [5] is better than the ones proposed in [26] and [28] in terms of area, while the proposed adders in [26] and [28] are better in terms of delay since they do the computations in parallel. Since activation functions have complex designs with large areas and they are highly power- and energy-consuming, in addition to the adders and multipliers, so many stochastic-based activation functions have been proposed in the literature. For instance, in [28], a new ReLU and softmax activation functions based on stochastic computing are proposed. The stochastic ReLU proposed in [28] is much more accurate than the previous stochastic ReLU activation functions and its functionality is much closer to the actual ReLU. In addition, the negative numbers are eliminated in this activation function and the inputs can be in integer format. The methodology used for the softmax implementation in [28] utilizes the least possible resources, which leads to the area reduction and has a very low delay. The design of [28] achieved better area overhead, critical path length, and power consumption than the previous works done in this sphere for implementing Lenet-5. However, it lost a bit of accuracy.

In [29], a new sigmoid activation function is proposed based on stochastic computing. This activation unit is called approximate stochastic computing activation unit (ASCAU), which can be used to implement various kinds of activation functions, and the correlation among the different stochastic bitstreams would not be considered an issue in this design. Thus, this activation function is much more energy-efficient than other existing stochastic activation functions. By using ASCAU, sigmoid as well as ReLU activation functions can be implemented. The sigmoid activation function can be implemented by assigning 0, 4, and 2 to p, r, and s, respectively, to the ASCAU. Besides the sigmoid activation function, ReLU can be implemented by ASCAU by setting p, r, and s to 0, 1, and 0, respectively. The stochastic-based deep belief neural network (DBN) proposed in this work is better than 8-bit fixed point and 32-bit floating point conventional DBNs in terms of all area, power, and latency. ASCAU linear function is shown in Eq. (15).

The conventional ReLU activation function uses the $f(x) = max(0, x)$. Since in stochastic computing numbers are represented in stochastic bitstreams with length m, the sign as well as the value of the number would be defined after m clock cycles. Hence, using this activation function with stochastic bitstreams would increase the delay and energy consumption. For solving this problem, a clipped version of the ReLU activation function is proposed in [26]. The clipped version uses this equation: $f(x) = min(max(0, x), 1)$. In this ReLU activation unit, first the input bitstreams are accumulated, and then, they are compared by the half of the clock cycles. If the accumulated number was less than the half of the clock cycles, 1 would be generated as the output. Otherwise, the output would be generated by the FSM that is embedded in the ReLU's architecture. This optimized ReLU decreases the delay and energy consumption compared to the conventional ReLU without hurting the accuracy.

In addition to using just approximate and stochastic computing methodologies in DL applications, using hybrid methods with imprecise computing methodologies has gained a vast popularity during the past recent years such as hybrid stochastic-binary designs. The reason for using these architectures is that the binary part would compensate for the inaccuracy caused by the stochastic methodology and the stochastic part would compensate for the large area and power consumption of the binary part so that a good balance would be achieved in these architectures. For instance, in [5], a new hybrid stochastic-binary neural network for near-sensor computing was proposed. In this architecture, the binary part of the neural network compensates for the accuracy loss of the stochastic section. As a result, this architecture is much more accurate than fully stochastic designs. In addition, this design is 9.8 times more efficient in terms of energy savings than fully binary

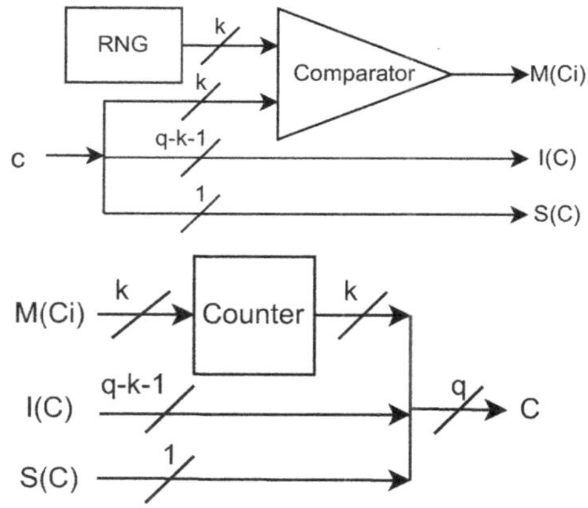

Fig. 7 Two-line stochastic elements. The upper one is the binary to stochastic bitstream converter and the lower one is the stochastic to binary converter. In these elements S is the sign bit, M is the mantissa part, and I is the integer part [30]

architectures, and it reaches the application-level accuracy within 0.05% compared to binary designs.

Another hybrid binary-stochastic design is proposed in [6] for the hardware implementation of a long-term short memory recurrent neural network. For evaluating this design, four different datasets (the Reder grammar, Japanese vowels, Texas Instruments, and Massachusetts Institute of Technology (TIMIT)) were used, and their results show that this design is much more noise-tolerant than its binary counterparts and it is highly efficient in terms of energy and area consumption. This design is better than both 8-bit fixed point and 32-bit floating point designs in terms of all area, power, and latency.

$$\Psi(x) = min(1, max(p, \frac{1}{r}x + s)) \tag{15}$$

Winograd algorithm [23] is one of the methods for reducing the hardware complexity of the CNNs by reducing the number of multiplications in the convolutional layers. Thus, it has gained popularity for being used in CNNs. For example, in [30], the WFCU is implemented using the stochastic adder and the two-line stochastic multiplier that was proposed in [31]. The stochastic elements used in [30] are shown in Figs. 7 and 8. According to the results, the number of LUTs in this work are reduced significantly compared to the binary Winograd fast convolution unit (WFCU). But since stochastic computing is used in this work, the range of the inputs are between -1 and 1 which has reduced the accuracy. The Winograd architecture proposed in [30] is shown in Fig. 9.

The resources needed for converting binary numbers into stochastic bitstreams are costly in terms of area. Another challenge that exist in stochastic computing is the correlation among the different stochastic bitstreams which contributes to inaccuracy. These problems are solved in [32] in CNN architecture. The results obtained in [32] show that their design shows a much better performance than binary

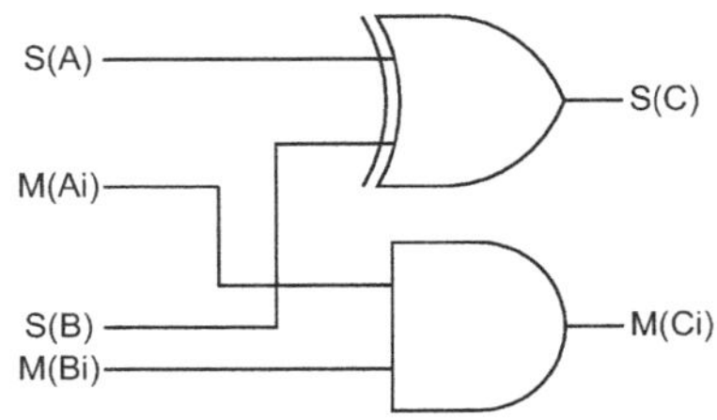

Fig. 8 Two-line stochastic multiplier. In this multiplier, S is the sign bit, I is the integer part, and M is the mantissa part [30]

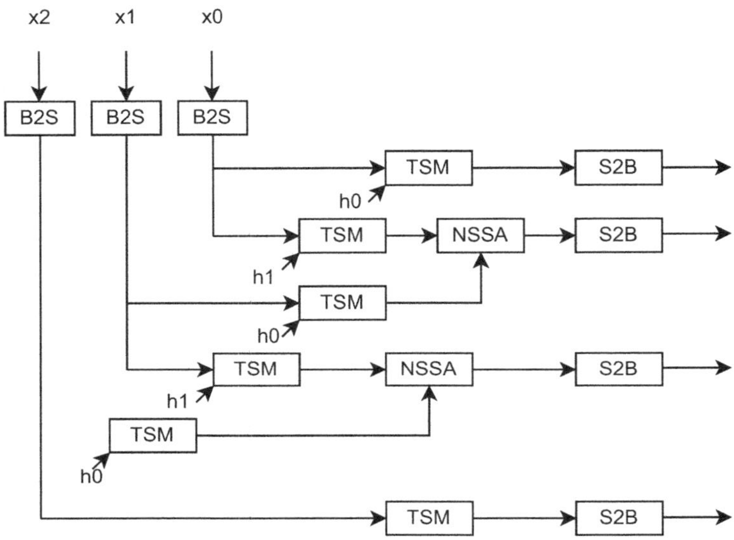

Fig. 9 Proposed 2 × 3 Winograd unit in [30]. In this unit B2S unit is the binary to stochastic converter, S2B is the stochastic to binary converter, TSM is the two-line stochastic computing multiplier, and NSSA is the high-accuracy non-scaled stochastic adder

and all other stochastic designs. The stochastic neuron in [32] uses both correlation and decorrelation in the stochastic computing methodology.

In [26], a highly efficient stochastic computing-based inference framework (HEIF) for large-scale deep convolutional neural networks (DCNNs) is proposed. In HEIF a modified APC and stochastic multiplier and an efficient way for storing the weights are proposed. In addition, this architecture reduced the energy consumption and bitstream length to their highest possible level with small loss of accuracy. The experimental results in this work showed that in HEIF, throughput has increased while the energy consumption and area reduced compared to the previous designs. Moreover, in [26], SRAM is used for storing the weights in the fully connected layers since it has high speed and small area while it is reliable.

Besides stochastic computing methodology, approximate computing has been used in several deep learning applications. Approximate computing eliminates the least significant elements in a design which would not hurt the accuracy of

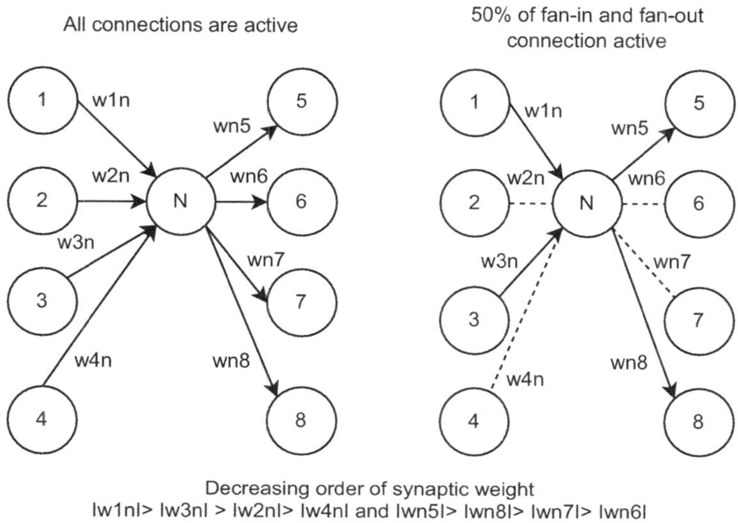

Fig. 10 The way that the neuron is approximated in [12]

the design significantly but it would reduce the area and power consumption. For example, in [12], AxSNN is proposed. AxSNN is an approximate computing methodology that is used for the implementation of a spiking neural network (SNN) in order to make its computations more efficient. This method eliminates the neurons that have the least impact on the network's output by using heterogeneous scales. The methodology which is used for eliminating the least significant neurons is shown in Fig. 10. This methodology is used for both software and hardware versions of this NN and for the hardware version a spiking neural network approximate processor (SNNAP) is also proposed in this paper, but the results are not compared to other works like exact SNN.

Using approximate computing in SNNs is widely popular because of their complex and large computations. In [33] a new approximate architecture based on FPGA is proposed for the spiking neural network accelerator. In this architecture, approximate adders are used to reduce area and power consumption. The proposed approximate adder in [33] is shown in Fig. 11 and its truth table is shown in Fig. 12. Furthermore, a new variable truncation method is proposed in [33] for the weights. In this methodology, different bit widths are dedicated to each weight instead of truncating all the weights uniformly. For instance, the weights that have more impact on the NN's output would have bigger bit widths, and the ones with the least importance are bound to have shorter bit widths. In that case, the number of ALUTs and the power consumption of the architecture are set to reduce significantly.

Approximate computing is also widely used in multilayer perceptron. For instance, in [34] exact multipliers in a multilayer perceptron and a convolutional neural network are replaced by deliberate design and Cartesian genetic program-

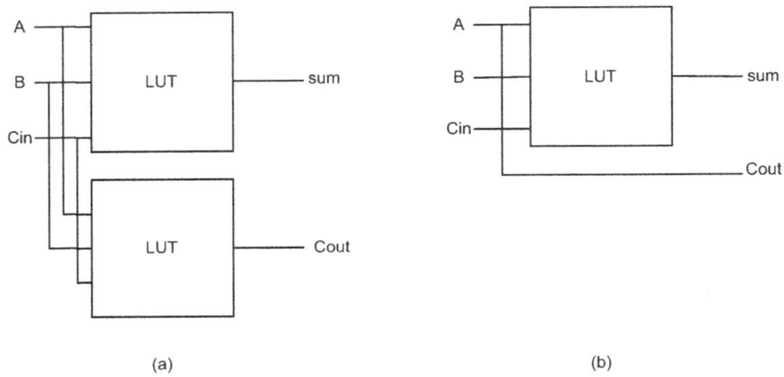

Fig. 11 (**a**) LUTs in exact full adder. (**b**) LUTs in the approximate adder proposed in [33]

Input operands			Exact addition		Approximate addition	
A	B	Cin	Cout	S	Cout	S
0	0	0	0	0	0	0
0	0	1	0	1	0	1
0	1	0	0	1	0	1
0	1	1	1	0	0	0
1	0	0	0	1	1	1
1	0	1	1	0	1	0
1	1	0	1	0	1	0
1	1	1	1	0	1	1

Fig. 12 Truth table of the exact full adder and the approximate adder proposed in [33]

ming (CGP)-based approximate multipliers. Deliberately designed multipliers are the ones made by making some slight changes in the exact multipliers' truth tables and the CGP-based multipliers are made by using the CGP heuristic algorithm. The five best approximate adders in terms of accuracy and hardware resource consumption are introduced in this paper. It should be noted that all these five multipliers are CGP-based.

4 Area- and Power-Efficient Hybrid Stochastic-Approximate Designs in Deep Learning Applications

Stochastic as well as approximate computing methodologies are widely used in deep learning applications in order to reduce their area and power consumption. In all of the previous works, stochastic and approximate methodologies have been used separately and the combination of these two methodologies have never been used in the literature. In this section of this chapter, we are going to propose two designs, in which we have used the combination of these two methodologies.

4.1 Low Complexity with High-Accuracy Winograd Convolution Based on Stochastic and Approximate Computing

Winograd algorithm is used in the convolutional layer of the convolutional neural network for reducing the area as well as the power consumption of the design. In this section we propose a new convolution architecture based on approximate and stochastic computing methodologies, which would reduce the hardware complexity compared to the exact designs and increase the accuracy compared to the stochastic architectures.

Convolutional neural network (CNN) is widely used in various applications such as image processing [1]. Imprecise computing methodologies can be used in Winograd algorithm like approximate and stochastic computing methodologies in order to reduce its area and power consumption even more. In the proposed architecture the range of the input numbers are between -10 and 10 for increasing the accuracy, but since the integer numbers cannot be used in stochastic computing methodologies, the multiplication of the integer part is done with an approximate multiplier which reduces the number of LUTs in FPGA.

In the proposed methodology a F(2×2) Winograd fast convolution unit is implemented using the stochastic and approximate computing methodologies. The matrices for F(2,2) are shown in Eq. (16) [35]:

$$g = \left[g_0\ g_1\right]^T\ d = \left[d_0\ d_1\ d_2\right]^T\ B^T = \begin{bmatrix} 1 & 0 & -1 \\ 0 & 1 & 1 \\ 0 & -1 & 1 \end{bmatrix}\ G = \begin{bmatrix} 1 & 0 \\ \frac{1}{2} & \frac{1}{2} \\ \frac{1}{2} & -\frac{1}{2} \end{bmatrix}\ A^T = \begin{bmatrix} 1 & 1 & 1 \\ 0 & 1 & -1 \end{bmatrix} \tag{16}$$

In this work, stochastic computing methodology is used for reducing the hardware complexity and for avoiding significant loss of accuracy; the range of the numbers are between -10 and 10. Since stochastic computing cannot be used for integer numbers, the numbers are divided into three sections: 1 sign bit (S), 4 bits for the

Table 1 Comparison of the hardware resource utilization between the convolutional operation without Winograd algorithm and different types of Winograd units

Hardware implementation results	Convolutional operation without Winograd algorithm	Binary Winograd unit	Stochastic Winograd unit	Approximate Winograd unit	Hybrid stochastic-approximate block
Multiplications (LUTs)	223	190	6	107	47
Additions (LUTs)	36	40	18	29	14
Power	0.051	0.044	0.01	0.029	0.014
Clock cycles	4	4	2N	4	N+1

integer part (I), and 2–10 bits for the mantissa (M). The multiplication is done, using Eq. (17) [31].

$$\begin{aligned}(I(A)+M(A)).(I(B)+M(B)) &= (I(A).I(B)) \\ &+(I(A).M(B))+(I(B).M(A))+(M(A).M(B))\end{aligned} \tag{17}$$

The M(A).M(B) is done by stochastic multiplier and as long as all three other multiplications contain integer numbers, their multiplications are done by approximate multipliers. For instance, the I(A).I(B) multiplication is done using a 4×4 approximate multiplier proposed in [19] and I(A).M(B) and I(B).M(A) are done using a 12×12 approximate multiplier which is proposed in this chapter. For reducing the number of LUTs in the multiplier, for example, when the mantissa part is 8 bits and the integer part is 4 bits, the partial products of the I(0)-I(7) with M multiplications and the partial products of the M(7)-M(11) with I multiplications are eliminated. This reduces the number of the multiplication's LUTs and the hardware complexity significantly.

The proposed design is implemented using Xilinx ISE and it is implemented on FPGA spartan-3e.

Table 1 shows a comparison between the number of LUTs, power consumption, and clock cycles in binary, stochastic, and proposed Winograd fast convolution units. As it can be seen, the number of LUTs in the proposed methodology are much less than all other designs but the stochastic one. The results shown in Table 1 are for the inputs with 1 sign bit, 4 bits for the integer part, and 8 bits for the mantissa. N is the number of bits in the stochastic bitstream.

To compare the accuracy, the mean squared errors (MSE) in two-line stochastic computing methodology, unipolar stochastic computing methodology, bipolar stochastic computing methodology, and the proposed design which is the hybrid stochastic-approximate computing methodology are compared. As it can be seen in Fig. 13, the MSE in the proposed design is less than all of the stochastic computing methodologies and it is approximately 62% less than the two-line stochastic one.

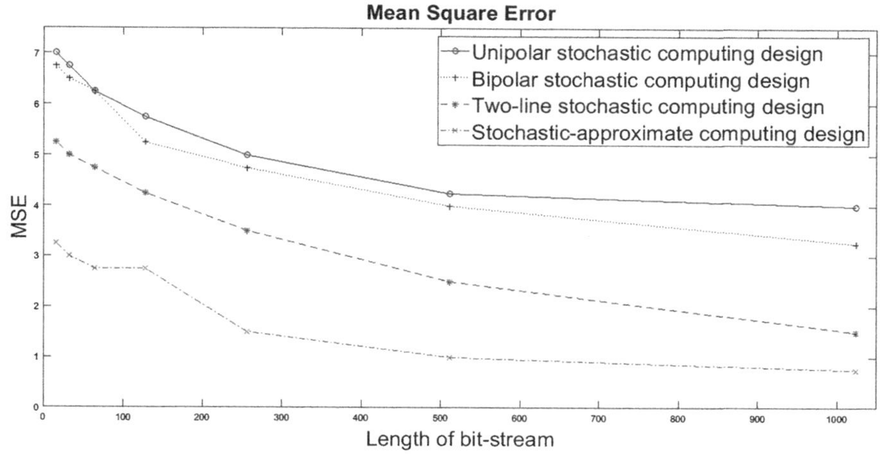

Fig. 13 Mean squared error (MSE) comparison between unipolar stochastic, bipolar stochastic, two-line stochastic, and proposed hybrid stochastic-approximate designs

4.2 A Hybrid Power- and Area-Efficient Stochastic-Approximate Neuron

In this section, we propose an efficient architecture in terms of speed, power consumption, and area for the hardware implementation of a neuron by sacrificing a small amount of accuracy. The proposed design uses an efficient combination of approximate and stochastic computing methodologies to achieve better results than the designs implemented by just stochastic and approximate computing methodologies. A stochastic multiplier is used instead of the approximate one for the following reasons:

- The stochastic multiplier's hardware complexity and power consumption are much less than approximate ones.
- The stochastic multiplier does not increase the bit widths after multiplication. Thus, we do not need to normalize the multiplier's output again to shorten the bit widths or use larger adders for adding the partial product's rows.

The approximate adders are also chosen instead of stochastic ones because they have a much higher speed and accuracy than stochastic ones, while their area and power consumption are less than stochastic adders. The architecture of the proposed hybrid stochastic-approximate neuron is shown in Fig. 14. In this design, we have 10-bit inputs with the fixed-point display. This design converts the neuron inputs and the binary weights to stochastic bitstreams and then the stochastic bitstreams multiplied by simple XNOR gates. XNOR gates are used because weights include negative numbers, and we should use bipolar stochastic numbers. Next, stochastic

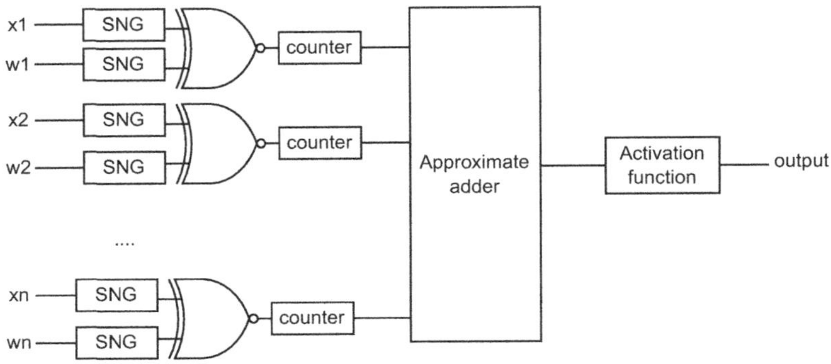

Fig. 14 Proposed hybrid stochastic-approximate neuron

numbers will be converted to binary ones to be added together using an approximate adder tree made by the approximate adders introduced in [20].

The adder tree used in the proposed methodology consists of three APAD4s [20] for the least three significant bits, one APAD3 [20] for the fourth least significant bit, and four full adders for the most four significant bits. The output of the adder tree feeds the activation function, which produces the output of the neuron block. The activation function used for the first layer is ReLU, and the Sigmoid activation function is used for the hidden layer. Since the hardware implementation of the Sigmoid activation function would be highly area-consuming, this activation function is implemented by approximate stochastic computing activation unit (ASCAU) [29].

Xilinx ISE is used to synthesize it, and it is implemented on FPGA Artix-7. Table 2 shows the power consumption, area, and the number of clock cycles for the proposed neuron block compared to the stochastic [3], approximate, and binary [3] ones. For comparison, we implemented an approximate neuron block in the same condition as the proposed methodology.

Table 2 shows that the number of clock cycles in the proposed hybrid approximate-stochastic neuron block is almost half of the number of clock cycles in the stochastic design. In terms of resource consumption, our proposed neuron block consumes 57% and 67% fewer LUTs than the approximate and binary designs, respectively. Our proposed methodology uses almost 1% fewer registers than the stochastic design. In terms of power, the proposed design can achieve 11.91%, 5.14%, and 17.78% less power consumption than the approximate, stochastic, and binary neurons, respectively.

Table 2 Comparison of the hardware resource utilization between the different types of neuron blocks

Hardware implementation results	Binary neuron block	Stochastic neuron block	Approximate neuron block	Hybrid stochastic-approximate neuron
Parallel inputs	16	16	16	16
Bit width	8	8	8	8
Frequency	250 MHz	286 MHz	250 MHz	286 MHz
LUTs	1268	416	982	418
Registers	23	299	18	295
IO	277	277	277	277
Power consumption	0.045 W	0.039 W	0.042 W	0.037 W
Clock cycles	3	258	3	130

5 Conclusion

In this chapter, stochastic and approximate computing-based deep learning applications are reviewed. We start with the introduction of the stochastic and approximate computing methodologies, including their computational circuits and their advantages and disadvantages. Then we go through the deep learning arithmetic units and accelerators. Finally, we propose two new designs for addressing the challenges that are faced while using stochastic and approximate computing methodologies such as low accuracy and computational speed in stochastic computing and the large area in the approximate multipliers. By reading this chapter it is expected that the readers become familiar with the basics of the stochastic and approximate computing and deep learning applications which will help them to propose new ideas for accommodating the challenges in these methodologies.

6 Future Research Directions

Stochastic and approximate computing methodologies reduce the hardware resource utilization in the deep learning applications [35, 36], though they have some downsides such as low accuracy and computational speed in stochastic computing and the large area in approximate multipliers. Hence, future works must mostly focus on tackling these issues. For example, in Sect. 5 two power- and area-efficient hybrid methodologies were proposed; in the future works these designs can be used in bigger dimensions and neural networks for getting better results compared to the previous designs.

For the future works, we have this intention to use the two proposed designs in a convolutional neural network. The area- and power-efficient hybrid approximate-stochastic neuron is going to be used in the fully connected layer, while the low complexity with high-accuracy Winograd unit based on stochastic and approximate computing is going to be used in the convolutional layer. By using these methodologies the accuracy of this convolutional neural network is expected to be improved compared to the fully stochastic design, while its area and power are predicted to be less than the fully binary and approximate designs.

References

1. Z. Jiang, H. Zhang, Y. Wang, S.B. Ko, Retinal blood vessel segmentation using fully convolutional network with transfer learning. Comput. Med. Imaging Graph. **68**, 1–15 (2018)
2. Q. Kong, Y. Cao, T. Iqbal, Y. Wang, W. Wang, M. Plumbley, PANNs: large-scale pretrained audio neural networks for audio pattern recognition. IEEE/ACM Trans. Audio Speech Lang. Process. **28**, 2880–2894 (2020)
3. D. Nguyen, H. Ho, D. Bui, X. Tran, An efficient hardware implementation of artificial neural network based on stochastic computing, in *5th Conference NAFOSTED on Information and Computer Science* (2018)
4. M.E. Nojehdeh, L. Aksoy, M. Altun, Efficient hardware implementation of artificial neural networks using approximate multiply-accumulate blocks, in *2020 IEEE Computer Society Annual Symposium on VLSI* (2020)
5. V. Lee, A. Alaghi, J. Hayes, V. Sathe, L. Ceze, Energy-efficient hybrid stochastic-binary neural networks for near-sensor computing, in *Design, Automation & Test in Europe Conference & Exhibition* (2017)
6. Y. Liu, L. Liu, F. Lombardi, J. Han, An energy-efficient and noise-tolerant recurrent neural network using stochastic computing. IEEE Trans. Very Large Scale Integr. Syst. **27**, 2213–2221 (2019)
7. Y. Liu, S. Liu, Y. Wang, F. Lombardi, J. Han, A survey of stochastic computing neural networks for machine learning applications. IEEE Trans. Neural Netw. Learn. Syst. **32**(7), 2809–2824 (2020)
8. W. Liu, F. Lombardi, *Approximate Computing* (Springer, Berlin, 2022)
9. H. Zhang, M. Asadikouhanjani, J. Han, D. Subbian, S.-B. Ko, Approximate computing for efficient neural network computation: a survey, in *Approximate Computing* (Springer, Cham, 2012), pp. 397–427
10. A. Alaghi, J.P. Hayes, Survey of stochastic computing. ACM Trans. Embed. Comput. Syst. **12**(2s), 1–19 (2013)
11. W. Liu, C. Gu, M. O'Neill, G. Qu, P. Montuschi, F. Lombardi, Security in approximate computing and approximate computing for security: challenges and opportunities. Proc. IEEE **108**(12), 2214–2231 (2020)
12. S. Sen, S. Venkataramani, A. Raghunathan, Approximate computing for spiking neural networks, in *Design, Automation & Test in Europe Conference & Exhibition (DATE)* (2017)
13. P. Kulkarni, P. Gupta, M. Ercegovac, Trading accuracy for power with an underdesigned multiplier architecture, in *2011 24th International Conference on VLSI Design* (2011)
14. M.S. Ansari, V. Mrazek, B.F. Cockburn, L. Sekanina, Z. Vasicek, J. Han, Improving the accuracy and hardware efficiency of neural networks using approximate multipliers. IEEE Trans. Very Large Integr. Syst. **28**, 317–128 (2019)
15. H. Zhang, H. Xiao, H. Qu, S.-B. Ko, FPGA-based approximate multiplier for efficient neural computation, in *2021 IEEE International Conference on Consumer Electronics-Asia (ICCE-Asia)* (IEEE, 2021), pp. 1–4
16. Q. Xu, T. Mytkowicz, N.S. Kim, Approximate computing: a survey. IEEE Design & Test **33**(1), 8–22 (2015)

17. S. Venkatachalam, S.-B. Ko, Approximate sum-of-products designs based on distributed arithmetic. IEEE Trans. Very Large Scale Integr. Syst. **26**(8), 1604–1608 (2018)
18. S. Venkatachalam, E. Adams, H.J. Lee, S.-B. Ko, Design and analysis of area and power efficient approximate booth multipliers. IEEE Trans. Comput. **68**(11), 1697–1703 (2019)
19. S. Ullah, S. Rehman, M. Shafique, A. Kumar, High-performance accurate and approximate multipliers for FPGA-based hardware accelerators. IEEE Trans. Comput.-Aided Design Integr. Circuits Syst. **41**(2), 211–224 (2021)
20. M.E. Nojehdeh, M. Altun, Systematic synthesis of approximate adders and multipliers with accurate error calculations. Integration **70**, 99–107 (2020)
21. L. Crespo, P. Tomás, N. Roma, N. Neves, Unified posit/IEEE-754 vector MAC unit for transprecision computing. IEEE Trans. Circuits Syst. II: Express Briefs **69**(5), 2478–2482 (2022)
22. S. Tang, J. Xia, L. Fan, X. Lei, W. Xu, A. Nallanathan, Dilated convolution based CSI feedback compression for massive MIMO systems. IEEE Trans. Veh. Technol. **71**(10), 11216–11221 (2022)
23. W. Shmuel, *Arithmetic Complexity of Computations*, vol. 33 (SIAM, 1980)
24. A. Lavin, S. Gray, Fast algorithms for convolutional neural networks, in *Proceedings of the IEEE Conference on Computer Vision and Pattern Recognition* (2016), pp. 4013–4021
25. K.S. Zaman, M.B.I. Reaz, S.H.M. Ali, A.A.A. Bakar, M.E.H. Chowdhury, Custom hardware architectures for deep learning on portable devices: a review. IEEE Trans. Neural Netw. Learn. Syst. **33**(11), 6068–6088 (2022)
26. Z. Li, J. Li, A. Ren, R. Cai, C. Ding, X. Qian, J. Draper et al., HEIF: Highly efficient stochastic computing-based inference framework for deep neural networks. IEEE Trans. Comput.-Aided Design Integr. Circuits Syst. **38**(8), 1543–1556 (2018)
27. N. Weste, D. Harris, *CMOS VLSI Design: A Circuits and Systems Perspective*, 4th edn. (Addison-Wesley, Boston, 2010)
28. M.H. Sadi, A. Mahani, Accelerating deep convolutional neural network base on stochastic computing. Integration **76**, 113–121 (2021)
29. Y. Liu, Y. Wang, F. Lombardi, J. Han, An energy-efficient stochastic computational deep belief network, in *2018 Design, Automation & Test in Europe Conference & Exhibition (DATE)* (2018)
30. H. Wang, Z. Zhang, X. You, C. Zhang, Low-complexity Winograd convolution architecture based on stochastic computing, in *In 2018 IEEE 23rd International Conference on Digital Signal Processing (DSP)* (IEEE, 2018), pp. 1–5
31. R. Xu, B. Yuan, X. You, C. Zhang, Efficient fast convolution architecture based on stochastic computing, in *2017 9th International Conference on Wireless Communications and Signal Processing (WCSP)* (IEEE, 2017), pp. 1–6
32. C.F. Frasser, P. Linares-Serrano, I.D. de Los Rios, A. Moran, E.S. Skibinsky-Gitlin, J. Font-Rossello, V. Canals, M. Roca, T. Serrano-Gotarredona, J.L. Rossello, Fully parallel stochastic computing hardware implementation of convolutional neural networks for edge computing applications. IEEE Trans. Neural Netw. Learn. Syst. (2022)
33. Y. Wang, H. Zhang, K.-I. Oh, J.-J. Lee, S.-B. Ko, Energy efficient spiking neural network processing using approximate arithmetic units and variable precision weights. J. Parallel Distrib. Comput. **158**, 164–175 (2021)
34. M.S. Ansari, V. Mrazek, B.F. Cockburn, L. Sekanina, Z. Vasicek, J. Han, Improving the accuracy and hardware efficiency of neural networks using approximate multipliers. IEEE Trans. Very Large Integr. Syst. **28**, 317–128 (2019)
35. J. Yepez, S.-B. Ko, Stride 2 1-D, 2-D, and 3-D Winograd for convolutional neural networks. IEEE Trans. Very Large Scale Integr. Syst. **28**(4), 853–863 (2020)
36. H. Zhang, D. Chen, S.-B. Ko, New flexible multiple-precision multiply-accumulate unit for deep neural network training and inference. IEEE Trans. Comput. **69**(1), 26–38 (2019)

Stochastic Computing Applications to Artificial Neural Networks

Josep L. Rosselló, Joan Font-Rosselló, Christiam F. Frasser, Alejandro Morán, Vincent Canals, and Miquel Roca

1 Introduction

The new generation of knowledge-based applications [1] is gaining increasing prominence in the field of information technologies. These kinds of applications are normally related to pattern recognition, data mining, and digital image/video synthesis. All these applications are characterized by their great complexity when tackled with conventional processing methods which entail large energy dissipation and processing times. Application-specific hardware designs for neural network implementation (hardware neural networks, HNN) are able to take advantage of the intrinsic parallelism of neural networks, thus optimizing energy dissipation and computation time. Therefore, many HNN designs and applications have been developed in recent years to provide a feasible solution to massive data mining, a sort of processing which is currently highly demanded by science and technology. This interest is reflected in Fig. 1, where a graph concerning the number of big data and HNN publications per year during this century is shown. As it can be observed, since 2012 there has been a surge in the interest in these two topics.

Two main artificial neuron models can be identified in the literature: non-spiking and spiking models. The first (and more common) model is characterized by providing at the neuron output a function (referred to as activation function) of the weighted sum of the inputs. The simplest model is the McCulloch-Pitts neuron, also called perceptron, that only provides a high or low value at the neuron output. This binary output is the result of applying the Heaviside function to the processed input signals. By means of perceptrons, any Boolean function can be reproduced in a simple way through a network with a single hidden layer [2]. Another type of

J. L. Rosselló (✉) · J. Font-Rosselló · C. F. Frasser · A. Morán · V. Canals · M. Roca
Industrial Engineering Department, Universitat de les Illes Balears, Palma de Mallorca, Spain
e-mail: j.rossello@uib.es

W. Liu et al. (eds.), *Design and Applications of Emerging Computer Systems*,
https://doi.org/10.1007/978-3-031-42478-6_12

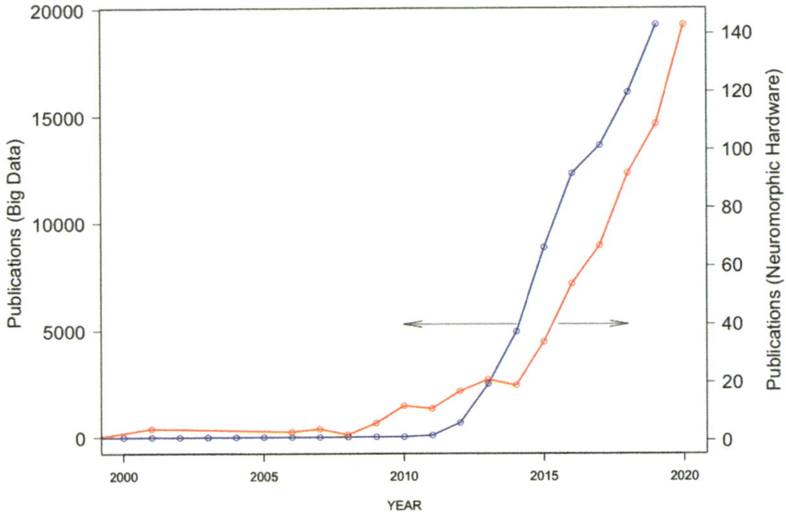

Fig. 1 Trends in the number of publications in big data and hardware implementation of neural networks (source: SCOPUS)

more complex non-spiking model is that which uses a smooth activation function so that the output is a continuous variable instead of a binary one. The main advantage provided by the use of continuous activation functions is their adaptability to backpropagation learning methods, not applicable when they are discontinuous. The high efficiency of backpropagation and its application to implement supervised learning processes has caused this type of ANN to be the most popular within the machine learning community.

Stochastic computing (SC) is an approximate computing technique that uses random bit sequences instead of precise arithmetic operations. Instead of representing data as precise numbers, SC uses the activation probability of a bit to perform calculations. This technique is particularly useful for tasks that do not require high accuracy but can benefit from the energy and space efficiency provided by SC. For these reasons, SC is gaining momentum in artificial intelligence implementations, such as neural network hardware design, since the high precision of the intermediate processing steps in an ANN is not a strict requirement for obtaining a high accuracy in the overall ANN behavior.

Stochastic computing has been extensively utilized in the implementation of machine learning systems [3], mainly for accelerating ANNs using hardware [4]. This is primarily attributed to the high energy efficiency and noise tolerance that can be achieved with this computational design technique [5]. Furthermore, the correlation/decorrelation between spiking signals within the SC circuit can be used to greatly compress the connections of neural networks. This effect is largely seen in the application of SC to the implementation of morphological neural networks (MNNs). These systems are a class of networks that are based on the

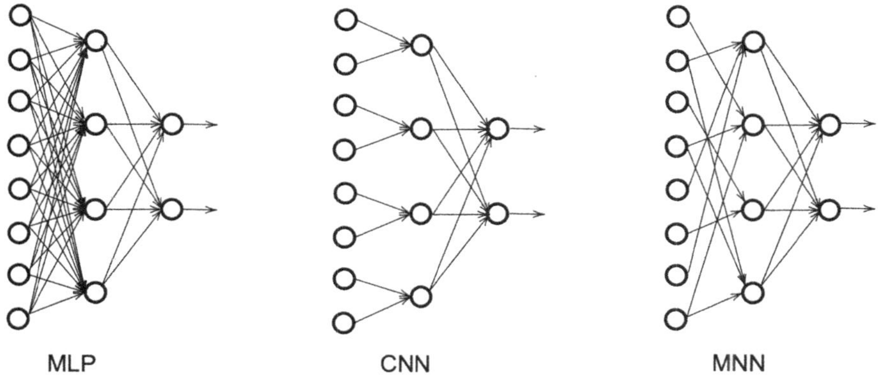

Fig. 2 Topological comparison between a multilayer perceptron (MLP), a convolutional neural network (CNN), and a morphological neural network (MNN)

theory of mathematical morphology, a technique for image analysis used to detect patterns and features in images. MNNs are based on basic arithmetic operations, such as addition, subtraction, maximum, and minimum, and avoid using complex activation functions used in classical neural networks schemes as convolutional neural networks. They are also characterized by their ability to maintain high accuracy in the morphological network despite the removal of a large number of connections within the min-max layers (known as tropical pruning [6]). In Fig. 2, we present a schematic comparison of MNN networks with two other classical topologies, MLPs (which use a fully connected topology between their layers) and CNNs where local connectivity is used. The complete topology of MLPs provides great generalization capability by being able to consider any type of correlation between any set of input signals. However, for networks with a large number of neurons or layers, such a topology is difficult to optimize using the backpropagation algorithms typically employed. To simplify training, CNNs take advantage of the fact that in image processing (one of the most tackled machine learning problems, if not the most), there will be a high correlation between adjacent pixels. This substantially reduces the number of connections to be trained, making the training process more optimal. The downside of CNNs is that they do not address correlations between distant pixels, requiring the use of multilayer systems to handle such possibilities (known as deep neural networks). Finally, MNNs enable a training process with pruning of connections that will not affect the overall accuracy of the network. In this way, MNNs can start with the training of a fully connected neural network (as is the case with MLPs) but eliminate those connections that are not necessary for global processing. Therefore, as with CNNs, we end up with a reduced set of connections between layers that are not restricted to specific areas of the image to be processed but are capable of connecting those regions that do have some relationship based on the patterns that need to be discerned from the inputs. Thus, MNNs are simpler in connectivity than MLPs (and similar to the simplicity of CNNs), but at the same time have similar generalization performance to MLPs.

In the next section we will demonstrate the fundamental principles of stochastic computing and its potential application in the implementation of ANNs, including networks like convolutional neural networks, radial basis functions, and the more recently established MNN networks.

2 Stochastic Computing Basic Principles

Stochastic computing (SC) was developed as a method to execute complex computing tasks with a low cost in terms of hardware and high noise tolerance capacities [7, 8]. Its primary characteristic is the use of the theory of probabilities instead of deterministic methods so that instead of fixed two-complement binary numbers, it uses variable random bit streams with a predefined switching activity. Therefore, each single bit signal is defined by a probability p of being in the high state, and the traditionally coded binary numbers must be converted to a stochastic one, firstly normalizing them to fit within the interval [0, 1]. Once these numbers are normalized, they are reconverted to unsigned binary depending on the bit precision. In this way, the number 1 would be assigned to the maximum binary number available, while 0 is settled to 0.

As an example, numbers π and $\frac{11}{3}$ would be firstly normalized to 0.856 and 1. And for the special case of a precision of 12 bits, they would be reconverted to $0xDB5$ and $0xFFF$. If we use 4-bit precision instead of 12 bits, then these numbers would be $0xD$ and $0xF$. All these codifications consider that number $\frac{11}{3}$ is the larger possible.

Once those conversions to simple unsigned binary are implemented, the next step is to convert them to a single switching bit signal. In this way, the resulting binary number U is randomized and converted to its stochastic equivalent, $u(t)$, by using a random number generator (RNG) $R(t)$ and a binary comparator:

$$u(t) = \begin{cases} 1 & U \geq R(t) \\ 0 & U < R(t) \end{cases} \tag{1}$$

Therefore, the binary number U is compared with a random binary number with the same bit size that is varying each clock cycle. Signal $R(t)$ must be as uniformly distributed as possible to have a fair conversion between the classical binary domain and the stochastic domain, so that the switching probability of the stochastic bit u ($u = \langle u(t) \rangle$) has to be proportional to U.

To convert from the stochastic domain to the classical binary domain, for each stochastic number $u(t)$ we have to count the specific number of high states that are present during a fixed number of clock cycles (what is called the evaluation time). This conversion is always subject to a fixed variability due to the random nature of stochastic signals. Defining $P_N(U^*)$ as the probability that signal $u(t)$ is converted to U^* given that the evaluation time is equal to N clock cycles, we have that:

$$P_N(U^*) = \binom{N}{U^*} u^{U^*} (1-u)^{N-U^*} \tag{2}$$

where u is the switching activity of stochastic signal $u(t)$. The expected value for U^* would be $\langle U^* \rangle = \sum U^* \cdot P_N(U^*) = Nu$. If stochastic signal $u(t)$ was initially generated using the n-bit precision binary number U, and that $N = 2^n - 1$, then we have that $\langle U^* \rangle = N \cdot \frac{U}{2^n - 1} = U$. Equation (2) is indeed the discrete probability distribution for the binomial distribution. The standard deviation, $\sigma^2(u) = Nu(1 - u)$, is related to the intrinsic error in the conversion process.

2.1 *Stochastic Signals and Correlation*

When working with stochastic signals, an important consideration is their cross-correlation, as noted by Alaghi [9]. The functional behavior of circuitry can be significantly affected by the correlation between signals [10, 11]. Consider two stochastic signals with probabilities p and q that are the inputs of an AND gate. We use symbol $\|$ to denote correlation between signals. For the case in which $p\|q$, the AND gate provides at its output the signal that presents the minimum switching activity, as shown in Fig. 3a. The maximum function would be obtained using an OR gate (not shown in the figure for simplicity). Conversely, if the signals are completely uncorrelated, $(p \perp q)$, the AND gate would perform the product of the switching probabilities, as shown in Fig. 3b.

This has implications such as the large area cost of operations requiring the use of different RNGs, which consume a large amount of logic elements or hardware area. These RNGs employed are usually linear-feedback shift registers (LFSRs) because of their relatively small size and low cost. An LFSR is a deterministic finite-state machine whose behavior is pseudo-random, meaning that it only approximates a true random source [12].

3 Classical Artificial Neural Networks

The first generation of neurons, which were developed in the 1940s and 1950s, were based on the McCulloch-Pitts model [13]. The internal state of this neuron, widely known as perceptron or threshold gate, is the weighted sum of its inputs. The neuron triggers when the internal state reaches a threshold voltage θ.

$$y_o = \text{sgn}(w_1 x_1 + \cdots + w_n x_n - \theta) = \text{sgn}\left(\sum_{i=1}^{n} w_i x_i - \theta\right) \tag{3}$$

where y_o is the output, x_i are the inputs, and w_i are the weights.

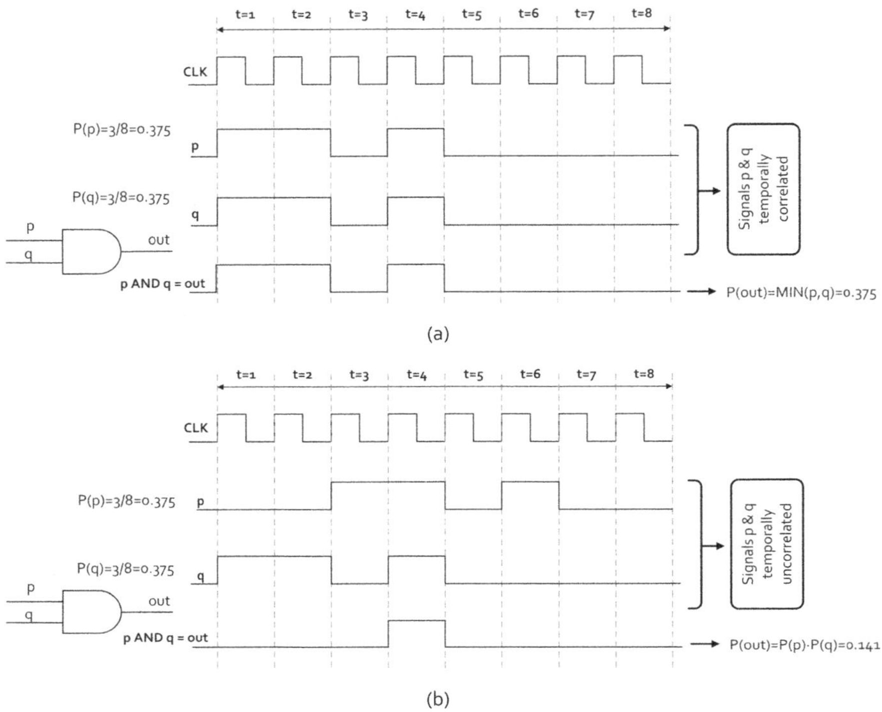

Fig. 3 (**a**) $p \| q$, that is, p and q are completely correlated signals. (**b**) $p \perp q$, that is, p and q are completely uncorrelated signals

For these first-generation neurons, the incoming inputs contribute to the variation of internal state given by $\sum_{i=1}^{n} w_i x_i$ and indirectly to the output y_o as well. Due to the simplicity of its activation function (the step function), these neural models are commonly used in hardware implementations [14–17], giving rise to the most popular neural network: the multilayer perceptron (MLP).

The second-generation neurons developed from the 1950s to 1990s intend to be more realistic and bioinspired. Now the activation function, which is applied to the same weighted sum of the inputs, is a continuous function and provides an analog output which can be understood as the average firing rate of the hypothetical postsynaptic neuron that is emulating. The most popular activation functions are the sigmoid and the hyperbolic tangent. However, other hardware-based activation functions have been implemented such as the ramp saturation, rectified linear, step or radial basis functions. All these activation functions must be continuous and also differential because they usually support learning algorithms that are based on gradient descent as backpropagation. These kinds of models are more powerful than the first-generation ones as they are capable of solving complex pattern recognition problems. Actually, they are also able to compute arbitrary Boolean functions, often with fewer gates than the nets from the first generation. Many works have been developed to reproduce these kinds of networks, using new devices and techniques as hybrid CMOS/memristor designs or spintronic devices [18–21]. Besides, these HNN designs normally use approximate functions with analog inputs and outputs. All these benefits make them powerful and suitable for implementing both feedforward and recurrent ANNs.

3.1 Application of Stochastic Computing to Artificial Neural Networks

Depending on the nature and purpose of an application, as well as the available data for training a system, it may be convenient to adopt a supervised or unsupervised learning method. This is independent of the computing paradigm used to implement a neural network in hardware. However, the internal architecture is different depending on whether the hardware is designed to include parameter optimization (on-chip learning) or not (off-chip learning). In the case of second-generation neural networks implemented using SC, it is more common and reasonable to maximize energy efficiency in edge applications by having the hardware-only handle inference. This is especially true in the case of deep learning, such as CNNs, particularly when a large amount of data is available, typically unstructured, resulting in high network generalization capability [22, 23]. This benefit can also be leveraged when addressing simpler tasks with structured data, as with shallow neural networks. In this context, it is worth noting that efforts have also been made to implement backpropagation using SC [24, 25], which could improve energy efficiency in servers dedicated to this purpose. However, integration capability and results are not yet competitive compared to those obtained using GPGPUs and other dedicated hardware based on floating-point arithmetic. On the other hand, in the case of unsupervised learning, as we do not rely on labeled data (ground truth), but on raw or preprocessed data, it can be beneficial to implement on-chip learning. This can be useful for anomaly detection [26], dimensionality reduction [27], or even for generative models [28].

Although implementing on-chip learning is a significant advantage for an intelligent system to adapt to changes in the external world, it also presents a more complex architecture, which makes it challenging to implement systems based entirely on SC. An example is the implementation of self-organizing maps (SOMs) based on SC [29], which requires a master node to update parameters using fixed-point arithmetic. There is a similar issue with [30], which proposes several SC building blocks to implement the autonomous data partitioning algorithm [31], but relies on a control block to update system parameters. Therefore, in this section, we will focus on describing the hardware architecture that performs inference. Specifically, we will explore the simulation, design, and implementation of hardware architectures that are optimized for efficient and accurate inference, including fully connected and CNN architectures [32, 33].

3.1.1 Fully Connected Neural Networks

A fully connected neural network or multilayer perceptron (MLP) is composed of multiple layers in which each neuron in one layer is connected to every neuron in the subsequent layer as depicted in Fig. 2 (MLP). The primary benefit of this type of network is its *structure agnosticism*, meaning that no assumptions need to be made about the input. Although this makes fully connected networks applicable to a wide range of problems, they usually have poorer performance than networks that are specifically designed for a particular class of problem, e.g., CNNs shine solving problems in which data present correlations in space and/or time (see Sect. 3.1.5).

The SC block diagram of a generic fully connected neural network with a single hidden layer is depicted in Fig. 4. The main difference compared to a regular fixed-point implementation is the fact that numbers need to be converted to stochastic bitstreams before performing equivalent arithmetic operations between these bitstreams. This figure tries to describe a generic high level block diagram, which can describe different types of fully connected networks depending on the random number correlation and the same concept can be extended to multiple hidden layers.

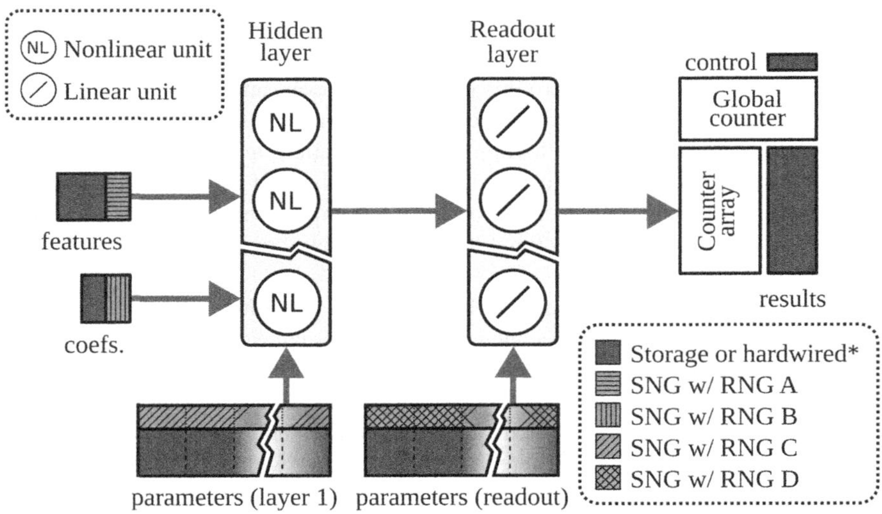

Fig. 4 Functional block diagram of a generic two-layer fully connected neural network. Different motifs indicate that SNGs use a different random sequence to generate the corresponding bit-streams. (*) Network parameters might be hardwired to save resources in FPGA implementations

3.1.2 Second-Generation MLP Configuration

For a classical second-generation MLP hidden unit, RNGs A and C generate uncorrelated random sequences. This generates maximally uncorrelated feature and hidden unit parameter bitstreams, so that SC multiplications can be implemented by simple logic gates. The same applies to the readout layer; RNGs C and D must generate uncorrelated random sequences. There are several SC options to encode signed numbers and the logic to perform arithmetic operations needs to be consistent:

- Using the (single-wire) bipolar representation, also known as sign-magnitude SC, each bitstream represents a number in the range $[-1, 1]$ and each multiplier is replaced by a single XNOR gate.
- Using the two-wire bipolar representation, each bitstream pair represents a number in the range $[-1, 1]$. The sign is represented by a bipolar bitstream, the magnitude is represented by a unipolar bitstream, and each multiplier is replaced by an XNOR gate to compute the output sign and an AND gate to compute the output magnitude. This approach requires some extra hardware but is more accurate and requires less evaluation cycles.

Even though these are the most common implementations, there are other original SC to extend the representation range, e.g., [34]. The bitstreams encoding multiplications need to be added together at each unit in order to approximate the dot product. This can be done by time multiplexing [8] (approximate addition) or an accumulative parallel counter (APC), which is more accurate. Moreover, notice the only difference between linear and nonlinear units is the activation function. Linear units do not have any activation, while nonlinear units can be configured with different functions and SC implementations usually have an ad-hoc design for that, e.g., SC sigmoid, hyperbolic tangent, or ReLU [35].

In the generic case with multiple hidden layers, let $\boldsymbol{z}^{(l)}$, $\boldsymbol{a}^{(l)}$, $\boldsymbol{W}^{(l)}$ be the bitstream pre-activation row vector, activation row vector, and weight matrix in the range $[-1, 1]$ for the l-th layer, respectively. Therefore, the pre-activation for the next layer is given by (4)

$$\boldsymbol{z}^{(l+1)} = s_l\, \boldsymbol{a}^{(l)} \boldsymbol{W}^{(l+1)} \tag{4}$$

where s is a scale factor and activations are obtained by applying an element-wise function f, i.e., $\boldsymbol{a}^{(l)} = f\left(\boldsymbol{z}^{(l)}\right)$. For the activation of the j-th unit in the l-th layer, we have:

$$z_j^{(l+1)} = s_l\, \boldsymbol{a}^{(l)} W_{:,j}^{(l+1)} = s_l \sum_i a_i^{(l)} W_{i,j}^{(l+1)} \tag{5}$$

Notice the scale factor is necessary to ensure that pre-activations are always in the range $[-1, 1]$,[1] so that they are in a valid bipolar range to be converted to stochastic bitstreams. For simplicity, we assume this scale factor is the same for all units in the same layer. It is reasonable to set these scales to powers of 2 in fully parallel designs to reduce area [33].

An equivalent SC model using the conventional bipolar representation for bitstreams is given by (6).

$$z_j^{(l+1)} \approx s_l \sum_{n=0}^{N-1} \sum_i \overline{\tilde{a}_i^{(l)}(nT) \oplus \tilde{W}_{i,j}^{(l+1)}(nT)} \tag{6}$$

Here $\tilde{a}_i^{(l)}(nT)$ and $\tilde{W}_{i,j}^{(l)}(nT)$ represent activation and parameter bitstream timeseries at time nT. So that if $\tilde{a}_i^{(l)}(nT)$ and $\tilde{W}_{i,j}^{(l)}(nT)$ are maximally uncorrelated bitstreams, then multiplications can be implemented by XNOR gates and additions with APCs. Similarly, for the two-wire bipolar representation, an equivalent SC model is given by:

$$z_j^{(l+1)} \approx s_l \sum_{n=0}^{N-1} \sum_i \left(2\left(\overline{\tilde{S}_{a;i}^{(l)}(nT) \oplus \tilde{S}_{W;i,j}^{(l+1)}(nT)}\right) - 1\right) \cdot \left(\tilde{M}_{a;i}^{(l)}(nT) \cdot \tilde{M}_{W;i,j}^{(l+1)}(nT)\right) \tag{7}$$

Here, $\tilde{S}_{a;i}^{(l)}(nT)$ and $\tilde{M}_{a;i}^{(l)}(nT)$ represent the bipolar sign and magnitude bitstream timeseries of $a_i^{(l)}$, respectively. Similarly, $\tilde{S}_{W;i,j}^{(l)}(nT)$ and $\tilde{M}_{W;i,j}^{(l)}(nT)$ represent the bipolar sign and magnitude bitstream timeseries of $W_{i,j}^{(l)}$. Notice operations between signs can be done with XNOR gates, unipolar multiplications between magnitudes can be done with AND gates, and additions can be implemented with APCs which can either add or subtract depending on the sign operation output. Notice this approach requires magnitude bitstreams to be maximally uncorrelated, so that the collision probability between two bitstreams is given by the product of probabilities.

3.1.3 Radial Basis Function Neural Network Configuration

A radial basis function neural network (RBFNN) shares some similarities with a MLP. In fact, Fig. 4 is representative for this network architecture too. Usually, RBFNNs have a single hidden layer in which nonlinear hidden units measure distances and the activations are radial basis functions (RBFs) applied to the distances and represent a measure of similarity. This network architecture makes

[1] It is worth highlighting other authors might apply the scale factor s after the activation function [36].

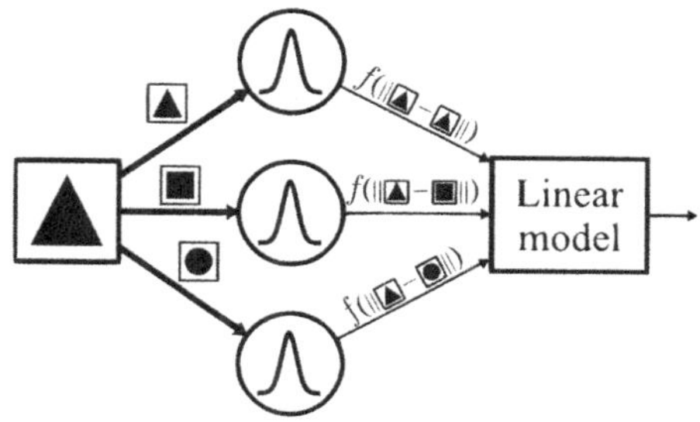

Fig. 5 Simple RBFNN example

the model decision more *explainable* compared to MLPs. Moreover, the core computations are the same in support vector machines [37] and similar ideas can be extended to enhance deep learning models in terms of accuracy and explainability [38].

Consider the example illustrated in Fig. 5. The hidden layer performs a form of template matching,[2] while the readout one is trained to fit the ground truth. Notice the parameters in the hidden layer can be interpreted as prototypes of the input data, and the more similar is the input to a prototype, the higher is the activation of the hidden unit. In the example, the input is almost equal to the prototype that corresponds to the top hidden unit. So, the corresponding distance is almost zero, which is not the case for the other two hidden units. As a result, the RBF activation from the top hidden unit must be much higher compared to the other two.

The core operations for the hidden layer are Euclidean distances and RBF activations. In contrast, the readout layer is the same as in the MLP case. So, returning to the block diagram description in Fig. 4, since the readout layer is the same as in the MLP configuration, then RNGs C and D must generate uncorrelated random sequences. It is, however, not the case for RNGs A and C. These two must generate the same random sequence or be substituted by a single RNG driving the SNGs for input features and hidden layer parameters.

A Euclidean distance, given by (8), involves four main arithmetic operations: difference, squaring, addition, and square root.

$$d(\boldsymbol{x}, \boldsymbol{y}) = \|\boldsymbol{x} - \boldsymbol{y}\| = \sqrt{\sum_i (x_i - y_i)^2} \tag{8}$$

However, typical RBF activations depend on the square of this distance and we do not need to worry about the SC implementation of the square root. The most common type of RBF unit is the Gaussian kernel, given by (9).

$$K(\boldsymbol{x}, \boldsymbol{y}) = f\left(d(\boldsymbol{x}, \boldsymbol{y})^2\right) = Ae^{-\gamma d(\boldsymbol{x}, \boldsymbol{y})^2} \tag{9}$$

[2] This part might be pretrained using unsupervised learning [39].

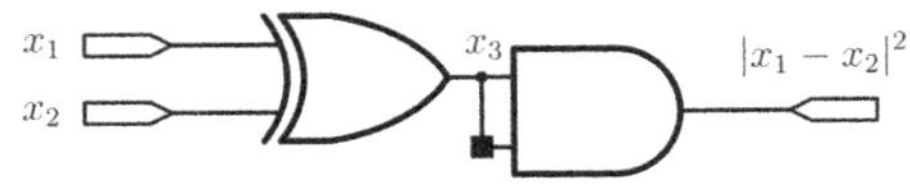

Fig. 6 SC square of the difference. The black square represents a time delay of one clock cycle or a D-type flip-flop. The variables x_1, x_2, x_3, and $|x_1 - x_2|^2$ indicate bitstream activation ratios

Consider the general case with multiple hidden layers. Pre-activations in the l-th layer are given by (10)

$$z_j^{(l+1)} = s_l\, d\left(\boldsymbol{a}^{(l)}, W_{:,j}^{(l+1)}\right)^2 = s_l \sum_i \left(a_i^{(l)} - W_{i,j}^{(l+1)}\right)^2 \tag{10}$$

which is followed by the RBF activation function $f(x) = Ae^{-\gamma x}$.

In order to perform absolute value subtraction by means of SC hardware, given two correlated unipolar bitstream timeseries $\tilde{x}_1(t)$ and $\tilde{x}_2(t)$, it can be approximated[3] by (11), as long as the corresponding activation ratios x_1 and x_2 are both in the unit range [0, 1].

$$|x_1 - x_2| \approx \sum_{n=0}^{N-1} \tilde{x}_1(nT) \oplus \tilde{x}_2(nT) \tag{11}$$

For the squaring operation, consider the bitstream timeseries $\tilde{x}(t)$ is generated from a quantity x in the unit interval, so that x^2 can be approximated by (12) and implemented using a time delay (D-type flip-flop) and an AND gate.

$$x^2 \approx \sum_{n=1}^{N} \tilde{x}(nT) \cdot \tilde{x}((n-1)T) \tag{12}$$

Notice this expression holds as long as the bitstream is maximally self-uncorrelated, which is true if the random sequence utilized to generate it does not have self-correlations. Both (11) and (12) can be combined to approximate the square of a difference, given by (13), and can be implemented using a XOR, a D-type flip-flop, and an AND gate as illustrated in Fig. 6.

$$|x_1 - x_2|^2 \approx \sum_{n=1}^{N} \tilde{x}_3(nT) \cdot \tilde{x}_3((n-1)T), \qquad \tilde{x}_3(t) \equiv \tilde{x}_1(t) \oplus \tilde{x}_1(t) \tag{13}$$

Therefore, a possible equivalent SC model for the hidden pre-activations using the unipolar representation, i.e., assuming activations and prototypes are in the unit

[3] This is actually an exact result if we are working with fixed-point quantities of $\log_2(N)$ bits.

range [0,1], is given by (14), where additions can be computed by an APC in a hardware implementation.

$$z_j^{(l+1)} \approx s_l \sum_{n=1}^{N} \sum_{i} \tilde{c}_{i,j}^{(l)}(nT) \cdot \tilde{c}_{i,j}^{(l)}((n-1)T), \qquad \tilde{c}_{i,j}^{(l)}(t) \equiv \tilde{a}_i^{(l)}(t) \oplus \tilde{W}_{i,j}^{(l+1)}(t) \tag{14}$$

Finally, the activation function can be computed using an ad-hoc design as in [40] or FSM implementation [41], and the readout layer is described by (6). It is also worth mentioning other approaches are also possible, e.g., [42] proposes a smart design to exploit the fact that the product of exponentials is the exponential of the arguments added together, but this approach does not scale well as the number of inputs increases due to undesired correlations among finite length bitstreams.

3.1.4 Applications

There is a wide range of applications in which fully connected networks outperform traditional methods as long as the amount of available training data is reasonable. However, here we discuss only SC implementations successfully implemented and/or applied to specific real-world applications or benchmarks consisting of structured data. Some examples are listed below:

- **Virtual screening acceleration with multiple parallel MLPs** [43]. This study demonstrates the potential use of MPE descriptors and artificial neural networks for virtual screening (VS). In addition, a novel approach to expedite the proposed VS process in a power-efficient manner using SC hardware design methods is presented, together with an FPGA-based implementation, reporting its accuracy, processing speed, and energy efficiency. In comparison to the current state of the art, the proposed model exhibits a processing speed increase of 44,670 times and an overall accuracy improvement of 2%. Furthermore, we have showcased the advantages obtained from utilizing unconventional hardware accelerators for ANN when handling large databases. They presented evidence of the advantages gained from utilizing SC hardware accelerators for neural networks when handling massive databases. Their design is based on the bipolar SC model in which each layer pre-activation is described by (6). Additionally, they introduced a novel approach where the ReLU activation function implementation leverages the signal correlation obtained through the use of APC isolation characteristics.
- **Benchmark datasets evaluated with a novel RBFNN architecture** [40]. This study highlights the significance of utilizing high-quality RNG to prevent self-correlation during the evaluation of Euclidean distances. Additionally, a new bitstream-to-bitstream APC architecture is introduced. This novel APC design can be conveniently adapted for the hardware implementation of other types of neural networks, such as MLP, CNN [44], or Haar-like feature extraction [45]. Nevertheless, the performance outcomes achieved for the proposed RBFNN

could be substantially enhanced with minor modifications, such as binary prototypes, varied kernel functions, or shorter evaluation times. The proposed solution proved very similar or same performance compared to the equivalent fixed-point model with an exact implementation of the activation function for a variety of benchmark datasets, including Iris [46], Banknote [47], Breast cancer [48], Digits [49], and MNIST [50]. Even though the energy efficiency deteriorates as the number of neurons increases due to elevated power consumption, the network's performance remains consistent. It is, however, more efficient than binary neural networks [51] for 265 neurons, but it is an order of magnitude less efficient than ReCA [52], with performance approximately 50% worse. When the number of neurons quadruples, the average power per unit or processing element experiences a negligible increase (2.6%). However, the power per RBF unit is 18 times less than that of BNN and 3.15 times that of ReCA. Finally, the average number of logical resources required to implement an RBFNN neuron is considerably less than those needed for a ReCA processing element, which includes a hardware multiplier block (DSP) and requires up to 18 times fewer logical resources than a BNN unit.

3.1.5 Convolutional Neural Networks

The sense of sight is essential for any human being. It provides us with more information about the physical world around us than any other sense. It allows us to navigate, identify, and manipulate objects and interpret emotions through facial expressions. Given its significance, the scientific community has invested significant effort for decades in understanding the mechanisms of vision and creating models that replicate its nature. In the 1960s, researchers' focus on the neural mechanisms underlying vision and the associated processes [53] prompted computer scientists to incorporate the latest findings in neuroscience into the development of artificial vision. During the 1990s, Y. LeCun et al. introduced and made popular a type of neural network called convolutional neural networks (CNNs) [54], building on earlier works such as the neocognitron [55]. In their research, they utilized a CNN to improve the recognition of handwritten digits, exceeding the performance of previous studies. Their initial implementation, called LeNet, found a specialized market in industries such as postal services and banking. However, due to power constraints at the time, CNNs remained relatively unexplored in visual computing. As transistor technology improved and processors became more robust and efficient, interest in CNNs was revived.

In 2012, a significant breakthrough occurred when A. Krizhevsky et al. introduced AlexNet, which is a type of CNN capable to identify intricate patterns in colored images from the ImageNet database [56]. The architecture of AlexNet was introduced in Krizhevsky et al.'s publication in 2012 [57]. Using a CNN pattern similar to LeCun's one but with five convolutional layers (as opposed to LeNet's with three), AlexNet managed to accomplish an error rate of 15.3% (for the top-5 ranking) in the ImageNet Large Scale Visual Recognition Challenge. This was a significant improvement over the runner-up's error rate of 26.1%. Since then, the field has experienced an enormous surge of research activity, and this trend continues to this day.

CNNs have found widespread applications in various fields, such as medical diagnosis [58], natural language processing [59], and autonomous navigation [60]. However, the efficiency of these applications heavily depends on the computational capacity of the processors that run them. CNNs require intensive computation and memory, with multiple layers, neurons per layer, filters, and filter sizes, which makes deploying them on embedded devices a challenge. While some progress has been made in this regard, most embedded solutions still rely on cloud computing, which brings its own limitations, such as data privacy, network latency, and security risks. With the proliferation of IoT devices, this challenge becomes even more pressing. Therefore, researchers have been exploring different approaches for running CNNs locally, on the edge. One promising optimization technique is model compression, which includes lower data precision [61], weight pruning [62], weight clustering sharing [63], and specific shrunken models [64]. An alternative research avenue seeks to minimize the cost of the most recurrent and costly operation in convolutional neural networks (CNNs), namely, multiplication. In this regard, stochastic computing (SC) stands out as a promising approach.

3.1.6 CNN Structure

At their core, CNNs consist of three main operations: multiplication, addition, and the max function. Similar to fully connected networks, convolutional neurons consist of a nonlinear activation function and a dot product. The most commonly used activation function in literature is the ReLU function, which involves determining the maximum value between 0 and the input (ReLU = $max(0, input)$).

The *pooling* layer is another essential component of CNNs that intends to decrease the dimensionality of the feature map. Its primary purpose is to alleviate the computational burden on the subsequent neural layers and mitigate the risk of overfitting during training.

An example of a simple CNN architecture is the LeNet-5 network, which was introduced by Lecun et al. [54]. As depicted in Fig. 7, the LeNet-5 comprises convolutional layers (designated as "Conv" in the figure) and Max-Pooling (MP) layers that extract the maximum value over a 2x2 pixel window. The network concludes with a fully connected (FC) layer, consisting of a stack of 128, 84, and 10 neurons, that serves as a categorizer.

3.1.7 Stochastic Computing Implementation

To implement the convolutional layer within the stochastic computing (SC) domain, a novel SC neuron design that leverages the correlation phenomenon has been recently proposed [65] (refer to Fig. 8). The multiplication is carried out by the

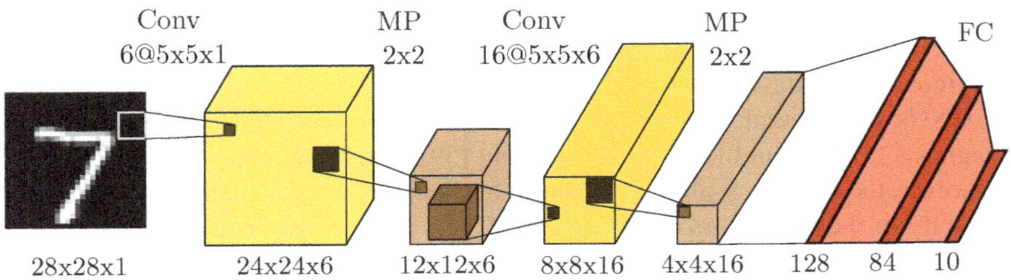

Fig. 7 The LeNet-5 architecture comprises a feature extraction block that includes two convolutional layers and two pooling layers. Following the feature extraction block, a fully connected layer is employed to classify the inputs into one of ten possible classes

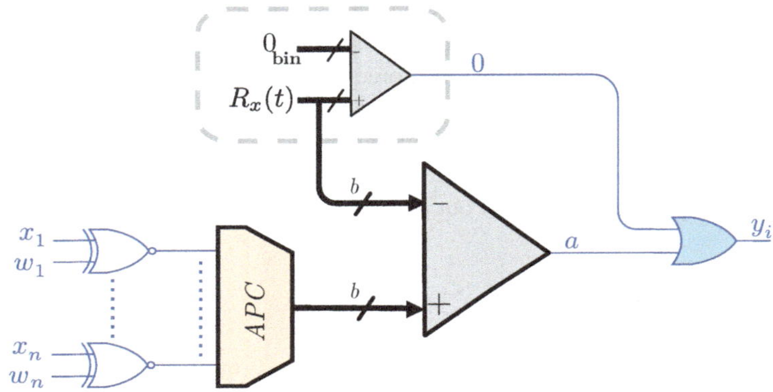

Fig. 8 SC-neuron leveraging the correlation phenomenon for optimal hardware implementation

XNOR gates array (using bipolar codification). Then, an accumulative parallel counter (APC) is used to add the ones present in the bitstreams multiplied, producing a b-bit binary output. The crucial step in this process occurs when the APC output is converted back into the SC domain. To ensure maximum correlation between the SC-APC signal and the zero reference signal, it is essential to employ the identical random number generator (R_x) used to generate the input values x. By doing so, it is possible to implement the SC-ReLU function with exactness using only a single two-input OR gate.

In the context of the MP layer, it is possible to utilize the correlated signals produced by the stochastic neurons to efficiently compute the maximum value within a pixel window. This can be accomplished by employing a four-input OR gate as shown in Fig. 9

3.1.8 Experiments and Results

To evaluate the SC design for the CNN, an implementation of LeNet-5 was carried out. The complete SC-CNN architecture was synthesized using the Cadence Genus

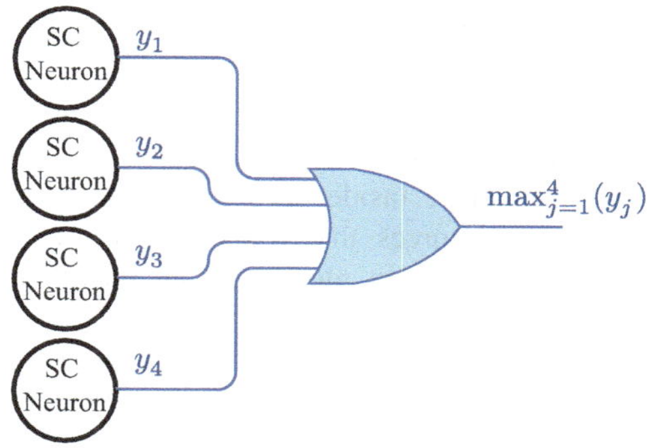

Fig. 9 Stochastic computing implementation of the Max-Pooling (MP) block

Table 1 Comparison with alternative VLSI implementations of LeNet-5

Methods	[66]	[67]	[36]	[68]	[69]	[65]
Year	2020	2017	2018	2019	2019	2021
Technology	40 nm	45 nm	45 nm	45 nm	40 nm	40 nm
Total operations synthesized	–	4.59M	4.59M	–	–	566640
Latency (μs)	–	1.28	0.31	–	37.50	0.03
Power (mW)	0.06	3530	2600	–	0.07	651
Area (mm^2)	0.124	36.4	22.9	–	0.006	2.01
Computational density (MOPs/mm^2)	0.155	0.126	0.201	–	–	**0.282**
Normalized throughput (TOPS)	$3 \cdot 10^{-4}$	3.59	14.72	–	0.01	**18.89**
Normalized energy efficiency (TOPS/W)	10.13	1.02	5.66	–	18.36	**29.01**
Normalized area efficiency (TOPS/mm^2)	$2.9 \cdot 10^{-3}$	0.098	0.64	–	2.47	**9.40**
Absolute throughput (images/μs)	–	0.78	3.20	–	0.03	**33.33**
Absolute energy efficiency (images/μJ)	–	0.22	1.23	0.66	24.47	**51.20**
Absolute area efficiency (images/μs · mm^2)	–	0.02	0.14	0.05	4.60	**16.58**

Tool in TSMC 40 nm CMOS technology. The full design consumes 651 mW and occupies 2 mm^2. The clock frequency was 200 MHz.

Table 1 presents a summary and comparison of the performance of the synthesized stochastic computing-based CNN LeNet-5 with other unconventional implementations reported in the literature. The proposed system achieves significantly higher computational density (1.4x in terms of $MOPS/mm^2$), throughput (1.28x in $TOPS$), energy efficiency (1.58x in $TOPS/W$), area efficiency (3.8x in $TOPS/mm^2$), throughput (10.4x in $Images/\mu s$), energy efficiency (2x in $Images/\mu J$), and area efficiency (3.6x in $Images/(\mu s \cdot mm^2)$) compared to the best case in each metric. This improvement is attributable to the compact implementation of the ReLU function and MP operation, which exploits signal correlations effectively.

4 Morphological Neural Networks

Two main reasons appear behind the renewed interest in morphological neural networks (MNNs). MNNs aim to meet the challenge of the complex functioning that takes place inside each neuron, a complexity that leads to a huge use of hardware resources that translates into a burden on energy efficiency. MNNs are aimed to obtain more efficient neural networks by reducing the number of neurons and parameters required in neural networks. The second reason behind the trendy interest in MNN in image processing (in fields as edge detection, pattern classification, denoising, image enhancement, or image segmentation) lies on the fact that the nonlinear operations used are able to better characterize the geometrical information of an image, such as borders, shapes, corners, and other spatial patterns. Current tools for image analysis still rely mainly on linear systems, which may not always be suitable or efficient for these tasks. As a powerful nonlinear methodology, mathematical morphology has become a viable alternative to the existing linear image analysis tools. Even though morphological operators are not so computationally efficient as the linear operators, they are more sensitive to capture different patterns compared to linear operators such as convolutional or dense linear layers.

The main morphological operations are dilation, erosion, closing, and opening [70]. Dilation and erosion are quite similar to a linear convolution where a structuring element (called a kernel in convolutions) composed of weights moves along and across the whole input image. But instead of the sum of products, a dilation computes the maximum value over sums, while an erosion computes the minimum. Indeed, instead of computing the overall addition of all the products between the value of each pixel inside the local zone of the image and the weight overlapping it, such as a linear convolution does, the dilation operation adds first the value of each pixel over the local region with its overlapping weight and then calculates the maximum over all these additions. Erosion performs in the same way but calculates the minimum. In the case of a morphological dense (or fully connected) layer, the structuring element is a 1x1 patch formed by a single weight, and the dilation or erosion operation takes place at each morphological neuron i, as described by Mondal et al. [71].

$$
\begin{aligned}
h_v^i &= \max_{j=1}^{Q}\{v_{ij} + x_j\} \\
h_w^i &= \min_{j=1}^{Q}\{w_{ij} + x_j\}
\end{aligned}
\tag{15}
$$

where v_{ij} represents the weight of the input x_j which excites the dilation neuron i. In turn, w_{ij} are the respective weights for the erosion neuron i. Q is the number of inputs for each neuron. The structuring elements v_{ij} and w_{ij} form matrices that will be learned during the training stage of the neural network. h_v^i is the outcome of the dilation neuron and h_w^i is the output of the erosion one.

From a visual standpoint, upon getting the maximum value over a local neighborhood given by the moving structuring element, dilation enlarges the bright zones and reduces the dark ones, thereby increasing the objects of the image. By contrast, erosion removes details and reduces the objects of the image [72]. Mathematically, the closing operation is a dilation followed by an erosion. This operation widens the boundaries of the foreground regions and shrinks background color holes. The opening operation is an erosion followed by a dilation and its visual effect is to erode the edges by eliminating its foreground pixels [72]. Other important morphological operations are the $top - hats$ and geodesic reconstructions [73].

Far beyond the aforementioned image processing advantages of the morphological operators, the other main motivation for the arising interest in MNN within the machine learning community is their ability of shrinking the amount of neurons and trainable parameters (weights, biases) in neural networks. Morphological neurons are characterized to encapsulate patterns within hyperboxes, unlike classical linear neurons that the simpler boundary is the hyperplane [74–80]. Therefore, the number of possible decision limits is relatively higher with respect to the parameters used. Consequently, for the same phase space separation, a smaller number of trainable parameters are needed in an MNN than in classical networks. Also, the shrinkage of the amount of neurons in the hidden layer leads to a decrease in the hardware resources needed. On top of that, the nonlinearity inherent to morphological neurons enables to remove the typical activation function of the linear neurons which provides them with the necessary element of nonlinearity. It is worth reminding that the implementation of digital blocks for activation functions such as the hyperbolic tangent or the sigmoid is very costly in terms of hardware resources as they require large numbers of logic gates.

Another crucial feature of the morphological layers is their integrability in classical neural networks by replacing entire layers or some blocks such as average or max pooling operators. This integrability into the conventional CNN architectures also extends to hardware, thus increasing the compactness of the full design because there is no encoding or decoding of data in between stacked layers of hybrid networks. This scenario allows to reduce the hardware complexity and also the energy consumption. Actually, both morphological convolutional layers and morphological dense layers are widely used in what is referred to as deep morphological networks [73, 81–83].

Another factor which helps morphological layers outperform linear layers when facing a reduction of hardware resources is the pruning capacity of the former ones. The pruning strategy derives from the intrinsic nature of morphological algebra. Let us focus on a specific dilation neuron i that computes the max-plus operation. Let us suppose that the layer in which this neuron is placed is the first hidden layer excited by an MNIST image (composed of $28 \times 28 = 784$ pixels) which configures the input stage of the network. As explained above, this neuron chooses as output only one input $v_{il} + x_l$ among the 784 possible inputs:

$$h_v^i = \max(v_{i1} + x_1, v_{i2} + x_2, \ldots, v_{i784} + x_{784}) = v_{il} + x_l \tag{16}$$

Since the other 783 inputs do not have any contribution to the i-the neuron output, there should be different weights v_{ik} that will never (or nearly never) be activated while the network is being trained through the 60,000 images that form the MNIST train dataset. Thus, these superfluous weights can be discarded or, to speak in neural networks' terms, "pruned." The same occurs with the other neurons of the morphological layer. Getting rid of these superfluous parameters doesn't impact on performance at all. It is a matter of course that the higher capacity of pruning of MNN comes from the very definition of the morphological operations which, unlike linear operations, only chooses one specific and single value (its maximum or minimum) over all the neighborhood, discarding the other values.

A drawback of MNN is that structuring elements take more time to be learned than regular layers. That is why, when applying the chain rule of the backpropagation algorithm, the gradients of the loss function with respect to the structuring element shall be always zero except for a single element. Therefore, just one weight (or one bias) is updated per each trained image; even some weights of the structuring element may not be updated throughout the training process [71]. As a result, the training stage for MNN lasts more than for the regular neural networks and may take up to hundreds of epochs.

To sum up, all these factors make MNN suitable for reducing the amount of hardware resources and the complexity of the hardware design and, as a consequence, also reducing the energy consumption.

4.1 Application of Stochastic Computing to the Implementation of Morphological Neural Networks

Among all hardware approaches to implement a morphological neural network (MNN), the stochastic computing (SC) technique seems to be the most suitable one because it seems tailor-made for it. Morphological neurons and morphological layers use four fundamental mathematical operations: the maximum, the minimum, the product, and the addition. Three of these operations, concretely the maximum, the minimum, and the product (improvement of the addition is still a challenge for the stochastic computing community), are highly straightforward when using SC as they can be implemented through simple logic gates which perform in a bitwise operation. Specifically, the multiplication can be implemented by a XNOR gate, a maximum by an OR gate, and a minimum by an AND gate. That is the main reason behind the use of the SC technique for implementing MNN in hardware. The elegant suitability of SC to MNN allows a huge shrinkage of the hardware resources and, as a consequence, a drastic reduction of the power dissipation.

In [84] we proposed a layered hybrid neural network composed of a morphological dense layer and a linear dense one without any activation function [71]. To the best of our knowledge, it was the first morphological neural network to be implemented in hardware, though not wholly morphological because it has an output layer which is linear. This network was trained through a backpropagation algorithm by gradient descent and implemented in hardware by mixing up classical digital circuitry with stochastic circuitry. The trainable parameters of this network were obtained by software for some traditional pattern recognition benchmarks like tic-tac-toe endgame, Iris, banknote authentication, Pen-based digit recognition, and optical digit recognition datasets.

A logical follow-up was to extend it to a more demanding dataset such as the MNIST benchmark [85]. We repeated the same network architecture, but now with an input layer of 784 inputs (as many as pixels per image), with a hidden and morphological layer composed of 100 dilation neurons and 100 erosion neurons and a 1-hot encoding output layer of 10 predicted outputs, as many as category labels (digits from 0 to 9). The criterion to choose the number of dilation and erosion neurons to work with was to compare our network with other networks with a similar number of total weights or trainable parameters we would like to compare with. This choice was set during the training process in order to maximize the figures of merit we took into consideration.

When using the MNIST dataset, the number of connections boosted dramatically. Therefore, in order to improve the overall performance, nonessential weights that have no impact on performance are removed. We managed to discard more than 92% of the original weights in a network with 200 neurons that start at the beginning of the backpropagation training with a total of $784 \times 200 + 200 \times 10 = 158{,}800$ parameters. After pruning, only 13,177 of these weights survived, a reduction of 91.7%. We have to take into account that the removed connections have a great impact in the overall area of the stochastic computing design, since the binary-to-stochastic converters (that convert the two-complement signals to the stochastic domain) and the binary adders are severely reduced (in the next section we show the details of the MNN hardware design).

4.1.1 SC-Based Hardware Implementation

The illustration shown in Fig. 10 depicts the hardware design of a morphological neural network that has been proposed to address the MNIST problem. To implement this network, a mixed digital circuitry has been chosen by combining classical 2s complement (C2) and stochastic bipolar (SCB) codifications. To convert between C2 and SCB signals, a random generator rnd_{data} is used for the data path and a different random generator $rnd_{weights}$ for the multipliers' weights. The addition operation in the binary realm is performed by using C2 adders, while the maximum, minimum, and product functions are computed in a compact form by using simple gates in SCB codification. As shown in Fig. 10, the C2 adder outputs are transformed into SCB signals through a random number (that changes with each clock cycle) and a C2 binary comparator. The outputs of SCB signals are therefore correlated between them since they use the same random number generator (RNG) in the conversion. These SCB signals pass through a chain of OR gates and another chain of AND gates in order to estimate their maximum and minimum values (because all

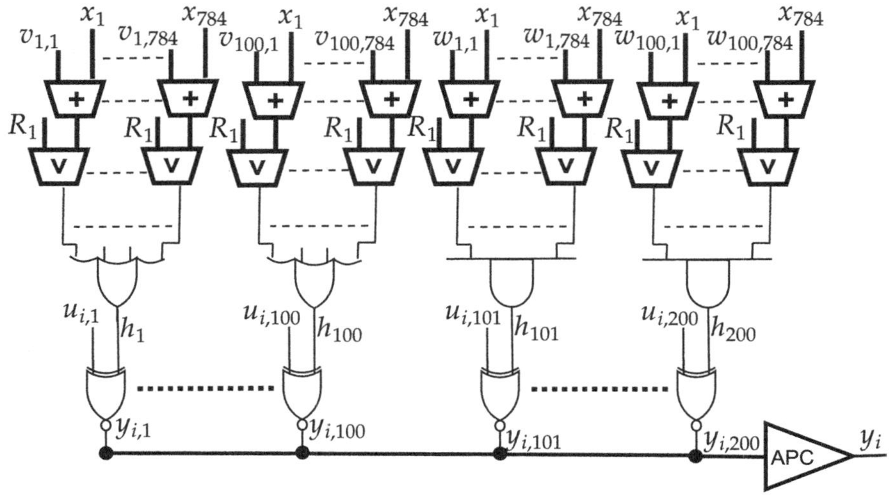

Fig. 10 Hardware design of the SC-based MNN for a single 1-hot encoding output (y_i) and a hidden layer with $M = 200$ neurons (100 dilation neurons and other 100 erosion neurons). The array of XNOR gates and the APC must be replicated for each one of the 10 outputs y_i, where $i = 1, 2..., 10$

SCB signals are correlated). After executing the morphological operations, the SCB outputs are multiplied by the weights of the linear dense layer u_{ij}. In this case, both the results of max and min functions along with the weights must be uncorrelated. Therefore, the RNG used to obtain the SCB weights signals must be different from the first RNG. Finally, an accumulative parallel counter (APC) calculates the overall addition. The APC is ideal to reconvert from the stochastic bipolar domain to the two-complement domain. It is worth noting that using maximum and minimum functions by means of OR and AND gates, along with XNOR gates for the products, allows for a high degree of compactness. Also, an APC block is used for the addition of the XNOR outputs, thus providing a C2 binary number at its output. The use of the APC block allows the possibility to connect different layers since it is providing to the output of the network the same coding used at its input (the two-complement).

Table 2 illustrates a comparison of various neural networks that address the MNIST problem. The last column of the table shows the performance of the proposed 7-bit SC-based MNN approach with a single hidden layer of 200 morphological neurons. This approach is compared with an SC-based solution for a classical CNN (Lenet-5, presented in the fourth column of the table) as well as with three optimized binary neural networks. The SC-MNN approach achieves competitive accuracy values and presents a similar overall performance (in terms of energy efficiency and normalized throughput) than the highly optimized binary neural networks.

Table 2 Comparison and results obtained for different FPGA-based implementations tackling the MNIST problem

	[86]	[87]	[65]	[88]	[6]
Year of introduction	2017	2019	2022	2022	2022
ML method	BMLP	BCNN	CNN	BCNN	MNN
Design method	Binary	Binary	Stochastic	Binary	Stochastic
# Parameters (millions)	2.91	1.11	0.044	0.0203	0.158
Test accuracy	98.4	**98.91**	97.4	**98.4**	96.29
Throughput (Inf/s/MHz)	**7805**	191.7	1960	5100	**8000**
Energy efficiency (Inf/J)	177386	36508	14000	**852367**	**594059**
Logic used K (LUT or ALM)	82.9	28.9	343	26.8	139.5
DSPs	–	–	0	0	0
BRAM (Mbits)	396	1.53	0	0	0

5 Conclusions

In recent years, stochastic computing (SC) has emerged as a popular technique for developing energy-efficient artificial neural networks (ANNs). In particular, SC has proven to be capable of compressing different layers of the widely used convolutional neural networks (CNNs), enabling a high degree of parallelism when implementing medium-complexity networks on a single chip, so that no auxiliary memory is needed to implement the full network algorithms. Therefore, the advantage of using SC is that it enhances the processing characteristics of the system by parallelizing the full system and removing any memory bottleneck that would degrade operational speed and energy efficiency. Therefore, SC represents a promising solution for implementing traditional deep learning algorithms (as CNNs) in hardware for edge computing applications due to its advantages of area shrinkage and low power consumption.

Recently, researchers have found that stochastic computing is particularly relevant for the implementation of morphological neural networks (MNNs), and that these hardware solutions provide better speed and energy efficiency characteristics than many other classical ANN implementations. MNNs have the additional advantage of not requiring activation functions, which can complicate the overall system's circuitry. Moreover, MNNs are constructed using basic arithmetic functions that are simple to implement using SC, such as multiplication, maximum, and minimum. The addition operation can also be used that is useful as a signal conversion tool so that they are properly correlated when needed. As a result, SC design methodologies are considered ideal for developing artificial neural networks. The results obtained using this technique are very excellent, with efficiency values similar to those provided by highly optimized networks such as binary neural networks. In summary, stochastic computing is a valuable design technique in the development of energy-efficient artificial neural networks that provide both operational speed and energy efficiency. The use of SC in developing ANNs can lead to significant benefits, including faster and more efficient processing, improved system design, and better overall performance.

References

1. A. Engelbrecht, *Computational Intelligence: An Introduction*, 2nd edn. (Wiley, London, 2007)
2. W. Maass, Networks of spiking neurons: the third generation of neural network models. Neural Netw. **10**(9), 1659–1671 (1997)
3. F. Niknia, Z. Wang, S. Liu, A. Louri, F. Lombardi, Nanoscale accelerators for artificial neural networks: Arithmetic design, analysis and ASIC implementations. IEEE Nanotechnol. Mag. **16**(6), 14–21 (2022)
4. Y. Liu, S. Liu, Y. Wang, F. Lombardi, J. Han, A survey of stochastic computing neural networks for machine learning applications. IEEE Trans. Neural Netw. Learn. Syst. **32**(7), 2809–2824 (2021)
5. Y. Liu, L. Liu, F. Lombardi, J. Han, An energy-efficient and noise-tolerant recurrent neural network using stochastic computing. IEEE Trans. Very Large Scale Integr. Syst. **27**(9), 2213–2221 (2019)
6. J.L. Rosselló, J. Font-Rosselló, C.F. Frasser, A. Moran, E.S. Skibinsky-Gitlin, V. Canals, M. Roca, Highly optimized hardware morphological neural network through stochastic computing and tropical pruning. IEEE J. Emer. Sel. Top. Circuits Syst. **13**(1), 249–256 (2022)
7. W.J. Poppelbaum, C. Afuso, J.W. Esch, Stochastic computing elements and systems, in *Proceedings of the November 14-16, 1967, Fall Joint Computer Conference. AFIPS '67 (Fall)* (Association for Computing Machinery, 1967), pp. 635–644
8. B.R. Gaines, in *Stochastic Computing Systems* (Springer, Boston, 1969), pp. 37–172
9. A. Alaghi, P. Ting, V.T. Lee, J.P. Hayes, in *Accuracy and Correlation in Stochastic Computing* (Springer, Cham, 2019), pp. 77–102
10. J.L. Rosselló, V. Canals, A. Oliver, A. Morro, Studying the role of synchronized and chaotic spiking neural ensembles in neural information processing. Int. J. Neural Syst. **24**(5), 1430003 (2014)
11. F. Galán-Prado, A. Morán, J. Font, M. Roca, J.L. Rosselló, Compact hardware synthesis of stochastic spiking neural networks. Int. J. Neural Syst. **29**(8), 1950004 (2019)
12. W. Qian, X. Li, M.D. Riedel, K. Bazargan, D.J. Lilja, An architecture for fault-tolerant computation with stochastic logic. IEEE Trans. Comput. **60**(1), 93–105 (2011)
13. W. McCulloch, W. Pitts, A logical calculus of the ideas immanent in nervous activity. Bull. Math. Biol. **52**(1–2), 99–115 (1990)
14. A. Ankit, T. Ibrayev, A. Sengupta, K. Roy, Trannsformer: clustered pruning on crossbar-based architectures for energy-efficient neural networks. IEEE Trans. Comput.-Aided Design Integr. Circuits Syst. **39**(10), 2361–2374 (2020)
15. F. Bayat, M. Prezioso, B. Chakrabarti, H. Nili, I. Kataeva, D. Strukov, Implementation of multilayer perceptron network with highly uniform passive memristive crossbar circuits. Nat. Commun. **9**(1), 2331 (2018)
16. M. Bavandpour, M. Mahmoodi, D. Strukov, Energy-efficient time-domain vector-by-matrix multiplier for neurocomputing and beyond. IEEE Trans. Circuits Syst. II: Express Briefs **66**(9), 1512–1516 (2019)
17. F. Silva, M. Sanz, J. Seixas, E. Solano, Y. Omar, Perceptrons from memristors. Neural Netw. **122**, 273–278 (2020)
18. R. Zand, K. Camsari, S. Pyle, I. Ahmed, C. Kim, R. DeMara, *Low-Energy Deep Belief Networks Using Intrinsic Sigmoidal Spintronic-Based Probabilistic Neurons* (2018), pp. 15–20
19. R. Zand, K. Camsari, S. Datta, R. Demara, Composable probabilistic inference networks using mram-based stochastic neurons. ACM J. Emer. Technol. Comput. Syst. **15**(2), 1–22 (2019)

20. F. Khanday, M. Dar, N. Kant, T. Zulkifli, C. Psychalinos, Ultra-low-voltage integrable electronic implementation of delayed inertial neural networks for complex dynamical behavior using multiple activation functions. Neural Comput. Appl. **32**(12), 8297–8314 (2020)
21. M. Bhardwaj, Aradhana, A. Kumar, P. Kumar, V. Nath, Digital implementation of sigmoid function in artificial neural network using VHDL. Lect. Notes Electr. Eng. **692**, 45–53 (2021)
22. G. Hinton, L. Deng, D. Yu, G.E. Dahl, A.R. Mohamed, N. Jaitly, A. Senior, V. Vanhoucke, P. Nguyen, T.N. Sainath, et al., Deep neural networks for acoustic modeling in speech recognition: the shared views of four research groups. IEEE Signal Process. Mag. **29**(6), 82–97 (2012)
23. A. Krizhevsky, I. Sutskever, G.E. Hinton, Imagenet classification with deep convolutional neural networks. Commun. ACM **60**(6), 84–90 (2017)
24. K. Kollmann, K.R. Riemschneider, H.C. Zeidler, On-chip backpropagation training using parallel stochastic bit streams, in *Proceedings of Fifth International Conference on Microelectronics for Neural Networks* (IEEE, 1996), pp. 149–156
25. S. Liu, H. Jiang, L. Liu, J. Han, Gradient descent using stochastic circuits for efficient training of learning machines. IEEE Trans. Comput.-Aided Design Integr. Circuits Syst. **37**(11), 2530–2541 (2018)
26. P. Angelov, *Autonomous Learning Systems: From Data Streams to Knowledge in Real-Time* (Wiley, London, 2012)
27. L. Van Der Maaten, E. Postma, J. Van den Herik, et al., Dimensionality reduction: a comparative. J. Mach. Learn. Res. **10**(66–71), 13 (2009)
28. I. Goodfellow, J. Pouget-Abadie, M. Mirza, B. Xu, D. Warde-Farley, S. Ozair, A. Courville, Y. Bengio, Generative adversarial networks. Commun. ACM **63**(11), 139–144 (2020)
29. A. Morán, J.L. Rosselló, M. Roca, E. Isern, V. Martínez-Moll, V. Canals, Self-organizing maps hybrid implementation based on stochastic computing, in *2019 XXXIV Conference on Design of Circuits and Integrated Systems (DCIS)* (IEEE, 2019), pp. 1–6
30. A. Morán, V. Canals, P.P. Angelov, C.F. Frasser, E.S. Skibinsky-Gitlin, J. Font, E. Isern, M. Roca, J.L. Rosselló, Stochastic computing co-processing elements for evolving autonomous data partitioning, in *2021 XXXVI Conference on Design of Circuits and Integrated Systems (DCIS)* (IEEE, 2021), pp. 1–6
31. X. Gu, P.P. Angelov, J.C. Príncipe, A method for autonomous data partitioning. Inform. Sci. **460**, 65–82 (2018)
32. A. Moran Costoya, *Compact Machine Learning Systems with Reconfigurable Computing* (2021)
33. C.C.F. Frasser, Hardware Implementation of Machine Learning and Deep-Learning Systems oriented to Image Processing. Ph.D. Thesis, Universitat de les Illes Balears (2022)
34. V. Canals, A. Morro, A. Oliver, M.L. Alomar, J.L. Rosselló, A new stochastic computing methodology for efficient neural network implementation. IEEE Trans. Neural Netw. Learn. Syst. **27**(3), 551–564 (2015)
35. J. Li, Z. Yuan, Z. Li, C. Ding, A. Ren, Q. Qiu, J. Draper, Y. Wang, Hardware-driven nonlinear activation for stochastic computing based deep convolutional neural networks, in *2017 International Joint Conference on Neural Networks (IJCNN)* (IEEE, 2017), pp. 1230–1236
36. Z. Li, J. Li, A. Ren, R. Cai, C. Ding, X. Qian, J. Draper, B. Yuan, J. Tang, Q. Qiu, Y. Wang, HEIF: highly efficient stochastic computing-based inference framework for deep neural networks. IEEE Trans. Comput.-Aided Design Integr. Circuits Syst. **38**(8), 1543–1556 (2019)
37. C. Cortes, V. Vapnik, Support-vector networks. Mach. Learn. **20**, 273–297 (1995)
38. P. Angelov, E. Soares, Towards explainable deep neural networks (XDNN). Neural Netw. **130**, 185–194 (2020)
39. B. Scholkopf, K.K. Sung, C.J. Burges, F. Girosi, P. Niyogi, T. Poggio, V. Vapnik, Comparing support vector machines with gaussian kernels to radial basis function classifiers. IEEE Trans. Signal Process. **45**(11), 2758–2765 (1997)

40. A. Morán, L. Parrilla, M. Roca, J. Font-Rosselló, E. Isern, V. Canals, Digital implementation of radial basis function neural networks based on stochastic computing. IEEE J. Emer. Sel. Top. Circuits Syst. **13**(1), 257–269 (2022)
41. Y. Liu, K.K. Parhi, Computing RBF kernel for SVM classification using stochastic logic, in *2016 IEEE International Workshop on Signal Processing Systems (SiPS)* (IEEE, 2016), pp. 327–332
42. Y. Ji, F. Ran, C. Ma, D.J. Lilja, A hardware implementation of a radial basis function neural network using stochastic logic, in *2015 Design, Automation & Test in Europe Conference & Exhibition (DATE)* (IEEE, 2015), pp. 880–883
43. C.F. Frasser, C. de Benito, E.S. Skibinsky-Gitlin, V. Canals, J. Font-Rosselló, M. Roca, P.J. Ballester, J.L. Rosselló, Using stochastic computing for virtual screening acceleration. Electronics **10**(23), 2981 (2021)
44. A. Ren, Z. Li, C. Ding, Q. Qiu, Y. Wang, J. Li, X. Qian, B. Yuan, SC-DCNN: highly-scalable deep convolutional neural network using stochastic computing. ACM SIGPLAN Notices **52**(4), 405–418 (2017)
45. P. Viola, M. Jones, Rapid object detection using a boosted cascade of simple features, in *Proceedings of the 2001 IEEE Computer Society Conference on Computer Vision and Pattern Recognition. CVPR 2001*, vol. 1 (IEEE, 2001)
46. R.A. Fisher, The use of multiple measurements in taxonomic problems. Ann. Eugenics **7**(2), 179–188 (1936)
47. Banknote dataset. https://archive.ics.uci.edu/ml/datasets/banknote+authentication
48. W.N. Street, W.H. Wolberg, O.L. Mangasarian, Nuclear feature extraction for breast tumor diagnosis, in *Biomedical Image Processing and Biomedical Visualization*, vol. 1905 (International Society for Optics and Photonics, 1993), pp. 861–870
49. E. Alpaydin, C. Kaynak, Cascading classifiers. Kybernetika **34**(4), 369–374 (1998)
50. L. Deng, The mnist database of handwritten digit images for machine learning research. IEEE Signal Process. Mag. **29**(6), 141–142 (2012)
51. S. Liang, S. Yin, L. Liu, W. Luk, S. Wei, FP-BNN: binarized neural network on FPGA. Neurocomputing **275**, 1072–1086 (2018)
52. A. Moran, C.F. Frasser, M. Roca, J.L. Rossello, Energy-efficient pattern recognition hardware with elementary cellular automata. IEEE Trans. Comput. **69**(3), 392–401 (2019)
53. T.N. Wiesel, D.H. Hubel, Single-cell responses in striate cortex of kittens deprived of vision in one eye. J. Neurophysiol. **26**(6), 1003–1017 (1963)
54. Y. LeCun, L. Bottou, Y. Bengio, P. Haffner, et al., Gradient-based learning applied to document recognition. Proc. IEEE **86**(11), 2278–2324 (1998)
55. K. Fukushima, S. Miyake, T. Ito, Neocognitron: a neural network model for a mechanism of visual pattern recognition. IEEE Trans. Syst. Man Cybern. **5**, 826–834 (1983)
56. J. Deng, W. Dong, R. Socher, L.J. Li, K. Li, L. Fei-Fei, Imagenet: a large-scale hierarchical image database, in *2009 IEEE Conference on Computer Vision and Pattern Recognition* (IEEE, 2009), pp. 248–255
57. A. Krizhevsky, I. Sutskever, G.E. Hinton, Imagenet classification with deep convolutional neural networks. Adv. Neural Inform. Process. Syst. **25**, 1097–1105 (2012)
58. B. Khagi, C.G. Lee, G.R. Kwon, Alzheimer's disease classification from brain MRI based on transfer learning from CNN, in *2018 11th Biomedical Engineering International Conference (BMEiCON)* (IEEE, 2018), pp. 1–4
59. E. Grefenstette, P. Blunsom, N. De Freitas, K.M. Hermann, A deep architecture for semantic parsing (2014). arXiv preprint arXiv:1404.7296
60. E. Ackerman, How drive. ai is mastering autonomous driving with deep learning. IEEE Spectr. Mag. **1** (2017)
61. M. Rastegari, V. Ordonez, J. Redmon, A. Farhadi, Xnor-net: imagenet classification using binary convolutional neural networks, in *European Conference on Computer Vision* (Springer, Berlin, 2016), pp. 525–542
62. H. Li, A. Kadav, I. Durdanovic, H. Samet, H.P. Graf, Pruning filters for efficient convnets (2016). arXiv preprint arXiv:1608.08710

63. S. Han, H. Mao, W.J. Dally, Deep compression: compressing deep neural networks with pruning, trained quantization and Huffman coding (2015). arXiv preprint arXiv:1510.00149
64. A.G. Howard, M. Zhu, B. Chen, D. Kalenichenko, W. Wang, T. Weyand, M. Andreetto, H. Adam, Mobilenets: efficient convolutional neural networks for mobile vision applications (2017). arXiv preprint arXiv:1704.04861
65. C.F. Frasser, P. Linares-Serrano, I.D. de los Ríos, A. Morán, E.S. Skibinsky-Gitlin, J. Font-Rosselló, V. Canals, M. Roca, T. Serrano-Gotarredona, J.L. Rosselló, Fully parallel stochastic computing hardware implementation of convolutional neural networks for edge computing applications. IEEE Trans. Neural Netw. Learn. Syst. 1–11 (2022)
66. A. Sayal, S. Nibhanupudi, S. Fathima, J. Kulkarni, A 12.08-TOPS/W all-digital time-domain CNN engine using bi-directional memory delay lines for energy efficient edge computing. IEEE J. Solid-State Circuits **55**(1), 60–75 (2020)
67. A. Ren, Z. Li, C. Ding, Q. Qiu, Y. Wang, J. Li, X. Qian, B. Yuan, SC-DCNN: highly-scalable deep convolutional neural network using stochastic computing. ACM SIGOPS Oper. Syst. Rev. **51**(2), 405–418 (2017)
68. H. Kung, B. McDanel, S.Q. Zhang, Packing sparse convolutional neural networks for efficient systolic array implementations: column combining under joint optimization, in *Proceedings of the Twenty-Fourth International Conference on Architectural Support for Programming Languages and Operating Systems. ASPLOS '19* (Association for Computing Machinery, 2019), pp. 821–834
69. Y. Zhang, X. Zhang, J. Song, Y. Wang, R. Huang, R. Wang, Parallel convolutional neural network (CNN) accelerators based on stochastic computing, in *2019 IEEE International Workshop on Signal Processing Systems (SiPS)* (2019), pp. 19–24
70. V. Charisopoulos, P. Maragos, A tropical approach to neural networks with piecewise linear activations (2018)
71. R. Mondal, S. Santra, B. Chanda, Dense morphological network: An universal function approximator (2019)
72. D. Mellouli, T.M. Hamdani, J.J. Sanchez-Medina, M. Ben Ayed, A.M. Alimi, Morphological convolutional neural network architecture for digit recognition. IEEE Trans. Neural Netw. Learn. Syst. **30**(9), 2876–2885 (2019)
73. K. Nogueira, J. Chanussot, M.D. Mura, J.A.D. Santos, An introduction to deep morphological networks. IEEE Access **9**, 114308–114324 (2021)
74. G. Ritter, P. Sussner, *An introduction to Morphological Neural Networks*, vol. 4 (1996), pp. 709–717
75. P. Sussner, Morphological perceptron learning, in *Proceedings of the 1998 IEEE International Symposium on Intelligent Control (ISIC) held jointly with IEEE International Symposium on Computational Intelligence in Robotics and Automation (CIRA) Intell.* (1998), pp. 477–482
76. P. Sussner, E.L. Esmi, An introduction to morphological perceptrons with competitive learning, in *2009 International Joint Conference on Neural Networks* (2009), pp. 3024–3031
77. H. Sossa, E. Guevara, Efficient training for dendrite morphological neural networks. Neurocomputing **131**, 132–142 (2014)
78. G. Hernandez, E. Zamora, H. Sossa, G. Tellez, F. Furlan, Hybrid neural networks for big data classification. Neurocomputing **390**, 327–340 (2020)
79. L. Pessoa, P. Maragos, Neural networks with hybrid morphological/rank/linear nodes: a unifying framework with applications to handwritten character recognition. Pattern Recogn. **33**(6), 945–960 (2000)
80. E. Zamora, H. Sossa, Dendrite morphological neurons trained by stochastic gradient descent. Neurocomputing **260**, 420–431 (2017)
81. G. Franchi, A. Fehri, A. Yao, Deep morphological networks. Pattern Recogn. **102**, 107246 (2020)
82. S.K. Roy, R. Mondal, M.E. Paoletti, J.M. Haut, A. Plaza, Morphological convolutional neural networks for hyperspectral image classification. IEEE J. Sel. Top. Appl. Earth Obs. Remote Sensing **14**, 8689–8702 (2021)

83. H. Zhang, Y. Chen, Y. Song, Z. Xiong, Y. Yang, Q.M. Jonathan Wu, Automatic cunninghamegreen for ct images using morphological cascade convolutional neural networks. IEEE Access **7**, 83001–83011 (2019)
84. J.L. Rosselló, J. Font-Rosselló, C.F. Frasser, A. Morán, E.S. Skibinsky-Gitlin, V. Canals, M. Roca, Hardware implementation of stochastic computing-based morphological neural systems, in *Proceedings International Symposium on Circuits and Systems (ISCAS)* (2022)
85. J.L. Rosselló, J. Font-Rosselló, C.F. Frasser, A. Morán, E.S. Skibinsky-Gitlin, V. Canals, M. Roca, Highly optimized hardware morphological neural network through stochastic computing and tropical pruning. IEEE J. Emer. Sel. Top. Circuits Syst. **13**(1), 249–256 (2022)
86. Y. Umuroglu, N.J. Fraser, G. Gambardella, M. Blott, P. Leong, M. Jahre, K. Vissers, Finn: a framework for fast, scalable binarized neural network inference, in *Proceedings of the 2017 ACM/SIGDA International Symposium on Field-Programmable Gate Arrays. FPGA '17* (ACM, New York, 2017), pp. 65–74
87. B. Liu, S. Chen, Y. Kang, F. Wu, An energy-efficient systolic pipeline architecture for binary convolutional neural network, in *2019 IEEE 13th International Conference on ASIC (ASICON)* (IEEE, 2019), pp. 1–4
88. Q. Vo, N. Le, F. Asim, L. Kim, C. Hong, A deep learning accelerator based on a streaming architecture for binary neural networks. IEEE Access **10**, 21141–21159 (2022)

Characterizing Stochastic Number Generators for Accurate Stochastic Computing

Yutao Gong, Heng Shi, and Siting Liu

1 Introduction

Stochastic computing (SC) was first proposed in the 1960s to reduce hardware cost and power consumption and re-emerges as the computing task becomes more and more complex in the past few decades. The primary advantages of SC are smaller hardware cost, lower power consumption, and higher fault tolerance compared to traditional binary-encoded computation. SC leverages a long random bit stream called stochastic number (SN) or stochastic sequence to represent a number. The occurrence frequency of 1s in the bit stream is used to encode the value. The commonly used methods to encode a number are shown in Table 1. We can directly use the probability that 1 occurs in the bit stream p to represent a number x. This is called the unipolar representation in SC. Figure 1a shows an example of the unipolar representation. There are 8 1s in the 16-bit sequence A, so it represents a 0.5. The unipolar representation is commonly used in SC due to its simplicity for binary-stochastic conversion and computation. However, the representation range of the unipolar presentation is limited to [0, 1]. Another SC representation expands the range to [−1,1] by using a linear mapping as shown in Table 1, which is called the bipolar presentation. As shown in Fig. 1b, the sequence represents a negative number of −0.5 using the bipolar representation.

Y. Gong · H. Shi
ShanghaiTech University, Shanghai, China
e-mail: gongyt2022@shanghaitech.edu.cn; shiheng2022@shanghaitech.edu.cn

S. Liu (✉)
ShanghaiTech University, Shanghai, China

Shanghai Engineering Research Center of Energy Efficient and Custom AI IC, Shanghai, China
e-mail: liust@shanghaitech.edu.cn

W. Liu et al. (eds.), *Design and Applications of Emerging Computer Systems*,
https://doi.org/10.1007/978-3-031-42478-6_13

Table 1 Stochastic number representations

	Mapping	Range	Example
Unipolar	$x = p$	[0,1]	$x = p = 0.5$, 01001101
Bipolar	$x = 2p - 1$	[−1,1]	$x = 0.5$, $p = 0.75$, 11011011
Sign magnitude	$\|x\| = p$	[−1,1]	$x = 0.5$, $p = 0.5$, 01001101, $sign = 0$
Nonlinear	$x = \frac{p}{1-p}$	[0,+∞)	$x = 0.5$, $p = \frac{1}{3}$, 1010000010

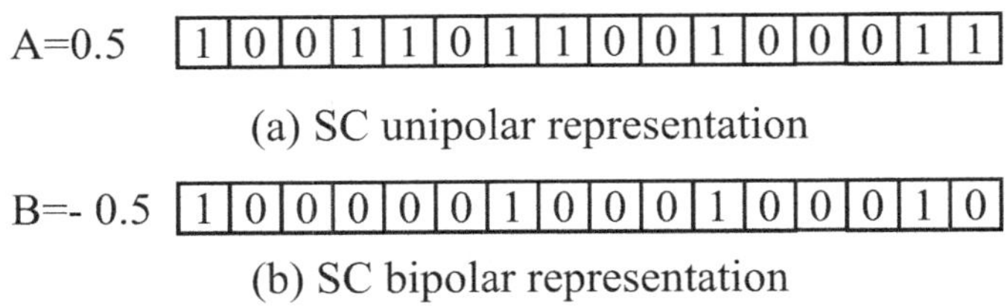

Fig. 1 The SC representations

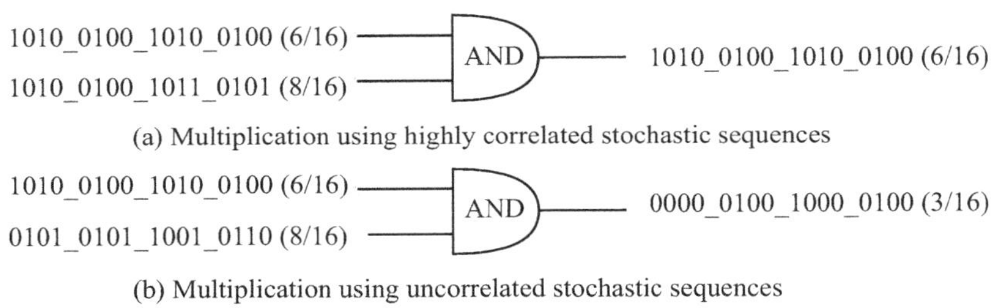

Fig. 2 Multiplication with (**a**) highly correlated and (**b**) uncorrelated sequences

Exploiting the probability theory, SC enables the implementation of complex arithmetic functions by means of simple logic gates. Figure 2b shows that a single AND gate realizes the multiplication in stochastic computing, under the condition that the input sequences are independent or uncorrelated. Similar simple SC circuit designs are observed in a wide range of applications such as neural networks [3], image processing [4, 15], and low-density parity check (LDPC) decoding [19].

A typical SC system consists of three components: the stochastic number generators (SNGs), the stochastic computation circuits, and the probability estimators (PEs), as shown in Fig. 3. Taking the unipolar representation as an example, the SNG can generate stochastic sequence by comparing the number to be encoded with a uniform random number between [0, 1], as shown in Fig. 4. It produces a 1 if the random number is smaller, otherwise, a 0. As a result, the probability of 1 in the generated sequence is the number to be encoded. Weighted binary generator (WBG) [12] is another way to implement an SNG without comparison, which will be covered in detail later.

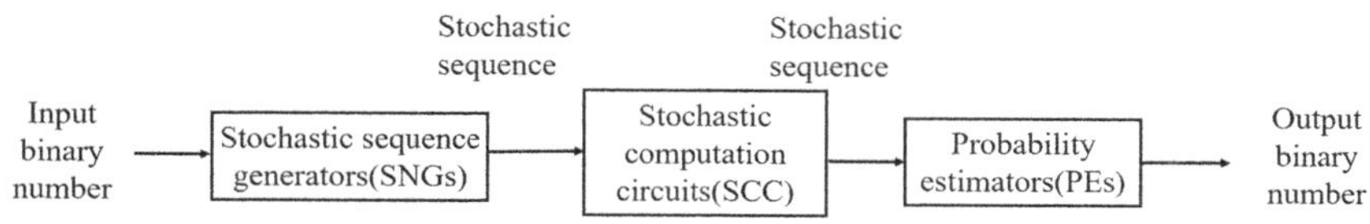

Fig. 3 Components of a stochastic computing system [12]

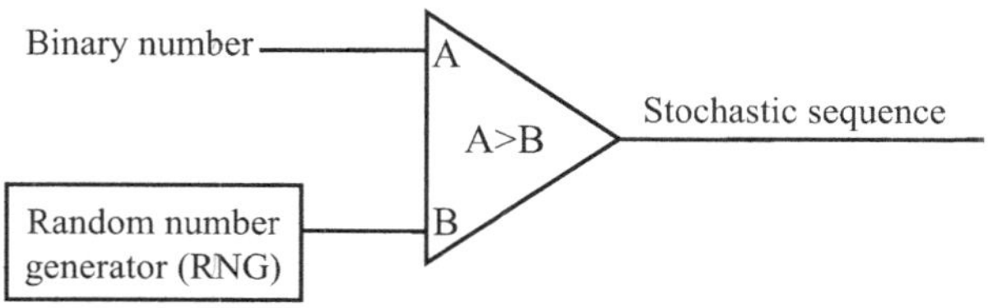

Fig. 4 A stochastic number generator [10]

The quality of the random numbers greatly impacts the accuracy of SC. Conventionally, the random numbers are generated by a pseudo-random number generator or a linear feedback shift register (LFSR). It reveals excellent randomness, but induces random errors in stochastic sequence generation, i.e., the probability of 1s in the sequence may not be exactly or even has a large variation compared with the number to be encoded. On the other hand, some of the stochastic circuits relies on independent or uncorrelated stochastic sequences to produce high accuracy such as in an AND gate (stochastic multiplier). Figure 2a shows an extreme case where the two inputs are strongly correlated with maximally overlapped 0s and 1s. The circuit executes a min(·) function instead of a multiplication.

Therefore, to reduce the random errors in sequence generation and the correlation for accurate stochastic computation, SNGs have been extensively studied and various strategies are proposed. In this work, we review some of the recently developed SNGs to evaluate and compare their efficacy in improving the accuracy of SC. The rest of this chapter is organized as follows. Section 2 introduces different types of SNGs; Sect. 3 discusses the accuracy metrics in SC; the experiment results are presented and analyzed in Sect. 4 with different sequence lengths and SNG settings; Sect. 5 concludes this work.

2 Stochastic Number Generators

Existing SNGs utilize various schemes to generate stochastic sequences. We start with the LFSR-based SNGs and then SNGs based on low-discrepancy sequences. The underlying theory and their hardware implementations are illustrated.

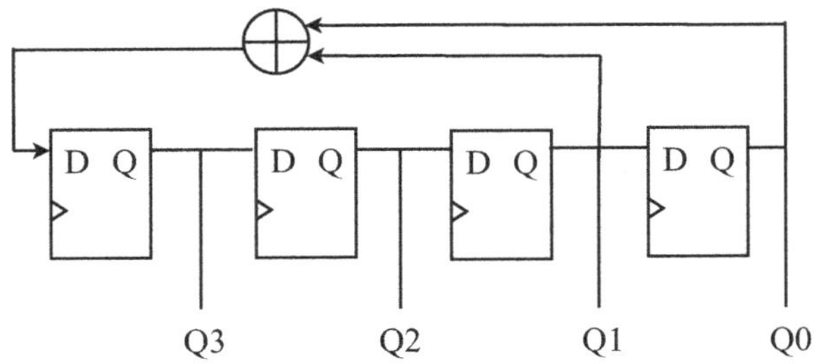

Fig. 5 A 4-bit LFSR

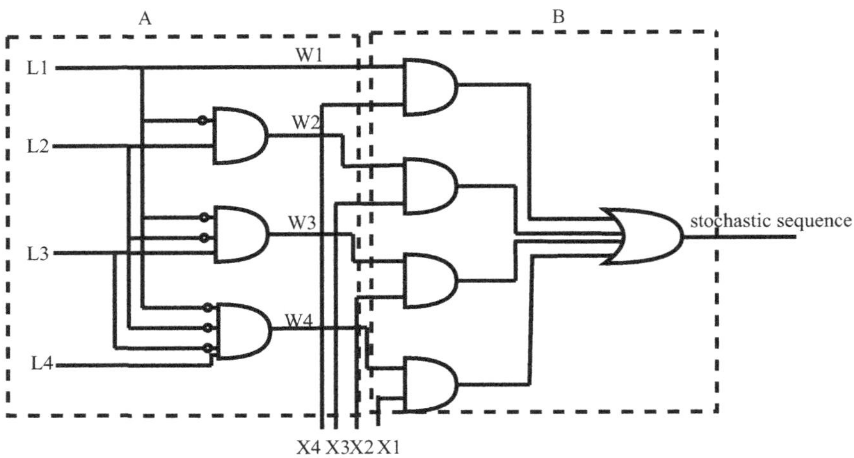

Fig. 6 A weighted binary generator (WBG)

2.1 Linear Feedback Shift Register (LFSR)-Based SNGs

As a most commonly used random number generator, an N-bit LFSR consists of N DFFs and several XOR gates for feedback signal generation. Figure 5 shows an example of a 4-bit LFSR. It works along with either a comparator or through a WBG to produce stochastic sequences. It generates 2^N-1 non-repeated patterns except the all-zero state. The output of the N DFFs is considered as an N-bit number. It is then normalized by 2^N and can be considered as a uniform random number between [0, 1]. Subsequently, it can serve as the random number generator as shown in Fig. 4. Since it traverses all the states between $[1, 2^N - 1]$, the sequence generation error is small especially when maximal-length LFSR is used, i.e., the sequence length is $2^N - 1$.

Other than a comparator, a WBG also implements an SNG with an LFSR, as shown in Fig. 6. WBG is proposed by P. K. Gupta and R. Kumaresan in 1988 [12]. The WBG is divided into two parts. In part A, it uses $L_1L_2L_3L_4$ to compute a weight vector $W_1W_2W_3W_4$. For part B, the number to be encoded, $X_1X_2X_3X_4$, is bit-wise ANDed with the weight vector $W_1 W_2 W_3 W_4$ to obtain an intermediate result. Finally, a 4-input OR gate takes the intermediate value to produce one bit in a stochastic sequence. In the WBG, each bit of the number to be encoded ($X*$) is weighted by the signal $W*$ using the AND gate and combined through the OR gate, thus the name "weighted" binary generator.

The above two LFSR-based designs both have a relatively simple structure. However, when multiple uncorrelated sequences are required, multiple SNGs with either different initial values (i.e., seeds) or feedback are required for the LFSRs. It leads to a linear increase of hardware cost with the number of uncorrelated sequences required. To reduce the hardware cost, Sayed proposed an area-efficient SNGs by sharing the output of a single LFSR for generating several stochastic sequences [18]. To avoid large correlation between the stochastic sequences, different permutations of the LFSR output bits are used as uncorrelated random numbers. For example, it is found that reversing the bit order of a random number generated by an LFSR de-correlates it with the original random numbers [18]. Thus, the original and reversed random numbers can be used to generate two uncorrelated stochastic sequences. To obtain three uncorrelated stochastic sequences using a single LFSR, an algorithm is proposed to search for the best permutations with minimum SC correlation (SCC). Figure 7 shows the algorithm.

Algorithm 1:

Initialization : $SM = 1$;
begin
  for $i = 3$ *to* $n!$ **do**
    for $j = 1$ *to* $i - 2$ **do**
      if $SCC_{avg}(PL_i, PL_j) < SM$ **then**
        $SP_1 = SCC_{avg}(PL_i, PL_j)$;
        for $k_1 = 1$ *to* $i - j - i$ **do**
          if $SCC_{avg}(PL_i, PL_{j+k_1}) < SM$ **then**
            $SP_{k_1} = SCC_{avg}(PL_i, PL_{j+k_1})$;
            if $SCC_{avg}(PL_j, PL_{j+k_1}) < SM$ **then**
              $SP_{h_1 h_2} = SCC_{avg}(PL_j, PL_{j+k_1})$;
              $P_1 = i$;
              $P_2 = j$;
              $P_3 = j + k_1$;
              $SM = max\{SP_1, SP_{k1}, SP_{h_1 h_2}\}$
        end for
    end for
  end for
end

Fig. 7 For an n-bit LFSR, the algorithm finds the optimal permutation so that the generated three stochastic sequences are of the lowest SCC. i, j, and k indicate different permutations that are encoded as unsigned numbers. $P1$, $P2$, and $P3$ are the optimal combination of the permutations [18]

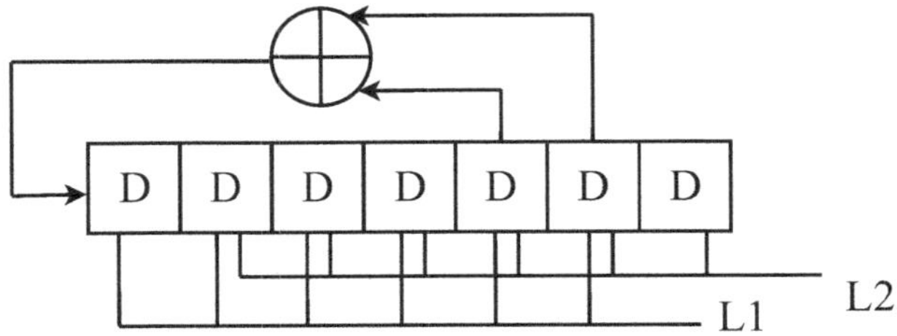

Fig. 8 A 7-bit LFSR is shared to generate two sequences of 6-bit random numbers [22]

Table 2 Logic function of a 4-bit S-box

Input	0	1	2	3	4	5	6	7	8	9	A	B	C	D	E	F
Output	6	B	5	4	2	E	7	A	9	D	F	C	3	1	0	8

The inputs and the outputs are in hexadecimal

Except reordering the LFSR's outputs, Zhong et al. proposed a structure of an efficient random number generator [22]. Similar to Sayed's proposal, it also reuses an LFSR to generate multiple stochastic sequences for hardware savings. To generate two uncorrelated sequences, it selects different combinations of the output bits to form a new sequence of random numbers. Figure 8 shows that two sequences of uncorrelated 6-bit random numbers can be generated by sharing one 7-bit LFSR. Compared to the previous case where the number of SNGs grows linearly with the number of uncorrelated sequences, [22] approximately halves the hardware cost. As can be seen from Fig. 8, L2 can be considered as a delayed version of L1. This lowers the correlation. To reduce the hardware cost, the DFFs that are neither selected to generate the random numbers nor used for the LFSR feedback can be removed.

In [17], an S-box number generator (SBoNG) is proposed, which combines LFSR with the basic components used in advanced encryption standard (AES). The 4-bit S-box performs [11] a nonlinear Boolean function in the SBoNG. The 4-bit input-output mapping is shown in Table 2. The logic function can be expressed as

$$\begin{aligned}
o_3 &= \overline{i_2 i_3} + i_0 \overline{i_1 i_3} + i_0 i_1 i_2, \\
o_2 &= i_0 \overline{i_1} i_2 \overline{i_3} + i_1 \overline{i_2} + \overline{i_0 i_2 i_3} + i_0 \overline{i_2 i_3} + \overline{i_0} i_1 \overline{i_3}, \\
o_1 &= \overline{i_0} i_1 \overline{i_2} i_3 + \overline{i_1 i_3} + i_2 \overline{i_3} + i_0 \overline{i_1} i_2, \\
o_0 &= i_0 \overline{i_1 i_2} + \overline{i_0} i_1 \overline{i_3} + \overline{i_1} i_3 + \overline{i_0 i_2} i_3,
\end{aligned}$$

where $(i_3 i_2 i_1 i_0)_2$ stands for the input and $(o_3 o_2 o_1 o_0)_2$ the output. The 4-bit S-box is constructed using simple logic gates. In addition to the S-box, the SBoNG, as shown in Fig. 9, also contains an internal state, S, of the same width as the output of the LFSR. When the width of LFSR is N bits, it generates non-repeating stochastic sequences with a length of 2^N-bit at most.

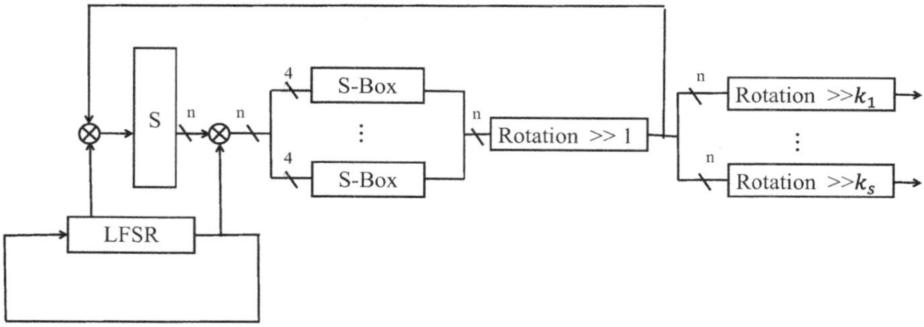

Fig. 9 An S-box number generator [17]

Algorithm 2:

Data: Internal state : $S = s_1 s_2 \ldots s_n$
Output: A random number : O
begin
  Choose maximum length LFSR sequence L randomly;
  $L = nextState(L)$;
  $S = S\ xor\ L$;
  for $i = 1\ to\ \frac{n}{4}$ **do**
    $[s_{4i-3} \ldots s_{4i}] = Sbox[s_{4i-3} \ldots s_{4i}]$
  end for
  $S = rotateToRight(S, 1bit)$;
  $O = S$;
  $S = S\ xor\ L$;
  Return(O)
end

Fig. 10 SBoNG procedure for generating a random number [23]

The internal state of S in SBoNG is a randomly selected binary number in the interval $[0, 2^N - 1]$. When SBoNG starts, the randomly selected S is XORed with the output of the LFSR. The result is split into bundles of 4 bits. They serve as the input to the S-boxes. After that, the output is rotated to the right by 1 bit. The internal state, S, is then updated by XORing the result produced by the LFSR and the rotated result. These operations are repeated to produce random numbers. To generate multiple uncorrelated random numbers in parallel, different numbers of bits are rotated based on the aforementioned rotated result. Therefore, the SBoNG is also able to generate multiple stochastic sequences using only one LFSR. Further correlation reduction can be achieved by adding delays to the rotation components in the last stage. Since the width of S-box is 4 bits, the whole SBoNG uses widths that are a multiple of 4 (Fig. 10).

2.2 Low-Discrepancy (LD) Sequence-Based SNGs

In order to improve the efficiency of numerical integration, the quasi-Monte Carlo method is proposed [8], which generates sampling points through specifically designed methodologies so that they are more evenly distributed in the integration space. SC can be considered as a Monte Carlo numerical integration problem; thus, LD sequences are applied in SC to accelerate the convergence process.

Discrepancy is a metric used to measure the uniformity of the random number distribution. A random point set P with length L can be considered as L points in an s-dimensional unit cube $\beta^*(\prod_{i=1}^{s}[0, 1])$. The star discrepancy $D^*(P)$ is given as

$$D^*(P) = \max_{B \in \beta^*} \left| \frac{A(B; P)}{L} - \lambda(B) \right|, \tag{1}$$

where B is any region in β^*. $\lambda(B)$ stands for the Lebesgue measure of the region, e.g., the volume of a region in three-dimensional space. $A(B; P)$ refers to the number of points contained in this region. In an "ideal" uniform distribution,

$$A(B; P) = \lambda(B)L \Rightarrow D^*(P) = 0. \tag{2}$$

Theoretically, the minimum possible star discrepancy asymptotically converges to $O(log(L)^s/L)$. When the convergence rate of the star discrepancy reaches $O(log(L)^{s-1}/L)$, the sequence is defined as an LD sequence. When it is applied to SC, higher precision can be achieved with shorter sequence length. Several types of LD sequence-based SNGs are described below.

2.2.1 Halton Sequence Generator

Halton sequences are constructed by a deterministic method based on prime numbers. A Halton sequence based on a prime number generates an infinite number of points. The base-b Halton sequence is given as [13]:

$$H_b = \{\Phi_b(0), \Phi_b(1), \Phi_b(2), \ldots, \Phi_b(i)\}. \tag{3}$$

$\Phi_b(i)$ is a function that converts i to base b and reverses the bit order, where b is a prime number. In Table 3, when the base-2 (or base-3) integers are normalized by 2^3 (or 3^3), we can obtain a Halton sequence within [0, 1].

Although the Halton sequence is evenly distributed in Fig. 11a, it suffers from the drawback that the projections are not as well distributed with large base. When the base is small, such as the points in Fig. 11a, the Halton random points are evenly distributed. However, when the base is large (as shown in Fig. 11b), the uniformity is undermined. It can be seen from Fig. 11b that the first several points form a straight line. As the base number increases, this linear alignment appears more frequently.

Table 3 Halton sequence generation

	i	0	1	2	3	4	5	6	7
Base-2	$(i)_2$	000	001	010	011	100	101	110	111
	$\Phi_2(i)$	000	100	010	110	001	101	011	111
	$(\Phi_2(i))_{10}$	0	4	2	6	1	5	3	7
Base-3	$(i)_3$	000	001	002	010	011	012	100	101
	$\Phi_3(i)$	000	100	200	010	110	210	001	101
	$(\Phi_3(i))_{10}$	0	9	18	3	12	21	1	10

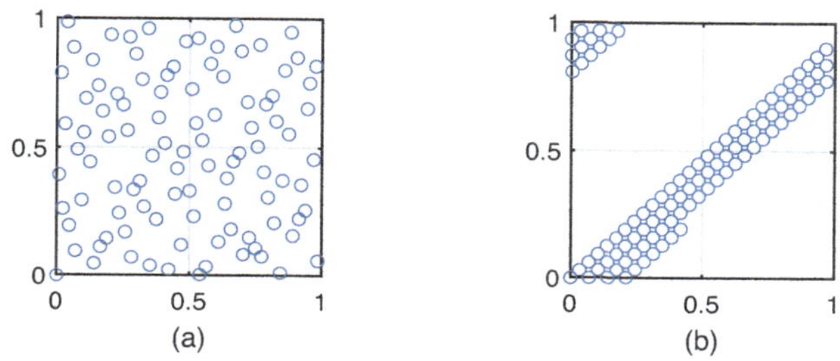

Fig. 11 Points of 2D Halton sequences with different bases. (**a**) Halton sequences base-(2,3). The x- and y-coordinates of the i-th point are the i-th Halton-2 and Halton-3 number, respectively. The combination of these two sets of Halton sequences is uniformly distributed on the 2D space. (**b**) Halton sequences base-(29,31). In the case of large prime numbers, the two sets of Halton sequences are strongly correlated, especially for the initial points

To solve this problem, the most efficient method is deterministic scrambling [14]. In SC, when multiple Halton sequences are required, scrambling modules can be added to the Halton sequence generator to prevent correlation.

As shown in Fig. 12, a Halton sequence generator is proposed in [2]. Binary-coded base-b counters can be used to convert i to base b. Then the order of the digits is reversed. The result is connected to a digit converter to turn it into a binary number. The digit converter transforms a base-b number to base-2, so that it fits in Fig. 4 as a binary random number generator. Alternatively, the number to be encoded can be converted to the same base as the Halton sequence for direct comparison. The digital converter then can be discarded to reduce hardware cost. When $b = 2$, the Halton sequence generator is an ordinary binary counter with reversed order.

2.2.2 Sobol Sequence Generator

Sobol sequence is an LD sequence in base-2. A Sobol sequence generation algorithm [9] is briefly summarized in Fig. 13, where $\{R_i\}$ is a Sobol sequence with length L and $\{V_k\}$ is a set of direction vectors (DVs) [6] that can be obtained from different primitive polynomials. Different polynomials lead to different directional vector arrays (DVAs), so that the multiple uncorrelated Sobol sequences are

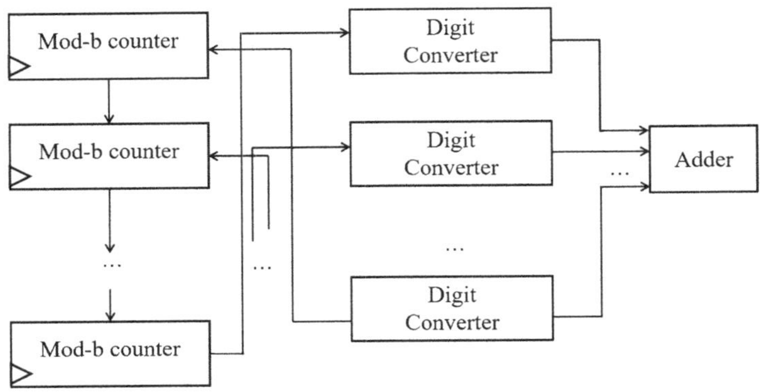

Fig. 12 A Halton sequence generator [2]

Algorithm 3:

Initaialization: Set the initial value of Sobol sequence $R_0 = 0$
begin
  for i = *0 to* $L - 2$ **do**
    Detection of LSZ: $k = LSZ\ position\ of\ i$;
    $R_{i+1} = R_i\ xor\ V_k$;
  end for
  Return($\{R_n\}, n = 0, 1, 2, \ldots, L - 1$)
end

Fig. 13 Sobol sequence generation algorithm [16]

Table 4 Example of a DVA

k	0	1	2	3	4	...	7
V_k	10000000	11000000	11100000	11110000	11111000	...	11111111

generated with different polynomials and DVAs. Table 4 shows an example of a DVA. "LSZ" stands for "least significant zero."

The initial value of a Sobol sequence is set to zero. To generate a Sobol sequence of length L, $L-1$ iterations are required. In each iteration, the newly generated R_{i+1} is obtained by XORing the previously generated R_i with V_k. A detailed example is shown in Table 5.

As shown in Fig. 14, a Sobol sequence generator [16] mainly consists of three parts: the LSZ detector, the DVA memory, and the Sobol sequence calculator. The LSZ detector includes a counter and a priority encoder. The counter is used to count the number of iterations. The priority encoder produces the LSZ position of the counter result. The LSZ position, k, is then used to index the direction vector in the DVA. The direction vector is XORed with the previous result to calculate a new Sobol number.

Table 5 Working example of a Sobol sequence generator

CLK	Reset	1	2	3	4	5	...
i	0	1	2	3	4	5	...
k	0	1	0	2	0	1	...
V_k	00000000	10000000	11000000	10000000	11100000	10000000	...
R_i	00000000	00000000	10000000	01000000	11000000	00100000	...
R_{i+1}	00000000	10000000	01000000	11000000	00100000	10100000	...

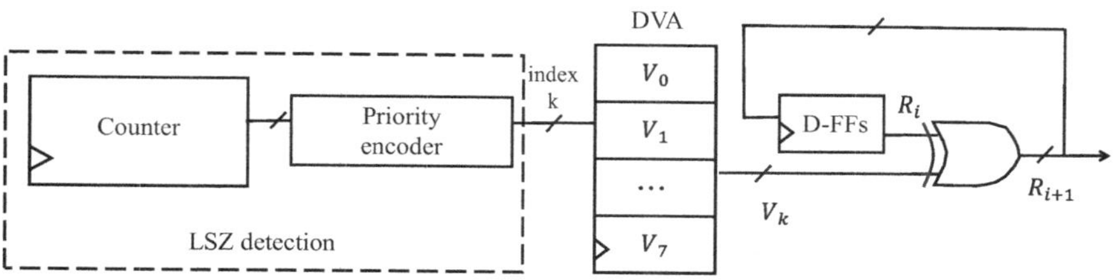

Fig. 14 Design of a Sobol sequence generator [9]

Table 6 LSZ position of continuous non-negative integers

i	0	1	2	3	4	5	6	7
LSZ position of i	0	1	0	2	0	1	0	3
i	8	9	10	11	12	13	14	15
LSZ position of i	0	1	0	2	0	1	0	4
i	...							
LSZ position of i	...							
i	$8j$	$8j+1$	$8j+2$	$8j+3$	$8j+4$	$8j+5$	$8j+6$	$8j+7$
LSZ position of i	0	1	0	2	0	1	0	$L(j)$+3

Moreover, by its unique properties, Sobol sequence generation can be highly parallelized, which reduces energy consumption and area compared to duplicating the RNGs. As described above, the generation of Sobol sequence requires the detection of LSZ. For continuous non-negative integers, the LSZ positions have a repeating pattern, as shown in Table 6. As can be seen from Table 6, the LSZ position is "nearly periodic" with any period of 2^m, $(m = 0, 1, \dots)$, which can be described by (4).

$$L(i) = \begin{cases} L(j) + m & i \equiv (2^m - 1) \mod 2^m \\ & \text{for } j = \lfloor i/2^m \rfloor \\ L(l) & i \equiv l \mod 2^m \\ & \text{for } l = 0, 1, \dots, 2^m - 2 \end{cases} \quad (4)$$

$L(j)$ stands for the LSZ position of j. $L(i) = 0$ when i is an even number. When $m = 1$, let $j = \lfloor i/2 \rfloor (i = 1, 3, 5, \dots)$, then $L(i) = L(j) + 1$. Therefore,

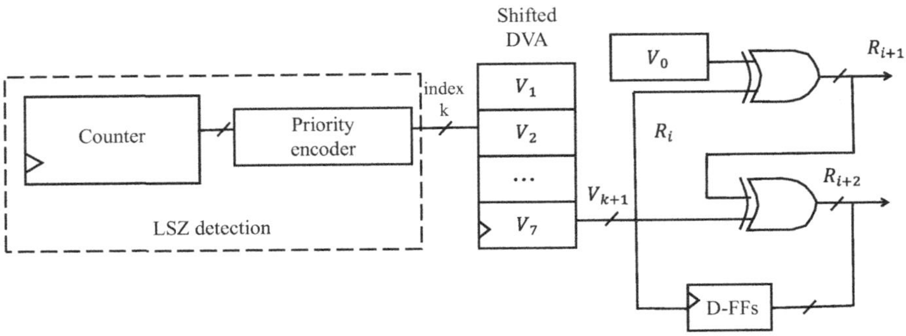

Fig. 15 Design of a 2× Sobol sequence generator [9]

to construct a 2× Sobol sequence generator, we can count j instead of i and detect $k = L(j)$, while for even numbers i, V_0 is directly used for XOR computation ($R_{i+1} = R_i \oplus V_0$). To compute the Sobol numbers corresponding to the odd i, $R_{i+2} = R_{i+1} \oplus V_{k+1}$ as per (4), where V_{k+1} is produced by the shifted DVA. Thus, two results can be computed simultaneously, namely, R_{i+1} and R_{i+2}, and the hardware implementation is shown in Fig. 15.

When $m = 2$, a 4× Sobol sequence generator can be constructed similarly. The LSZ position for i can be expressed as

$$L(i) = \begin{cases} 0 & i \equiv 0 \mod 4 \\ 1 & i \equiv 1 \mod 4 \\ 0 & i \equiv 2 \mod 4 \\ L(j) + 2 & i \equiv 3 \mod 4 \\ & j = \lfloor i/4 \rfloor \end{cases} \tag{5}$$

V_0 and V_1 can be separately stored from the DVA. They are used to compute R_{i+1}, R_{i+2}, and R_{i+3} given R_i. In Fig. 16, the counter counts j, where $j = \lfloor i/4 \rfloor$ and the shifted DVA produces V_{k+2} to obtain R_{i+4}, which is later used to compute R_{i+5}, R_{i+6}, and R_{i+7} in the next clock cycle. Similarly, an arbitrarily $2^m\times$ Sobol sequence generator can be realized by continuing to exploit this regular pattern. Besides, it is also possible to generate uncorrelated Sobol sequences by sharing the LSZ detection with different DVAs.

2.2.3 Finite State Machine (FSM)-Based SNG

LD sequences such as Halton and Sobol sequences improve the accuracy of SC and reduce the computation time of SC. However, these LD sequence generators generally take more hardware compared to the pseudo-random number generators

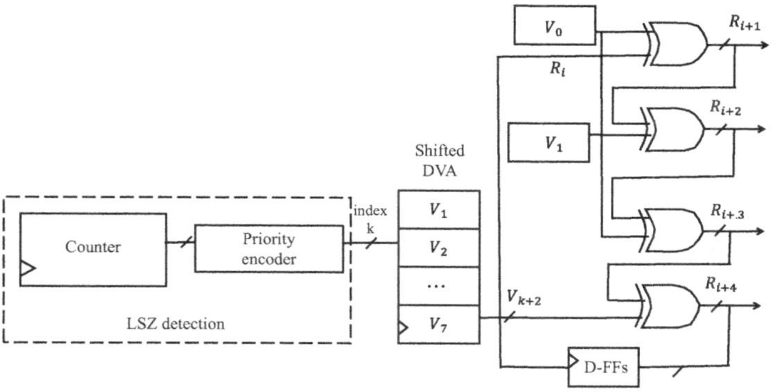

Fig. 16 Design of a 4× Sobol sequence generator [9]

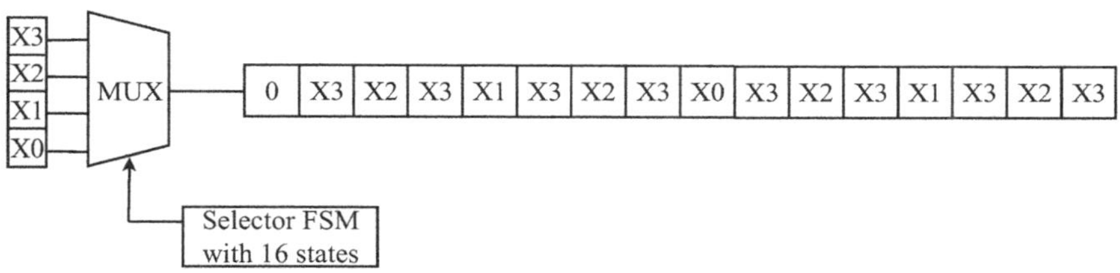

Fig. 17 An FSM-based SNG. $(X3\ X2\ X1\ X0)_2$ is the number to be encoded [20]

such as an LFSR. To reduce the hardware cost while maintaining a low discrepancy, Sim et al. proposed an FSM-based LD bit stream generator [20]. In [20], the output of the FSM-based SNG, which is a stochastic sequence, is ensured a low discrepancy, so it directly improves the accuracy of the computing results. While in conventional comparator-based LD SNG designs, the low discrepancy is only ensured for the random number generation, and the comparison after the random number generation might increase the discrepancy of the resulting stochastic sequences. Figure 17 shows the FSM-based SNG. Different bits of the input binary number are selected by the state of the FSM. It is found that this strategy results in less computation error compared to the other conventional SNGs and has zero bias.

However, this FSM-based design is only able to produce one stochastic sequence. Sina et al. proposed an FSM-based generator that produces multiple uncorrelated LD stochastic sequences [5]. The FSMs for generating multiple LD stochastic sequences are pre-calculated by the algorithm in Fig. 18. Since the FSMs are obtained based on different Sobol sequences, the independence of the stochastic sequences is guaranteed by using uncorrelated Sobol sequences.

Algorithm 4:

Input: Sobol sequence(-num$[0 : 2^n - 1]$),data-width (n)
Output: A 2^n-state FSM
begin
  for $k = 1$ *to* $2^n - 1$ **do**
    if $0 \leq Sobol - num(k) < \frac{1}{2}$ **then**
      FSM output = n-1;
    else if $\frac{1}{2} \leq Sobol - num(k) < \frac{3}{4}$ **then**
      FSM output = n-2;
      ...
    else if $\frac{2^{n-1}-1}{2^{n-1}} \leq Sobol - num(k) < \frac{2^n-1}{2^n}$ **then**
      FSM output = 0;
    else
      FSM output = n;
  end for
end

Fig. 18 An algorithm constructing an FSM-based on a Sobol sequence. Sobol-num(k) is the kth number of a Sobol sequence [5]

3 Accuracy Metrics

Random errors affect both the generation and the computation of stochastic sequences. The probability of 1s in a stochastic sequence may not exactly be the number to be encoded during generation. This is largely affected by the SNGs used. How precise a number can be represented by SC is also determined by the sequence length. For a true random stochastic sequence, the generation error is proportional to $1/\sqrt{L}$, where L is the sequence length, according to probability theory. For the other types of stochastic sequences, the precision is also limited by the sequence length since a 2^N-length stochastic sequence represents at most 2^N different values. Thus, to evaluate the generation errors of different types of SNGs, the RMSEs between the probability of a stochastic sequence and its real value encoded are compared with different sequence length. RMSE is obtained by

$$\text{RMSE} = \frac{\sum_n (y - \hat{y})^2}{n}, \tag{6}$$

where y is the real value and $\hat{y}$ is the probability of the corresponding stochastic sequence. n is the total number of test cases.

The SC computation error mainly stems from the correlation between multiple stochastic sequences. The computation model of the SC circuits is changed when correlation is taken for full consideration. For example, an AND gate, or stochastic multiplier, intends to compute

$$P(A = 1, B = 1) = P(A = 1)P(B = 1),$$

where A and B are the input stochastic sequences. It only holds true when A and B are independent. However, when we consider the correlation between A and B, the computation model is modified to [21]

$$P(A = 1, B = 1) = P(A = 1)P(B = 1|A = 1)$$

according to conditional probability theory. Therefore, the correlation between A and B should be minimized, so that $P(B = 1) = P(B = 1|A = 1)$. Although correlation can be manipulated for efficient stochastic divisor [7] and sorting [16], and correlated stochastic sequences are used in these circuits, we focus on the common cases where correlation degrades the accuracy. To evaluate the computation error induced by correlation using different types of SNGs, SCC [1] is often employed. The SCC of two stochastic sequences A and B is defined as

$$\mathrm{SCC}(A, B) = \begin{cases} \dfrac{ad - bc}{L \times min(a+b, a+c) - (a+b)(a+c)}, & \text{if } ad > bc; \\ \dfrac{ad - bc}{(a+b)(a+c) - L \times max(a-d, 0)}, & \text{otherwise.} \end{cases} \tag{7}$$

a stands for the number of bit pairs where both sequence A and B are 1 at the same bit position, b is the number of (1,0) pairs, c the number of (0,1) pairs, and d the number of (0,0) pairs. L is the sequence length. SCC $= 0$ means that the two sequences are independent, while either -1 or 1 indicates a high correlation. Besides, RMSEs of the SC multiplication results are also measured along with SCC to evaluate the results.

4 Experimental Methods and Results

Three metrics are measured to evaluate the aforementioned SNGs: the accuracy generating a single stochastic sequence, the SCC of two sequences, and the accuracy of multiplication by an AND gate. The numbers within [0.01, 0.99] with an interval of 0.01 are used as the numbers to be encoded or the multipliers/multiplicands. The RMSEs and SCCs produced by the usage of different SNGs are measured. The sequence lengths are 2's powers and vary from 2^3 to 2^{12}. In addition, since SBoNG uses a 4-bit S-box, we only measure the sequence length of 2^4, 2^8, and 2^{12}. Since it is generally considered that SC has little or no advantage over binary computation when the sequence length is greater than 2^{12}, especially when considering speed and energy consumption, they are not considered for comparison.

As shown in Fig. 19, the RMSE measures the accuracy of sequence generation. For LD-based SNGs, the Sobol-based ones use two different sets of DVAs, the Halton-based ones use Halton(2) (denoting base-2 Halton sequence, the same

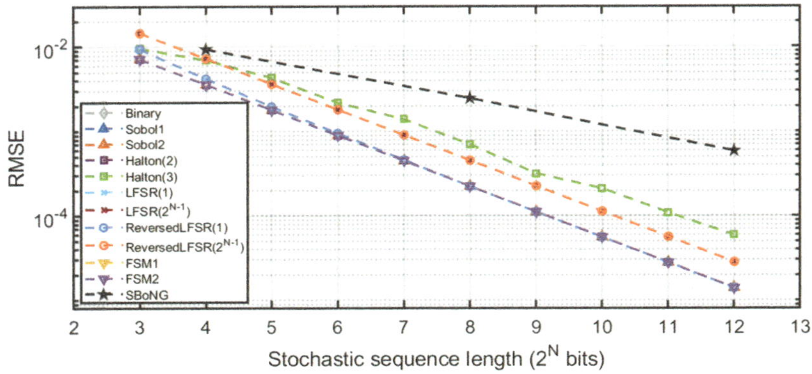

Fig. 19 Accuracy for single sequence generation

notation is used below) and Halton(3) as the random numbers, and the FSM-based ones are constructed from two Sobol sequences. For the LFSR-based SNGs, they use different seeds (1 and 2^{N-1}) and feedback polynomials. The bits are then reversed to form a new random number sequence (it is equivalent to the design proposed in [18] and marked as "ReversedLFSR"). The maximal-length LFSRs are used for better accuracy. In SBoNG, the internal states S are randomly selected in the interval $[1, 2^N - 1]$.

The horizontal axis represents the length of the stochastic sequences, and the vertical axis shows the RMSE. For all types of SNGs, the RMSEs of the stochastic sequences decrease with the sequence length. This can be attributed to the property of SC which is called progressive precision (PP). N-bit width binary numbers are used as references. Among these SNGs, Halton(2), Sobol-based, and FSM-based designs produce similarly low RMSE. SBoNG-generated sequences show the highest RMSE. The other SNGs such as the LFSR-based ones using seed 1 also show a high accuracy. However, with a different seed, the accuracy of the LFSR-based one is slightly degraded. So, the selection of seed affects the accuracy of the LFSR-based SNG.

The SCCs of two stochastic sequences are shown in Fig. 20, and the RMSE of the stochastic multiplier is shown in Fig. 21. The two operands for SCC and multiplication measurements vary from 0.01 to 0.99, with an interval of 0.01, and thus a total of 9801 pairs of operands is measured at different sequence lengths. The two Sobol-, Halton-, and FSM-based SNGs use the same settings as in the single-sequence generation accuracy evaluation. To evaluate the correlation and multiplication accuracy for the LFSR-based designs, the reversed version is used for generating one of the stochastic sequences, and the original LFSR generates another. SBoNG, on the other hand, chooses a random initial internal state for each computation. The second stochastic sequence of SBoNG is obtained by rotating the first sequence by 2^{N-1} bits. Figure 20 shows the SCCs of the stochastic sequences by different SNGs. Except LFSR-based SNGs, the other SCCs decrease with

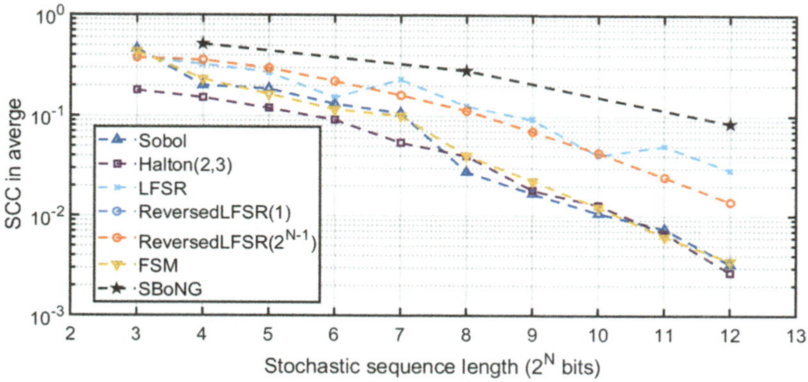

Fig. 20 Average SCC of two stochastic sequences. "LFSR" refers to the SCC of the stochastic sequence generated by two LFSRs with different seeds. "ReversedLFSR" refers to the SCC of the stochastic sequence generated by an LFSR and its reversed version

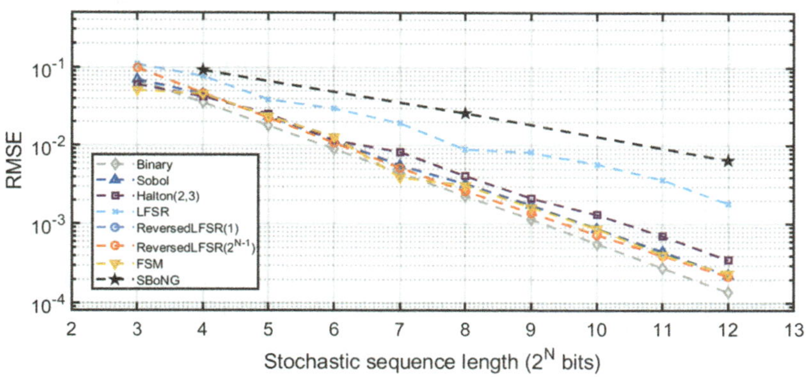

Fig. 21 RMSE of multiplication by an AND gate. "LFSR" refers to the multiplication accuracy of the stochastic sequences generated by two LFSRs with different seeds. "ReversedLFSR" refers to the multiplication accuracy of the stochastic sequence generated by an LFSR and its reversed version

the sequence length. However, the SCC of the LFSR-based sequences fluctuates between the sequence length of $[2^5, 2^{10}]$. Generally, SNGs using Halton(2,3) produce the lowest and SBoNGs produce the highest SCC. When the sequence length is between 2^8 and 2^{10}, the one using Sobol sequence has the lowest SCCs compared to the others. The FSM-based designs show a similar trend with the Sobol-based ones.

For the RMSEs of the stochastic multiplication, the reversed LFSR-, FSM-, and Sobol-based designs produce the highest accuracy, while the Halton-based design produces a slightly higher RMSE, generally. The SBoNGs produce a lower accuracy mainly because that it is designed for higher randomness. It is shown that the

SBoNG outperforms LFSR in many standard random tests. Also, it shows better auto-correlation and cross-correlation properties than its counterparts [17].

5 Conclusion

Stochastic computing reduces both the hardware cost and power consumption compared to their binary counterparts. However, its computation relies on the generation of stochastic sequences, which is achieved by the SNGs. The use of various SNGs affects the accuracy of the stochastic sequence generation and stochastic computation results. In this chapter, we review several recently developed SNGs and evaluate their sequence generation accuracy, SCC, and stochastic multiplication accuracy. Generally speaking, the LD-sequence-based SNGs perform the best. When it comes to stochastic multiplication, however, it is a surprise to see that the results produced by an LFSR and its reversed version match or even outperform the ones using LD sequences. This is explained in [18] by using a similarity function. The more the index changes between two permutations of a single LFSR, the less the similarity function is and thus the less correlation between the two random numbers. However, the theory underlying may need a further exploration.

References

1. A. Alaghi, J.P. Hayes, Exploiting correlation in stochastic circuit design, in *2013 IEEE 31st International Conference on Computer Design (ICCD)* (2013), pp. 39–46. https://doi.org/10.1109/ICCD.2013.6657023
2. A. Alaghi, J.P. Hayes, Fast and accurate computation using stochastic circuits, in *2014 Design, Automation I& Test in Europe Conference I & Exhibition (DATE)* (2014), pp. 1–4. https://doi.org/10.7873/DATE.2014.089
3. A. Alaghi, W. Qian, J.P. Hayes, The promise and challenge of stochastic computing. IEEE Trans. Comput.-Aided Design Integr. Circuits Syst. **37**(8), 1515–1531 (2018). https://doi.org/10.1109/TCAD.2017.2778107
4. A. Ardakani, F. Leduc-Primeau, N. Onizawa, T. Hanyu, W.J. Gross, Vlsi implementation of deep neural network using integral stochastic computing. IEEE Trans. Very Large Scale Integr. Syst. **25**(10), 2688–2699 (2017). https://doi.org/10.1109/TVLSI.2017.2654298
5. S. Asadi, M.H. Najafi, M. Imani, A low-cost fsm-based bit-stream generator for low-discrepancy stochastic computing, in *2021 Design, Automation I & Test in Europe Conference I & Exhibition (DATE)* (2021), pp. 908–913. https://doi.org/10.23919/DATE51398.2021.9474143
6. P. Bratley, B.L. Fox, Algorithm 659: implementing sobol's quasirandom sequence generator. ACM Trans. Math. Soft **14**(1), 88–100 (1988)
7. T.H. Chen, J.P. Hayes, Design of division circuits for stochastic computing, in *2016 IEEE Computer Society Annual Symposium on VLSI (ISVLSI)* (2016), pp. 116–121. https://doi.org/10.1109/ISVLSI.2016.48
8. B.J. Collings, H. Niederreiter, Random number generation and quasi-monte carlo methods. J. Amer, Stat. Assoc. **88**(422), 699 (1993)

9. I.L. Dalal, D. Stefan, J. Harwayne-Gidansky, Low discrepancy sequences for monte carlo simulations on reconfigurable platforms, in *Application-Specific Systems, Architectures, and Processors* (2008)
10. B.R. Gaines, Stochastic computing systems. Adv. Informat. Syst. Sci. **2**, 37–172 (1969)
11. M. Gay, J. Burchard, J. Horácek, A.S. Messeng Ekossono, T. Schubert, B. Becker, M. Kreuzer, I. Polian, *Small Scale AES Toolbox: Algebraic and Propositional Formulas, Circuit-Implementations and Fault Equations* (2016). http://hdl.handle.net/2117/99210
12. P. Gupta, R. Kumaresan, Binary multiplication with PN sequences. IEEE Trans. Acoust. Speech Signal Process. **36**(4), 603–606 (1988). https://doi.org/10.1109/29.1564
13. A. Keller, Quasi-monte carlo image synthesis in a nutshell, in *Monte Carlo and Quasi-Monte Carlo Methods 2012*, ed. by J. Dick, F.Y. Kuo, G.W. Peters, I.H. Sloan (Springer, Berlin, 2013), pp.213–249
14. L. Kocis, W.J. Whiten, Computational investigations of low-discrepancy sequences. ACM Trans. Math. Softw. **23**(2), 266–294 (1997). https://doi.org/10.1145/264029.264064
15. P. Li, D.J. Lilja, W.Qian, K. Bazargan, M.D. Riedel, Computation on stochastic bit streams digital image processing case studies. IEEE Trans. Very Large Scale Integr. Syst. **22**(3), 449–462 (2014). https://doi.org/10.1109/TVLSI.2013.2247429
16. S. Liu, J. Han, Toward energy-efficient stochastic circuits using parallel sobol sequences. IEEE Trans. Very Large Scale Integr. Syst. **26**(7), 1326–1339 (2018). https://doi.org/10.1109/TVLSI.2018.2812214
17. F. Neugebauer, I. Polian, J.P. Hayes, S-box-based random number generation for stochastic computing. Microprocess. Microsyst. **61**, 316–326 (2018). https://doi.org/10.1016/j.micpro.2018.06.009
18. S.A. Salehi, Low-cost stochastic number generators for stochastic computing. IEEE Trans. Very Large Scale Integr. Syst. **28**(4), 992–1001 (2020). https://doi.org/10.1109/TVLSI.2019.2963678
19. S. Sharifi Tehrani, W. Gross, S. Mannor, Stochastic decoding of LDPC codes. IEEE Commun. Lett. **10**(10), 716–718 (2006). https://doi.org/10.1109/LCOMM.2006.060570
20. H. Sim, J. Lee, A new stochastic computing multiplier with application to deep convolutional neural networks, in *2017 54th ACM/EDAC/IEEE Design Automation Conference (DAC)* (2017), pp. 1–6. https://doi.org/10.1145/3061639.3062290
21. D. Wu, J. Li, R. Yin, H. Hsiao, Y. Kim, J.S. Miguel, uGEMM: unary computing for gemm applications. IEEE Micro **41**(3), 50–56 (2021). https://doi.org/10.1109/MM.2021.3065369
22. K. Zhong, Z. Li, H. Jin, W. Qian, Exploiting uniform spatial distribution to design efficient random number source for stochastic computing, in *Proceedings of the 41st IEEE/ACM International Conference on Computer-Aided Design, ICCAD '22* (Association for Computing Machinery, New York, 2022). https://doi.org/10.11453/508352.3549396
23. F. Neugebauer, I. Polian, J.P. Hayes, Building a better random number generator for stochastic computing. in *2017 Euromicro Conference on Digital System Design (DSD)*, pp. 1–8 (2017). https://doi.org/10.1109/DSD.2017.29

Part III
Inexact/Approximate Computing

Automated Generation and Evaluation of Application-Oriented Approximate Arithmetic Circuits

Ao Liu, Yong Wu, Qin Wang, Zhigang Mao, Leibo Liu, Jie Han, and Honglan Jiang

1 Introduction

The high requirements for performance and energy efficiency due to the development of big data processing and artificial intelligence have motivated alternative computing paradigms and techniques. As many applications such as image processing and machine learning are error-tolerant [1, 2], approximate computing (AC) is a potential technique to trade off between the quality of result (QoR) and computational efficiency. A common approach to employ AC in an application is to replace the original exact arithmetic units with the corresponding approximate arithmetic circuits (AACs). To this end, various AACs have been manually designed, e.g., approximate adders [3, 4], approximate multipliers [5–7], and approximate dividers [8]. Readers interested in the manually designed AACs can refer to the recent surveys [9, 10]. However, manual approximation is difficult and time-consuming to satisfy the requirements of a complex application, such as the deep neural network (DNN). Hence, automated generation methodologies for AACs have recently been investigated.

A. Liu · Q. Wang · Z. Mao · H. Jiang (✉)
Department of Micro-Nano Electronics, Shanghai Jiao Tong University, Shanghai, China
e-mail: 2017la@sjtu.edu.cn; qinqinwang@sjtu.edu.cn; maozhigang@sjtu.edu.cn; honglan@sjtu.edu.cn

Y. Wu · L. Liu
School of Integrated Circuits, Tsinghua University, Beijing, China
e-mail: wuyong20@mails.tsinghua.edu.cn; liulb@tsinghua.edu.cn

J. Han
Department of Electrical and Computer Engineering, University of Alberta, Edmonton, AB, Canada
e-mail: jhan8@ualberta.ca

W. Liu et al. (eds.), *Design and Applications of Emerging Computer Systems*,
https://doi.org/10.1007/978-3-031-42478-6_14

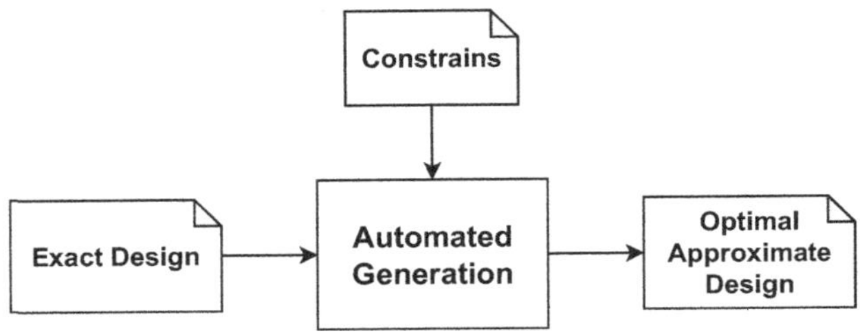

Fig. 1 Overview of the automated generation of AACs

As shown in Fig. 1, the automated generation of AACs typically takes an exact circuit and user-defined constraints as inputs and iteratively modifies the circuit to obtain well-performing approximate implementations satisfying the constraints. During the generation process, the quality metrics of the candidate AACs are often evaluated and compared with the constraints. The comparison results are then used to guide the evolution direction, which is an essential step but may be time-consuming. The commonly used quality metrics include output error, power, delay, and area; their relatively accurate values are usually obtained from circuit simulation and synthesis.

While generic AACs may not be efficient for some applications [11], methodologies aiming to generate application-oriented AACs have been proposed [12–14]. An obvious difference between the automated generation of generic and application-oriented AACs is the inputs. Generally, the inputs of the generic AAC generation are the exact arithmetic circuit and the error constraints of the AAC, whereas they are the exact application and the QoR constraints of the application for the generation of the application-oriented AACs. Therefore, the complexity of the evaluation for the application-oriented AACs is much higher than that for the generic AACs and highly depends on the specific application. In the cases where exact metric results are not required (e.g., for comparison only), the power, delay, and area of an AAC-based application can be intuitively or mathematically estimated from the metrics of each integrated AAC [15]. For example, integrating an AAC with larger power, delay, or area is more likely to result in a larger corresponding metric for the application circuit. However, the relationship between the error metrics of AACs and the QoRs of AAC-based applications is rather difficult to discover; integrating AACs with larger output errors may even slightly improve the QoR [16, 17]. Thus, the QoR evaluation is necessary for an efficient generation of application-based AACs.

Recent research on automated approximation mainly focuses on generic AACs or applications constructed from generic AACs [18] and rarely considers the QoR evaluation or recovery. This chapter aims to provide an overview and discussion of the automated generation methodologies for generic and application-oriented AACs, methods to construct AAC-based applications, as well as the techniques for accelerating the QoR evaluation and mitigating the QoR degradation of AAC-based applications.

Due to the error resilience of compute-intensive DNNs, AC has widely been employed into DNNs and achieved significant gains. Detailed AC techniques for DNN accelerators can be seen in a recent survey [19]. Moreover, the automated generation of DNN-oriented AACs is also increasingly investigated in recent years [20, 21]. As a typical complex application, DNN is utilized in this chapter as the application example to discuss the QoR evaluation and recovery methodologies.

This chapter is organized as follows. Section 2 reviews the automated generation methodologies of AACs and the related error constraints. Section 3 presents the acceleration methodologies for the QoR evaluation of AAC-based applications. Section 4 describes the techniques for QoR recovery. Finally, Sect. 5 concludes the chapter and prospects the possible techniques that can facilitate the automated generation of application-oriented AACs.

2 Automated Generation Methodologies for AACs

For the purpose of flexibility and quick delivery, many dedicated methodologies have been proposed to generate AACs automatically. In the generation, the error characteristics of an AAC are commonly provided as the basic constraints. For most methodologies, statistical error metrics of AACs are duly evaluated as the constraints to guide the approximation degree. Therefore, in this section, we first introduce some commonly used statistical error metrics. The automated generation methodologies for generic AACs are then reviewed by classifying them into three categories. Next, state-of-the-art automated generation methodologies for application-oriented AACs are discussed. Finally, a brief summary is made for the mentioned methodologies.

2.1 Statistical Error Metrics

To evaluate the error characteristics of AACs, various error metrics have been utilized. Given the approximate and accurate results M' and M, one of the basic error metrics is the error rate (ER), which shows the probability of producing an erroneous result, i.e., the probability of $M' \neq M$. The other basic error metric is the error (E) calculated as

$$\mathrm{E} = M' - M \ , \tag{1}$$

which shows the difference between the approximate and accurate results. To evaluate the relative deviation from the accurate result, the relative error (RE) is calculated as

$$\mathrm{RE} = \mathrm{E}/M \ , \tag{2}$$

Table 1 Statistical error metrics of AACs

Metrics	Description	Computation formula
ER	Error rate	$p(M_i' \neq M_i), i \in [1, N]$
ME	Mean error	$\sum_{i=1}^{N} p(\mathrm{E_i})\mathrm{E_i}$
MRE	Mean relative error	$\sum_{i=1}^{N} p(\mathrm{RE_i})\mathrm{RE_i}$
MED	Mean error distance	$\sum_{i=1}^{N} p(\mathrm{ED_i})\mathrm{ED_i}$
MRED	Mean of relative error distance	$\sum_{i=1}^{N} p(\mathrm{RED_i})\mathrm{RED_i}$
VarE	Variance of error	$\sum_{i=1}^{N} p(\mathrm{E_i})(\mathrm{E_i} - \mathrm{ME})^2$
VarRE	Variance of relative error	$\sum_{i=1}^{N} p(\mathrm{RE_i})(\mathrm{RE_i}-\mathrm{MRE})^2$
VarED	Variance of error distance	$\sum_{i=1}^{N} p(\mathrm{ED_i})(\mathrm{ED_i}-\mathrm{MED})^2$
VarRED	Variance of relative error distance	$\sum_{i=1}^{N} p(\mathrm{RED_i})(\mathrm{RED_i}-\mathrm{MRED})^2$
NMED	Normalized mean error distance	MED/M_{max}
WCE	Worst-case error	$\max_{i\in[1,N]} \mathrm{ED_i}$
WCRE	Worst-case relative error	$\max_{i\in[1,N]} \mathrm{RED_i}$
MSE	Mean-square error	$\sum_{i=1}^{N} p(\mathrm{ED_i})\mathrm{ED_i}^2$
RMSE	Root-mean-square error	$\sqrt{\mathrm{MSE}}$

which is the error normalized by the accurate result. To show the error magnitude and relative error distance, the absolute values of E and RE are commonly used, denoted as the error distance (ED) and the relative error distance (RED), respectively. In addition, to evaluate the errors from the perspective of statistic, the mean and variance of E, RE, ED, and RED can be obtained through statistical methods such as Monte Carlo simulation. Table 1 shows the statistical error metrics of AACs as well as their computation formulas, where N is the total number of the test cases and $p(\mathrm{E_i})$ is the probability of generating an error of $\mathrm{E_i}$. Moreover, the normalized mean error distance (NMED) has been proposed to compare the errors between two AACs with different bit widths; it is calculated as the normalization value of the MED by the maximum accurate result [7, 22]. The maximum values of ED and RED are also included in Table 1 to indicate the worst-case error (WCE) and the worst-case relative error (WCRE).

The mean-squared error (MSE) and root-mean-square error (RMSE) are closely related to the signal-to-noise ratio and thus are widely used in the evaluation for signal processing applications [23] and neural networks (NNs) [24]. The computation formulas of MSE and RMSE are shown in Table 1.

For simple AACs, high-precision statistical error metrics can be obtained by exhaustive simulation, i.e., traversing all possible inputs [17, 25]. For example, $256 \times 256 = 65536$ cases are required for 8×8 unsigned approximate multipliers. For more complex AACs, traversal method is no longer applicable since the number of cases grows exponentially with the bit width of inputs, and Monte Carlo-based methods are widely employed [26].

2.2 Automated Generation of Generic AACs

An AAC can be obtained through voltage overscaling (VOS) or functional approximation. VOS aims to reduce the supply voltage of a circuit that allows acceptable output errors, thereby reducing power consumption. However, the supply voltage reducing possibly leads to larger delay and induces unpredictable timing errors [9]. Hence, functional approximation is more commonly used for AAC generation, which modifies an exact circuit via netlist transformation, Boolean rewriting, or high-level approximation [18]; details of these three categories of methodologies will be discussed later.

While VOS does not perform well on its own, it can be employed along with functional approximation to achieve further energy efficiency gains. For instance, Lee et al. [27] proposed an approximate high-level synthesis tool that jointly uses bit rounding, operation elimination, and VOS. The experimental results show that 21% additional energy gains on average are obtained when considering VOS. Similarly, Zervakis et al. [28] also utilized VOS as a circuit-level approximation method after a higher-level functional approximation. Also, [28] proposed the voltage island grouping (VIG) to group approximate components into different voltage islands and limited the total number in a power efficient manner. By employing the VIG with three voltage islands, about 2% additional power gains are obtained compared to the design with only one voltage island.

2.2.1 Netlist Transformation

The netlist transformation aims at the gate-level netlists and obtains AACs by modifying, removing, or adding some gates or wires.

Hrbacek et al. [12] proposed a Cartesian genetic programming (CGP)-based methodology for generating AACs automatically. This methodology represents a candidate circuit as a fixed-sized 2D Cartesian grid of nodes interconnected by a feed-forward network, each node representing a two-input Boolean function. Thus, a candidate solution can be represented by the function and the input connections of each node as well as the output connections of the circuit. The number of active nodes determines the circuit size. For trading off between output error (characterized by MRED), power consumption, and delay of AACs, [12] utilized non-dominated sorting genetic algorithm II (NSGA-II) [29] to solve the multi-objective optimization problem and obtained the Pareto-optimal AACs automatically. Based on this CGP-based methodology, Mrazek et al. [30] built an open-source AAC library called EvoApprox8b, which includes 430 generic 8-bit approximate adders and 471 generic 8×8 approximate multipliers. To adapt to more complex applications and introduce application-oriented AACs, EvoApprox8b has been extended [21]. The latest EvoApproxLib contains various (un)signed approximate adders, $N \times N$ (un)signed approximate multipliers, and $M \times N$ unsigned approximate multipliers.

Schlachter et al. [31] represented a circuit netlist as a directed acyclic graph and utilized the automatic gate-level pruning (GLP) according to the *significance* and *activity* of each node. The *significance* shows the impact of the node on the output and is computed as

$$\sigma_i = \sum \sigma_{desc(i)} , \tag{3}$$

where σ_i is the *significance* of the node i and $\sigma_{desc(i)}$ is the *significance* of the direct descendants of node i. The *significance* of each primary output is predefined, and the *significance* of each node can be obtained through reverse topological graph traversal. The *activity* is obtained through a gate-level hardware simulation and extracted from the toggle count and the time spent at the logic levels 0 and 1. After calculating the significance-activity product (SAP), the GLP methodology iteratively prunes nodes with the lowest SAP until reaching the error threshold constrained by MRED and ER. Similarly, Zervakis et al. [20] generated AACs by identifying the wires that have to be approximated, but replacing the wire with a switch that selects between the original value and the approximate constant instead of pruning it. Thus, the AACs with runtime reconfigurable accuracy are obtained. Moreover, GLP can also be used to generate AACs with variable delay by reducing the critical path [32].

AxLS [33] is an open-source framework for an automated generation of AACs based on netlist transformation. It converts a Verilog-based register-transfer level (RTL) circuit into an XML file that includes the gate-level netlist obtained through synthesis and other information, such as the switching activity from post-synthesis simulation. Approximation techniques such as GLP [31] can be implemented in Python and used in the XML file to generate AACs. AxLS can be used as a general framework to test various methodologies that based on netlist transformation.

2.2.2 Boolean Rewriting

The Boolean rewriting aims at the Boolean representation of the circuits, which obtains AACs by rewriting the input-output mapping.

Wu and Qian [34] proposed a methodology based on the simplification of Boolean expression to automatically generate min-area AACs under the ER constraint. This methodology iteratively selects one most effective node of the circuit and replaces its factored-form original expression with the best simplified one obtained by deleting one or more literals. For example, replacing the expression $n = (a + b)(c + d)$ with $n = a(c + d)$ for node n, when this substitution leads to lower ER and smaller circuit. Furthermore, to speed up the simplification for more complex circuits, the whole process of an iteration was formulated as a knapsack problem and multiple nodes are selected and simplified simultaneously.

Beyond Boolean expression simplification-based methodologies, some works employ Boolean matrix factorization (BMF) to generate AACs automatically [35–

37]. For a $k \times m$ Boolean matrix $\mathbf{M}$ where all elements are restricted to Boolean values, it can be factored into two Boolean matrices, a $k \times f$ matrix $\mathbf{B}$ and an $f \times m$ matrix $\mathbf{C}$, such that $\mathbf{M} \approx \mathbf{BC}$ [35]. For example,

$$\begin{pmatrix} 1\,1\,1\,0\,1 \\ 1\,0\,0\,1\,0 \\ 1\,0\,1\,0\,1 \\ 0\,1\,0\,0\,1 \\ 0\,0\,1\,1\,1 \end{pmatrix} \approx \begin{pmatrix} 1\,1\,0 \\ 0\,0\,1 \\ 1\,0\,0 \\ 0\,1\,0 \\ 1\,0\,1 \end{pmatrix} \begin{pmatrix} 1\,0\,1\,0\,1 \\ 0\,1\,0\,0\,1 \\ 0\,0\,0\,1\,0 \end{pmatrix} . \tag{4}$$

For Boolean values, the multiplications are performed using the logical AND; the additions are performed using the logical OR or XOR [36]. Therefore, a simplified circuit can be obtained by replacing the truth table $\mathbf{M}$ of the exact circuit with $\mathbf{BC}$. The degree of simplification is negatively related to f (the number of columns of $\mathbf{B}$). Moreover, the BMF can be formulated as an optimization problem as follows.

$$\underset{\mathbf{B},\mathbf{C}}{\operatorname{argmin}} \|\mathbf{M} - \mathbf{BC}\|_2 , \tag{5}$$

which means minimizing the Hamming distance between the exact and approximate truth tables. In BLASYS [35], (5) is modified to take the bit significance into account, which is

$$\underset{\mathbf{B},\mathbf{C}}{\operatorname{argmin}} \|(\mathbf{M} - \mathbf{BC})\mathbf{w}\|_2 , \tag{6}$$

where $\mathbf{w}$ is a constant weight vector, indicating the impact of each bit on the output error. BLASYS does not directly solve this optimization problem. It decomposes a large exact circuit into smaller subcircuits and obtains the optimal BMF plan of each subcircuit via design space exploration (DSE), i.e., finding the optimal design from a search space that includes all possible designs.

2.2.3 High-Level Approximation

High-level approximation aims at the behavior-level descriptions of circuits, which obtains AACs by replacing some building blocks.

Similar to high-level synthesis (HLS) [38], the flow of high-level approximation can be abstracted as that shown in Fig. 2. First, with an exact design and the pre-built approximate unit library as the inputs, some candidate parts of the exact design are replaced with approximate units, resulting in a set of approximate designs. Second, the QoR and hardware evaluations are performed on each approximate design to obtain the quality metrics. Finally, an optimization solver is employed to find the Pareto-optimal approximate designs with the best quality metrics under user-defined error constraints (generally characterized by statistical error metrics).

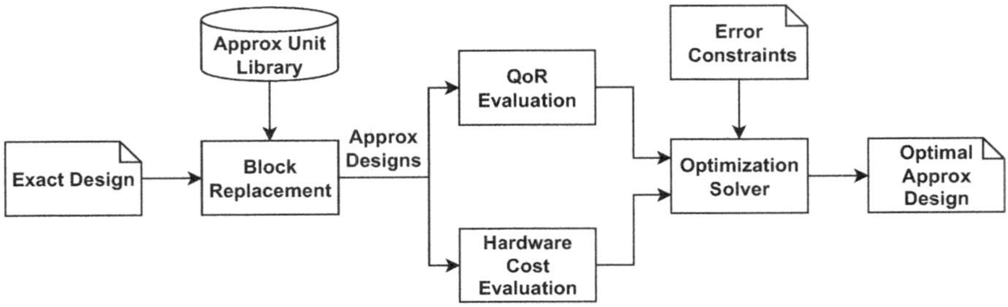

Fig. 2 Overview of high-level approximation

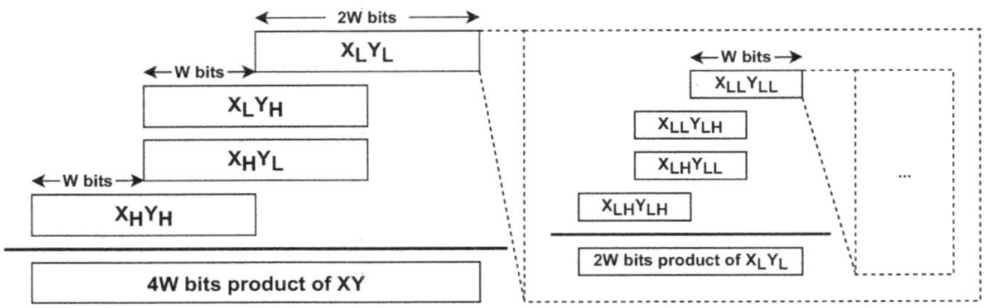

Fig. 3 Recursively breaking down a large multiplication into four small multiplications, where X_L and X_H denote the W-bit LSBs and MSBs of X, respectively

As shown in Fig. 3, A $2W \times 2W$ multiplication can be broken down into four $W \times W$ multiplications, and this process can be recursively performed on each $W \times W$ multiplication until a set of 2×2 multiplications are required[1] [5]. Based on this technique, Rehman et al. [39] proposed three approximation methods to generate approximate multipliers, approximating in the elementary 2×2 multipliers, in the adders that compress the partial products, and in the LSBs of the adder tree. To implement these methods, a library containing various approximate 2×2 multipliers and 1-bit full adders is built in [39]. Assuming a need for N 2×2 multipliers, with M optional multipliers and D adders in the library, and offering B options for LSB approximation, a design space of size $D \times B \times M^N$ is then constructed. Therefore, the approximate multiplier generation is transformed into a DSE problem, which can be solved according to the flow shown in Fig. 2. In [39], the depth-first search (DFS) algorithm is used to find the design with the smallest area and power consumption under user-defined error constraints.

In addition to AAC generation, many high-level approximation methodologies aim to automatically select AACs from a library to construct complex applications, such as accelerators [15, 27, 28, 40, 41]. These can be considered as the automated generation methodologies for AAC-based applications. In [28], each arithmetic unit

[1] Assuming that W is a power of 2, i.e., $W = 2^n$, where n is an integer.

of an accelerator is processed respectively to generate its corresponding Pareto-optimal approximate designs that form a library. Therefore, various approximate configurations of the accelerator can be obtained by randomly selecting and combining AACs from this library. The remaining task is to iteratively evaluate each approximate configuration to find the best-performing one under the user-defined QoR constraints of the accelerator, as shown in Fig. 2. To accelerate the QoR evaluation, a machine learning-based predictor is introduced, which will be discussed in Sect. 3.2. Mrazek et al. [40] built a general library of fully characterized AACs that will be pruned to reduce the size of a specific application. Based on this library, various approximate designs can be obtained through arithmetic unit replacement. An iterative heuristic algorithm is then employed to construct the Pareto front of these designs. ALWANN [41] is an automated approximation framework for DNN accelerators. In ALWANN, a pre-trained DNN is firstly quantized, and all convolutional layers of this DNN are implemented by using approximate units. Then, the approximate DNN is fine-tuned and evaluated for obtaining the accuracy and energy characteristics. The NSGA-II algorithm is employed in ALWANN to select the Pareto-optimal designs from various approximate DNNs. AxHLS [15] introduces two techniques for fast automated generation of accelerators, referred to as AxME and DSEwam. AxME uses mathematical calculations to estimate area, power, and delay for a specific design. DSEwam uses AxME and the error propagation rules proposed in [42] for a fast evaluation and employs an optimization solver based on tabu search (TS) to find the Pareto-optimal approximate designs.

2.3 *Automated Generation of Application-Oriented AACs*

In the last section, we discussed how to generate a generic AAC according to its own error constraints and how to select generic AACs from a library to construct an AAC-based application. These two processes are combined to generate application-oriented AACs, as shown in Fig. 4. Rather than selecting existing generic AACs to implement error-tolerant applications, in this flow, the AAC library in Fig. 2 is replaced with the AAC generation process. After the evaluation and optimal design selection, a judgment according to the user-defined application QoR constraints is performed. When the constraints are not satisfied, the parameter adjustment module is activated to adjust the parameters for the AAC generation as per the quality of the current design. In some automated generation methodologies for application-oriented AACs, error modeling is involved. The error modeling process analyzes the error tolerance of the application and obtains some application-specific suggestions to guide the AAC generation.

The generation methodologies for generic AACs can also be utilized to generate application-oriented AACs by introducing application-specific constraints. For example, similar to [12], Mrazek et al. [16] adopted the CGP-based automated generation methodology, but for generating NN-oriented approximate multipliers. In this methodology, exact multipliers in a pre-trained NN are replaced with the

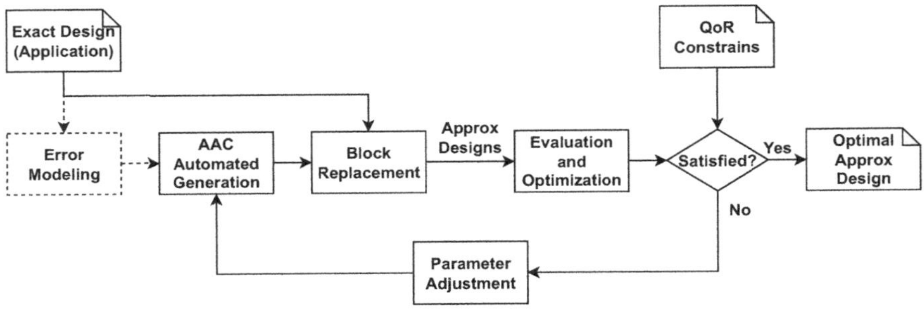

Fig. 4 Overview of the automated generation of application-oriented AACs

approximate multipliers generated through CGP. The approximate NN is then retrained to refine its QoR. The QoR obtained via simulation is compared with the user-defined accuracy constraint. If the constraint is not met, the WCE that used as the error constraint in CGP will be reduced; this process will be repeated until the constraint is met. Since an NN usually contains a massive amount of zero-valued data, a zero-error constraint is also involved in CGP to ensure that the output of the multiplier must be zero when one of the inputs is zero.

Vasicek et al. [43] proposed a new error metric referred to as the weighted mean error distance (WMED) to generate approximate application circuits with high quality. The WMED is an extension of MED, which takes the distribution of inputs into account. For a $w \times w$ signed approximate multiplier, the WMED obtained by traversing all possible inputs is defined as

$$\mathrm{WMED} = \frac{1}{2^{2w}} \sum_{i=-2^{w-1}}^{2^{w-1}-1} \sum_{j=-2^{w-1}}^{2^{w-1}-1} \alpha_{i,j} \left| M'_{i,j} - M_{i,j} \right| , \tag{7}$$

where $M'_{i,j}$ and $M_{i,j}$ are the outputs of the approximate and accurate multiplier for inputs i and j, respectively, and $0 \leq \alpha_{i,j} \leq 1$ is the weight determined by the probability mass function of the inputs. The experimental results show that WMED performs better than the conventional error metrics when used in CGP-based NN-oriented approximate multiplier generation.

Soares et al. [13] proposed a hybrid approximate adder for constructing image and video processing accelerators. Operations in these applications mainly include additions and left shifts, and the two operands of an addition may require different bit numbers of left shift. Hence, the hybrid approximate addition is divided into four regions as shown in Fig. 5. The region of bit width k_1 is the overlapped shift region where a truncated adder (i.e., the output is set to zero) is used, the region of bit width k_2 is the excessive shift region where a copy adder (i.e., the output is the same as one of the inputs) is used, the region of bit width k_3 is the approximate part of no-shift region where an approximate adder is used, and the rest MSBs are calculated by an

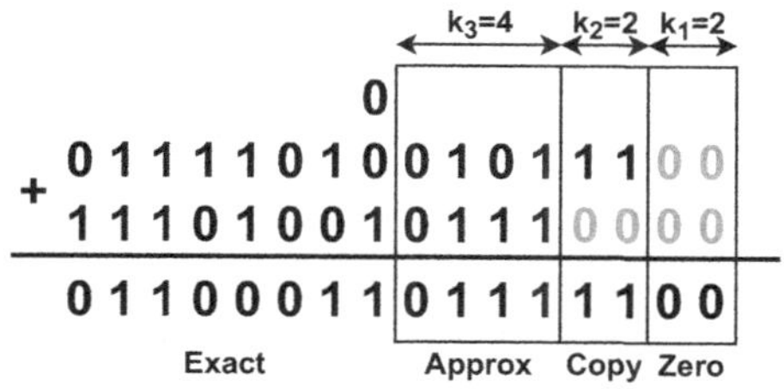

Fig. 5 Example of the hybrid approximate adder [13], where the first and second operands are left shifted for two and four bits, respectively, before addition, and the approximate adder used in the approximate part is ETAI [3]

exact adder. When the approximate adder for the approximate part is specified, the hybrid approximate adder can be defined by three parameters, k_1, k_2, and k_3. While k_1 and k_2 are determined by the shift bit number of the operands and obtained by a traversal of all operations in the application, k_3 is determined through iterative evaluation and optimization according to the application quality.

For some methodologies, the error modeling is critical. Shah et al. [44] proposed a framework for automatically generating low-precision probabilistic inference circuits, referred to as ProbLP. The probabilistic inference operations in Bayesian networks are regular and only contain multiplications and additions. Therefore, ProbLP first analyzes the single multiplier and single adder with bit width reduction to obtain the WCE as well as the error propagation rules. Then, the minimum bit width of each arithmetic unit is recursively obtained by limiting the accumulated WCE on the output. Both fixed-point and floating-point designs are considered in ProbLP, and the one with lower energy consumption will be selected.

Some methodologies focus on the field-programmable gate array (FPGA), which utilize look-up tables (LUTs) and carry-chains to design AACs [11, 14, 45, 46]. Typically, Ullah et al. [11] proposed a methodology for automatically generating application-oriented AACs on FPGA, referred to as AppAxO. AppAxO represents an FPGA-based AAC by a binary string denoting a subset of LUTs to be disabled in the exact implementation. Two DSE methods for finding the Pareto-optimal designs are included in AppAxO, referred to as AppAxO_MBO and AppAxO_ML. AppAxO_MBO is based on the multi-objective Bayesian optimization, to generate only the Pareto-optimal AACs that satisfy the application quality constraints. AppAxO_ML is based on the genetic algorithm to explore a large design space of AACs, and machine learning models are introduced to predict the power, area, delay, and QoR of an AAC-based application, thereby accelerating the exploration process. Details of these machine learning models will be discussed in Sect. 3.2. SyFAxO-GeN [14] uses the same AAC representation method as AppAxO, and generative adversarial networks (GANs) are introduced to generate candidate AACs for better accuracy-performance trade-offs. Similar to AppAxO_ML, SyFAxO-GeN also adopts machine learning models for a fast DSE.

2.4 Summary

The abovementioned automated generation methodologies of AACs can be further classified as follows.

- From the perspective of the circuit representation: [12, 16, 20, 31–33, 43] are based on the gate-level netlists and generate AACs by using netlist transformation. [34–37] are based on the Boolean representation and generate AACs by using Boolean rewriting. [13, 15, 27, 28, 39–41, 44] are based on the behavior-level descriptions and generate AACs or AAC-based applications by using high-level approximation. In [11, 14], the FPGA-based AACs are represented by LUTs and carry-chains, which are generated using a method similar to the netlist transformation.
- From the perspective of the purpose: [12, 20, 31–37, 39] aim to generate generic AACs. [15, 27, 28, 40, 41] select the proper generic AACs to implement error-tolerant applications. [11, 13, 14, 16, 43, 44] directly generate application-oriented AACs.

For generating the generic AACs, each methodology has its own pros and cons. The netlist transformation is the closest to the actual structure of the circuit and thus can involve some specific optimizations, but requires the detailed circuit description. The Boolean rewriting, especially the BMF-based methodologies, can explore a wider range of possible designs, but the corresponding netlist of each design is unknown before synthesis. High-level approximation can construct larger designs based on the provided approximate units and is essential for devising complex approximate applications; however, it highly depends on the pre-built approximate unit library.

The pros and cons discussed above also apply to methodologies generating application-oriented AACs, as these methodologies can also be classified into the three categories according to the approximation level. In addition, the techniques for error modeling as well as the evaluation and optimization processes greatly affect the overall generation efficiency.

The main difference between the generic AAC generation and application-oriented AAC generation is the objective of the optimization. The optimization objective of the former is the quality of the AAC itself, while the optimization objective of the latter is the quality of the AAC-based application. Moreover, the optimization of a generic AAC is generally constrained by a statistical error metric of the AAC, while the optimization of an application-oriented AAC is generally limited by a QoR of the application. Details of the QoR evaluation for AAC-based applications will be discussed in the next section.

3 QoR Evaluation of the AAC-Based Applications

For some simple applications based on AACs, the process of error propagation and accumulation can be modeled mathematically, e.g., analyzing the error generation and propagation of each possible arithmetic unit and then using these to recursively model the output error of a given application [42]. The error propagation modeling is similar to the error modeling in Fig. 4, but analyzes approximate designs and aims to predict the QoR. Based on the error propagation model, the QoR evaluation of the automated generation for application-oriented AACs can be accelerated, because the QoR can be estimated without performing simulation. However, for complex applications such as DNN accelerators, the error propagation process is hard to be modeled through mathematical analysis due to the extremely high complexity of the calculation. Therefore, the circuit simulation is inevitable to obtain the QoRs of these applications. As the simulation can be very time-consuming, the QoR evaluation needs to be optimized to effectively generate application-oriented AACs and use them to construct complex applications. Generally, there are three categories of techniques for the acceleration of the QoR evaluation.

- *Simulation acceleration.* To accelerate an evaluation, the most basic technique is to increase the simulation efficiency, e.g., using high-performance simulation platforms.
- *Prediction model.* To avoid massive repeated time-consuming simulations, some prediction models have been introduced based on the machine learning techniques to predict the QoRs of the AAC-based applications.
- *Functional abstraction.* Although integrating an AAC into the application circuit is an accurate and reliable way to evaluate the QoR for the AAC-based application, it is usually time-consuming and hardly to be generalized to other AACs. Thus, some methodologies ignore the specific structure of AACs and only consider the errors introduced by them, which can obtain the impact of AAC errors on the application QoR more efficiently.

3.1 Simulation Acceleration

For obtaining the QoR of an AAC-based application, an intuitive method is to perform an RTL simulation. However, the structure of the circuit can be very complex, and the RTL simulation based on conventional central processing unit (CPU) is time-consuming. In addition, some QoR recovery methods (will be discussed in Sect. 4), such as DNN retraining, are much more time-consuming when performed by using RTL simulation [19]. Therefore, for the purpose of acceleration, the conventional RTL simulation is commonly replaced with high-level functional simulation. Some open-source AAC libraries, e.g., lpACLib [47], SMApproxLib [45], and EvoApproxLib [21], provide both hardware and software implementations of AACs, which facilitate the software modeling for AAC-based applications. AxDNN [48] is a systematical framework for approximate DNN design, which integrates a pre-RTL simulator for evaluation. This simulator represents the dataflow of the hardware as a dynamic data dependence graph based on C code, achieving a 20× speedup in power estimation compared with conventional RTL method. Zervakis et al. [20] implemented a gate-level to C converter to translate the netlist of an AAC to a C function, so as to evaluate the error and power of the AAC at C-level. Based on this converter, an average speedup of 120× for AAC simulation is achieved compared to RTL simulation. Similarly, [21] and [32] also employed a netlist to C-code converter for software-based AAC simulation. MAxPy [49] builds a cycle-accurate Python model for an AAC described in RTL code (e.g., Verilog). This model can then be integrated into the Python-based applications as a module to perform high-level simulation.

The mainstream standard machine learning frameworks, such as TensorFlow and PyTorch, implement high-level descriptions of algorithms/models and support graphics processing unit (GPU)-based fast simulation. However, these frameworks lack support for approximate computing. To integrate user-defined approximate operations into these frameworks, some LUT-based methodologies have been proposed. TFApprox [50] is a TensorFlow-based approximate DNN simulation framework that supports GPU-based simulation. In TFApprox, the outputs for the exhaustive inputs of an AAC are pre-written into the GPU memory in the form of an LUT. Thus, the approximate operation of the AAC is implemented by accessing the LUT as per the inputs. By replacing the exact multiplications with the LUT-based approximate ones, TFApprox implements the approximate Conv2D (2D convolution) operation of TensorFlow. Hence, approximate DNNs can be easily constructed based on the TensorFlow framework. Moreover, in TFApprox, the AAC can be easily replaced by just changing the LUT. The experimental results show that the GPU-based TFApprox can reduce the simulation time by 200× with respect to an optimized CPU version on the approximate inference of ResNet [51]. Similarly, ProxSim [52] and AdaPT [53] also utilize LUT-based functional simulation for approximate DNNs involving approximate multiplications. ApproxTrain [54] approximates not only the forward, but also the backward propagations of DNNs, which enables an approximate training. As the intermediate data in the training phase changes in a large range and is precision-sensitive [55], the floating-point (FP) data representation is essential. For implementing approximate FP multiplication, ApproxTrain computes the mantissa using the same LUT-based method as TFApprox, while the sign and exponent remain accurate. In ApproxTrain, both forward and backward propagations of both Conv2D and Dense (densely connection) operations in the TensorFlow framework are modified based on this method. Detailed differences between the above four LUT-based simulation frameworks are concluded in Table 2. It can be seen that due to the limitation of the memory size, an AAC with a large input bit width cannot be converted to an LUT, which is an obvious bottleneck of these LUT-based frameworks.

Table 2 LUT-based approximate DNN simulation frameworks

Framework	Basic Python framework	Platforms	Open source	Data type[a]	Backward propagation
TFApprox [50]	TensorFlow	CPU, GPU	Yes	8-bit integer	Accurate
ProxSim [52]	TensorFlow	CPU, GPU	No	≤8-bit integer	Accurate
AdaPT [53]	PyTorch	CPU	Yes	≤15-bit integer	Accurate
ApproxTrain [54]	TensorFlow	CPU, GPU	Yes	(1,8,1-11) FP[b]	Approximate

[a] Supported input data types of approximate multiplications
[b] Floating-point data with 1-bit sign, 8-bit exponent, and from 1- to 11-bit mantissa

3.2 Prediction Model

As shown in Fig. 4, multiple iterations are required to find the optimal application-oriented AAC, where the QoR evaluation based on simulation needs to be performed at every iteration. Note that the difference between two candidate approximate designs for the same application is only the approximation strategies, i.e., the design and integration strategies of AACs. Therefore, it may be possible to predict the QoR empirically without performing simulation at every iteration.

The similar problem also exists in the field of neural architecture search (NAS) that aims to automatically generate an optimal NN architecture by selecting and combining some basic operations from a predefined search space [56]. Multiple NNs are generated and evaluated during the searching process. As training an NN from scratch and evaluating its performance is time-consuming, some NAS methodologies build machine learning models to predict the accuracy of the generated NNs [56–58]. For example, Moons et al. [58] defined an NN consisting of blocks and built a block library as the search space, and a set of architectures is then constructed by sampling and combining blocks from this library. These architectures are then fine-tuned and evaluated to obtain the accuracy. As per the architecture accuracy and the characteristics of the constituent blocks, a linear accuracy predictor was trained. This predictor is finally used to predict the accuracy of all architectures in the search space.

Similarly, the approximate operations can be integrated into the search space, and the automated generation of approximate NNs can be treated as a NAS problem [59]. Thus, the predictor-based evaluation method of NAS can also be adopted for the evaluation of AAC-based applications. Zervakis et al. [28] trained an NN model to fit the relationship between the approximate configurations and the QoRs of the approximate designs, where the approximate configuration is defined by the introduced error of each node in the data flow graph. Mrazek et al. [40] built a regression model to predict the QoR of the accelerator according to the WMED (defined in (7)) of all employed arithmetic units. [11] and [14] used the configuration vector (i.e., positions of the utilized LUTs) of an FPGA-based AAC as the input of machine learning models to predict the QoR as well as other quality metrics of the application based on the AAC. These methodologies require a small subset of possible configurations and the corresponding simulation results for the application as the data set, which are used to train different machine learning models. The trained models are then tested to select the one with the highest prediction accuracy. Although the process of training and selection brings extra workload, it is much faster than performing the simulation at every iteration for a large search space. For example, traversing the search space with 10^{23} possible solutions would take 10^{17} years, whereas [40] reduced the time to a few hours by employing a predictor and a Pareto construction algorithm.

3.3 Functional Abstraction

The above-introduced two types of QoR evaluation methodologies are based on functional simulation, where integrating the modeled AACs into the application circuit is an inevitable step, e.g., describing the behavior of an AAC through C code or using an LUT. This step may be time-consuming when the structure of the AAC or application is complex. To further accelerate the evaluation process, some methodologies describe AACs at an abstract level, i.e., ignoring the specific structure and only considering the introduced errors.

As functional abstraction generally involves mathematical analysis, we take DNN as an example to introduce the common approximation strategies. In contrast to the traditional neuron-wise NN approximation methodologies such as AxNN [60] and ApproxANN [61], most state-of-the-art works adopt layer-wise approximation [41, 59, 62], i.e., using the same approximation strategy for each arithmetic unit in the same layer. The basic operation in a DNN layer can be described as

$$y = \sum_i (x_i \times w_i) \ , \tag{8}$$

where y is one of the output features, x_i is the input activation, and w_i is the weight. Equation (8) can represent the operation in both a convolutional layer and a fully connected layer by changing the number and order of accumulation. The most widely used DNN-oriented AAC is the approximate multiplier (AM). The operation of an approximate layer where the multipliers are replaced with the same AM $\widetilde{M}$ can be described as

$$\begin{aligned} y' &= \sum_i \widetilde{M}(x_i, w_i) \\ &= \sum_i (x_i \times w_i + \epsilon(x_i, w_i)) \ , \end{aligned} \tag{9}$$

where y' is one of the approximate output features and ϵ is the operand-dependent multiplication error. The purpose of the functional abstraction denoted by (9) is using ϵ to represent $\widetilde{M}$.

A typical functional abstraction approach is to treat the errors introduced by AACs as noise and simulate the effect of AACs by injecting noise signals. Hanif et al. [63] assumed that the error sources introduced by quantization and approximate hardware components are independent and identically distributed. Hence, [63] introduced white Gaussian noise individually at the outputs of convolutional layers and tested the inference accuracy to analyze the error resilience of a DNN. The experimental results show that a network has different levels of error tolerance for the same error in different convolutional layers, and the error with a distribution of zero mean is more tolerable. Hammad and El-Sankary [2] simulated the impact of AMs on the accuracy of VGG [64] by introducing an error matrix $\mathbf{E}$ to element-wise multiply with the layer input $\mathbf{Y}$ during inference, i.e.,

$$\mathbf{Y}' = \mathbf{Y} \odot \mathbf{E} \,, \tag{10}$$

where $\mathbf{Y}'$ is the modified layer input and $\odot$ is the element-wise multiplication operator. The error matrix $\mathbf{E}$ is generated randomly using the probability density function of the uniform distribution or Gaussian distribution according to the MRED of the AM and the user-defined standard deviation. Simulation results in [2] show that using AMs leads to very little impact on the accuracy of VGG. Ansari et al. [17] treated the error introduced by the AM in an NN layer as the noise injected into the weights, i.e., transforming (9) to

$$\begin{aligned} y' &= \sum_i (x_i \times (w_i + \frac{\epsilon(x_i, w_i)}{x_i})) \\ &= \sum_i (x_i \times (w_i + \delta_i)) \,, \end{aligned} \tag{11}$$

where δ_i is the equivalent noise. Hence, [17] built an analytical AM whose error ϵ is replaced with Gaussian noise, and discovered that adding small amounts of noise followed by a retraining operation can achieve a higher accuracy for the NN than the one without noise. This phenomenon can be explained by the fact that an appropriate amount of noise helps to mitigate the overfitting problem. Additionally, [17] also found that the unbiased (i.e., with zero mean) noise performs better, which is consistent with the experimental results in [63]. For obtaining the correlation between the outputs of each DNN layer and the errors introduced by AMs, [65] calculated the Pearson correlation coefficient and observed a linear correlation. Following this, a neuron-wise error model based on Gaussian distributed signals was built and verified [65]. This model mimics the impacts of AMs on the outputs of the DNN layer and can be added to the outputs of each layer. The experiments show that employing this model can speed up the retraining process by about 10× compared to the functional simulation based on ProxSim [52].

Another functional abstraction approach is to represent an AAC by its statistical error metrics. As described in Sect. 2, most automated generation methodologies of AACs require statistical error metrics as the constraints, and the error metrics can be regarded as the most basic characteristics of an AAC. Some error modeling methods are also based on error metrics, e.g., Li et al. [66] treated the error of an AAC as a random variable characterized by its mean and variance, and [2] used the MRED to generate the error matrix for the simulation. Therefore, it can be assumed that there are correlations between the statistical error metrics of AACs and the QoRs of AAC-based applications. Ansari et al. [17] built a machine learning-based classifier to classify AMs according to their impact on the accuracy of an AM-based NN. The input features of this classifier are several statistical error metrics of the AM. Thus, the importance of each error metric can be obtained by feature ranking, i.e., selecting features that can result in higher classification accuracy. [17] finally concluded that the two most important error metrics are VarE and RMSE. [62] and [67] analyzed the error accumulation in an AM-based approximate convolutional layer during inference. They demonstrated that the mean of the output errors can be zero after compensating the layer outputs; in this case, the accuracy of the convolution is only subject to the variance of the errors. This conclusion illustrates that the variance-related error metrics is essential for the evaluation of DNN-oriented AMs. In [68], the operation in an AM-based layer-wise approximate layer denoted by (9) is described as

$$\begin{aligned} y' &= \sum_i ((x_i \times w_i) \times (1 + \frac{\epsilon(x_i, w_i)}{x_i \times w_i})) \\ &= \sum_i ((x_i \times w_i) \times (1 + \epsilon_r)) , \end{aligned} \tag{12}$$

where ϵ_r is the relative error defined by (2). For an AM with a low enough variance of relative error (VarRE), ϵ_r can be treated as an operand-independent constant and (12) can be transformed to

$$\begin{aligned} y' &= (1 + \epsilon_r) \times \sum_i (x_i \times w_i) \\ &= (1 + \epsilon_r) \times y , \end{aligned} \tag{13}$$

where y is the accurate output feature defined in (8). Equation (13) shows that the errors introduced by the AM lead to a constant multiplication on each output feature. [68] demonstrated that features are not identified by their absolute values but by the relatively higher values within each feature map. Therefore, the feature change shown in (13) does not affect the layer function. This conclusion illustrates that the AM with lower VarRE is more suitable for the layer-wise DNN approximation.

In addition to accelerate the QoR evaluation, functional abstraction methods can also be adopted to build an error model. As shown in Fig. 4, the error model is used to analyze the exact application and guide the AAC generation. By injecting noise to the exact design or investigating the critical error metrics, more constraints for generating a better application-oriented AAC can be obtained. For example, while concluding the importance of the variance of errors, [62] generated low-variance AMs based on [20] for constructing NN accelerators. Additional AAC constraints can reduce the searching space for the automated generation of the application-oriented AACs, achieving an overall speedup.

3.4 Summary

The QoRs of the AAC-based simple applications may be mathematically modeled, while the QoR evaluation of complex applications inevitably requires time-consuming simulations. The reviewed acceleration techniques for the QoR evaluation are classified into three categories, as shown in Fig. 6.

For the simulation acceleration, the high-level functional simulation is more general, as the functions of majority AACs can be modeled by software code (e.g., Python and C), while LUT-based simulation is only available for AACs with relatively low input bit width due to the memory size limitation. However, the LUT-based simulation is faster and more flexible than software-based simulations. The machine learning-based predictors generally use the characteristics of the employed AACs as the inputs to predict the QoR, where the statistical error metrics are usually selected. They demonstrate that the error metrics can be used to characterize the impact of AACs on the QoRs of the AAC-based applications. The functional abstraction methods can not only accelerate the QoR evaluation but also be used to build error models for specific applications. It is noteworthy that most state-of-the-art functional abstraction methods cannot accurately describe the impact of AAC errors on the QoR; thus, the evaluation results are more suitable for comparison and analysis. The statistical error metrics of AACs are more abstract and diverse than noise and thus harder to be used directly but better for analysis.

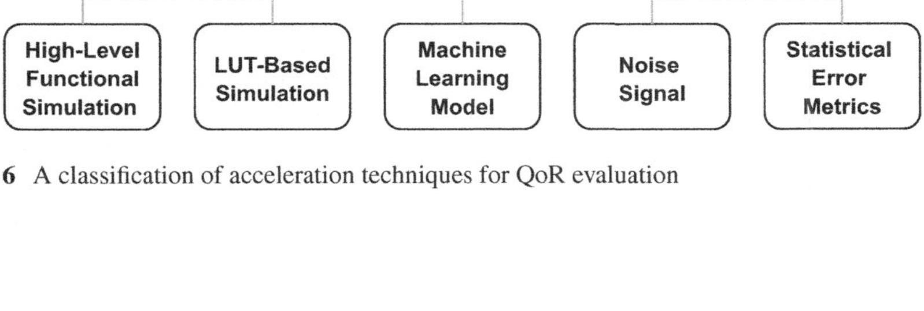

Fig. 6 A classification of acceleration techniques for QoR evaluation

4 QoR Recovery

Directly integrating AACs into an application circuit may lead to unexpected QoR loss. Therefore, QoR recovery is an essential complementary step for devising AAC-based applications. Generally, three categories of techniques have been investigated for QoR recovery.

- *Self-adjustment.* It is a simple QoR recovery approach for the applications with learning properties or constant parameters, e.g., adaptive filter and DNN. By adjusting the parameters of the application itself, the errors due to AACs can be complemented.
- *Error-aware adjustment.* Manually adjusting the approximate design according to the errors introduced by AACs is a fast QoR recovery method. As the AAC error is input-dependent, the statistical analysis is usually required.
- *Robustness enhancement.* Error tolerance can be considered when building the original exact application circuit. If the exact circuit is robust enough (i.e., resilient to noise and errors), lower QoR degradation will be caused by the integration of AACs.

4.1 Self-Adjustment

The self-adjustment methods use the same parameter obtaining approach or algorithm as the original exact application to update the parameters of the AAC-based application. Take DNN as an example; training from scratch is the most basic approach to obtain parameters; however, there are two challenges for training an approximate DNN. One challenge is the time and computing resource consumption. Training a DNN to convergence is extremely expensive even without approximation [56], and the implementation of AACs further worsens the situation. The other challenge is the accuracy loss due to error fluctuations. As the output error of an AAC commonly varies with its inputs, the parameters of the DNN may be frequently updated in a large range when training from scratch. Thus, the varying errors may affect the convergence speed of the parameters.

As approximate designs aim to mimic the original exact design as close as possible with simpler circuits, the optimal parameters for approximate and exact circuits are also likely to be similar. Thus, a basic idea is to obtain the parameters of an AAC-based application by fine-tuning the parameters of the exact application. For approximate accelerators performing the inference of a DNN, a common fine-tuning approach is to employ retraining based on the parameters of a pre-trained model. Venkataramani et al. [60] showed that retraining can recover a large amount of quality loss due to approximation with very few iterations (two iterations). Similarly, ApproxANN [61] and [16] applied retraining for the accuracy recovery with less than ten iterations. Based on the pre-trained parameters, retraining is much faster than training from scratch that requires tens or hundreds of iterations. Also, retraining only fine-tunes the parameters, which avoids frequent changes in a large range. Thus, retraining overcomes the above two challenges.

To improve the retraining performance, in [69], the learning rate for the retraining of an approximate circuit is iteratively reduced until the accuracy improvement of the retraining saturates. Ansari et al. [17] found that retraining an approximate NN not only can mitigate the accuracy degradation, but also can achieve higher accuracy than the original NN without approximation. [70] introduced the knowledge distillation (KD) [71] into retraining to achieve a better accuracy recovery. KD is a process of transferring the knowledge from a large model to a smaller one, i.e., using the inference results of a large model to guide the training of a smaller model. [70] employed two-stage KD that transfers the knowledge from a pre-trained exact model to a quantized one and then to an approximated one, achieving a higher accuracy than normal retraining.

4.2 Error-Aware Adjustment

Although the self-adjustment methods are simple and effective, they recover the accuracy by updating the parameters as per the QoRs of the AAC-based applications rather than the errors due to the AACs. Considering the errors caused by the AACs, some error-aware adjustment methods have been proposed to recover the QoRs of the AAC-based applications via statistical analyses.

ALWANN [41] fine-tunes the weights of the approximate convolutional layers based on the weight mapping function $map_{\widetilde{M}}$. For an approximate layer whose multiplications are implemented by the same AM $\widetilde{M}$, the $map_{\widetilde{M}}$ is defined as

$$\forall w \in W : map_{\widetilde{M}}(w) = \underset{w' \in W}{\text{argmin}} \sum_{x \in X} \left| \widetilde{M}(x, w') - x \times w \right| , \tag{14}$$

where W and X are the sets of possible values for the weights and input activations, respectively. Equation (14) indicates that for each weight w, the fine-tuned weight w' is determined by minimizing the sum of the error distances for all possible input activations. The experiments show that this method can result in an average of 5% accuracy improvement for an AM-based DNN. Similarly, Tasoulas et al. [62] assumed that the weights are fixed and the input activations are varied in a convolutional layer during inference, which works for most accelerators based on the systolic array. To mitigate the output error of an AM-based layer-wise approximate convolutional layer, [62] biased the approximate output feature y' (defined in (9)) as

$$y'' = y' - \sum_{i} \mu(\epsilon_{w_i}) , \tag{15}$$

where y'' is the biased output feature and $\mu(\epsilon_{w_i})$ is the mean error of performing the multiplication $x_j \times w_i, \forall x_j$. With the bias, the output error e is given by

$$\begin{aligned} e &= y'' - y \\ &= \sum_i \epsilon(x_i, w_i) - \sum_i \mu(\epsilon_{w_i}) \,, \end{aligned} \tag{16}$$

where y is the accurate output feature defined in (8). It can be easily obtained that the mean of the output errors $\mu(e)$ is equal to zero and the variance $\sigma^2(e)$ is given by

$$\sigma^2(e) = \sum_i \sigma^2(\epsilon_{w_i}) \,. \tag{17}$$

Therefore, $\sigma^2(e)$ can be reduced by employing multipliers with a low error variance. Based on [62], Zervakis et al. [67] focused on partial product perforation-based multipliers [6] and further reduced the $\sigma^2(e)$. For the multiplier perforating the m least partial products, [67] biased the approximate output feature y' as

$$\begin{aligned} y'' &= y' - \mu(w_i) \sum_i \frac{\epsilon(x_i, w_i)}{w_i} \\ &= y' - \mu(w_i) \sum_i (x_i \bmod 2^m) \,, \end{aligned} \tag{18}$$

where $\mu(w_i)$ is the mean of the weights. This method biases each multiplication according to the input activation x_i as well as the mean of all weights. After biasing as (18), $\mu(e)$ is equal to zero and $\sigma^2(e)$ depends on whether the distribution of the weights is centralized (i.e., concentrated close to $\mu(w_i)$). Since the weights of a DNN layer tend to follow a Gaussian-like distribution [43, 67, 72], $\sigma^2(e)$ will be close to zero.

4.3 Robustness Enhancement

The robustness of a circuit can be enhanced at both the hardware and application levels. At the hardware level, the circuits are modified to enable an error compensation mechanism. At the application level, the parameters or architecture are adjusted to gain the error tolerance.

An intuitive hardware-level approach is to compensate the erroneous output, resulting in an accurate result. Hanif et al. [73] proposed an equivalent accurate system and employed it into a systolic array. The error generated by an approximate multiply-accumulate (MAC) unit is completely compensated in the subsequent

MAC unit, thereby limiting the final output error. However, this approach is not suitable for the AAC generation, because the error generation mechanism of AACs is not fixed and hard to be compensated by a generic method. To mitigate the input-dependent errors of various AACs, Masadeh et al. [25] implemented a machine learning-based compensation module that can dynamically compensate the output errors according to the inputs. Although some extra hardware resources are required, adding this module can achieve about 9% QoR improvement for the image blending application.

Without changing the circuits, the robustness enhancement methods at the application level are easier to be employed. AxTrain [74] contains an active (AxTrain-act) and a passive (AxTrain-pas) methods for obtaining robust NNs based on training. AxTrain-act introduces a regularization term representing how sensitive the network is to the noise. The regularization term is then incorporated into the cost function for training; thus, the noise resilience can be improved along with the network accuracy during training. In AxTrain-pas, numerical models of the errors introduced by approximate operations (e.g., stochastic noise model discussed in Sect. 3.3) are incorporated into the forward propagation, so that the robustness of the network can be enhanced by learning the noise distribution during training. The regularization-based method similar to AxTrain-act is also included in ProxSim [52], while the error modeling-based method similar to AxTrain-pas is also utilized in [65]. Kim et al. [68] demonstrated that the batch normalization layer can mitigate the accumulated errors in an approximate DNN by redistributing the feature maps, which provides an architecture-level method to improve the error tolerance of the DNNs. Moreover, [68] demonstrated that the depthwise separable convolution is more error-sensitive than conventional convolution, because the accumulated errors on different output features are more independent and harder to recover, which provides a guidance for the NN architecture design.

4.4 Summary

The self-adjustment methods are simple and general for the QoR recovery of AAC-based applications; however, they are extremely time-consuming for complex applications. The self-fine-tuning methods, e.g., retraining for DNNs, can greatly reduce the time requirement, but highly depend on the original parameters. When the approximate design is far from the exact implementation, the parameters obtained from fine-tuning may not be optimal. Without the AAC simulation, the error-aware adjustment methods are significantly faster than those of the self-adjustment; however, the application-specific analyses are required. The robustness enhancement methods aim at the exact application, which are orthogonal to the above two adjustment techniques. Generally, the robustness enhancement at hardware level requires extra circuits, whereas the application-level methods are easier to employ, yet more difficult to optimize for specific circuits.

5 Conclusions and Prospects

By using AACs, the error-tolerant applications can be implemented with high performance and energy efficiency. To automatically generate the application-oriented AACs, the evaluation of the AAC-based applications is a critical step that consumes massive time and computing resources. In this chapter, the automated generation and evaluation methodologies for application-oriented AACs are reviewed, characterized, classified, and compared. Moreover, the QoR recovery techniques for AAC-based applications are also discussed.

As per the approximation level, we classified the automated generation methodologies for AACs into three categories: netlist transformation, Boolean rewriting, and high-level approximation. With the approximation level goes from low to high, the design space increases, leading to an increased difficulty for optimizing the details. As the high-level approximation depends on the library of approximate units, it can be improved by using the optimized results derived from the netlist transformation and Boolean rewriting methods. Moreover, the high-level approximation is essential for generating large AACs or employing AACs to construct approximate applications. Thus, applying the netlist transformation and Boolean rewriting to the library extension and optimization for the high-level approximation are potential to be investigated.

For evaluating the QoRs of AAC-based applications, the high complexity of conventional RTL simulation has motivated various acceleration techniques, including high-level functional simulation, LUT-based simulation, machine learning-based prediction, and functional abstraction. The high-level functional simulations are suitable for most AACs but specific for each AAC, while LUT-based simulations can be generalized to specific operations but are limited by the employed framework and the input bit width of the operation. The machine learning-based predictions avoid repeated time-consuming simulations by mapping the characteristics of the ACCs to the QoRs of the ACC-based applications; however, its prediction accuracy depends on various factors. The functional abstraction methods can effectively estimate the application QoR by describing the errors due to AACs with noise or statistical error metrics; however, the estimation results are generally less accurate and more suitable for comparison and analysis.

The QoR recovery is a complementary step for the QoR evaluation. For the applications with variable parameters, updating or fine-tuning the original parameters based on the algorithm or principle of the application itself is a general approach; however, it is usually time-consuming due to the requirement for executing complex algorithms. Manually adjusting the approximate design based on the statistical analysis of the errors due to AACs is much faster, but the outcomes are limited to specific applications. As the robustness of an application affects the degree of QoR degradation, enhancing the robustness through hardware compensation and application optimization can improve the QoR of an ACC-based application.

It is noteworthy that the statistical error metrics of AACs are critical to the automated generation of application-oriented AACs, which play three roles, as the constraints for AAC generation (Sect. 2), as the inputs of predictors for QoR prediction (Sect. 3.2), and representing AACs for error modeling and evaluation acceleration (Sect. 3.3). Thus, the statistical error metrics are not only key characteristics of an AAC, but also indicators showing the impacts of an AAC on the application. Since the relationship between the error metrics of an AAC and the QoR of the AAC-based application is hard to be mathematically modeled, the machine learning techniques can be useful, which is similar to the QoR predictor for evaluation acceleration. Hence, the predictor based on the statistical error metrics can also be used to guide the generation of application-oriented AACs. Moreover, new statistical error metrics for AACs can be devised considering the error effects on the applications, which would facilitate the QoR evaluation.

References

1. V.K. Chippa, S.T. Chakradhar, K. Roy, A. Raghunathan, Analysis and characterization of inherent application resilience for approximate computing, in *Proceedings of the 50th Annual Design Automation Conference* (2013), pp. 1–9
2. I. Hammad, K. El-Sankary, Impact of approximate multipliers on VGG deep learning network. IEEE Access **6**, 60438–60444 (2018)
3. N. Zhu, W.L. Goh, W. Zhang, K.S. Yeo, Z.H. Kong, Design of low-power high-speed truncation-error-tolerant adder and its application in digital signal processing. IEEE Trans. Very Large Scale Integr. Syst. **18**(8), 1225–1229 (2009)
4. P. Balasubramanian, R. Nayar, D.L. Maskell, N.E. Mastorakis, An approximate adder with a near-normal error distribution: design, error analysis and practical application. IEEE Access **9**, 4518–4530 (2020)
5. K. Bhardwaj, P.S. Mane, J. Henkel, Power-and area-efficient approximate Wallace tree multiplier for error-resilient systems, in *Fifteenth International Symposium on Quality Electronic Design* (IEEE, Piscataway, 2014), pp. 263–269
6. G. Zervakis, K. Tsoumanis, S. Xydis, D. Soudris, K. Pekmestzi, Design-efficient approximate multiplication circuits through partial product perforation. IEEE Trans. Very Large Scale Integr. Syst **24**(10), 3105–3117 (2016)
7. R. Pilipović, P. Bulić, U. Lotrič, A two-stage operand trimming approximate logarithmic multiplier. IEEE Trans. Circuits Syst. I: Regul. Pap. **68**(6), 2535–2545 (2021)
8. Y. Wu, H. Jiang, Z. Ma, P. Gou, Y. Lu, J. Han, S. Yin, S. Wei, L. Liu, An energy-efficient approximate divider based on logarithmic conversion and piecewise constant approximation. IEEE Trans. Circuits Syst. I: Regul. Pap. **69**(7), 2655–2668 (2022)
9. H. Jiang, F.J.H. Santiago, H. Mo, L. Liu, J. Han, Approximate arithmetic circuits: a survey, characterization, and recent applications. Proc. IEEE **108**(12), 2108–2135 (2020)
10. Y. Wu, C. Chen, W. Xiao, X. Wang, C. Wen, J. Han, X. Yin, W. Qian, C. Zhuo, A survey on approximate multiplier designs for energy efficiency: from algorithms to circuits (2023). arXiv preprint arXiv:2301.12181
11. S. Ullah, S.S. Sahoo, N. Ahmed, D. Chaudhury, A. Kumar, Appaxo: designing application-specific approximate operators for FPGA-based embedded systems. ACM Trans. Embed. Comput. Syst. **21**(3), 1–31 (2022)
12. R. Hrbacek, V. Mrazek, Z. Vasicek, Automatic design of approximate circuits by means of multi-objective evolutionary algorithms, in *2016 International Conference on Design and Technology of Integrated Systems in Nanoscale Era (DTIS)* (IEEE, Piscataway, 2016), pp. 1–6

13. L.B. Soares, M.M.A. da Rosa, C.M. Diniz, E.A.C. da Costa, S. Bampi, Design methodology to explore hybrid approximate adders for energy-efficient image and video processing accelerators. IEEE Trans. Circuits Syst. I: Regul. Pap. **66**(6), 2137–2150 (2019)
14. R. Ranjan, S. Ullah, S.S. Sahoo, A. Kumar, Syfaxo-gen: synthesizing FPGA-based approximate operators with generative networks, in *Proceedings of the 28th Asia and South Pacific Design Automation Conference* (2023), pp. 402–409
15. J. Castro-Godínez, J. Mateus-Vargas, M. Shafique, J. Henkel, Axhls: design space exploration and high-level synthesis of approximate accelerators using approximate functional units and analytical models, in *2020 IEEE/ACM International Conference On Computer Aided Design (ICCAD)* (IEEE, Piscataway, 2020), pp. 1–9
16. V. Mrazek, S.S. Sarwar, L. Sekanina, Z. Vasicek, K. Roy, Design of power-efficient approximate multipliers for approximate artificial neural networks, in *2016 IEEE/ACM International Conference on Computer-Aided Design (ICCAD)* (ACM, Berlin, 2016), pp. 1–7
17. M.S. Ansari, V. Mrazek, B.F. Cockburn, L. Sekanina, Z. Vasicek, J. Han, Improving the accuracy and hardware efficiency of neural networks using approximate multipliers. IEEE Trans. Very Large Scale Integr. Syst. **28**(2), 317–328 (2019)
18. I. Scarabottolo, G. Ansaloni, G.A. Constantinides, L. Pozzi, S. Reda, Approximate logic synthesis: a survey. Proc. IEEE **108**(12), 2195–2213 (2020)
19. G. Armeniakos, G. Zervakis, D. Soudris, J. Henkel, Hardware approximate techniques for deep neural network accelerators: a survey. ACM Comput. Surv. **55**(4), 1–36 (2022)
20. G. Zervakis, H. Amrouch, J. Henkel, Design automation of approximate circuits with runtime reconfigurable accuracy. IEEE Access **8**, 53522–53538 (2020)
21. V. Mrazek, L. Sekanina, Z. Vasicek, Libraries of approximate circuits: automated design and application in CNN accelerators. IEEE J. Emer. Sel. Top. Circuits Syst. **10**(4), 406–418 (2020)
22. W. Liu, L. Qian, C. Wang, H. Jiang, J. Han, F. Lombardi, Design of approximate radix-4 booth multipliers for error-tolerant computing. IEEE Trans. Comput. **66**(8), 1435–1441 (2017)
23. D. Esposito, A.G.M. Strollo, E. Napoli, D. De Caro, N. Petra, Approximate multipliers based on new approximate compressors. IEEE Trans. Circuits Syst. I: Regul. Pap. **65**(12), 4169–4182 (2018)
24. M. Nagel, R.A. Amjad, M. Van Baalen, C. Louizos, T. Blankevoort, Up or down? Adaptive rounding for post-training quantization, in *International Conference on Machine Learning* (PMLR, 2020), pp. 7197–7206
25. M. Masadeh, O. Hasan, S. Tahar, Machine learning-based self-compensating approximate computing, in *2020 IEEE International Systems Conference (SysCon)* (IEEE, Piscataway, 2020), pp. 1–6
26. S. Su, Y. Wu, W. Qian, Efficient batch statistical error estimation for iterative multi-level approximate logic synthesis, in *2018 55th ACM/ESDA/IEEE Design Automation Conference (DAC)* (IEEE, Piscataway, 2018), pp. 1–6
27. S. Lee, L.K. John, A. Gerstlauer, High-level synthesis of approximate hardware under joint precision and voltage scaling, in *Design, Automation & Test in Europe Conference & Exhibition (DATE), 2017* (IEEE, Piscataway, 2017), pp. 187–192
28. G. Zervakis, S. Xydis, D. Soudris, K. Pekmestzi, Multi-level approximate accelerator synthesis under voltage island constraints. IEEE Trans. Circuits Syst. II: Express Briefs **66**(4), 607–611 (2018)
29. K. Deb, A. Pratap, S. Agarwal, T. Meyarivan, A fast and elitist multiobjective genetic algorithm: NSGA-II. IEEE Trans. Evol. Comput. **6**(2), 182–197 (2002)
30. V. Mrazek, R. Hrbacek, Z. Vasicek, L. Sekanina, Evoapprox8b: library of approximate adders and multipliers for circuit design and benchmarking of approximation methods, in *Design, Automation & Test in Europe Conference & Exhibition (DATE), 2017* (IEEE, Piscataway, 2017), pp. 258–261
31. J. Schlachter, V. Camus, K.V. Palem, C. Enz, Design and applications of approximate circuits by gate-level pruning. IEEE Trans. Very Large Scale Integr. Syst. **25**(5), 1694–1702 (2017)

32. K. Balaskas, F. Klemme, G. Zervakis, K. Siozios, H. Amrouch, J. Henkel, Variability-aware approximate circuit synthesis via genetic optimization. IEEE Trans. Circuits Syst. I: Regul. Pap. **69**(10), 4141–4153 (2022)
33. J. Castro-Godinez, H. Barrantes-Garcia, M. Shafique, J. Henkel, Axls: a framework for approximate logic synthesis based on netlist transformations. IEEE Trans. Circuits Syst. II: Express Briefs **68**(8), 2845–2849 (2021)
34. Y. Wu, W. Qian, An efficient method for multi-level approximate logic synthesis under error rate constraint, in *2016 53nd ACM/EDAC/IEEE Design Automation Conference (DAC)* (IEEE, Piscataway, 2016), pp. 1–6
35. S. Hashemi, H. Tann, S. Reda, Blasys: approximate logic synthesis using Boolean matrix factorization, in *2018 55th ACM/ESDA/IEEE Design Automation Conference (DAC)* (IEEE, Piscataway, 2018), pp. 1–6
36. S. Hashemi, S. Reda, Generalized matrix factorization techniques for approximate logic synthesis, in *2019 Design, Automation & Test in Europe Conference & Exhibition (DATE)* (IEEE, Piscataway, 2019)
37. J. Ma, S. Hashemi, S. Reda, Approximate logic synthesis using boolean matrix factorization. IEEE Trans. Comput.-Aided Design Integr. Circuits Syst. **41**(1), 15–28 (2021)
38. B.C. Schafer, Z. Wang, High-level synthesis design space exploration: past, present, and future. IEEE Trans. Comput.-Aided Design Integr. Circuits Syst. **39**(10), 2628–2639 (2019)
39. S. Rehman, W. El-Harouni, M. Shafique, A. Kumar, J. Henkel, J. Henkel, Architectural-space exploration of approximate multipliers, in *2016 IEEE/ACM International Conference on Computer-Aided Design (ICCAD)* (IEEE, Piscataway, 2016), pp. 1–8
40. V. Mrazek, M.A. Hanif, Z. Vasicek, L. Sekanina, M. Shafique, autoax: an automatic design space exploration and circuit building methodology utilizing libraries of approximate components, in *2019 56th ACM/IEEE Design Automation Conference (DAC)* (IEEE, Piscataway, 2019), pp. 1–6
41. V. Mrazek, Z. Vasícek, L. Sekanina, M.A. Hanif, M. Shafique, Alwann: automatic layer-wise approximation of deep neural network accelerators without retraining, in *2019 IEEE/ACM International Conference on Computer-Aided Design (ICCAD)* (IEEE, Piscataway, 2019), pp. 1–8
42. J. Castro-Godínez, S. Esser, M. Shafique, S. Pagani, J. Henkel, Compiler-driven error analysis for designing approximate accelerators, in *2018 Design, Automation & Test in Europe Conference & Exhibition (DATE)* (IEEE, Piscataway, 2018), pp. 1027–1032
43. Z. Vasicek, V. Mrazek, L. Sekanina, Automated circuit approximation method driven by data distribution, in *2019 Design, Automation & Test in Europe Conference & Exhibition (DATE)* (IEEE, Piscataway, 2019), pp. 96–101
44. N. Shah, L.I.G. Olascoaga, W. Meert, M. Verhelst, Problp: a framework for low-precision probabilistic inference, in *Proceedings of the 56th Annual Design Automation Conference 2019* (2019), pp. 1–6
45. S. Ullah, S.S. Murthy, A. Kumar, Smapproxlib: library of FPGA-based approximate multipliers, in *2018 55th ACM/ESDA/IEEE Design Automation Conference (DAC)* (IEEE, Piscataway, 2018), pp. 1–6
46. S. Ullah, H. Schmidl, S.S. Sahoo, S. Rehman, A. Kumar, Area-optimized accurate and approximate softcore signed multiplier architectures. IEEE Trans. Comput. **70**(3), 384–392 (2020)
47. M. Shafique, W. Ahmad, R. Hafiz, J. Henkel, A low latency generic accuracy configurable adder, in *2015 52nd ACM/EDAC/IEEE Design Automation Conference (DAC)* (IEEE, Piscataway, 2015), pp. 1–6
48. Y. Fan, X. Wu, J. Dong, Z. Qi, Axdnn: towards the cross-layer design of approximate DNNs, in *Proceedings of the 24th Asia and South Pacific Design Automation Conference* (2019), pp. 317–322
49. Y. Arbeletche, G. Paim, B. Abreu, S. Almeida, E. Costa, P. Flores, S. Bampi, Maxpy: a framework for bridging approximate computing circuits to its applications. IEEE Trans. Circuits Syst. II: Express Briefs (2023)

50. F. Vaverka, V. Mrazek, Z. Vasicek, L. Sekanina, Tfapprox: towards a fast emulation of DNN approximate hardware accelerators on GPU, in *2020 Design, Automation & Test in Europe Conference & Exhibition (DATE)* (IEEE, Piscataway, 2020), pp. 294–297
51. K. He, X. Zhang, S. Ren, J. Sun, Deep residual learning for image recognition, in *Proceedings of the IEEE Conference on Computer Vision and Pattern Recognition* (2016), pp. 770–778
52. C. De la Parra, A. Guntoro, A. Kumar, Proxsim: GPU-based simulation framework for cross-layer approximate DNN optimization, in *2020 Design, Automation & Test in Europe Conference & Exhibition (DATE)* (IEEE, Piscataway, 2020), pp. 1193–1198
53. D. Danopoulos, G. Zervakis, K. Siozios, D. Soudris, J. Henkel, Adapt: fast emulation of approximate DNN accelerators in pytorch. IEEE Trans. Comput.-Aided Design Integr. Circuits Syst. (2022)
54. J. Gong, H. Saadat, H. Gamaarachchi, H. Javaid, X.S. Hu, S. Parameswaran, Approxtrain: fast simulation of approximate multipliers for DNN training and inference (2022). arXiv preprint arXiv:2209.04161
55. Y. Zhao, C. Liu, Z. Du, Q. Guo, X. Hu, Y. Zhuang, Z. Zhang, X. Song, W. Li, X. Zhang, et al., Cambricon-q: a hybrid architecture for efficient training, in *2021 ACM/IEEE 48th Annual International Symposium on Computer Architecture (ISCA)* (IEEE, Piscataway, 2021), pp. 706–719
56. X. He, K. Zhao, X. Chu, Automl: a survey of the state-of-the-art. Knowl.-Based Syst. **212**, 106622 (2021)
57. W. Wen, H. Liu, Y. Chen, H. Li, G. Bender, P.-J. Kindermans, Neural predictor for neural architecture search, in *European Conference on Computer Vision* (Springer, Berlin, 2020), pp. 660–676
58. B. Moons, P. Noorzad, A. Skliar, G. Mariani, D. Mehta, C. Lott, T. Blankevoort, Distilling optimal neural networks: rapid search in diverse spaces, in *Proceedings of the IEEE/CVF International Conference on Computer Vision* (2021), pp. 12229–12238
59. M. Pinos, V. Mrazek, L. Sekanina, Evolutionary approximation and neural architecture search. Genet. Program. Evolvable Mach. **23**(3), 351–374 (2022)
60. S. Venkataramani, A. Ranjan, K. Roy, A. Raghunathan, Axnn: energy-efficient neuromorphic systems using approximate computing, in *2014 IEEE/ACM International Symposium on Low Power Electronics and Design (ISLPED)* (IEEE, Piscataway, 2014), pp. 27–32
61. Q. Zhang, T. Wang, Y. Tian, F. Yuan, Q. Xu, Approxann: an approximate computing framework for artificial neural network, in *2015 Design, Automation & Test in Europe Conference & Exhibition (DATE)* (IEEE, Piscataway, 2015), pp. 701–706
62. Z.-G. Tasoulas, G. Zervakis, I. Anagnostopoulos, H. Amrouch, J. Henkel, Weight-oriented approximation for energy-efficient neural network inference accelerators. IEEE Trans. Circuits Syst. I: Regul. Pap. **67**(12), 4670–4683 (2020)
63. M.A. Hanif, R. Hafiz, M. Shafique, Error resilience analysis for systematically employing approximate computing in convolutional neural networks, in *2018 Design, Automation & Test in Europe Conference & Exhibition (DATE)* (IEEE, Piscataway, 2018), pp. 913–916
64. K. Simonyan, A. Zisserman, Very deep convolutional networks for large-scale image recognition (2014). arXiv preprint arXiv:1409.1556
65. C. De la Parra, A. Guntoro, A. Kumar, Efficient accuracy recovery in approximate neural networks by systematic error modelling, in *Proceedings of the 26th Asia and South Pacific Design Automation Conference* (2021), pp. 365–371
66. C. Li, W. Luo, S.S. Sapatnekar, J. Hu, Joint precision optimization and high level synthesis for approximate computing, in *Proceedings of the 52nd Annual Design Automation Conference* (2015), pp. 1–6
67. G. Zervakis, O. Spantidi, I. Anagnostopoulos, H. Amrouch, J. Henkel, Control variate approximation for DNN accelerators, in *2021 58th ACM/IEEE Design Automation Conference (DAC)* (IEEE, Piscataway, 2021), pp. 481–486
68. M.S. Kim, A.A. Del Barrio, H. Kim, N. Bagherzadeh, The effects of approximate multiplication on convolutional neural networks. IEEE Trans. Emer. Top. Comput. **10**(2), 904–916 (2021)

69. S.S. Sarwar, S. Venkataramani, A. Ankit, A. Raghunathan, K. Roy, Energy-efficient neural computing with approximate multipliers. ACM J. Emer. Technol. Comput. Syst. **14**(2), 1–23 (2018)
70. C. De la Parra, X. Wu, A. Guntoro, A. Kumar, Knowledge distillation and gradient estimation for active error compensation in approximate neural networks, in *2021 Design, Automation & Test in Europe Conference & Exhibition (DATE)* (IEEE, Piscataway, 2021), pp. 679–684
71. G. Hinton, O. Vinyals, J. Dean, et al., Distilling the knowledge in a neural network (2015). arXiv preprint arXiv:1503.02531
72. C. Guo, L. Zhang, X. Zhou, W. Qian, C. Zhuo, A reconfigurable approximate multiplier for quantized CNN applications, in *2020 25th Asia and South Pacific Design Automation Conference (ASP-DAC)* (IEEE, Piscataway, 2020), pp. 235–240
73. M.A. Hanif, F. Khalid, M. Shafique, CANN: curable approximations for high-performance deep neural network accelerators, in *Proceedings of the 56th Annual Design Automation Conference 2019* (2019), pp. 1–6
74. X. He, L. Ke, W. Lu, G. Yan, X. Zhang, Axtrain: hardware-oriented neural network training for approximate inference, in *Proceedings of the International Symposium on Low Power Electronics and Design* (2018), pp. 1–6

Automatic Approximation of Computer Systems Through Multi-objective Optimization

Mario Barbareschi, Salvatore Barone, Alberto Bosio, and Marcello Traiola

1 Introduction

In this chapter, we foster an application-independent, unified methodology able to automatically explore the impact of different approximation techniques on a given application while resorting to the Approximate Computing (AxC) design paradigm and Multi-objective Optimization Problem (MOP)-based Design-Space Exploration (DSE). We also devote particular relevance to all the phases and steps of the proposed methodology which can be automated. Our methodology is tailored neither to a specific application nor to an approximation technique, it does not require the designer to specify which part(s) of the application should be approximated and how, and it only requires the definition of the acceptable output degradation from the user. Moreover, it addresses the design problem as a MOP, which allows optimizing different figures of metrics, e.g., error and hardware requirements, at the same time, providing the designer with a set of equally good

Authors are listed in alphabetical order.

M. Barbareschi · S. Barone
Department of Electrical Engineering and Information Technologies, University of Naples Federico II, Naples, Italy
e-mail: Mario.Barbareschi@unina.it; Salvatore.Barone@unina.it

A. Bosio (✉)
University of Lyon, ECL, INSA Lyon, CNRS, UCBL, CPE Lyon, INL, UMR5270, Villeurbanne, France
e-mail: alberto.bosio@ec-lyon.fr

M. Traiola
University of Rennes, CNRS, Inria, IRISA - UMR6074, Rennes, France
e-mail: marcello.traiola@inria.fr

W. Liu et al. (eds.), *Design and Applications of Emerging Computer Systems*,
https://doi.org/10.1007/978-3-031-42478-6_15

Pareto-optimal solutions, leaving the designer free to choose the one that, according to his experience, best suits the context or the requirements of the application considered. We also discuss the application of the presented methodology to relevant applications in the scope of the AxC paradigm.

This chapter is organized as follows: Sect. 2 provides the reader with a brief introduction concerning the AxC design paradigm, including issues and challenges to be addressed to exploit the AxC full potential. Section 3 discusses the automatic design methodology exploiting AxC and MOP-based design methodology, including the main steps of the method that we propose, and how the method helps in addressing the challenges that the AxC paradigm poses to the designer. Section 4 applies the methodology to the design of combinational logic circuits, i.e., those that typically constitute building blocks for larger, more complex designs. Section 5 presents the design of hardware accelerators for image processing, while Sect. 6 discusses the approximation of two of the most common classification models in the machine-learning domain, namely, Deep Neural Networks (DNNs) and Decision Tree based Multiple Classifier Systems (DT MCSs). These applications are even more challenging, since hardware accelerators are utterly resource intensive, and reducing the amount of induced error is very critical, because machine-learning systems process a huge amount of data.

2 The Approximate Computing Design Paradigm and Its Application

This section provides the reader with a brief introduction concerning the AxC design paradigm, including issues and challenges to be addressed to exploit the AxC full potential.

2.1 Overview

The scientific literature demonstrated that inexact computation can be selectively exploited to enhance computing system performance, defining the AxC paradigm [84]. It is based on the intuitive observation that, while performing exact computation, or maintaining peak-level service performance, requires a high number of resources, selective approximation or occasional violation of the specification can provide quite interesting gains in efficiency. In other words, the AxC paradigm exploits the gap between the level of accuracy required by the application or the end-users, and that provided by the computing system—with the former being often far lower than the latter—for achieving diverse optimizations. Thus, this design paradigm has the potential to benefit a wide range of applications, including data analytic, scientific computing, multimedia and signal processing, machine learning, etc. [54].

Anyway, exploiting AxC requires coping with (i) the characterization of parts of the considered software or hardware component, identifying those that are suitable to be approximate; (ii) the approach to introduce actual approximation; (iii) the selection of appropriate error metrics, which generally depend on the particular application; (iv) the actual error-assessment procedure, to guarantee output quality constraints are met [27]; and finally, (v) the DSE, to select the best approximate configurations among those generated by a certain approximation technique. As for the first two of the aforementioned issues, pinpointing approximable code or data portions may require the designer to have profound insights into the application. Error injection is quite common as an approach to find the data or operation that can be approximated with little impact on the quality of result [66, 68]. Once portions, or data, to be approximate have been identified, being able to introduce approximation is not a straightforward matter and may require coping with several technical challenges [18]. Indeed, a naive approach—e.g., using uniform approximation—is unlikely to be efficient. Moreover, no method can be universally applied to all approximable applications. Therefore, the approximation strategy needs to be determined on a per-application basis.

One of the most commonly adopted techniques to introduce approximation is *precision scaling*, also referred to as *bit-width reduction*. Essentially, it reduces the number of bits used for representing input data and intermediate operands [75, 87]. Precision scaling basically combines close values into a single value, which paves the way for the *memoization* technique [43]. Memoization is based on storing the results of functions for later reuse with similar inputs. Finally, another quite widespread approach is *loop perforation* that is based on skipping some iterations of a loop to reduce computational overhead. It has proven to be effective when applied to several computational patterns, such as the Monte Carlo simulation, iterative refinement, and search space enumeration [72].

Concerning the approximation of circuits, the scientific literature distinguishes between timing and functional techniques [67]. The former consists of forcing the circuit to operate on reduced voltage or higher frequency than nominal ones, while the latter includes altering the logic being implemented. Technology-independent functional approximation currently represents the most popular technique to introduce approximations within hardware components, and many libraries consisting of thousands of elementary approximate circuits have been proposed in the scientific literature, supplying hundreds of implementations of even a single arithmetic operation [42, 56].

As for error assessment, it typically requires the simulation of both exact and approximate applications. Nevertheless, Bayesian inference [76] or machine-learning-based approaches [57] have been proposed to reduce computation-demanding simulations. However, these approaches can provide only an estimate of the error, meaning that they do not offer any guarantees on the maximum value that the error can yield. Hence, various analytical and formal approaches have been proposed and applied for the exact quantification of the error. They do not make any assumption on the structure of the approximate circuits, and, albeit potentially time-consuming, they can be applied to determine almost every error metric [78].

Finally, concerning DSE, initial approaches either combine multiple design objectives in a single-objective optimization problem or optimize a single parameter while keeping the others fixed. Therefore, the resulting solutions are centered around a few dominant design alternatives [30]. Recently published studies address the approximate design problem by using MOP to search for Pareto-optimal approximate circuit implementations [13]. Unfortunately, such approaches did not focus on complex systems, rather on arithmetic components, such as adders and multipliers, since they are building blocks for more complex designs. Conversely, in the following section we foster an application-independent, unified methodology able to automatically explore the impact of different approximation techniques on a given application while resorting to the AxC design paradigm and MOP-based DSE.

3 Automatic Application-Driven, Multi-Objective Approximate Design Methodology

This section addresses the automatic approximation of computer systems through multi-objective optimization. We describe the main steps of the methodology and how the method helps in addressing the challenges that the AxC paradigm poses to the designer. As discussed in Sect. 2.1, there are several challenges to be addressed to effectively exploit the AxC design paradigm. Although diverse research articles in the scientific literature proposed well-founded approaches addressing the abovementioned challenges, there are still plenty of open ones holding AxC back from wider employment. In particular, one of the key points is the lack of a general and automatic DSE methodology. Indeed, existing AxC design tools consider specific transformations and domains, and they are not fully automatic, providing only a guided approach for approximation. Therefore, in the following we discuss a generic, MOP-based, and fully automatic methodology to design hardware accelerators for error-resilient applications.

In particular, we break down our methodology into different phases: (i) how to identify which part of the application is amenable for approximation and (ii) a suitable approximation technique, (iii) how MOP-based DSE can be defined, and, finally, (iv) how to pinpoint suitable fitness functions for error assessment and performance estimation to effectively drive the DSE toward Pareto-optimal approximate configurations.

3.1 Identifying Approximable Portions and Suitable Approximation Techniques

The first challenge to be addressed when dealing with the AxC is identifying error-resilient—i.e., approximable—data or portions of a given algorithm/application and, consequently, a suitable approximation technique. Although it seems trivial, this step of the methodology is rather quite crucial. Indeed, as we discuss in the following, an improper design choice concerning either parts to be approximate, or the technique to be adopted, impacts all the subsequent phases.

Despite many of the methods from the scientific literature claiming to be generic, they actually require the designer to have an in-depth knowledge of the target application to choose a suitable approximation technique. Unfortunately, this may not be trivial or even not possible: there are plenty of applications for which, albeit conceptually simple, having profound understanding is very difficult indeed, e.g., DNNs and DT MCSs. Furthermore, once portions to be approximated have been correctly pinpointed, and a suitable approximate technique selected, the actual approximation has to be performed. However, manual introduction of approximation within applications is definitely inconvenient, due to their complexity or due to the amount of data/operations amenable for approximation.

Conversely, the methodology we are bound to discuss requires only minimal knowledge of the target application and provides the designer with a systematic approach to automatically generate approximate variants. An approximation variant is an implementation of a given application where approximable parts are implemented by approximate components. In general, the goal is to automatically generate approximate variants while having control on the error. Therefore, it is needed to collect information on the operations suitable for approximation. The gathering process can be automated by analyzing the Abstract Syntax Tree (AST) of the given an algorithm implementation. Then, AST manipulation using *mutators* [16] allows automatic generation of approximate variants.

Mutators are defined as a set of search-and-modify rules on the AST; the rule definition is generally application-independent and does not require the designer to know the algorithm or its specific implementation. Furthermore, mutators do not depend on the specific approximation technique being adopted, and they effectively allow introducing a suitable tuning knobs for approximation, replacing exact operations within the AST using their approximate counterparts. Consider, for instance, an approximate multiplier designed using the precision-scaling technique: let the Number of Approximate Bit (NAB) be the parameter for such approximation and suppose the approximate operation truncates the least *nab* significant bits of operands, with *nab* being configurable. Setting a value for the *nab* parameters tunes the approximation degree, resulting in an *approximate configuration* of the algorithm. Mutators can be exploited, for instance, to implement the *inexact-component* technique. Indeed, exact multiplications can be automatically replaced using a mutator that allows selecting which implementation to be adopted among those provided by a given library, e.g., the EvoApproxLib library [56]. In this case, the configuration parameter would allow selection of an optimal multiplier implementation regarding a given error metric, required silicon area, and power dissipation.

3.2 *Optimization and Design-Space Exploration*

The number of approximate variants and, consequently, the number of approximate configurations grow quickly with the number of parts suitable for approximation. Consider, for instance, an algorithm implementation with n approximable operations, each allowing k different degrees of approximation: $\binom{n}{j}$ different approximate variants can be defined by simultaneously approximating j operations, and k^j different approximate configurations can be defined for each of the variants. Therefore, the total number of approximate configurations is $\sum_{i=1}^{n} k^i \times \binom{n}{i}$. At this point, the main challenge is to find values for the approximation parameters leading to the Pareto-optimal trade-offs between performance gains and accuracy losses.

In fact, each one of the introduced approximation parameters impacts both accuracy and performance. Hence, the automated design of approximate applications is inherently a MOP in which variants satisfying user-defined constraints and showing the desired trade-off between the quality and other performance-related parameters are sought within all possible implementations [79]. As we mentioned, most of the approximation approaches either combine multiple design objectives in a single-objective optimization problem or optimize a single parameter while keeping the others fixed. Therefore, the resulting solutions are centered around a few dominant design alternatives [30]. We propose to find Pareto-optimal configurations for approximation parameters through an automatic MOP-based DSE, which is not only tailored to the target application, yet it considers the latter target as a whole. Indeed, recent works addressing the circuit design problem as MOP, e.g., [71], did not focus on complex systems, rather on arithmetic components, such as adders and multipliers, since they are building blocks for more complex designs.

In the following, we provide the reader with the required knowledge concerning MOP.

3.2.1 Multi-objective Optimization Problems

Basically, a MOP is an optimization problem involving multiple objectives. More formally, given the set of *fitness functions* (1), or *objective functions*, and the set of *constraints* (2), a MOP can be formulated as in Eq. (3). While the functions of the former set assume values in $\mathbb{R}$, or its subset, the constraint functions assume either the value *1* or *0* to indicate that the constraint is or is not met, respectively.

$$\Gamma = \{\gamma_i : X \to \mathbb{R},\ i = 1 \cdots k\},\ X \subseteq \mathbb{R}^n \tag{1}$$

$$\Psi = \{\psi_j : A \to \{0, 1\},\ j = 1 \cdots l\},\ A \subseteq \mathbb{R}^n \tag{2}$$

$$\begin{aligned} &min/max\ \{\gamma \in \Gamma\} \\ &s.t.\ \psi \in \Psi \end{aligned} \tag{3}$$

The $X \subseteq \mathbb{R}^n$, which is defined by constraints (2), is the set of feasible solutions to the MOP, or the *feasible set*. An element $x \in X$ is a *feasible solution*, while its image through Γ, i.e., $z = \{\gamma(x), \gamma \in \Gamma\}$, is called the *outcome* of x.

Let us consider two solutions, $x, y \in X : x \neq y$, x is said to *dominate* y i.f.f. $x \prec y \iff \gamma_i(x) \leq \gamma_i(y) \, \forall i \in [1, k] \ \wedge \ \exists j \in [1, k] : \gamma_j(x) < \gamma_j(y)$ holds, i.e., x shows better or equally good objective values than y in all objectives and at least better in one objective. If a solution is not dominated by any others, it is called a *Pareto-optimal* solution.

Due to the rapid growth of the size of the solution space as the number of decision variables, fitness functions, and constraints increases, using exact solving algorithms for MOPs turns out to be very computation-intensive and time-consuming. Consequently, a variety of (meta-)heuristics aiming at producing an approximation of the Pareto-front have been proposed in the scientific literature: Genetic Algorithm (GA) [51], Simulated Annealing (SA) [44], and particle swarm [29] are just some of the most commonly adopted ones, and among the heuristics belonging to the mentioned families, the Non-dominated Sorting Genetic Algorithm-II (NSGA-II) [31] and Archived Multi-Objective Simulated Annealing (AMOSA) [7] are the most common ones.

3.2.2 MOP Modeling: Identifying Decision Variables and Suitable Fitness Functions

In the context of AxC, modeling a specific optimization problem is not trivial, and no general rules exist. Anyway, considering the technique used to generate the approximate variants definitely helps in at least identifying the decision variables of the problem. Indeed, the latter find natural correspondence in the configuration parameters introduced to govern the degree of approximation.

As already discussed, the identification of suitable decision variables is only the first step to complete in order to define a MOP-based DSE. Indeed, we need to also define fitness function driving the DSE. In particular, we must assess the error entailed by the approximations. Hence, we need to pinpoint an appropriate error metric to define a suitable error fitness function to minimize. Unfortunately, when using the AxC paradigm, defining an appropriate error metric is of major concern, and it is usually not a trivial task. Therefore, the error metric is usually selected case by case. Anyway, for some applications, the choice of error metric is obvious, if not outright forced, by the application domain. The classification accuracy loss, for instance, is a meaningful error metric for either DNNs or DT MCSs applications, while the Peak Signal-to-Noise Ratio (PSNR) and the Structural SIMilarity (SSIM) are common error metrics in the image-processing field [82].

For what pertains to performances in terms of either computational time, power consumption, or hardware overhead, to accurately consider the resource savings in the DSE, we should measure area, power consumption, and maximum operating frequency of the explored approximate configurations. Unfortunately, this would require the synthesis and simulation of each approximate configuration explored during the DSE, which is definitely a time-consuming process. Therefore, we propose a model-based estimation of performance increases to drive the DSE. This has to consider the impact of the selected approximation technique on the final hardware implementation, to provide a faithful estimation, albeit not accurate. Defining such a model is not straightforward. Although removing some parts of an arithmetic circuit, for instance, undoubtedly leads to specific gains in terms of area/energy, model-based hardware requirement estimation becomes trickier when the approximation has to be tailored to the application and performance to be evaluated in the application's context, since they depend on the specific implementation.

3.3 Summary

For the reader's convenience, Fig. 1 summarizes the proposed methodology. Starting from the model of the application to be approximated, an automatic approximation engine generates configurable approximate variants. Variants may either be generated from scratch starting from the model or result from alterations of the model itself. Furthermore, for each approximate portion, approximate variants allow to selectively adjust the degree of introduced approximation, through the use of convenient configuration parameters. The value for such parameters leading to optimal trade-offs between quality of results and performance gains is searched through a MOP-based DSE. The latter is performed using a suitable heuristic—e.g., either NSGA-II or AMOSA—minimizing the error entailed by the approximation and, at the same time, a figure of merit that correlates to performances, e.g., computational time, power consumption, or hardware overhead. At the end of the DSE, resulting non-dominated approximate configurations are adopted to suitably shape a configurable implementation of the target application.

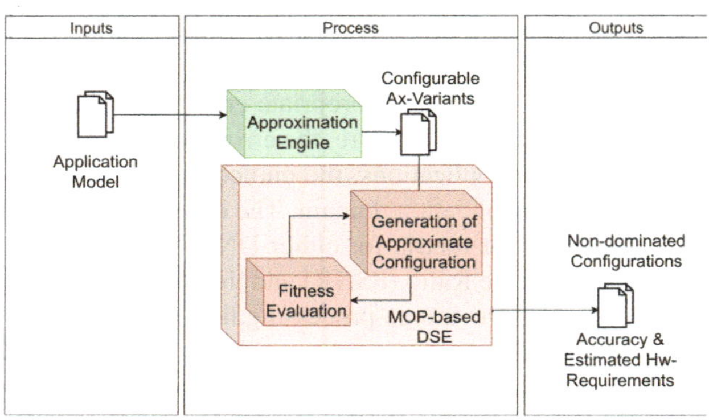

Fig. 1 Workflow of the discussed automatic design methodology

In the next sections we apply the proposed methodology to several applications from different domains, including logic, image processing, and machine-learning applications.

4 Automatic Approximation of Combinational Circuits

In this section we discuss the design of combinational logic circuits that is particularly relevant since combinational circuits typically constitute the building blocks for larger, more complex designs. We resort to the catalog-based And-Inverter Graph (AIG)-rewriting technique from [12] and to the *pyALS* framework [15] that implements it. Anyway, the method we discuss is not constrained to a specific approximation technique.

Figure 2 sketches the overall flow of the *pyALS* framework [15]: starting from HDL source code describing the design under study, the approach goes through a first phase of circuit analysis and synthesis, then, it deals with the DSE by defining a MOP to obtain Pareto-optimal configurations in terms of error and hardware requirements.

Since the methodology is based on non-trivial local rewriting of AIGs and MOP, in the following we provide the reader with essentials concerning these building blocks.

4.1 Approximate Variant Generation

The method from [12] exploits the AIG representation of digital circuits to introduce approximation. An AIG is a structural representation, based on directed acyclic

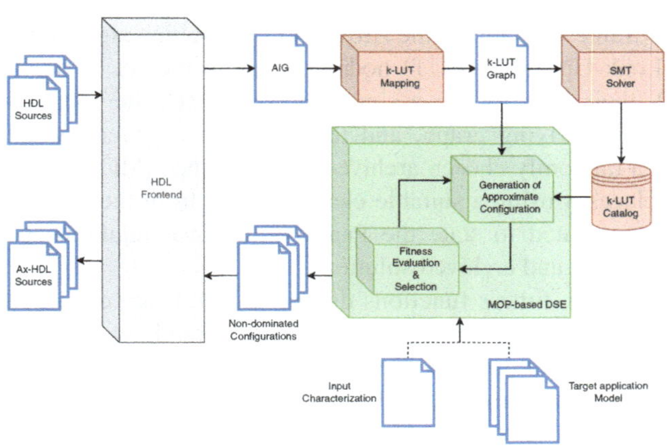

Fig. 2 Overview of the automatic workflow

graphs, for Boolean networks. In an AIG, nodes can be either Primary Input (PI) nodes that have no *fan-in*, meaning that they have no incoming edges and hence, there is no node driving them, or logic-AND nodes, which have two incoming edges. Furthermore, nodes having no *fan-out*, i.e., nodes that do not drive any other node, are called Primary Outputs (POs). For what pertains to edges, they represent physical connections between nodes, and they can be indicated as complemented or not. Conventionally, the polarity of complemented edges is 0. The AIG is non-canonical as a representation, which means that the same Boolean function can be realized by multiple, different AIGs. Briefly, given the AIG of a combinational circuit, the approach from [12] first enumerates k-feasible cuts and then introduces approximation by superseding k-cuts with approximate ones. A k-cut of a node n, called *root*, is a set of at most K nodes of the AIG such that each path from a PI to n passes through at least one node of the set. Approximate k-cuts are generated by exploiting the Exact Synthesis (ES). Being f the Boolean function implemented by a k-cut, ES searches for those k-cuts implementing a function $f'4 \neq f$ that requires less resources w.r.t. f, and yet complies with error constraints. The latter are expressed in terms of the Hamming distance between f and f'. The main challenge is hence to find replacements leading to Pareto-optimal trade-offs between the quality of results and hardware requirements. This requires coping with two major concerns: (i) the number of approximate configurations grows exponentially with the size of the concerned Boolean functions and (ii) preserving the quality of results while pursuing a reduction in hardware requirements is a conflicting design goal.

4.2 Design-Space Exploration

Once the catalog has been built, the main challenge is to find the combination of cut replacements leading to Pareto-optimal trade-offs between error and performance, i.e., to perform a DSE.

The *pyALS* framework adopts the AMOSA [7] searching algorithm to orchestrate the DSE: k-Look-Up Table (LUT) nodes constitute the set of decision variables of the MOP, their indexes are assigned according to the topological ordering defined by the underlying graph, and their domain is given by catalog entries. Starting from a randomly chosen archived solution, the AMOSA selects a random LUTs and replaces it using a suitable element taken from the catalog, then fitness functions are computed to state the Pareto-dominance relationship between the altered configuration and archived solutions.

As we mentioned, fitness functions driving the DSE are error and silicon-area minimization. As far as the silicon area is concerned, we resort to a model-based gain estimation to drive the DSE. In particular, we estimate the silicon-area requirements from the number of AIG nodes, since the relationship between the latter and hardware requirements—in terms of critical path and LUTs or standard cells—has been empirically proven in [52]. We evaluate both the number of nodes and the depth for a given approximate configuration on Functionally-Reduced And-Inverter Graphs (FRAIGs) [53].

For what pertains to error metrics, the one to be used strictly depends on the final target application. In the following, we discuss experimental results while targeting generic combinational logic circuits and arithmetic circuits. In the former case, we adopt the Error Probability (EP) metric, while in the latter case we resort to the Absolute Worst-Case Error (AWCE) metric.

4.3 Experimental Results

To evaluate the proposed methodology, we first considered a subset of the LGSynt91 benchmark [85], including both logic and arithmetic circuits. We also considered arithmetic circuits from [38] for further evaluation. At the end of the DSE, the AMOSA heuristic provided several approximate configurations for each of the considered benchmark circuits. We synthesized resulting Hardware Description Language (HDL) implementations targeting the 45 nm standard-cell library, to measure actual hardware requirements.

Figure 3 plots silicon area against EP for circuits from the LGSynth91 benchmark: the red star denotes the exact circuits, while blue dots denote approximate configurations resulting from our workflow. As the reader can observe, a general decreasing trend for silicon area can be observed while the error increases, and, depending on the specific error resiliency of the circuit being considered, our technique allows achieving significant savings. Furthermore, although we do not report plots for brevity's sake, a similar trend can be observed for power consumption.

Figure 4, instead, plots the silicon area against AWCE for various 8-bit and 16-bit arithmetic circuits, including array-tree multiplier (ATM), Dadda-tree multiplier (DTM), Wallace-tree multiplier (WTM), carry-skip adder (CSkA), ripple-carry adder (RCA) Han-Carlson adder (HCA), and carry-lookahead adder (CLA). Note that the x-axis for Fig. 4 is in semilogarithmic scale. Once again, the red star is the reference—i.e., the exact circuits—while the blue dots denote approximate configurations resulting from our workflow. A general decreasing trend for silicon area can be observed in these results too, denoting the overall approach allows effectively introducing approximation also within arithmetic circuits, resulting in significant savings as the allowed error increases.

5 Approximation of Image-Processing Applications

In this section, we discuss the application of our methodology to the design of hardware accelerators for image processing. Image processing is one of the main fields of application for AxC, since imperceptible reduction of image quality can lead to important computational resources savings [27]. Specifically, we address the

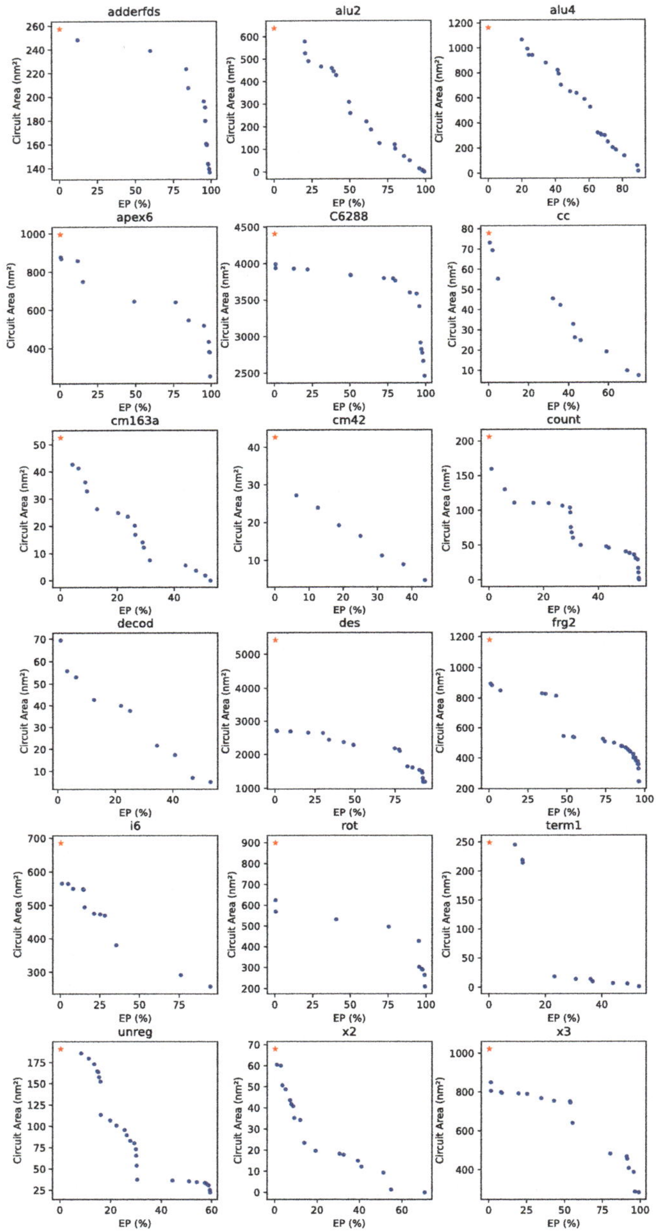

Fig. 3 Experimental results for LGSynt91 benchmark circuits

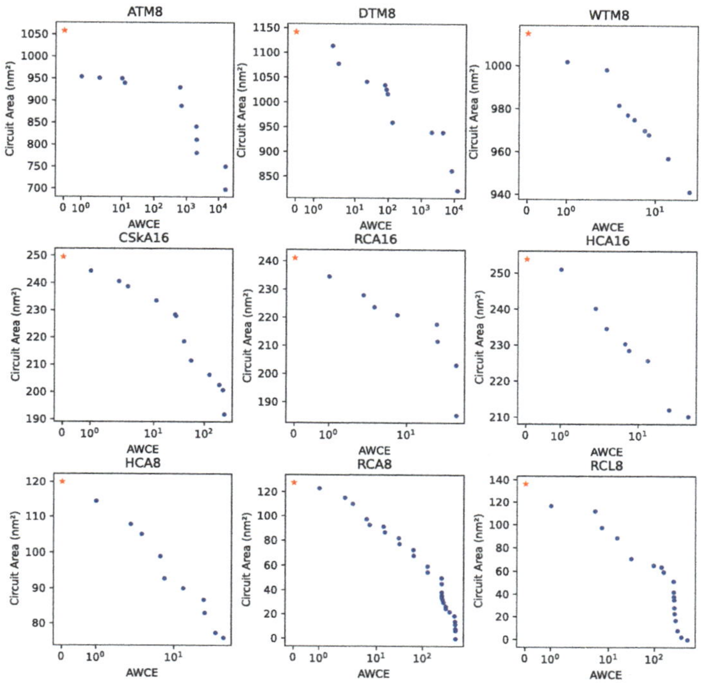

Fig. 4 Experimental results for arithmetic circuits. Kindly note that the x-axis is in semilogarithmic scale

design of an accelerator for Discrete Cosine Transform (DCT), which is the most resource-demanding step of the JPEG, one of the most commonly adopted lossy image and video compression algorithms.

Before discussing the mentioned case studies, we briefly describe the implementation of the methodology from Sect. 3 as a state-of-the-art approximation framework, i.e., we present the Evolutionary-IIDEAA Is a Design Exploration tool for Approximate Algorithm (E-𝕀DE𝔸) framework [16].

5.1 *The E-𝕀DE𝔸 Framework*

Evolutionary-𝕀DE𝔸 (IIDEAA is a design exploration tool for approximate algorithm) is an approximation framework for C/C++ applications. Figure 5 sketches the overall flow of E-𝕀DE𝔸. It consists of two main components, i.e., (i) a source-to-source manipulation tool, *the Clang-Chimera*, and (ii) an evolutionary search engine, *the Bellerophon*. E-𝕀DE𝔸 requires:

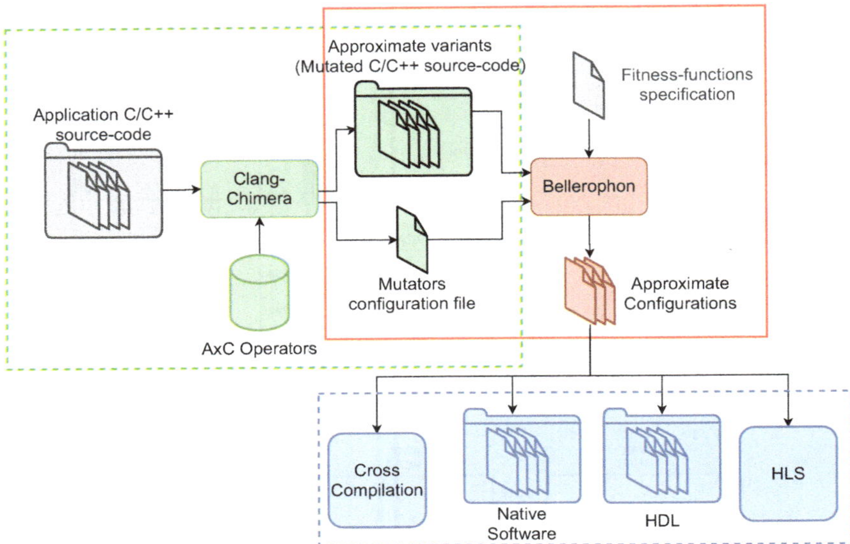

Fig. 5 E-IDEA flow, which includes Clang-Chimera and Bellerophon tool

1. The original application, described as C/C++ code
2. The set of approximate operators, i.e., mutators
3. The fitness functions to select the appropriate approximation outcomes

The green dotted box in Fig. 5 highlights Clang-Chimera flow. Clang-Chimera is a mutation engine for C/C++ code. It is based on the Clang compiler [6], used to rapidly develop source-to-source C/C++ compilers. Clang-Chimera applies the set of mutators (i.e., AxC operators) to the input application code. It exploits Low Level Virtual Machine (LLVM)/Clang facilities—e.g., *ASTMatcher* and *Rewriter*—to apply a given mutator and to make systematic modifications to the code. This process generates a set of mutated files that are the configurable approximate variants of the input application code. Clang-Chimera analyzes and manipulates the input application source code through its AST, which is a tree-based representation of the application code, where each node of the tree denotes a language construct of the analyzed code. A set of AST nodes defines an AST pattern, which corresponds to a specific structure of the code. Altering the AST allows introducing constructs that enable the approximation degree tuning. Clang-Chimera is already provided with a set of mutators implementing common approximation techniques, such as (i) two loop-perforation mutators, namely, LOOP1 and LOOP2; (ii) two precision-scaling mutators for the floating-point arithmetic, namely, Variable Precision Arithmetic (VPA) and FLexible Arithmetic Precision (FLAP); (iii) a precision-scaling mutator for the integer arithmetic, namely, TRUNC; and (iv) a mutator supporting approximate arithmetic operator models of circuits being part of the EvoApproxLib [56] and EvoApproxLib-Lite [60] libraries.

The solid red box in Fig. 5 depicts the Bellerophon flow. The tool analyzes the set of mutated files generated by Clang-Chimera and explores the different possible configurations of mutator tunable parameters, i.e., decision variables for the MOP driving the DSE. Therefore, Bellerophon explores the different approximate variants while "moving" toward the *Pareto front* of the solutions, in terms of the defined fitness functions. Fitness functions might be defined accordingly to the particular exploited AxC technique. In the case of precision-scaling technique, for instance, a feasible fitness function could be based on the NAB, since it translates in less hardware resources. Bellerophon is based on the NSGA-II [31], but, since implementing a full-featured NSGA-II may be cumbersome, it exploits the *ParadisEO* framework [49]. Furthermore, to be evaluated, each individual has to be compiled and executed. To speed up the execution time, the compilation strategy adopted by Bellerophon allows compiling just what it is necessary to retrieve information about approximate variants. Bellerophon uses the just-in-time engine provided by Clang-LLVM: each time the software needs to be altered to test a new variant, Bellerophon does not invoke the system loader; rather it alters the program image which is already loaded into the memory.

5.2 The DCT Case Study

Most of the research work concerning image-processing applications focuses on the JPEG compression, either considering the algorithms as a whole or its individual computational steps. Concerning the design of hardware accelerators, researchers focused on the approximation of DCT accelerators, mainly targeting figures of merit such as circuit complexity, delay, area, and power dissipation. Unfortunately, the effects of the different approximation techniques and relative configurations (i.e., approximation degrees) are only analyzed individually and without a supporting methodology.

In [3], for instance, a framework relying on inexact computing to perform the DCT computation for the JPEG has been proposed. The framework acts on three levels: (i) At the application level, it exploits human insensitivity to high-frequency variation to use a filter and discard high-frequency components. (ii) At the algorithmic level, multiplier-less fast algorithms are employed for the actual DCT computation on integer coefficients. (iii) At the hardware level, rather than using a simple truncation for adder circuits, the authors used Inexact-Adder Cells (IACs) to compute less significant bits instead of the Full-Adder Cells (FACs). Therefore, firstly, the JPEG quantization step is performed only on low-frequency components of an image block; thus the high-frequency filter implementation comes down to simply setting some DCT coefficients to zero. Then, at the algorithmic level, since the DCT is the most effort-demanding step in JPEG, fast DCT algorithms have been used, reducing complexity from $O(N^2)$ to $O(N)$ and requiring only integer additions. Finally, at the hardware level, different families ofIAC are considered to further reduce power consumption.

The framework in [3] mainly aims at assessing the joint impact of those three levels of approximation. However, it presents one rather important shortcoming, i.e., approximation is introduced by manually tuning the individual approximation parameters. Conversely, in this case study, we assess the impact of approximation on the DCT computation by performing a fully automated DSE. Applying our methodology, we start from the DCT algorithm, and we perform an AST analysis to gather information on the operations suitable for approximation. Then, we generate parametric approximate versions which allow the approximation degree to be tuned through approximation parameters. Finally, we build a MOP to find the Pareto-optimal values for the aforementioned approximation parameters, using the NSGA-II to converge toward the Pareto front.

First, we provide the reader with an overview on computing the DCT using fast algorithms in Sect. 5.2.1, as done for all the above-discussed case studies. Then, we discuss the generation of approximate variants in Sect. 5.2.2 and several aspects concerning MOP-based DSE in Sect. 5.2.3, including MOP-modeling and fitness functions to drive the DSE. Finally, in Sect. 5.2.4 we present experimental results.

5.2.1 Toward Approximate DCT

As we already mentioned, the DCT computation is known to have $O(N^2)$ complexity and requires resource-intensive functional units, such as floating-point arithmetic modules. Cosine coefficients required by the DCT can be scaled and rounded such that floating-point operations can be superseded by integer ones: the resulting algorithms are significantly faster, and they find extensive use in practical applications. However, integer multiplication is still complex and resource intensive; thus, many low-complexity multiplier-less algorithms have been proposed [17, 19–21, 28, 63, 64]. They all avoid computing DCT transformed coefficients separately or iteratively; rather they extensively resort to matrix algebra and its properties.

The DCT transform of an input image tile X, which is an 8×8 matrix, is given by $F = C \cdot X \cdot C'$, where C contains the cosine function values at the needed frequencies, and it is referred to as *DCT matrix*. Multiplier-less algorithms first split the C matrix into two sub-matrices, namely, T and D, as reported in Eq. (4). T contains only the values $\{0, \pm\frac{1}{2}, \pm 1, \pm 2\}$ and it is orthogonal—i.e., $T' = T^{-1} \Rightarrow TT' = T'T = I$, where I is the identity matrix—while D is a diagonal matrix consisting of values in the $[-1, 1]$ range, with $\{\frac{1}{2}, \frac{1}{\sqrt{2}}, \frac{1}{\sqrt{8}}\}$ being typical values.

$$F = C \cdot X \cdot C' = D \cdot (T \cdot X \cdot T') \cdot D \tag{4}$$

The multiplication of D in (4) still requires floating-point operations; nevertheless, resorting to properties of diagonal matrices, the integer null-multiplicative part $T \cdot X \cdot T'$ can be isolated from floating-point operations required by D, as reported in Eq. (5), where $\circ$ is the Hadamard product, i.e., an element-wise multiplication.

$$F = T \cdot X \cdot T' \circ (diag(D) \cdot diag(D)') \tag{5}$$

Afterward, floating-point operations can be performed outside the DCT and embedded into the JPEG quantization step, as shown in Eq. (6), where $\hat{Q}$ is the *complete quantization matrix* and the $\oslash$ operator is the Hadamard division, i.e., an element-wise division.

$$\begin{aligned} F_Q &= \lceil F \oslash Q \rfloor = \lceil T \cdot X \cdot T' \circ (diag(D) \cdot diag(D)') \oslash Q \rfloor \\ &= \lceil T \cdot X \cdot T' \circ \hat{Q} \rfloor = \lceil (T \cdot (T \cdot X')') \circ \hat{Q} \rfloor \\ \hat{Q} &= (diag(D) \cdot diag(D)') \oslash Q \end{aligned} \tag{6}$$

Besides allowing the use of integer arithmetic for the calculation of coefficients, Eq. (6) also allows computing the two-dimensional DCT using the one-dimensional DCT transform twice, reducing the computational complexity from quadratic to linear.

5.2.2 Generating of Approximate Variants

Once the addition-based equations for the DCT coefficients are defined, simple implementations for the DCT computation algorithm can be derived. Within those, we introduce further approximation by replacing exact sums by configurable approximate ones. Such approximate sums allow setting two parameters, i.e., the NAB and the type of adder cell to use (namely, a classic FAC or an IAC). We consider three different IAC families, i.e., the Approximate Mirror Adder (AMA) [35], the Approximate XOR-based Adder (AXA) [86], and the IneXact Adder (InXA) [2].

As mentioned, this is the same approach adopted in [3]. However, while in [3] the approximation was manually introduced, we automate the replacement process by considering the AST of the algorithm implementation, resorting to E-𝕀DE𝔸 [16]. Thus, we model each of the abovementioned multiplier-less DCT algorithms [17, 19–21, 28, 63, 64] by using C/C++ implementations, and starting from such implementations, the generation of approximate variants is performed using the Clang-Chimera tool [16]. For each DCT algorithm, the tool produces mutated sources which allow configuring, for each of the sums, both the NABs and type of adder hardware cell to use (i.e., either FAC or IAC).

5.2.3 Design-Space Exploration

Decision variables of the MOP are parameters introduced during the generation step for approximate variants, i.e., the NAB value and the type of adder hardware cell to be used for each of the approximate operations. Thus, if N_{op} is the number of addition required by a given algorithm, each approximate configuration is identified through the use of a vector, i.e., a chromosome, which is composed of $2 \cdot N_{op}$ different elements, or *genes*. Chromosomes are provided with an additional gene representing the approximation degree for the high-frequency filter. Hence, each chromosome is composed of $2 \cdot N_{op} + 1$ genes.

As for error fitness function, we resort to Structural DiSSIMilarity (DSSIM) [82] to evaluate differences among images. The higher the DSSIM, the higher the error between X and Y sets. We compute the DSSIM between a standard JPEG compressed image X and an image Y which is obtained by using a certain approximate configuration of a given approximate algorithm, with both X and Y originating from the same non-compressed source image. We considered the SIPI image data set [73] that consists of 44 different images, covering a wide set of common features, including among others a flat gray scale, foreground subject with a messy background, and high contrast images. Concerning silicon-area requirements, we again resort to model-based estimation to drive the DSE, estimating the hardware overhead based on the number of transistors required to implement one-bit adder cells, using the data from [3].

5.2.4 Experimental Results

To be able to measure the final gains, we encoded all the abovementioned DCT algorithms in VHDL. Such implementations guarantee high flexibility: they handle the configuration of both the types of adder cells to use for each addition and the number of bits to approximate (NABs). This allows the synthesis of any solution to be eventually found as a result of the DSE process. VHDL implementations follow Eq. (6), and each of the partial sums is performed using a configurable approximate adder: it allows computing the least significant bits of the sum using IACs, while the most significant bits are computed by classical FACs. The number of approximate bits, i.e., IACs, is configurable through the NAB parameter.

As we mentioned above, seven different DCT algorithms and ten types of IACs are considered during this case study. As for the DCT algorithms, we considered BAS08 [19], BAS09 [20], BAS11 [21], BC12 [17], CB11 [28], PEA12 [63], and PEA14 [64]. As for the IACs families, we considered AMA [35], AXA [86], and InXA [2]. All of these are encoded in the C++ language, and the generation of approximate variants is performed through the Clang-Chimera tool. The latter variants are, then, evolved using the Bellerophon tool. After the DSE, to correctly evaluate the final gains, we synthesized the obtained approximate configurations to both Application-Specific Integrated Circuit (ASIC) and Field Programmable Gate Array (FPGA) technologies.

Concerning ASIC, we synthesized all the obtained non-dominated approximate configurations while targeting the 65 nm Fin Field-Effect Transistor (FinFET) technology through the Cadence *Genus Synthesis Solution* tool. We resorted to the synthesis reports for the silicon-die area of the approximate configurations. In Fig. 6, we report the results. Pertaining to the power consumption, to determine

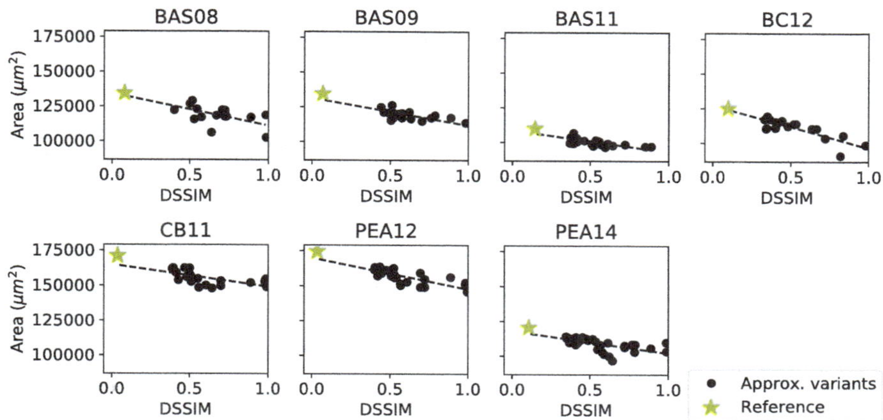

Fig. 6 Silicon-die area requirements (in μm^2)DCT hardware-resource requirements while targeting the 65 nm FinFET technology

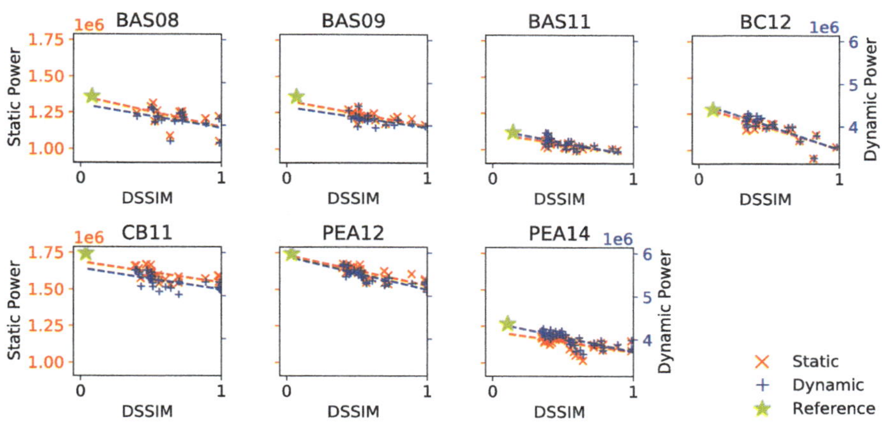

Fig. 7 Power consumption requirements (in nW) DCT hardware-resource requirements while targeting the 65 nm FinFET technology

whether the synthesis power report provides a satisfying accuracy, we simulated the whole workload for two algorithms (BAS08 and BAS09) and collected the resulting power consumption. As a result, we realized that the difference between the power consumption resulted from the workload simulation and that coming from the synthesis tool only differed by 5%, on average. We considered the synthesis report accuracy sufficient; thus, in Fig. 7 we show the power results from the synthesis report. Kindly note that the scale on the left axis (static power) is different from the scale on the right axis (dynamic power). Power savings are achieved due to both the reduced area and the lower switching activity that IACs exhibit w.r.t FACs, as also reported in [3].

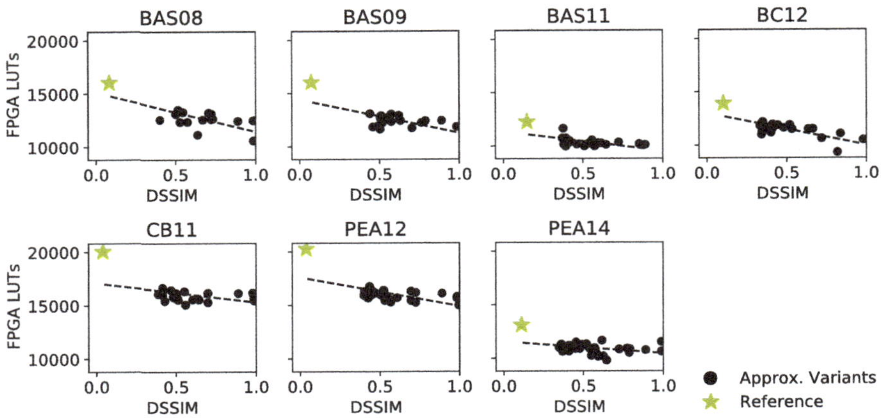

Fig. 8 LUTs requirements while targeting a Xilinx Zynq-7020 FPGA

Regarding FPGA synthesis, we synthesized all the obtained non-dominated approximate configurations to a Xilinx Zynq-7020 MPSoC, using only its embedded FPGA and inhibiting Digital Signal Processings (DSPs) usage, to get a fair estimation of hardware requirements. Figure 8 reports the synthesis result in terms of the number of LUTs for all the considered algorithms. As expected, approximate solutions require fewer resources than the precise implementation, as highlighted by the decreasing general trend. To correctly evaluate energy savings, we performed a post-synthesis timing simulation, using the Dynamic Power Analysis tool provided by the Xilinx Vivado. In this case, since the synthesis report has a very low confidence level for power consumption estimation, we resorted to a workload simulation for all the solutions the DSE provided, for all the algorithms. In this way, we achieved a high confidence level of power estimation. Figure 9 shows static and dynamic power consumption for all the algorithms. The static power of the FPGA is largely caused by the fabric of the device and does not directly depend on used resources, while the dynamic one is directly linked to the user design, due to the input data pattern and the design internal activity. Being our hardware implementations of approximate DCT characterized by low overhead, i.e., device resource usage falls between 6 and 13%, it is necessary to split power consumption in static and dynamic since the former turned out to be about an order of magnitude greater than the latter one for the target FPGA device. Also, in this case, power savings are achieved thanks to both the reduced total area and the logical structure of IACs: FPGA LUTs implementing IACs have a lower switching activity than those implementing FACs, as reported in [3].

Visual Test

Since JPEG belongs to the image processing domain, we also provide a visual test: Fig. 10 shows, from left to right, the standard JPEG-compressed image of Baboon, as taken from the SIPI image database [73], the same image compressed

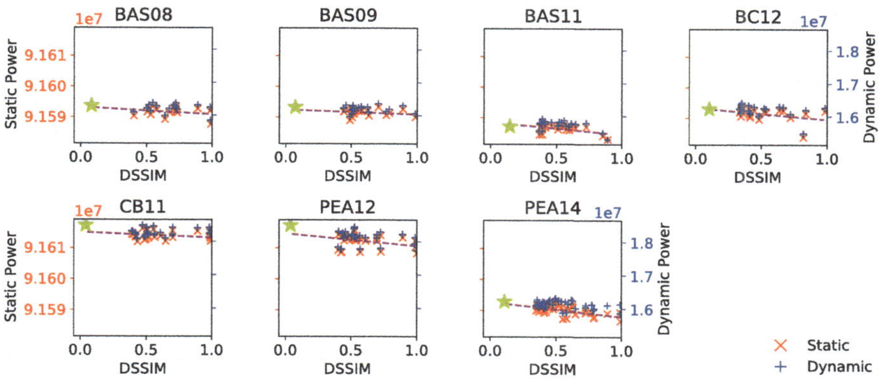

Fig. 9 Power consumption (in nW) while targeting a Xilinx Zynq-7020 FPGA

Fig. 10 Visual test

using the exact version of the BC12 algorithm [17]—which exhibits a DSSIM of 0.10 and requires 125,473.92 μm^2 and 5,691,946 μW when implemented on ASIC, or 5902 LUTs and 107,933,980 μW while targeting FPGA—and, finally, the ones compressed with its approximate variant having 0.33 as DSSIM value, which correspond to 8362.64 μm^2 and 352.711 μW saved for ASIC and 1846 LUTs and 94,506.744 μW saved for FPGA. As the reader can easily figure out, the quality differences are barely perceivable.

6 Automatic Approximation of Artificial Intelligence Applications

This section discusses the automatic approximation of two common Artificial Intelligence (AI) applications, namely, DNNs and DT MCSs. These case studies are particularly relevant since the state-of-the-art hardware accelerators targeting both these models are tremendously resource intensive, due to the massive amount of processing elements needed to effectively accelerate computations [26, 55]. In either case, the high demands in terms of both silicon area and power consumption utterly hinder the spread of commercial devices. The desire to reduce the hardware overhead by resorting to the AxC, however, cannot jeopardize more than half a century of efforts to achieve the accuracy that modern models exhibit. Luckily, the scientific literature demonstrated that MOP-based DSE can provide a full Pareto front consisting of several trade-offs between the considered fitness functions to optimize [11, 14, 60].

In the following, we first provide the reader with a brief background concerning DNNs and DT MCSs before discussing how to design approximate systems based on those models.

6.1 Neural Networks

Recently, Artificial Neural Network (ANN) have won numerous contests in pattern recognition, classification, object recognition, and so forth, imposing themselves on everyone's attention as one of the most successful learning techniques. The basic processing element is the *artificial neuron*: input signals are multiplied with *learned weights* at *synapses*, while *dendrites* carry the weighted input signals to the *neuron body*, where partial products are summed and *biased*. An output is produced along the *axon*—i.e., the neuron "fires"—if the weighted sum is greater than a threshold, defined by the neuron's *activation function*. Synaptic weights, as well as biases, are learned by exploiting the backpropagation algorithm [46].

It is worth mentioning that the model of artificial neurons is quite different from that of biological ones: biological dendrites do not simply carry signals, but, as well as biological synapses, they perform very complex—and still partially unknown—nonlinear functions and the exact instant in which the neuron "fires" encodes the information, not the frequency of firing [70].

Instead of being modeled as an amorphous blob of connected neurons, DNNs are organized in distinct *layers*, the number of which defines the network's depth. Three main types of layers are peculiar to Convolutional Neural Networks (CNNs), namely, the Convolutional Layer (CL), the Pooling Layer (PL), and the Fully–Connected Layer (FCL). While PLs perform a single function, i.e., subsampling, CLs and FCLs perform computation that is not just function of the inputs, but also of learned synaptic weights and biases. PLs are placed in between CLs, to progressively reduce the spatial size of the intermediate representation. In FCLs, neurons of the layer are connected to all the preceding layer, while in CLs, neurons in a layer are connected only to a small region of the previous layer. In any case, there is no connection between neurons within the same layer. In Recurrent Neural Networks (RNNs), cycles are allowed, while they are strictly forbidden in feed-forward ANNs, such as CNNs. Moreover, CNN architecture is constrained to be arranged in three dimensions, and explicit assumptions are made on the input, which is images, unveiling a more efficient forward function implementation and a reduction in the amount of learned parameters.

6.1.1 Approximate DNNs

As we mentioned, the inner error resiliency of ANN makes them the ideal field of application for AxC; consequently, a significant amount of research focused on both the training and the inference phase, attempting to further reduce resource requirements of hardware accelerators. In the following, we briefly report some of the most relevant contributions.

One of the approaches to identify approximation-resilient neurons in ANNs is discussed in [80]. It leverages the backpropagation algorithm [46] to obtain a measure of the sensitivity of the output of the considered DNNs, to the output of each neuron. The latter are sorted based on the magnitude of their average error contribution. Depending on whether the latter falls below a predetermined threshold, they are labeled as resilient or sensitive. Resilient neurons are replaced using approximate ones, which are designed using the precision-scaling technique, and allow modulating the bit widths of both inputs and weights based on resilience. A subsequent retraining step suitably adjusts the learned parameters, alleviating the impact of approximation-induced errors and allowing further approximation.

As mentioned, multipliers are recognized as the most demanding component within neurons. Therefore, as foreseeable, several contributions focus precisely on them. The authors of [5] investigated on properties that an approximate multiplier should exhibit to maintain acceptable classification accuracy and, at the same time, reduce the use of silicon area. They observed that multipliers having low values for the variance σ_{ED} and Root Mean Squared Error (RMSE) do not deteriorate the classification accuracy. Furthermore, they noted that when a multiplier underestimates or overestimates the exact product with equal probability, the classification accuracy tends to increase, since such a multiplier prevents the errors from accumulating. However, this is a necessary-but-not-sufficient condition.

To alleviate approximation-induced error, the mentioned contributions resort to network retraining. While, on one hand, this even allows for further approximation, on the other hand, it increases the overall design time. Moreover, retraining might not be possible, e.g., the data set on which the network has been trained may not be available. A way to overcome this limitation while using approximate multipliers has been proposed in [58]: a weight-tuning algorithm adapts the learned weights to the employed multipliers, allowing accuracy recovering. The proposed algorithm exploits the fact that, for each of the multiplications, the value of the one operand—the one holding the synaptic weight—is constant, while the second operand varies with the input data. Thus, a *map function* can be computed offline and exploited during approximation to determine the suitable weight update.

In [59], the EvoApprox8b library of arithmetic components [56], designed through Cartesian Genetic Programming (CPG) as in [71], is further evolved and employed to conduct a resiliency analysis targeting CNNs, specifically different networks belonging to the ResNet [36] family whereby 8-bit quantization is exploited to preemptively lower resource requirements. Further savings are pursued using approximate hardware components, selected as in [57]. In this regard, the EvoApprox8b is expanded considering both standard $n \times n$ and $m \times n$ approximate multipliers, and CNNs are approximated considering either a single layer at a time or the network as a whole. As for the former, all multiplications of all layers are replaced using one particular implementation taken from the mentioned library, regardless of layers' resiliency.

6.1.2 Automatic Approximation of DNN Applications

Reviewing the reported contributions, the following drawbacks can be easily recognized: (i) they are related to a specific approximation technique; (ii) they typically require a retraining step to alleviate the impact of approximation on the classification accuracy, which undoubtedly increases the design time; (iii) they either optimize a single parameter (silicon area, for instance) under quality constraints; and (iv) although they recognize that each neuron may contribute to error and performances differently, depending on the layer it belongs to, the degree of introduced approximation does not consider these differences.

Conversely, the approach we discussed (i) supports different approximation techniques, (ii) does neither leverage the backpropagation algorithm nor require retraining, to avoid lengthening the design time and make the method applicable even when retraining is not possible, (iii) allows for the different degree of error resilience exhibited by different parts of the same application to be considered, and (iv) is based on multi-objective optimization, which allows for solutions that simultaneously optimize multiple figures of merit, e.g., classification accuracy, ASIC silicon area or FPGA LUTs, power consumption, etc.

In the following, we discuss a case study concerning the automatic approximation of a DNN. To perform approximate variants generation and DSE, we use the E-𝕀DE𝔸 framework that we extensively discussed in Sect. 5.1; hence, through a MOP-based DSE, we find the correct approximation degrees leading to non-dominated solutions exhibiting near-Pareto trade-offs between accuracy loss and hardware efficiency. Indeed, E-𝕀DE𝔸 allows specifying multiple fitness functions; for our scenario, we define the accuracy loss and the hardware requirements to be both minimized. Furthermore, this approach allows to selectively introduce approximation within layers while considering the network as a whole, contextually analyzing error resiliency of layers.

The precision-scaling technique, for instance, can be applied by carefully manipulating the AST to supersede precise multiplications and/or additions in CL or FCL. As discussed, such approximate operations should allow selecting the appropriate degree of approximation to be introduced, through tunable parameters. Nevertheless, the amount of such parameters can easily explode, since the number of operations within layers. Structural properties of layers, however, can be exploited to reduce the amount of introduced parameters while effectively introducing approximation. In CLs, for instance, *weights sharing*, which reduces the number of parameters to be learned during the training phase by sharing synaptic weights among neurons within the same layer, allows applying the same approximation degree to all neurons belonging to the same CL. However, operations in different

CLs must have their approximation degree. Neurons belonging to FCLs usually do not share synaptic weights, yet they process the same input volume. Thus, neurons belonging to the same FCL can share the same approximation degree. PLs can also be subject to approximation. Their contribution in terms of calculation burden is, however, negligible w.r.t. CLs and FCLs. Moreover, subsampling utilizing stride in convolutions is progressively supplanting PLs. Therefore, we do not apply any approximation to such types of layers. We configured Clang-Chimera to truncate input operands and results of multiplications in CLs and FCLs. Thus, the Clang-Chimera tool produces an approximate version which allows configuring, for each of the approximate layers, the approximation degree for the multiplications and additions involved in the weighted sum, depending on the considered layer.

To estimate the error introduced by the approximation, we configured Bellerophon to execute the approximate CNN on the training test data set, to assess the classification-accuracy loss. Concerning hardware requirements, we estimate savings by considering several parameters that definitely have some impact on the former, such as (i) the input volume of a neuron, which impacts the number of operations performed within it and (ii) the number of neurons within a layer, i.e., the output volume size of a layer, which impacts the hardware requirements of the whole layer.

In the following, we consider the LeNet5 network [47]. LeNet5 is a CNN that performs well in large-scale image processing: it is basic CNN when compared to the state-of-the-art architectures, but it is a common reference point. The network being considered has been trained to classify images from the Modified National Institute of Standards and Technology (MNIST) test data set [48], which consists of a training set of 60,000 examples, and a test set of 10,000 examples of handwritten digits. The network has been trained using 64-bit floating point and exhibits a 99.07% accuracy when classifying images from the mentioned data set. We performed 8-bit quantization, without accuracy loss.

We configured the Clang-Chimera tool to supersede exact multiplications within the three CLs and two FCLs of LeNet5 using approximate multipliers designed while resorting to the precision-scaling technique. The latter allow selectively introducing approximation through configurable parameters. As discussed in Sect. 3, such configuration parameters constitute decision variables for MOP-based DSE; therefore, the Bellerophon tool encodes each approximate configuration, i.e., each individual, using a five-element-long vector, i.e., using a chromosome composed of five genes. Each of the latter governs the NAB for multiplications within a given layer. Concerning fitness functions, we resort to simulations performed on the test data set to assess the classification accuracy loss due to approximation. As far as hardware requirements are concerned, we estimate hardware requirements from the number of bits being kept after the precision scaling is applied. Furthermore, we set our GA parameters as follows: initial population equals to 300 individuals, mutation, and crossover probabilities both set at 0.9 and 31 generation epochs. Finally, we set a maximum error threshold to 1% accuracy loss.

After the DSE, to correctly evaluate the final gains, we performed FPGA synthesis of non-dominated configurations while targeting a Xilinx Virtex Ultrascale+.

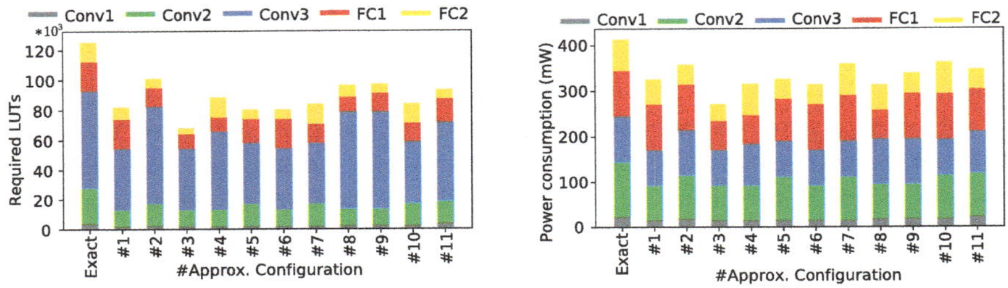

(a) (b)

Fig. 11 Hardware requirements for 8-bit LeNet5 while targeting a Xilinx Virtex Ultrascale+ FPGA. (**a**) Required LUTs for 8-bit LeNet5 while targeting a Xilinx Virtex Ultrascale+ FPGA. (**b**) Estimated power consumption for 8-bit LeNet5 while targeting a Xilinx Virtex Ultrascale+ FPGA

These syntheses involve only one single neuron, to provide a fair estimation of hardware requirements, i.e., as independent as possible from configuration parameters governing the structure of the accelerator. Furthermore, for the same reason, we disabled advanced FPGA features, e.g., DSPs, during syntheses.

Figure 11a reports synthesis results: these stacked bar graphs show, for each of the layers, the amount of FPGA LUTs required by a single neuron. A significant reduction, actually up to 45%, of required resources in terms of FPGA LUTs can be observed. As foreseeable, savings achieved due to precision scaling do not only concern hardware requirements, but also energy consumption. Trivially, the less hardware a circuit requires, the less energy is spent to power it, and since the least significant bits of inputs, weights, and biases are always set to zero, the whole approximate circuit is also expected to exhibit a lower switching activity w.r.t. its exact counterpart. To evaluate potential power savings, we performed simulations on the exact neuron and on approximate ones lying on the Pareto front, resulting from the DSE. Simulations involve 10,000 input combinations, each consisting of an appropriate number of inputs, weights, and bias vectors, depending on the input volume size of the considered neuron. Figure 11b reports simulation results. Again, up to 35% savings in terms of power consumption can be observed.

6.2 *Decision-Tree-Based Multiple Classifier Systems*

Decision Trees (DTs) stand out for their simplicity and high interpretability level, placing them as one of the most widely used classifier models [83]. A DT is a white-box classification model representing its decisions through a tree-like structure composed of an internal set of nodes containing *test conditions*, and leaf nodes which represent *class labels* [25]. Nodes are joined by arcs symbolizing possible outcomes of each test condition. Classes can be either categorical or numerical. In the former case, we refer to classification trees, while in the latter case, we refer to regression trees.

According to the number of attributes evaluated in each test condition—i.e., each internal node—two DT types can be induced: *univariate* and *multivariate* [25]. In the former, each test condition evaluates a single attribute to split the training set, while a combination of attributes is used in multivariate DTs. Some advantages of univariate DTs are their comprehensibility and the simplicity of their induction algorithms; however, they may include many internal nodes when the training data set instance distribution is complex. In such trees, test conditions are defined as $x_i \lesseqgtr c$, where x_i is the i-th attribute value and c is a threshold value used to define a partition. Therefore, test conditions represent axis-parallel hyperplanes dividing the instance space, so they are also known as axis parallel DTs. Anyway, when a categorical attribute is evaluated, the training set is split into as many subsets as values exist in the attribute domain.

In its essence, the procedure to construct DT MCS consists in using the training data set—that is constituted of historical, labeled data—to determine its best splits. These splits reduce the data set into smaller and smaller pieces while aiming at splits that best emphasize the differences between data points belonging to the different partitions. The most widely adopted algorithms to construct DTs are CART [25] and C4.5 [65]. Significant improvements in classification accuracy have resulted from growing an ensemble of trees and letting them vote for the most popular class. Besides *bagging* predictors [22], several other techniques have been proposed. Some examples are *random split selection* [33], *random feature selection* [37], or searches for over a random selection of features for the best split at each node [4] and *random error injection* [23]. According to [24], all these procedures generate *random forest classifiers*.

6.2.1 Hardware Accelerators Targeting Decision-Tree-Based Classifiers

As mentioned, their simplicity and understandability make DTs one of the most popular machine-learning algorithms [83]. However, at the beginning, they were not considered for hardware accelerator, since they need only comparison to be performed during the inference phase; thus, they were not viewed as being computationally expensive. Nevertheless, the need to accelerate the inference phase emerged inherently in some application fields. Many real-time tasks require high prediction speed. However, the software implementation of the inference phase cannot meet the requirement even if the multi-threading technology is adopted. Hence, attention has been paid to the hardware-based accelerators [41].

As pinpointed in [61], General Purpose - Graphic Processing Unit (GP-GPU) is an ill-fated choice for accelerating DT-based predictors since (i) high-precision arithmetic circuit—e.g., single/double precision Floating-Point Units (FPUs)—is inefficient in the amount of both hardware and power consumption, since, during the inference phase, each node in the tree is evaluated using an *if-then-else* statements, which compare input values with constant values; (ii) DT-based predictors may consist of trees with a different size and depth that causes an unbalanced computation; and (iii) all-to-all communication with a DT-based predictor requires evaluating all the trees will certainly lead to performance bottleneck, since communication between near processing cores with the same local memory can be performed at a relatively high speed, while communication penalty is large for the all-to-all communications. The authors of [77] compared performances of different implementations of hardware accelerators for DT-based predictors, including CPU, GP-GPUs, and FPGAs while using offline generated models. They empirically proved that FPGA implementations provide the highest performance, but may require a multi-chip/multi-board system to execute even modest sized predictors.

According to [50], architectures of FPGA-based accelerators for DT-based predictors can be categorized as comparator-centric and memory-centric. The former implements the model as a threshold network that consists of a layer of threshold logic units and a layer of combinational logic units, while the latter accelerates the model by introducing the pipeline in layers of the tree [1, 34, 62, 69, 74]. On the other hand, comparator-based accelerators reduce the reliance on memory elements by using custom comparators at each internal level [32, 40]. Comparator-centric accelerators could achieve high throughput and low latency as well; however, since designing such an accelerator requires all information of the model for the logic design, the architecture is tightly coupled with the underlying model, meaning any change in the model requires the accelerator to be reconfigured.

6.2.2 Approximate DTMCSs

Concerning DT MCSs, the widespread adoption of hardware-based accelerators is actually hindered by scalability issue, as reported in [77]. Nevertheless, there is very little research on both efficient architectures and methods to reduce their hardware-resource consumption, as the effort is mainly devoted to other classification systems, e.g., DNNs.

One of the few contributions from the scientific literature concerning the improvement of energy efficiency of DT MCSs has been proposed in [81]. The authors noticed that, usually, only a part of the data of a given data set really needs the full computational power of a classifier. Therefore, they dynamically configure the classifier, making it more or less accurate, according to the difficulty in classifying the inputs. Hence, rather than building a single complex model, during the training phase they construct a set of models with progressively increasing complexity. Then, during the testing phase, the number of decision models applied to a given input varies depending on the difficulty of the considered input instance. To estimate the difficulty of a certain input, a confidence level for each classification is computed. If the latter confidence falls above a certain threshold, the classification process is terminated; otherwise, a more accurate classifier is used.

6.2.3 Automatic Approximation of DT MCSs Applications

In this section, we discuss a case study concerning the approximation of a DT MCSs. In particular, we discuss the automatic approximation of some hardware accelerators trained to tackle with the Spambase data set from the UCI Machine Learning Repository [39], which contains 4601 emails, 1813 of which are spam. This data set is freely available and makes use of 57 different features, expressed in the floating-point notation, to characterize elements that are part of the data set. Each of the features specifies how often a word or a character appears in each element of the data set, i.e., in an email. During the training phase, conducted using the KNIME [45] tool, 40 different random-forest classifiers with a number of DTs ranging from 1 to 40 are trained.

6.2.3.1 Reference Architectures

In the case studies discussed in this section, we consider the hardware implementation from [9]. In order to speed up DTs visiting, the authors of [9] adopt a speculative approach, which consists in a DT flattening so that the visiting is performed over every possible path. Predicates are performed concurrently, regardless of the position and depth at which nodes are located, and a Boolean decision variable, which indicates whether a condition is fulfilled, is produced for each one of the evaluated predicates. To determine which leaf of the DT is reached, i.e., which class the input belongs to, a Boolean function, called *assertion*, is defined for each different class. Since a path that leads to a specific leaf is obtained by computing the logic AND between the Boolean decision variables along that path, and since it is possible to compute the logic OR between the conditions related to different paths leading to leaves belonging to the same class, assertions can be defined as a sum of products of Boolean functions. A majority voter [10] combines the outcome of each tree to produce the outcome. The scalability of this approach has been formally demonstrated in [8]. In particular, the number of literals in each assertion is always less or equal to twice the size of the features set.

To generate approximate variants and to perform DSE, the authors of [11] resort to the E-IDEA framework, which we discussed in Sect. 5.1. Hence, through a MOP-based DSE, they find the correct approximation degrees leading to non-dominated solutions exhibiting near-Pareto trade-offs between accuracy loss and hardware efficiency, with accuracy loss and the hardware requirements to be both minimized.

6.2.3.2 Generating Approximate Variants

Several opportunities inherently arise, for instance, from the hardware implementation which is discussed above. Both comparators and assertion functions are excellent candidates for approximation, albeit the contributions of the former seem much more substantial w.r.t the latter, meaning their approximation can lead to significant savings. Furthermore, to introduce approximation, a wide plethora of techniques can be used regarding comparators, including precision scaling, inexact hardware, and functional approximation. Since the latter is well suited to Boolean functions, it is also ideal for introducing approximation within assertion functions.

As mentioned, DT MCSs may consist of hundreds or even thousands of nodes, each comparing one of the many features considered by the predictive model with their corresponding threshold value. Consider introducing approximation through approximate comparators designed using the precision-scaling technique: such approximate comparators allow, through a configuration parameter, to govern the degree of introduced approximation by tuning the amount of neglected mantissa bits at each comparison. Hence, the value to be assigned to each one of such parameters constitutes the decision variables of the MOP. Furthermore, the domain of such variables depends on the particular data type being adopted for representing features. However, albeit straightforward, this naive MOP definition may result in an utterly infeasible DSE. Indeed, hardware implementations of DT MCSs may consist of hundreds or even thousands of comparators, resulting in an enormous design space. Nevertheless, we can exploit the fact that, albeit against different threshold values, the same feature can be taken into consideration for comparison several times while visiting trees. This allows setting the same degree of approximation to all the comparators processing the same feature, reducing the number of decision variables and, consequently, the size of the solution space. Thus, being F the features set, each approximate configuration—i.e., individual in the GA context—can be represented using a vector consisting of $|F|$ elements, i.e., genes, each governing the degree of approximation for the corresponding feature.

As we mentioned, in [11], the actual generation of approximate variants is performed while resorting to the E-$\mathbb{IDE}\mathbb{A}$ framework, applying the precision-scaling technique on comparators when using the mentioned double-precision floating-point representation [16].

6.2.3.3 Design Space Exploration

For what pertains to hardware requirements, to accurately consider the resource savings in the DSE, we should measure area, power consumption, and maximum clock speed of the explored approximate variants. This would require the hardware synthesis—and also simulations, in the case we are interested in measuring power consumption, for instance—of each variant explored in the DSE. This is utterly a time-consuming process. Hence, again, we resort to a model-based gain estimation to drive the DSE.

Since the purpose of the model is determining the $<$, $=$, or $>$ relation between two different approximate configurations, it is not necessary to focus on its accuracy. We rather focus on its fidelity, i.e., how often the estimated values are in the same relation as the real values for each pair of configurations. Concerning comparators, it is easy to recognize the fewer bit to compare the fewer hardware requirements.

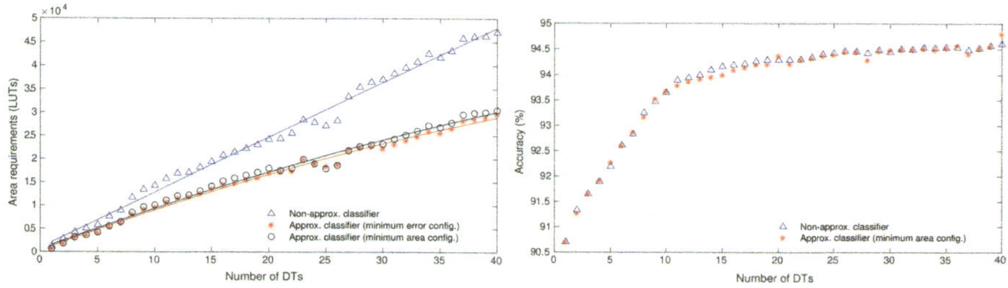

(a) (b)

Fig. 12 Resource requirements and accuracy of approximate DT MCSs for Spambase. (**a**) Required LUTs. (**b**) Accuracy

6.2.3.4 Results

The AxC exploration phase found, for each of the 40 classifiers, a certain number of approximate configurations on the Pareto frontier, but for each of them only the configuration with minimum error and the one that requires less silicon area has been reported.

Figure 12a shows the area requirements in terms of LUTs, as the number of DTs used by the classifier increases. For all the measured quantities, an increasing trend, as the number of trees grows, is shown for area requirements. The growth, however, is clearly sublinear. In addition, it can be seen that the difference between the requirements of the exact classifier and the approximate one increases as the number of trees grows. This is because even if the complexity of single DTs—i.e., the number of nodes of which they consist of and the height of DTs themselves—decreases significantly as the number of trees used by the classifier increases, the total number of nodes increases, providing more approximation opportunities. This behavior can be observed also for the number of FPGA slices and registers and both when considering solutions providing minimum error and those requiring the minimum silicon area. Furthermore, it can be noted that the difference in terms of area requirements between the minimum error and the minimum area solutions always remains negligible.

Figure 12b compares the levels of classification accuracy, as the number of trees used by the classifier increases, provided by the precise version—without approximation—and by the approximate version that has minimum area requirements. It is evident, from the graph, that there is only a small difference in accuracy between the configurations. Moreover, it remains minimal as the number of trees used for classification varies. On the other hand, the increase in the number DTs used in the classification process makes a smaller contribution as the number of DTs grows. This asymptotic behavior can be seen in exact and approximate classifiers, and it is because, by increasing the number of models, data sets involved for training turn out simpler and the corresponding DTs get less branched, which leads to a saturation of the accuracy level provided by the classifier model.

7 Conclusion

In this work, we discussed a unified methodology able to automatically explore the impact of different approximation techniques on a given application while resorting to the AxC design paradigm and MOP-based DSE. We discussed the steps the methodology breaks into while devoting particular relevance to all those that can be automated. In particular, we discussed how to effectively select parts of an application to be approximated and how to choose a well-suited approximation technique. Then, we discussed how approximate variants can be generated automatically and how to identify decision variables and suitable fitness functions to define the MOP driving the DSE. The methodology is not dependent on a particular application; rather it can be generally applied to all applications.

To evaluate the proposed methodology, we selected some significant and relevant applications in the scope of the AxC paradigm, including generic combinatorial logic circuits, image-processing applications, and artificial intelligence applications. For what pertains to generic logic, we propose local rewriting of AIG to reduce the number of nodes, resulting in lower hardware-resource requirements, while resorting to MOP-based DSE to carefully introduce approximation. We evaluate our approach using different benchmarks, and experimental results show our method allows performing a meaningful exploration of the design space to find the best trade-offs in a reasonable time, thus resulting in approximate circuits exhibiting lower requirements and restrained error. Concerning image processing applications, the discussed case study concerns the design of hardware accelerators for the discrete cosine transform (DCT), which is the most demanding step of the JPEG compression algorithm. We analyzed and modeled several algorithms from the literature to compute a fast and lightweight version of the DCT, and, for each algorithm, we applied approximation by substituting full-precise adders with several approximate ones from the literature having configurable approximation degree. For each algorithm, we performed a DSE to find the non-dominated approximate designs in terms of trade-off between inaccuracy and gains. We modeled the DSE as a MOP and we used a GA to solve it, and after the DSE, we synthesized the obtained designs by targeting both FPGA and ASIC. Experimental results clearly showed that, with the proposed approach, it is possible to perform a meaningful DSE to find the best trade-offs between output accuracy and resource gains in a reasonable time. Finally, the comparison performed with previous work clearly showed the advantages of the proposed approach.

Finally, we applied our methodology to two of the most promising classification models in the machine-learning domain, namely, deep neural networks (DNNs) and decision-tree-based multiple classifier systems (DT MCSs). Leveraging the AxC design paradigm, a very limited quantity of classification accuracy is traded off for a reduction in the silicon-area requirements and power consumption of hardware-implemented DT MCS and CNN. Concerning DNNs, we exploited our methodology to investigate the impact of approximation on the classification accuracy, and experimental results prove the validity and efficiency of our methodology, even in applications in which the error has to be minimized as possible, providing savings up to 75% for silicon area and 50% for power consumption. Pertaining to DT MCSs, to prove the validity of the proposed approach of several classifiers, the optimal number of bits to be used to represent each of the features of the model is searched using NSGA-II. Among all Pareto-optimal hardware configurations, the one providing minimum classification error configuration and the one requiring the minimum amount of silicon area were considered for further consideration. Experimental results show a significant reduction in area requirements, for both of them. Furthermore, since the classification is very resistant to error, those configurations are very similar both in terms of area requirements and classification error.

References

1. A. Alcolea, J. Resano, FPGA accelerator for gradient boosting decision trees. Electronics **10**(3), 314 (2021). Publisher: Multidisciplinary Digital Publishing Institute
2. H.A. Almurib, T.N. Kumar, F. Lombardi, Inexact designs for approximate low power addition by cell replacement, in *2016 Design, Automation & Test in Europe Conference & Exhibition (DATE)* (2016), pp 660–665. ISSN: 1558–1101
3. H.A. Almurib, T.N. Kumar, F. Lombardi, Approximate DCT image compression using inexact computing. IEEE Trans. Comput. **67**(2), 149–159 (2018). https://doi.org/10.1109/TC.2017.2731770. Conference Name: IEEE Transactions on Computers
4. Y. Amit, D. Geman, Shape quantization and recognition with randomized trees. Neural Comput. **9**(7), 1545–1588 (1997). https://doi.org/10.1162/neco.1997.9.7.1545. https://direct.mit.edu/neco/article/9/7/1545-1588/6116
5. M.S. Ansari, V. Mrazek, B.F. Cockburn, L. Sekanina, Z. Vasicek, J. Han, Improving the accuracy and hardware efficiency of neural networks using approximate multipliers. IEEE Trans. Very Large Scale Integr. Syst. **28**(2), 317–328 (2020). https://doi.org/10.1109/TVLSI.2019.2940943. Conference Name: IEEE Transactions on Very Large Scale Integration (VLSI) Systems
6. Apple Inc, Clang C Language Family Frontend for LLVM (2023). https://clang.llvm.org/
7. S. Bandyopadhyay, S. Saha, U. Maulik, K. Deb, A simulated annealing-based multiobjective optimization algorithm: AMOSA. IEEE Trans. Evolut. Comput. **12**(3), 269–283 (2008). https://doi.org/10.1109/TEVC.2007.900837. Conference Name: IEEE Transactions on Evolutionary Computation
8. M. Barbareschi, Implementing hardware decision tree prediction: A scalable approach, in *2016 30th International Conference on Advanced Information Networking and Applications Workshops (WAINA)* (IEEE, Crans-Montana, Switzerland, 2016), pp. 87–92. https://doi.org/10.1109/WAINA.2016.171. http://ieeexplore.ieee.org/document/7471178/
9. M. Barbareschi, S. Del Prete, F. Gargiulo, A. Mazzeo, C. Sansone, Decision tree-based multiple classifier systems: An FPGA perspective, in *International Workshop on Multiple Classifier Systems* (2015). https://doi.org/10.1007/978-3-319-20248-8
10. M. Barbareschi, S. Barone, A. Mazzeo, N. Mazzocca, Efficient Reed-Muller implementation for fuzzy extractor schemes, in *2019 14th International Conference on Design & Technology of Integrated Systems in Nanoscale Era (DTIS)* (2019), pp. 1–2. https://doi.org/10.1109/DTIS.2019.8735029
11. M. Barbareschi, S. Barone, N. Mazzocca, Advancing synthesis of decision tree-based multiple classifier systems: an approximate computing case study. Knowl. Informat. Syst. **63**, 1–20 (2021). https://doi.org/10.1007/s10115-021-01565-5. https://link.springer.com/article/10.

1007/s10115-021-01565-5. Company: Springer Distributor: Springer Institution: Springer Label: Springer Publisher: Springer London
12. M. Barbareschi, S. Barone, N. Mazzocca, A. Moriconi, A catalog-based AIG-rewriting approach to the design of approximate components. IEEE Trans. Emerg. Topics Comput. (2022). (In press) https://doi.org/10.1109/TETC.2022.3170502
13. M. Barbareschi, S. Barone, N. Mazzocca, A. Moriconi, Design space exploration tools, in ed. by Bosio, A., Ménard, D., Sentieys, O., *Approximate Computing Techniques: From Component- to Application-Level* (Springer International Publishing, Cham, 2022), pp. 215–259. https://doi.org/10.1007/978-3-030-94705-7_8
14. S. Barone, Catalog of approximate LUTs for pyALS (2022). https://github.com/SalvatoreBarone/pyALS-lut-catalog. Original-date: 2021-12-29T10:31:12Z
15. S. Barone, pyALS (2022). https://github.com/SalvatoreBarone/pyALS. Original-date: 2021-06-30T11:20:07Z
16. S. Barone, M. Traiola, M. Barbareschi, A. Bosio, Multi-objective application-driven approximate design method. IEEE Access **9**, 86975–86993 (2021). https://doi.org/10.1109/ACCESS.2021.3087858. Conference Name: IEEE Access
17. F.M. Bayer, R.J. Cintra, DCT-like transform for image compression requires 14 additions only. Electron. Lett. **48**(15), 919 (2012). https://doi.org/10.1049/el.2012.1148. http://arxiv.org/abs/1702.00817. arXiv:1702.00817 [cs, stat]
18. A. Bosio, D. Ménard, O. Sentieys (eds.), *Approximate Computing Techniques: From Component- to Application-Level* (Springer International Publishing, Cham, 2022). https://doi.org/10.1007/978-3-030-94705-7. https://link.springer.com/10.1007/978-3-030-94705-7
19. S. Bouguezel, M.O. Ahmad, M.N.S. Swamy, Low-complexity 8 × 8 transform for image compression. Electron. Lett. **44**(21), 1249–1250 (2008). Publisher: IET
20. S. Bouguezel, M.O. Ahmad, M.N.S. Swamy, A fast 8 × 8 transform for image compression, in *2009 International Conference on Microelectronics - ICM* (2009), pp. 74–77. https://doi.org/10.1109/ICM.2009.5418584. ISSN: 2159-1679
21. S. Bouguezel, M.O. Ahmad, M. Swamy, A low-complexity parametric transform for image compression, in *2011 IEEE International Symposium of Circuits and Systems (ISCAS)* (2011), pp. 2145–2148. https://doi.org/10.1109/ISCAS.2011.5938023. ISSN: 2158-1525
22. L. Breiman, Bagging predictors. Mach. Learn. **24**(2), 123–140 (1996). https://doi.org/10.1007/BF00058655. http://link.springer.com/10.1007/BF00058655
23. L. Breiman, Randomizing Outputs to Increase Prediction Accuracy. Mach. Learn. **40**(14), (1998)
24. L. Breiman, Random forests. Mach. Learn. **45**(1), 5–32 (2001). Publisher: Springer
25. L. Breiman, J.H. Friedman, R. Olshen, C.J. Stone, *Classification and Regression Trees* (Routledge Publisher, Wadsworth, 1984)
26. M. Capra, B. Bussolino, A. Marchisio, M. Shafique, G. Masera, M. Martina, An updated survey of efficient hardware architectures for accelerating deep convolutional neural networks. Fut. Int. **12**(7), 113 (2020). https://doi.org/10.3390/fi12070113. https://www.mdpi.com/1999-5903/12/7/113. Publisher: Multidisciplinary Digital Publishing Institute
27. V.K. Chippa, S.T. Chakradhar, K. Roy, A. Raghunathan, Analysis and characterization of inherent application resilience for approximate computing, in *Proceedings of the 50th Annual Design Automation Conference on - DAC '13* (ACM Press, Austin, 2013)), p. 1. https://doi.org/10.1145/2463209.2488873. http://dl.acm.org/citation.cfm?doid=2463209.2488873
28. R.J. Cintra, F.M. Bayer, A DCT approximation for image compression. IEEE Signal Process. Lett. **18**(10), 579–582 (2011). https://doi.org/10.1109/LSP.2011.2163394. Conference Name: IEEE Signal Processing Letters
29. C. Coello Coello, M. Lechuga, MOPSO: A proposal for multiple objective particle swarm optimization, in *Proceedings of the 2002 Congress on Evolutionary Computation*. CEC'02 (Cat. No.02TH8600), vol. 2 (2002), pp. 1051–1056. https://doi.org/10.1109/CEC.2002.1004388
30. I. Das, J.E. Dennis, A closer look at drawbacks of minimizing weighted sums of objectives for Pareto set generation in multicriteria optimization problems. Struct. Optim. **14**(1), 63–69 (1997). https://doi.org/10.1007/BF01197559. http://link.springer.com/10.1007/BF01197559

31. K. Deb, A. Pratap, S. Agarwal, T. Meyarivan, A fast and elitist multiobjective genetic algorithm: NSGA-II. IEEE Trans. Evolut. Comput. **6**(2), 182–197 (2002). https://doi.org/10.1109/4235.996017. Conference Name: IEEE Transactions on Evolutionary Computation
32. I.A. Dávila-Rodríguez, M.A. Nuño-Maganda, Y. Hernández-Mier, S. Polanco-Martagón, Decision-tree based pixel classification for real-time citrus segmentation on FPGA, in *2019 International Conference on ReConFigurable Computing and FPGAs (ReConFig)* (2019), pp. 1–8. https://doi.org/10.1109/ReConFig48160.2019.8994792. ISSN: 2640-0472
33. T.G. Dietterich, An experimental comparison of three methods for constructing ensembles of decision trees: bagging, boosting, and randomization. Mach. Learn. **40**(2), 19 (1998)
34. M. Elnawawy, A. Sagahyroon, T. Shanableh, FPGA-based network traffic classification using machine learning. IEEE Access **8**, 175637–175650 (2020). https://doi.org/10.1109/ACCESS.2020.3026831. Conference Name: IEEE Access
35. V. Gupta, D. Mohapatra, A. Raghunathan, K. Roy, Low-power digital signal processing using approximate adders. IEEE Trans. Comput.-Aided Design Integr. Circuits Syst. **32**(1), 124–137 (2013). https://doi.org/10.1109/TCAD.2012.2217962. Conference Name: IEEE Transactions on Computer-Aided Design of Integrated Circuits and Systems
36. K. He, X. Zhang, S. Ren, J. Sun, Deep residual learning for image recognition, in *2016 IEEE Conference on Computer Vision and Pattern Recognition (CVPR)* (IEEE, Las Vegas, 2016), pp. 770–778. https://doi.org/10.1109/CVPR.2016.90. http://ieeexplore.ieee.org/document/7780459/
37. T.K. Ho, The random subspace method for constructing decision forests. IEEE Trans. Pattern Analy. Mach. Intell. **20**(8), 832–844 (1998). https://doi.org/10.1109/34.709601. Conference Name: IEEE Transactions on Pattern Analysis and Machine Intelligence
38. N. Homma, T. Aoki, Arithmetic Module Generator (2022). https://www.ecsis.riec.tohoku.ac.jp/topics/amg/
39. M. Hopkins, E. Reeber, G. Forman, J. Suermondt, Spambase Data Set (1999). https://archive.ics.uci.edu/ml/datasets/spambase
40. T. Ikeda, K. Sakurada, A. Nakamura, M. Motomura, S. Takamaeda-Yamazaki, Hardware/algorithm Co-optimization for fully-parallelized compact decision tree ensembles on FPGAs, in ed. by Rincón, F., Barba, J., So, H.K.H., Diniz, P., Caba, J., *Applied Reconfigurable Computing. Architectures, Tools, and Applications*. Lecture Notes in Computer Science (Springer International Publishing, Cham, 2020), pp. 345–357. https://doi.org/10.1007/978-3-030-44534-8_26
41. W. Jiang, V.K. Prasanna, Large-scale wire-speed packet classification on FPGAs, in *Proceedings of the ACM/SIGDA International Symposium on Field Programmable Gate Arrays* (2009), pp. 219–228
42. H. Jiang, F.J.H. Santiago, H. Mo, L. Liu, J. Han, Approximate arithmetic circuits: a survey, characterization, and recent applications. Proc. IEEE **108**, 1–28 (2020). https://doi.org/10.1109/JPROC.2020.3006451. https://ieeexplore.ieee.org/document/9165786/
43. G. Keramidas, C. Kokkala, I. Stamoulis, Clumsy value cache: An approximate memoization technique for mobile GPU fragment shaders, in *Workshop on Approximate Computing (WAPCO'15)* (2015), p. 6
44. S. Kirkpatrick, Optimization by simulated annealing: quantitative studies. J. Statist. Phys. **34**(5–6), 975–986 (1984). https://doi.org/10.1007/BF01009452. http://link.springer.com/10.1007/BF01009452
45. A.G. Knime, KNIME | Open for Innovation (2022). https://www.knime.com/
46. Y. LeCun, B. Boser, J.S. Denker, D. Henderson, R.E. Howard, W. Hubbard, L.D. Jackel, Backpropagation applied to handwritten zip code recognition. Neural Comput. **1**(4), 541–551 (1989). https://doi.org/10.1162/neco.1989.1.4.541
47. Y. Lecun, L. Bottou, Y. Bengio, P. Haffner, Gradient-based learning applied to document recognition. Proc. IEEE **86**(11), 2278–2324 (1998). https://doi.org/10.1109/5.726791. Conference Name: Proceedings of the IEEE
48. Y. LeCun, C. Cortes, C. Burges, MNIST Handwritten digit database (1998). http://yann.lecun.com/exdb/mnist/

49. A. Liefooghe, L. Jourdan, T. Legrand, J. Humeau, E.G. Talbi, ParadisEO-MOEO: A software framework for evolutionary multi-objective optimization, in ed. by Kacprzyk, J., Coello Coello, C.A., Dhaenens, C., Jourdan, L., *Advances in Multi-Objective Nature Inspired Computing*, vol. 272 (Springer, Berlin, 2010), pp. 87–117. https://doi.org/10.1007/978-3-642-11218-8_5. http://link.springer.com/10.1007/978-3-642-11218-8_5. Series Title: Studies in Computational Intelligence
50. X. Lin, R.S. Blanton, D.E. Thomas, Random forest architectures on FPGA for multiple applications, in *Proceedings of the on Great Lakes Symposium on VLSI 2017*, GLSVLSI '17 (Association for Computing Machinery, New York, 2017) pp. 415–418. https://doi.org/10.1145/3060403.3060416
51. M. Melanie (1998) *An Introduction to Genetic Algorithms* (MIT Press, Cambridge, 1998), p. 162
52. A. Mishchenko, S. Cho, S. Chatterjee, R. Brayton, Combinational and sequential mapping with priority cuts, in *2007 IEEE/ACM International Conference on Computer-Aided Design* (2007), pp. 354–361. https://doi.org/10.1109/ICCAD.2007.4397290. ISSN: 1558-2434
53. A. Mishchenko, S. Chatterjee, R. Jiang, R. Brayton, FRAIGs: A Unifying Representation for Logic Synthesis and Verification. ERL Technical Report (2015), p. 7
54. S. Mittal, A survey of techniques for approximate computing. ACM Comput. Surv. **48**(4), 1–33 (2016). https://doi.org/10.1145/2893356, https://dl.acm.org/doi/10.1145/2893356
55. S. Mittal, A survey of FPGA-based accelerators for convolutional neural networks. Neural Comput. Appl. **32**(4), 1109–1139 (2020). https://doi.org/10.1007/s00521-018-3761-1
56. V. Mrazek, R. Hrbacek, Z. Vasicek, L. Sekanina, EvoApprox8b: Library of approximate adders and multipliers for circuit design and benchmarking of approximation methods, in *Design, Automation Test in Europe Conference Exhibition (DATE)* (2017), pp. 258–261. https://doi.org/10.23919/DATE.2017.7926993. ISSN: 1558-1101
57. V. Mrazek, M.A. Hanif, Z. Vasicek, L. Sekanina, M. Shafique, autoAx: An automatic design space exploration and circuit building methodology utilizing libraries of approximate components, in *2019 56th ACM/IEEE Design Automation Conference (DAC)* (2019), pp. 1–6. ISSN: 0738-100X
58. V. Mrazek, Z. Vasicek, L. Sekanina, M.A. Hanif, M. Shafique, ALWANN: Automatic layer-wise approximation of deep neural network accelerators without retraining, in *2019 IEEE/ACM International Conference on Computer-Aided Design (ICCAD)* (2019), pp. 1–8. https://doi.org/10.1109/ICCAD45719.2019.8942068. http://arxiv.org/abs/1907.07229. arXiv: 1907.07229
59. V. Mrazek, L. Sekanina, Z. Vasicek, Libraries of approximate circuits: automated design and application in CNN accelerators. IEEE J. Emerg. Sel. Topics Circuits Syst. **10**(4), 406–418 (2020). https://doi.org/10.1109/JETCAS.2020.3032495. Conference Name: IEEE Journal on Emerging and Selected Topics in Circuits and Systems
60. V. Mrazek, L. Sekanina, Z. Vasicek, Using libraries of approximate circuits in design of hardware accelerators of deep neural networks, in *2020 2nd IEEE International Conference on Artificial Intelligence Circuits and Systems (AICAS)* (2020), pp. 243–247. https://doi.org/10.1109/AICAS48895.2020.9073837
61. H. Nakahara, A. Jinguji, S. Sato, T. Sasao, A random forest using a multi-valued decision diagram on an FPGA, in *2017 IEEE 47th International Symposium on Multiple-Valued Logic (ISMVL)* (2017), pp. 266–271. https://doi.org/10.1109/ISMVL.2017.40. ISSN: 2378-2226
62. M. Owaida, A. Kulkarni, G. Alonso, Distributed inference over decision tree ensembles on clusters of FPGAs. ACM Trans. Reconfig. Technol. Syst. **12**(4), 1–27 (2019). https://doi.org/10.1145/3340263. https://dl.acm.org/doi/10.1145/3340263
63. U.S. Potluri, A. Madanayake, R.J. Cintra, F.M. Bayer, N. Rajapaksha, Multiplier-free DCT approximations for RF multi-beam digital aperture-array space imaging and directional sensing. Measure. Sci. Technol. **23**(11), 114003 (2012). https://doi.org/10.1088/0957-0233/23/11/114003. https://iopscience.iop.org/article/10.1088/0957-0233/23/11/114003
64. U.S. Potluri, A. Madanayake, R.J. Cintra, F.M. Bayer, S. Kulasekera, A. Edirisuriya, Improved 8-point approximate DCT for image and video compression requiring only 14 additions. IEEE Trans. Circuits Syst. I Regular Papers **61**(6), 1727–1740 (2014). https://doi.org/10.1109/TCSI.

2013.2295022. Conference Name: IEEE Transactions on Circuits and Systems I: Regular Papers
65. J.R. Quinlan, *C4.5: Programs for Machine Learning* (Morgan Kaufmann Publishers., San Francisco, 1993)
66. A. Raha, S. Venkataramani, V. Raghunathan, A. Raghunathan, Quality configurable reduce-and-rank for energy efficient approximate computing, in *2015 Design, Automation & Test in Europe Conference & Exhibition (DATE)* (2015), pp. 665–670. https://doi.org/10.7873/DATE.2015.0569. ISSN: 1558-1101
67. A. Ranjan, S. Venkataramani, S. Jain, Y. Kim, S.G. Ramasubramanian, A. Raha, K. Roy, A. Raghunathan, Automatic synthesis techniques for approximate circuits, in ed. by Reda, S., Shafique, M., *Approximate Circuits: Methodologies and CAD* (Springer International Publishing, Cham, 2019), pp. 123–140. https://doi.org/10.1007/978-3-319-99322-5_6
68. P. Roy, R. Ray, C. Wang, W.F. Wong, ASAC: Automatic sensitivity analysis for approximate computing, in *Proceedings of the 2014 SIGPLAN/SIGBED Conference on Languages, Compilers and Tools for Embedded Systems* (ACM, Edinburgh United Kingdom, 2014), pp. 95–104. https://doi.org/10.1145/2597809.2597812. https://dl.acm.org/doi/10.1145/2597809.2597812
69. F. Saqib, A. Dutta, J. Plusquellic, P. Ortiz, M.S. Pattichis, Pipelined decision tree classification accelerator implementation in FPGA (DT-CAIF). IEEE Trans. Comput. **64**(1), 280–285 (2015). https://doi.org/10.1109/TC.2013.204. Conference Name: IEEE Transactions on Computers
70. J. Schmidhuber, Deep learning in neural networks: an overview. Neural Netw. **61**, 85–117 (2015). https://doi.org/10.1016/j.neunet.2014.09.003. https://linkinghub.elsevier.com/retrieve/pii/S0893608014002135
71. L. Sekanina, Z. Vasicek, V. Mrazek, Automated search-based functional approximation for digital circuits, in ed. by Reda, S., Shafique, M., *Approximate Circuits* (Springer International Publishing, Cham, 2019), pp. 175–203. https://doi.org/10.1007/978-3-319-99322-5_9. http://link.springer.com/10.1007/978-3-319-99322-5_9
72. S. Sidiroglou-Douskos, S. Misailovic, H. Hoffmann, M. Rinard, Managing performance vs. accuracy trade-offs with loop perforation, in *Proceedings of the 19th ACM SIGSOFT Symposium and the 13th European Conference on Foundations of Software Engineering - SIGSOFT/FSE '11* (ACM Press, Szeged, 2011), p. 124. https://doi.org/10.1145/2025113.2025133. http://dl.acm.org/citation.cfm?doid=2025113.2025133
73. SIPI Image Database (1977). https://sipi.usc.edu/database/
74. D. Tong, Y.R. Qu, V.K. Prasanna, Accelerating decision tree based traffic classification on FPGA and multicore platforms. IEEE Trans. Parall. Distrib. Syst. **28**(11), 3046–3059 (2017). https://doi.org/10.1109/TPDS.2017.2714661. Conference Name: IEEE Transactions on Parallel and Distributed Systems
75. J.Y.F. Tong, D. Nagle, R.A. Rutenbar, Reducing power by optimizing the necessary precision/range of floating-point arithmetic. IEEE Trans. Very Large Scale Integr. Syst. **8**(3), 273–286 (2000). https://doi.org/10.1109/92.845894. Conference Name: IEEE Transactions on Very Large Scale Integration (VLSI) Systems
76. M. Traiola, A. Savino, M. Barbareschi, S.D. Carlo, A. Bosio, Predicting the impact of functional approximation: From component- to application-level, in *2018 IEEE 24th International Symposium on On-Line Testing And Robust System Design (IOLTS)* (2018), pp. 61–64. https://doi.org/10.1109/IOLTS.2018.8474072. ISSN: 1942-9401
77. B. Van Essen, C. Macaraeg, M. Gokhale, R. Prenger, Accelerating a random forest classifier: Multi-core, GP-GPU, or FPGA? in *2012 IEEE 20th International Symposium on Field-Programmable Custom Computing Machines* (2012), pp. 232–239. https://doi.org/10.1109/FCCM.2012.47
78. Z. Vasicek, Formal methods for exact analysis of approximate circuits. IEEE Access **7**, 177309–177331 (2019). https://doi.org/10.1109/ACCESS.2019.2958605. Conference Name: IEEE Access
79. Z. Vasicek, L. Sekanina, Circuit approximation using single- and multi-objective cartesian GP. Europ. Confer. Genetic Programm. **9025**, 229 (2015). https://doi.org/10.1007/978-3-319-16501-1_18

80. S. Venkataramani, A. Ranjan, K. Roy, A. Raghunathan, AxNN: Energy-efficient neuromorphic systems using approximate computing, in *2014 IEEE/ACM International Symposium on Low Power Electronics and Design (ISLPED)* (2014), pp. 27–32. https://doi.org/10.1145/2627369.2627613
81. S. Venkataramani, A. Raghunathan, J. Liu, M. Shoaib, Scalable-effort classifiers for energy-efficient machine learning, in *Proceedings of the 52nd Annual Design Automation Conference* (ACM, San Francisco California, 2015), pp. 1–6. https://doi.org/10.1145/2744769.2744904. https://dl.acm.org/doi/10.1145/2744769.2744904
82. Z. Wang, A. Bovik, H. Sheikh, E. Simoncelli, Image quality assessment: from error visibility to structural similarity. IEEE Trans. Image Process. **13**(4), 600–612 (2004). https://doi.org/10.1109/TIP.2003.819861. http://ieeexplore.ieee.org/document/1284395/
83. X. Wu, V. Kumar, J. Ross Quinlan, J. Ghosh, Q. Yang, H. Motoda, G.J. McLachlan, A. Ng, B. Liu, P.S. Yu, Z.H. Zhou, M. Steinbach, D.J. Hand, D. Steinberg, Top 10 algorithms in data mining. Knowl. Informat. Syst. **14**(1), 1–37 (2008). https://doi.org/10.1007/s10115-007-0114-2
84. Q. Xu, T. Mytkowicz, N.S. Kim, Approximate computing: a survey. IEEE Design Test **33**(1), 8–22 (2016). https://doi.org/10.1109/MDAT.2015.2505723. Conference Name: IEEE Design Test
85. S. Yang, *Logic Synthesis and Optimization Benchmarks User Guide: Version 3.0* (Citeseer, 1991)
86. Z. Yang, A. Jain, J. Liang, J. Han, F. Lombardi, Approximate XOR/XNOR-based adders for inexact computing, in *2013 13th IEEE International Conference on Nanotechnology (IEEE-NANO 2013)* (2013), pp. 690–693. https://doi.org/10.1109/NANO.2013.6720793. ISSN: 1944-9399
87. T. Yeh, P. Faloutsos, M. Ercegovac, S. Patel, G. Reinman, The art of deception: Adaptive precision reduction for area efficient physics acceleration, in *40th Annual IEEE/ACM International Symposium on Microarchitecture (MICRO 2007)* (2007), pp. 394–406. https://doi.org/10.1109/MICRO.2007.9. ISSN: 2379-3155

Evaluation of the Functional Impact of Approximate Arithmetic Circuits on Two Application Examples

Jordi Fornt, Leixin Jin, Josep Altet, Francesc Moll, and Antonio Rubio

1 Introduction

Basic arithmetic operations (add, subtract, multiply, divide) are essential in computing systems. Arithmetic units are part of the Arithmetic Logic Unit (ALU) in processor systems to perform these operations. High-performance processors include several specialized units to support various flavors of number representations: fixed-point or floating-point, with varying number of bits. Each representation is selected based on the application. Simple processors may support only integer operations with a fixed number of bits. Accelerators in a computing system also usually include their own arithmetic units, either integer or floating-point. Integer adders are also essential in control units, for example, to compute jump addresses in a program.

Some applications require as much precision as possible and are not at all tolerant of errors in the computation. One example is the address calculation in the program control flow, in which an error may cause a catastrophic result in most cases. Many scientific computing applications also demand high precision and accuracy in the results. However, other applications are error tolerant, such as, for example, those in which the result has to be interpreted by a human: image and audio processing, among others.

In the early 2000s, a number of authors considered the advantage of tolerating imperfect circuits. Breuer introduced the concept of error-tolerance with respect

J. Fornt · F. Moll
Department of Electronic Engineering, Universitat Politècnica de Catalunya, Barcelona, Spain

Barcelona Supercomputing Center, Barcelona, Spain

L. Jin · J. Altet · A. Rubio (✉)
Department of Electronic Engineering, Universitat Politècnica de Catalunya, Barcelona, Spain
e-mail: antonio.rubio@upc.edu

W. Liu et al. (eds.), *Design and Applications of Emerging Computer Systems*,
https://doi.org/10.1007/978-3-031-42478-6_16

to an application or circuit which contains defects that cause internal errors that may cause punctual functional errors, and still the system that incorporates this application or circuit produces acceptable results to the end user [2]. Palem and his group advocated for a trade-off between accuracy of the hardware circuit and thermodynamic energy of a device [16], considering the random errors produced by noise on the logic operation. Shanbhag et al. [6] studied how input-dependent errors due to voltage overscaling can be compensated at the algorithmic level.

These early works considered random errors or physical defects as the cause of errors and how those errors still could be tolerated at the application level for certain types of applications. These approaches and conclusions raised a question: can we purposefully introduce errors by design to obtain some benefit and still obtain error tolerance? More specifically, can we design arithmetic units which provide purposefully inexact results? If the errors thus introduced can be tolerated at the algorithmic level, the use of inexact or approximate arithmetic units has several advantages:

- A reduction in the number of logic gates. This reduction has an impact on area and power, potentially timing as well.
- A reduction in the timing critical path. Some approximate unit designs have the explicit goal of reducing the critical path, which can potentially lead to an increased throughput (by increasing the clock frequency) or a decreased voltage operation and thus power consumption.

Several methods have been proposed to evaluate the numerous proposals for approximate arithmetic unit designs [7]. Some of them are based on the quantification of the error by calculating a statistical measurement: mean error distance (MED) and its relative version (MRED); mean squared error (MSE) and its root (RMSE), as well as the worst-case error. Circuit parameters are also used to assess the approximate circuits: power-delay product, area-delay product, energy-delay product.

These metrics are useful to compare different designs. However, they do not provide information on what the impact on the application is when these units are part of the system.

This chapter presents some examples in which exact arithmetic units are substituted by approximate units, and characteristics at the application level are presented in order to evaluate the impact of this substitution. First, Section 2 enumerates the different approximate adders and multipliers considered in the subsequent sections. Section 3 uses approximate units in the design of a filter for signal processing, calculating their impact using filter parameters. Section 4 shows the results of the YOLO neural network when different approximate units are considered, using the mean average precision in object identification as the metric. Finally, in Sect. 5, some conclusions are drawn.

2 Description of Approximate Arithmetic Units

In this section, we introduce the arithmetic units (adders and multipliers) that will be considered in the application examples. They have been selected as good candidates based on their timing and area savings according to the literature. It is not the purpose of this chapter to present an exhaustive list of approximate arithmetic units, which are available in the literature presented in the reference section of this chapter.

2.1 *Approximate Adders*

Adders are basic blocks in arithmetic units and also key components in multipliers. We introduce in this section three approximate strategies: the lower-OR adder (LOA), the generic accuracy configurable adder (GeAr), and the truncated adder (TruA).

2.1.1 Lower-OR Adder (LOA)

The lower-OR adder (LOA) [12] modifies the *"a"* least significative bit stages of an exact *N*-bit adder, substituting these *"a"* stages by OR gates instead of full adder blocks, thus cancelling the carry propagation that could exist between these bits, as depicted in Fig. 1. The effect is a moderate hardware reduction and mainly a faster calculation (the carry propagation chain is reduced to *N*-*a*) reducing critical path requirements.

2.1.2 Generic Accuracy Configurable Adder (GeAr)

Proposed in Shafique et al. [19], the GeAr adder splits the global addition into several sub-additions with independent and unconnected carry chains. This strategy reduces drastically the propagation delay of the adder to the propagation time of

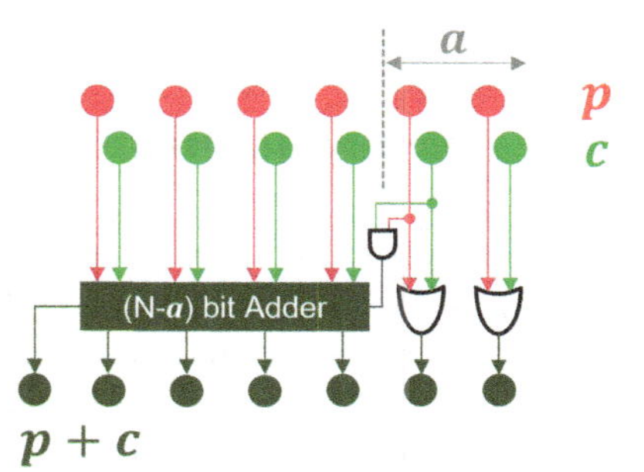

Fig. 1 Principle of approximation in the lower-OR adder

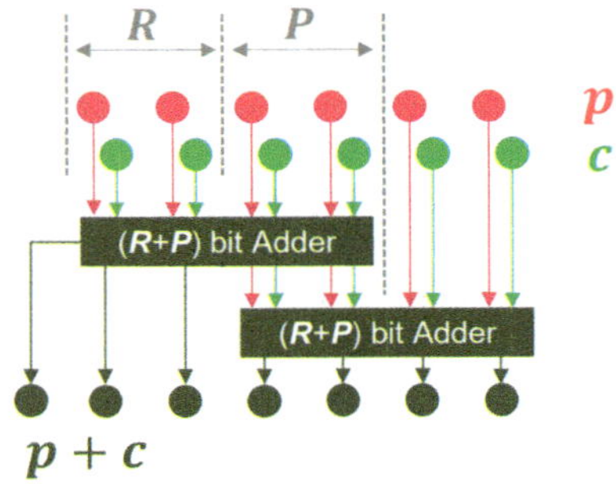

Fig. 2 Approximation strategy for the GeAr adder

the longest sub-addition. Figure 2 illustrates how a 6-bit adder can be spited into two 4-bit adders. The four least significant bits of the final addition are directly the result of the addition of the 4 least significant bits of the operands with one adder. The remaining adders (just one in this example) add up the Previous (P) bits and the Result (R) bits. Previous bits are simultaneously connected both adders in Fig. 2, and they are used to perform the addition of the Result bits with higher accuracy: thanks to them, partial previous carry propagation is considered. The total number of adders depends on the size of the operands and the size of the adders. For instance, if the operands are 8 bits and $R = P = 2$, a total of 3 adders are needed, whereas if $R = 1$ and $P = 3$, the number of adders required is 5. The GeAr multiplier is also known as ACA1 when $R = 1$ and ACA2 when $R = P$.

2.1.3 Truncated Adder (TruA)

The truncated adder takes "a" number of bits of the lower significant adder and sets them to a fixed value, reducing the length of the adder to N-a bits, and consequently the adder propagation time and power consumption. The fix value of the lower "a" bits can be 0, giving as result the truncated adder (TruA [7]), or 1, called the set-one adder (SOA). In fact, any combination of values for that fixed section would be considered a truncated adder. Moreover, another design decision is the value of the input carry for the (N-a) bit adder, which can be either 1 or 0. Figure 3 shows the basic scheme of truncated adders.

2.2 *Approximate Multipliers*

The multiplier unit is typically the most area- and power-consuming unit in arithmetic circuits and the cause of the main power and performance limitations. In this section, we will introduce five approximate multiplier units: the under-designed multiplier, the broken array multiplier, the approximate booth multiplier, the carry-in prediction multiplier, and the approximate logarithmic multiplier.

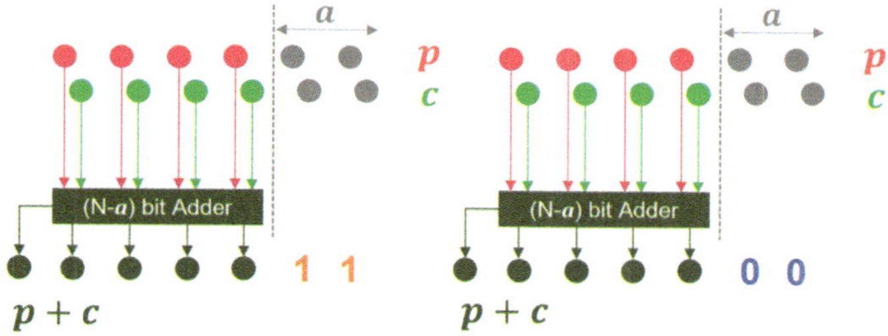

Fig. 3 Concept of truncated adder (left) and set-one adder (right)

2.2.1 Under-Designed Approximate Multiplier (UDM)

The under-designed multiplier (UDM) [10] is a modification of a recursive multiplier simplifying its structure. Recursive multipliers use smaller multipliers to generate a complete wider operands multiplier. A 2 × 2 bits multiplier block is the smallest block used to generate wider operand multipliers following the recursive strategy. For instance, if A and B are two positive numbers encoded with 4 bits, they can be written as:

$$A = A_H \cdot 2^2 + A_L \tag{1}$$

$$B = B_H \cdot 2^2 + B_L$$

Where A_H and B_H are respectively the two most significative bits of A and B, and A_L and B_L are the two least significative bits of A and B. Then, the product A·B can be expressed as an addition of the partial products given by four 2 × 2 bits multipliers:

$$A \cdot B = A_H \cdot B_H \cdot 2^4 + A_H \cdot B_L \cdot 2^2 + A_L \cdot B_H \cdot 2^2 + A_L \cdot B_L \tag{2}$$

The under-designed approximate multiplier simplifies the basic block multiplier, thereby saving time and power following the principle of recursive multiplier. Figure 4 shows a simplification of a basic 2 × 2 bits multiplier block. Note that the number and complexity of logic gates decrease significantly compared with the exact circuit (8 gates in the exact case and 5 in the inexact, the XOR gates of the exact circuit are replaced by simpler OR gates). Despite the simplification, the approximate multiplier yields high accuracy, causing only one error out of 16 possible combinations, being very close results to the exact ones. Using the inexact 2 × 2 unit, an N-bit approximate multiplier can be constructed applying the recursive principle. In the design of this N-bit multiplier, one can use all 2 × 2 bits units as approximate, or only a part of them, resulting in different levels of area saving and accuracy.

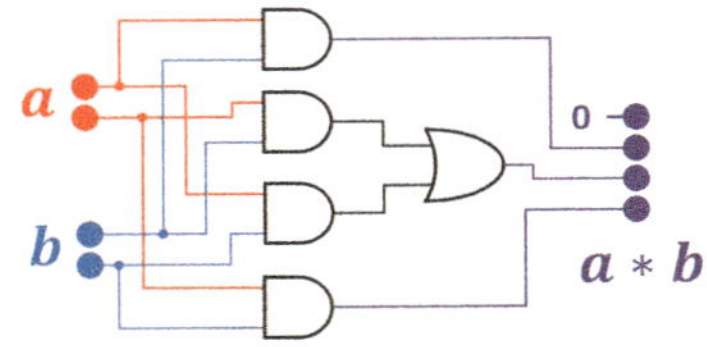

a/b	00	01	10	11
00	0000	0000	0000	0000
01	0000	0001	0010	0011
10	0000	0010	0100	0110
11	0000	0011	0110	0111

Fig. 4 Simplified 2 × 2 multiplier basic block, reducing hardware and delay. Bottom: Karnaugh map of the simplified multiplier showing the error combination causing approximate results when used in a recursive circuit

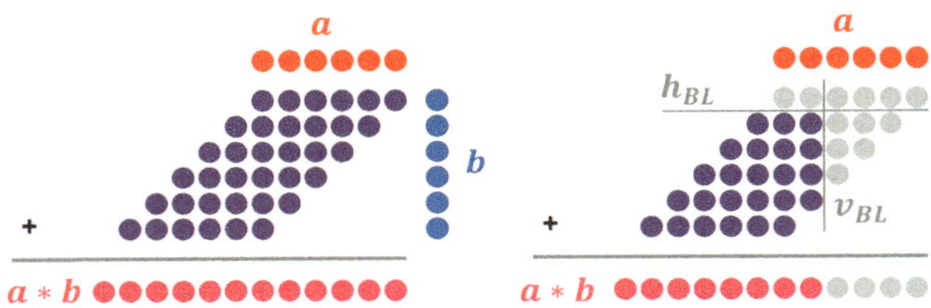

Fig. 5 Exact and BAM multiplier using partial product bits strategy

2.2.2 Broken Array Multiplier (BAM)

The broken array multiplier (BAM) [12] simplifies an exact array multiplier by removing some of the cells that generate the partial product bits. Figure 5 shows the scheme of an exact multiplier and a BAM, for comparison.

This architecture defines two parameters that allow tuning the precision and complexity of the multiplier: the horizontal break line and the vertical break line (h_{BL} and v_{BL} respectively, BAM (h_{BL}, v_{BL}) in Fig. 5).

2.2.3 Approximate Booth Multiplier (ABM)

Approximate booth multipliers (ABM) apply some simplifications on the encoding logic of a booth multiplier. There are many examples of ABMs in the literature, and here two proposed versions [22] are presented: the ABM-M1 and the ABM-M3. Both multipliers are based on the radix-4 booth encoding.

The ABM-M1 simplifies the partial product generation by omitting the 2x factor present for some input combinations of the radix-4 encoding, as well as some further circuit modifications. Figure 6 shows the modified partial product generation circuit. The multiplier is constructed by combining this approximate circuit with its exact counterpart, with the number of simplified partial products being defined by

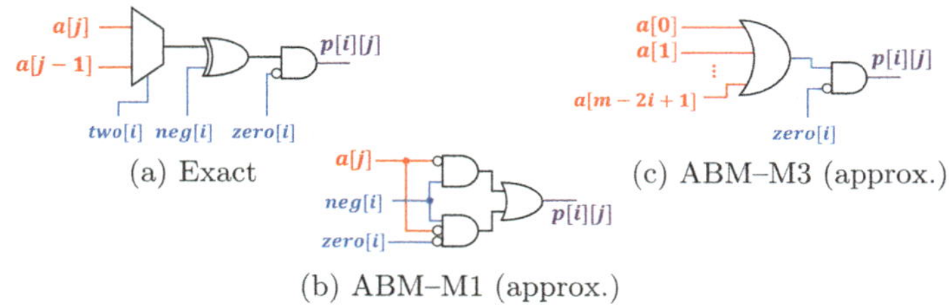

Fig. 6 Exact and approximate partial product generation circuits used in ABM-M1 and ABM-M3. (Adapted from Ref. [22])

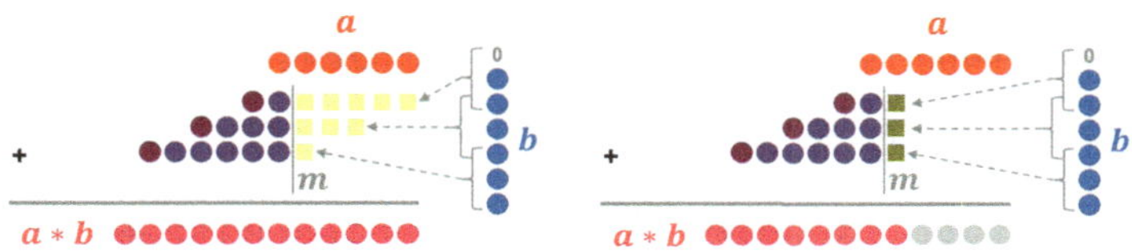

Fig. 7 Main idea behind ABM-M1 and ABM-M3. Purple: bits generated by the exact circuit. Yellow: generated by the ABM-M1 circuit. Brown: generated by the ABM-M3 circuit

a parameter *m*. By tuning *m*, the circuit can be configurated for different levels of hardware saving and output error magnitude.

The ABM-M3 applies a more aggressive simplification by only taking into account the case in radix-4 encoding in which the partial product is zero (see Fig. 7). Furthermore, all partial product bits with indices lower than *m* use an OR gate, so if any of the simplified partial product bits was non-zero, the most significant simplified bit is set to high. This effectively reduces the number of partial product bits to sum, as well as the active bits in the multiplier output.

ABM multipliers can achieve a good accuracy-complexity trade-off. For instance, in the case of ABM-M3, metrics of area and power can be reduced up to 31% and 23%, respectively, for m = 12 (TMSC 65 nm 16-bit operands [22]).

2.2.4 Carry-In Prediction Multiplier

This multiplier is based on the work from Kartikeya et al. [8], whose strategy consists of the separation of the multiplication into three parts: the multiplication of high significant bits part (H), the low significant bits part (L), and the mid-low significant part (ML). This multiplier operates accurately on the high significant bits part and on the low significant bits part, while an inaccurate multiplication will be applied in the mid-low bits part. The approximation of that part is based on putting the results by 1's and bringing a carry to the left column.

The complexity of the partial products increases with the number of bits. For example, the multiplication of two operands of 4 bits will have 4 partial products, while in case of 8 bits operands, the number of partial products will be 8. The

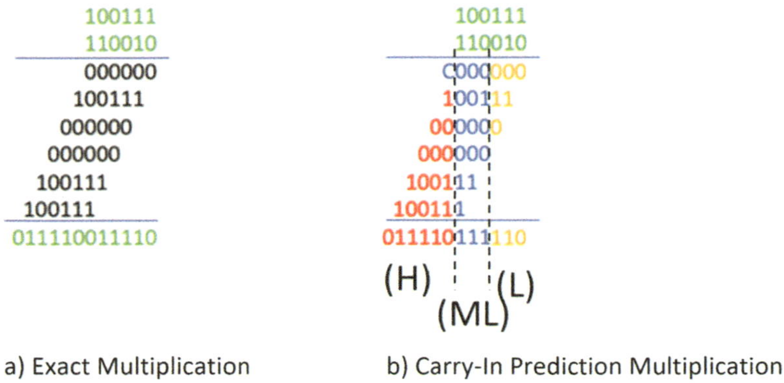

Fig. 8 Numeric example of carry-in prediction multiplier

advantage of this multiplier is that we can reduce the computation complexity of the partial products.

Figure 8 shows an example of approximate multiplication, and the partial products are separated into three parts: yellow for low significant bits (L), blue for mid-low bits (ML), and red for more significant bits part (H). The blue products are the ones calculated approximately; this means the result has been forced to 1's bringing a carry (C) to the left column. In the example, the exact result is 39*50 = 1950 while the approximate is 1982, giving a relative error of 1.64%. All the adders needed in the exact multiplier to obtain the ML bits of the result are suppressed in the approximated version.

2.2.5 Logarithmic Multiplier (LM)

This approximate logarithmic multiplier is based on the work of Mitchell [13], where the properties of logarithms are used to perform the multiplication, reducing the multiplication to an addition and saving time, energy, and logic resources in comparison with other multipliers. The basic principle is to make multiplications (divisions) by using adders (subtractors), with the only use of shift registers and adder units.

$$Log_2(P) = Log_2\,(A\cdot B) = Log_2(A) + Log_2(B) \tag{3}$$

While the multiplication has been simplified to an addition, the complexity now resides in how to obtain the logarithms of A and B and how to calculate the antilogarithm of the result of the addition. Computing the exact logarithm can be very costly, so Mitchell proposed an approximation to calculate the logarithm by defining a fixed-point number with the index of the leading 1 forming the integer part, and all the bits to the right of the leading one forming the decimal part. For

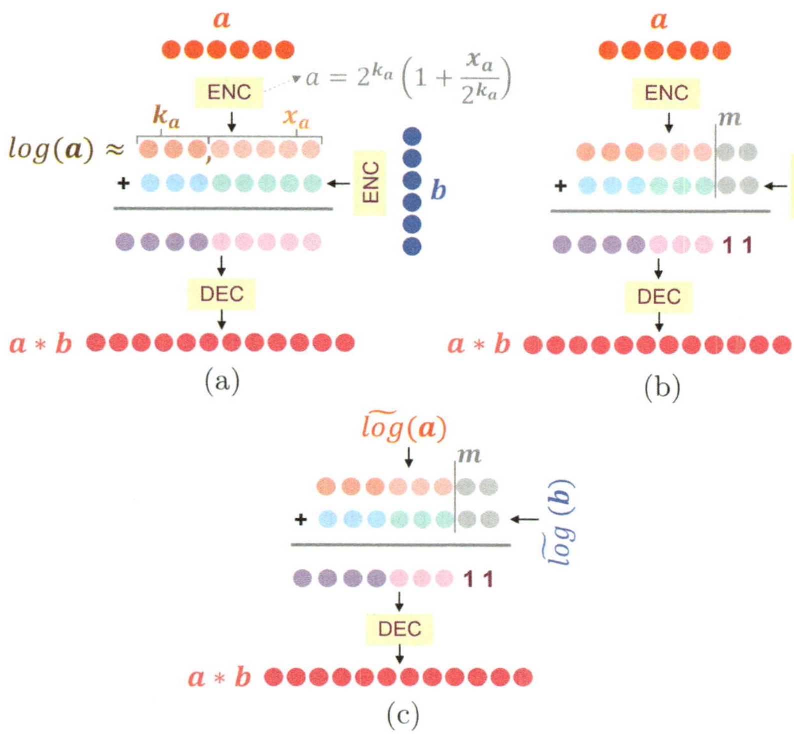

Fig. 9 Main idea behind LM, ALM-SOA, and encoder-less ALM-SOA

instance, the binary number A = b10100011 has the leading one (i.e., the 1 in the number with the most significant weight) in the seventh position, whereas the number B = b01001000 has it in the sixth position. Following Mitchell's algorithm, $\text{Log}_2(A)$ would be approximated by b111. 0100011 (7.27 instead of the exact value 7.34) and $\text{Log}_2(B)$ would be approximated by b110.001000 (6.125 instead of the exact value 6.169).

An LM, as depicted in Fig. 9, consists of two encoder modules that perform the logarithmic approximations, a single adder that sums the logarithm and a decoder that performs the complementary antilogarithmic approximation. In these architectures, the encoder and decoder modules are usually the most area- and power-consuming sections.

Approximate logarithmic multipliers (ALMs) typically apply simplifications to the addition in log space. Note that even though ALMs are called approximate, regular LMs are not exact either. A popular ALM architecture is the ALM with set-one adder (SOA) [11], which skips the lowest bits of the addition and sets the result bits to 1 (see Fig. 9b, c). Basic LMs have the feature that they always underestimate the value of the product if compared to the exact solutions. When an adder like SOA is introduced (that overestimates additions), the average error can be reduced.

Additionally, an LM can be further simplified by taking the encoders out of the multiplier [15] (see Fig. 9c). Given that the log-encoding process depends only on the operand values, we can compute the approximate logarithms beforehand and pass them to the LM as inputs. This makes sense in the context of an application where the input signal has to be multiplied by a set of constants (e.g., 1D or 2D digital filters). The logarithmic transformation of all these constants can be computed beforehand and stored in the hardware in the logarithmic form.

2.3 Comparison of Approximate Multiplier Approaches

Approximate multipliers have to be compared analyzing their implementation performances (area, power, delay) as well as their accuracy figures of merit. Table 1 shows [4, 23] the relative area, power, and delay (with respect to the exact version implementation) reduction (negative mean increment) and the normalized mean error distance (NMED, ten million input pairs) for a specific implementation of four type of multipliers (40 nm CMOS and 16-bit operands). UDM shows a low accuracy in terms of error distance and a relative high circuit overhead. BAM(8) multipliers show good performances set with high area and power reduction and reduced NMED. BAM(12) exhibits even higher performance improvements but at a higher mean error distance. LM shows a moderate NMED but with a high error rate and a reasonable improvement of circuit characteristics. The selection of the type of multiplier will be very influenced by the circuit requirements, target application (image, filters, object detection, etc.) concerned with the functional characteristics, its cost, its error metrics, and its fitness to the goal algorithms. Table 1 shows a comparison between the exposed multipliers.

Table 1 Comparison of circuit characteristics of power, delay, and area as well as NMED and ER error metrics for UDM, BAM, and LM multipliers

	2×2 UDM	BAM(8)	BAM(12)	LM
Delay (ns) ([a])	80.50	40.31	60.47	80.56
Power (μW) ([a])	21.26	25.05	37.56	36.37
Area (μm^2) ([a])	125.22	174.03	261.04	223.18
Relative area reduction ([b])	−20%	+50%	+70%	+30%
Relative power reduction ([b])	+20%	+60%	+65%	+50%
Relative delay reduction ([b])	−10%	+10%	+50%	+20%
NMED ([b])	0.06	0.02	0.08	0.04
ER (%) ([b])	85	98.60	99.99	1

[a]Goswami et al. [4], [b]Wu et al. [23]

3 Application to Digital Filters

3.1 Filter Description and Specifications

The work by Navarro [14] analyzed the effects of approximate operators when used in a finite impulse response (FIR) filter. Some results are shown in this section. Figure 10 shows the structure of a discrete-time finite impulse response (FIR) filter or order N, where $x[n]$ is the input signal, $y[n]$ is the output signal, and b_i is the ith filter coefficient. Then, the nth sample of the output can be calculated as:

$$y[n] = \sum_{i=0}^{N} b_i \cdot x[n-i] \tag{4}$$

As exemplary purposes, a 12th-order low-pass filter has been designed [14] with the purpose of suppressing an interfering 25 kHz signal superimposed to a 1 kHz signal. Figure 11 shows the spectrum of the signal applied to the input of the filter. To get the signal-to-noise ratio of Fig. 11, the input samples are encoded with 12 bits binary, and the sampling frequency is 100 kHz.

With the goal of achieving an attenuation of the interfering signal of 64 dB, the filter coefficients have been calculated and are encoded with 18 bits. This implies that, at the output of the multipliers, the result of the operation is encoded with 30 bits. To get the sample y[n], the output of the 13 multipliers has to be added up. The reduction of all these terms can be done with 4 levels of adders. As a binary addition (all numbers are positive, binary encoded) gives a result encoded with an extra bit respect to the number of bits of the summands, the final result is encoded

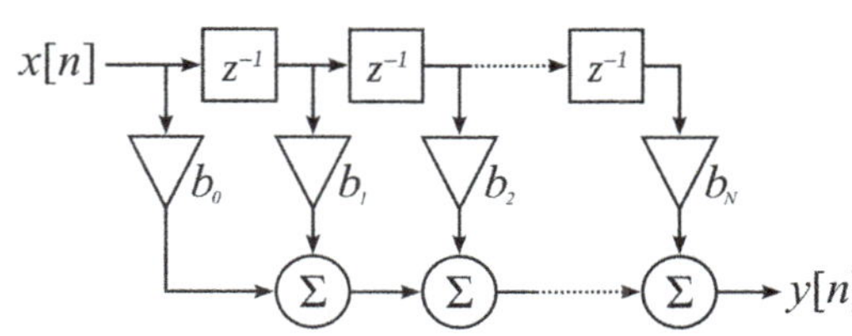

Fig. 10 Structure of FIR filter of order N. (Source: Wikipedia)

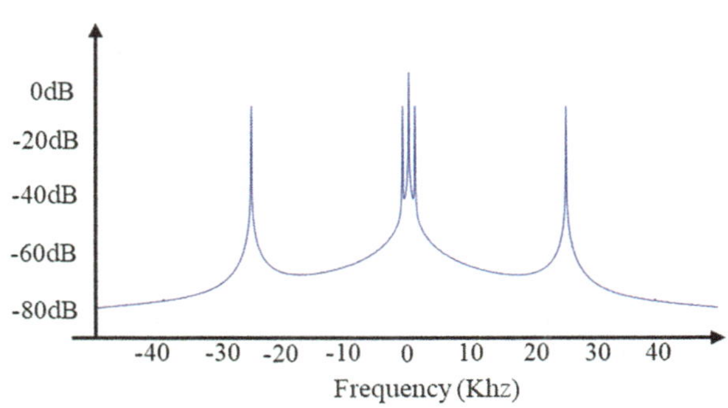

Fig. 11 Spectrum of the signal applied to the input of the FIR filter. Goal of the filter: to suppress the interfering 25 kHz signal

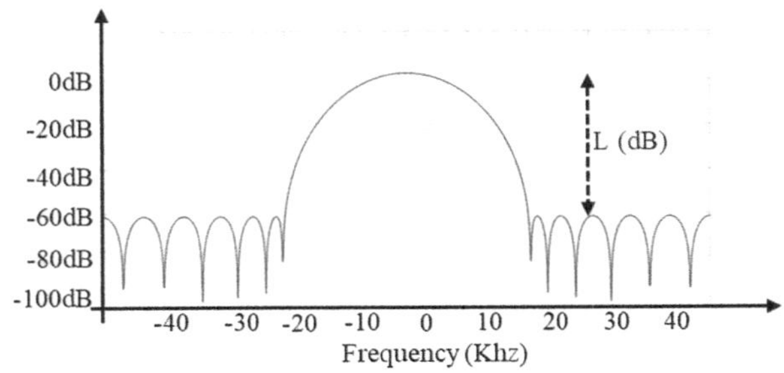

Fig. 12 Impulsional response of the exact FIR filter. Attenuation of $L = 64$ dB

Table 2 Characteristics of the impulsional response shown in Fig. 12

	LOA		GEAR	
Size of adder	BW(kHz)	L(dB)	BW	L(dB)
16	12.15	64.1	12.15	64.1
6	11.3	42.34	12.04	60
2	11.9	18.9	13.74	19.3

with 34 bits, which is later on truncated to get the original 12 bits/sample. The total hardware requirements of the filter are: 12 flip-flops, 13 multipliers and 12 adders. Figure 12 shows the impulse response of the filter. The attenuation between the main lobe and the secondary lobes is $L = 64$ dB. The bandwidth of the first lobe is 12.1 kHz. We are going to denominate this filter as exact FIR filter. Table 2 shows the characteristics of the impulsional response.

3.2 *Effects of Approximate Operators in the Filter Specs*

The first experiment conducted by Navarro [14] analyzed how the transfer function of the filter is affected when the adders of the filter are replaced by approximated adders. Figure 13 shows the Impulsional Response of 6 different circuits, where the approximate adders used are either lower-OR adder (LOA) or generic accuracy configurable adder (GEAR). For each adder, three different levels of aggressiveness have been considered, depending on the size of the adders used: 16 bits (LOA (16), GEAR (8 8)), 6 bits (LOA (6), GEAR (3 3)), or 2 bits (LOA (2) or GEAR (1 1)). For the LOA adders, the number in parenthesis indicates the size of the adder, not the truncation parameter (a in section 2.1.1). When the goal is to replace a full precision 30 bits adder, LOA(16) implies a truncation parameter $a = 30–16$. LOA(6) implies a truncation parameter $a = 30–8$ whereas LOA(2) implies a truncation parameter of $a = 28$.

When the exact adders are replaced by LOA(16) or GEAR(8,8), the impulsional response of the original circuit does not experience major degradations. Stronger reduction of the attenuation and loss of secondary lobes are observed when the approximation is more aggressive. The higher complexity of the GEAR (3 3) approximate adder explains the good performance of this version of the filter when

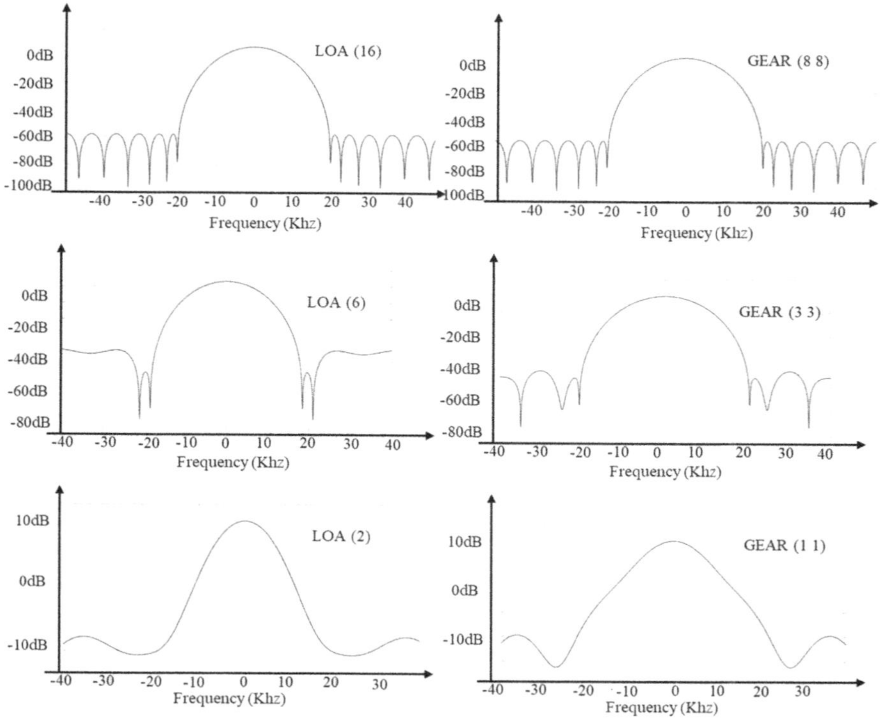

Fig. 13 Impulsional response of the FIR filter as a function of the approximate adder used

compared to the LOA (6) version: to replace the exact adder with the GEAR(3,3), the 30 bits adder is replaced by 9 adders of 6 bits (54 full adders!). On the other hand, replacing the original exact 30 bits adder by the LOA (6) adder requires just one 6 bits adder and 24 OR gates (the truncation factor is a = 24). Figure 14 (right) shows the spectrum of the signal at the output of the filter for the exact case and when the adder is the LOA. In the approximated case LOA (16), although the attenuation between the 25 kHz signal and the 1 kHz signal is the same as in the exact case, the noise level of the output signal has increased. When the LOA (6) adder is used, the attenuation L decreases and the noise level increases with respect to the LOA (16). The effect of this change in noise level is observable in the output signal, when represented in the time domain (Fig. 14 left).

Figure 15 shows the effects on the impulsional response and the spectrum of the signal at the output of the filter when the multipliers in the exact filter have been replaced with exact ones (adders are exact). Two different approximate multipliers are used: under-design multiplier (UDM) and the logarithmic multiplier (LM). The attenuation L obtained with the UDM is higher than the one obtained with the LM. However, the noise level is higher in the UDM version than in the LM version. The 3 dB bandwidth is 13.7 for the LM and 11.5 for the UDM.

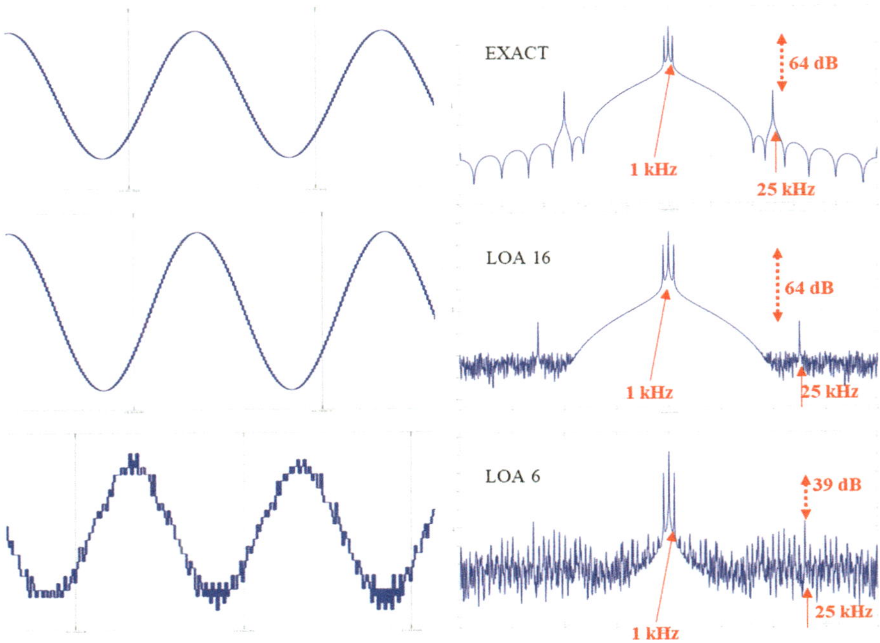

Fig. 14 Signal at the output of the filter for the exact case and when two LOA adders are used. Spectrum of the output signal. The amplitude of the signal at 1 kHz and 25 kHz (interfering signal) is marked. At the input, both signals had the same amplitude

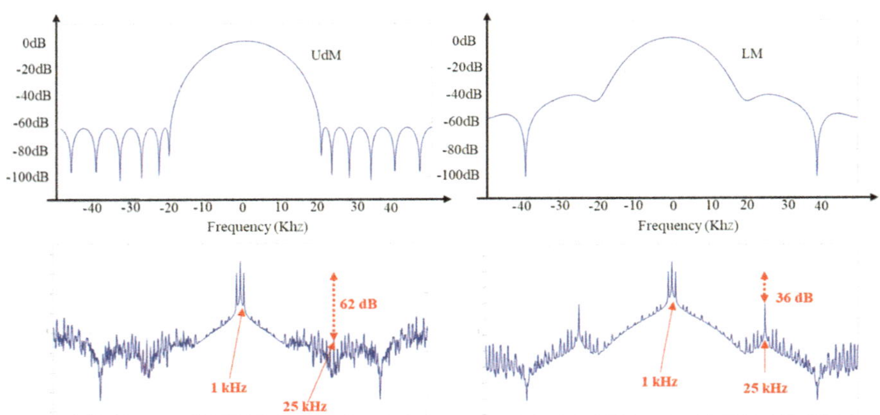

Fig. 15 Impulsional response of the FIR filter and spectrum of the signal at the output of the filter when the multipliers are replaced by under-design multipliers (UdM-left) and logarithmic multipliers (LM)

In conclusion, approximate arithmetic modules are an interesting option in filters. The choice of the particular approximate module depends on the requirements of the final application, the particular nature of the input signal (audio, video, instrumentation), and the hardware resources available. In any case, results in Figs. 14 and 12 indicate that approximate adders and multipliers can be used in digital filters without remarkable performance degradation.

4 Application to Deep Neural Networks

4.1 Neural Network Basics

Artificial neural networks (often simply called neural networks) are machine-learning models loosely based on the idea of neuron activity in a biological brain. They are composed of different layers in which neurons are connected via learnable weights. Inference (also known as a *forward pass*) consists of feeding input values to the first layer (the *input layer*) and propagating the computation through the network. As depicted in Fig. 16, on a particular layer, this propagation consists of a linear combination of the previous layer's values ($a_i^{(l)}$) and the weight of the connection ($w_{i,j}^{(l)}$), plus a per-neuron bias value ($b_j^{(l)}$). The result of this sum is passed to a nonlinear function ($\sigma(\bullet)$) that decouples linear operations between layers.

Neural networks are trained by adjusting the weight and bias values according to experiences gathered from data. Depending on the application, training or learning can be categorized as supervised or unsupervised. *Supervised learning* happens when we have labeled data, i.e., when we know what the output of the network should be at any given moment. Many tasks can be framed as supervised learning, from image classification to depth perception, speech recognition, etc. *Unsupervised*

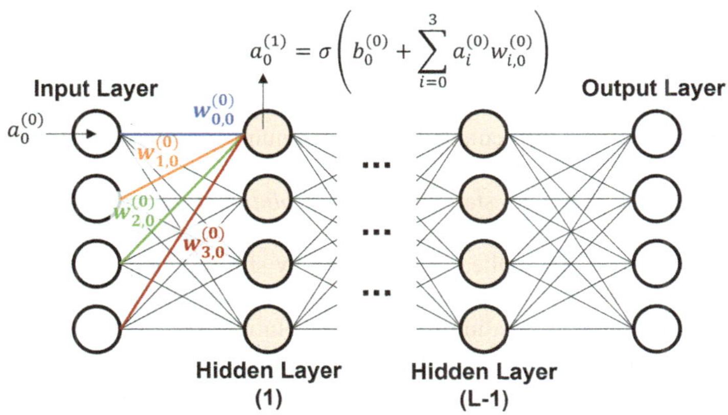

Fig. 16 Sketch of a fully connected ANN

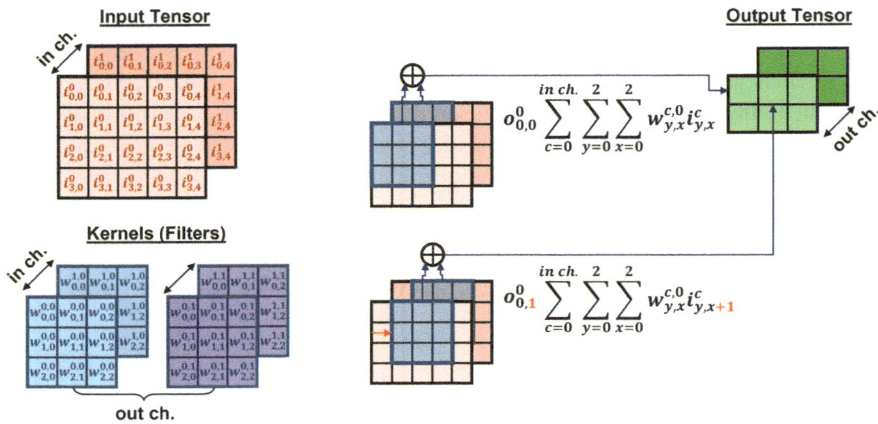

Fig. 17 Example of a convolution operation in a CNN convolutional layer

learning is used when the outputs or labels are unknown and is usually used in clustering applications.

In supervised learning, we define a *loss function* that penalizes deviations of the network output from the ideal values (the labels) and we try to minimize it by computing the gradient of the loss function with respect to every weight and bias and by updating (learning) the parameters in the gradient direction. *Stochastic gradient descent* (SGD) is the most used technique, in which we approximate the gradient by considering only a handful of training examples (a *batch*) and constantly nudge the weight and bias parameters in the estimated gradient direction.

Image recognition and computer vision tasks are some of the applications that have benefited the most from the advent of neural networks. The first networks used for these applications relied on the classical multilayer structure shown in Fig. 16. However, the full connectivity of the neuron layers proved to be problematic in terms of execution and learning. As every pixel in the image was a standalone input neuron and was connected to *all* the neurons in the subsequent layer, the number of connections (and weights) was extremely large, so processing was only possible for very small resolutions. Regarding training, the huge number of weights per layer made convergence more difficult and limited the number of layers that the networks could have.

To solve this, researchers started using *convolutional neural networks* (CNNs) to process images, see Fig. 17. In a CNN, the weights are represented as a set of *kernels* or *filters* that are applied to the image by shifting them through its spatial dimensions. This idea decouples the input resolution with the size of the model and its connections, allowing medium- to large-resolution images to be processed by these models without exploding in complexity.

Since their popularization in 2012 [9], CNNs have become the dominant model for computer vision tasks and have also been used in other applications such as speech recognition and temporal series analysis. Moving on from Krizhevsky's

AlexNet, many techniques have been proposed to improve the accuracy and generalization capabilities of CNNs. One of the most important trends that have arisen, motivated by successful networks such as VGG [20] and later ResNet [5], is the use of networks with a very large number of layers, known as deep convolutional neural networks (DCNNs) or just deep neural networks (DNNs).

In terms of DNN execution, most of the computational intensity resides in matrix-vector multiplications (for fully connected layers) and convolution operations (for convolutional layers). For the sake of discussion, it is useful to just consider convolutions, as matrix-vector operations can be formulated as a special case of a convolution operation. A convolution operation can be described as a series of multiply-add (MAD) operations inside several levels of recursion along different dimensions, as shown below:

```
for h in Tensor Height:
    for w in Tensor Width:
        for k in Output Channels:
            for c in Input Channels:
                for y in Kernel Height:
                    for x in Kernel Width:

                        O[k][h][w] += I[c][h+y][w+x]*W[k][c][y][x]
```

where O, I, and W are the tensors of outputs, inputs, and weights, respectively. This loop representation gives a good intuition about why DNNs can be so expensive to execute. Having so many nested loops, if the tensor dimensions or channel sizes are even moderately large, the number of multiplications and additions (which results from multiplying all these sizes) can easily get very big. For this reason, using general-purpose consumer-grade CPUs for executing these tasks usually results in unacceptable inference times. Hence, using specific hardware built for DNN execution is critical for many applications. In this setting, the use of approximations on this hardware systems can help achieve lower inference times and lower power consumption while having a negligible effect on the network accuracy, thanks to the inherent resilience and redundancy of these models.

4.2 *YOLO – DCNN for Object Detection*

While deep neural networks in general present some degree of resilience to noise and errors, their precise behavior and accuracy degradation under arithmetic approximations depend on the target application and network architecture. Tasks that are very simple, like the classification of MNIST handwritten digits, may be able to easily achieve high accuracy with extremely aggressive approximations that do not work in most other applications. Hence, when evaluating these circuits, one must be mindful of choosing a task and neural network that is nontrivial enough to

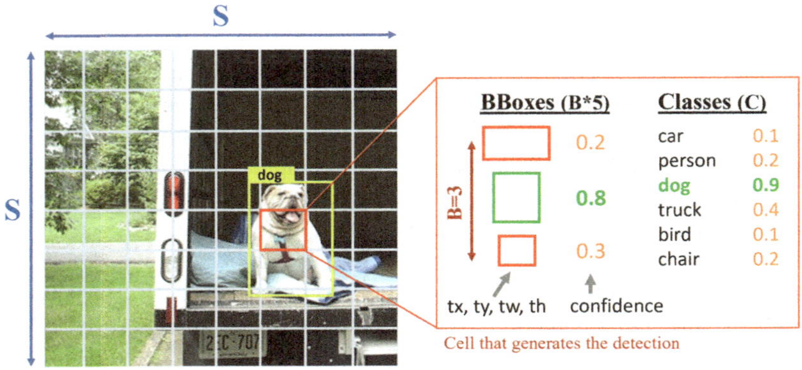

Fig. 18 Depiction of how YOLO generates a detection. Each possible bounding box (BBox) is defined by its position coordinates (tx, ty), size values (tw, th), and confidence score

ensure that the results are meaningful in general. For our discussion in the following pages, we will use the task of object detection (finding out the location and size of interest objects in a 2D image) as a means of evaluation and the YOLO deep neural network as our benchmark model.

You Only Look Once (YOLO) is a real-time object detection DCNN [17] that has been very successful in both academia and industry. It takes a color image as input and, after a single forward pass, it outputs a set of *bounding boxes* that mark the location and shape of the detected objects in the image. Each bounding box is defined by four coordinates (two describing its position and two for its size) plus a class value that defines the class that the particular object belongs to.

Before the proposal of YOLO, prior object detection networks worked by first generating *detection proposal regions*, zones of the image where there could be an object, and then analyzing each proposal region with a classification CNN. Hence, if N proposal regions were generated, these algorithms needed N forward passes of a whole CNN for analyzing a single image. The innovation of YOLO was to generate all the detections with only one forward pass of the network (hence the name).

The main idea is to partition the input image into a grid of $S \times S$ cells (see Fig. 18) and, for each of these cells, solve a regression task to find the most likely bounding box shape and object class. To do so, on each grid cell, a set of B possible bounding boxes, each one with different aspect ratio and scale, is checked. Besides the four coordinates that define each bounding box, a *confidence score* is evaluated to assess the probability of each box being the best fit. Additionally, for each cell, a *class probability* value is generated for every class, representing the probability of each object class to be present in the grid cell. In a post-processing step after the CNN forward pass, only cells with high-class probabilities are considered real detections. Hence, each grid cell is associated with $B_* 5 + C$ numeric values representing these concepts. These dimensions are encoded as the output channels of the network's output layer.

Regarding the evaluation of the object detection bounding boxes, many different metrics can be used to determine the fitness of the detections compared to the ideal result. The most used metric in the literature is the mAP (mean average precision), and to understand it, we need to define some intermediate metrics that are used to compute it.

To quantify the fitness of the bounding boxes, the intersection over union (IoU) metric is typically used. This metric is defined as the ratio between the overlap of the predicted and the ground truth bounding boxes, and the union of these two regions, ranging between 0 (no overlap) and 1 (perfect fit). To accept a detection, its IoU is calculated and compared with a threshold (e.g., 0.5). If the IoU is above the threshold and the object class is correct, the detection is considered a true positive (TP). On the other hand, if the detection has an IoU below the threshold or the object class is incorrect, it is considered a false positive (FP). When no detection is associated with an existing object, we call the non-detection a false negative (FN).

Based on these detections, to fully describe the fitness of the model, we need two metrics: *precision* and *recall*. These can be defined as follows:

$$\text{Precision} = \frac{\text{TP}}{\text{TP} + \text{FP}}; \text{Recall} = \frac{\text{TP}}{\text{TP} + \text{FN}} \tag{5}$$

The precision and recall of a detector give us different yet equally important insight on how good it is. Precision gives information on how valid the detections are (*how confident we are that a detection is correct*), while recall tells us how complete the detections are (*how confident we are that all objects were detected*). Note that we can have perfect precision or recall in arbitrarily bad detectors; e.g., consider a dummy detector that outputs all possible bounding boxes at every pixel as detections. As, by definition, it will never miss an object present in the image, it is a perfect detector in terms of recall. However, to see that this is no detector at all we can check its precision, which will be almost zero. Hence, a truly good detector should have both high precision and recall. Nevertheless, these two metrics are at odds with each other, and real detectors deal with this trade-off by adjusting the detection threshold to the desired precision-recall point. A low threshold results in a very sensitive detector, with high recall but low precision, and a high threshold results in a reliable detector with high precision, but low recall. In YOLO, this threshold would be applied to the class probability of the bounding boxes of each grid cell.

To take all these factors into account, detectors often create a precision-recall curve that depicts the whole extension of this trade-off. This curve, exemplified in Fig. 19 for different detectors, is obtained from varying the detection threshold value from the minimum to the maximum possible and shows the whole range of operating points that a particular detector can have.

From Fig. 19, it is clear that detectors with precision-recall curves on the top-right part of the plot are the best since they have both high precision and high recall in the intermediate regions. To quantify this as a metric, the *average precision* (AP) is defined as the area under the precision-recall curve of a detector and is typically

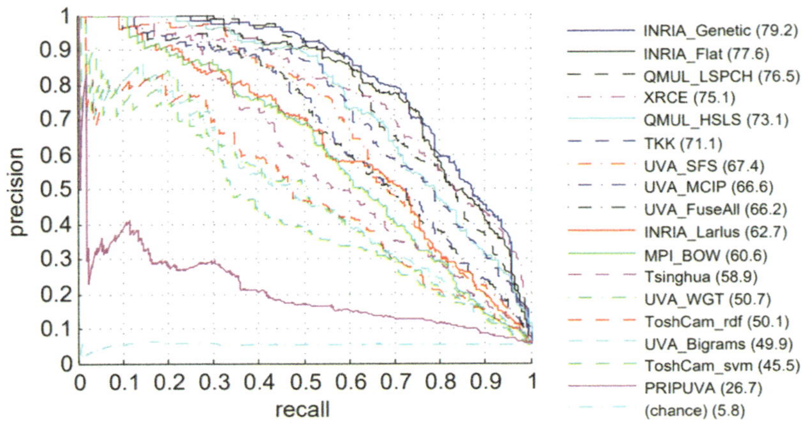

Fig. 19 Precision-recall curve of various object detection models and their associated AP, for a single class (train) of the PASCAL VOC challenge. (Source: Everingham et al. [3])

written as a percentage (see the numbers in the legend on Fig. 18). An AP of 100% (area of 1) would be a perfect detector, while an AP of 0% would be the worst detector possible.

When considering applications with different classes, the *mean average precision* (mAP) is used as a global metric including all object classes, and it is defined as the mean of the average precision values for all different classes. In real-time object detectors as YOLO, the mAP metric defined in the PASCAL VOC challenge is used in most works [3], and it is the metric we will consider in the following sections when discussing the effects of using approximate arithmetic units to execute the YOLO network.

4.3 Approximate FP MAD Units

For the evaluation of how the use of approximate adders and multipliers affects the functional performance of the YOLO network, a software implementation of this net can be used. An example is the open-source code written in C for the YOLOv3 network [18], which is available in Acton [1]. This code was developed to analyze the performances of YOLOv3 when numbers within the net are encoded with a float point of 16 bits, instead of the common 32 bits float point representation. To perform the evaluation of how the net behaves with approximate arithmetic operators, either the multiplication or the addition primitives (or both) that appear within the part of the code that implement convolution functions have to be replaced by functions that emulate the behavior of the approximate operators. An example is in Fig. 20, where the lines of the original code using the C-standard multiplier and adder operator are in green as they are commented and not compiled. The functions half_add,

```
for(k = 0; k < K; ++k){
    //sum += ALPHA*A[i+k*lda]*B[k+j*ldb];
    sum = half_add(sum, half_mul(ALPHA, half_mul2(A[i+k*lda], B[k+j*ldb])));
}
//C[i*ldc+j] += sum;
C[i*ldc+j] = half_add(C[i*ldc+j], sum);
```

Fig. 20 Example of how primitive multiplications and operators have been replaced within the C code by functions that emulate the behavior of approximate operators. Functions half_add, half_mult, half_mult2 substitute the original C operators

half_mult, half_mult2 replace the commented ones and emulate the behavior of approximate arithmetic operators.

This procedure has the advantage of providing flexibility as it allows to evaluate the effect of several operators, but as a drawback, instructions that were before executed with just one or two clock periods now can take up to several thousand, as the emulation of the approximated operator must be executed.

4.4 Effects of Approximate FP16 in YOLOv3

Using the C code depicted in the previous section, in this section, we are going to evaluate the performance of the YOLOv3 network in 5 different situations:

(i) *Exact_32*. Variables within the C code are defined as the C-default float type, i.e., float point numbers encoded with 32 bits. All additions and multiplications within the net are exact.
(ii) Exact_16. Variables within the C code are defined using a custom 16 bits float point type (10 bits of mantissa and 6 bits of exponent). All additions and multiplications within the net are exact.
(iii) Aprox_1. All operands are 16 bits float point numbers. Adders within the net are exact. Float point multipliers are approximated: the carry-in prediction multiplier is used to perform the product of the two 10 bits mantissa. In this implementation, the bits of the result product fixed to one (ML bits in Fig. 8) are from the ninth down to the 5th. No approximation is introduced in the processing of the exponent within the multiplier.
(iv) Aprox_2. Float numbers are encoded with 16 bits. The multipliers that perform the multiplication of the two mantissas are replaced by logarithmic multipliers. All float numbers are assumed to be normal. No approximation is introduced for the processing of the exponent or in the adders within the net.
(v) Aprox_3. Almost the same situation as Aprox_2. Here, an additional approximation has been introduced in the logarithmic multiplier for the mantissas: the addition needed in the logarithmic domain to perform the multiplication in the linear domain is performed with a set-one adder. Parameter $S = 4$ (6 bits set to 1). The carry input of this adder is set to 1.

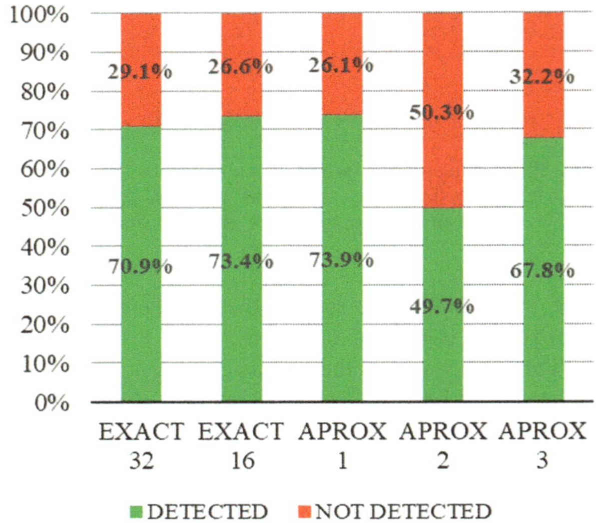

Fig. 21 Percentage of detected and undetected objects for each YoloV3 version analyzed. Total number of objects: 199. Total number of images: 25

Figures 21 and 22 show how the net performs when 25 different images were analyzed. The images contain a total of 199 different objects susceptible of being detected by the net. Although the number of images is not high enough to extract reliable statistical data about the network performance, it allows us to trace tendencies about the net behavior. Figure 21 shows the percentage of total object detected by the YOLOv3 network depending on the size of the internal variables and on the nature of the operators: exact or approximate. Figure 22 shows the number of images whose outcome, when analyzed by each net version, provides a 100% detection or a detection ratio higher than 80%. Surprisingly, the Exact_16 version provides the highest number of correct detections, even surpassing the Exact_32 version. Most likely, this is one of the statistical errors that is result of the small size population. But in any case, it indicates that reduced operand sizes give competitive results. Focusing on the three approximated versions, Aprox_1 version provides similar detection ratio than the exact version, proving that approximate computing in the mantissa can be used to further simplify the hardware (in addition to reduce from 32 float to 16 float) with minor losses in detection accuracy.

Logarithmic approximations are much more aggressive approximation approaches. This fact has repercussions in the percentage of objects detected when compared with the other network versions. One of the issues of logarithmic multipliers following Mitchel's algorithm is the fact that the result of the multiplication is always an underestimation of the exact result. This constant bias, when constantly accumulated in a large network, has clear implications in its performance. However, when a logarithmic multiplier is modified in such a way that the accumulative error of this multiplier has a mean value closer to zero, the performance of the net increases. Precisely, this is what corrects the additional approximation introduced to the logarithmic multiplier used in the

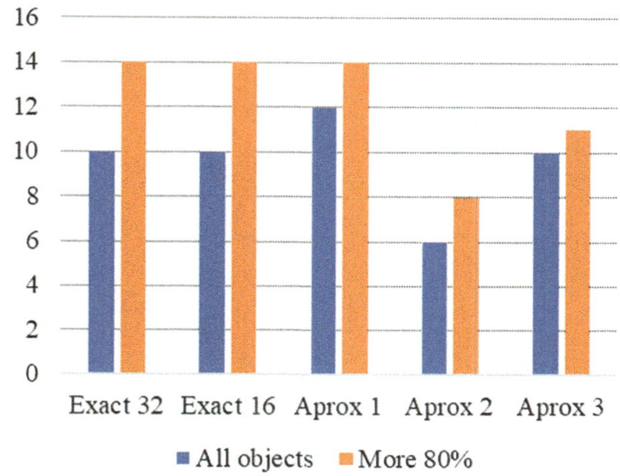

Fig. 22 Number of images (maximum would be 25) where either all objects or more than 80% of the objects have been detected

version Approx_3. This correction is exposed in Fig. 23: it shows the histogram of the relative error produced when the multipliers used in the Aprox_2 and Aprox_3 versions perform one million of multiplications of two numbers of 11 bits randomly generated with uniform distribution. There, clearly, the LM alone always underestimates the result of the multiplication, which is not observed in the LM + SOA multiplier, where both overestimations and underestimations of the exact result are produced. Therefore, if we analyze further the results in Fig. 21: the increase of accuracy of version Aprox_3 respect Aprox_2 is, at some point, counterintuitive, as the modification that we have introduced to the logarithmic multiplier used in Aprox_2 to generate the one used in Aprox_3 is precisely the introduction of another approximate operator, which a priori should provide worse results. However, the results indicate that when the approximated added compensates for the constant bias of the error produced by the multiplier, the net performance can increase.

Interestingly, the additional approximation introduced to the ALM used in the version Aprox_3 of the net has an optimal point. This is analyzed in Fig. 24, where the performance of the net is plotted as a function of the approximation aggressivity of the SOA adder: the horizontal axis shows the number of least significative bits set to one in the result of the SOA adder. When the number is 6, the maximum number of detected objects is achieved: 67.5%. For values higher than 6, the number of erroneously identified objects increases. For the extreme situation of 8 bits (just the 2 most significative bits from the mantissa are added up!), from the 199 objects, 42 are misidentified, and what is worse, 309 areas are wrongly marked as detected objects (false positives). This behavior does not appear when the approximation introduced by the SOA is smaller than 6 bits.

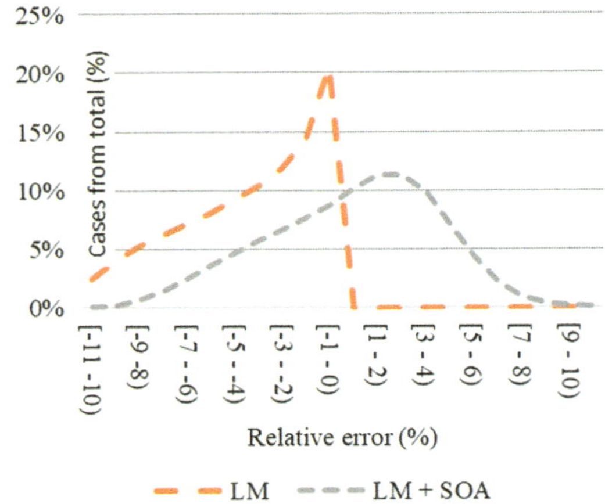

Fig. 23 Histogram of the relative error of the results after one million operations of two numbers of 11 bits randomly generated. *LM* logarithmic multiplier, *SOA* set-on adder

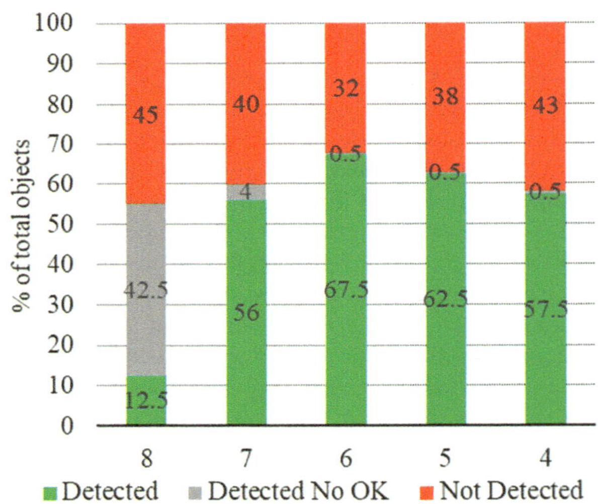

Fig. 24 Performance of the net Aprox_3 as a function of the aggressivity of the SOA adder used

4.5 *Approximate Accelerator Application: Approximate Systolic Array*

To understand the impact of replacing exact FP16 arithmetic by approximate units in a real hardware system, let us consider a DCNN accelerator architecture based on a systolic array structure. A systolic array is a parallel computing architecture in which many simple processing elements (PE) are distributed in a regular structure through which the operands flow to perform the computation. Figure 25 depicts a systolic array built as a 2D mesh, in which each PE shares its input operands with its bottom and right neighbors via simple pipelining registers. On every clock cycle, all PEs perform a multiplication and an accumulation (adding the product to the

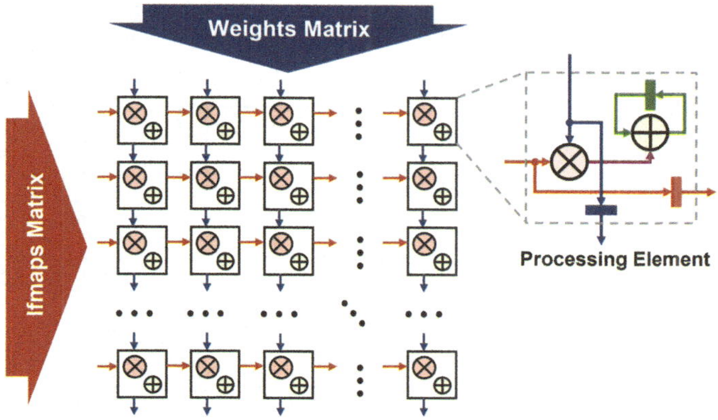

Fig. 25 Sketch of a 2D systolic array for DNN acceleration

previous value), and the product operands pass from neighbor to neighbor until they have propagated to the whole array.

Systolic arrays have seen widespread success in recent years due to their energy efficiency and potential for very high throughput. When the pipeline is full, a systolic array with X columns and Y rows performs $2XY$ operations (a product and a sum per PE) per clock cycle. However, the array only needs to be fed X weight values and Y inputs per cycle to operate, since the structure reuses the incoming values internally. This data reuse is a key factor for achieving high-energy efficiency, since reading data from memory is often as energy expensive as performing arithmetic operations, if not more.

Systolic array architectures can be used to compute most of the operations in convolutional and fully connected layers of DNNs. Mapping a convolution operation to a systolic array, as depicted in Fig. 26, consists of transforming the input and weight tensors into matrices that can be fed to the array and result in a computation that is equivalent to the convolution. Reshaping the weight tensor is straight-forward, since it only requires some reordering of the values and applying some latency between neighboring PE columns to account for the compute pipeline. On the other hand, adapting the input tensor is a bit more challenging, since one must account for the overlap between consecutive positions of the weight kernel. Transforming the input tensor of the convolution into an equivalent matrix is known as *convolution lowering* or the *im2col* algorithm [24].

To make a systolic array accelerator faster or more energy-efficient, one can consider introducing approximate arithmetic in its processing elements. From the point of view of an accelerator system, the most important decision is to determine which degree of approximation can be introduced without degrading the target DNN accuracy too much. As in the previous sections, we use the YOLOv3 object detection network as an example for assessing this. Regarding the definition of "too

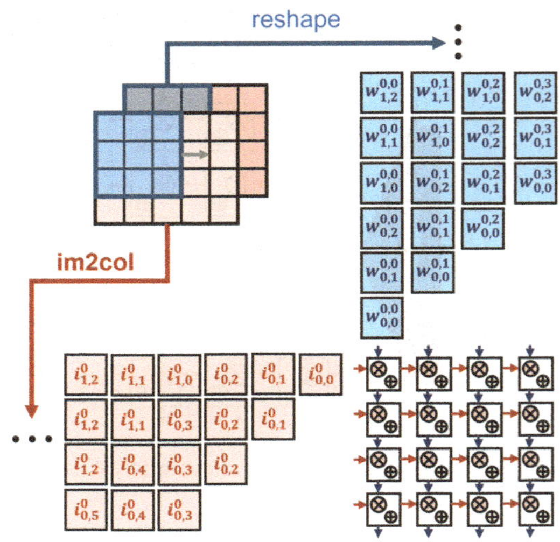

Fig. 26 Mapping a convolution operation into a systolic array accelerator

much accuracy loss," for our explanatory purposes, we can set a limit of a 2% decrease in the mAP.

Studying the hardware effects of approximate arithmetic, unlike pure functional analysis like in Sect. 4.3, requires an analysis of the resulting electrical circuit at the silicon level. To do so, a physical implementation of the digital accelerator system is required in order to obtain estimations of the silicon area, power consumption, and maximum frequency of a design. We are particularly interested in an ASIC implementation of the accelerator, so this physical design process can get very complex. However, for the sake of discussing the effect of the approximate units, we can make two simplifications. First, instead of performing the full physical design (including place-and-route, power planning, clock tree synthesis, etc.), we can use the results of a logic synthesis to analyze area, power and delay. Even though the absolute values of these metrics will not be exact (a 20–30% difference with respect to the full flow is expected), the tendencies and relative differences between approximate units will be consistent with the actual values. Second, as the systolic array is built from many identical processing elements, we can study a single PE instance in isolation and get a good approximation on how much savings we can get from the approximate circuits.

As an example of this, we performed a logic synthesis of a processing element featuring a fused FP16 multiply-add unit with one pipeline stage, an accumulator register and two systolic pipelining registers (as sketched in the right part of Fig. 25). Then, we applied different approximate circuits for the computation of the mantissas in the FP16 unit, relaunched the synthesis, and compared the results to the exact version. The synthesis process was performed using standard cells from a 22 nm technology library and considering operating conditions of 0.8 V, 25 °C, and 200 MHz. In terms of hardware metrics, we report the total area of all synthesized cells, the delay of the critical path, and a power consumption estimation. For the power consumption estimate, which includes static (leakage) and dynamic power, the latter depending on the input data, we performed a gate-level simulation of the synthesized circuit to annotate the switching activity of all cells and used a power estimation tool to generate our total power estimate based on this activity data. To be faithful to the application, the input stimuli of the simulation are feature maps and weights from the YOLOv3 network.

Finally, in order to contextualize the hardware metrics of each approximate FP16 multiply-add version, we need to analyze the impact of the approximation in YOLOv3. Gate-level simulations, and even pre-synthesis RTL simulations, are too time-consuming to consider for simulating a DNN forward pass, let alone a whole dataset. Therefore, as in Sect. 4.3, the evaluation of YOLO is done by executing the C model of the network and replacing the multiplication or the addition by a model of the approximate circuit. In this section, we will use the PASCAL VOC mAP metric (see Sect. 4.2) to report the object detection accuracy. To ensure our metrics are meaningful, we analyze the impact of each circuit with a set of 900 images from the COCO dataset [21].

Tables 3 and 4 present the main figures of merit of our example study, for different approximate multipliers and adders with different parametric configuration. When considering approximate multipliers, we use an exact adder and vice versa. Figure 27 shows the object detection results on a single image, for different units at different parameters, to give a visual intuition on how the accuracy degradation happens and its correlation to the mAP metric.

From these results, we see several trends that are expected from the concept of approximate computing. In general, higher values in the approximation parameters of multipliers and adders yield smaller area, lower power consumption, and less delay, at the cost of some degradation on the object detection mAP. This is the fundamental trade-off of approximate arithmetic: giving up some accuracy to gain efficiency.

However, we can also observe some surprising facts. First, clearly in our case, approximating the mantissa adders provides in general better delay and power reduction than doing so for the mantissa multipliers. This may seem counterintuitive

Table 3 Hardware metrics and YOLO mAP of several approximate FP16 units, replacing the mantissa multiplier by an approximate circuit

Multiplier type	m Parameter	Area [μm]	Delay [ps]	Power [μW]	mAP [%]
Exact	–	1013	3449	1251	57.35
BAM	0, 6	951	3463	1264	57.37
BAM	0, 12	781	3472	1153	56.17
BAM	0, 16	718	3388	1100	51.60
ALM-SOA	3	950	3449	1220	44.63
ALM-SOA	6	916	3449	1200	47.50
ALM-SOA	9	861	3449	1165	56.19
ALM-SOA	10	838	3449	1130	1.13

For the BAM: $m = h_{BL}, v_{BL}$

Table 4 Hardware metrics and YOLO mAP of several approximate FP16 units, replacing the mantissa adder by an approximate circuit

Adder type	*a* Parameter	Area [μm]	Delay [ps]	Power [μW]	mAP [%]
Exact	–	1013	3449	1251	57.35
GeAr	8, 8	963	2584	968	0.0
TruA	8	819	3085	874	57.54
TruA	12	784	2768	787	57.31
TruA	18	690	2625	685	48.91
LOA	8	964	3167	1002	57.75
LOA	20	952	3124	905	57.48
LOA	22	929	3118	897	51.16

For the GeAr: $a = R, P$

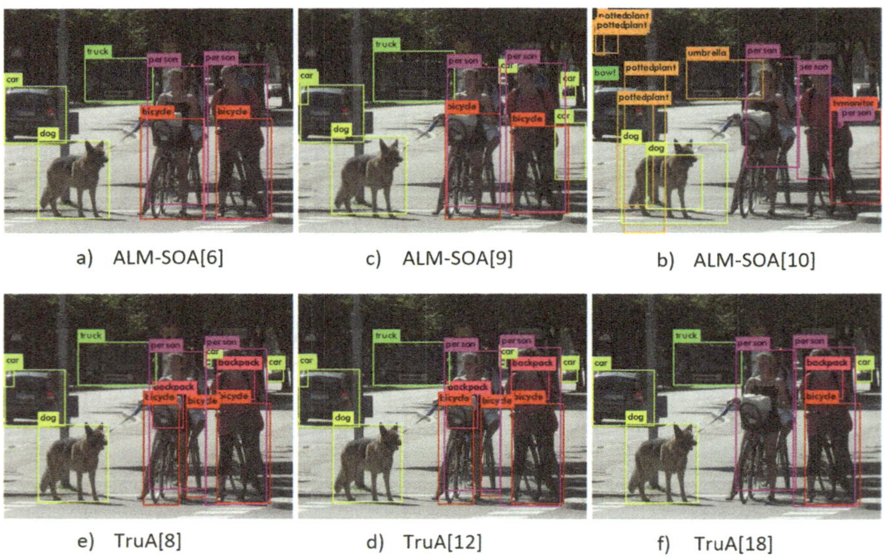

Fig. 27 Qualitative results of the YOLOv3 network executed with an approximate multiplier (ALM-SOA) and an approximate adder (TruA) at different configurations

since multipliers are more expensive than adders, in general. However, due to the nature of the FP16 fused multiply-add unit we are evaluating, the adder has quite a lot of impact. Considering that the floating-point inputs have 10-bit mantissas, the multiplier deals with two 11-bit operands (adding the implicit bit) and outputs a 22-bit product. On the other hand, the adder deals with two 33-bit operands, since the product and the input addend must be renormalized. Therefore, even though multipliers are in general more complex operators than adders, a big enough adder can be quite expensive as well, especially in terms of delay and power consumption.

The second interesting trend is the relation of the approximating parameter in the ALM-SOA and the mAP, where more aggressive approximations yield better object detection accuracy, up to a certain point. This effect, explained further in Sect. 4.4, creates an interesting situation in which the optimal point in terms of mAP also has significant hardware gains, unlike the other approximate units in which the accuracy monotonically decreases with the approximation parameter.

Lastly, it is worth commenting on the GeAr adder and how using it results in disastrous degradation in the network outputs. The reason is that this circuit approximates the sum by cutting its carry chain (see Sect. 2.1.2). Even though missing the carry is a low-probability event, it can result in very large deviations from the exact sum. In the context of neural networks, due to their high connectivity, a single low-probability and high-impact error can propagate through the whole network and completely alter the results. Moreover, due to the very large number of additions that must be performed on a single forward pass of a DNN, low-probability events are almost guaranteed to happen. In contrast, on Sect. 4.4, we saw how applying a GeAr for filtering applications results in very good behaviors. This showcases how important it is to consider the application when studying approximate arithmetic circuits: a terrible circuit for one application can work wonders in another one.

5 Conclusions

There are many applications that can afford the usage of approximate arithmetic units, with the target to enhance minimum one of the following circuit parameters: delay, area, and power consumption with minimum compromise of the circuit functionality. In the specialized literature, there are many approximate adders, dividers, multipliers that individually analyzed, isolated from any context, might offer good error metrics. However, the real impact on the functionality and performance when inserted in complex processing circuits can only be asserted when specific ad-hoc analyses are done. We have seen examples in this chapter where an approximate adder, whose specific error metric such as the mean error distance (MED) can be known beforehand, can provide when inserted in a circuit either almost negligible impact on the circuit functionality or catastrophic effects. In this chapter, we have presented two case studies that allow us to study the impact of approximate arithmetic circuits: an FIR-digital filter and a convolutional neural network (YOLOv3). We have studied how their functionality and, in some cases, its electrical performance (power, delay, area) are modified by the insertion of approximate adder and multiplier circuits. The methodology presented here can be extended to any kind of digital circuit.

References

1. R. Acton, *Branch-Free Implementation of Half-Precision (16 Bit) Floating Point* (2006). https://cellperformance.beyond3d.com/articles/2006/07/update-19-july-06-added.html. [Online; Accessed 23 May 2022]
2. M.A. Breuer, *Multi-media Applications and Imprecise Computation* (8th Euromicro Conference on Digital System Design (DSD'05), 2005), pp. 2–7. https://doi.org/10.1109/DSD.2005.58
3. M. Everingham, L. Gool, C.K. Williams, J. Winn, A. Zisserman, The Pascal Visual Object Classes (VOC) challenge. Int. J. Comput. Vision **88**(2), 303–338 (2010). https://doi.org/10.1007/s11263-009-0275-4
4. S.P. Goswami, B. Paul, S. Dutt, G. Trivedi, *Comparative Review of Approximate Multipliers* (30th Int. Conf. Radioelectronika, Bratislava, 2020). https://doi.org/10.1109/RADIOELEKTRONIKA49387.2020.9092370
5. K. He, X. Zhang, S. Ren, J. Sun, *Deep Residual Learning for Image Recognition* (2016 IEEE Conference on Computer Vision and Pattern Recognition (CVPR), Las Vegas, 2016), pp. 770–778. https://doi.org/10.1109/CVPR.2016.90
6. R. Hegde, N.R. Shanbhag, Soft digital signal processing. IEEE Trans. Very Large Scale Integr. (VLSI) Syst. **9**(6), 813–823 (2001). https://doi.org/10.1109/92.974895
7. H. Jiang, F.J.H. Santiago, H. Mo, L. Liu, J. Han, Approximate arithmetic circuits: A survey, characterization, and recent applications. Proc. IEEE **108**(12), 2108–2135 (2020). https://doi.org/10.1109/JPROC.2020.3006451
8. B. Kartikeya, P.S. Mnae, J. Henkel, *Power and Area Efficient Approximate Wallace Tree Multiplier for Error-Resilient Systems* (Fifteen International Symposium on Quality Electronic Systems, 2014). https://doi.org/10.1109/ISQED.2014.6783335
9. A. Krizhevsky, I. Sutskever, G.E. Hinton, in *Image Net Classification with Deep Convolutional Neural Networks, in Advances in Neural Information Processing Systems*, ed. by F. Pereira, C. Burges, L. Bottou, K. Weinberger, vol. 25, (Curran Associates, Inc, 2012)
10. P. Kulkarni, P. Gupta, M. Ercegovac, *Trading Accuracy for Power with an Underdesigned Multiplier Architecture* (2011 24th Internatioal Conference on VLSI Design, 2011), pp. 346–351. https://doi.org/10.1109/VLSID.2011.51
11. W. Liu, J. Xu, C. Wang, P. Montuschi, F. Lombardi, Design and evaluation of approximate logarithmic multipliers for low power error tolerant applications. IEEE Trans. Circuit. Syst.: I Regul. Pap. **65**(9), 2856–2868 (2018). https://doi.org/10.1109/TCSI.2018.2792902
12. H.R. Mahdiani, A. Ahmadi, S.M. Fakhraie, C. Lucas, Bio-inspired imprecise computational blocks for efficient VLSI implementation of soft-computing applications. IEEE Trans. Circuit. Syst. I Regul. Pap. **57**(4), 850–862 (2010). https://doi.org/10.1109/TCSI.2009.2027626
13. J.N. Mitchell, Computer multiplication and division using binary logarithms. IEEE Trans. Electron. Comput. **EC-11**(4), 512–517 (1962). https://doi.org/10.1109/TEC.1962.5219391
14. C. Navarro, *Anàlisi i implementació d'operadors aproximats per al processat digital* (Polytechnical University of Catalonia. Bachelor's Degree in Electronic Engineering, 2022) https://upcommons.upc.edu/pfc
15. C. Ni, J. Lu, J. Lin, Z. Wang, *LBPF: Logarithmic Block Floating Point Arithmetic for Deepe Neural Networks* (ITTT Asia Pacific Conference on Circuits and Systems, 2020), pp. 201–204. https://doi.org/10.1109/APCCAS50809.2020.9301687
16. K. Palem, A. Lingamneni, C. Enz, C. Piguet, *Why Design Reliable Chips When Faulty Ones Are Even Better* (2013 Proceedings of the ESSCIRC (ESSCIRC), 2013), pp. 255–258. https://doi.org/10.1109/ESSCIRC.2013.6649121
17. J. Redmon, S. Divvala, R. Girshick, A. Farhadi, *You Only Look Once: Unified, Real-Time Object Detection* ((cite arxiv:1506.02640), 2015)
18. J. Redmon, A. Farhadi, *YOLOv3: An Incremental Improvement* (CoRR, vol. abs/1804.02767, 2018, 2018)

19. M. Shafique, W. Ahmad, R. Hafiz, J. Henkel, A low latency generic accuracy configurable adder. *Proceedings of the 52nd Annual Design Automation Conference*, pp. 1–6 (2015) https://doi.org/10.1145/2744769.2744778
20. K. Simonyan, A. Zisserman, Very deep convolutional networks for large-scale image recognition. *3rd International Conference on Learning Representations (ICLR 2015)*, pp. 1–14 (2015)
21. L. Tsung-Yi, M. Maire, S.J. Belongie, L.D. Bourdev, R.B. Girshick, J. Hays, et al., *Microsoft COCO: Common Objects in Context* (CoRR, abs/1405.0312, 2014) Retrieved from http://arxiv.org/abs/1405.0312
22. S. Venkatachalam, E. Adams, H.J. Lee, S.-B. Ko, Design and analysis of area and power efficient approximation booth multipliers. IEEE Trans. Comput. **68**(11), 1697–1703 (2019). https://doi.org/10.1109/TC.2019.2926275
23. Y. Wu, C. Chen, W. Xiao, X. Wang, C, Wen, J. Han, X. Yin, W. Qian, C. Zhuo, arXiv: 2301.12181v1[cs.AR], 28 Jan 2023
24. Y. Zhou et al., *Characterizing and Demystifying the Implicit Convolution Algorithm on Commercial Matrix-Multiplication Accelerators* (2021 IEEE International Symposium on Workload Characterization (IISWC), Storrs, 2021), pp. 214–225

A Top-Down Design Methodology for Approximate Fast Fourier Transform (FFT) Design

Chenyi Wen, Ying Wu, Xunzhao Yin, and Cheng Zhuo

1 Introduction

The fast Fourier transform (FFT) is a crucial algorithm in digital communication and sensor signal processing, making it an essential component of computer systems [1–4]. Recently, with the popularity of human perception-related tasks (e.g., audio/image processing, machine learning), it has become apparent that full precision and exactness are not always necessary for FFT-based applications [5, 6]. As a result, approximate FFT designs have emerged as a promising alternative to trade computational accuracy for energy efficiency and high performance, which have received increasing interest from both academia and industry [7–16]. Hence, it is a natural idea to explore the approximate design of FFT to achieve sufficient instead of excessively accurate computational precision while obtaining maximum hardware benefits.

Despite decades of research on architectures for FFT, two key questions remain in the deployment of approximate computing in such a complex IP design. The first issue is *the lack of a clear link between the precision of the FFT algorithm and the introduced approximation*. Prior work has mostly focused on directly approximating the underlying arithmetic circuits, such as adders and multipliers, to replace the original exact units [9–12]. However, it is unclear how such approximations affect the precision of the FFT algorithm, leading to ineffective optimization. The second issue is *the limited flexibility to support versatile applications*. Given the diverse range of FFT-based applications, the approximate design for FFT is desired to have a large dynamic range and be configurable for different accuracy specifications. However, most existing approximate FFT designs are specific to a particular fixed

C. Wen · Y. Wu · X. Yin · C. Zhuo (✉)
School of Micro-Nano Electronics, Zhejiang University, Hangzhou, China
e-mail: wwency@zju.edu.cn; ying.wu@zju.edu.cn; xzyin1@zju.edu.cn; czhuo@zju.edu.cn

W. Liu et al. (eds.), *Design and Applications of Emerging Computer Systems*,
https://doi.org/10.1007/978-3-031-42478-6_17

point with limited dynamic range or freeze the approximation precision at the design stage [13–18].

Thus, in this work, we propose *a top-down design methodology* for approximate FFT design to address the abovementioned issues[19]. Instead of introducing approximations from the bottom-up, which may result in over-design, or relying on fixed point/word length at the cost of dynamic range and performance degradation, [9–15, 17, 18, 20], the top-down design methodology fully exploits the error-tolerance nature of the FFT algorithm and automatically determine the appropriate approximation levels of the underlying circuits. The methodology consists of three processes:

- **An FFT approximation error model** is proposed to bound the impact of circuit approximation to the FFT algorithm precision by modeling the introduced error with a relative error measure.
- **An FFT approximation optimization flow** is formulated to configure the desired approximate FFT, maximizing the hardware benefits while meeting the design specifications.
- **A configurable floating-point FFT structure** is implemented to achieve a wide precision range and high performance while supporting different approximation levels for versatile application scenarios.

The experimental results demonstrate the effectiveness of the proposed top-down design methodology for approximate FFT. The implemented approximate FFT design using a 40 nm UMC library achieves up to 52% improvement in area-delay product and 23% energy saving when compared to the conventional exact FFT. The proposed design also outperforms other approximate FFT designs by covering almost 2× wider precision range with low hardware cost for high precision constraints. The practicality of the proposed design is further demonstrated through its application in an audio classification system, where a loss in accuracy of around 1% is observed.

The chapter is organized as follows. Section 2 provides a review of the background on FFT hardware implementation and approximate multiplier techniques. Sections 3 and 4 present the proposed top-down design methodology, providing an overview and detailed information, respectively. In Sect. 5, we provide experimental results and analysis. Finally, Sect. 6 concludes this chapter.

2 Background

2.1 FFT Hardware Implementation

FFT architectures can be typically divided into two classes: pipeline and parallel architectures [21, 22]. In general, pipeline FFT requires fewer hardware resources with moderate control overhead, while its latency is higher due to its sequential process. On the other hand, parallel FFT is faster at the cost of huge resources,

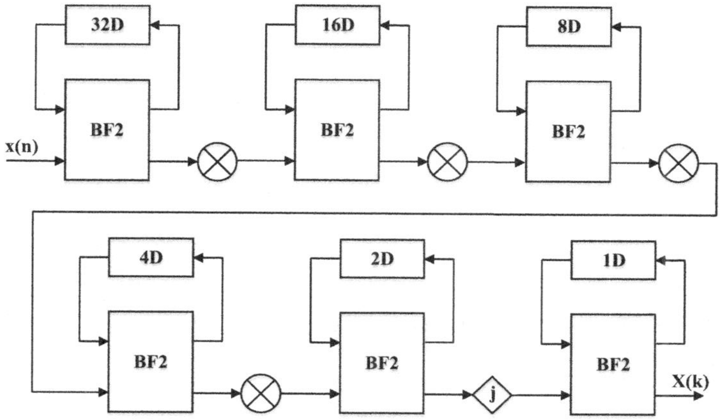

Fig. 1 An example of a 64-point FFT using R2SDF pipeline structure

which makes it less efficient in hardware implementation. Thanks to the popularity of mobile and portable devices, pipeline architecture is then widely used and can be further categorized to multi-path delay commutator (MDC) and single-path delay feedback (SDF) structures for different control policies. Although MDC has a simpler data controller, it has a higher area overhead than SDF and hence lower hardware utilization efficiency. Thus, most prior works employ SDF structure, whose control complexity depends on the radix of the FFT algorithm [23].

A commonly used radix-2 SDF (R2SDF) pipeline structure is presented in Fig. 1 and will be used as the baseline in the rest of this chapter due to its popularity. However, it is noted that the techniques presented in our work are general and applicable to other architectures. The R2SDF pipeline FFT typically employs radix-2 decimation-in-frequency (R2DIF) FFT algorithm. Thus, 64-point pipelined FFT in Fig.1 consists of six stages, each of which involves a radix-2 butterfly (BF2) unit, a delay buffer, and a complex number multiplier. The last two stages can be completed without using the actual multiplier, where multiplicands are just $+/-1$ or $+/-j$, with j as the imaginary unit. For R2SDF structure, a larger number of FFT points indicate more pipeline stages, e.g., eight stages for 256-point and ten stages for 1024-point. Each stage then has different approximation tolerance and hence can be exploited for approximation optimization.

2.2 Configurable Floating-Point Approximate Multiplier

According to Fig. 1, FP multiplier is a critical and probably the most resource-consuming component in the hardware implementation of FFT. Consequently, many prior works focus on designing approximate FP multipliers to replace the exact multipliers [24–26]. Recently, a configurable FP approximate multiplier, called the

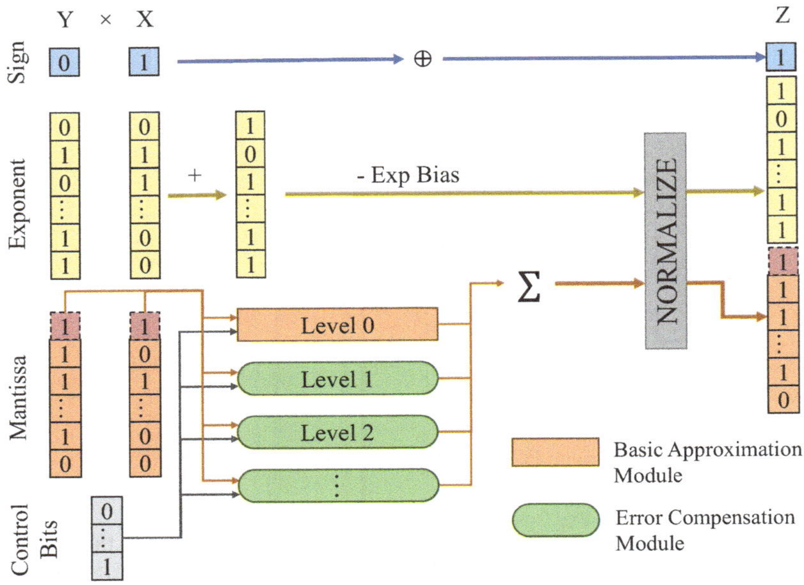

Fig. 2 Structure of the approximate FP multiplier PAM from [28]

PAM, has been proposed in [27, 28]. The *PAM* employs multi-level piece-wise linear approximations to converge to the exact result and has been found to achieve both unbiasedness and runtime configurability while maintaining state-of-the-art energy efficiency (up to 1 order of magnitude improvement) when compared to prior works. Therefore, the *PAM* will serve as the baseline for our approximate FP FFT design.

The ranges of two mantissas in the FP multiplication are partitioned into multi-level hierarchical sub-domains, and linear functions are used to approximate each sub-domain. The basic approximation module, denoted as Level 0 in Fig. 2, provides an initial estimation f^0_{approx}:

$$f = xy \approx f^0_{approx} = ax + by + c \tag{1}$$

where x and y are the input operands and a, b, and c are the coefficients. The deeper levels then act as error compensation to gradually improve the overall accuracy. Thus, the configurability can be flexibly realized by specifying the desired depth. Then, as shown in Fig. 3 [28], with one level deeper, each domain (or sub-domains) can be partitioned into four identical smaller ones. For a rectangular domain $[x_1, x_2] \times [y_1, y_2]$, the optimal coefficients to minimize the errors are $a = \frac{y_1+y_2}{2}$, $b = \frac{x_1+x_2}{2}$, and $c = -ab$ [28]. Then the multi-level configurable multiplication can be written as:

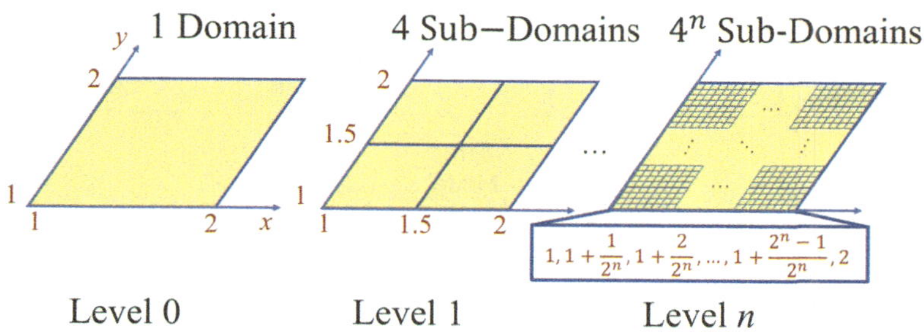

Fig. 3 Procedure of domain partitioning for PAM from [28]

$$f^n_{approx} = f^0_{approx} + \sum_{i=1}^{n} \Delta f_i \tag{2}$$

where n is the selected configuration level and $\Delta f_i = f^i_{approx} - f^{i-1}_{approx}$ is the additional accuracy improvement by going to the next approximation level. The equation above can be further simplified to:

$$f^n_{approx}(x, y) = -k_x k_y + k_y x + k_x y \tag{3}$$

where $k_x = [0.5 + floor\ (x \times 2^n)] \times 2^{-n}$ and $k_y = [0.5 + floor\ (y \times 2^n)] \times 2^{-n}$; $floor(.)$ is the round-down function.

3 Overview

Figure 4 provides an overview of the proposed top-down design methodology, which consists of three phases:

- **Error modeling**: This phase maps the algorithm-level accuracy requirement to the circuit-level specification by analyzing the relationship between the error characteristics of configurable approximate multipliers and the accuracy of FFT execution. In this chapter, we use peak signal-to-noise ratio (PSNR) as the metric for algorithm-/application-level accuracy, which is commonly used in human perception-related tasks [5]. PSNR is defined as:

$$PSNR = 10log_{10}\left[\frac{\max_i \mathbf{X}^2_{\mathbf{apr}}}{\frac{1}{N}\|\mathbf{X}_{\mathbf{apr}}} - \mathbf{X}_{\mathbf{exa}}\|^2\right] \tag{4}$$

where $\mathbf{X}_{\mathbf{apr}}$ and $\mathbf{X}_{\mathbf{exa}}$ represent the vectors obtained from the approximate FFT and the exact FFT, respectively, while N denotes the number of entries in the vector, and $|.|$ refers to the Euclidean norm. It is important to note that the

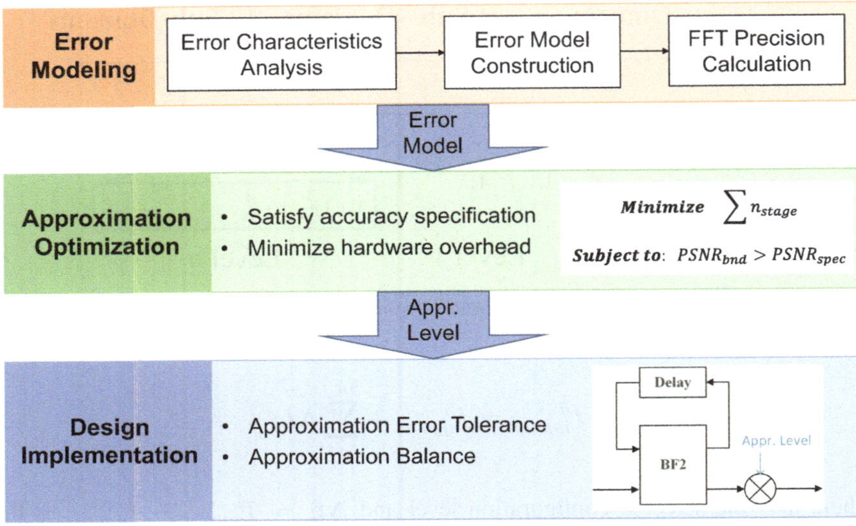

Fig. 4 Overview of the proposed design methodology

proposed error modeling method is general and hence not limited to a specific error measure or approximate multiplier.

- **FFT approximation optimization**: Once the algorithm-/application-level accuracy in terms of PSNR is mapped to the circuit-level specification of the entire FFT hardware, we need to further assign the appropriate approximation margin to each stage of a common floating-point FFT in Fig. 1. An optimization flow is formulated in this phase to minimize the hardware overhead while satisfying the overall accuracy specification for the FFT hardware implementation.
- **Design implementation**: In this phase, we implement the approximate FFT hardware based on the optimized approximation level assigned to each FFT pipeline stage. If there are multiple groups of valid assignments, we can follow the approximation balance and approximation tolerance principles to maximize the overall performance.

4 Method

4.1 Error Modeling

In this section, we explore the relationship between the algorithm-level PSNR and the circuit-level error specification. We here employ a configurable approximate multiplier PAM [28] as a demonstration vehicle to showcase the modeling techniques. Then, we investigate the error tolerance of different stages for the FFT pipeline structure to establish an error model link between the two different design layers.

4.1.1 Error Characteristics Analysis

Without loss of generality, an FP number can always be represented as below according to IEEE 754 standard:

$$F_x = sign_x \times x \times 2^{E_x} \tag{5}$$

where $sign_x$, x, and E_x refer to the sign, mantissa, and exponent of the FP number F_x, with $x \in [1, 2)$. Thus, for the product of two FP numbers F_x and F_y, we have:

$$F_x F_y = sign_x sign_y \times xy \times 2^{E_x + E_y} \tag{6}$$

According to Eqs. (2) and (3), when using the approximate multiplier PAM [28] at level n to approximate the above multiplication, we have:

$$F_x F_y \approx sign_x sign_y \times f^n_{approx}(x, y) \times 2^{E_x + E_y} \tag{7}$$

where $f^n_{approx}(x, y) = -k_x k_y + k_y x + k_x y$ and n is the configured approximation level.

The error introduced by the approximate multiplier is then:

$$error^n = -sign_x sign_y \times (x - k_x)(y - k_y) \times 2^{E_x + E_y} \tag{8}$$

We can see from Eq. (8) that the magnitude of $error^n$ is affected by F_x and F_y and decreases as the approximate level n increases. Therefore, deriving an analytical formulation for $error^n$ is challenging. To simplify the analysis process, we use the bound $|error^n_{bnd}|$ as a conservative measure of error, which is easier to model.

Now assume both F_x and F_y are in the range of $[-bnd, bnd]$. Then we can formulate the estimation of absolute error bound $|error^n_{bnd}|$ as:

$$|error^n_{bnd}| = (bnd - k_{bnd})^2 \times 2^{2 \times E_{bnd}} \tag{9}$$

where $k_{bnd} = [0.5 + floor\ (bnd \times 2^n)] \times 2^{-n}$. If bnd happens to be the power of 2, we can further simplify Eq. (9) to:

$$|error^n_{bnd}| = 4^{E_{bnd} - n - 1} \tag{10}$$

which indicates:

- The bound of the absolute error only depends on *bnd* and *n*, and it exponentially decreases with a growing *n* in a factor of 4 for a given *bnd*.
- When *bnd* is increased, e.g., larger dynamic range causing a larger E_{bnd}, the absolute error bound also grows. However, since both E_{bnd} and *n* are on the exponent, the impact of larger dynamic range on accuracy can be cancelled out by using a larger *n*.

For the convenience of the following analysis, we further denote the absolute error bound as $\theta^2_{bnd,n}$.

4.1.2 Error Model Construction

Since the error measure of FFT is PSNR, in addition to the absolute error bound, it is necessary to further explore the relative error of the approximate FP multiplier PAM [28]. Similar as Eq. (8), we can define the relative error for the product of $F_x F_y$ as:

$$rel\_err^n(F_x, F_y) = \frac{error^n}{F_x F_y} = \frac{-(x - k_x)(y - k_y)}{xy} \tag{11}$$

According to Eq. (10), we have:

$$rel\_err^n(F_x, F_y) \leq \frac{1}{4^{n+1}} \tag{12}$$

We can reasonably assume that the mantissa *x* or *y* is uniformly selected from the range [1, 2), as defined by the IEEE standard for FP number representation. The distribution of rel_err^n for a given *n* is essentially a symmetric normal product-like distribution, as shown in Fig. 5a. To measure the relative approximation, we can still use the worst-case relative error bound as in Eq. (12), but this can be overly pessimistic. Instead, we propose to use the error at a particular percentile, such as σ, to measure the approximation induced by the approximate multiplier. This approach is similar to the practices in statistical yield or timing analysis [29]. This can be pre-characterized from Eqs. (8)–(11) and denoted as $\beta^2_{\sigma,n}$ in the rest of the chapter.

Figure 5b shows the proposed percentile-σ error measure of the relative error with respect to different *n*, where σ takes the values 0.85, 0.9, and 0.95. A larger σ value corresponds to a more conservative measure. As observed from the figure, $\beta^2_{\sigma,n}$ depends exponentially on the approximate level *n*. Furthermore, based on the IEEE FP number standard, the mantissa range is always [1, 2), indicating that $\beta^2_{\sigma,n}$ is universal for any FP number multiplication, and can thus be pre-characterized for subsequent optimization.

Based on the discussions above, *for a configured approximation level n and a selected σ, we can use $\beta^2_{\sigma,n}$ to measure the relative approximation degree:*

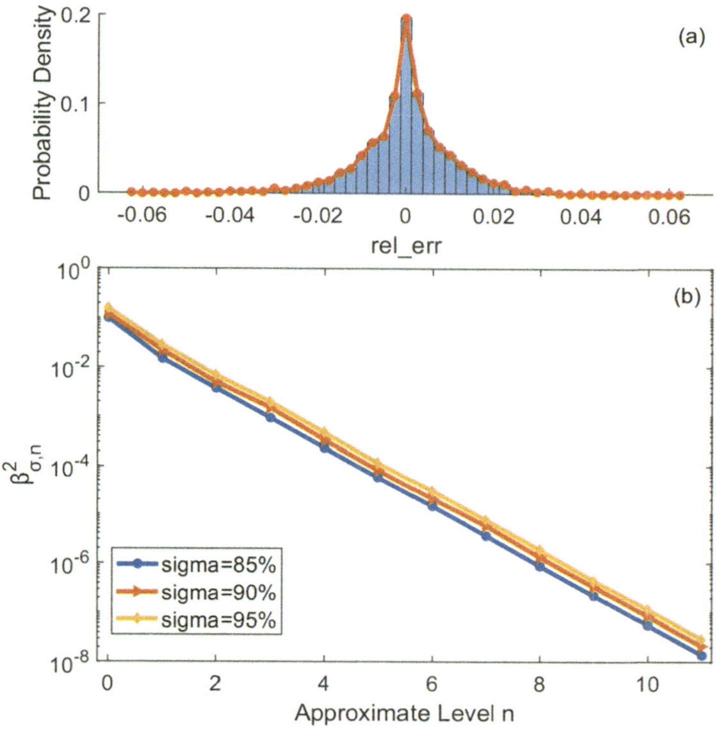

Fig. 5 (**a**) An example of the distribution for relative error $rel\_err^1(F_x, F_y)$ when both F_x and F_y are uniformly sampled. (**b**) The proposed percentile-σ error measure $\beta^2_{\sigma,n}$ for different approximation levels n when σ is 0.85, 0.90, and 0.95

$$F_xF_y + error^n = F_xF_y(1 + rel\_err^n(F_x, F_y)) \approx F_xF_y\left(1 + \beta^2_{\sigma,n}\right) \tag{13}$$

Based on Eq. (13), the complex multiplication in FFT will be simplified. The two operands are the entry F_x in the input signal vector $\mathbf{X}$ and the rotation factor W. The output of the complex multiplication is:

$$O_r \approx F_{x,r}W_r\left(1 + \beta^2_{\sigma,n}\right) - F_{x,i}W_i\left(1 + \beta^2_{\sigma,n}\right) = (F_{x,r}W_r - F_{x,i}W_i)\left(1 + \beta^2_{\sigma,n}\right) \tag{14}$$

$$O_i \approx F_{x,i}W_r\left(1 + \beta^2_{\sigma,n}\right) + F_{x,r}W_i\left(1 + \beta^2_{\sigma,n}\right) = (F_{x,i}W_r + F_{x,r}W_i)\left(1 + \beta^2_{\sigma,n}\right) \tag{15}$$

$$O = F_xW\left(1 + \beta^2_{\sigma,n}\right) \tag{16}$$

where O denotes the output and the subscripts r and i denote real and imaginary parts. Equation (16) greatly improves the efficiency of error estimation and FFT precision analysis.

4.1.3 FFT Precision Calculation

To demonstrate how we calculate FFT algorithm-level precision with lower-level circuit approximation specifications, a 64-point R2DIF algorithm is used as a demonstration example. We focus on how approximation errors are introduced and propagated along the first four stages of the FFT pipeline structure shown in Fig. 1. Note that the last two stages do not require actual multipliers, so we exclude them from our analysis.

According to Eq. (8), the introduced approximation from the approximate multiplier can be considered as additive noise. With the discussion in the last two subsections, we can use $\theta^2_{bnd,n}$ as a conservative estimate to analyze the error propagation, as below:

$$F_x F_y + error^n \approx F_x F_y + \theta^2_{bnd,n} \tag{17}$$

Since the approximation in the pipelined FFT is only introduced by the approximate multiplier, the stage error considering both real and imaginary parts in Eqs. (14)–(16) can then be written as θ_s^2 for the s_{th} stage:

$$\theta_s^2 = 2\theta^2_{bnd,n} + i2\theta^2_{bnd,n} \tag{18}$$

In each stage of FFT, the butterfly cell only involves additive and subtraction operations, which does not impact the error bound. When using the R2DIF algorithm to compute the 64-point FFT, the i_{th} output data for the s_{th} stage is:

$$O_s[i] = \left(D_{s-1}[i] \pm D_{s-1}[i \pm 2^{6-(s-1)}] + \theta^2_{s-1}\right) W_s[i] + \theta_s^2 \tag{19}$$

where $D_s[i]$ and $O_s[i]$ are the i_{th} entry of the input and output signal vector for the s_{th} stage and $W_s(i)$ is the ith rotation factor in the s_{th} stage. Since the rotation factors are always smaller than 1, with the error propagated to the next stage, θ_1^2 may get reduced by multiplying the rotation factor. Thus, this simply indicates that:

- **Approximation error tolerance**: The prior stages in the pipeline FFT may have larger error tolerance and hence can be assigned larger approximations.
- **Approximation balance**: If one stage already introduces a large error and hence dominates the entire FFT, deploying very accurate multipliers in other pipeline stages will not improve the overall accuracy.

Figure 6a and b validates the above findings. The input signal for the FFT in these experiments is a randomly selected normally distributed signal. The x-axis

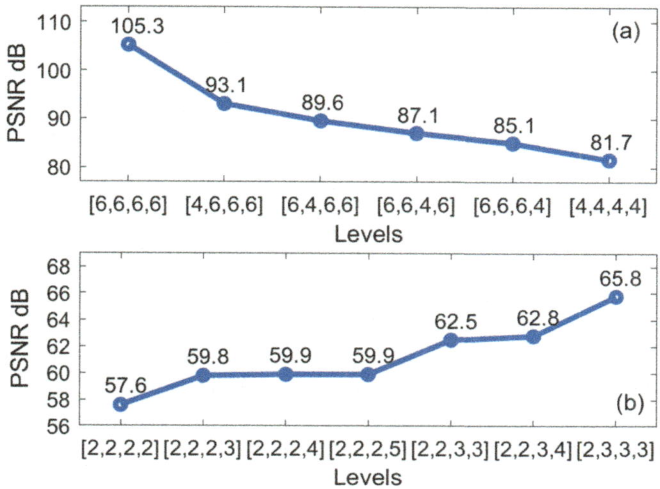

Fig. 6 (**a**) Impact of approximation-level assignment to different stages of a pipelined FFT. (**b**) Impact of approximation balance by assigning similar approximation levels across the stages of a pipelined FFT

represents the assigned approximation levels for the four stages of the FFT. For example, [6, 6, 6, 6] denotes that all the approximate multipliers in stages 1 to 4 are assigned with an approximate level 6. As shown in Fig. 6a, it is clear that PSNR decreases less when the earlier stages are assigned with a higher approximate level. In the first four data points of Fig. 6b, increasing the accuracy in the last stage does not improve the overall accuracy.

4.2 Approximation Optimization

According to the general methodology in Fig. 4, the precision optimization flow of FFT is to assign the appropriate approximate levels to each stage in order to maximize the benefits of approximate computing. This can be formulated as a mixed-integer nonlinear optimization problem (MINLP) as shown below:

$$\begin{aligned} &\textbf{Minimize} \quad \sum n_{stage} \\ &\textbf{Subject to}: \ PSNR_{bnd} > PSNR_{spec} \\ &P(x \mid x \in Set_m,\ x \geq PSNR_{bnd}) > Prob_{spec} \\ &n_{stage} \in [0, 11] \end{aligned} \tag{20}$$

where n_{stage} is the approximate level of complex multiplication unit and needs to be an integer, $PSNR_{spec}$ is the PSNR specification required by the application, Set_m is the set of PSNR obtained by calculating approximate FFT for m input sequences, $PSNR_{bnd}$ is the estimated PSNR, and $Prob_{spec}$ is the specified percentile to locate the $PSNR_{bnd}$ in Set_m. To speed up the optimization, $Prob_{spec}$ can be set to a value less than 1.

The sum of approximate levels is the most straightforward and effective way to represent the optimization objective, as higher levels come with increased hardware overhead. To ensure that the optimized design meets the precision specification, the first constraint requires that the $PSNR_{bnd}$ remains above a certain threshold set by the application requirement. The second constraint employs the error model to ensure the accuracy of error estimation and further speed up the optimization process. This optimization algorithm is versatile and extensible, allowing for the incorporation of findings from the previous section to ensure approximation balance and make use of approximation error tolerance.

4.3 Design Implementation

In Sect. 2, we introduce the R2SDF FFT pipeline and the PAM approximate multiplier [28], which serve as the baseline for this work. For hardware implementation, all multipliers in the complex multiplication unit are replaced with PAMs. We adjust the approximate level of PAMs to balance hardware overhead and the FFT algorithm-level precision.

Figure 7 depicts two complex multiplication unit architectures. In Fig. 7a, four multipliers and two adders are used, resulting in a shorter critical path and a higher clock frequency. However, this comes at the cost of higher hardware overhead. In contrast, Fig. 7b uses three multipliers and five adders, reducing the number of

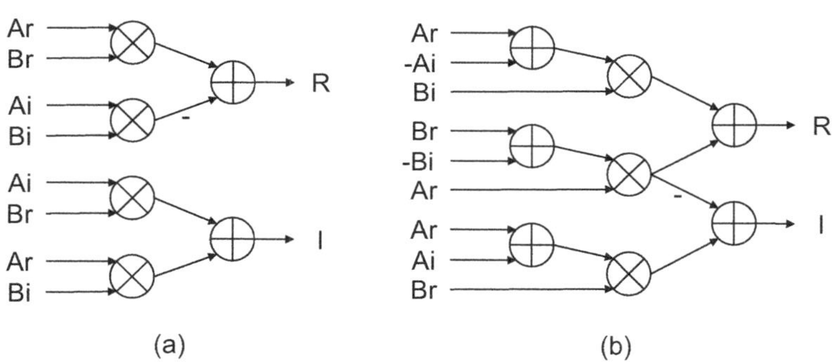

Fig. 7 Different architectures of the complex multiplication unit: (**a**) the complex multiplication unit consisting of four multipliers and two adders; (**b**) the complex multiplication unit consisting of three multipliers and five adders

multipliers but increasing the number of adders. This lengthens the critical path and is not conducive to increasing the clock frequency. Given that this work will use approximate multipliers instead of exact multipliers, resulting in significantly reduced multiplier hardware overhead, the (a) structure is more conducive to further enhancing the operation rate of the FFT hardware. Therefore, we choose the (a) structure for the complex multiplication units in our hardware implementation.

To finely control the accuracy of FFT operations and reduce hardware overhead, it may seem beneficial to use different approximate levels for the four multipliers in a complex multiplication unit. However, we believe that such an approach is unnecessary. From an algorithmic perspective, each of these four multipliers has an equal impact on the accuracy of a complex multiplication operation. Using different approximate levels in one complex multiplication unit could actually prevent us from accurately estimating the introduced approximate error. This would hinder our ability to predict the impact of approximation behavior on the accuracy of FFT operations. Therefore, we propose using the same approximate level for all four multipliers in a complex multiplication unit, which would also characterize the approximate level of that FFT pipeline stage.

For hardware implementation, the approximate levels of FFT pipeline stages are given by the optimization flow described in Sect. 4.2. If the optimization gives multiple groups of level assignments, the principles (approximation balance and approximation tolerance) obtained in Sect. 4.1.3 will guide the implementation.

5 Experimental Results

We conduct comprehensive experiments to validate the performance of our design. Our experiment infrastructure, as shown in Fig. 8, includes both simulation and verification. We use MATLAB to build an FFT accuracy simulation tool and then

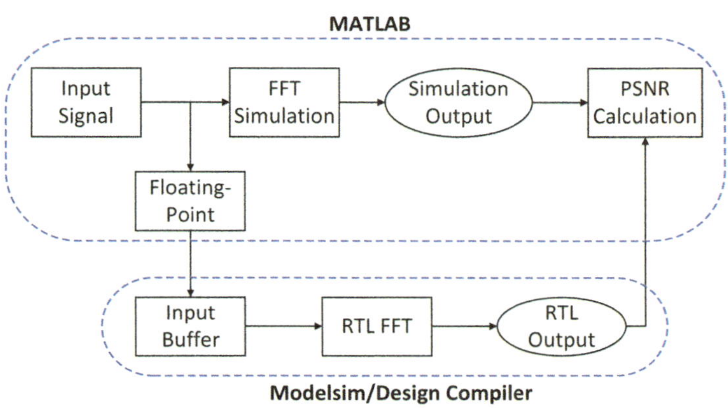

Fig. 8 Experiment infrastructure to validate the performance of the optimized FFT design

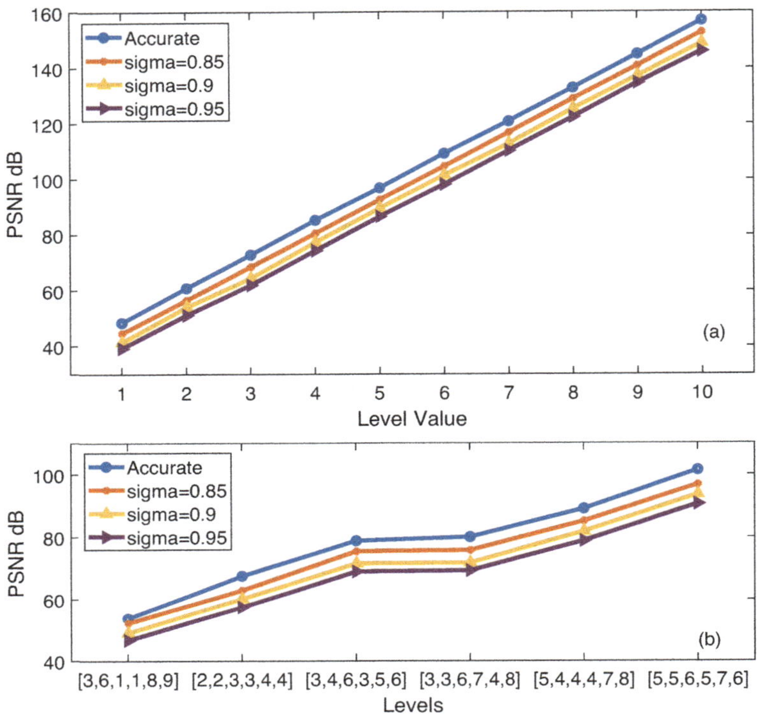

Fig. 9 Comparison on error modeling using an accurate error estimation and the proposed relative error with different σ. (**a**) All the stages are assigned with the same approximation level. (**b**) Each stage is randomly assigned with an approximation level

complete the RTL implementation of the optimized FFT to perform functional verification and timing analysis. We use Design Compiler to synthesize and analyze the design on a UMC 40 nm library. Furthermore, we implement an audio classification system using the proposed design to investigate the effect of approximation on the accuracy of practical system applications.

5.1 Performance of Error Model

We conducted multiple experiments to validate the accuracy of the proposed error modeling on a 256-point 8-stage FP FFT using the R2DIF algorithm. The experiments are conducted using two settings: In Experiment 1, it traverses the approximate levels from 1 to 10 assigned to the first 6 stages of the pipeline FFT. Figure 9a shows the PSNR for the proposed relative errors using different values of σ and accurate error estimation. It is clear that all the lines have very small

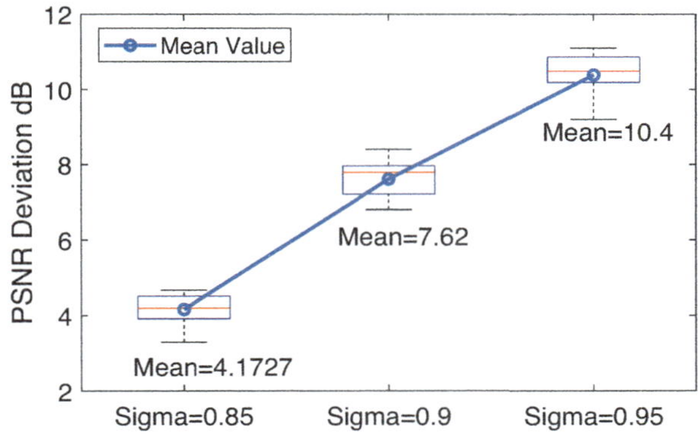

Fig. 10 Statistics of PSNR deviation between the relative error model and accurate error estimation using different σ

Table 1 Comparison on precision optimization results for different PSNR specifications

$PSNR_{spec}$	Optimal	Optimization found	$PSNR_{bnd}$	Runtime
60	[2,2,2,3,3,3]	[2,2,3,3,3,4]	62.3	1m13 s
70	[3,3,3,3,4,4]	[3,3,3,4,4,5]	72.7	55 s
80	[4,4,4,4,4,4]	[4,4,4,4,5,5]	82.4	59 s
90	[5,5,5,5,5,5]	[5,5,5,5,5,6]	90.8	45 s
100	[5,6,6,6,6,6]	[5,6,6,6,6,6]	101	1m23 s

deviation and follow consistent trends. In Experiment 2, we randomly assign an approximation level to each stage of the pipelined FFT and then conduct the same comparison as in the first experiment. Figure 9b shows similar findings to Fig. 9a. Figure 10 further plots the output statistics of the two settings. It can be seen that the difference increases with a growing σ, indicating the inclusion of more conservativeness.

5.2 *Performance of Optimization*

Table 1 summarizes the precision optimization results of a 256-point 8-stage FP FFT using R2DIF algorithm, with each stage assigned an approximation level in the range [0, 11]. We validate the proposed optimization flow with different PSNR specifications with $\sigma = 0.85$. The second column lists the best approximate designs obtained through an exhaustive search that takes several hours to complete, while columns 3–5 list the results from the proposed optimization. It is evident that the proposed optimization flow can complete within around 1 minute for different PSNR specifications and the solution found is very close to the optimal combinations.

Table 2 Comparison of the proposed approximate FFT and an exact FFT without approximation

No.	1	2	3	4	5
$PSNR_{spec}$ (dB)	60	70	80	90	150
Levels	[2,2,2,3,3,3]	[3,3,3,3,4,4]	[4,4,4,4,4,4]	[5,5,5,5,5,5]	exact FFT
$PSNR_{act}$ (dB)	62.8	73.8	85	95.6	145.7
Area (μm^2)	214502	216401	221176	230821	302902
Freq. (MHz)	188	178.6	179.5	174.2	127.2
Power (mW)	149	150	152	155	131
Energy per FFT (nJ)	202	215	216	227	263
Norm. ADP	0.48	0.50	0.51	0.55	1

5.3 Approximate FFT Design Comparison

Table 2 presents a comparison between the approximate FFT using the proposed design methodology and an exact FFT. We implement both designs using UMC40 nm and report the results from Design Compiler. The second row $PSNR_{spec}$ in this table specifies the PSNR requirement at the application level. The third row shows the optimized approximation assignment achieved from the proposed methodology, while the fourth row $PSNR_{act}$ presents the actual PSNR of the implemented FFT hardware. All the optimized designs can satisfy the PSNR requirements with less than 6 dB difference, which is only 6% away from the PSNR requirement. The last five rows of the table report the area, clock frequency, power, energy per FFT operation, and normalized area-delay product (ADP). The optimized design achieves almost 50% speed improvement with around 23% energy saving for each FFT operation. Moreover, we can achieve up to 52% ADP improvement to achieve much higher energy efficiency than the exact FFT design.

Finally, we compare the proposed approximate FFT with prior state-of-the-art approximate FFT designs [17] in Fig. 11. It is noted that other prior approximate designs do not account for their approximation impact on the overall FFT algorithm precision, which then demands multiple design iterations or non-trivial design overhead to meet the design target. It can be seen from the figure that the proposed approximate FFT has the following advantages over [17]:

- **Wider precision range:** The PSNR range of [17] is just below 70 dB, while the PSNR coverage of this work can reach 150 dB. This indicates our design can adapt to higher application accuracy requirements.
- **Higher energy efficiency with tight PSNR constraint:** While the ADP of [17] rises rapidly with a growing PSNR, the ADP of our design maintains at a similar value. Our design has almost the same ADP at $PSNR > 80$ dB as [17] at $PSNR = 50$ dB, which indicates the large hardware benefits of the proposed design.

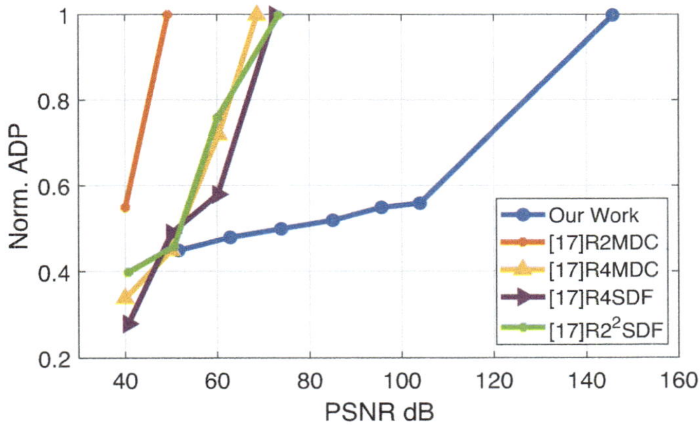

Fig. 11 Comparison of normalized area-delay products between our work and other approximate FFT designs [17]

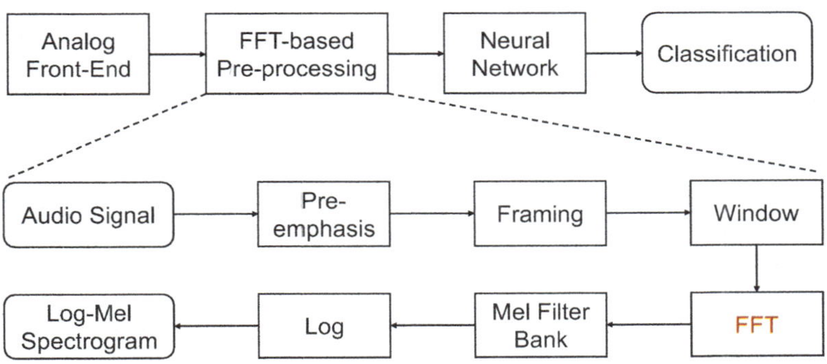

Fig. 12 An example of edge AI-based audio classification system

5.4 System Application

Figure 12 displays a typical architecture of an edge AI-based audio classification system. It employs FFT to create log-mel spectrograms of speech signals and convolutional neural networks for speech classification. To investigate the effect of approximate FFT on the accuracy of practical applications, we replace the original exact FFT with the proposed approximate FFT for speech pre-processing. Then, we check the accuracy of CNN for classification tasks using approximate log-mel spectrograms.

The audio signals are processed using a 256-point FFT with a 50% overlap between each frame. The UrbanSound8K dataset [30] is selected as the input for our application, which contains 8732 digital audio clips divided into 10 categories. To create the log-mel spectrograms, we use 26 mel bands and a logarithmic function.

Table 3 Architecture of the convolutional neural network

Layer	Input	Output	Kernel
2D Conv	26×694×1	24×692×32	3×3
2D Conv	24×692×32	22×690×32	3×3
Max pooling	22×690×32	11×345×32	2×2
2D Conv	11×345×32	9×343×64	3×3
2D Conv	9×343×64	7×341×64	3×3
Global average pooling	7×341×64	64	–
Fully connected	64	10	–

Table 4 Comparison of classification accuracy on the proposed FFTs using different approximation levels

No.	1	2	3	4	5
FFT PSNR (dB)	60	70	80	90	Exact
Accuracy (%)	89.16	89.28	89.28	89.29	90.26

The structure of the CNN model used for classification is presented in Table 3. The model is a 2D-CNN [31], and each convolutional layer is followed by a ReLU activation function, batch normalization, and a dropout layer. To train the model, we used a batch size of 128 and an Adam optimizer with a learning rate of 0.0001 and trained it for 360 epochs.

We randomly selected 732 records from the UrbanSound8K dataset as prediction samples, which were not used in the CNN training process. These samples were processed using FFT with different approximation accuracies, and then the trained CNN was used to classify them. Only the remaining 8000 speech data processed by exact FFT were used to train the CNN.

Table 4 presents the classification prediction accuracy for approximate FFT and exact FFT processing. The table indicates that the overall impact of using approximate FFT processing on the application is minimal, with a reduction in accuracy of around 1%. Meanwhile, approximate FFT offers significant advantages over exact FFT in terms of area, speed, and energy consumption. Therefore, this experiment highlights the potential of the proposed design methodology for enhancing the performance of FFT-based applications in various domains.

6 Conclusions

In this work, we propose a top-down methodology for designing approximate floating-point FFTs. By using the proposed error model and optimization flow, the design methodology can fully exploit the error-tolerance nature of a pipeline FFT and maximize energy efficiency with limited hardware overheads. Experimental results show that our design can achieve almost 2× wider precision range and higher energy efficiency compared to prior approximate FFT designs [17]. Additionally, the system application that uses the proposed design experiences a loss in accuracy of around 1%, further demonstrating the practical potential of this approach.

References

1. S. Liu, D. Liu, A high-flexible low-latency memory-based fft processor for 4g, wlan, and future 5g. IEEE Trans. Very Large Scale Integr. VLSI Syst. **27**(3), 511–523 (2019)
2. J. Deng, Z. Shi, C. Zhuo, Energy-efficient real-time uav object detection on embedded platforms. IEEE Trans. Comput. Aided Des. Integr. Circuits Syst. **39**(10), 3123–3127 (2020)
3. J. Bhattacharya, S. De, S. Bose, I. Banerjee, T. Bhuniya, R. Karmakar, K. Mandal, R. Sinha, A. Roy, A. Chaudhuri, Implementation of OFDM modulator and demodulator subsystems using 16 point FFT/IFFT pipeline arhitecture in FPGA, in *2017 8th IEEE Annual Information Technology, Electronics and Mobile Communication Conference (IEMCON)* (2017), pp. 295–300
4. C. Zhuo, S. Luo, H. Gan, J. Hu, Z. Shi, Noise-aware DVFS for efficient transitions on battery-powered iot devices. IEEE Trans. Comput. Aided Des. Integr. Circuits Syst. **39**(7), 1498–1510 (2020)
5. W. Liu, F. Lombardi, M. Shulte, A retrospective and prospective view of approximate computing [point of view. Proc. IEEE **108**(3), 394–399 (2020)
6. C. Zhuo, K. Unda, Y. Shi, W.-K. Shih, From layout to system: early stage power delivery and architecture co-exploration. IEEE Trans. Comput. Aided Des. Integr. Circuits Syst. **38**(7), 1291–1304 (2019)
7. S. Mittal, A survey of techniques for approximate computing. ACM Comput. Surv. **48**(4) (2016). https://doi.org/10.1145/2893356
8. D. Gao, D. Reis, X.S. Hu, C. Zhuo, Eva-cim: A system-level performance and energy evaluation framework for computing-in-memory architectures. IEEE Trans. Comput. Aided Des. Integr. Circuits Syst. **39**(12), 5011–5024 (2020)
9. J. Du, K. Chen, P. Yin, C. Yan, W. Liu, Design of an approximate FFT processor based on approximate complex multipliers, in *2021 IEEE Computer Society Annual Symposium on VLSI (ISVLSI)* (2021), pp. 308–313
10. A.K.Y. Reddy, S.P. Kumar, Performance analysis of 8-point FFT using approximate radix-8 booth multiplier, in *2018 3rd International Conference on Communication and Electronics Systems (ICCES)* (2018), pp. 42–45
11. V. Arunachalam, A.N. Joseph Raj, Efficient vlsi implementation of FFT for orthogonal frequency division multiplexing application. IET Circuits Devices Syst. **8**, 526–531 (2014)
12. A. Lingamneni, C. Enz, K. Palem, C. Piguet, Highly energy-efficient and quality-tunable inexact fft accelerators, in *Proceedings of the IEEE 2014 Custom Integrated Circuits Conference* (2014), pp. 1–4
13. X. Han, J. Chen, B. Qin, S. Rahardja, A novel area-power efficient design for approximated small-point FFT architecture. IEEE Trans. Comput. Aided Des. Integr. Circuits Syst. **39**(12), 4816–4827 (2020)
14. J. Emmert, J. Cheatham, B. Jagannathan, S. Umarani, An FFT approximation technique suitable for on-chip generation and analysis of sinusoidal signals, in *Proceedings 18th IEEE Symposium on Defect and Fault Tolerance in VLSI Systems* (2003), pp. 361–368
15. V. Ariyarathna, D.F.G. Coelho, S. Pulipati, R.J. Cintra, F.M. Bayer, V.S. Dimitrov, A. Madanayake, Multibeam digital array receiver using a 16-point multiplierless dft approximation. IEEE Trans. Antennas Propag. **67**(2), 925–933 (2019)
16. J. Park, J.H. Choi, K. Roy, Dynamic bit-width adaptation in DCT: an approach to trade off image quality and computation energy. IEEE Trans. Very Large Scale Integr. VLSI Syst. **18**(5), 787–793 (2010)

17. W. Liu, Q. Liao, F. Qiao, W. Xia, C. Wang, F. Lombardi, Approximate designs for fast fourier transform (FFT) with application to speech recognition. IEEE Trans. Circuits Syst. I Regul. Pap. **66**(12), 4727–4739 (2019)
18. B. Liu, X. Ding, H. Cai, W. Zhu, Z. Wang, W. Liu, J. Yang, Precision adaptive MFCC based on R2SDF-FFT and approximate computing for low-power speech keywords recognition. IEEE Circuits Syst. Mag. **21**(4), 24–39 (2021)
19. C. Wen, Y. Wu, X. Yin, C. Zhuo, Approximate floating-point FFT design with wide precision-range and high energy efficiency, in *Proceedings of the 28th Asia and South Pacific Design Automation Conference*, ser. ASPDAC '23 (Association for Computing Machinery, New York, 2023), pp. 134–139. https://doi.org/10.1145/3566097.3567885
20. S.L.M. Hassan, N. Sulaiman, I.S.A. Halim, Low power pipelined FFT processor architecture on FPGA, in *2018 9th IEEE Control and System Graduate Research Colloquium (ICSGRC)* (2018), pp. 31–34
21. M. Garrido, A survey on pipelined fft hardware architectures. J. Signal Process. Syst. **94**(11), 1345–1364 (2021)
22. J.H. Bahn, J. Yang, N. Bagherzadeh, Parallel FFT algorithms on network-on-chips, in *Fifth International Conference on Information Technology: New Generations (ITNG 2008)* (2008), pp. 1087–1093
23. Z. Wang, X. Liu, B. He, F. Yu, A combined SDC-SDF architecture for normal I/O pipelined radix-2 FFT. IEEE Trans. Very Large Scale Integr. VLSI Syst. **23**(5), 973–977 (2015)
24. M. Ha, S. Lee, Multipliers with approximate 4–2 compressors and error recovery modules. IEEE Embed. Syst. Lett. **10**(1), 6–9 (2018)
25. H. Jiang, J. Han, F. Qiao, F. Lombardi, Approximate radix-8 booth multipliers for low-power and high-performance operation. IEEE Trans. Comput. **65**(8), 2638–2644 (2016)
26. L. Qian, C. Wang, W. Liu, F. Lombardi, J. Han, Design and evaluation of an approximate wallace-booth multiplier, in *2016 IEEE International Symposium on Circuits and Systems (ISCAS)* (2016), pp. 1974–1977
27. C. Chen, S. Yang, W. Qian, M. Imani, X. Yin, C. Zhuo, Optimally approximated and unbiased floating-point multiplier with runtime configurability, in *2020 IEEE/ACM International Conference On Computer Aided Design (ICCAD)* (2020), pp. 1–9
28. C. Chen, W. Qian, M. Imani, X. Yin, C. Zhuo, Pam: a piecewise-linearly-approximated floating-point multiplier with unbiasedness and configurability. IEEE Trans. Comput. **71**(10), 2473–2486 (2022)
29. M. Hashimoto, J. Yamaguchi, T. Sato, H. Onodera, Timing analysis considering temporal supply voltage fluctuation, in *Proceedings of the 2005 Asia and South Pacific Design Automation Conference*, ser. ASP-DAC '05 (Association for Computing Machinery, New York, 2005), pp. 1098–1101. https://doi.org/10.1145/1120725.1120833
30. Urbansound8k (2020). https://www.kaggle.com/datasets/chrisfilo/urbansound8k
31. Urban sounds classification with convolutional neural networks (2019). https://github.com/GorillaBus/urban-audio-classifier

Approximate Computing in Deep Learning System: Cross-Level Design and Methodology

Yu Gong, You Wang, Ke Chen, Bo Liu, Hao Cai, and Weiqiang Liu

1 Introduction

The edge computing systems in artificial neural network (ANN) applications own massive numbers of devices as user interfaces. Thus, ultra-low power consumption and high accuracy become essential factors while designing ANN accelerators. The approximate computing paradigm has been treated as a promising candidate to improve the power/area efficiency of computing hardware for ANNs [1–3].

For the highly efficient ANN optimization approaches, quantizing the weights to single bit is the commonly used way to meet the area-power constraint conditions. Such networks are nominated as binary-weight neural networks (BWNNs). With single-bit weights, the vast majority of the operations in BWNN processing are additions. The multiplications in ANN are almost substituted by bit operations, which makes BWNN the most used low-power ANN variant. Meanwhile, with smart training and quantization approaches, the accuracies of BWNNs may satisfy the requirements of specific low-power recognition applications [4, 5] with tolerable accuracy loss.

With single-bit weights and the loss of BWNN accuracy, the designs and optimizations of the BWNN computing units are significant for accelerating systems, especially for the facts that the approximate computing technologies are radically replacing full-precision computing. By using approximate computing, system precision and consumption can be balanced for different scenarios. The system accuracy constrains the deployment of approximation techniques, and the impacts on the

Y. Gong (✉) · Y. Wang · K. Chen · W. Liu (✉)
Key Laboratory of Aerospace Integrated Circuits and Microsystem, Nanjing University of Aeronautics and Astronautics, Nanjing, China
e-mail: gongyu@nuaa.edu.cn; liuweiqiang@nuaa.edu.cn

B. Liu · H. Cai
College of Integrated Circuits, Southeast University, Nanjing, China

W. Liu et al. (eds.), *Design and Applications of Emerging Computer Systems*,
https://doi.org/10.1007/978-3-031-42478-6_18

system accuracy by introducing approximate computing are usually evaluated by computing qualities. For the basic computing units, a series of approximate multipliers with error tolerate techniques [6–9], for different applications under varying requirements. Also, the implementation of approximate addition units [10–15] has been widely used in signal processing systems. Some of the significant approximate adders, such as ETA-I/II [10, 11], the AMAs [12], GeAr [13], QuAd [14], and the InXAs [15], were integrated as silicon-based works [16].

For ANN accelerators, approximation techniques were also assigned at the system level. Yang et al. [17] proposed an analog circuit-based adder array, which only needs a small BWNN with single-bit data. By using analog modulation, the computing energy consumptions were significantly reduced, whereas the system accuracy remains guaranteed. Yin et al. [18] used processing elements with reconfigurable bit width to accelerate different scales of BWNNs. By the reconfigurable units, the BWNN engine can effectively deal with multiple applications. They also implemented a processor with precision configurable approximate adder cluster for both binary-weight and ternary-weight neural networks [19]. The adder cluster can accelerate almost all processes of the networks (excludes the activation and batch normalization layers), which leads to the enhanced power/area efficiency.

For NN systems deploying approximate computing in different levels, including algorithm, architecture, and circuit, the control of computing quality is significant. Several types of approaches are innovated to quantify the impacts on different levels of approximation and further to achieve high accuracy and low hardware consumption. The mathematical models, multi-level simulations, and validations are the main methods to estimate and evaluate the output qualities in approximate computing systems. For example, the statistical approaches were obtained by simulations [15, 20], and the Monte Carlo (MC) simulations can be used to obtain a certain computation quality [21]. Due to the tremendous runtime of MC simulation, the probabilistic and analytical models or frameworks for approximate units were proposed to speed up the evaluation of the output qualities. Liu et al. [22] conducted evaluation approaches using statistical methods and systematically integrated the approximate computing designs in a low-power system [1]. A statistical error model [23] as a design methodology was proposed for the approximate addition cluster. MACACO [24] was an analytical model of approximate computing circuits. Another error model for a selected class of approximate adders is conducted by Mazahir et al. [25]. Ayub et al. [26] analyzed approximate adders using statistical error analysis under specific low-power scenarios.

2 Related Work

In this section, the features of BWNN and typical SOTA BWNN-inspired and approximate computing-enabled NN engines are introduced. Then previous works on approximation quality estimation approaches are also addressed to find innovation for evaluating NN systems.

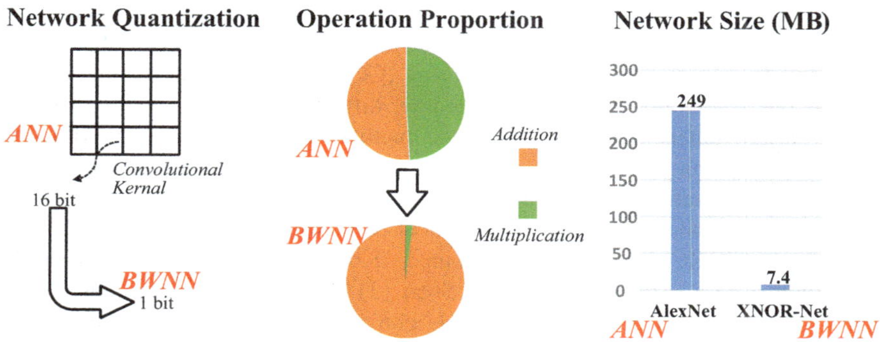

Fig. 1 Comparisons between conventional ANNs and BWNNs

2.1 *Binary-Weight Neural Networks*

For NNs, the massive synapses lead to large consumption of memory accessing, caching, and allocating. Thus, the "Memory Wall" becomes very critical, and the most promising way to solve this issue is the quantization of NNs. With quantization approaches, the topologies of the original NNs and the quantized NNs are the same, which means the datapath in the corresponding acceleration system for original NNs can still work with the quantized network, only with reduced data/weight bit width. As the reduced bit width leads to the reduction of operation complexities, the quantization of ANNs is one of the most used approaches to achieve tradeoff between accuracy and hardware costs. The BWNNs, owning only 1 bit weight with multi-bit data choices, are therefore widely deployed in low-power systems with high accuracy and low hardware consumption. Figure 1 compares ANNs to BWNNs and shows the almost retained accuracy of BWNN, with 32× less memory occupation than ANN. As the weight bit width is quantized from 16 bits to 1 bit, which removes almost all of the multiplication operations, the design of BWNN acceleration focused on the addition units, and thus, the designs of approximate adders are significant.

2.2 *Low-Power BWNN System with Approximation*

In recent years, approximate computing technology has been widely used in different arithmetic operation circuits as well as neural network accelerators. Jiang [2] provided a comprehensive research and a comparative evaluation of recently developed approximate computing circuits and accelerators under different design constraints. Then the approximate computing tolerance between these accelerators is assessed. In widespread application situations of neural network accelerators, different kinds of approximate computing [27–32] have been used to improve the whole system performance and to reduce the power consumption. The works [27–29] integrated various levels of approximate adders and multipliers in a coarse-grained reconfigurable array (CGRA) for different configurations so that the accelerator can be optimized by different performances or different energy improvements. Yin [30] took the method of lower-5-bit approximation to optimize 16-bit RCA (ripple carry adder) and realized a BNN accelerator. With the optimization of the carry chain in the approximate part, they realized an approximate adder with a shorter critical path, which is 49.28% less than an accurate 16-bit RCA's. They also obtained 48.08% power-delay product (PDP) with only 1.27% error rate increase. Shan [31] applied an approximate MAC unit (AP-MAC) in a depthwise separable convolution neural network (DSCNN) accelerator and saved 37.7%~42.6% computing power compared to the traditional MAC units within 0.8% accuracy loss. Liu [32] proposed a BWNN accelerator with both approximate computing mode and standard computing mode. The accelerator realized an addition unit which is precision self-adaptive due to various situations with the use of lower-bit OR gate adder (LOA). The error is compensated by a positive-and-negative error accumulating method. Benefited from the approximate computing mode, the accelerator obtained 1.4× power consumption reduction with almost no deviation.

To realize the low-power NN systems, the approximate computing techniques are contributing a lot, including network quantization, approximate unit design, and systematical tradeoff between approximation and hardware complexity. Thus, the tradeoff approach is significant for such approximate systems. The estimation model or framework is gradually introduced as one of the most effective approaches to do tradeoffs fast and accurately.

2.3 Quality-Driven Approximate Computing System

For the quality estimation approaches and models, many metrics are employed for quantification, such as the mean error, the mean error distance (MED), the mean relative error distance (MRED), the mean squared error (MSE), the error rate (ER), and so on [33]. Deploying different metrics, various models and frameworks are proposed. MACACO [24] is an analytical model of approximate computing circuits by Boolean analysis which can analyze a range of approximate designs for datapath building blocks. Liu et al. [23] proposed a statistical error model using MSE and ER as a design methodology for approximate accumulators and can reduce the simulation effort by over 1.5× speedups with 8.56% accuracy degradation. The error model [25] for a selected class of approximate adders is conducted and validated by Mazahir et al., and the work shows the model generalization possibility for one type of massive approximate adders. Ayub et al. [26] analyzed low-power approximate adders using statistical error analysis and showed how to extract the error from iteration. For complex approximate unit arrays, Masoud [33] presented SNR as the estimation metric to evaluate the output quality between inexact and exact results. The works on approximate adders and the error models have shown great progress in the past years, and the analytical, probabilistic, and statistical approaches are progressing and focusing on more accurate estimations of the output quality. The circuit or architecture optimization is the secondary use of the models or frameworks. Referring to the estimation of the output quality on a full acceleration system for BWNN applications, it is still critically required as the adders are jointed and coupled tightly. Toward this end, we focus on presenting a probabilistic model for quality estimation of BWNN applications using massive approximate addition units. The optimization will also be considered following the model.

3 Estimation and Evaluation of Low-Power Approximate NN System

In this section, the low-power approximate NN system estimation and evaluation approach is introduced, including the binarization methods for NNs, the estimation of approximate units, the approximation noise abstraction, and datapath-related noise propagation analysis. The key of the estimation approach is the analytical model for evaluating approximate arrays, which is composed of the approximate adder quality model and the approximate noise in datapath model. By the analytical model, the most time-consuming work, which is the design space of approximate computing arrays, can be explored fast. The system computing quality is then evaluated via simulation.

3.1 Estimation of Approximate Units

Approximate adders can be categorized as low-latency and low-power adders. For low-latency approximate adders (LLAA), to achieve high performance, they are mostly block-based. Multiple overlapped sub-adders with approximate unit inside are usually consisted. For the most significant bits (MSBs), they are calculated accurately in exact part, and the less significant bits (LSBs) are calculated inaccurately in approximate sub-adder [10, 13, 14].

Approximate adders (AxAdders) are mostly with controllable precision, which means we can know the error probability when we decide the carry-in and adding mechanism. Taking the OR gate-based addition unit as an example, the design of a single approximate adder is demonstrated in Fig. 2a, and the quality configurable approximate addition units are set with both full addition (FA)- and OR gate (OR)-based addition circuit. The details for FA and OR unit for the ith bit addition are presented in Fig. 2b and c, respectively. The truth tables of them are also given in Fig. 2d. It can be seen that the OR unit introduces computing quality loss, but the critical path is significantly reduced. Similarly, other approximate addition circuits

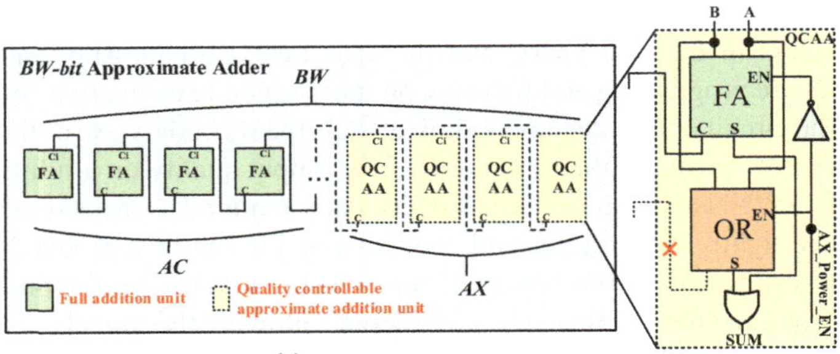

(a)

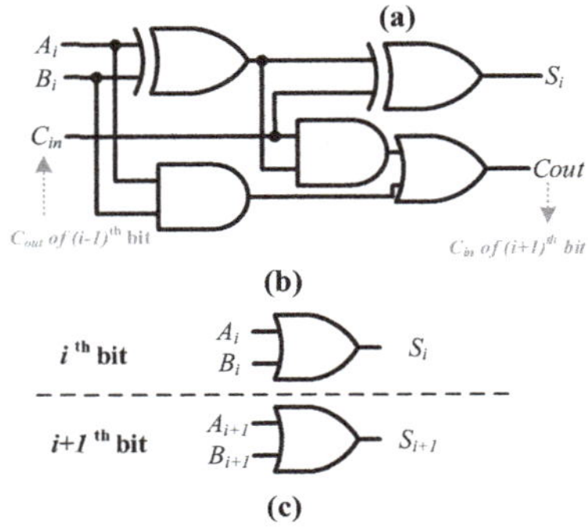

	A	B	C_{in}	S	C_{out}
FA	0	0	0	0	0
	0	0	1	1	0
	1	0	0	1	0
	1	0	1	0	1
	1	1	0	0	1
	1	1	1	1	1
OR	0	0	N/A	0	N/A
	1	0	N/A	1	N/A
	1	1	N/A	1	N/A

(d)

Fig. 2 Design of approximate adder with quality control: (**a**) the design of proposed approximate adder; (**b**) the circuit of the FA unit for the ith bit; (**c**) the circuit of the OR circuit for the ith bit; (**d**) the truth table of the FA and OR circuits

can also be adopted in the unit of QCAA. The signal of "AX_Power_EN" is used to configure the QCAA function as FA or OR. For the approximate adder, the design parameters are defined as follows: the bit width of AxAdder is BW and AC bits are implemented with full-precision adder blocks, with full addition circuit as accurate cell, and AX bits ($AX \leq 8$ and they are all in fraction blocks) are adopting approximate addition circuit as an approximate adder block. By the definitions, for a BW-bit AxAdder, AC-bit full-precision blocks and AX-bit approximate blocks are contained. Thus, the quality is impacted by the LSBs and the output carry of the AX bits. Since the approximate blocks in one AxAdder are in the same structure, we can assume that the AX-bit blocks own the same independent error probability, P_e, and it is determined by the truth table of the block. Based on previous works [34, 35], the following formulas can be obtained. Firstly, the AX bits of AxAdder own the MSE of:

$$MSE = E\{e^2[n]\} = \frac{1}{N}\sum_{n=0}^{N-1} e^2[n] \tag{1}$$

in which $N = BW$. With the definition of error distance of each bit (ED_i), and the P_e of a certain AxAdder circuit which can be calculated from the truth table, we can get:

$$MSE = \sum_{i=0}^{k-1} ED_i^2 \times P(ED_i) \tag{2}$$

in which k=AC.

Assuming every bit in the AxAdder processes error, $ED_i = 2i$, and the error rate is P_e. Using the equation $\sum_{i=0}^{n-1} x^i = \frac{1-x^n}{1-x}$, we can get the final form of MSE:

$$MSE = P_e \times \frac{2^{2j}}{3} \tag{3}$$

in which $j = AX$. To be clear, the deduction approaches are following the principles in work [35].

3.2 *Approximation Noise in Datapath*

As noise is the extra signal mixed in the needed information, the computing error is similar to the noise during the computing process, and it is called approximation noise [33]. In this chapter, approximation noise is introduced to the key module for BWNN acceleration, which is the approximate adder cluster. Considering $x[n]$ and $y[n]$ as the n-th bit of input signal x and signal y, the precise addition result is $s[n]$, as illustrated in the right top of Fig. 3.

With the noise signal $e[n]$ shown in the right bottom, the precise result is in the middle bottom of Fig. 3 with detailed components. Compared with the approximate results in the middle up of Fig. 3, the similar figures prove that without the details of

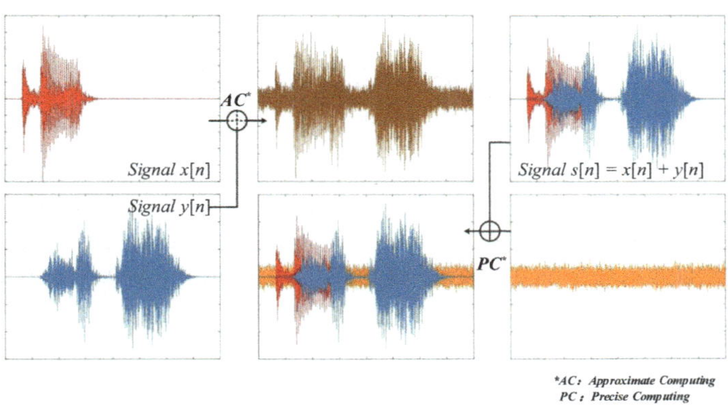

Fig. 3 Approximation noise propagation in system

the final results components, the approximation noise theory can show the impacts of the noise introduced by approximation. For the error is the noise signal and the accurate result is the required signal, following the definition of SNR, which is:

$$SNR = 10log_{10}\left(\frac{P_s}{P_n}\right) \tag{4}$$

in which P_s is the signal power and P_n is the noise power.

In this chapter, input signal power is defined as A_s^2, and the noise signal power is n^2; then the approximation noise is:

$$SNR = 10\log_{10}\frac{A_s^2}{\hat{n}^2} \tag{5}$$

Generally, in an approximate system, the noise is mainly from the noise in the input signal, the noise in the quantization for signal scaling or multiplication, and the noise from approximate computing. The formula is modified:

$$SNR = 10\log_{10}\frac{A_s^2}{\hat{n}_i^2 + \hat{n}_q^2 + \hat{n}_a^2} \tag{6}$$

in which the noise power of $\hat{n}^2$ comes from three parts, which are the noise $\hat{n}_i^2$ in the input signal, the noise introduced by approximate multipliers $\hat{n}_q^2$, and the noise introduced by approximate adders $\hat{n}_a^2$.

Since, in BWNN, the key module we use is only adder cluster, $\hat{n}_q^2$ can be ignored. The noise is propagating in an ordered chain as $\hat{n}_i^2$ can be defined as the [23] of input noise signal, and $\hat{n}_a^2$ is the *MSE* of approximation noise for the adder cluster. Considering one convolutional computation with m weights, and w_i is the i-th weight, then the power of the output noise in one convolutional computation is given:

$$\hat{n}_o^2 = \hat{n}_i^2 \times \sum_{i=1}^{m} w_i^2 + \hat{n}_a^2 \times m \tag{7}$$

For the BWNNs, the input data and the weights can be seen as two independent distributed signals; thus, the computation can be seen as an addition of two signals. And the power of the result is the sum of the powers from the two signals, just as the formula above. If n^2 is the power of input signal, the output power is $2n^2$.

3.3 System Evaluation Approach and Mechanism

For the recognition accuracy evaluation, the analytical models for approximate arrays are integrated in MATLAB models to evaluate the whole BWNN system. The design of the approximate addition arrays is taken under the quality model, and a large design space can be explored fast. With the design parameters of the arrays, the circuits for other functions, such as the pooling layers, the activation layers, and the batch normalization layers, are all processed with conventional 16-bit fixed-point designs. And thus, for the systematical recognition accuracy estimation, we take MATLAB to calculate the accuracy based on the design parameters of architectures. First, we compare the different approximate adders to select which is the better structure. Then, we take the selected structure to ensure the approximate full adder operation logic and build the accumulative adder model. The model is transformed into approximate addition unit model in MATLAB, in which the approximate bits of each layer are decided. Afterward, we embed the approximate addition unit model into the MATLAB BWNN model for keyword spotting and calculate the accuracy of recognition based on the test dataset. The simulation result is compared to the recognition accuracy of full-precision BWNN to tell the accuracy change by introducing approximate computing. The analytical model-based approach can significantly reduce the evaluation time consumption, compared to adder function models.

The designs that meet the accuracy requirements will be taken to realizations and further simulation and verification to get the power and area specifications. We take Synopsys Design Compiler (DC) synthesis tool and PrimeTime PX (PTPX) power consumption evaluation tool to evaluate the power consumption benefit brought by the introduction of the approximate addition unit in the BWNN network. Due to the approximate bits decided before in MATLAB, we make the RLT realization of the approximate addition structure based on Verilog and make the functional simulation and verification to make sure the validity. After this, we make DC synthesis based on an industrial 22-nm COMS process library and PTPX simulation on the approximate addition structure to measure the power consumption and area specifications. Referring to the power and area distribution of full-precision BWNN, the BWNN specifications are calculated with the synthesis and simulation results, and the benefits are preliminarily evaluated. The design with balanced power consumption and system accuracy will be chosen as the final architecture. Besides, the quality-power co-optimization considers the hardware overheads to the designs, which can improve the energy efficiency of the system.

By the evaluations, we further optimized the system using the estimation approaches, with the optimization flow in Fig. 4. First of all, the flow can be divided into two parts, the systematical recognition accuracy evaluation and the power consumption optimization evaluation.

4 Quality Configurable Approximate Computing

In this section, a quality configurable approximate computing architecture is proposed after the design approach as discussed in Sect. 3. Firstly, the array of low-power approximate computing units is evaluated to prove the effectiveness of the

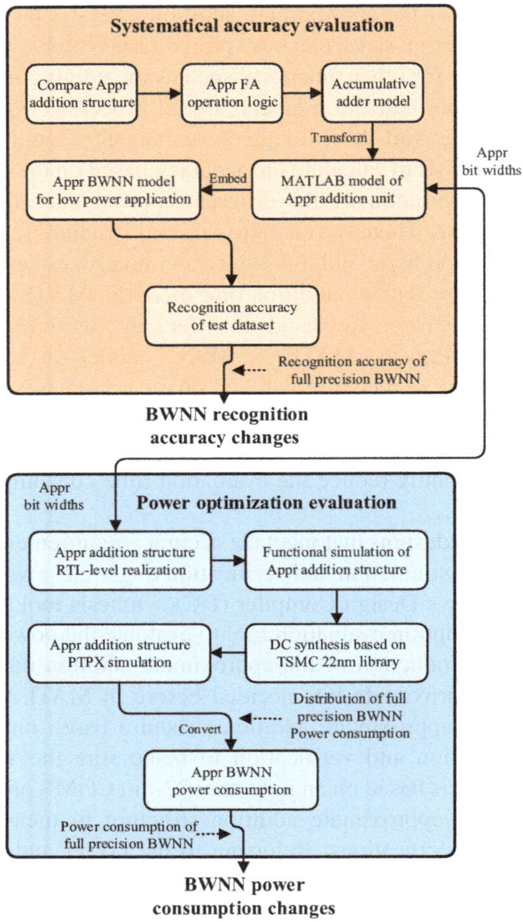

Fig. 4 Systematical evaluation-based design with quality-power consumption co-optimization

proposed approach. Then the design of a heterogeneous approximate computing array for BWNN is introduced and compared with conventional function circuits. As discussed in Sect. 2, the LLAAs are block-based, for which is modeled block by block and then jointed. In this chapter, we focused on the modeling of LLAAs with 8-bit fraction computing blocks and configurable bits of integer blocks. Besides, to ensure the real-time response, the mainstream BWNN accelerators mainly use the types of LLAAs too.

4.1 Evaluation of Low-Power Approximate Array

By the approximation noise deduction and the MSE estimation of AxAdders, we can estimate the approximation error of AxAdder clusters. To verify the efficiency of the approach, we conducted simulation and estimation for six types of approximate addition circuits, including AMA1, AMA2, AMA3, AMA4, TGA1, and TGA2. The AxAdder clusters are with the same configuration of nine stages.

The analysis of the results in Fig. 5 is as follows. Figure 5a~f shows the experimental results for one AxAdder with the AX of LSB 8 bits, 10 bits, 12 bits, and 14 bits. The comparisons show for one AxAdder, the estimation accuracy is almost the same as the simulation results, which proves the efficiency of the estimation approach for AxAdders. Figure 5g presents the experimental results for AxAdder clusters, also with AX of 8 bits, 10 bits, 12 bits, and 14 bits. Referring to AMA2, AMA3, AMA3, and TGA, compared to simulated MSE, the differences between estimated MSE for AMA2, AMA3, AMA3, and TGA are within 6%. The differences between the two factors of AMA1 and TGA2 are also tolerable.

4.2 Design of Hierarchical Adder Cluster

For the BWN blocks, the most used computation operation is the summation operation. As BWNNs are usually used in low-power and real-time application, the latency becomes critical, and the structure of approximate addition clusters is used. The design of block-based approximate full adder is presented in Fig. 2. It is a BW_i-bit full adder with AC_i-bit accurate addition block and AX_i-bit approximate addition block. We can describe the approximate adder (AxAdder) in a parameter pair of $\{BW_i, AX_i, AC_i\}$. In Fig. 2, the LSB OR gate [36]-based AxAdder (LOA) is used as an example.

The example of a conventional approximate adder cluster in tree form is given in Fig. 6. To state the model clearly, we take OR gate-based addition block (ORA) as an example, and the parameters and the structure of the AxAdder are also listed in Fig. 6. The proposed signal SIG_EN and the multiplexer MUX are designed for power gating and precision reconfiguration and will be discussed later. The left part of the figure is the adder cluster, with the design of each addition unit. For each unit, an AxAdder and a multiplexer are consisted, and the signal of "SIG_EN" is used to control the dataflow. The control method is shown on the right figure of the circuit. While "SIG_EN" = 1, the unit works as an adder; and while the signal is 0, the input data is just passed by without processing, and the data is transferred to the next addition unit.

For the adder cluster in tree form, the parameter pair is defined as N, BW_i, AX_i, and AC_i . The parameters are described as follows: i represents the stage order, ranging from 1, as the last stage, to N, as the first stage; 2^{i-1} stands for the total number of adders used at stage i; at each stage, the bit width of all AxAdders is BW_i;

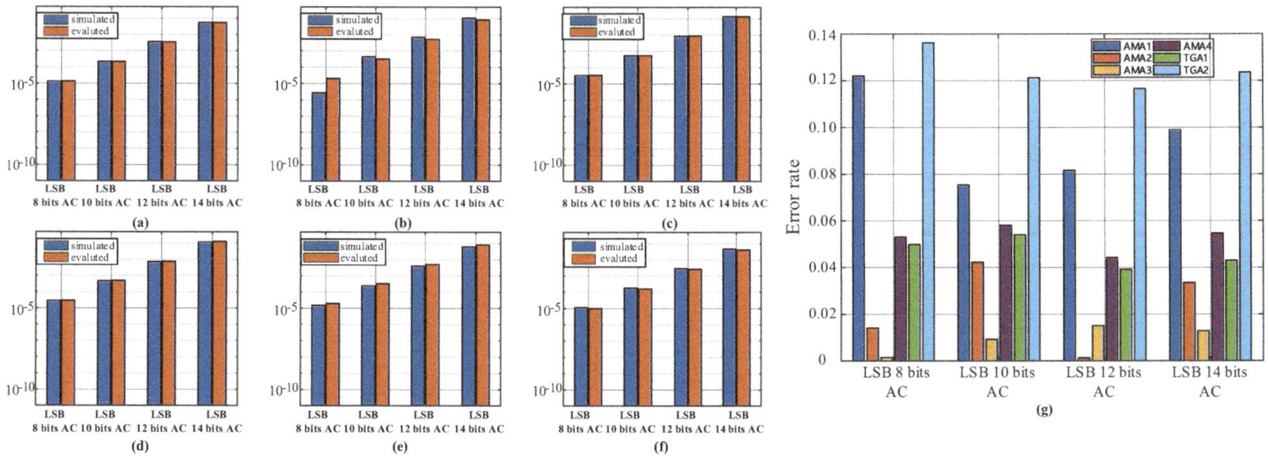

Fig. 5 Simulation and evaluation results for different types of AxAdder trees: MSE (mean squared error) for AxAdders integrating (**a**) AMA1; (**b**) AMA2; (**c**) AMA3; (**d**) AMA4; (**e**) TGA1; (**f**) TGA2,=; and (**g**) ER (error rate) for previous AxAdders

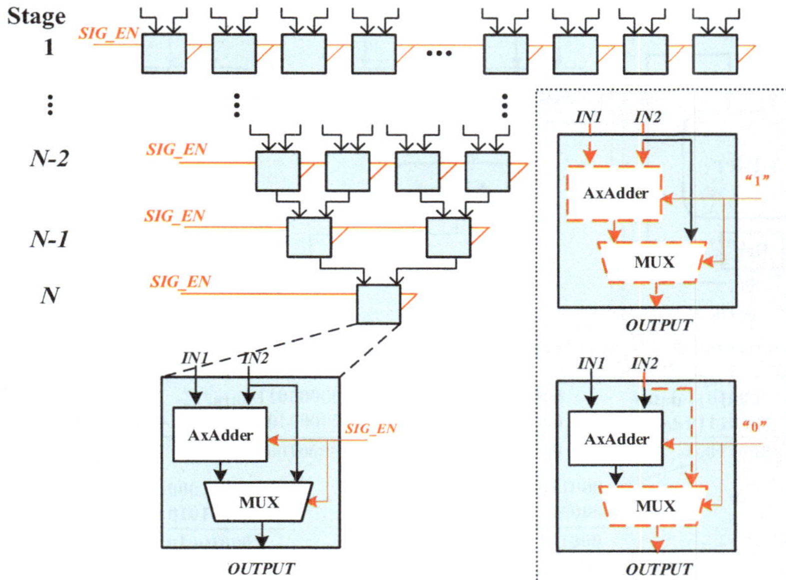

Fig. 6 Architecture of the proposed AxAdder cluster in tree form

and for the adders, AC_i bits are implemented with full adders, with full addition circuit as accurate cell, and AX_i bits are adopting approximate addition circuit as approximate cell. At all stages, the AxAdders are fixed pointed with 8-bit fractional blocks and $(BW_i - 8)$-bit integer blocks as stated in the former section.

As accumulators have many forms, we compare the adder cluster design with the conventional accumulator in Fig. 7, to show the advantages of the AxAdder cluster design in tree form. The accumulator in Fig. 7a consists of the adder and an extra register, and it is coupled with the clock. It may be time-consuming in low-frequency systems and cannot satisfy the requirements of real-time response of ANN acceleration. The tree form shown in Fig. 7b is used in this chapter. Despite the high static power and extend critical circuit path, it can deal with real-time applications. And the static power can be relieved by the high threshold voltage and power-gating technology in the IC design flow. Also, the example in Fig. 7c shows that the tree form can help control the significant bits by using heterogeneous AxAdder in different computation stages. In Fig. 8, the experiments are to compute 32-bit accumulations by using the two types of designs. The results show the error distance of them each to the accurate accumulation. The one in tree form shows superiority over the conventional approximate accumulation design.

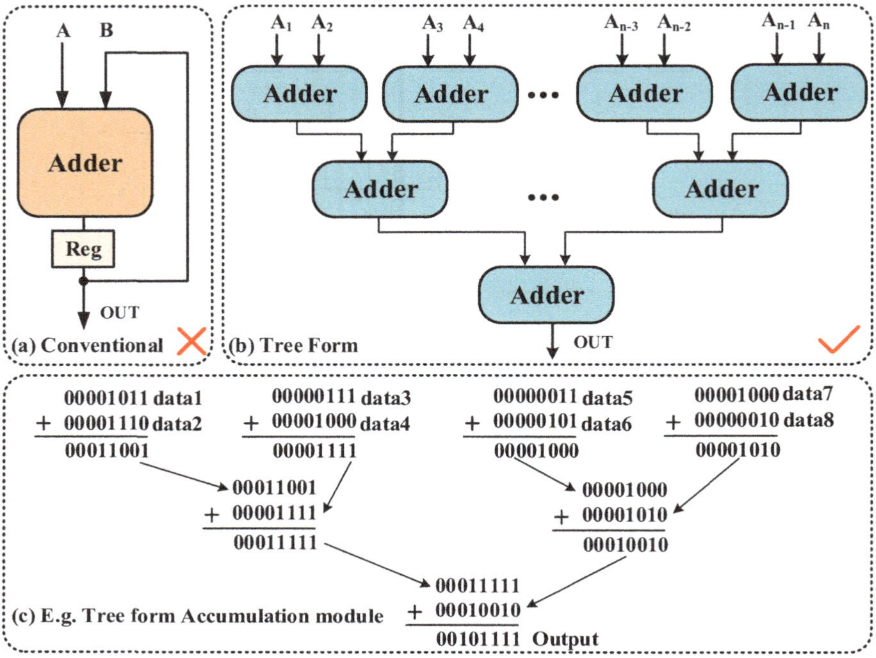

Fig. 7 Comparison between conventional form and tree form

4.3 Pre-analysis for Estimating BWNN Acceleration System

For further estimation and design work on the hardware consumption and the computing quality of AxAdders [12, 36], the hardware specifications are listed as follows: for MA, AMA1, AMA2, AMA3, AMA4, and ORA, the transistor costs in one block are 24, 20, 14, 11, 15, and 6 in order.

4.3.1 Adaptability for Convolutional/Pooling/Activation Layers

As for BWNN accelerations, the main topology of BWNN consists of convolutional, pooling, and activation layers. For the testing architecture in this chapter, the computing components used are pooling and activation layers are all 16-bit full-precision units, which can be regarded as no accuracy loss inner the layers. For convolutional layers in BWNNs, the addition scale is changeable, relating to the data reuse strategy, the kernel size, the channel numbers, and so on. But the calculation operation is the same, which is summation, since multiplication is eliminated in convolutional layers. Thus, the proposed approach can be directly used in BWNN systems.

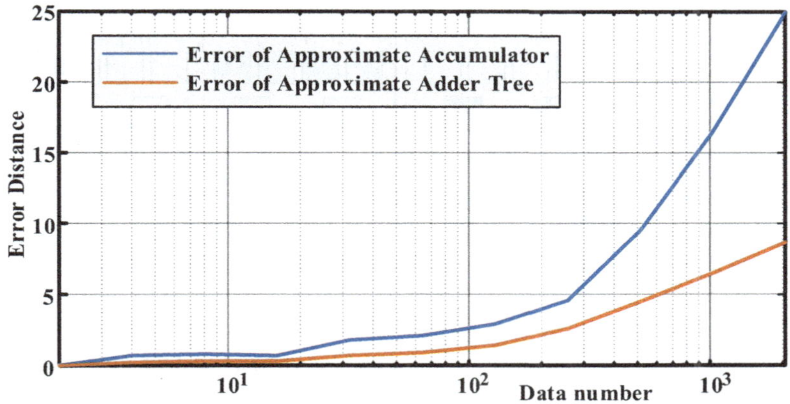

Fig. 8 Comparison between two types of accumulation designs

4.3.2 Parameterized Adder Cluster Design

According to different application accuracy requirements, the tolerable maximum error is usually known, and thus, different combinations of $\{N, BW_i, AX_i, AC_i\}$ can be made to test out the best one for implementation. The tradeoff between accuracy and hardware cost can also be made. The following strategies show the design methodology based on the error model.

Definitions System accuracy, as the output quality, is set as Sys_acc; the tolerable error rate is $err = (1 - Sys\_acc)$; and the maximum and minimum numbers to be accumulated are Sum_No_max and Sum_No_min.

The Input Analysis For the input feature maps, the last 8 bits of data are taken to do statistical analysis to show the distribution of "1" appears in each bit position.

Adder Clusters The depth of adder cluster N is decided due to Sum_No_max and Sum_No_min and should be ranging from $INT(log_2Sum\_No\_min) + 1$ to $INT(log_2Sum\_No\_max) + 1$, which is determined by the transistor numbers also.

AX_i: The 1st stage AxAdders should keep most of the information of input data, so AX_1 considers the input data distributions. If there are more than 2^x of "1" in the position of 2^{-x} bit, which means carry-in may occur, AX_1 would be set to $(8 - x)$. The following stages are evaluated the same way from the output of the previous stages.

AC_i: $(=BW_i - AX_i)$ Based on the quality model, different AC_i can be tested fast and efficiently.

BW_i: When AC_i is confirmed, BW_i is also set.

Tradeoff The number of transistors can be calculated when N, BW_i, AX_i, and AC_i are set. As one of the N-stage adder clusters can be replaced by two of the $(N-1)$-stage adder clusters with different settings of $\{N, BW_i, AX_i, AC_i\}$, tradeoff

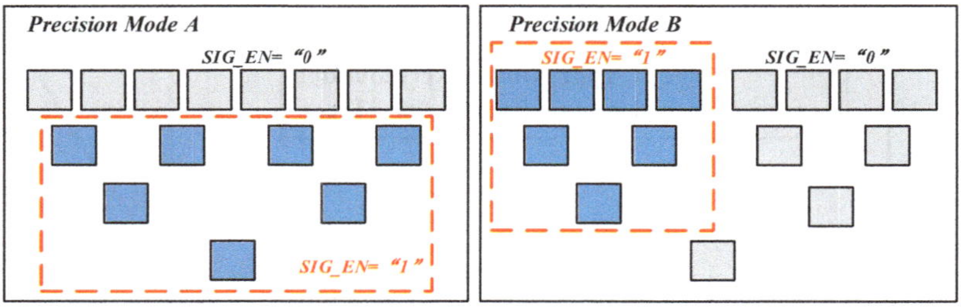

Fig. 9 Example of precision scaling by gating scheduling

between accuracy and hardware cost can be conducted with the proposed error model.

4.3.3 Evaluation-Based Gating for Scheduling Mechanism

With the signal of "SIG_EN" and the help of power gating, when we process an M-stage ($M < N$) accumulation on an N-stage adder cluster, we can use different parts of the adder cluster for different precision requirements, by putting M to the model. The energy consumption can also be improved by using only part of the adder cluster.

In a certain BWNN, the scales of accumulation tasks may change from layer to layer. For example, for a four-numbered accumulation, a three-stage adder cluster is enough; for an eight-numbered accumulation, a four-stage adder cluster is needed. To process such tasks on the same hardware, the adder cluster can make use of the enable signal "SIG_EN" and consider the accuracy needs at the same time. If we process a three-stage accumulation on a four-stage adder cluster as in Fig. 9, we can use either the two to four stages or half of the one to three stages. Since the AxAdders may not be the same settings, different processing accuracy can be evaluated by the proposed quality model during the acceleration procedures.

The configuration workflow is established in Fig. 10. Firstly, the application and the corresponding NN topology are analyzed. From the application input features, the computing precision requirements are mentioned and used as the control of system computing quality. Within the minimal computing quality requirements, the configurations of AxAdders, such as the approximation degree of the fraction part, can be achieved. By the NN topology analysis, the computing scale of the kernel steps in the network, such as the convolutional kernel sizes, is considered to compare with the adder cluster scale, to see if partial clusters should be gated. By the proposed analytical computing quality evaluation approaches, the finally adder cluster configurations for the specific application can be achieved to process the task on the proposed system, with high energy efficiency.

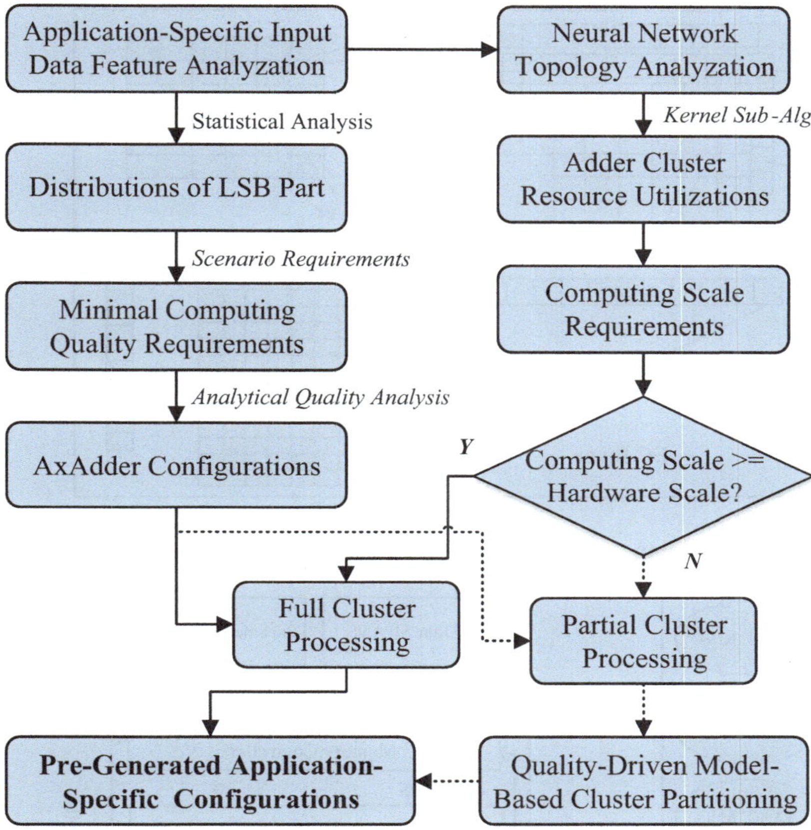

Fig. 10 Workflow for gating-enabled AxAdder cluster configuration

5 Reconfigurable Lower-Power NN System for Keyword Spotting Application

For previous sections, we proposed an efficient approach for low-power NN system design and deployed it in a quality-controllable approximate computing architecture. In this section, we use the previous approach and architecture to implement an MFCC [37]-based low-power keyword spotting (KWS) system.

5.1 Deployment of Approximate BWNN

The architecture consists of the computing engine to process the BWNN as illustrated in Fig. 11, the memory modules to keep the weights and data, and the

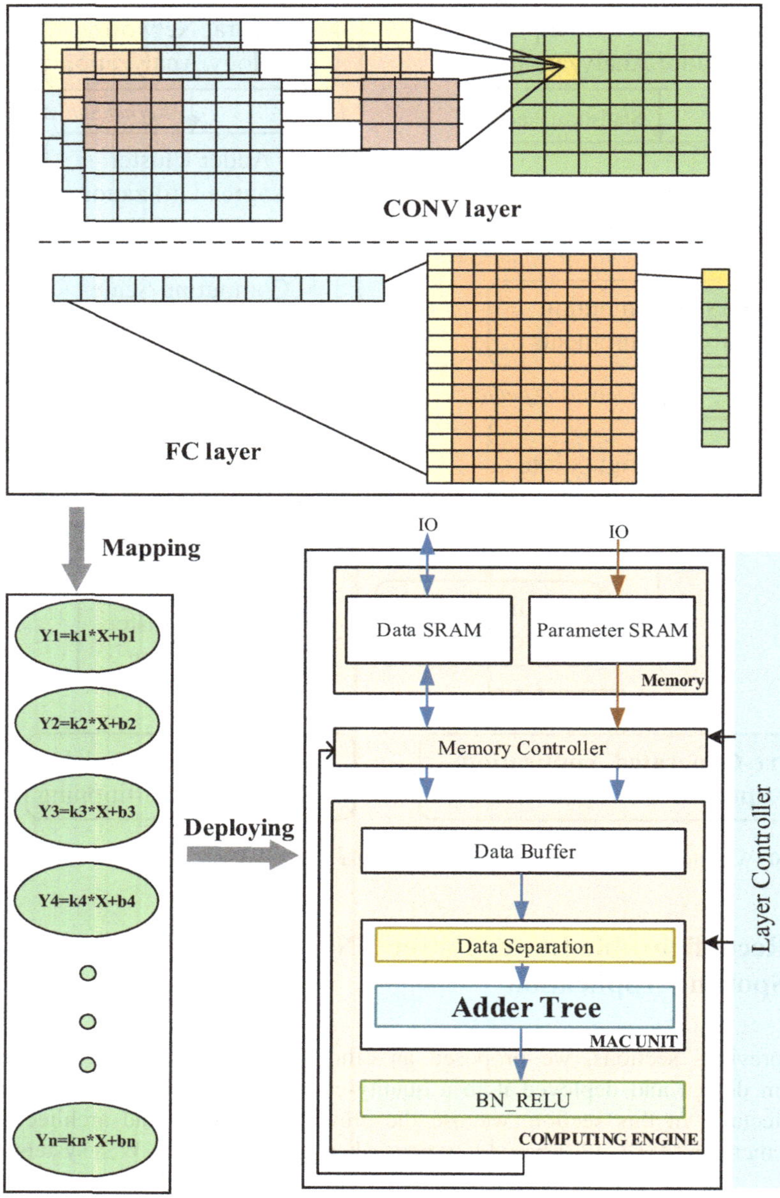

Fig. 11 System design: from algorithm to architecture

layer control module to make the engine work in the right order. The computing engine is the approximate adder cluster we proposed in this chapter, and it can be reconfigured into different precision referring to the scheduling mechanism.

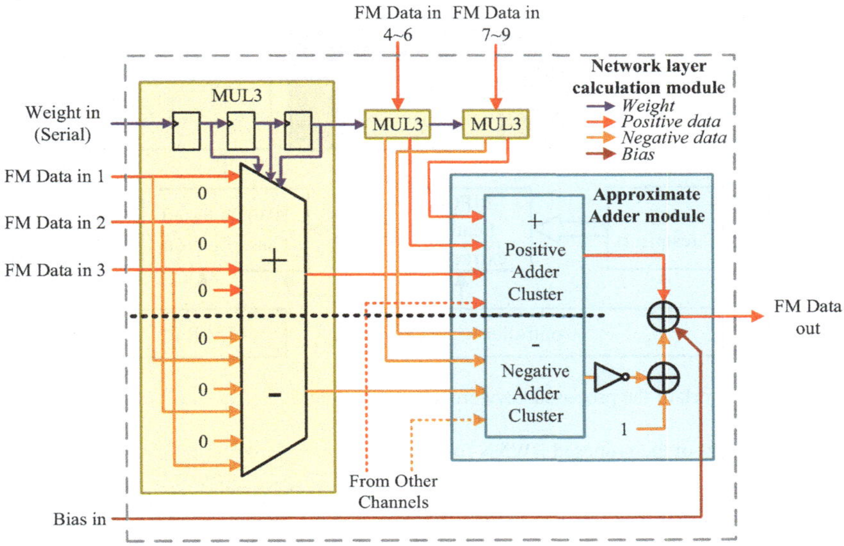

Fig. 12 The computing module for one neural channel

The detailed design of the computing module for one channel is presented in Fig. 12. Since BWNN owns only binary weights, the multiplication operation in ANNs becomes a complement operation. We design a positive-negative circuit to avoid the operation. While the weight is "+1," the data is sent to the positive adder cluster, and while the weight is "−1" ("0" in the circuit design), the negative adder cluster is enabled. The two modules are identically mirrored. By taking the inverted result of the negative modules and plus one, the two modules can be added to achieve the final results.

5.2 *A 22-nm Low-Power System for Always-On KWS System*

For the keyword spotting designs, the hidden Markov model [38] and the Gaussian mixture model [39] are usually used for high recognition accuracy. But in recent years, the ANN structures and ANN-based accelerators [40–44] are proposed for high energy efficiency.

In this section, to design a real-time low-power ANN system, we chose the keyword spotting system as application target, and the framework we proposed is presented in Fig. 13. It consists of the front-end feature extractor, MFCC module, and the back-end feature classifier, BWNN accelerator. To be specific, we use ten keywords ("yes, no, up, down, left, right, on, off, stop, go"), chosen from the Google Speech Commands Dataset (GSCD) as the training and validation data.

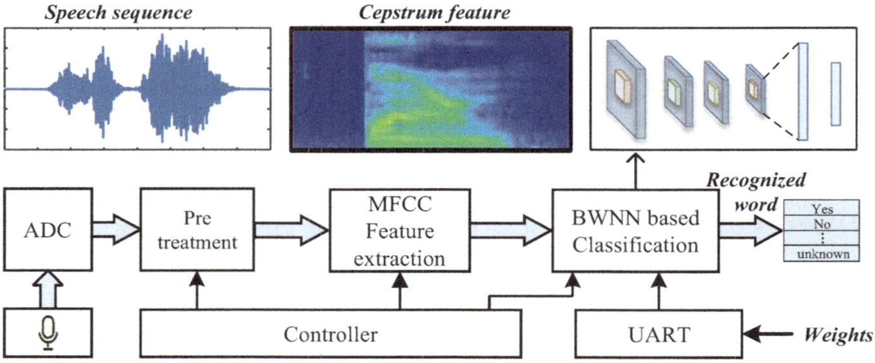

Fig. 13 Framework of the proposed keyword spotting system

Table 1 Settings of the proposed BWNN for KWS system

Layer	Type	Kernel size (Stride x, y)	Channels	Neurons
1	CONV	3 × 3 × 1 (1, 2)	28	–
2	CONV	3 × 3 × 28 (1, 1)	24	–
3	CONV	3 × 3 × 24 (2, 1)	16	–
4	CONV	3 × 3 × 16 (1, 2)	12	–
5	FC	–	–	30
6	FC	–	–	6

The proposed BWNN in this case is designed for a ten-keyword spotting system, and it consists of four binary convolutional layers (CONV) and two binary fully connected layers (FC). Between the six layers, batch normalization layers followed by $ReLU$ operations are conducted for five times. The detailed network descriptions are presented in Table 1.

To implement the approximate circuit and architectures for the keyword spotting system, the design flow is illustrated in Fig. 14. Firstly, the network weights are quantized, and the input data is considered to determine the maximum adder integer bit width. Then, the fraction bit width is decided also by the input data sample distributions. With all the input data features and weight features, we can use the estimation approach to determine the appropriate approximation structure and design parameters. After the design is fixed, the further tradeoff between accuracy and hardware costs can be conducted. For the implementation approach, the algorithm estimation, the approximate architecture design, and the detailed circuit optimization are all involved and considered, and with the help of statistical methods, analytical models, and functional simulations, the systematic approximation can be finally deployed. Figure 11 also shows the final architecture of the proposed neural engine, and we use the Design Compiler and the PrimeTime PX to evaluate the profits by adopting approximation to such system. The network layers are mapped to vector computing to ensure the datapath is simple. Then the

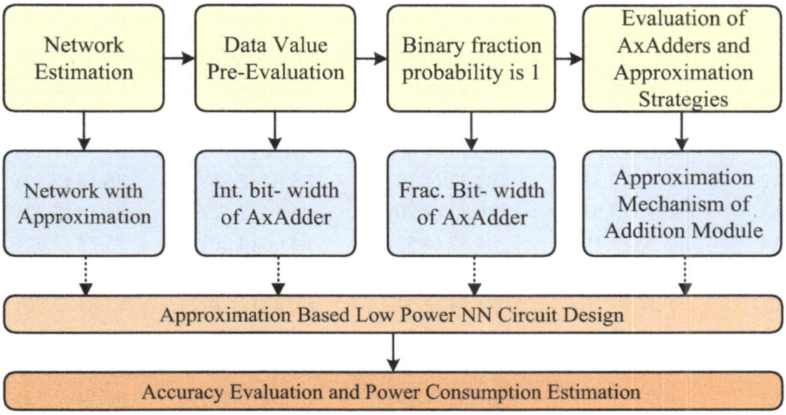

Fig. 14 Implementation from algorithm to circuit

computing engine with approximate AxAdder cluster is adopted. In this case, the data buffer is set to 0.42 KB, and the data and weight SRAM are set to 8.74 KB and 4.99 KB, respectively.

6 Experimental Results

The implementation and evaluation of the proposed keyword spotting system are on an industrial 22-nm ULL UHVT CMOS technology. The power consumption is evaluated with Synopsys PTPX at 25°C TT corner.

The system accuracies by adopting different types of AxAdders are illustrated in Table 2. We applied four parameter pairs for the AxAdders, and the parameters are representing the AX_is in the stages of 1–8, which are [0,4,4,6,6,6,6,8], [0,4,4,4,6,6,8,8], [0,4,4,4,6,6,6,8], and [0,2,4,4,4,4,6,8]. The accuracy relative loss is also demonstrated in the table following every accuracy. It can be seen that for the first configuration, only the proposed LOA can keep less than 1% relative loss of accuracy (0.35%) and the others are introducing more than three times higher. For the four configurations, the approximation degrees are decreasing. It can be seen that the accuracy is higher, while the approximation degree is smaller. The AMA2 and InXA1 appear irregular rules, and it may be because the parameter pairs impact their circuit features. In general, when the rule is with higher approximation degrees, the system accuracy is decreasing. In conclusion, under the four configurations, the approximate degrees are decreasing, which means the hardware resources are increasing. For LOA design, it not only owns the highest accuracy under the first configuration but also shows high energy efficiency.

Table 3 presents the detailed parameter comparison between different AxAdders such as the power consumption of a single AxAdder, addition cluster configuration,

Table 2 The system accuracy comparison for different AxAdder configurations

Types	System accuracy (%) (Relative accuracy Loss (%))[a]			
	Config#1[b]	Config#2[b]	Config#3[b]	Config#4[b]
LOA	88.06 (0.35)	85.48 (3.27)	86.78 (1.80)	86.55 (2.06)
AMA1	85.84 (2.86)	86.73 (1.86)	87.32 (1.19)	87.32 (1.19)
AMA2	86.14 (2.52)	87.32 (1.19)	85.84 (2.86)	86.14 (2.52)
AMA3	85.25 (3.53)	86.43 (2.20)	86.14 (2.52)	86.14 (2.52)
AMA4	85.55 (3.19)	84.37 (4.53)	84.66 (4.20)	85.25 (3.52)
AXA1	86.73 (1.86)	87.02 (1.53)	86.14 (2.52)	86.43 (2.20)
AXA2	86.37 (2.26)	86.14 (2.52)	86.43 (2.20)	85.84 (2.86)
AXA3	86.14 (2.52)	86.43 (2.20)	85.84 (2.86)	85.84 (2.86)
InXA1	84.31 (4.59)	84.37 (4.53)	83.19 (5.86)	85.25 (3.53)
InXA2	86.71 (1.88)	87.32 (1.19)	87.61 (0.86)	86.14 (2.52)
InXA3	86.14 (2.52)	87.32 (1.19)	85.84 (2.86)	86.14 (2.52)
FA	88.37 (0)			

[a] The relative accuracy loss is compared to the FA design, which is non-approximated, following the equation $\frac{Accuracy\_of\_FA\_System - Accuracy\_of\_AxAdder\_System)}{Accuracy\_of\_FA\_System}$

[b] Config#1~4 are presenting the AX_i configurations of AxAdders in stages 1~8, and the configurations are [0,4,4,6,6,6,6,8], [0,4,4,4,6,6,8,8], [0,4,4,4,6,6,6,8], and [0,2,4,4,4,4,6,8], respectively

Table 3 The comparison between different types of AxAdder clusters

Type	$Power_b$ (nW)[a]	Accu (%)[b]	Area (μm^2)	Re_{area} (%)[c]	Power (μW)	Re_{power} (%)[d]
FA	–	88.4	3498.11	0	0.418	0
AMA1	45.8	85.8	3216.77	8	0.399	4.6
AMA2	28.9	86.1	3005.76	14.1	0.387	7.4
AMA3	27.1	85.3	2900.25	17.1	0.385	7.8
AMA4	36.8	85.5	3040.93	13.1	0.394	5.8
InXA1	14.5	84.3	2794.41	20.1	0.357	14.6
InXA2	18.3	86.7	2794.75	20.1	0.369	11.6
InXA3	16.7	86.1	2794.41	20.1	0.362	13.5
LOA	13.88	88.1	2769.38	20.8	0.343	18.0

[a] $Power_s$ is the power of single-bit AxAdder block under the same simulation settings

[b] Accu is the accuracy of recognition for the proposed system integrating corresponding type of AxAdder clusters

[c] Re_{area} is the ratio of area reduction for system integrating different AxAdder clusters compared to full-precision standard cell

[d] Re_{power} is the ratio of power reduction for system integrating different AxAdder clusters compared to full-precision standard cell

network structure, network bit width, network accuracy, number of transistors saved in one approximate calculation unit, saved area, saved power consumption, etc. For the proposed AxAdder cluster in tree form, the power consumption and area overhead are reduced in two ways: (1) during the design procedure, we used the analytical model to use as less hardware resources as possible while keeping the

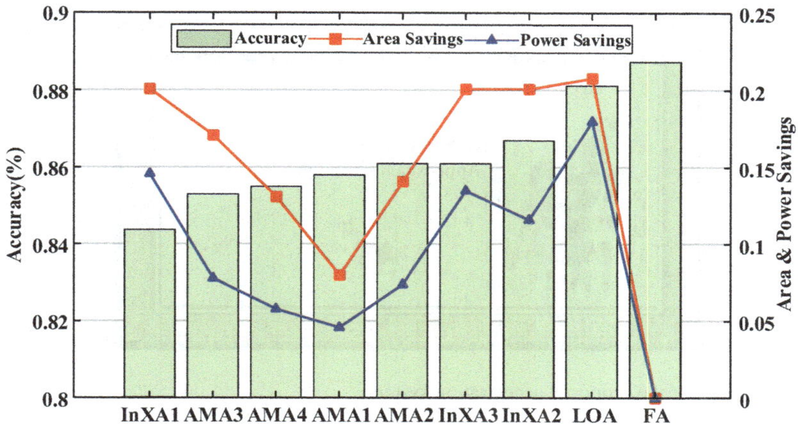

Fig. 15 Accuracy and area/power consumption comparisons between AxAdders

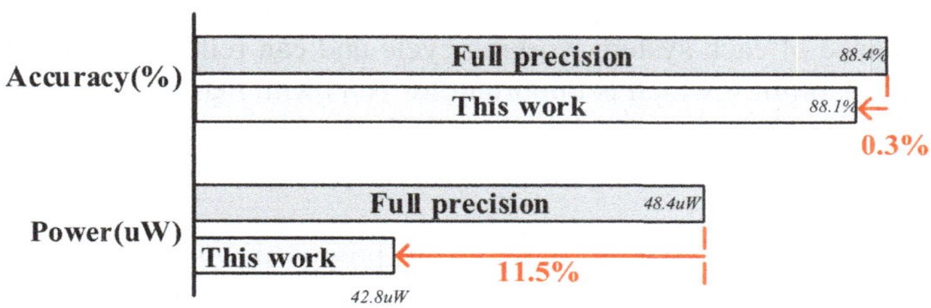

Fig. 16 Accuracy and power reduction by introducing approximation in BWNN system

computing quality; and (2) we compared the area and power between several types of different approximate adders.

The more intuitive comparison result of accuracies, power-saving rates, and area-saving rates for different types of AxAdder clusters are compared in Fig. 15. It can be seen that the LOA AxAdder can achieve the best balance between the system accuracy and the hardware consumption. With the least transistors in the LOA circuit, the system accuracy is still in a high level, due to the quality model which can help control the computing quality loss.

For the whole keyword spotting system, with the proposed approximate computing, the power consumption has changed from 48.4 μW to 42.8 μW, which is a decrease of 11.5%, while the accuracy of network recognition only decreases by 0.3%. The schematic diagram of the power consumption and accuracy of the keyword recognition system before and after the introduction of approximate calculation is shown in Fig. 16. The LOA is the design we finally use, and the layout is shown in Fig. 17 with an area of 3.499 mm^2.

Table 4 compared the proposed design with the state-of-the-art designs for keyword spotting processors. We use operations per system cycle (OPC) as the

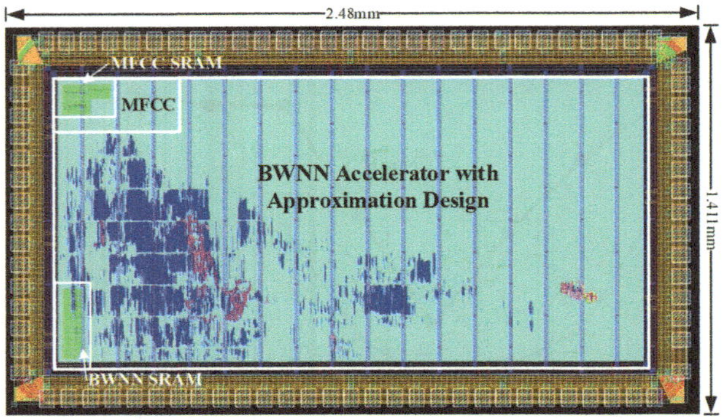

Fig. 17 Layout of the proposed acceleration system

key performance metric [45] and the power/area efficiency as the normalized metrics to compare different processors. For the metric of OPC, it can represent the performance of each system working cycle and can reflect the system parallel computing capabilities, which is important for NNs with massive parallel computation requirements. The normalized power efficiency and normalized area efficiency are the ratio of (OPC/μW and OPC/mm^2) to baseline (the baseline is set to TCAS-I'20 [32] in this chapter).

Among the designs listed in Table 4, the proposed work is with the highest OPC. For the computation features, works [30, 32, 46–48] are using approximate computing techniques, while works [48–50] are processing with full-precision computing cells. All the works except [50] deploy quantized neural networks but with different quantization degrees. The works [30, 32, 48] including this chapter are adopting BWNNs. Compared to work [30] with on-chip self-learning circuit, our work is near 90× better in power efficiency at the cost of 8% accuracy loss. Work [48] only supports one to two keyword recognition, while our work supports ten keywords with much more complex system design. For comparison, we can improve the area efficiency by 9.18×, 0.11×, 1.49×, 123.10×, 7.50×, 3.44×, and 4.48×, than works [30, 32, 46–50], respectively. Compared to work [49] deploying complex and memory-consuming LSTM network with 4/8-bit weight bit width and precise computing, the proposed work is 123.10× better in area efficiency, benefitting from BWNN and quality-driven approximate computing. The work [30] can support both KWS and speaker verification with 1-bit weight and data, but it only supports one keyword recognition, while our work supports ten keywords. Our work can achieve 81× better area efficiency at the cost of only 1.7× larger area than work [30]. Due to the complexity of multi-bit LSTM network and precise computing design in [49], our work can achieve over 40× better power efficiency than work [49], with accuracy loss within 3% for GSCD dataset.

Compared to work [32], we can achieve 6.4× OPC performance and maintain high accuracy with similar BWNN topology settings of (4CONV+2FC). The work [32] can process BWNN with only 17% area occupation of our work, but the

Table 4 Comparisons with the state-of-the-art energy-efficient keyword spotting processors

	VLSI'18 [30]	TCAS-I'20 [32]	ACCESS'19 [46]	JSSC'20 [49]	ESSCIRC'18 [50]	DATE'21 [47]	ISSCC'20 [48]	This work
Application[a]	KWS+SV	KWS	KWS	KWS	KWS	KWS	KWS	KWS
Process (*nm*)	28	22	22	65	65	22	28	22
Voltage (V)	0.57	0.6	0.55	0.6	0.57	0.6	0.41	0.72
Frequency (KHz)	2500	250	250	250	250	250	40	50
Network topology[b]	4CONV+3FC	4CONV+2FC	3CONV+2FC	1LSTM+1FC	1LSTM+1FC	5CONV+1LSTM+1FC	CONV+DSC+FC	4CONV+2FC
Dataset[c]	TIDIGITS	GSCD	GSCD	GSCD/TIDIGITS	GSCD	GSCD	GSCD	GSCD
Keyword number	1	10	10	10	4	1~5	1~2	10
On-chip SRAM size	52 KB	11 KB	44 KB	36 KB	34 KB	46 KB	2 KB	13.73 KB
Data bit width (bit)	1	16	8	10	16	9	1	16
Weight bit width (bit)	1	1	7	4/8	16	8	1	1
Computing mode[d]	Approximate	Approximate	Approximate	Precise	Precise	Approximate	Precise	Approximate
OPs per inference (OPI)[e]	11074048	3036456	2101824	115380	681240	1790250	150496	3914280
Recognition accuracy	96.11%	87.9%	90.51%	90.87/98.5%	91.2%	93.2%	94.6%	88.1%
System power (μW)	141	10.8	52	10.8	5	2.7	0.51	42.8
Area (mm^2)	1.29	0.602	0.75	2.56	1.035	0.758	0.23	3.499
Throughput (MOP/s)	442.96	189.78	105.09	7.21	42.58	59.62	2.35	244.64
OPs per cycle (OPC)[f]	177.18	759.11	420.36	28.85	170.31	238.46	58.69	4892.85
Power efficiency (OPC/μW)	1.26	70.29	8.08	2.67	34.06	88.32	115.09	114.32
Norm. power efficiency[g]	0.02	1.00	0.12	0.04	0.48	1.26	1.64	1.63
Area efficiency (OPC/mm^2)	137.35	1260.99	560.49	11.27	164.55	314.59	255.19	1398.36
Norm. area efficiency[g]	0.11	1.00	0.44	0.01	0.13	0.25	0.20	1.11

[a] KWS is short for keyword spotting, and SV is short for speaker verification
[b] CONV for convolutional layer, FC for fully connected layer, LSTM for long short-term memory layer, and DSC for depthwise separable CONV
[c] GSCD is the Google Speech Commands Dataset collected by Google Inc., and TIDIGITS is collected by Texas Instruments Inc.
[d] Two computing modes are compared, approximate computing with approximate units and precise computing with standard computing cells
[e] OPI is the total number of operations in one-time inference, which equals the operations of the corresponding neural network
[f] OPC is the operations processed by the hardware system in one system cycle, which equals to 1/frequency
[g] Normalized (norm.) power/area efficiency is the ratio of the work's power efficiency to work TCAS-I'20

area efficiency is 10% lower, which is close to the proposed work. Besides, based on the parameter setting of AxAdder clusters by quality estimation approaches, the proposed work is 63% better than work [32] in power efficiency. Compared to the work [32], it can be seen that by the proposed quality-driven design approaches, we can further improve power and area efficiency while optimizing approximate BWNN systems. The overall comparison results show that the proposed work owns the highest OPC performance of over 6× than SOTA works. Compared with SOTA KWS processors, our work can achieve over 60% improvement in power efficiency and 1.1× area efficiency while achieving similar recognition accuracy.

7 Conclusion

In this chapter, we innovated an approach to design low-power BWNN acceleration system by approximation estimation. Quality model is based on the metric of MSE, and the parametric forms help guide the architecture optimization flow. The estimation approach for a single approximate adder is first analyzed, and then with the help of approximate noise, we can also estimate complex approximate adder arrays, which are the key modules used for BWNN acceleration. By taking several approximate adders into experiments, we validate the effectiveness of the estimation approach. In addition, we adopt the proposed design by taking a ten-word keyword spotting system as an example, and detailed design flows are presented. At last, the proposed architecture is implemented under an industrial 22-nm ULL UHVT process, and the results show the superiority of the proposed design compared to SOTA keyword spotting processors, which can achieve 1.6× power efficiency while maintaining high recognition accuracy.

References

1. W. Liu, J. Xu, D. Wang, C. Wang, P. Montuschi, F. Lombardi, IEEE Trans. Circuits Syst. I Regul. Pap. **65**(9), 2856 (2018)
2. H. Jiang, F.J.H. Santiago, H. Mo, L. Liu, J. Han, Proc. IEEE **108**(12), 2108 (2020)
3. A. Sohrabizadeh, J. Wang, J. Cong, in *Proceedings of the 2020 ACM/SIGDA International Symposium on Field-Programmable Gate Arrays* (2020), pp. 133–139
4. R. Andri, L. Cavigelli, D. Rossi, L. Benini, IEEE Trans. Comput. Aided Des. Integr. Circuits Syst. **37**(1), 48 (2017)
5. M. Rastegari, V. Ordonez, J. Redmon, A. Farhadi, in *European Conference on Computer Vision* (Springer, Berlin, 2016), pp. 525–542
6. W. Liu, L. Qian, C. Wang, H. Jiang, J. Han, F. Lombardi, IEEE Trans. Comput. **66**(8), 1435 (2017)
7. W. Liu, F. Lombardi, M. Shulte, Proc. IEEE **108**(3), 394 (2020)
8. W. Liu, T. Cao, P. Yin, Y. Zhu, C. Wang, E.E. Swartzlander, F. Lombardi, IEEE Trans. Comput. **68**(6), 804 (2018)

9. C. Chen, S. Yang, W. Qian, M. Imani, X. Yin, C. Zhuo, in *Proceedings of the 39th International Conference on Computer-Aided Design* (2020), pp. 1–9
10. N. Zhu, W.L. Goh, W. Zhang, K.S. Yeo, Z.H. Kong, IEEE Trans. Very Large Scale Integr. VLSI Syst. **18**(8), 1225 (2009)
11. N. Zhu, W.L. Goh, K.S. Yeo, in *Proceedings of the 2009 12th International Symposium on Integrated Circuits* (IEEE, New York, 2009), pp. 69–72
12. V. Gupta, D. Mohapatra, A. Raghunathan, K. Roy, IEEE Trans. Comput. Aided Des. Integr. Circuits Syst. **32**(1), 124 (2012)
13. M. Shafique, W. Ahmad, R. Hafiz, J. Henkel, in *2015 52nd ACM/EDAC/IEEE Design Automation Conference (DAC)* (IEEE, New York, 2015), pp. 1–6
14. M.A. Hanif, R. Hafiz, O. Hasan, M. Shafique, in *Proceedings of the 54th Annual Design Automation Conference 2017* (2017), pp. 1–6
15. H.A. Almurib, T.N. Kumar, F. Lombardi, in *2016 Design, Automation & Test in Europe Conference & Exhibition (DATE)* (IEEE, New York, 2016), pp. 660–665
16. A. Madanayake, R.J. Cintra, V. Dimitrov, F. Bayer, K.A. Wahid, S. Kulasekera, A. Edirisuriya, U. Potluri, S. Madishetty, N. Rajapaksha, IEEE Circuits Syst. Mag. **15**(1), 25 (2015)
17. M. Yang, C.H. Yeh, Y. Zhou, J.P. Cerqueira, A.A. Lazar, M. Seok, in *2018 IEEE International Solid-State Circuits Conference-(ISSCC)* (IEEE, New York, 2018), pp. 346–348
18. S. Zheng, P. Ouyang, D. Song, X. Li, L. Liu, S. Wei, S. Yin, IEEE Trans. Circuits Syst. I Regul. Pap. **66**(12), 4648 (2019)
19. S. Yin, P. Ouyang, J. Yang, T. Lu, X. Li, L. Liu, S. Wei, IEEE J. Solid State Circuits **54**(4), 1120 (2018)
20. Z. Yang, J. Han, F. Lombardi, in *Proceedings of the 2015 IEEE/ACM International Symposium on Nanoscale Architectures (NANOARCH 15)* (IEEE, New York, 2015), pp. 145–150
21. N. Zhu, W.L. Goh, G. Wang, K.S. Yeo, in *2010 International SoC Design Conference* (IEEE, New York, 2010), pp. 323–327
22. W. Liu, L. Chen, C. Wang, M. O'Neill, F. Lombardi, IEEE Trans. Comput. **65**(1), 308 (2015)
23. C. Liu, X. Yang, F. Qiao, Q. Wei, H. Yang, in *The 20th Asia and South Pacific Design Automation Conference* (IEEE, New York, 2015), pp. 237–242
24. R. Venkatesan, A. Agarwal, K. Roy, A. Raghunathan, in *2011 IEEE/ACM International Conference on Computer-Aided Design (ICCAD)* (IEEE, New York, 2011), pp. 667–673
25. S. Mazahir, O. Hasan, R. Hafiz, M. Shafique, J. Henkel, IEEE Trans. Comput. **66**(3), 515 (2016)
26. M.K. Ayub, O. Hasan, M. Shafique, in *Proceedings of the 54th Annual Design Automation Conference 2017* (2017), pp. 1–6
27. M. Brandalero, A.C.S. Beck, L. Carro, M. Shafique, in *2018 55th ACM/ESDA/IEEE Design Automation Conference (DAC)* (IEEE, New York, 2018), pp. 1–6
28. O. Akbari, M. Kamal, A. Afzali-Kusha, M. Pedram, M. Shafique, in *2018 Design, Automation & Test in Europe Conference & Exhibition (DATE)* (IEEE, New York, 2018), pp. 413–418
29. M.S. Ansari, V. Mrazek, B.F. Cockburn, L. Sekanina, Z. Vasicek, J. Han, IEEE Trans. Very Large Scale Integr. VLSI Syst. **28**(2), 317 (2019)
30. S. Yin, P. Ouyang, S. Zheng, D. Song, X. Li, L. Liu, S. Wei, in *2018 IEEE Symposium on VLSI Circuits* (IEEE, New York, 2018), pp. 139–140
31. Y. Lu, W. Shan, J. Xu, in *2019 IEEE Asia Pacific Conference on Circuits and Systems (APCCAS)* (IEEE, New York, 2019), pp. 309–312
32. B. Liu, H. Cai, Z. Wang, Y. Sun, Z. Shen, W. Zhu, Y. Li, Y. Gong, W. Ge, J. Yang, et al., IEEE Trans. Circuits Syst. I Regul. Pap. **67**(12), 4733 (2020)
33. M. Pashaeifar, M. Kamal, A. Afzali-Kusha, M. Pedram, IEEE Trans. Circuits Syst. I Regul. Pap. **66**(1), 327 (2018)
34. B. Widrow, IRE Trans. Circuit Theory **3**(4), 266 (1956)
35. Y. Wu, Y. Li, X. Ge, Y. Gao, W. Qian, IEEE Trans. Comput. **68**(1), 21 (2018)
36. M.S. Ansari, B.F. Cockburn, J. Han, IEEE Trans. Emerg. Top. Comput. **10**(1), 500–506 (2020)
37. B. Liu, X. Ding, H. Cai, W. Zhu, Z. Wang, W. Liu, J. Yang, IEEE Circuits Syst. Mag. **21**(4), 24 (2021). https://doi.org/10.1109/MCAS.2021.3118175

38. M.L. Seltzer, D. Yu, Y. Wang, in *2013 IEEE International Conference on Acoustics, Speech and Signal Processing* (IEEE, New York, 2013), pp. 7398–7402
39. G.E. Dahl, D. Yu, L. Deng, A. Acero, IEEE Trans. Audio Speech Lang. Process. **20**(1), 30 (2011)
40. S. Bang, J. Wang, Z. Li, C. Gao, Y. Kim, Q. Dong, Y.P. Chen, L. Fick, X. Sun, R. Dreslinski, et al., in *2017 IEEE International Solid-State Circuits Conference (ISSCC)* (IEEE, New York, 2017), pp. 250–251
41. S. Yin, P. Ouyang, S. Tang, F. Tu, X. Li, S. Zheng, T. Lu, J. Gu, L. Liu, S. Wei, IEEE J. Solid State Circuits **53**(4), 968 (2017)
42. M. Price, J. Glass, A.P. Chandrakasan, in *2017 IEEE International Solid-State Circuits Conference (ISSCC)* (IEEE, New York, 2017), pp. 244–245
43. M. Shah, S. Arunachalam, J. Wang, D. Blaauw, D. Sylvester, H.S. Kim, J.s. Seo, C. Chakrabarti, J. Signal Process. Syst. **90**(5), 727 (2018)
44. J.P. Giraldo, S. Lauwereins, K. Badami, H. Van Hamme, M. Verhelst, in *2019 Symposium on VLSI Circuits* (IEEE, New York, 2019), pp. C52–C53
45. J. Cong, Z. Fang, M. Lo, H. Wang, J. Xu, S. Zhang, in *2018 IEEE 26th Annual International Symposium on Field-Programmable Custom Computing Machines (FCCM)* (IEEE, New York, 2018), pp. 93–96
46. B. Liu, Z. Wang, W. Zhu, Y. Sun, Z. Shen, L. Huang, Y. Li, Y. Gong, W. Ge, IEEE Access **7**, 186456 (2019)
47. B. Liu, Z. Shen, L. Huang, Y. Gong, Z. Zhang, H. Cai, in *2021 Design, Automation & Test in Europe Conference & Exhibition (DATE)* (IEEE, New York, 2021), pp. 495–500
48. W. Shan, M. Yang, J. Xu, Y. Lu, S. Zhang, T. Wang, J. Yang, L. Shi, M. Seok, in *2020 IEEE International Solid-State Circuits Conference-(ISSCC)* (IEEE, New York, 2020), pp. 230–232
49. J.S.P. Giraldo, S. Lauwereins, K. Badami, M. Verhelst, IEEE J. Solid State Circuits **55**(4), 868 (2020)
50. J.S.P. Giraldo, M. Verhelst, in *ESSCIRC 2018-IEEE 44th European Solid State Circuits Conference (ESSCIRC)* (IEEE, New York, 2018), pp. 166–169

Adaptive Approximate Accelerators with Controlled Quality Using Machine Learning

Mahmoud Masadeh, Osman Hasan, and Sofiène Tahar

1 Introduction

The ongoing scaling in feature size has caused integrated circuit (IC) behavior vulnerable to soft errors as well as process, voltage, and temperature variations. Thus, the challenge of assuring strictly exact computing is increasing [1]. On the other hand, present-age computing systems are pervasive, portable, embedded, and mobile, which led to an ever-increasing demand for ultra-low power consumption, small footprint, and high-performance systems. Such battery-powered systems are the main pillars of the internet of things (IoT), which do not necessarily need entirely accurate results.

Approximate computing (AC), known as best-effort computing, is a nascent computing paradigm that allows us to achieve these objectives by compromising the arithmetic accuracy [2]. Nowadays, many applications, such as image processing, multimedia, recognition, machine learning, communication, big data analysis, and data mining, are error-tolerant and thus can benefit from approximate computing. These applications exhibit *intrinsic error resilience* due to the following factors [3]: (i) redundant and noisy input data, (ii) lack of golden or single output, (iii) imperfect

M. Masadeh (✉)
Computer Engineering Department, Yarmouk University, Irbid, Jordan
e-mail: mahmoud.s@yu.edu.jo

O. Hasan
Electrical Engineering Department, National University of Sciences and Technology, Islamabad, Pakistan
e-mail: osman.hasan@seecs.nust.edu.pk

S. Tahar
Department of Electrical and Computer Engineering, Concordia University Montreal, Montreal, QC, Canada
e-mail: tahar@ece.concordia.ca

W. Liu et al. (eds.), *Design and Applications of Emerging Computer Systems*,
https://doi.org/10.1007/978-3-031-42478-6_19

perception in the human sense, and (iv) the usage of implementation algorithms with self-healing and error attenuation patterns.

Different approximation strategies, which fall under the umbrella of approximate computing, e.g., the voltage over scaling [4], algorithmic approximations [5], and approximation of basic arithmetic operations [6], have gained a significant research interest, in both academia and industry, such as IBM [7], Intel [8], and Microsoft [9]. However, approximate computing is still immature and does not have standards yet, which poses severe bottlenecks and main challenges. Thus, future work of AC should be guided by the following general principles to achieve the best efficiency [3]:

1. Significance-driven approximation: Identifying the approximable parts of an application or circuit design is a great challenge. Therefore, it is critical to distinguish the approximable parts with their approximation settings.
2. Measurable notion of approximation quality: Quality specification and verification of approximate design are still open challenges, where quality metrics are application and user-dependent. To quantify approximation errors, various quality metrics are used.
3. Quality configuration: Error resiliency of applications depends on the applied inputs and the context in which the outputs are consumed.
4. Asymmetric approximation benefits: It is essential to identify the approximable components of the design, which reduces the quality insignificantly while improving efficiency considerably.

For a static approximate design, the approximation error continues during its operational lifetime. It restricts approximation versatility and results in under- or over-approximated systems for dynamic input data, causing excessive power usage and insufficient accuracy, respectively. Given the dynamic nature of the applied inputs into static approximate designs, errors are the norm rather than the exception in approximate computing, where the error magnitude depends on the user inputs [10]. On the other hand, the defined tolerable error threshold, i.e., target output quality (TOQ), can be dynamically changed. In both cases, errors with a high value produced by approximate components in an approximate accelerator, even with a low error rate, have a more significant impact on the quality than those caused by approximate parts with a small magnitude. This is in line with the notion of fail-small, fail-rare, or fail-moderate approaches, [11], where error magnitudes and rates should be restricted to avoid high loss in the output quality. The fail-small technique allows approximations with low error magnitudes, while the fail-rare technique allows approximations with low error rates. On the other hand, the fail-moderate technique allows approximations with moderate error magnitude and moderate error rate [12]. Thus, the approaches mentioned above limit the design space to prevent approximations with high error rates and high error magnitudes, where such a combination degrades the quality loss significantly.

Quality assurance of approximate computing is still missing a mathematical model for the impact of approximation on the output quality [3]. Toward this goal, in this chapter, we develop a runtime adaptive approximate accelerator. For that, we utilize a set of energy-efficient approximate multipliers which we designed in [13]. The adaptive design is based on fine-grained input data to satisfy a user-defined target output quality (TOQ) constraint. Design adaptation uses a machine learning-based design selector to dynamically choose the most suitable approximate design for runtime data. The target approximate accelerator is implemented with configurable levels and types of approximate multipliers.

1.1 Approximate Computing Error Metrics

Approximation introduces accuracy as a new design metric. Thus, several application-dependent error metrics are used to quantify approximation errors and evaluate design accuracy [14]. For example, considering an approximate design with two inputs, i.e., X and Y, of n-bit each, where the exact result is (P) and the approximate result is (P'), these error metrics include:

- Error Distance (ED): The arithmetic difference between the exact output and the approximate output for a given input, which is presented by $ED = |P - P'|$.
- Error Rate (ER): Also called error probability, which is the percentage of erroneous outputs among all outputs.
- Mean Error Distance (MED): The average of ED values for a set of outputs obtained by applying a set of inputs. MED is a useful metric for measuring the implementation accuracy of multiple-bit circuit design.
- Normalized Error Distance (NED): The normalization of MED by the maximum result that an exact design can have (P_{Max}). NED is an invariant metric independent of the size of the circuit. Therefore, it is used for comparing circuits of different sizes, and it is expressed as:
- Relative Error Distance (RED): The ratio of ED to the accurate output, which equals $RED = ED/P$.
- Mean Relative Error Distance (MRED): The average value of all possible relative error distances (RED).
- Mean Square Error (MSE): It is defined as the average of the squared ED values.
- Peak Signal-to-Noise Ratio (PSNR): The peak signal-to-noise ratio is a fidelity metric used to measure the quality of the output images; it indicates the ratio of the maximum pixel intensity to the distortion.

The presented metrics are not mutually exclusive, where one application may use several quality metrics.

1.2 Approximate Accelerators

Hardware accelerators are special hardware, which is devoted for executing frequently called functions. Accelerators are more efficient than software running on general-purpose processors. Generally, they are constructed by connecting multiple simple arithmetic modules. The existing literature has proposed the design of approximate accelerators using neural networks [15] or approximate functional units, particularly approximate adders [16] and multipliers [17]. Moreover, several functionally approximate designs for basic arithmetic modules, including adders [6], dividers [18], and multipliers [19], have been investigated for their pivotal role in various applications. These individually designed components are rarely used alone, especially in computationally intensive error-tolerant applications, which are amenable to approximation. The optimization of accuracy performance at the accelerator level has received little or no attention in the previous literature. Generally, hardware accelerators are constructed by connecting multiple simple arithmetic modules. For example, discrete Fourier transform (DFT) and discrete cosine transform (DCT) modules are used in signal and image processing. Approximate multipliers and multiply-accumulate units (MACs) are intensively used to build approximate accelerators.

Multipliers are one of the most foundational components for most functions and algorithms in classical computing. However, they are the most energy-costly units compared to other essential CPU functions such as register shifts or binary logical operators. Thus, their approximation would introduce an enhancement in their performance and energy, which automatically induces crucial benefits for the whole application. Approximate multipliers have been mainly designed using three techniques:

(i) Approximation in partial product generation: For example, Kulkarni et al. [20] proposed an approximate 2×2 binary multiplier at the gate level by changing a single entry in the Karnaugh map with an error rate of $1/16$.
(ii) Approximation in partial product tree: For example, error-tolerant multipliers (ETM) [21] divide the input operands into two parts, i.e., the multiplication part for the MSBs and the non-multiplication part for the LSBs, thus omitting the generation of some partial products [19].
(iii) Approximation in partial product summation: Approximate full adder (FA) cells are used to form an array multiplier, e.g., in [22], the approximate mirror adder has been used to develop a multiplier.

We focus on array multipliers, which are not the fastest neither the smallest. Their short wiring gives them a periodic structure with a compact hardware layout. Thus, they are one of the most used in embedded system on chip (SoC). In [23] and [24], we designed various 8- and 16-bit approximate array multipliers based on approximation in partial product summation.

1.3 Quality Control of Approximate Accelerators

Managing the quality of approximate hardware designs for dynamically changing inputs has substantial significance to guarantee that the obtained results satisfy the required target output quality (TOQ). To the best of our knowledge, there are very few works targeting the assurance of the accuracy of approximate systems compared to designing approximate components. While most prior works focus on *error prediction*, we propose to overcome the approximation error through an input-dependent *self-adaptation of design*.

Mainly, there are two approaches for monitoring and controlling the accuracy of the results of approximate accelerators at runtime. The first approach suggests to periodically, through *sampling techniques*, measure the error of an accelerator through comparing its outcome with the exact computation performed by the host processor. Then, a re-calibration and adjustment process is performed to improve the quality in subsequent invocations of the accelerator if the error is found to be above a defined range, e.g., Green [25] and SAGE [26]. However, the quality of unchecked invocations cannot be ensured, and the previous quality violations cannot be compensated. The second approach relies on implementing lightweight pre-trained error predictors to expect if the invocation of an approximate accelerator would produce an unacceptable error for a particular input dataset [27, 28].

However, the works [25–27], and [28] mainly target controlling software approximation, i.e., loops and functions approximation, through program re-execution and thus are not applicable for hardware designs. Moreover, they ignore input dependencies and do not consider choosing an adequate design from a set of design choices. Overall, none of these state-of-the-art techniques exploits the potential of different settings of approximate computing and their adaptations based on a user-specified quality constraint to ensure the accuracy of the individual outputs, which is the main idea we propose. Design adaptation could be implemented in software-based systems by having different versions of the approximate code, while hardware-based systems rely on having various implementations for the functional units. However, concurrently having such functional units diminishes approximation benefits. Thus, dynamic partial reconfiguration (DPR) could be used to have only a single implementation of the design at any instance of time.

2 Proposed Methodology

We aim to assure the quality of approximation by design adaptation by predicting the most suitable settings of the approximate design to execute the inputs. The proposed method predicts the *design settings* based on the applied input data and user preferences, without losing the gains of approximations. We mostly consider the case of approximate accelerators built with approximate functional units such as approximate multipliers.

We propose a comprehensive methodology that handles the limitations of the current state of the art in terms of fine-grained input dependency, suitability for various approximate modules (e.g., adders, dividers, and multipliers), and applicability to both hardware and software implementations. Figure 1 provides a general overview of the proposed methodology for design adaptation. As shown in

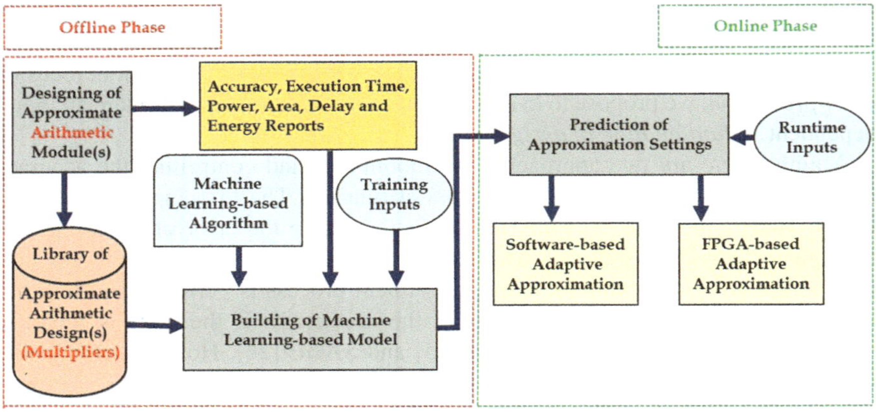

Fig. 1 General overview of the proposed methodology

the figure, the methodology includes two phases: (1) The first is an offline phase, which is executed once for building a machine learning-based model. Such a model predicts the design settings. (2) The second is an online stage, where the machine learning-based model constantly accepts inputs and predicts accordingly based on the runtime inputs. The proposed methodology encompasses the following main steps:

(1) *Building a library of approximate designs*: The first step is designing the library of basic functional units, such as adders, multipliers, and dividers, with different settings, which will be integrated into a quality-assured approximate design. The characteristics of each design, e.g., accuracy, area, power, delay, and energy consumption, should be evaluated to highlight the benefits of approximation.
(2) *Building a machine learning-based model*: In the offline phase, we use supervised learning and employ decision trees (DT) and neural network (NN) algorithms to build a model to predict the unseen data, e.g., the design settings. This step incorporates generating and pre-processing the training data, such as quantization, sampling, and reduction. The training inputs are applied exhaustively to an approximate design to create the training data. For n-bit designs with two inputs, the size of the input combinations is 2^{2n}.
(3) *Predicting the approximation settings*: In the online phase, the user-specified runtime inputs, i.e., the target output quality (TOQ) and the inputs of the approximate design, are given to the ML-based models to predict approximation-related output, i.e., setting of the adaptive design. The implemented ML-based model should be lightweight, i.e., have a high prediction accuracy with fast execution.
(4) *Integrating the approximate accelerator into error-resilient applications*: For adaptive design, the approximate accelerator, which has been nominated by the ML-based model, is adapted within an error-resilient application. Such design

could be implemented in software (off-field-programmable gate array (FPGA), as explained in Sect. 3) or in hardware (on-FPGA, as described in Sect. 4).

Approximation approaches demand a *quality assurance* to adjust approximation settings/knobs and monitor the quality of fine-grained individual outputs. There are two approaches to adjusting the settings of an approximate program to ensure the quality of results:

(i) *Forward design* [29], which sets the design knobs and then observes the quality of results. However, the output quality of some inputs may reach unacceptable levels.
(ii) *Backward design* [30], which tries to locate the optimal knob setting for a given bound of output quality; this requires examining a large space of knob settings for a given input, which is unmanageable.

We present an *adaptive approximate design* that allows altering the settings of approximation, at runtime to meet the preferred output quality. The principal idea is to generate a machine learning-based input-aware *design selector*, which can adjust the approximate design based on the applied inputs, to meet the required quality constraints. Our technique is general in terms of quality metrics and supported approximate designs. It is primarily based on a library of 8- and 16-bit approximate multipliers with 20 different configurations and well-known power dissipation, performance, and accuracy profiles [13]. Moreover, we utilize a backward design approach to dynamically adjust the design to satisfy the desired target output quality (TOQ) based on machine learning (ML) models. The TOQ is a user-defined quality constraint, which represents the maximum permissible error for a given application. The proposed design flow is adaptable, i.e., applicable to approximate functional units other than multipliers, e.g., approximate multiply-accumulate units [31] and approximate meta-functions [32].

2.1 Machine Learning-Based Models

ML-based algorithms find solutions by learning through training data [33]. Supervised learning allows for a rapid, flexible, and scalable way to develop accurate models that are specific to the set of application inputs and TOQ. The error for an approximate design with particular settings can be predicted based on the applied inputs. In [34], we designed and evaluated various ML-based models, based on the analyzed data and several algorithms, developed in the statistical computing language R. These models express the design selector for the adaptive design. *Linear regression* (LR) models were found to be the simplest to develop; however, their accuracy is the lowest, i.e., around 7%. Thus, they are not suitable for our proposed methodology. On the other hand, *decision tree* (DT) models based on both *C5.0* and *rpart* algorithms achieve an accuracy of up to 64%, while *random forest* (RF) models, with an overhead of 25 decision trees, achieve an accuracy of up to

68%. The most accurate models are based on *neural networks*, but they suffer from long development time, design complexity, and high-energy overhead [27]. In this work, we implement and evaluate two versions of the *design selector*, based on decision tree and neural network models. Accordingly, we identify and select the most suitable one to implement in our methodology.

2.1.1 Decision Tree-Based Design Selector

The DT algorithm uses a flowchart-like tree layout to partition data into various predefined classes, thereby providing the description, categorization, and generalization of the given datasets [35]. Unlike the linear model, it models non-linear relationships quite well. Thus, it is used in a wide range of applications, such as credit risk of loans and medical diagnosis [36]. Decision trees are usually employed for *classification* over datasets, through recursively partitioning the data, such that observations with the same label are grouped [36].

Generally speaking, a decision tree model could be replaced by a lookup table (LUT) which contains all the training data that are used to build the DT model [34]. When searching the LUTs, we could use the first matched value, i.e., design settings that satisfy the TOQ, which could be a better solution obtained with a little search effort. For DT-based models, we do not need to specify which value to retrieve. However, it is possible to obtain a result which is closer to the TOQ by changing the settings of the tree such as (1) the maximum depth of any node of the tree, (2) the minimum number of observations that must exist in a node in order for a split to be attempted, and (3) the minimum number of observations in any terminal node. In general, for embedded and limited resource systems, a lookup table is not a viable solution if the number of entries becomes very large [37]. In fact, for a circuit with two 16-bit inputs, we need to generate 2^{32} input patterns to cover all possible scenarios of a circuit.

2.1.2 Neural Network-Based Design Selector

We implemented a two-step NN-based design selector by predicting the design *Degree* first (how much to approximate) and then the *Type* (which approximate full adder to use). The model for *Degree* prediction has an accuracy of 82.17%, while the four models for *Type* prediction have an average accuracy of 67.3%. These models have a single hidden layer with a *sigmoid* activation function.

3 Software-Based Adaptive Design of Approximate Accelerators

We present a detailed description of the proposed methodology for designing adaptive approximate accelerators, where the proposed design can be implementable in both software and hardware. This section shows the aspects of software-based implementation. Section 4 is devoted to the FPGA-based hardware implementation.

3.1 Adaptive Design Methodology

As shown in Fig. 2, the proposed methodology contains two phases: (1) an offline stage, where we build an ML-based model, and (2) an online stage, where we use the ML-based model based on the inputs to anticipate the settings of the adaptive design. The detailed steps of the presented methodology are:

(1) *Generating of Training Data*: Inputs are applied exhaustively to the approximate library to create the training data for building the ML-based model (design selector). For 8- and 16-bit designs, the size of the input combinations is 2^{16} and 2^{32}, respectively. Thus, a sampling of the training data could be used because it is impossible to generate an exhaustive training dataset for large circuits.

(2) *Clustering/Quantizing of Training Data*: Evaluating the design accuracy for a single input can provide the error distance (ED) metric only. However, mean error metrics (e.g., mean square error (MSE), peak signal-to-noise ratio (PSNR), and normalized error distance (NED)) are evaluated over a set of successively applied data rather than a scalar input. Thus, inputs with a specific distance from each other are considered a single cluster with the same estimated error metric. We propose to cluster every 16 consecutive input values. Based on that, each input for an 8-bit multiplier encompasses 16 clusters rather than 256 inputs. Similarly, for the 16-bit multiplier design, the number of clustered inputs is reduced to 2^{24} rather than 2^{32}.

(3) *Pre-processing of Training Data*: Inputs could be applied exhaustively for small circuits, e.g., 8-bit multipliers. However, the size of the input combinations for 16- and 32-bit designs is significant. Therefore, we have to reduce the size of the training data through sampling approaches to design a smaller and more efficient ML-based model. Moreover, for 16-bit designs, we prioritize the training data based on their area, power, and delay as well as accuracy and then reduce the training data accordingly.

(4) *Building of Machine Learning-Based Model*: We built decision trees and neural network-based models, which act as design selectors, to predict the most suitable settings of the design based on the applied inputs.

(5) *Selection of Approximate Design*: In the online phase, the user inputs, i.e., TOQ and inputs of the multiplier, are given to the ML-based models to predict the setting of the approximate design, i.e., *Type* of approximate components and *Degree* of approximation, which is utilized within an error-resilient application, e.g., image processing, in a software-based adaptive approximate execution.

The flow of the proposed methodology is depicted in Fig. 2. The main steps are done once offline. During the online phase, the user specifies the TOQ, where we build our models based on normalized error distance (NED) and peak signal-to-

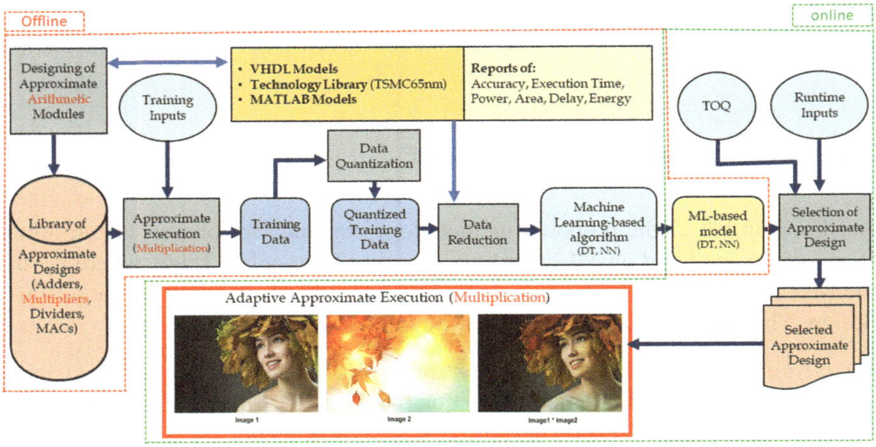

Fig. 2 A detailed methodology of software-based adaptive approximate design

noise ratio (PSNR) error metrics. An important design decision is to determine the configuration granularity, i.e., how much data to process before re-adapting the design, which is termed the window size (N). For example, in image processing applications, we select N to be equal to the size of colored components of an image. Then based on the length of inputs, i.e., L and N, we determine the number of times to reconfigure the design such that the final approximation benefits are still significant. After N inputs, a design adaptation is done, if any of the inputs or TOQ changes. The first step in such adaptation is input quantization, i.e., specifying the corresponding cluster for each input based on its magnitude, since design adaptation for every scalar input is impractical. To evaluate the inputs of an approximate design, various metrics, such as median, skewness, and kurtosis, have been used [38]. Thus, the input magnitude is the most suitable characteristic of design selection.

3.2 Machine Learning-Based Models

We developed a forward design-based model, as shown in Fig. 3a. The obtained accuracy for this model is 97.6% and 94.5% for PSNR and NED error metrics, respectively. Such high efficiency is due to the straightforward nature of the problem. However, we target the inverse design of finding the most suitable design settings (degree and type) for given inputs (C1 and C2) and error threshold, as shown in Fig. 3b.

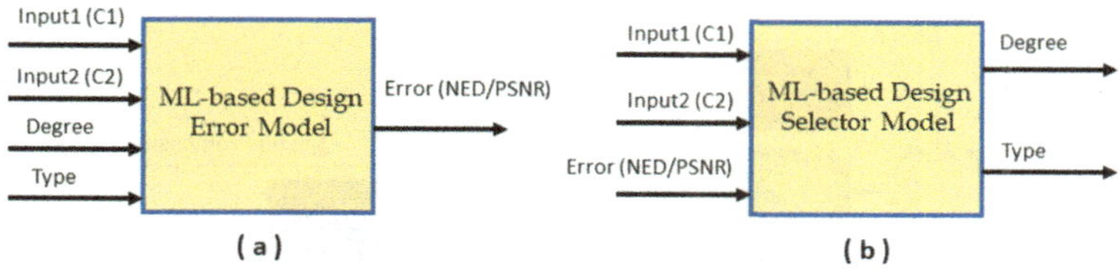

Fig. 3 Models for AC quality manager, (**a**) forward design and (**b**) inverse design

Table 1 Accuracy and execution time of DT- and NN-based design selectors

Model		Accuracy		Execution time (ms)	
Inputs	Output	DT	NN	DT	NN
C1, C2, PSNR	Degree	77.8%	82.17%	8.87	18.9
C1, C2, PSNR, s2=D1	Type	75.5%	66.52%	25.03	18.0
C1, C2, PSNR, s2=D2	Type	76.1%	70.21%	19.3	9.0
C1, C2, PSNR, s2=D3	Type	71.3%	73.22%	11.94	18.7
C1, C2, PSNR, s2=D4	Type	74.1%	59.08%	6.61	7.4

3.2.1 Decision Tree-Based Design Selector

Based on the error analysis of the approximate designs [39], we noticed that the error magnitude is correlated to the approximation *Degree* in a more significant manner than the design *Type*. Such correlation is evident in the accuracy of the models, where these models have an average accuracy of 77.8% and 74.3% for predicting the design *Degree* and *Type*, respectively, as shown in Table 1. The time for executing the *software* implementation of these models is very short, i.e., 24.6 ms in total with 8.87 ms to predict the design *Degree* and 15.72 ms to predict the design *Type*. This time is negligible compared to the time of running an application, such as image blending.

3.2.2 Neural Network-Based Design Selector

As shown in Table 1, the model for *Degree* prediction has an accuracy of 82.17%, while the four models for *Type* prediction have an average accuracy of 67.3%. The time for executing the *software* implementation of these models is short, i.e., 32.18 ms in total with 18.9 ms to predict the design *Degree* and 13.28 ms to predict the design *Type*. This time is negligible compared to the running time of an application, such as image processing. Compared to the DT-based model, the NN-based model has an execution time, which is 1.31× higher than the DT, while its average accuracy is almost 0.98× of the accuracy achieved by the DT-based model. Next, we evaluate the *software* implementation of the proposed methodology, which utilizes the DT-based design selector. We discard the NN-based design selector due to the absence of advantages over DT.

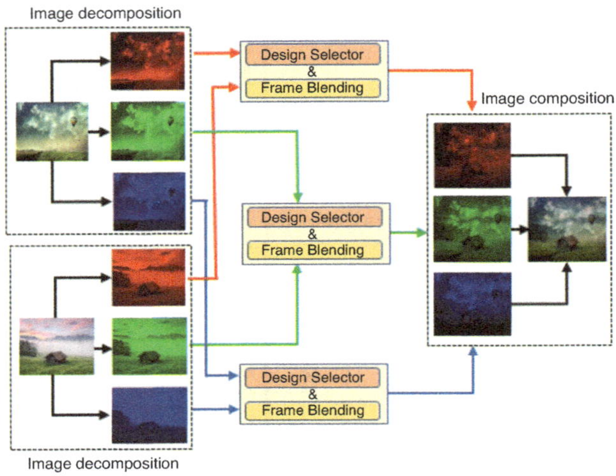

Fig. 4 Adaptive image/video blending at component level

3.3 Experimental Results of Image Blending

Here, we evaluate the effectiveness of the *software* implementation of the fully automated proposed system. We run MATLAB on a machine with 8 GB DRAM and an i5 CPU with a speed of 1.8 GHz. We assess the proposed methodology based on an image blending application, where we use a set of images. The *execution time* is a quality metric, where its overhead is relatively small compared to the original applications, as shown in the sequel.

Image blending in multiplication mode multiplies numerous images to look like a single image. For example, blending two-colored videos, each with N_f frames of size N_r rows by N_c columns per image, involves a total of $3 \times N_f \times N_r \times N_c$ pixels. Each image has three colored components/channels, i.e., red, green, and blue, where the values of their pixels are expected to differ. A static configuration uses a single design of an 8-bit multiplier to perform all multiplications, even when their pixels are different. Therefore, for improved output quality, we propose to adapt the approximate design per channel as shown in Fig. 4. However, for a video with a set of successive frames, e.g., 30 frames per second, the proposed methodology can be run for the first frame only since the other frames have very close pixel values. This way, the design selector continuously monitors the inputs and efficiently finds the most suitable design for each colored component to meet the required TOQ.

Various metrics, e.g., median, skewness, and kurtosis, have been used in the literature to represent the inputs of approximate designs [38]. However, their proposed approximate circuits heavily depend on the training data used during the approximation process. Since the error magnitude depends on the user inputs, we rely on pixel values to select a suitable design. However, setting the configuration

Table 2 Characteristics of the blended images

			Input 1 (Image1)			Input2 (Image2)		
Example	(Image1, Image2)	Frame characteristic	Red	Green	Blue	Red	Green	Blue
1	(Frame, City)	Average	131	163	175	172	153	130
		Cluster	9	11	11	11	10	9
2	(Sky, Landscape)	Average	121	149	117	160	156	147
		Cluster	8	10	8	11	10	10
3	(Text, Whale)	Average	241	241	241	48	156	212
		Cluster	16	16	16	4	10	14
4	(Girl, Beach)	Average	177	158	140	168	176	172
		Cluster	12	10	9	11	12	11
5	(Girl, Tree)	Average	102	73	40	239	193	118
		Cluster	7	5	3	16	13	8

granularity at the pixel level is impractical. On the other hand, the design selection per colored component is more suitable.

We compute the average of the pixels of each colored component to determine the most suitable design. Two completely different images may have the same average of their pixels. Unfortunately, this could result in the same selected approximate design. To avoid this scenario, we reduce the configuration granularity by dividing the colored component into multiple segments, e.g., four segments. Thus, we use various designs, rather than a single design, for each colored component. Next, we analyze the results of applying the proposed methodology on a set of ten images. The photos of each set are then blended at the component level, as shown in Fig. 4, to evaluate the efficiency of the proposed methodology.

We use a set of ten different images, each of size $N_r \times N_c = 250 \times 400 = 10^5$ pixels, and each image is segmented into three colored components. Table 2 shows the average values of the pixels of each colored component and the associated input cluster, which are denoted as *Average* and *Cluster*, respectively.

We target 49 different values of TOQ, i.e., PSNR ranges from 17 dB to 65 dB, for each blending example. Thus, we run the methodology 245 times, i.e., 5 × 49. For every invocation, based on the corresponding cluster for each input, i.e., $C1$ and $C2$, and the associated target PSNR, 1 of the 20 used designs is selected and used for blending. For illustration purposes, in the sequel, we explain *Example5* in detail. As shown in Table 2, the *Girl* image has a red component with an average of 102, which belongs to *Cluster 7*, i.e., $C1_R = 7$. Similarly, the *Tree* image has a red component with an average of 239, which belongs to *Cluster 16*, i.e., $C2_R = 16$. The green components belong to *Clusters* 5 and 13 ($C1_G = 5$, $C2_G = 13$), while the blue components belong to *Clusters* 3 and 8 ($C1_B = 3$, $C2_B = 8$). Then, we adapt the design by calling the design selector thrice, i.e., once for every colored component, assuming TOQ= 17 dB. The selected designs are used, and the obtained quality is 16.9 dB, which is insignificantly less than the TOQ.

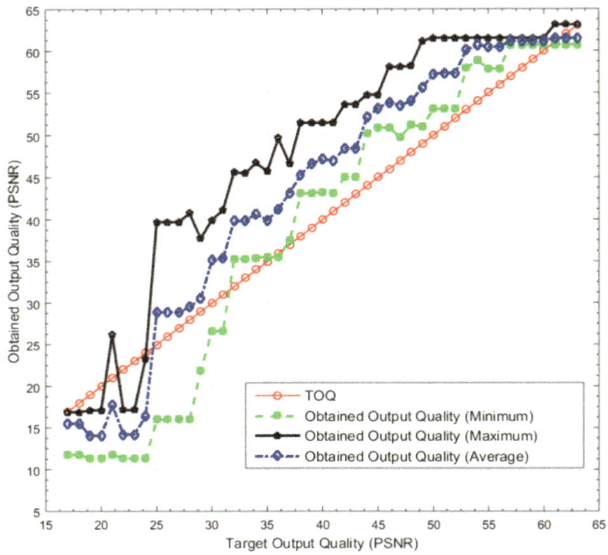

Fig. 5 Obtained output quality for image blending of *Set-1*

Accuracy Analysis of Adaptive Design Figure 5 shows the minimum, maximum, and average curves of the obtained output quality, each evaluated over five examples of image blending. Out of the 245 selected designs, 49 predicted designs are violating the TOQ, even insignificantly, i.e., the obtained output quality is below the red line. The unsatisfied output quality is attributed mainly to model imperfection. The best achievable prediction accuracy is based on the accuracy of the two models executed consecutively, i.e., *Degree* model with 77.8% and *Type* model with 76.1%. The accuracy of our model prediction is 80%, which is in agreement with the average accuracy of the DT-based models, as shown in Table 1.

Execution Time Analysis of Adaptive Design Figure 6 displays the average execution time of the 5 examples of image blending evaluated over 20 static designs. The shown time is normalized for the execution time of the exact design. All designs have a time reduction ranging from 1.8% to 13.6% with an average of 3.96%. For the five examples of image blending, we assessed the execution time of the adaptive design, where the target PSNR ranges between 17 dB and 65 dB for each case. Figure 7 shows the execution time for the 5 examples using the exact design, the adaptive design averaged over 49 different TOQ, and the static design averaged over 20 approximate designs. Design adaptation overhead, which represents the time for running the ML-based design selector, is 30.5 ms, 93.9 ms, 164.6 ms, 148.6 ms, and 42.1 ms, for the five examples, respectively. Moreover, the five examples have a data processing time based on three selected designs per example of 50.90 s, 50.91 s, 51.10 s, 51.69 s, and 51.04 s, respectively. Thus, for these five examples, the design

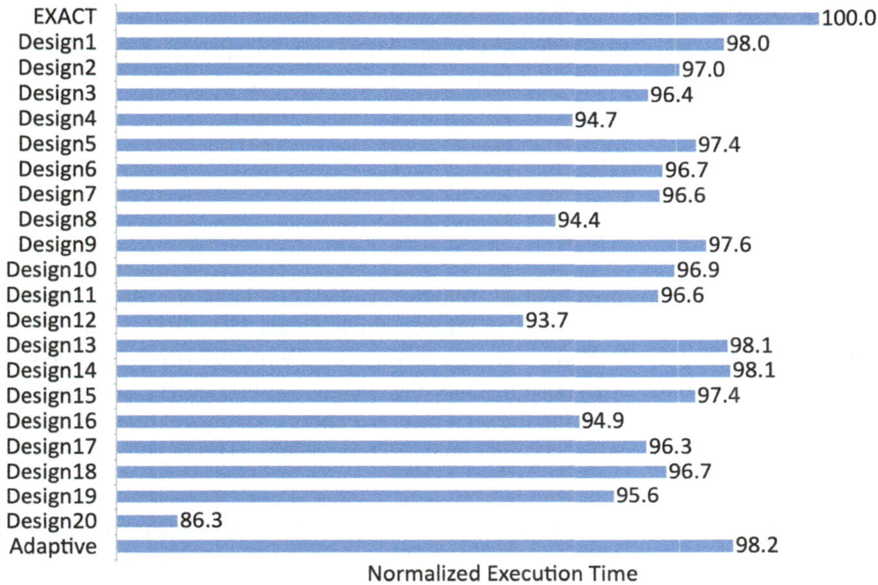

Fig. 6 Normalized execution time for image blending using 20 static designs

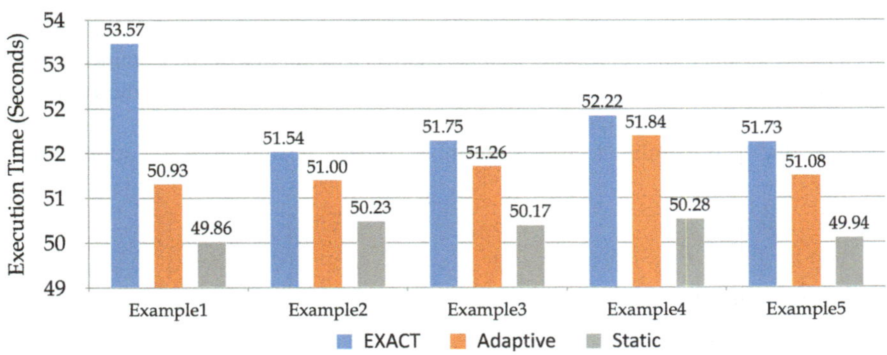

Fig. 7 Execution time of the exact, static, and adaptive design

adaptation time represents 0.06%, 0.18%, 0.32%, 0.28%, and 0.08% of the total execution time, respectively, which is a negligible overhead.

Energy Analysis of Adaptive Design Designing a library of approximate arithmetic modules aims to enhance the energy efficiency [13]. To calculate the energy consumed by the approximate multiplier to process an image, we use the following equation:

$$Energy = Power \times Delay \times N \tag{1}$$

Table 3 Obtained accuracy (PSNR) for various approximate designs

Application		KUL [20]	ETM [21]	ATCM [40]	Adaptive design (proposed)
Blending	*Set-1*, Ex. 1	24.8	27.9	41.5	61
	Set-1, Ex. 2	29.2	29.1	43.7	61.1
	Set-1, Ex. 3	20.3	24.8	33.1	63
	Set-1, Ex. 4	23	28	38.2	60.7
	Set-1, Ex. 5	27.6	29.4	40.3	61.5

where *Power* and *Delay* are obtained from the synthesis tool and N is the number of multiplications required to process an image, which equals $250 \times 400 = 10^5$ pixels. *Design9* multiplier has the highest energy consumption with 2970 pj and a saving of 896 pj compared to the exact design. Thus, the design adaption overhead of 733.7 pj is almost negligible compared to the total minimal energy savings of 89.6 μj (896 pj $\times 10^5$) obtained by processing a single image. These results validate our lightweight *design selector*.

Comparison with Related Work We now compare the output accuracy achieved by our adaptive design with the precision of two static approximate models based on approximate multipliers proposed by Kulkarni et al. [20] and Kyaw et al. [21] that have *similar structures* as the used approximate array multipliers. Moreover, we compare the accuracy of our work with a third approximate design based on the approximate tree compressor multiplier (ATCM), proposed by Yang et al. [40], which is a Wallace tree multiplier. Table 3 shows a summary of the obtained PSNR for image blending based on KUL [20], ETM [21], ATCM [40], and the proposed adaptive design. The proposed model achieves better output quality than static designs due to the ability to select the most suitable design from the approximate library.

3.4 Summary

For dynamic inputs, an approximate static design may lead to substantial output errors for changing data. Previous work has ignored the consideration of the changing inputs to assure the quality of individual outputs. We proposed a novel fine-grained input-dependent adaptive approximate design, based on machine learning models. Then, we implemented a fully automated toolchain utilizing a DT-based design selector. The proposed solution considers the inputs in generating the training data, building ML-based models, and then adapting the design to satisfy the *TOQ*. The "software" implementation of the proposed methodology, developed, provided a negligible delay overhead and was able to satisfy an output accuracy of 80% to 85.7% for image blending applications. Such quality-assured results come at the one-time cost of generating the training data and deploying and evaluating the design selector, i.e., a machine learning-based model. With runtime design adaptation, the model always identifies and selects the most suitable design for controlling the quality loss.

4 Hardware-Based Adaptive Design of Approximate Accelerators

The *software* implementation of the proposed adaptive approximate accelerate was able to satisfy the required TOQ with a minimum accuracy of 80%. Now, we present a hardware implementation of the adaptive approximate accelerator based on a field-programmable gate array (FPGA) and utilizing the feature of dynamic partial reconfiguration (DPR), with a database of 21 reconfigurable modules.

An essential advantage of FPGAs is their flexibility, where these devices can be configured and reconfigured on-site and at runtime by the user. In 1995, Xilinx introduced the concept of *partial reconfiguration* (PR) in its XC6200 series to increase the flexibility of FPGAs by enabling re-programming parts of design at runtime while the remaining parts continue operating without interruption [41]. The basic assumption of PR is that the device hardware resources can be time-multiplexed, similar to the ability of a microprocessor to switch tasks. PR eliminates the need to reconfigure and re-establish links fully and dramatically improves the flexibility that FPGAs offer. PR enables adaptive and self-repairing systems with reduced area and dynamic power consumption.

We propose to dynamically adapt the functionality of the FPGA-based approximate accelerators using machine learning (ML) and dynamic partial reconfiguration (DPR). We utilize the previously proposed DT- and NN-based *design selector* that continually monitors the input data and determines the most suitable approximate design and then, accordingly, partially reconfigures the FPGA with the chosen approximate design while maintaining the whole error-tolerant application intact. The proposed methodology applies to any error-tolerant application where we demonstrate its effectiveness using an image processing application. As FPGA vendors announced the technical support for the runtime partial reconfiguration, such systems are becoming feasible. To our best knowledge, the design framework for adaptively changeable approximate functional modules with input awareness does not exist.

4.1 Dynamic Partial Reconfiguration (DPR)

Field-programmable gate array (FPGA) devices conceptually consist of [42] (i) hardware logic (functional) layer which includes flip-flops, lookup tables (LUTs), block random-access memory (BRAM), digital signal processing (DSP) blocks, routing resources, and switch boxes to connect the hardware components and (ii)

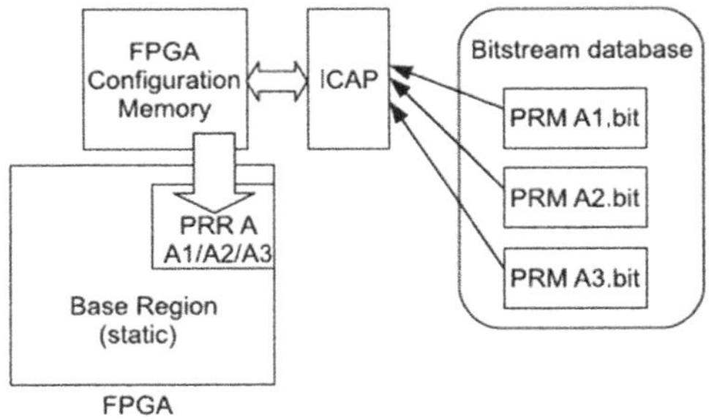

Fig. 8 Principle of dynamic partial reconfiguration on Xilinx FPGAs

configuration memory which stores the FPGA configuration information through a binary file called *configuration file* or *bitstream* (BIT). Changing the content of the *bitstream* file allows us to improve the functionality of the hardware logic layer. Xilinx and Intel (formerly Altera) are the leading manufacturing companies for FPGA devices. We use the VC707 evaluation board from Xilinx, which provides a hardware environment for developing and evaluating designs targeting the Virtex-7 XC7VX485T-2FFG1761C FPGA.

Partial reconfiguration (PR) is the ability to modify portions of the modern FPGA logic by downloading partial *bitstream* files while the remaining parts are not altered [43]. PR is a hierarchical and bottom-up approach and is an essential enabler for implementing *adaptive* systems. It can be static or dynamic, where the reconfiguration can occur while the FPGA logic is in the reset state or running state, respectively [42]. The DPR process consists of two phases: (i) fetching and storing the required bitstream files in the flash memory, which is not time-critical, and (ii) loading bitstreams into the reconfigurable region through a controller, i.e., internal configuration access port (ICAP). Implementing a partially reconfigurable FPGA design is similar to implementing multiple non-partial reconfiguration designs that share a common logic. Since the device is switching tasks in hardware, it has the benefit of both flexibility of software implementation and the performance of hardware implementation. However, it is not commonly employed in commercial applications [43].

Logically, the part that will host the reconfigurable modules (dynamic designs) is the dynamic *partial reconfigurable region* (PRR), which is shared among various modules at runtime through multiplexing. Figure 8 illustrates a reconfigurable design example on Xilinx FPGAs, with a partially reconfigurable region (PRR) A, which is reserved in the overall design layout mapped on the FPGA, with three possible partially reconfigurable modules (PRM). During PR, a portion of the FPGA needs to keep executing the required tasks, including the reconfiguration process. This part of the FPGA is known as the *static region*, which is configured only once at the boot time with a full *bitstream*. This region will also host static parts of the system, such as I/O ports as they can never be physically moved. When a hardware (signal) or a software (register write) trigger event occurs, the Partial Reconfiguration Controller (PRC) fetches/pulls partial *bitstreams* from the memory/database and delivers them to a configuration port.

4.2 Machine Learning-Based Models

Here, we describe the FPGA-based implementation of the design selector based on DT and NN models.

4.2.1 Decision Tree-Based Design Selector

As described previously, the DT-based models have an average accuracy of 77.8% and 74.3% for predicting the design *Degree* and *Type*, respectively, as shown in Table 1. Time overhead for executing the *software* implementation of these models is around 24.6 ms in total, with 8.87 ms to predict the design *Degree* and 15.72 ms to predict the design *Type*. This section evaluates the power, area, delay, and energy of the *FPGA-based* implementation of the DT-based *design selector*. We utilize the XC6VLX75T FPGA, which belongs to the Virtex-6 family. The configurable logic block (CLB) comprises 2 slices, each containing 4 6-input LUTs and 8 flip-flops, for a total of 8 6-input LUTs and 16 flip-flops per CLB. We use Mentor Graphics ModelSim [44] for functionality verification. We use Xilinx XPower Analyzer for the power calculation based on exhaustive design simulation [45], while for logic synthesis, we use the Xilinx Integrated Synthesis Environment (ISE 14.7) tool suite [46].

The obtained characteristics of the DT-based model are shown in Table 4, where the power consumption of the model ranges between 35 mW and 44 mW. This value is insignificant compared to the power consumption of approximate multipliers, where these multipliers being selected are used for *N* inputs. Similarly, the introduced area, delay, and energy overhead are amortized by running the approximate design for *N* inputs. The area of the model, represented in terms of the number of slice LUTs, is 1099, at maximum. Also, the number of occupied slices could reach 452 slices. The worst-case frequency that the model could run is 43.65 MHz, with a period of 22.91 ns. The designed model could consume a maximum energy of 733.7 pj.

The *design selector*, which is synthesized only once, is specific for the considered set of approximate designs. However, the proposed methodology is applicable to other approximate designs as well. The implementation overhead, i.e., power, area, delay, and energy, for the DT-based model is insignificant compared to the approximate accelerator since it is a simple nesting of if-else statements with a maximum depth of 12 to reach a node of a final result.

4.2.2 Neural Network-Based Design Selector

Neural networks (NNs) have typically been implemented in software. However, recently with the exploding number of embedded devices, the hardware implementation of NNs is gaining substantial attention. FPGA-based implementation of NN is complicated due to a large number of neurons and the calculation of complex equations such as activation function [47]. We use the sigmoid function $f(x)$ as an activation function. A piecewise second-order approximation scheme for the implementation of the sigmoid function is proposed in [48] as provided by Eq. (2). It has inexpensive hardware, i.e., one multiplication, no lookup table, and no addition.

$$f(x) = \begin{cases} 1 & x > 4.0 \\ 1 - \frac{1}{2}(1 - \frac{|x|}{4})^2, & 0 < x \leq 4.0 \\ \frac{1}{2}(1 - \frac{|x|}{4})^2, & -4.0 < x \leq 0 \\ 0, & x \leq -4.0 \end{cases} \tag{2}$$

As shown in Table 1, we implemented a two-step design selector by predicting the design *Degree* first and then the *Type*, with an accuracy of 82.17% and 67.3%, respectively. The execution time of the NN-based model ranges between 37.6 ms and 26.3 ms, with an average of 32.7 ms.

We implemented the NN-based model on FPGA, and its characteristics, including dynamic power consumption, slice LUTs, occupied slices, operating frequency, and consumed energy, are shown in Table 4. These values are insignificant when compared to the characteristics of approximate multipliers, where these multipliers are used for N inputs. However, compared to the DT-based model, the NN-based model has an execution time, which is 1.31× higher than the DT, while its average accuracy is almost 0.98× of the accuracy achieved by the DT-based model. Moreover, regarding other design metrics, including power, slice LUTs, occupied slices, period, and energy, the NN-based model has a value of 8.06×, 13.93×, 11.74×, 1.61×, and 6.8×, consecutively, compared with the DT-based model. Unexpectedly, the DT-based model is better than the NN-based model in all design characteristics, including accuracy and execution time.

4.3 Adaptive Design Methodology

Figure 9 shows the FPGA-based methodology for quality assurance of approximate computing through design adaptation, inspired by the general methodology shown in Fig. 1. In order to utilize the available resources of the FPGA and show the benefits of design approximation, we integrate 16 multipliers into an accelerator to be used altogether. Figure 10 shows the internal structure for the approximate accelerator with 16 multipliers. Each input, i.e., A_i and B_i where $16 \geq i \geq 1$, is 8-bit wide.

The implemented ML-based models (design selectors) are DT-based only, where model training is done once offline, i.e., off-FPGA. Then, the VHDL implementation of the obtained DT-based model, which is the output of the offline phase, is integrated as a functional module within the online phase of the FPGA-based

Table 4 Power, area, delay, frequency, and energy of DT- and NN-based design selectors

Model		Dynamic power (mW)		Slice LUTs		Occupied slices		Period (ns)		Frequency (MHz)		Energy (pJ)	
Inputs	Output	DT	NN	DT	NN	DT	NN	DT	NN	DT	NN	DT	NN
C1, C2, PSNR	Degree	16	155	602	7835	231	2683	22.910	31.504	43.65	31.74	366.6	4883.1
C1, C2, PSNR, s2=D1	Type	19	164	497	8427	189	2791	18.596	31.746	53.78	31.50	353.3	5206.3
C1, C2, PSNR, s2=D2	Type	23	153	449	6625	221	2309	15.962	31.718	62.65	31.53	367.1	4852.8
C1, C2, PSNR, s2=D3	Type	23	159	390	5360	149	1731	15.494	29.164	64.54	34.29	356.4	4637.0
C1, C2, PSNR, s2=D4	Type	28	170	298	4549	134	1420	11.838	28.678	84.47	34.87	331.5	4875.3

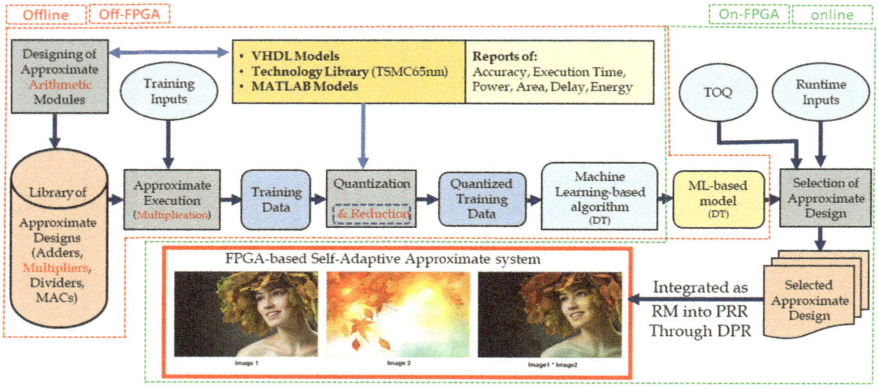

Fig. 9 Methodology of hardware-based adaptive approximate design

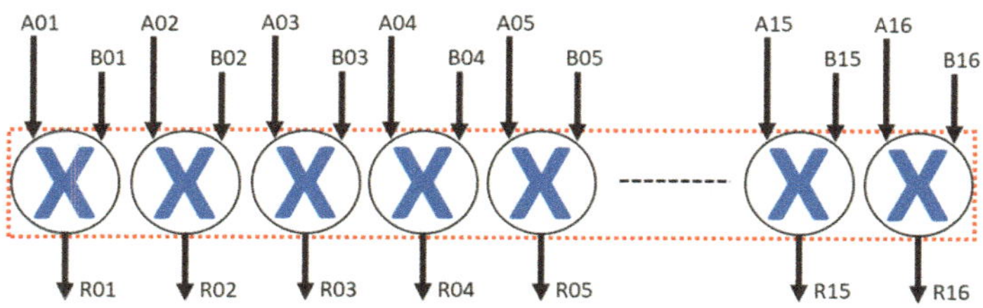

Fig. 10 An accelerator with 16 identical approximate multipliers

adaptive system, as shown in Fig. 11. The proposed FPGA architecture contains a set of intellectual property (IP) cores, connected through a standard bus interface. The developed approximate accelerator core is with the capability of adjusting processing features as commanded by the user to meet the given TOQ. For the parallel execution, we utilize the existing block RAM in the Xilinx 7 series FPGAs, which have 1030 blocks of 36Kbits. Thus, we store the input data (images) in a distributed memory, e.g., save each image of size 16 KByte into 16 memory slots each of 1 KByte. Other configurations of the memory are also possible and can be selected to match the performance of the processing elements within the accelerator.

The *online* phase of the adaptive design, based on the decision tree, is presented in Fig. 11, where the annotated numbers, i.e., ① to ⑧, show the flow of its execution for image blending application. The target device is xc7vx485tffg1761-2, and the evaluation kit is Xilinx Virtex-7 VC707 Platform [49]. The main components are the reconfiguration engine, i.e., DT-based design selector, and the reconfigurable core (RC), i.e., approximate accelerator. The RC is placed in a well-known partially reconfigurable region (PRR) within the programmable logic.

We evaluate the effectiveness of the proposed methodology for an FPGA-based adaptive approximate design utilizing DPR. For that, we select an *image blending* application due to its computationally intensive nature and its amenability

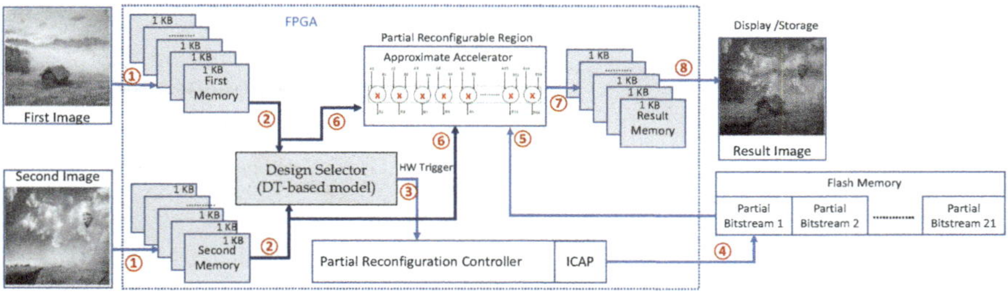

Fig. 11 Methodology of FPGA-based adaptive approximate design—online phase

to approximation. As a first step, to prove the validity of the proposed design adaptation methodology, we evaluate a design without the DPR feature, utilizing the exact accelerator as well as 20 approximate accelerators that exist simultaneously, based on the proposed methodology. Thus, 21 different accelerators evaluate the outputs. Next, based on the inputs and the given TOQ, the design selector chooses the output of a specific design, which has been selected based on the DT model. Finally, the selected result will be forwarded as the final result of the accelerator. The evaluated area and power consumption of such a design are 15× and 24× more significant than the exact implementation, respectively.

We use MATLAB to read the images, re-size them to 128 × 128 pixels, convert them to grayscale, and then write them into coefficient (.COE) files. Such files contain the image pixels in a format that the Xilinx CORE Generator can read and load. We store the images in an FPGA block RAM (BRAM). The design evaluates the average of the pixels of each image retrieved from the memory; then, the *hardware selector* decides which reconfigurable module, i.e., *bitstream* file, to load into the reconfigurable region. The full bitstream is stored in flash memory to be booted up into the FPGA at power-up. Moreover, the partial bitstreams are stored in well-known addresses of the flash memory.

4.4 Experimental Results

In the following, we discuss the results of our proposed methodology when evaluated on image processing applications. In particular, we present the obtained accuracy results along with reports of the area resources utilized by the implemented system.

Accuracy Analysis of the Adaptive Design We evaluate the accuracy of the proposed design over 55 examples of image blending. For each example, our TOQ (PSNR) ranges from 15 dB to 63 dB. The images we use are from the database of "8 Scene Categories Dataset" [50], which is downloadable from [51]. Figure 12 shows the minimum, maximum, and average curves of the obtained output quality,

Fig. 12 Obtained output quality for FPGA-based adaptive image blending

each evaluated over 55 examples. Generally, for image processing applications, the quality is typically considered acceptable if PSNR is 30 dB at least and otherwise unacceptable [52]. Based on that, the design adaptation methodology has been executed 1870 times, while the TOQ has been satisfied 1530 times. Thus, the accuracy of our obtained results in Fig. 12 is 81.82%.

Area Analysis of the Adaptive Design Table 5 shows the primary resources of the XC7VX485T-2FFG1761 FPGA [53]. Moreover, it shows the resources required for the image blending application utilizing an approximate accelerator, both static and adaptive implementation. Design checkpoint files (.DCP) are a snapshot of a design at a specific point in the flow, which includes the current netlist, any optimizations made during implementation, design constraints, and implementation results. For the static implementation, the .DCP file is 430 KByte only, while for the dynamic implementation, it is 17411 KByte. This increase in the file size is due to the logic which has been added to enable DPR, as well as the 20 different implementations for the reconfigurable module (RM). Moreover, the overhead of such logic is shown in the increased number of occupied slice LUTs and slice registers. However, both static and dynamic implementations have the same size of the *bitstream* file (692 KByte), which is to be downloaded into the FPGA. DPR enables downloading the partial *bitstream* into the FPGA rather than the full *bitstream*. Thus, downloading 692 KByte rather than 19799 KByte would be 28.6 × faster. Since different variable-size reconfigurable modules will be assigned to the same reconfigurable region,

Table 5 Area/size of static and adaptive approximate accelerator

Design	.DCP file KByte	Slice LUTs	Slice registers	RAMB36	RAMB18	Bonded IOB	DSPs	Bitstream size (KByte)
XC7VX485T-2FFG1761 FPGA	–	303600	607200	1030	2060	700	2800	–
Static design	430	1472	357	235	51	65	0	19799
Adaptive–Top	17411	12876	15549	235	51	65	0	19799
Adaptive–Exact RM	770	1287	0	0	0	0	0	692
Adaptive–Max Approx RM	647	800	0	0	0	0	0	692
Adaptive–Min Approx RM	458	176	0	0	0	0	0	692

it must be large enough to fit the biggest one, i.e., the exact accelerator in our methodology.

Table 5 shows the main features of the Xilinx XC7VX485T-2FFG1761 device, including the number of slice LUTs, slice registers, and block RAM. The total capacity of block RAM is 37080 Kbit, which could be arranged as 1030 blocks of size 36Kbit each or 2060 blocks of size 18Kbit each. The reconfigurable module (RM) with exact implementation occupies 1287 slice LUTs. However, the number of slice LUTs occupied by the RM with approximate implementation varies from 800 to 176 LUTs. Thus, the area of the approximate RM varies from 62.16% to 13.68% of the area of the exact RM. Despite all of that, all 21 RMs have the same *bitstream* size of 692 KB.

4.5 Summary

To ensure the quality of approximation by design adaptation, we described the proposed methodology to adapt the architecture of the FPGA-based approximate design using dynamic partial reconfiguration. The proposed design with low power, reduced area, small delay, and high throughput is based on runtime adaptation for changing inputs. For this purpose, we utilized a lightweight and energy-efficient *design selector* built based on decision tree models. Such input-aware *design selector* determines the most suitable approximate architecture which satisfies user-given quality constraints for specific inputs. Then, the partial *bitstream* file of the selected design is downloaded into the FPGA. Dynamic partial reconfiguration allows quickly reconfiguring the FPGA devices without having to reset the complete device. The obtained analysis results of the image blending application showed that it is possible to satisfy the TOQ with an accuracy of 81.82%, utilizing a partial *bitstream* file that is 28.6× smaller than the full *bitstream*.

5 Conclusions

Approximate computing has re-emerged as an energy-efficient computing paradigm for error-tolerant applications. Thus, it is promising to be within the architecture and algorithms of brain-inspired computing, which has massive device parallelism and the ability to tolerate unreliable operations. However, there are essential questions to be answered before approximate computing can be made a viable solution for energy-efficient computing, such as [54] (1) how much to approximate at the component level to be able to observe the gains at the system level, (2) how to measure the final quality of approximation, and (3) how to maintain the desired output quality of an approximate application.

Toward addressing these challenges, we proposed a methodology that assures the quality of approximate computing through design adaptation based on fine-grained inputs and user preferences. For that, we designed a lightweight machine learning-based model, which functions as a design selector, to select the most suitable approximate designs to ensure the final quality of the approximation.

We proposed a novel methodology to generate an adaptive approximate design that satisfies user-given quality constraints, based on the applied inputs. For that, we have built a machine learning-based model (that functions as a design selector) to determine the most suitable approximate design for the applied inputs based on the associated error metrics. To solve the *design selector* model, we used decision tree and neural network techniques to select the approximate design that matches the closest accuracy for the applied inputs.

We realized the *software* and *hardware* implementations of the proposed methodology, with negligible overhead. The obtained analysis results of the image processing application showed that it is possible to satisfy the TOQ with accuracy ranging from 80% to 85.7% for various error-resilient applications. The *FPGA-based* adaptive approximate accelerator with constraints on size, cost, and power consumption relies on dynamic partial reconfiguration to assist in satisfying these requirements. In summary, the general proposed design adaptation methodology can be seen as a basis for automatic quality assurance. It offers a promising solution to reduce the approximation error while maintaining approximation benefits.

References

1. B. Moons, M. Verhelst, Energy-efficiency and accuracy of stochastic computing circuits in emerging technologies. IEEE J. Emerging Sel. Top. Circuits Syst. **4**(4), 475–486 (2014)

2. J. Han, M. Orshansky, Approximate computing: An emerging paradigm for energy-efficient design, in *European Test Symposium* (2013), pp. 1–6
3. S. Venkataramani, S.T. Chakradhar, K. Roy, A. Raghunathan, Approximate computing and the quest for computing efficiency, in *Design Automation Conference* (2015), pp. 1–6
4. R. Ragavan, B. Barrois, C. Killian, O. Sentieys, Pushing the limits of voltage over-scaling for error-resilient applications, in *Design, Automation Test in Europe* (2017), pp. 476–481
5. P. Roy, R. Ray, C. Wang, W.F. Wong, ASAC: automatic sensitivity analysis for approximate computing. SIGPLAN Not. **49**(5), 95–104 (2014)
6. V. Gupta, D. Mohapatra, A. Raghunathan, K. Roy, Low-power digital signal processing using approximate adders. IEEE Trans. Comput. Aided Des. Integr. Circuits Syst. **32**(1), 124–137 (2013)
7. R. Nair, Big data needs approximate computing: technical perspective. Commun. ACM **58**(1), 104–104 (2014)
8. A. Mishra, R. Barik, S. Paul, iACT: A software-hardware framework for understanding the scope of approximate computing, in *Workshop on Approximate Computing Across the System Stack* (2014), pp. 1–6
9. J. Bornholt, T. Mytkowicz, K. McKinley, UnCertain: a first-order type for uncertain data. SIGPLAN Not. **49**(4), 51–66 (2014)
10. M. Laurenzano, P. Hill, M. Samadi, S. Mahlke, J. Mars, L. Tang, Input responsiveness: using canary inputs to dynamically steer approximation, in *Programming Language Design and Implementation* (ACM, New York, 2016), pp. 161–176
11. V.K. Chippa, S.T. Chakradhar, K. Roy, A. Raghunathan, Analysis and characterization of inherent application resilience for approximate computing, in *Design Automation Conference* (2013), pp. 1–9
12. E. Nogues, D. Menard, M. Pelcat, Algorithmic-level approximate computing applied to energy efficient hevc decoding. IEEE Trans. Emerg. Top. Comput. **7**(1), 5–17 (2019)
13. M. Masadeh, O. Hasan, S. Tahar, Comparative study of approximate multipliers, in *ACM Great Lakes Symposium on VLSI* (2018), pp. 415–418
14. M. Masadeh, O. Hasan, S. Tahar, Machine-learning-based self-tunable design of approximate computing. IEEE Trans. Very Large Scale Integr. VLSI Syst. **29**(4), 800–813 (2021)
15. H. Esmaeilzadeh, A. Sampson, L. Ceze, D. Burger, Neural acceleration for general-purpose approximate programs, in *International Symposium on Microarchitecture* (2012), pp. 449–460
16. M. Shafique, R. Hafiz, S. Rehman, W. El-Harouni, J. Henkel, Invited: cross-layer approximate computing: From logic to architectures, in *Design Automation Conference* (2016), pp. 1–6
17. S. Ullah, S. Rehman, B.S. Prabakaran, F. Kriebel, M.A. Hanif, M. Shafique, A. Kumar, Area-optimized low-latency approximate multipliers for FPGA-Based hardware accelerators, in *Design Automation Conference* (2018), pp. 1–6
18. M. Imani, R. Garcia, A. Huang, T. Rosing, Cade: configurable approximate divider for energy efficiency, in *Design, Automation Test in Europe Conference* (2019), pp. 586–589
19. G. Zervakis, K. Tsoumanis, S. Xydis, D. Soudris, K. Pekmestzi, Design-efficient approximate multiplication circuits through partial product perforation. IEEE Trans. Very Large Scale Integr. Syst. **24**(10), 3105–3117 (2016)
20. P. Kulkarni, P. Gupta, M. Ercegovac, Trading accuracy for power with an underdesigned multiplier architecture, in *International Conference on VLSI Design* (2011), pp. 346–351
21. K.Y. Kyaw, W.L. Goh, K.S. Yeo, Low-power high-speed multiplier for error-tolerant application, in *International Conference of Electron Devices and Solid-State Circuits* (2010), pp. 1–4
22. K.M. Reddy, Y.B.N. Kumar, D. Sharma, M.H. Vasantha, Low power, high speed error tolerant multiplier using approximate adders, in *VLSI Design and Test* (2015), pp. 1–6
23. M. Masadeh, O. Hasan, S. Tahar, Comparative study of approximate multipliers, in *Great Lakes Symposium on VLSI* (ACM, New York, 2018), pp. 415–418
24. M. Masadeh, O. Hasan, S. Tahar, Comparative study of approximate multipliers, in *CoRR*, vol. abs/1803.06587 (2018)
25. W. Baek, T. Chilimbi, Green: a framework for supporting energy-conscious programming using controlled approximation. SIGPLAN Not. **45**(6), 198–209 (2010)

26. M. Samadi, J. Lee, D. Jamshidi, A. Hormati, S. Mahlke, SAGE: self-tuning approximation for graphics engines, in *International Symposium on Microarchitecture* (2013), pp. 13–24
27. T. Wang, Q. Zhang, N. Kim, Q. Xu, On effective and efficient quality management for approximate computing, in *International Symposium on Low Power Electronics and Design* (2016), pp. 156–161
28. X. Chengwen, W. Xiangyu, Y. Wenqi, X. Qiang, J. Naifeng, L. Xiaoyao, J. Li, On quality trade-off control for approximate computing using iterative training, in *Design Automation Conference* (2017), pp. 1–6
29. M. Shafique, W. Ahmad, R. Hafiz, J. Henkel, A low latency generic accuracy configurable adder, in *Design Automation Conference* (ACM, New York, 2015), pp. 86:1–86:6
30. X. Sui, A. Lenharth, D. Fussell, K. Pingali, Proactive control of approximate programs, in *International Conference on ASPLOS* (ACM, New York, 2016), pp. 607–621
31. M. Masadeh, O. Hasan, S. Tahar, Input-conscious approximate multiply-accumulate (MAC) unit for energy-efficiency. IEEE Access **7**, 147129–147142 (2019)
32. D. Mohapatra, V.K. Chippa, A. Raghunathan, K. Roy, Design of voltage-scalable meta-functions for approximate computing, in *Design, Automation Test in Europe* (2011), pp. 1–6
33. S. Shalev-Shwartz, S. Ben-David, *Understanding Machine Learning: From Theory to Algorithms* (Cambridge University Press, Cambridge, 2014)
34. M. Masadeh, O. Hasan, S. Tahar, Controlling approximate computing quality with machine learning techniques, in *Design, Automation and Test in Europe* (2019), pp. 1575–1578
35. L. Breiman, J. Friedman, R. Olshen, C. Stone, *Classification and Regression Trees* (Chapman and Hall, Wadsworth, 1984)
36. R.C. Barros, A.C. de Carvalho, A.A. Freitas, *Automatic Design of Decision-Tree Induction Algorithms* (Springer, Berlin, 2015)
37. A. Raha, V. Raghunathan, qLUT: Input-Aware quantized table lookup for energy-efficient approximate accelerators. ACM Trans. Embed. Comput. Syst. **16**(5s), 130:1–130:23 (2017)
38. S. Xu, B.C. Schafer, Approximate reconfigurable hardware accelerator: adapting the micro-architecture to dynamic workloads, in *International Conference on Computer Design* (IEEE, New York, 2017), pp. 113–120
39. M. Masadeh, O. Hasan, S. Tahar, Error analysis of approximate array multipliers, in *CoRR* (2019). https://arxiv.org/pdf/1908.01343.pdf
40. T. Yang, T. Ukezono, T. Sato, Low-power and high-speed approximate multiplier design with a tree compressor, in *International Conference on Computer Design* (2017), pp. 89–96
41. Partial Reconfiguration User Guide (2013). https://www.xilinx.com/support/documentation/sw_manuals/xilinx14_7/ug702.pdf. Last accessed on 2023-02-24
42. K. Vipin, S.A. Fahmy, FPGA dynamic and partial reconfiguration: a survey of architectures, methods, and applications. ACM Comput. Surv. **51**(4), 72:1–72:39 (2018)
43. D. Koch, *Partial Reconfiguration on FPGAs: Architectures, Tools and Applications* (Springer, Berlin, 2012)
44. Mentor Graphics Modelsim (2019). https://www.mentor.com/company/higher_ed/modelsim-student-edition. Last accessed on 2023-02-24
45. Xilinx XPower Analyser (2019). https://www.xilinx.com/support/documentation/sw_manuals/xilinx11/ug733.pdf. Last accessed on 2023-02-24
46. Xilinx Integrated Synthesis Environment (2019). https://www.xilinx.com/products/design-tools/ise-design-suite/ise-webpack.html. Last accessed on 2023-02-24
47. S. Ngah, R. Abu Bakar, A. Embong, S. Razali, Two-steps implementation of sigmoid function for artificial neural network in field programmable gate array. ARPN J. Eng. Appl. Sci. **11**(7), 4882–4888 (2016)
48. M. Zhang, S. Vassiliadis, J.G. Delgado-Frias, Sigmoid generators for neural computing using piecewise approximations. IEEE Trans. Comput. **45**(9), 1045–1049 (1996)
49. VC707 Evaluation Board for the Virtex-7 FPGA: User Guide (2019). https://www.xilinx.com/support/documentation/boards_and_kits/vc707/ug885_VC707_Eval_Bd.pdf. Last accessed on 2023-02-13

50. A. Oliva, A. Torralba, Modeling the shape of the scene: a holistic representation of the spatial envelope. Int. J. Comput. Vis. **42**(3), 145–175 (2001)
51. Modeling the shape of the scene: a holistic representation of the spatial envelope (2020). http://people.csail.mit.edu/torralba/code/spatialenvelope/. Last accessed on 2023-02-04
52. M. Barni, *Document and Image compression* (CRC Press, New York, 2006)
53. 7 Series FPGAs Data Sheet: Overview (2020). https://www.xilinx.com/support/documentation/data_sheets/ds180_7Series_Overview.pdf. Last accessed on 2023-02-13
54. J. Han, Introduction to approximate computing, in *VLSI Test Symposium* (2016), pp. 1–1

Design Wireless Communication Circuits and Systems Using Approximate Computing

Chenggang Yan, Ke Chen, and Weiqiang Liu

1 Introduction

In recent years, the communication circuits and systems have become more complicated and more power hungry. However, due to the channel noise and forward error correction (FEC) module, the communication systems can inherently tolerate certain errors at the receiving end. Approximate computing schemes have been explored in many communication systems to achieve low complexity and low power design, such as multi-in multi-out (MIMO) detector [1, 2] and polar decoders [3, 4]. When approximation computing is applied to wireless communication systems the following two benefits can be obtained: (1) approximate architecture can achieve low hardware consumption without signal-to-noise ratio (SNR) performance deterioration; (2) when system performance requirements are determined, the hardware efficiency can approach the limit through more fine-grained quantization.

The block diagram of a 4 × 32 MIMO Orthogonal Frequency Division Multiplexing (OFDM) wireless communication system without feedforward error correction code is shown in Fig. 1. Floating point representation is used in plenty of scientific and engineering applications because of its large dynamic range. In 2011, the fixed-point and floating-point units in MIMO-OFDM detectors have been systematically compared [5]. The results showed that a 12-bit width floating-point system can achieve the same bit error rate (BER) performance as a 16-bit width fixed-point system which has smaller circuit area and power consumption. It also has been demonstrated in the literature [6] that the wireless communication system supporting floating-point arithmetic units can obtain nearly 30% performance and energy efficiency compared to fixed-point only arithmetic units.

C. Yan (✉) · K. Chen · W. Liu
College of Electronic and Information Engineering/College of Integrated Circuits, Nanjing University of Aeronautics and Astronautics, Nanjing, China
e-mail: yancg@nuaa.edu.cn

W. Liu et al. (eds.), *Design and Applications of Emerging Computer Systems*,
https://doi.org/10.1007/978-3-031-42478-6_20

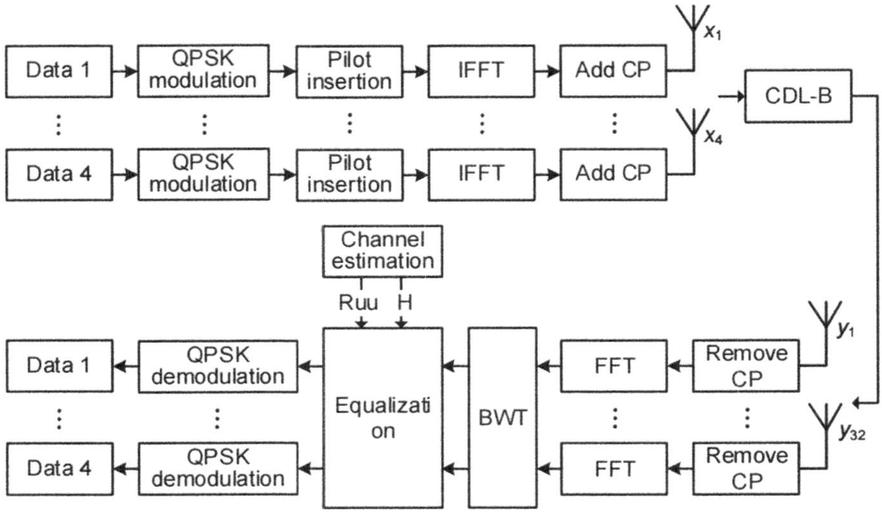

Fig. 1 The block diagram of a 4 × 32 MIMO wireless communication system

In beam weight transformation (BWT) and channel estimation module, there are large number of floating point (FP) multiplication and addition and fast Fourier transform (FFT) operations, which occupy many hardware resource and power consumption. Thus, this chapter will introduce several approximate FP arithmetic circuits (such as adders and multipliers) for wireless communication system. Furthermore, this chapter will present approximate baseband processing units (such as FFT processor and polar decoder) and their evaluation.

2 Approximate FP Arithmetic Unit in Wireless Communication System

Currently, more and more researchers are conducting research on floating-point units in wireless communication systems. In the literature [7], the authors propose a floating-point block-enhanced MMSE filter matrix computation architecture for MIMO-OFDM communication systems, which significantly reduces the circuit area without degrading the BER performance. Large and complex matrix inversion operations are often involved in wireless communication systems, and QR decomposition is one of the common solutions for matrix inversion. A floating-point 4 × 4 matrix design is proposed to increase the data throughput in the QR decomposition process [8]. In the literature [9], researchers proposed an approximate floating-point compression strategy to accelerate MPI (message passing interface) communication, which greatly increases the throughput performance within a certain. It can be seen that the application of floating-point computing unit in wireless communication system has achieved good results. While the wireless communication system is fault-tolerant, this section focuses on the research of approximate floating-point arithmetic unit.

2.1 Approximate FP Adder

The floating-point adder, as a key component in the computing systems, and its energy efficiency have been widely concerned. Accurate floating-point adder operations include exponential difference, pair order, mantissa adder, special value processing, and rounding operations, as shown in Fig. 2. The approximate floating-point adder is usually designed with the approximate mantissa adder block. The mantissa adder can be seen as a fixed-point adder, and the architectures of approximate fixed-point adder have been extensively explored, which can be well extended to the approximate floating-point adders design.

According to the characteristics of approximate fixed-point adders, the existing approximate adders are mainly classified into four categories [10]: approximate speculative adders, approximate segmented adders, approximate carry-select adders, and approximate full adder adders. The main idea of approximate speculative adders [11, 12] is to take the advantage of the characteristic that the bit propagation chain is relatively short in most cases. When calculating the carry bits, only the nearest k lower bits are used to speculate the carry bits of each stage in a n-bit adder ($k < n$) instead of calculating all the bits of the operand. The length of carry propagation chain has been effectively cutting off which has much lower latency than the conventional ripple carry adder. The approximate segmented adder mainly divides the n-bit adder into k-bit adders ($k < n$) in equal ratios. The multiple k-bit adders are calculated parallelly to reduce the critical path delay by cutting off the carry propagation chain between the segments adders. The approximate refinement adder [13], dithered adder [14], and error-tolerant adder [15] are all approximate segmented adders. The approximate carry-select adder combines the features of approximate speculative adder and approximate segmented adder by first dividing the n-bit adder uniformly into a number of m-bit adders ($m < n$) and then using the k-bit carry speculative in each m-bit adder ($k < m$). The random carry select adder [16] and carry speculative adder [17] are all approximate carry select adders. The approximate full adder can reduce the overall delay and power consumption by applying to the low weighted bits of the exact adder with a small loss of accuracy. The approximate transmission XOR adder [18], the approximate mirror adder [19], and the low-weight OR gate adder [20] all belong to the approximate full adder designs.

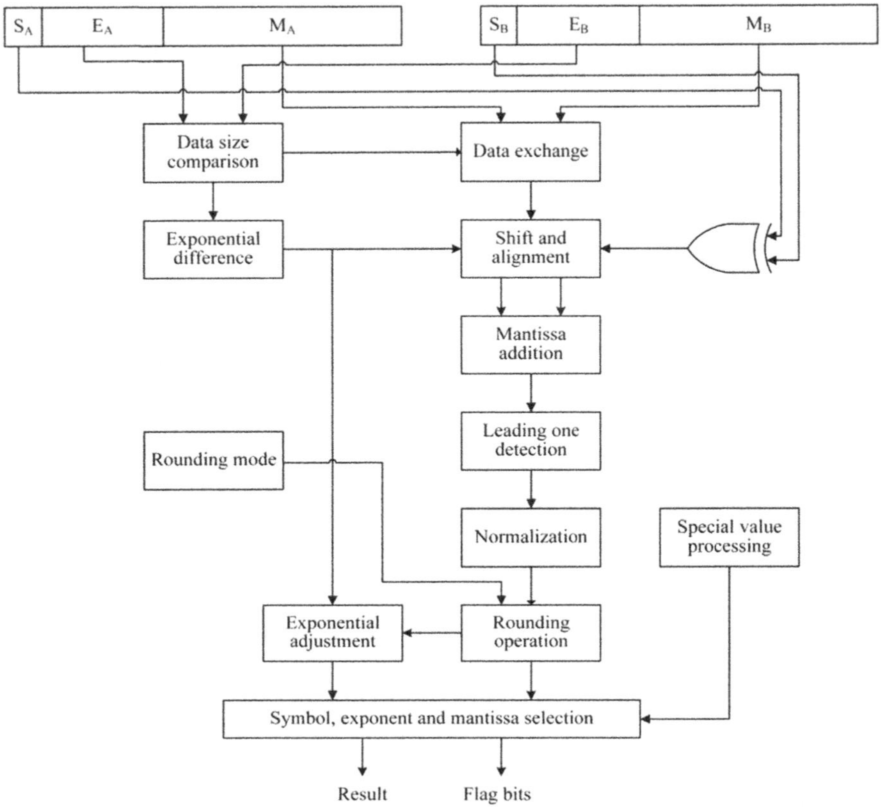

Fig. 2 The block diagram of standard precision floating-point multiplier

2.1.1 Data Distribution Characteristics in Wireless Communication Systems

In the built 4 × 32 MIMO-OFDM system based on half-precision floating-point presentation, the data at the receiver side are classified. The statistical results show that the proportion of normal and subnormal numbers in the real and imaginary parts of the complex data at the receiver side is consistent. The percentage of subnormal numbers is only about 1.2%. The data distribution characteristic at the receiving end is shown in Fig. 3. From the figure, it can be seen that both the real and imaginary parts of the complex data at the receiving end of the system follow the Gaussian distribution.

The probability of each bit being 1 in the mantissa of a floating-point number that follows the Gaussian distribution is calculated, and the related results are shown in Table 1. From the table, the probability that the highest significant bit (implied bit) of the floating-point number is 1 is as high as 97%, which indicates that the majority of

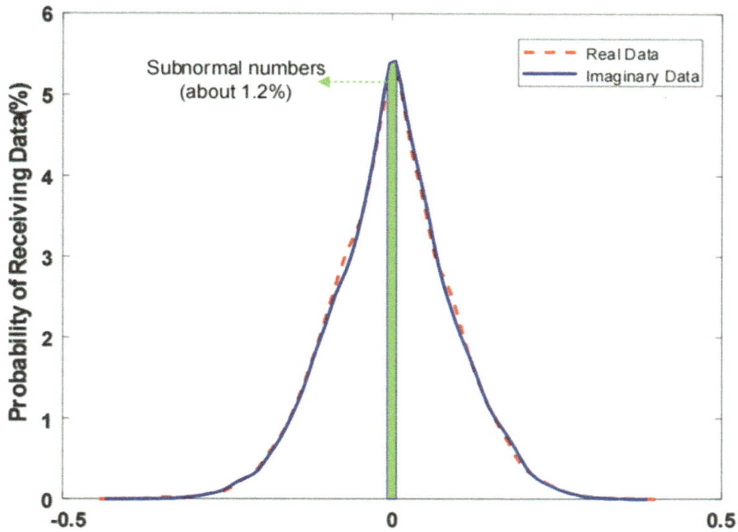

Fig. 3 The data distribution characteristic at the receiving end

Table 1 The probability of each bit being 1 in the mantissa

Mantissa bits	A [10]	A [9]	A [8]	A [7]	A [6]	A [5]	A [4]	A [3]	A [2]	A [1]	A[0]
Probability of being 1	0.97	0.42	0.46	0.48	0.49	0.50	0.50	0.50	0.50	0.50	0.50

the data are normal numbers, in accordance with the two kinds of data percentage patterns shown in Fig. 3. Except for the highest significant bit, the probability of the other bits being 1 increases gradually from 0.42 (or 0.41) to 0.5 as the weight decreases, and finally follows a uniform distribution. This is consistent with the fact that zero-mean Gaussian distributed data are mainly concentrated around zero.

2.1.2 Truncation-Based Approximate FP Adders

The schematic of the approximate floating-point adder based on the standard algorithm is shown in Fig. 4. Firstly, the least significant bits (LSBs) of the operands after the shift pair order are truncated in the approximate mantissa adder, which will result in the approximate sum result. Therefore, only the maximum significant bits are detected by the leading 1 processing. Secondly, considering the small percentage of subnormal numbers in the wireless communication system and special values are not allowed, the specification module can approximate the subnormal number to zero and does not require the corresponding operation. Similarly, since the approximate mantissa sum result is no longer exact, the rounding module is directly discarded. The control signals for the four rounding modes are also no longer needed. The effect after discarding the rounding module is essentially similar

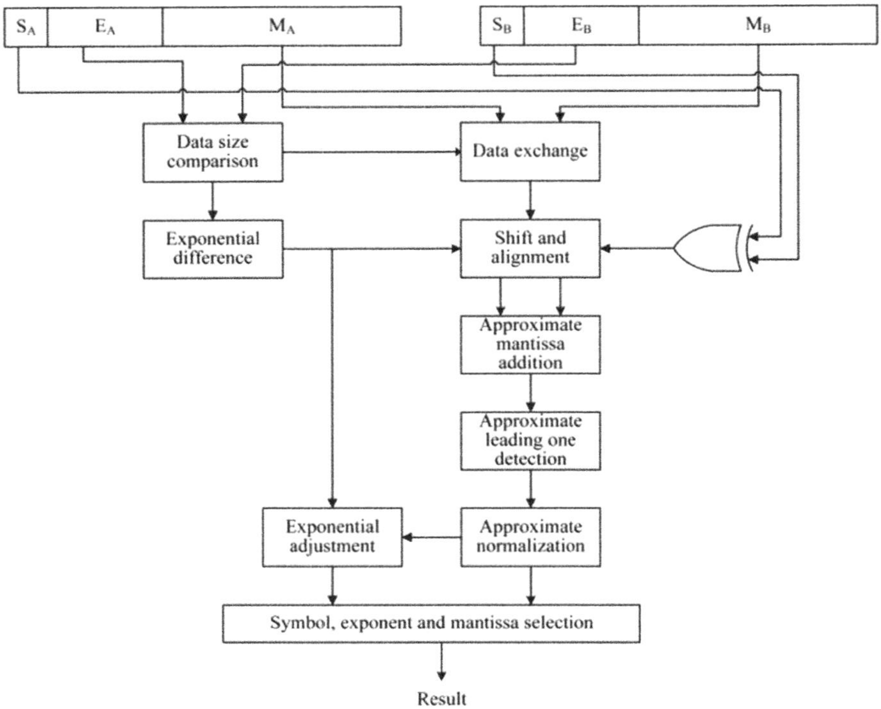

Fig. 4 The block diagram of approximate floating-point adder

to that of rounding to zero mode which directly truncates the redundant data in the mantissa. Since the rounding module consumes a large amount of resources, it is better to discard it directly. Finally, the input of the floating-point addition operation in wireless communication systems only need two operands. The flag bits from the higher level are not required, so the result output flag bits are also discarded in the approximate floating-point adder design, and its corresponding operational logic circuit is also no longer needed.

Truncating LSBs of the operands is an efficient approximation method, and the schematic of proposed half-precision floating-point adder is shown in Fig. 5. The existing floating-point mantissa approximate adder scheme LOA compression is shown in Fig. 6 as a comparison, where A is the mantissa of the larger operand after position swapping, and B′ is the mantissa of the smaller operand after the pairwise order operation.

In the mantissa addition, the bits of data B′ which shifted out of orders are not involved in the addition operation, and they will be discarded after the specification. The truncation of the lower 3 bits of the operand in Fig. 5 effectively shortens the critical path of the circuit. Compared to the LOA approximation as shown in Fig. 6, the direct truncation strategy saves three OR gates, but it also results in more errors

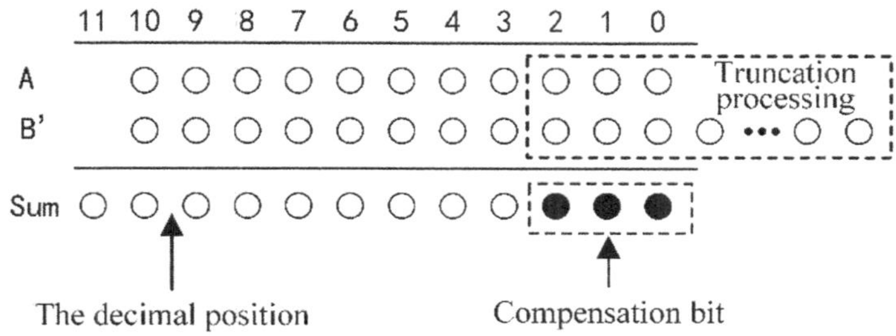

Fig. 5 The operand truncation approximation scheme

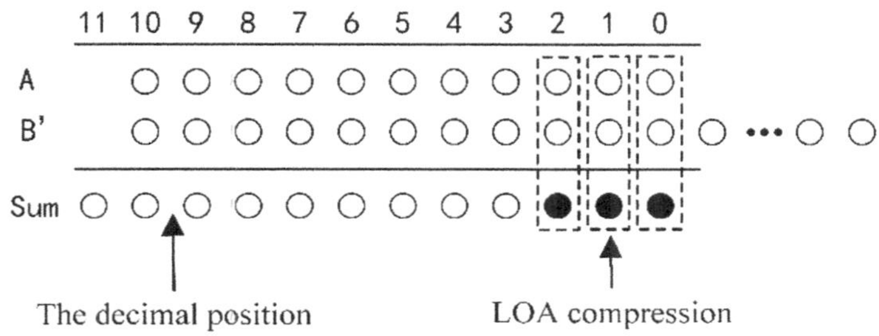

Fig. 6 The LOA approximate scheme

in the final result. Therefore, we add compensation after the truncation to improve the accuracy. The truncation operands will inevitably result in an error in the final result. By setting all the truncation bits to 1, the error of the final result will be zero mean, which can achieve a higher SNR compared to the errors with DC offset. At the same time, by adjusting the number of truncated bits, the approximate design with different accuracy can be achieved.

2.2 Approximate FP Multiplier

Multiplication is one of the most common and expensive FP operations, slowing down the computation in many applications, such as image processing, neural network, and wireless communication [21, 22]. The general FP multiplier structure is shown in Fig. 7.

The application of approximate computing to FP multipliers has been focused a lot in academic researches. Zhang et al. [23] propose a low-power accuracy-configurable FP multiplier by applying Mitchell's algorithm to the mantissa calculation. The maximum error margins of the log path and the full path provided in this work are 11% and 2.04%, respectively. In [24], an FP multiplier with configurable accurate and approximate modes is proposed. After adaptive selection checks for a

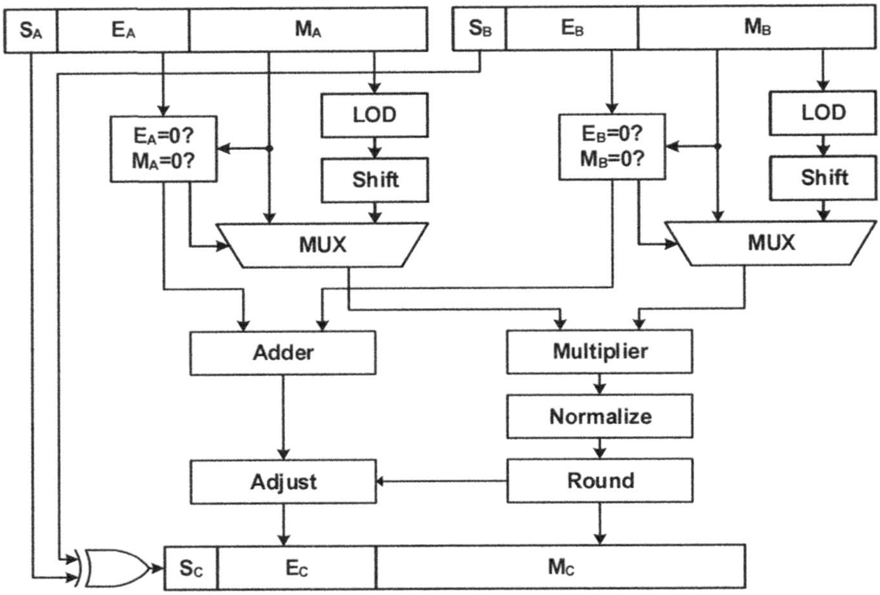

Fig. 7 General structure of the floating-point multiplier

mantissa which will produce an exact output when possible, one of the two modes is selected according to the accuracy requirements. In the approximate mode, one of the mantissas is directly used as the output mantissa to avoid costly multiplication operation. RMAC [25] also has exact and approximate mode; however, approximate result is computed first. The approximate mode only uses bit addition for all bits to replace multiplication. RMAC inspects the result and decides whether exact computation is necessary for the specific input data. Hashemi et al. [26] propose a dynamic unbiased multiplier called DRUM, which dynamically extracts shorter fixed bit-width data for multiplier by leading one detection. On the contrary, in [27], the authors discarded the leading one detection and directly divided the data into two segments statically. Which segment of data to choose for multiplication is determined by judging whether the high-weight segment is zero. These two works are approximate fixed-point multipliers, but their principles can also be extended to FP multipliers. In [28], the input mantissa is divided into a short exactly processed part and a remaining approximately processed part by leading one detection. Higher bits are exactly calculated to ensure high accuracy and lower bits are approximated for simplification to reduce the hardware consumption. A floating-point logarithmic multiplier applied to neural network training is proposed in [29]. It realizes mantissa multiplier by converting mantissa multiplier into logarithmic domain addition. The area can be reduced 10.7 times compared with the accurate floating-point multiplier.

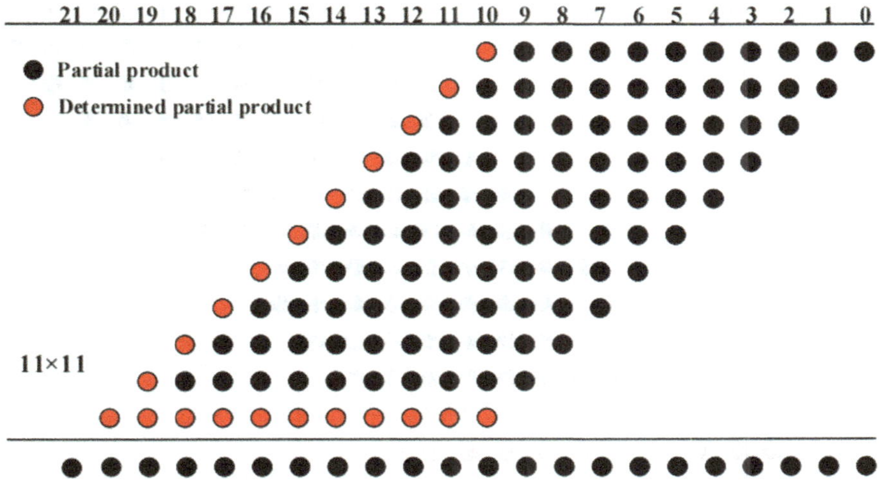

Fig. 8 The traditional 11 × 11 array multiplier

2.2.1 Low-Complexity Exact Mantissa Multiplier

Among the many proposed multiplier designs, the Booth multiplier at wide bit widths is widely used. However, the partial product saved by the Booth algorithm is not enough to make up for the additional hardware overhead caused by the Booth encoder with narrow bit widths. Therefore, the mantissa multiplier design of our work is based on an array multiplier.

With the subnormal data approximated to zero, the mantissa feature that the implicit leading significant bit is always 1 for non-zero data. Based on this feature, a partial product array of 11-bit multipliers is obtained, as shown in Fig. 8. Since the MSB of the multipliers A and B are always 1, the leftmost and bottom-most partial products are fixed (i.e., the red partial products). Therefore, these partial products can be generated directly without any hardware consumption. In this case, the original 11-bit multiplier can be split, with the black partial product compressed with a 10-bit multiplier and the white partial product compressed with a 10-bit adder, and the result of both parts treated as input to the 11-bit adder to produce the final result as shown in Fig. 9. In this proposed mantissa multiplier, 21 "AND" gates are saved and the overall degree of parallelism is improved, which can reduce certain power and area consumption without losing accuracy. At the same time, there's still a lot of approximate space for the 10-bits multiplier.

2.2.2 Truncation and Compensation Scheme for Mantissa Multiplication

Mantissa multiplier in the FP multiplier accounts for more than 80% of the total power consumption [30]. Therefore, reducing the power consumption of this part is

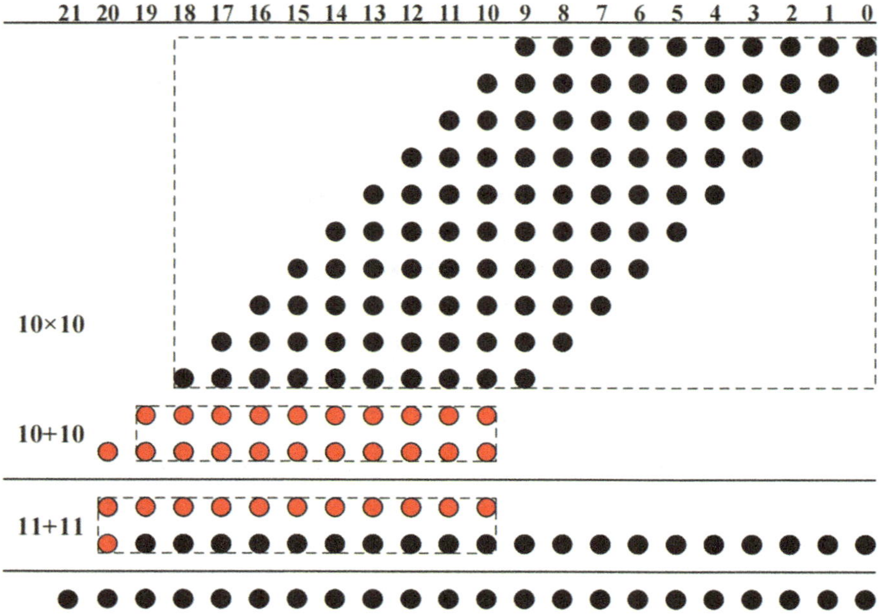

Fig. 9 The proposed low-complexity 11 × 11 multiplier

important and efficient to improve the energy efficiency of the overall FP multiplier. Limited by the floating-point number format, half of the exact multiplier output should be cut off when normalizing the mantissa. Therefore, it is a highly profitable approximation method for an FP multiplier to directly truncate the partial product of the low-weight bits. After truncation, a constant determined by probability analysis is compensated to make up for accuracy loss. The distribution characteristics of the data at the receiving end are shown in Fig. 2. The result shows that the distributions of real and imaginary parts of the received complex data are consistent and obey the Gaussian distribution. Therefore, the input data following the Gaussian distribution is fully considered in our approximate design. And the data of many applications in nature obey Gaussian distribution, and thus the FP multiplier proposed in this work has application scalability. Through MATLAB simulation, the probability of each bit in the mantissa of the receiving data being 1 is shown in Table 1. The probability of MSB being 1 is as high as 97%, which indicates that most data are normal numbers, consistent with the data distribution characteristic in Fig. 2. Except for the MSB, the probability of other bits in mantissa being 1 increases gradually from 0.42 to 0.5 with the decrease of weight, and finally tends to be a uniform distribution. This is also consistent with the fact that Gaussian-distributed data is concentrated in the central region (i.e., near-zero).

In the wireless communication system, the two operands obey the Gaussian distribution. To explore the approximate topology, the probability of being 1 in the

20	19	18	17	16	15	14	13	12	11	10	9	8	7	6	5	4	3	2	1	0
										0.5000	0.2100	0.2300	0.2400	0.2450	0.2450	0.2500	0.2500	0.2500	0.2500	0.2500
									0.5000	0.2100	0.2300	0.2400	0.2450	0.2450	0.2500	0.2500	0.2500	0.2500	0.2500	
								0.5000	0.2100	0.2300	0.2400	0.2450	0.2450	0.2500	0.2500	0.2500	0.2500	0.2500		
							0.5000	0.2100	0.2300	0.2400	0.2450	0.2450	0.2500	0.2500	0.2500	0.2500	0.2500			
						0.5000	0.2100	0.2300	0.2400	0.2450	0.2450	0.2500	0.2500	0.2500	0.2500	0.2500				
					0.4900	0.2058	0.2254	0.2352	0.2401	0.2401	0.2450	0.2450	0.2450	0.2450	0.2450					
				0.4900	0.2058	0.2254	0.2352	0.2401	0.2401	0.2450	0.2450	0.2450 (adjusted bit)	0.2450	0.2450						
			0.4800	0.2016	0.2208	0.2304	0.2352	0.2352	0.2400	0.2400	0.2400	0.2400	0.2400							
		0.4600	0.1932	0.2116	0.2208	0.2254	0.2254	0.2300	0.2300	0.2300	0.2300	0.2300								
	0.4200	0.1764	0.1932	0.2016	0.2058	0.2058	0.2100	0.2100	0.2100	0.2100	0.2100									
1.0000	0.4200	0.4600	0.4800	0.4900	0.4900	0.5000	0.5000	0.5000	0.5000	0.5000	1 (compensated bit)									

truncated bits (columns 7–0)

Fig. 10 The probability distribution of the mantissa multiplier partial product being 1 and the approximate design of truncated 8 columns

partial products array of mantissa multiplier and the approximate design of truncated 8 columns is shown in Fig. 10. The probability of each bit being 1 in the mantissa is different, resulting in a non-uniform probability distribution of the partial product being 1. Thus, after the mantissa multiplication is truncated, the error compensation according to the partial product probability distribution can effectively reduce the error. The truncated partial products are multiplied by the corresponding weight, and then the weighted probability sum is divided by the lowest reserved bit weight to obtain the compensation value. For example, when the last four columns are truncated, the corresponding probability compensation value (CPCV) is calculated as shown in Eq. (1). The result 0.766 is closer to 1, so a constant "1" is compensated in the fifth column after truncating 4 columns. As the number of truncated columns increases, the introduced error increases gradually. Thus, the FP multiplier with different precision can be realized by truncating different partial product column numbers.

$$\mathrm{CPCV} = \frac{1}{2^4}\left(\begin{array}{c} 0.25 \times 2^0 + 0.25 \times 2 \times 2^1 \\ + 0.25 \times 3 \times 2^2 + 0.25 \times 4 \times 2^3 \end{array}\right) = 0.766 \tag{1}$$

However, when the CPCV deviates greatly from the integer, large error will be generated by directly compensating "1" on the corresponding partial product column. For example, the CPCV for a design truncated 8 columns is 1.725, which is closer to the integer 2. If we directly compensate one "1" in the tenth column, the error due to overcompensation is $(2–1.725) \times 2^8 = 70.4$, which is equivalent to a 27.5% error in the column with a weight of 2^8. In order to further reduce the error, it's important to make full use of the advantages of probability analysis. We can adjust CPCV elaborately to balance the error caused by overcompensation. As shown in Fig. 10, for the approximate design that truncates the low-weight 8 columns partial product, one "1" is compensated in column 10, while a partial product with a probability of 0.245 in column 9 is discarded, that is, its probability is artificially adjusted to zero. Then, the error introduced by the overcompensation at this time becomes $(2–1.725 - 0.245) \times 2^8 = 7.68$, and the error generated on the column with a weight of 2^8 is reduced from 27.5% to 3%.

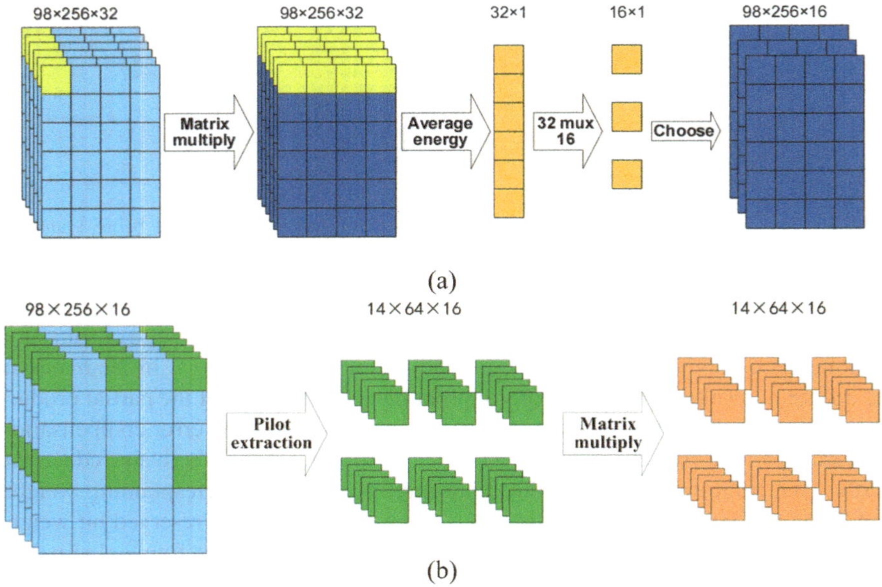

Fig. 11 The block diagram of the (**a**) BWT module and (**b**) channel estimation module

2.3 Application in Wireless Communication System

In this chapter, the proposed approximate design is applied to the BWT and channel estimation module in the wireless communication receiver. As shown in Fig. 11a, the BWT module selects half of the antenna data after the beam weight transformation with the highest energy for subsequent processing to reduce the complexity of receivers. The main operation in the BWT module is matrix multiplication. The channel estimation also contains large number of matrix multiplication as shown in Fig. 11b.

Therefore, the proposed approximate FP multiplier FPmul-T10 is applied to this module. The simulation results are shown in Fig. 12. With the adoption of the proposed approximate FP multiplier, the bit error rate (BER) of the whole communication system hardly deteriorated, about increasing 0.2 dB when SNR is 35 dB, proving that the proposed approximate design has high reliability and practical applicability.

3 Approximate FP FFT Processor

The FFT is a typical DSP unit and it can be realized with approximate computing for low-power consumption. Many approximate FFTs have been proposed to reduce

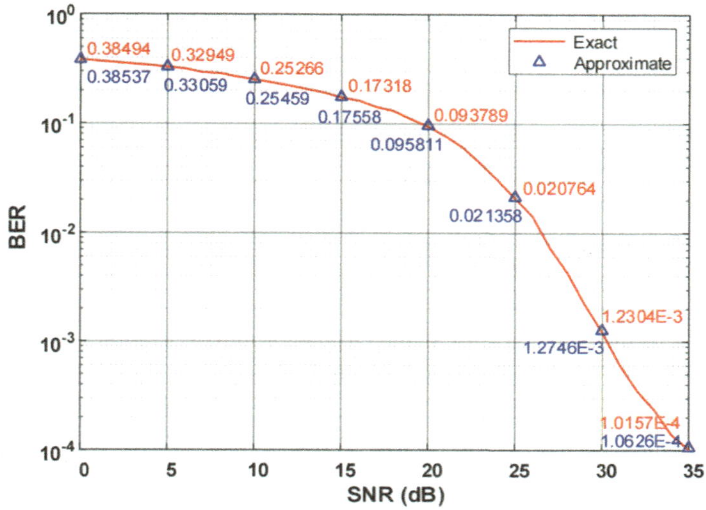

Fig. 12 The BER performance of exact and approximate multipliers adopted in MIMO-OFDM system

hardware complexity. Approximate Booth multipliers with different degrees of approximation are proposed and applied to low-power FFT designs [31]. Twiddle factors fusion architecture is proposed to reduce the number of multipliers in FFT, and the application of common sub-expression sharing architecture is proposed to reuse hardware resources [32]. In [31, 32], the approximate FFTs are realized by optimizing the basic units such as compressors or partial product generators. Nevertheless, the accuracy of these FFTs only can be obtained by simulations.

To design an FFT with a specific accuracy, the error sensitivity of the FFT is analyzed and two algorithms of bit-width selection are proposed [33]. However, the bit-width selection algorithms also require extensive simulation to obtain the final results. Moreover, previous most bit-width selection algorithms are mainly focused on the fixed-point FFT. The floating-point FFT which is widely used in DSPs and wireless communication systems has not been sufficiently studied for its more complex error characteristics. The standard floating point format is fixed and in most cases is redundant for accuracy requirements. In this section, the error sensitivity of FP FFT is analyzed and a mantissa bit-width adjustment algorithm is proposed for specific accuracy requirements. The proposed algorithm can significantly reduce the area and hardware resources of an approximate FFT circuit without extensive iterations.

3.1 DFT and FFT

For a sequence $x(n)$ with N points, its DFT transform [34] is given as follows:

$$X(k) = \sum_{n=0}^{N-1} x(n) W_N^{nk}, k = 0, 1, \ldots, N - 1 \tag{2}$$

$$W_N^{nk} = e^{-j\frac{2\pi nk}{N}}, k = 0, 1, \ldots, N - 1 \tag{3}$$

where $W_N{}^{nk}$ is the twiddle factor. In the DFT algorithm, N^2 complex multiplications and $N(N\text{-}1)$ complex additions are required to calculate the N-point DFT. The number of calculations increases significantly with increasing N. The FFT was proposed by Cooley and Turkey to reduce the complexity of the DFT algorithm [35]. It reduces the multiplication number of the DFT from N^2 to $N2 \log_2 N$. Firstly, In Eq. (2), $x(n)$ can be divided into two groups:

$$X(k) = \sum_{n=0}^{\frac{N}{2}-1} x(n) W_N^{nk} + \sum_{n=\frac{N}{2}}^{N-1} x(n) W_N^{nk}$$

$$= \sum_{n=0}^{\frac{N}{2}-1} x(n) W_N^{nk} + \sum_{n=0}^{\frac{N}{2}-1} x\left(n + \frac{N}{2}\right) W_N^{nk} W_N^{\frac{Nk}{2}} \tag{4}$$

$$= \sum_{n=0}^{\frac{N}{2}-1} \left(x(n) + x\left(n + \frac{N}{2}\right) W_N^{\frac{Nk}{2}}\right) W_N^{nk}$$

Thus, an N-point DFT is decomposed into two N/2-point DFTs. Similarly, A(n) and B(n) can be decomposed, as shown in Fig. 13. In terms of circuit implementation, the FFT architectures can be classified as the iterative FFT, the parallel FFT, and the pipelined FFT. The iterative FFT usually only includes a butterfly unit, a control unit, and memories. The parallel FFT is directly mapped from the butterfly scheme. The pipelined FFT includes multi-path delay commutator (MDC), single-path delay feedback (SDF), single-path delay commutator (SDC), and other structures. Among them, MDC and SDF are widely used for ease of control and medium throughput. Figures 14 and 15 show the 16-point radix-2 SDF (R2SDF) DIF-FFT and the 64-point radix-2^2 SDF (R2^2SDF) DIF-FFT.

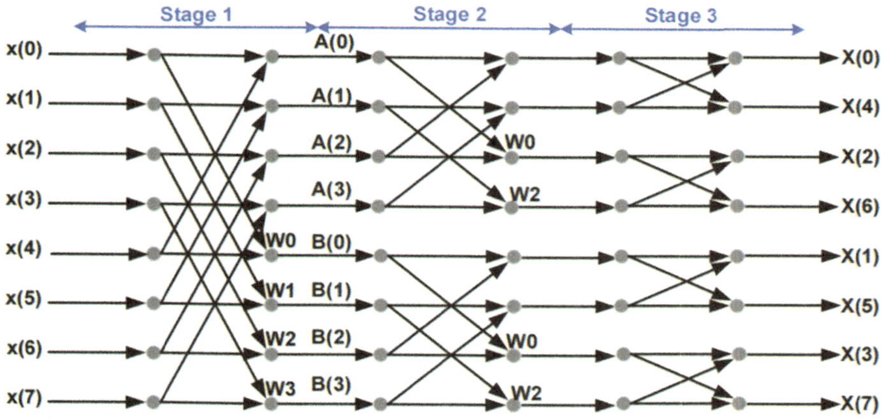

Fig. 13 8-point radix-2 DIF-FFT algorithm

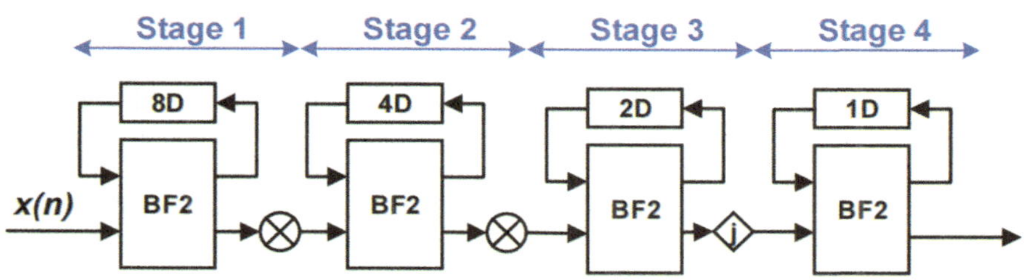

Fig. 14 16-point R2SDF DIF-FFT

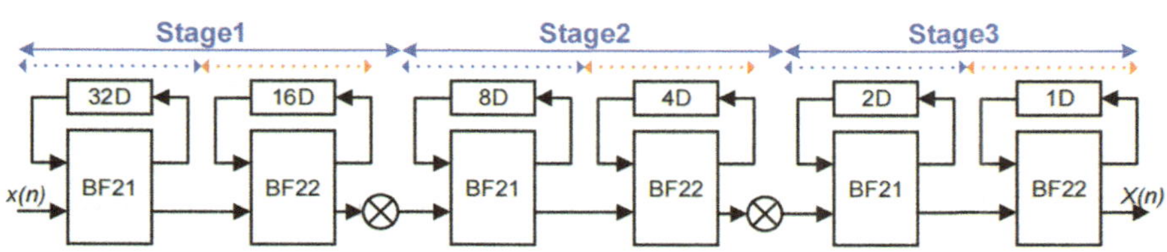

Fig. 15 64-point R2^2SDF DIF-FFT

3.2 *Mantissa Bit-Width Adjustment Algorithm*

For a floating-point number, the exponent determines the dynamic range, and the mantissa affects the precision. In addition, the mantissa bit-width is usually larger than the other parts and its operation requires more hardware resources. Truncation is a common and easy-to-implement approximation, which is often used in the design of arithmetic units, especially in bit-width optimization designs [33]. Therefore, the mantissa bit-width in floating-point FFT can be properly truncated to achieve the trade-off between precision and hardware resources.

Table 2 The error transfer characteristic in 8-point FFT

Output port	Stage 2	Stage 3
0	(e0 + e2) + (e1 + e3)	e0 + e1
1	(e0 + e2) – (e1 + e3)	e0 – e1
2	(e0-e2) + (e1-e3)W2	e2 + e3
3	(e0-e2) – (e1-e3)W2	e2 – e3
4	(e4 + e6) + (e5 + e7)	e4 + e5
5	(e4 + e6) – (e5 + e7)	e4 – e5
6	(e4-e6) + (e5-e7)W2	e6 + e7
7	(e4-e6) – (e5-e7)W2	e6 – e7

3.2.1 The Error Sensitivity of FFT

The error introduced by truncation is assumed to be additive white noise in this work. As shown in Fig. 13, the error introduced by the mantissa approximation for each port in Stage 3 will transfer to two output nodes, while the error introduced in Stage 2 and Stage 1 affect 4 and 8 output nodes, respectively. Thus, the error introduced in the previous stage affects more outputs. An approximate method with a zero-mean noise for the mantissa of a floating-point number, PAM, has been proposed to analyze the FP errors, and the conclusion is also suitable for the FP truncation [36].

In our work, the error sensitivity of the FP FFT is analyzed based on PAM approximation. Since the mean of the errors introduced by PAM is 0, we focus on the magnitude of the errors. The errors introduced by PAM in each stage are named as e0 e7 in order. After introducing errors at the input ports in Stage 2 and Stage 3, the errors at the output ports are given in Table 2. Obviously, in the case of an equal amount of error introduced in Stage 2 and Stage 3, the errors introduced in Stage 2 have a greater impact on the results. Similarly, the errors introduced in Stage 1 have the greatest impact. The error sensitivity of the N-point FFT is also the same as for the 8-point FFT. Therefore, the error sensitivity of each stage can be presented as follows: Stage1 > Stage2 > $\cdots$ > StageN. Under the condition of introducing an equal amount of error, stages with lower error sensitivity can truncate more bits and save more hardware resources.

3.2.2 The Mantissa Bit-Width Adjustment Algorithm

In terms of the accuracy, complexity, and maximum operating frequency of the FFT circuit, a high-efficiency mantissa bit-width adjustment algorithm for FP FFT design is proposed in this section, which is shown as Algorithm 1. According the noise propagation model of FP arithmetic operation units proposed in [36], the cumulative noise normalized to the mantissa LSB in FFT, CN, can be obtained with Eq. (5), where mult and add donate the operation name, and α and β denote the noise factors of the two operands. The initial noise factor is 1.This design uses a simplified complex multiplier architecture with 3 multipliers and 5 adders.

$$CN\,(\text{mult}, \alpha, \beta) = 1.18\,(\alpha + \beta) + 1$$

$$CN\,(\text{add}, \alpha, \beta) = (\alpha + \beta)\,/2 + 1 \tag{5}$$

For example, the CN of the 16-point R2SDF DIF-FFT is 18.2187, while the CN of the 64-point R2^{2}SDF DIF-FFT is 15.4696. The relationship between the SNR and the mantissa bit-width, λ, is shown in Eq. (6), where the mean value of the mantissa, M, is 1.5. Thus, the λ can be directly obtained with Eq. (6) for a specific SNR requirement. However, λ is an estimated value within 2 bits error range. Therefore, the default value of the mantissa bit-width is set to λ + 2 in step 1 of Algorithm 1 to ensure sufficient margin of accuracy at initial station. In step 2, the bit-width of each stage is optimized finely from the last stage according to the error sensitivity. Compared to the adjustment from the standard format, Algorithm 1 can reduce the time complexity by about half.

$$\text{SNR} = 10\log_{10}\frac{M^2}{3\frac{CN}{2^{2\lambda}}} \tag{6}$$

To better express the proposed algorithm, two single precision format FFTs with different points and different accuracy requirements are designed. For the 64-point R2^{2}SDF DIF-FFT with the quality restriction set to larger than 60 dB, [23 23 23] is the stander bit-width of the mantissa. [13 13 13] is derived from Eq. (5). [12 12 12] is chosen in step 1 and [12 12 11] is finally obtained in step 2. Similarly, [11 11 11 10] is designed for the 16-point R2SDF DIF-FFT by employing the proposed algorithm. Obviously, by employing Algorithm 1, the mantissa bit-width of the FFT is significantly reduced on the promise of ensuring the accuracy requirements.

Algorithm 1: The Mantissa Bit-Width Adjustment Algorithm

```
 Input:
Required accuracy: SNR > α dB
mant_width_model = λ
s: Number of stages of N-point FFT
 Output:
 Mant_adj: The bit-width combination after adjustment
 SNR_curr: The accuracy of current output
step 1:
Set Mant_adj = λ + 2;
Generate Gaussian distributions as inputs to the FFT
for Mant_adj > 0 do
  FFT SNR calculation (SNR_curr)
  if SNR_curr ≥ αthen
   Mant_adj = Mant_adj - 1;
  else
```

(continued)

```
      Mant_adj = Mant_adj + 1; break
    end if
  end for
  step 2:
  for s > 0 do
    for Mant_adj[s] ≥ 0 do
      FFT SNR calculation (SNR_curr)
     if SNR_curr ≥ αthen
      Mant_adj[s] = Mant_adj[s] - 1;
     else
      Mant_adj[s] = Mant_adj[s] + 1; break
     end if
    end for
    s = s - 1;
  end for
```

3.3 Approximate FP FFT Processor in Channel Estimation

Due to the channel noise and interference, the wireless communication system is inherently error-tolerant, which means that approximate computing can be employed to reduce its hardware complexity. The 64-point FFTs with SNR not less than 60 dB are adopted in the channel estimation module of 5G wireless communication receivers without forward error correction code Forward Error Correction. The architecture of the channel estimation module is shown in Fig. 16 [37].

The channel estimation module first estimates the channel information using the least squares algorithm with pilots' data. Secondly, the estimated coefficients will be converted to the time domain and truncated its channel impulse response for noise reduction. Then, the FFT module will transform the time domain information back to the frequency domain for interpolation to obtain the complete channel coefficients

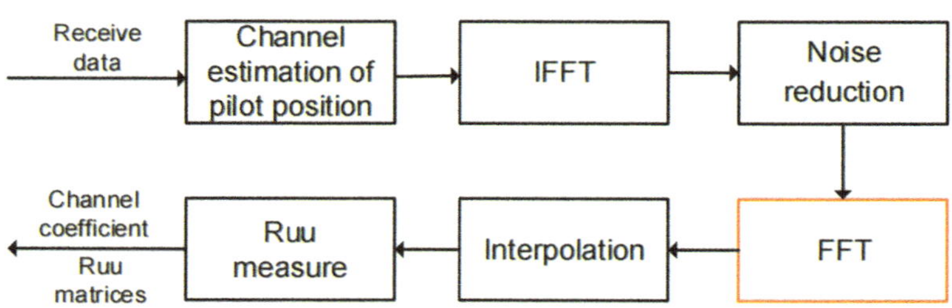

Fig. 16 Channel estimation with time-domain noise reduction module

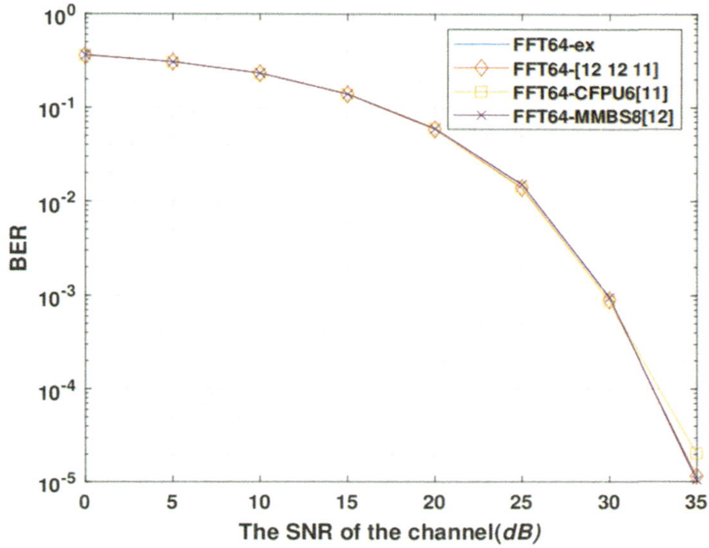

Fig. 17 The BER performance of exact and approximate FFTs in communication system

H. In our work, approximate design is adopted in this step. Finally, the Ruu_measure module calculates the interference and noise covariance matrix Ruu from H and sends H and Ruu to the equalization module. Figure 17 shows the simulation results with different SNRs of the channel. Obviously, the proposed approximate design with 60 dB SNR has a neglectable impact on the BER performance compared to the exact design. The difference in BER between FFT64-[12 12 11] and FFT64-ex is less than 10%. And in most required SNRs, the difference is less than 1%. However, FFT64-CFPU6 with a better SNR has a higher BER. This is mainly because the Gaussian distribution data is used as input for FFT design, while the input of the FFT in the system simulation is the output data of the noise reduction module. Therefore, the proposed design can be applied to a wireless communication system. And other modules can also take advantage of the idea of approximate computation to reduce hardware complexity.

4 Approximate Polar Decoder

Polar codes, utilizing channel polarization for binary input discrete memoryless channels (B-DMCs) [38], have been proved to be one of the first capacity-achieving codes and has received significant attention. In the fifth generation (5G)- enhanced mobile broadband (eMBB) scenarios [39], the polar codes are finally identified as the coding scheme for the control channel. Successive cancellation (SC) decoding and belief propagation (BP) decoding are two main decoding algorithms for the polar codes. Due to its serial architecture, the hardware complexity of the SC algorithm is low, leading to limited throughput. On the other hand, the BP algorithm [40] can achieve much higher throughput while costing more hardware consumption. However, the performance of the two decoding algorithms also depends on the SNR scenarios. In a high SNR scenario, which has fewer iterations in BP decoding, the performance of the BP decoding algorithm is better than the SC decoding, while in a low SNR scenario [41], the SC decoding delivers better performance.

From the performance perspective of the SC decoding design, the throughput improvement is the primary target. The successive cancellation list (SCL) algorithm [42] was proposed to improve the throughput of the SC decoder at the expense of power and area consumption. Based on the SCL algorithm, the successive cancellation stack algorithm [43] and hybrid successive cancellation algorithm [44] were developed to improve error-correction performance by further increasing the complexity and reducing the throughput. The error-correction performance of cyclic redundancy check (CRC)-SCL scheme can outperform LDPC codes with a comparable code length.

Due to the noise in the wireless channel and its inherent error correction function, the forward error correction decoder can be regarded as an error-tolerant system, and the full computing accuracy is not always necessary. Consequently, applying approximate computing to the decoder design has been considered a promising solution for trade-offs between decoding performance and hardware cost. In prior works, approximate computing has already been successfully explored in LDPC decoders [45], BP polar decoders [46], and fast SSC polar decoders [47] respectively. However, most approximations in the decoders are realized by directly truncating the lower significant bits, which hinder further hardware efficiency improvement with specific accuracy requirement.

4.1 Approximate SC Polar Code Decoder

Polar codes are based on the theory of channel polarization when transmitted over binary input discrete memoryless channel (B-DMC). N identical B-DMCs can be divided into mostly reliable and mostly unreliable channels through channel combination and channel splitting. The K channels with "1" capacity transmit information bits, and the N − K unreliable ones transmit frozen bits which are known, usually zeros. The indices of bit-channels to transmit information bits are in set $I \in \{1, 2, \ldots, N\}$. Correspondingly, those bit channels whose indices are in the complement set I^c are used to transmit the frozen bits [48]. Generally, PC(N,K) is used to represent polar codes in the text, and the rate of polar SC decoding of polar codes estimates transmitted bits u_0 to u_{N-1} sequentially at step i when i is not in the frozen set.

The root of the tree contains the log-likelihood ratio (LLR) from the channel observation. In Fig. 18, the LLR values (denoted by α) contained in the parent nodes

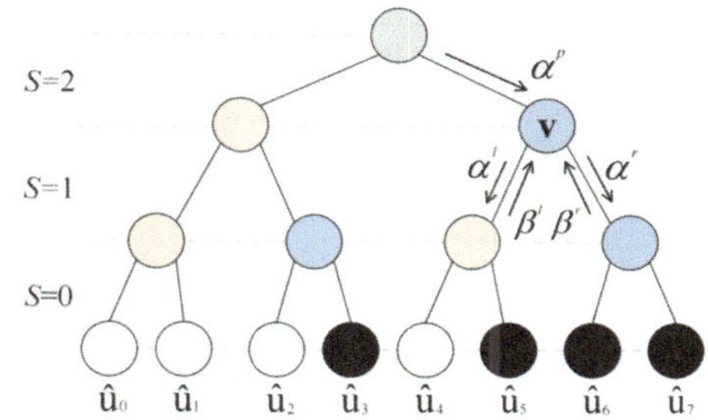

Fig. 18 The SC decoding tree for PC (8,4)

pass to the child node. In return, child nodes pass partial sum values (denoted by β) to their parent nodes. For a parent node p that contains LLR values, the values passed to left, and right child nodes are calculated as:

$$\alpha_i^l = \text{sgn}\left(\alpha_i^p\right) \text{sgn}\left(\alpha_{i+2^{S-1}}^p\right) \min\left(\left|\alpha_i^p\right|, \left|\alpha_{i+2^{S-1}}^p\right|\right) \tag{7}$$

$$\alpha_i^r = \alpha_{i+2^{S-1}}^p + \left(1 - 2\beta_i^l\right) \alpha_i^p \tag{8}$$

The operations of Eqs. (7) and (8) are well known as the operations of f and g nodes, where *sgn* is the sign operation and β represents the partial sum calculated by bitwise XOR operation or wired-and from the left β^l and right β^r child nodes during the backward process.

LLR values are passed from the root node to the leaf nodes and partial sum values are fed back all the way between different stages. The conventional SC decoding tree butterfly structure of PC (8, 4) is depicted in Fig. 19, which shows the feedback signals of f and g nodes in detail [49]. It requires 14 clock cycles to finish the decoding operation, where f and g functions are along forward trace, and the dotted lines indicate the feedback signals from partial sums to the previous stage. The labels attached to each node indicate the index of clock cycle when the responding stage S is activated.

4.1.1 Processing Element for Decoding Intermediate Stages

The PE unit processes two input LLRs with *f* and *g* function to generate output LLR in each stage. There are three types of data formats used in previous PE designs, which are redundant LLR representation (RR) [50], two's complement (2CR) representation [51], and sign-magnitude representation (SMR) [52]. The sg control signal decides the g function value, and the *fg* control signal decides the final LLR output in all three cases in Fig. 20.

RR: In the RR format, the first bit is one additional bit called negating bit and the other bits are the specific value encoded according to two's complement format.

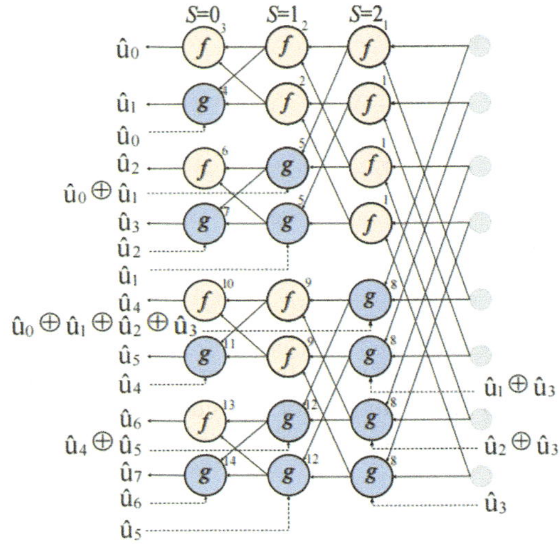

Fig. 19 The conventional butterfly structure for PC (8,4)

As shown in Fig. 20 (a), the negating bit and the value of LLR are represented by x.neg and x.val. The value of f function can be obtained by a.val without any negations, whereas the results of g function can be obtained by b.val ± a.val. Final output is computed based on fg control signals. Its hardware complexity and critical path delay are *2Cadder + 5Cmul + 5Cxor* and *Tadder + 3Tmul + 2Txor*, respectively. *Cadder, Cmul, Cxor* and *Tadder, Tmul, Txor* represent the complexity and logic delay of adder/subtractor, 2:1 multiplexer, and xor gate, respectively. It is noteworthy that the RR decoder costs more registers for the redundant LLRs in architecture which needs more PEs. This is not friendly for pipelining architecture, in which area is proportional to the increased registers [51].

2CR: In the case of 2CR, the first bit is sign bit, and the basic arithmetic operation can be processed directly. Both the received LLRs (LLR (a) & LLR (b)) are first added to determine whether it is needed to negate them or not (Fig. 20b). If this sum is greater than or equal to zero, then the minimum value of input LLRs is calculated for f function. Otherwise, we should negate them and determine the minimum one. The g function and the final LLR output are computed based on sg and fg control signals. The critical path depends on the f function since the absolute value should be obtained by the additional operation. The complexity and critical path delay are *2Cadder + 5Cmul + 2CIAOU* and Tadder +3Tmul, respectively. The function of inversion-and-adding-one unit (IAOU) is for negation process by inverting each bit of the operand and adding '1' at the bit of the lowest weight position. The size of adders used in 2CR architecture [51] is also 1 bit larger than SMR scheme [52] to achieve the same quantitative accuracy.

SMR: In the case of SMR, the first bit is sign bit and other bits are magnitude bits. The f function can be operated directly for the magnitude of received LLRs using comparator; however, it needs a conversion circuit before the adder or subtractor, which means the g function calculation path determines the critical path. The g result

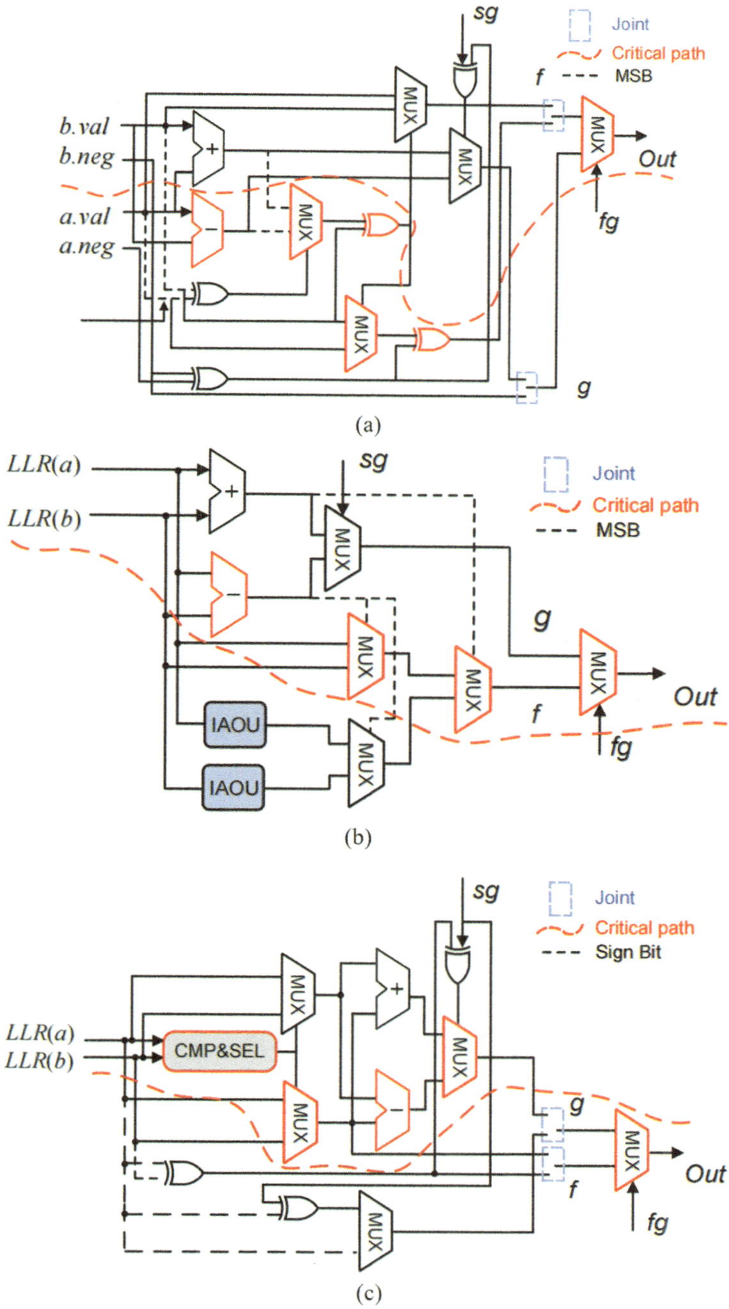

Fig. 20 The previous different architecture of PEs: (**a**) RR [50] (**b**) 2CR [51] (**c**) SMR [52]

is decided by sg and the sign of inputs, and final LLR output is computed based on fg control signals. As shown in Fig. 20c, it consists of 2 adders, 1 comparator, 5 2:1 multiplexers, and 3 XOR gates. The complexity can be represented as *2Cadder + Ccomparator + 5Cmul + 3Cxor.* Moreover, the critical path delay is *Tcomparator + Tadder + 3Tmul.*

4.1.2 The Proposed Low-Complexity Approximate PE

This chapter employs the sign-magnitude format for its simple f operation, which is friendly for approximate circuit design for g node [46, 47]. In our design, the operation of the g node is divided into two categories according to the partial sum control signal sg and the algorithm of the g node. Sa, Sb, Ss represent the sign of LLR (a), LLR (b), and the output of g node, respectively. Ma, Mb, Ms. represent the magnitude of LLR (a), LLR (b), and the output of g node, respectively.

As shown in Fig. 22, the proposed low-complexity PE merges addition and subtraction operations into an adder-subtractor. The subtract operation occurs when only one or three of Sa, Sb, sg are 1. Thus, we derive a control signal cin to determine whether the g node executes subtraction. The architecture of the proposed adder-subtractor is shown in Fig. 23. If cin $=$ 1, the inversion of Mb is selected, for Ma $-$ Mb $=$ Ma $+$ ($-$Mb). The adder-subtractor has two outputs. One is the sum or difference of the magnitude of LLR (a) and LLR (b), and the other is a carry output called cout to show whether the magnitude of LLR (a) is greater than LLR (b). Meanwhile, the sign of g result is determined by the Eq. (9), which is demonstrated as the module I in Fig. 21.

$$
\begin{aligned}
S_S = {} & \overline{\Big(\overline{S_a \bullet S_b} \bullet \overline{S_a \bullet \overline{S_b} \bullet cout} \bullet \overline{\overline{S_a} \bullet S_b \bullet \overline{cout}}} \bullet \overline{sg} \\
& + \overline{\Big(\overline{S_a \bullet \overline{S_b}} \bullet \overline{\overline{S_a} \bullet \overline{S_b} \bullet \overline{cout}} \bullet \overline{S_a \bullet S_b \bullet cout}} \bullet sg \\
= {} & \overline{\overline{op1 \bullet \overline{sg}} \bullet \overline{op2 \bullet sg}}
\end{aligned} \tag{9}
$$

It consists of 4 inverters and 11 NAND gates. Note that the area of an XOR gate is about 2.3 times that of a NAND gate, and the delay of an XOR gate is about 3 times that of a NAND gate. Compared with the design in Fig. 20c, the proposed low-complexity PE replaces 1 XOR gate with 1 AND gate and 1 inverter. Meanwhile, the number of 2:1 multiplexers reduces from five to three, including a multiplexer in the adder-subtractor. The overflow judgment part in subtraction is also omitted as shown in Fig. 22. The IAOU is needed in the design; however, it can be approximated to 4 inverters. This design can significantly reduce the complexity of the PE units compared to previous designs.

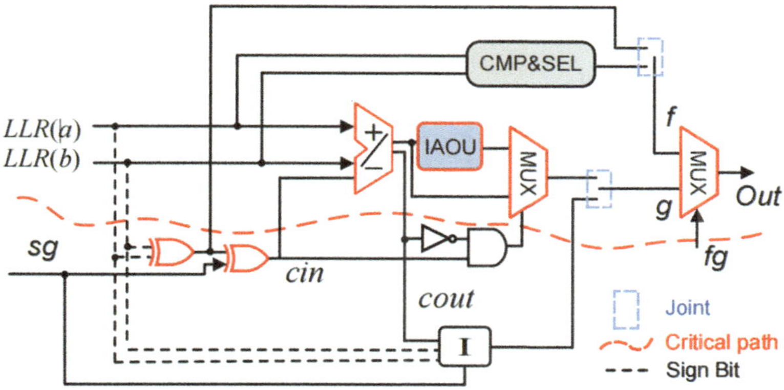

Fig. 21 The proposed low-complexity PE

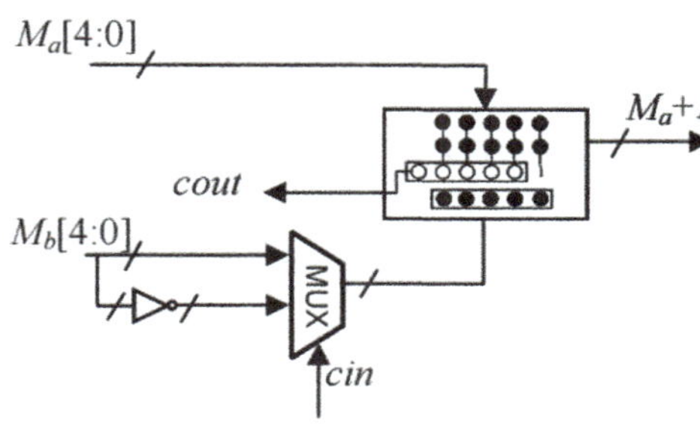

Fig. 22 The proposed low-complexity adder-subtractor

1. Approximate comparator: In conventional comparators, operands are processed by NOT gates and NAND gates to find which bits are not equal, and the operand with the first different bit that equals 1 is the bigger number. There are several approximate comparators that have been proposed with direct truncation [46]. However, a large error will be introduced by truncation directly. The accuracy of truncating 1 bit is acceptable for its error distance is only 1 with 25% probability. However, truncation of 2 bits will cause a non-negligible error [47]. Thus, the entries for lower two or three bits that need to be compared in the Karnaugh map are listed in Fig. 23a. When Ma < Mb, the output of comparator is 1, and LLR(a) will be chosen directly for the output of f node. Changing specific bit from "1" to "0" or vise versa can simplify logical expression as shown in the shaded part in the figure. For example, when 2 bits are approximated, only two entries will generate false results, as shown in the shaded part in Fig. 23a. Meanwhile, only four entries are changed when three bits are approximated, as shown in Fig. 23b. The error probability is 2/16 = 12.5% for two approximate bits and 4/64 = 6.25% for three approximate bits. Most importantly, the maximum error distances for both schemes are only 1.

$$\text{Out1} = \overline{M_a\,[1]}$$

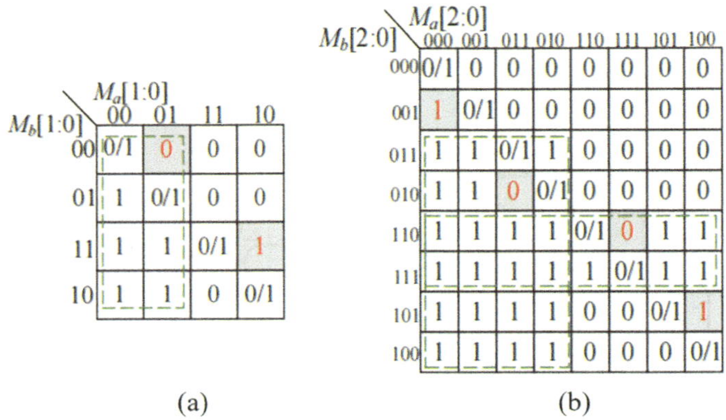

Fig. 23 The Karnaugh map of proposed approximate comparators: (**a**) 2 bits (**b**) 3 bits

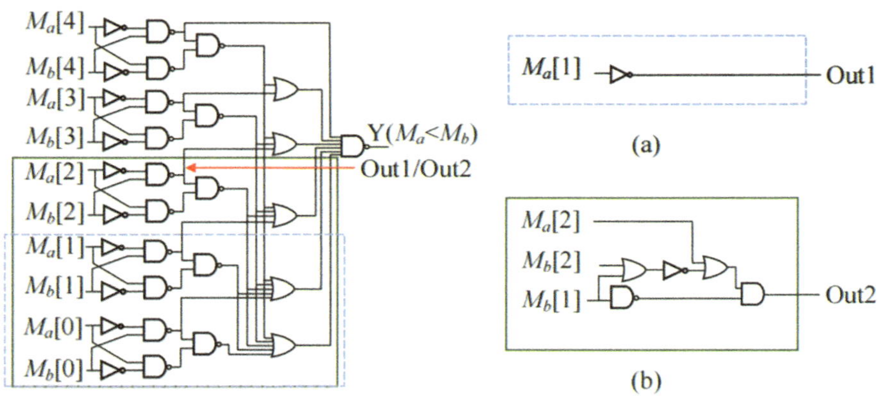

Fig. 24 The circuit of the proposed comparators for approximating (**a**) 2 and (**b**) 3 bits

$$\text{Out2} = \left(M_a\,[2] + \overline{M_b\,[2] + M_b\,[1]}\right) \bullet \overline{(M_b\,[2] \bullet M_b\,[1])} \tag{10}$$

The schematics of the proposed approximate comparators are shown in Fig. 24. The approximate 2 bits schemes shown in Fig. 24a only need an inverter to replace the circuit in dotted line in the exact ones. The approximate 3 bits scheme needs 1 NAND gate, 1 inverter, 1 AND gate, and 2 OR gates to replace the circuit in a solid box shown in Fig. 24b. The Boolean algebras from the Karnaugh maps and circuits are shown in Eq. (10).

Both the two proposed approximate comparators have smaller area and delay. The proposed approximate 3 bits comparator has achieved 37.15% area and 29.41% delay improvement compared with direct truncation 1 bit design, and the area

Table 3 The probability of amplitude bits being 1

Weights	M_a [4]	M_a [3]	M_a [2]	M_a [1]	M_a[0]
Probability	0.7502	0.6329	0.6149	0.5655	0.5098
Weights	M_b [4]	M_b [3]	M_b [2]	M_b [1]	M_b[0]
Probability	0.7508	0.6320	0.6139	0.5642	0.5082

and delay improvements of approximate 2 bits design are 12.36% and 11.76%, respectively.

2. Approximate adder and subtractor: In this work, we propose an approximate adder and subtractor by truncating the less significant bits of the operands and by compensating based on the input data distribution. The probabilities of the truncating bits are obtained by analyzing the distribution characteristics of the input data of the g node. Table 3 shows the probability of per bit being 1 with SNR equation to 4.5. It has been demonstrated that truncating 3 bits in the adder or subtractor will negatively impact the BER performance. In our design, only the lower two bits are considered for approximation design to guarantee performance. From Table 3, it can be seen that the probabilities of the lower two bits in the two operands are all a little bigger but close to 0.5. Therefore, the lower two bits in adder and subtractor can be directly set to "11" to save hardware consumption. With few hardware consumption overheads, the MSE can be significantly reduced through constant compensation. The MSE of adder with truncating only and the proposed approximate adder are 0.3597 and 0.19, respectively.

4.1.3 Overall Decoder Architecture

The tree-based architecture has low latency and high achievable throughput among SC decoders [53]. In this work, the 2b-SC polar decoder with $N = 1024$ based on tree architecture using the proposed exact and approximate PEs is implemented as shown in Fig. 25. p node is adopted to replace the common PEs at the last stage to enhance the throughput by decoding two bits in one cycle with critical path delay reducing to $Tcomparator + 2T_{AND} + 2T_{OR}$. The comparator in the p node can also be designed with approximate scheme. In the decoder, the partial sum accumulator (PSA) accumulates the partial sums from decoded bits generated by the p node by XOR operations. There are nine stages of intermediate PEs (a total of 1022 intermediate PEs). The last stage consists of the p node that computes two LLRs in each cycle. The selection of f and g functions is controlled by fg and sg, respectively. The controller determines whether the decoded bits are accumulated or not and selects one of the outputs of each PE. The number of registers of the 2b-SC decoder is proportional to the length of polar code N and the total number of decoding stages Ns (i.e., 10 in our decoder). The size of each register depends on the bit-width of received LLRs, which is 6 in our design.

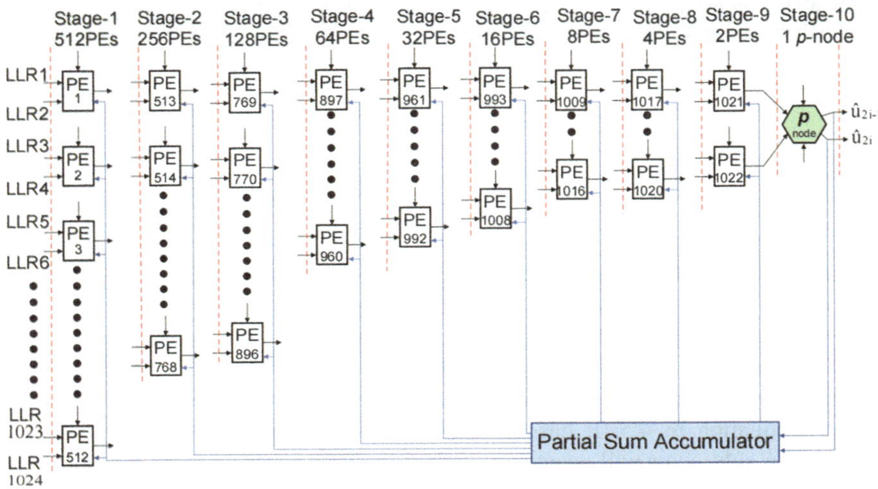

Fig. 25 The proposed overall design of 2b-SC polar decoder based on tree architecture of $N = 1024$

The BER and FER performances using approximate PEs designs are shown in Fig. 26. It can be found that the BER and FER performances with approximate PEs are almost same as the schemes with the exact fixed-point 5-bit for $N = 1024$, which is close to the floating-point scheme.

From the simulation result after synthesized, that the approximate design has effectively reduced the delay, area, and power consumption. The power-delay-product (PDP) and area of the proposed approximate design are reduced by 46.70% and 29.95%.

4.2 Approximate BPF Polar Code Decoder

The BP algorithm decodes in parallel, which is a good choice for low-latency and high-throughput applications. However, its error correction ability cannot reach the performance of SC list (SCL) algorithm. To solve this problem, the academic researches mainly include two directions: modifying the channel construction algorithm and improving the decoding scheme. In recent years, the genetic algorithm-tailored construction [54] for explicit decoding schemes and information bits selection method based on Monte Carlo simulation [55] are two encoding construction algorithms with significant error correction performance improvements. For the decoding scheme, influenced by the SC flip (SCF) [56] decoder, [57] introduces bit-flip into BP decoder. Such strategy can approach the performance of SCL-16 decoding through additional hundreds of decoding attempts. However, it does not reflect the concept of flipping for trying to assign

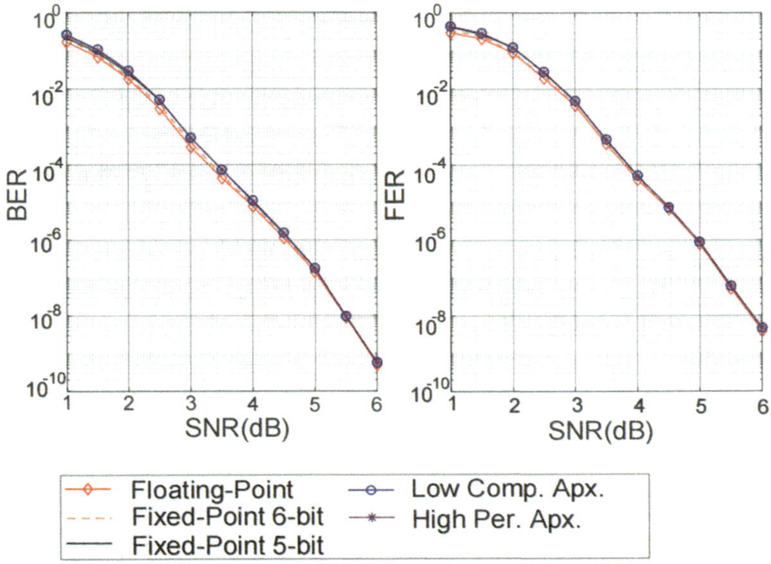

Fig. 26 The error-correction performance using the proposed approximate designs of SC decoder for PC (1024,512)

each error-prone bit to both "0" and "1." [58] proposed generalized BPF (GBPF) decoding with a redefinition of flip set construction based on the absolute value of LLR. The enhanced BP flip (EBPF) decoding halves the searching range for lower complexity with error correction performance deterioration. Zhang et al. [59] proposed an advanced BPF (A-BPF) scheme that reduces the decoding latency with the help of one critical bit and improves the error correction performance by offset min-sum (OMS) function and joint detection criterion. An optimized sorting network was also proposed in A-BPF to improve area efficiency. However, the genetic algorithm used in A-BPF does not measure the channel differences between its internal sub channels, and its complexity is very high.

1. The proposed approximate LLR sorter

In the BPF decoder as shown in Fig. 27, LLR sorter is used to select the smallest T disordered LLRs in the form of absolute values. Conventionally, bitonic merge sorting algorithm and multiplexing scheme are commonly used methods. A 256-in-16-out sorter can be realized by multiplexing a 64-in-16-out sorter five times and uses shift registers to store decision LLRs, flip signs, and bit indices within the search set. In our sorter, the parallel inputs pipelined sorter is adopted to reduce latency. It consists of two compare layers, including four in the first layer and one in the second layer, and each layer contains 6 stages. It is worth noting that the collection of flip set is an error-tolerant process inherently for the uncertainty of bit selection. Therefore, the approximate computing can be explored in the LLR sorter

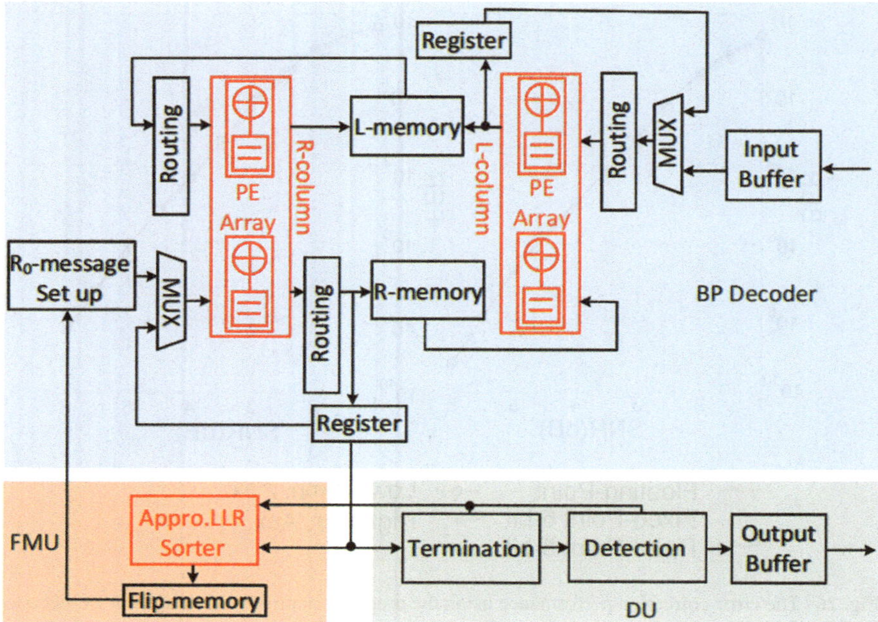

Fig. 27 The hardware architecture of BPF decoder

design. Since data in the second layer will be reordered, the first layer just needs to ensure that a minimum of 16 values are selected, which leads to more headroom for approximation in the first layer design. The approximate comparator truncated 1 least significant bit (LSB) to further reduce the latency. For example, for a 7 bit data if A[6: 1] > = B[6, 1], B is chosen as the minor one.

Table 4 shows the effect of approximate comparators on the accuracy of bitonic merge sorting networks. Per1 and Per2 are the probability that the 16 values selected are in the correct order for the results of the first layer and for the overall output of the second layer, respectively. In the table, Appro.1 Exa.2 denotes that the four 64-in-16-out sorters in the first layer are approximated and the sorter in the second is exact and vice versa. Appro.123 and Appro.6 denote only applying approximation to the first three stages and the last stage of the second layer. This comparison shows that errors can be compensated in the subsequent propagation process both in inter-layer and intra-layer based on the analysis using the same number of approximate comparators. So, in this chapter, we adopt the approximation of first layer and first three stages of second layer.

2. PE Arrays

Two kinds of PE arrays (Type 1 and Type 2) calculating g(LLR(b), LLR(c)) + LLR(a)), and g(LLR(b) + LLR(c), LLR(a)), are employed in the L-column and R-column with different offset factors β^L, β^R = *0, 0.25*

Table 4 The effect of approximate comparators on LLR sorter

		Per1	Per2
Inter-layer	All exact	1	1
	All Appro.	0.7094	0.5583
	Appro.1Exa.2	0.7102	0.9975
	Exa.1Appro.2	1	0.7012
Inter-layer	Appro.123	1	0.9922
	Appro.6	1	0.8109

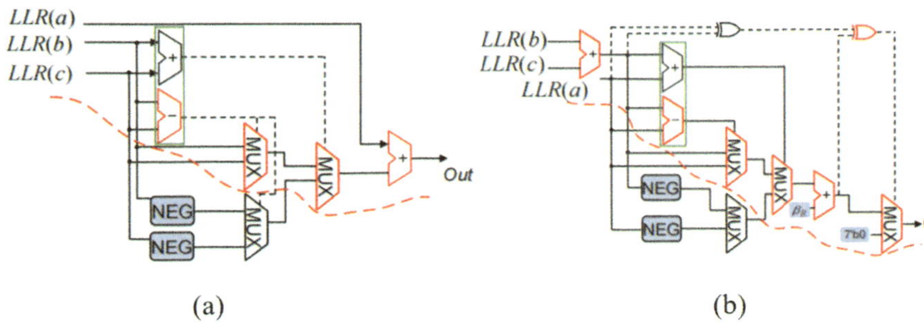

Fig. 28 Schematic of the proposed PEs: (**a**) Type 1 and (**b**) Type 2

[8]. The g function approximates the sum-product, which can be denoted as $g(a,b) = sign(a)sign(b)max(min(|a|,|b|)\text{-}\beta, 0)$. However, the previous PEs have two conversions of number representation between 2's complement and sign-magnitude (C2S/S2C) form in the critical path leading to longer latency. In this chapter, we propose a new low-latency Type 1 and Type 2 PEs by removing conversions. Take the Type I PE for example, when the inputs LLR (b) $= -2$ and LLR(c) $= 1$, the most significant bit (MSB) of their summation is 1, so we select the minimum of their negation to add with LLR (a). The schematic of the proposed PEs is shown in Fig. 28.

The adder and subtractor in green box only calculate carry bit resulting in saving two XOR gates for one bit. Totally, 27 and 21 nand gates are saved in proposed Type I and Type II PE, respectively. Table 2 presents the hardware performance comparison of the proposed PEs with previous work. The proposed Type I and Type II PEs improve critical path delay by 22.4% and 18.4%, respectively. More importantly, the proposed Type I PE also saves 15.4% area.

5 Conclusion

This chapter introduces the application of approximate computing in wireless communication systems. It mainly includes approximate floating point adder design with mantissa truncation and compensation method, low-complexity floating-point multiplier design with partial product location adjustment and method based on partial product probability compensation; a top-down FP FFT processor design strategy for determined accuracy requirement. All the above approximate circuits are verified in a 4 $\times$ 32 MIMO OFDM system, which achieves significant hardware efficiency enhancement with negligible BER performance deterioration. At last, two approximate polar code decoders are introduced. With approximate PE units or approximate sorter design, the hardware efficiency increased more than 30%. The application of approximate computing in wireless communication systems can be further studied in terms of joint system and circuit approximation design and the approximation errors transfer function in top level systems.

References

1. J. Chen, J. Hu, G.E. Sobelman, Stochastic MIMO detector based on the Markov chain Monte Carlo algorithm. IEEE Trans. Signal Process. **62**(6), 1454–1463 (2014)
2. J. Yang, C. Zhang, S. Xu, et al., *Efficient stochastic detector for large-scale MIMO* (IEEE International Conference on Acoustics, Speech and Signal Processing, 2016), pp. 6550–6554
3. C. Yan, Y. Cui, K. Chen, et al., *Hardware Efficient Successive-Cancellation Polar Decoders Using Approximate Computing[J]* (IEEE Journal on Emerging and Selected Topics in Circuits and Systems, 2023)
4. Z. Zhou, J.L. Chen, Z. Wang, A high-speed successive cancellation decoder for polar codes using approximate computing. IEEE Trans. Circuits Syst. II, Exp. Briefs **66**(2), 227–231 (2019)
5. J. Janhunen, T. Pitkanen, O. Silven, et al., Fixed-and floating-point processor comparison for MIMOOFDM detector[J]. IEEE J. Sel. Top. Signal Proc. **5**(8), 1588–1598 (2011)
6. S.Z. Gilani, N.S. Kim, M. Schulte, *Energy-Efficient Floating-Point Arithmetic for Software-Defined Radio Architectures[C]* (Proceedings of ASAP 2011-22nd IEEE International Conference on Application-specific Systems, Architectures and Processors. IEEE, 2011), pp. 122–129
7. C. Senning, A. Burg, *Block-floating-point enhanced MMSE filter matrix computation for MIMOOFDM communication systems[C]* (Proceedings of 2013 IEEE 20th International Conference on Electronics, Circuits, and Systems (ICECS). IEEE, 2013), pp. 787–790
8. S. Amin-Nejad, K. Basharkhah, T.A. Gashteroodkhani, Floating point versus fixed point trade offs in FPGA Implementations of QR Decomposition Algorithm[J]. Eu. J. Elect. Eng. Comput. Sci. **3**(5) (2019)
9. Y. Hu, M. Koibuchi, *Accelerating MPI Communication Using Floating-point Compression on Lossy Interconnection Networks[C]* (Proceedings of 2021 IEEE 46th Conference on Local Computer Networks (LCN). IEEE, 2021), pp. 355–358
10. H. Jiang, J. Han, F. Lombardi, *A Comparative Review and Evaluation of Approximate Adders[C]* (Proceedings of Proceedings of the 25th edition on Great Lakes Symposium on VLSI, 2015), pp. 343–348
11. S.L. Lu, Speeding up processing with approximation circuits[J]. Computer **37**(3), 67–73 (2004)
12. A.K. Verma, P. Brisk, P. Ienne, *Variable Latency Speculative Addition: A New Paradigm For Arithmetic Circuit Design[C]* (Proceedings of Proceedings of the Conference on Design, Automation and Test in Europe, 2008), pp. 1250–1255
13. A.B. Kahng, S. Kang, *Accuracy-Configurable Adder for Approximate Arithmetic Designs[C]* (Proceedings of Proceedings of the 49th Annual Design Automation Conference, 2012), pp. 820–825

14. J. Miao, K. He, A. Gerstlauer, et al., *Modeling and synthesis of quality-energy optimal approximate adders[C]* (Proceedings of Proceedings of the International Conference on Computer-Aided Design, 2012), pp. 728–735
15. N. Zhu, W.L. Goh, G. Wang, et al., *Enhanced low-power high-speed adder for error-tolerant application[C]* (Proceedings of 2010 International SoC Design Conference. IEEE, 2010), pp. 323–327
16. K. Du, P. Varman, K. Mohanram, *High performance reliable variable latency carry select addition[C]* (Proceedings of 2012 Design, Automation & Test in Europe Conference & Exhibition (DATE). IEEE, 2012), pp. 1257–1262
17. C. Lin, Y.M. Yang, C.C. Lin, High-performance low-power carry speculative addition with variable latency[J]. IEEE Trans Very Large Scale Integr VLSI Syst **23**(9), 1591–1603 (2014)
18. Z. Yang, A. Jain, J. Liang, et al., *Approximate XOR/XNOR-based adders for inexact computing[C]* (Proceedings of 2013 13th IEEE International Conference on Nanotechnology (IEEE-NANO 2013). IEEE, 2013), pp. 690–693
19. V. Gupta, D. Mohapatra, A. Raghunathan, et al., Low-power digital signal processing using approximate adders[J]. IEEE Trans. Comput.-Aided Des. Integr. Circuits Syst. **32**(1), 124–137 (2012)
20. H.R. Mahdiani, A. Ahmadi, S.M. Fakhraie, et al., Bio-inspired imprecise computational blocks for efficient VLSI implementation of soft-computing applications[J]. IEEE Trans. Circuits Syst. I: Regul. Pap. **57**(4), 850–862 (2009)
21. S. Venkataramani, A. Ranjan, K. Roy, A. Raghunathan, Axnn: Energy-efficient neuromorphic systems using approximate computing, in 2014 IEEE/ACM International Symposium on Low Power Electronics and Design (ISLPED), pp. 27–32 (2014)
22. M. Imani, S. Patil, T. S. Rosing, Masc: Ultra-low energy multipleaccess single-charge tcam for approximate computing, in 2016 Design, Automation Test in Europe Conference Exhibition (DATE), pp. 373–378 (2016)
23. H. Zhang, W. Zhang, J. Lach, A low-power accuracy-configurable floating point multiplier, in 2014 IEEE 32nd International Conference on Computer Design (ICCD), pp. 48–54 (2014)
24. M. Imani, D. Peroni, T. Rosing, Cfpu: Configurable floating point multiplier for energy-efficient computing, in 2017 54th ACM/EDAC/IEEE Design Automation Conference (DAC), pp. 1–6 (2017)
25. M. Imani, R. Garcia, S. Gupta, T. Rosing, Rmac: Runtime configurable floating point multiplier for approximate computing, in Proceedings of the International Symposium on Low Power Electronics and Design. Association for Computing Machinery, pp. 12–17 (2018)
26. S. Hashemi, R. I. Bahar, S. Reda, Drum: A dynamic range unbiased multiplier for approximate applications, in 2015 IEEE/ACM International Conference on Computer-Aided Design (ICCAD), pp. 418–425 (2015)
27. S. Narayanamoorthy, H.A. Moghaddam, Z. Liu, T. Park, N.S. Kim, Energy-efficient approximate multiplication for digital signal processing and classification applications. IEEE Trans. Very Large Scale Integ. (VLSI) Syst. **23**(6), 1180–1184 (2015)
28. J. Li, Y. Guo, S. Kimura, Accuracy-configurable low-power approximate floating-point multiplier based on mantissa bit segmentation, in 2020 IEEE Region 10 Conference (TENCON), pp. 1311–1316 (2020)
29. Z. Niu, H. Jiang, M.S. Ansari, B.F. Cockburn, L. Liu, J. Han, *A Logarithmic Floating-Point Multiplier for the Efficient Training of Neural Networks* (Association for Computing Machinery, New York, 2021), pp. 65–70
30. J. Tong, D. Nagle, R. Rutenbar, Reducing power by optimizing the necessary precision/range of floating-point arithmetic. IEEE Trans. Very Large Scale Integ. (VLSI) Syst. **8**(3), 273–286 (2000)
31. J. Du, K. Chen, P. Yin, C. Yan, W. Liu, Design of an approximate fft processor based on approximate complex multipliers, in 2021 IEEE Computer Society Annual Symposium on VLSI (ISVLSI), pp. 308–313 (2021)
32. X. Han, J. Chen, B. Qin, S. Rahardja, A novel area-power efficient design for approximated small-point FFT architecture. IEEE Trans. Comput.-Aided Design Integ. Circuits Syst. **39**(12), 4816–4827 (2020)

33. W. Liu, Q. Liao, F. Qiao, W. Xia, C. Wang, F. Lombardi, Approximate designs for fast fourier transform (fft) with application to speech recognition. IEEE Trans. Circuits Syst. I: Regul. Pap. **66**(12), 4727–4739 (2019)
34. J.G. Proakis, *Digital Signal Processing: Principles Algorithms and Applications* (Pearson Education India, 2001)
35. J.W. Cooley, J.W. Tukey, An algorithm for the machine calculation of complex fourier series. Math. Comput. **19**(90), 297–301 (1965)
36. Y. Xiang, L. Li, S. Yuan, W. Zhou, B. Guo, Metrics, noise propagation models, and design framework for floating-point approximate computing. IEEE Access **9**, 71039–71052 (2021)
37. H. Al-Salihi, M. R. Nakhai, T. A. Le, DFT based channel estimation techniques for massive MIMO systems, in 2018 25th International Conference on Telecommunications (ICT), pp. 383–387 (2018)
38. E. Arikan, Channel polarization: A method for constructing capacity achieving codes for symmetric binary-input memoryless channels. IEEE Trans. Inf. Theory **55**(7), 3051–3073 (2009)
39. "Evaluation on channel coding candidates for embb control channel," document 3GPP TSG RAN WG1, #87,R1–1611109, Reno, USA, Nov. 2016
40. E. Arikan, A performance comparison of polar codes and Reed-Muller codes. IEEE Commun. Lett. **12**(6), 447–449 (2008)
41. O. Dizdar, E. Arıkan, A high-throughput energy-efficient implementation of successive cancellation decoder for polar codes using combinational logic. IEEE Trans. Circuits Syst. I: Reg. Pap. **63**(3), 436–447 (2016)
42. C. Zhang, Z. Wang, X. You, B. Yuan, Efficient adaptive list successive cancellation decoder for polar codes, in Proc. 48th Asilomar Conference on Signals, Systems and Computers, 2014, pp. 126–130
43. W. Song, H. Zhou, K. Niu, Z. Zhang, L. Li, X. You, C. Zhang, Efficient successive cancellation stack decoder for polar codes. IEEE Trans. Very Large Scale Integ. (VLSI) Syst. **27**(11), 2608–2619 (2019)
44. K. Niu, K. Chen, Hybrid coding scheme based on repeat-accumulate and polar codes. Electron. Lett. **48**(20), 1273–1274 (2012)
45. Y. Zhou, J. Lin, and Z. Wang, Efficient approximate layered LDPC decoder, in Proc. IEEE International Symposium on Circuits and Systems (ISCAS), 2017, pp. 1–4
46. M. Xu, S. Jing, J. Lin, W. Qian, Z. Zhang, X. You, C. Zhang, Approximate belief propagation decoder for polar codes, in Proc. IEEE International Conference on Acoustics, Speech and Signal Processing (ICASSP), 2018, pp. 1169–1173
47. Y. Zhou, Z. Chen, J. Lin, Z. Wang, A high-speed successive cancellation decoder for polar codes using approximate computing. IEEE Trans. Circuits Syst. II: Exp. Brief **66**(2), 227–231 (2019)
48. D. Wu, Y. Li, Y. Sun, Construction and block error rate analysis of polar codes over AWGN channel based on Gaussian approximation. IEEE Commun. Lett. **18**(7), 1099–1102 (2014)
49. C. Leroux, I. Tal, A. Vardy, W. J. Gross, Hardware architectures for successive cancellation decoding of polar codes, in Proc. IEEE International Conference on Acoustics, Speech and Signal Processing (ICASSP), 2011, pp. 1665–1668
50. H.-Y. Yoon, T.-H. Kim, Efficient successive-cancellation polar decoder based on redundant LLR representation. IEEE Trans. Circuits Syst. II: Exp. Brief **65**(2), 1944–1948 (2018)
51. R. Shrestha, A. Sahoo, High-speed and hardware-efficient successive cancellation polar-decoder. IEEE Trans. Circuits Syst. II: Exp. Brief. **66**(7), 1144–1148 (2019)
52. B. Yuan, K.K. Parhi, Low-latency successive-cancellation polar decoder architectures using 2-bit decoding. IEEE Trans. Circuits Syst. I: Regul. Pap. **61**(4), 1241–1254 (2014)
53. C. Leroux, A.J. Raymond, G. Sarkis, W.J. Gross, A semi-parallel successive-cancellation decoder for polar codes. IEEE Trans. Signal Process. **61**(2), 289–299 (2013)
54. A. Elkelesh, M. Ebada, S. Cammerer, S.T. Brink, Decoder-tailored polar code design using the genetic algorithm. IEEE Trans. Commun. **67**(7), 4521–4534 (2019)

55. J. Liu, J. Sha, Frozen bits selection for polar codes based on simulation and BP decoding. IEICE Electron. Expr. **14**(6), 20170026 (2017)
56. A. Elkelesh, S. Cammerer, M. Ebada, S. ten Brink, *Mitigating Clipping Effects on Error Floors Under Belief Propagation Decoding of Polar Codes* (2017 International Symposium on Wireless Communication Systems (ISWCS), 2017), pp. 384–389
57. O. Afisiadis, A. Balatsoukas-Stimming, and A. Burg, "A low-complexity improved successive cancellation decoder for polar codes," in 2014 48th Asilomar Conference on Signals, Systems and Computers, 2014, pp. 2116–2120
58. Y. Yu, Z. Pan, N. Liu, X. You, Belief propagation bit-flip decoder for polar codes. IEEE Access **7**, 10937–10946 (2019)
59. Y. Shen, W. Song, Y. Ren, H. Ji, X. You, C. Zhang, Enhanced belief propagation decoder for 5g polar codes with bit-flipping. IEEE Trans. Circuits Syst. II: Exp. Brief **67**(5), 901–905 (2020)

Logarithmic Floating-Point Multipliers for Efficient Neural Network Training

Tingting Zhang, Zijing Niu, Honglan Jiang, Bruce F. Cockburn, Leibo Liu, and Jie Han

1 Introduction

Due to its high accuracy over a very wide dynamic range, floating-point (FP) arithmetic is commonly used in applications where the relative accuracy of intermediate values, as well as the final results, must be preserved [1]. For example, an FP representation is usually adopted in the training process of neural networks (NNs) for its more accurate encoding of the large number of weights and activations [2]. However, compared with fixed-point units, FP arithmetic units require substantially more hardware area and power. Moreover, as Dennard scaling comes to an end, the increase in power density has become a principal limitation to further performance improvement.

As an emerging computing paradigm, approximate computing improves the efficiency of circuits and system for error-tolerant applications, such as digital signal processing, data mining, and NNs, that can accept a certain loss in accuracy [3, 4]. As a basic operation, multiplication is interesting for approximation due to its relatively high circuit implementation cost [5–7]. Approximate FP multiplier designs have been investigated by using approximate Booth encoding [8], reconfiguration [5, 9], truncation techniques [10], and approximate adders

T. Zhang · Z. Niu · B. F. Cockburn · J. Han (✉)
University of Alberta, Edmonton, AB, Canada
e-mail: ttzhang@ualberta.ca; zijing2@ualberta.ca; cockburn@ualberta.ca; jhan8@ualberta.ca

H. Jiang
Shanghai Jiao Tong University, Shanghai, China
e-mail: honglan@sjtu.edu.cn

L. Liu
Tsinghua University, Beijing, China
e-mail: liulb@tsinghua.edu.cn

W. Liu et al. (eds.), *Design and Applications of Emerging Computer Systems*,
https://doi.org/10.1007/978-3-031-42478-6_21

[6]. However, these methods lead to limited performance improvement since the expensive multiplication hardware is still present.

Logarithmic computing which converts conventional multiplication to addition provides an energy-efficient alternative for implementing multiplication. Since its number representation is naturally suited to FP numbers, efficient logarithmic FP multipliers (LFPMs) have been investigated to reduce circuit area and improve computational efficiency [11–15]. Moreover, Sun et al. have recently shown the superiority of using logarithmic computation in training deep NNs [16]. It was previously shown that applying logarithmic multiplication in Lognet led to higher classification accuracy at low resolutions [17]. Therefore, in this chapter, we consider the design of LFPMs for efficient NN training. Due to the unknown error characteristics in NNs, however, it is difficult to exploit the trade-off between the accuracy of NNs and hardware efficiency. Hence, additional approximate designs with a diverse range of accuracies and circuit characteristics must be considered experimentally to gain a better understanding of their roles in the implementation of NNs.

A large number of approximate fixed-point multipliers can be found in the literature [18, 19]; however, existing LFPM designs are not sufficient for analyzing the use of LFPMs in different error-tolerant applications. This chapter presents a design framework to generate a library of LFPMs [20]. These LFPMs are developed by using two FP representation formats, the IEEE 754 Standard FP Format (FP754) and the Nearest Power-of-Two FP Format (NPFP2), in the logarithm and anti-logarithm approximation. In both processes, the applicable regions of inputs are first evenly divided into several sub-regions. Then, rather than using a single approximation method, two candidate designs that, respectively, introduce negative and positive errors are considered for each sub-region. Numerous LFPMs are generated by configuring different approximation methods throughout the regions based on a piecewise function. Example designs with optimization strategies are synthesized and used in NN applications as case studies and for evaluation [15].

The remainder of this chapter is organized as follows. Section 2 presents basic concepts and theory. Section 3 discusses the approximation framework for building LFPMs. The generic hardware implementation of LFPMs is introduced in Sect. 4. Case studies and NN applications are presented in Sects. 5 and 6. Section 7 concludes this chapter.

2 Preliminaries

2.1 FP Representation

2.1.1 IEEE 754 Standard FP Format (FP754)

As defined in the IEEE 754 Standard, the FP number is represented as a string that consists of a 1-bit sign S, a w-bit exponent E, and a q-bit normalized mantissa M,

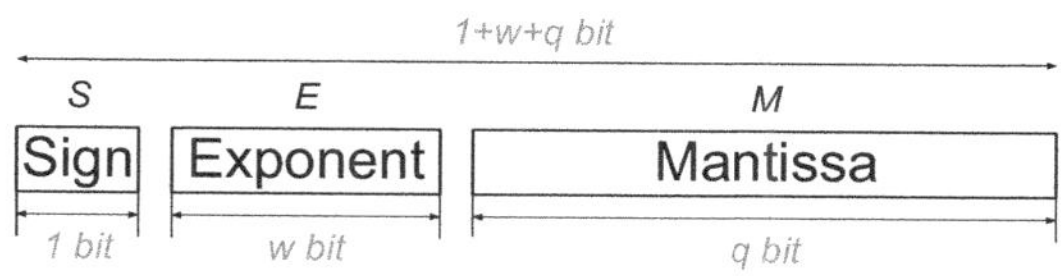

Fig. 1 IEEE 754 Standard FP Format [21]

as shown in Fig. 1. A normalized FP number N is expressed as [21]:

$$N = (-1)^S \cdot 2^{E-bias} \cdot (1+x), \quad (1)$$

where S takes either 0 or 1 for a positive or negative number, respectively. The exponent has a $bias$ of $2^{w-1} - 1$ to ensure that $E \geq 0$; thus, the actual exponent is given by $E - bias$. With the hidden "1," x is the fractional part of the FP number, and hence, $0 \leq x < 1$. The hidden "1" does not need to be presented in the FP754 encoding of a FP number.

2.1.2 Nearest Power-of-Two FP Format (NPFP2)

The FP754 format conventionally provides the largest power of two smaller than N in the exponent E. Based on it, another FP format, named NPFP2, was considered in an LFPM design [15, 22]. It employs the exponent E' that is the nearest power of two to N by comparing the mantissa x with 0.5, as:

$$N = (-1)^S \cdot 2^{E'-bias} \cdot (1+x'), \quad (2)$$

where

$$E' = \begin{cases} E, & x < 0.5 \\ E+1, & x \geq 0.5 \end{cases}, \quad (3)$$

$$X' = 1 + x' = \begin{cases} 1+x, & x < 0.5 \\ \frac{1+x}{2}, & x \geq 0.5 \end{cases}, \quad (4)$$

where $x' \in [-0.25, 0.5)$ and $X' \in [0.75, 1.5)$.

2.2 *Logarithmic FP Multiplication*

Consider $P = A \times B$. Performing this FP multiplication involves sign bit generation, the addition of the exponents, and the multiplication of the mantissas [15]:

$$S_p = S_A \oplus S_B, \quad (5)$$

$$X_{AB} = X_A X_B = (1 + x_A) \times (1 + x_B), \tag{6}$$

$$E_P = \begin{cases} E_A + E_B - bias, & if\ X_{AB} < 2 \\ E_A + E_B - bias + 1, & if\ X_{AB} \geq 2 \end{cases}, \tag{7}$$

$$X_P = \begin{cases} X_{AB}, & if\ X_{AB} < 2 \\ \frac{X_{AB}}{2}, & if\ X_{AB} \geq 2 \end{cases}, \tag{8}$$

where the sign bit, exponent, and mantissa of A, B, and P are denoted with the corresponding subscripts. The exponent and mantissa of product P are determined by the comparison of the obtained mantissa X_{AB} with 2.

To ease the complexity of computing X_{AB}, logarithmic multiplication in base-2 scientific notation converts the multiplication in (6) to addition as:

$$log_2 X_{AB} = log_2(1 + x_A) + log_2(1 + x_B). \tag{9}$$

Then, the logarithmic result $log_2 X_{AB}$ is interpolated back into the original result (X_{AB}) using a base-2 anti-logarithm evaluation. Finally, the values of X_P and E_P are obtained by using (7) and (8).

Mitchell's logarithm approximation (LA) and a corresponding anti-logarithm approximation (ALA) are commonly used [23]:

$$log_2(1 + k) \cong k, \tag{10}$$

where $k \in [0, 1)$, and

$$2^l \cong l + 1, \tag{11}$$

where $l \in [0, 1)$.

3 Piecewise Approximation Design Framework

3.1 Logarithm Approximation

The input operands for the logarithm approximation use either FP754 (given by x_A and x_B) or NPFP2 (given by x'_A and x'_B) formats. When using FP754, the LA for $log_2(1 + k)$ is considered for $k \in [0, 1)$; when using NPFP2, as in (2), the LA is considered for $k \in [-0.25, 0.5)$.

Mitchell's LA (10) always underestimates the logarithm when $k \in [0, 1)$, which introduces negative errors. However, beyond the applicable region, (10) overestimates the logarithm. Hence, by applying Mitchell's LA method outside of the original applicable domain, positive errors are introduced.

First, we extend Mitchell's LA in (10) into the region outside of $k \in [0, 1)$. Considering that the multiplication or division of powers of two can be easily performed by bit shifts, we identify a factor n of the given number such that n is a power of two. Let $k \in (-\infty, +\infty)$; the LA can be obtained from (10) with different applicable regions, as:

$$\begin{aligned} log_2(1+k) = &\quad log_2(n \times \tfrac{1+k}{n}) \\ &= log_2(n) + log_2(1 + \tfrac{1+k}{n} - 1) \\ \cong &\quad log_2(n) + \tfrac{1+k}{n} - 1, \end{aligned} \tag{12}$$

where $0 \leq \frac{1+k}{n} - 1 < 1$ and where $k \in [n-1, 2n-1)$.

Consider both FP754, where $k \in [-0.25, 0.5)$, and NPFP2, where $k \in [0, 1)$. The LA methods in which the applicable domains are close to the interval $[-0.25, 1)$ are considered. Let $n = \frac{1}{2}$, 1, and 2; then

$$log_2(1+k) \cong \begin{cases} 2k, & -0.5 \leq k < 0 \\ k, & 0 \leq k < 1 \\ \frac{k+1}{2}, & 1 \leq k < 3 \end{cases} . \tag{13}$$

Figure 2 indicates that these LA methods introduce negative errors within their respective applicable domains; otherwise, they overestimate the logarithm results. The LA method is developed by considering three candidates (i.e., k, $\frac{k+1}{2}$, and $2k$) for $k \in [-0.25, 1)$ to introduce double-sided errors. When using FP754, i.e., $k \in [0, 1)$, $2k$ and $\frac{k+1}{2}$ overestimate the true $log_2(1+k)$ value, whereas k

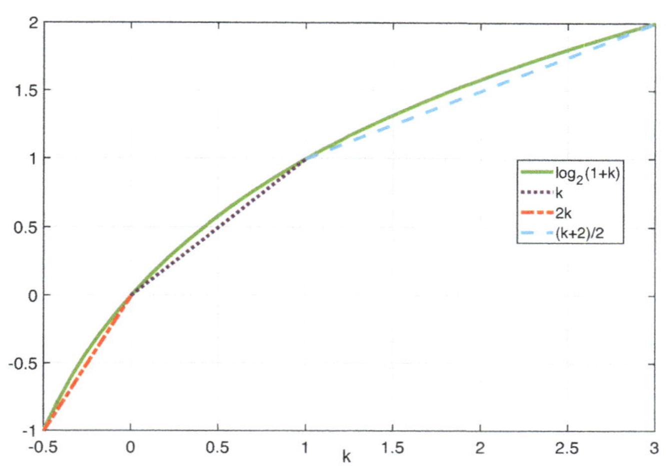

Fig. 2 LA in (13)

underestimates that value. When using NPFP2, i.e., $k \in [-0.25, 0.5)$, k, $2k$, and $\frac{k+1}{2}$ overestimate the $log_2(1+k)$ value in the regions $[-0.25, 0)$, $[0, 1)$, and $[0, 1)$, respectively. However, $2k$ and k underestimate the $log_2(1+k)$ value in the regions $[-0.25, 0)$ and $[0, 0.5)$, respectively.

The LAs for $log_2(1+k)$ when $k \in [0, 1)$ based on FP754 and $k \in [-0.25, 0.5)$ based on NPFP2 are both developed by using a piecewise function with different configurations in these sub-regions, respectively. First, the region is divided into several sub-regions with a fixed width. Then, within each sub-region, two candidate methods with a high accuracy are considered: one that overestimates the logarithmic result with positive errors and another that underestimates the logarithmic results with negative errors. An LA method can be independently selected for each sub-region. For example, the LA methods based on FP754 can be found in Fig. 3.

A trade-off is assessed when selecting an appropriate length of each sub-region. An excessively small length leads to an increase of hardware complexity, whereas an excessively large length reduces the variety of the accuracy for those generated designs. If 0.25 is taken as the length of each sub-region, the piecewise logarithm approximation (PWLA) based on FP754 and NPFP2 are presented in Tables 1 and 2, respectively.

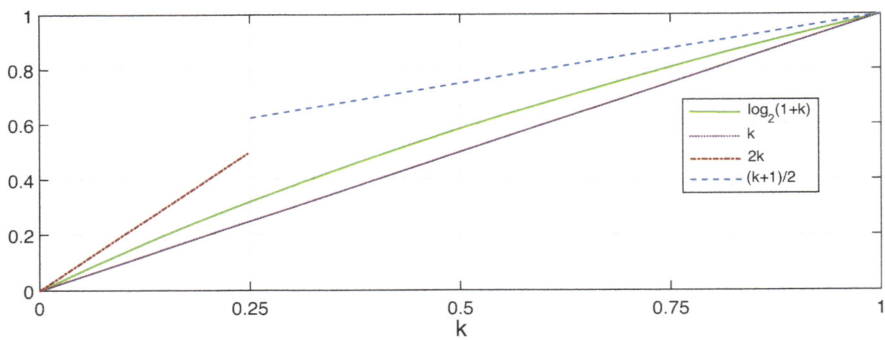

Fig. 3 Piecewise logarithm approximation methods

Table 1 Piecewise logarithm approximation using the IEEE 754 Standard FP Format

k		$[0, 0.25)$	$[0.25, 0.5)$	$[0.5, 0.75)$	$[0.75, 1)$
$log_2(1+k)$	Positive	$2k$	$\frac{k+1}{2}$	$\frac{k+1}{2}$	$\frac{k+1}{2}$
	Negative	k	k	k	k

Table 2 Piecewise logarithm approximation using the Nearest Power-of-Two FP Format

k		$[-0.25, 0)$	$[0, 0.25)$	$[0.25, 0.5)$
$log_2(1+k)$	Positive	k	$2k$	$\frac{k+1}{2}$
	Negative	$2k$	k	k

3.2 Anti-logarithm Approximation

Consider that two logarithm results in the LA process can be obtained from the approximation based on either FP754 or NPFP2, as in Tables 1 and 2. Then the sum of two logarithm results, denoted by l, ranges from -1 to 2, and so the ALA method is considered for $l \in [-1, 2)$. The ALA method, as per (11), always introduces positive errors in the applicable region. Similar as the methods in Sect. 3.1, the ALA methods can be derived from (11), as:

$$2^l = n \times 2^{l - log_2 n} \cong n \times (l + 1 - log_2 n), \tag{14}$$

where $0 \leq l - log_2 n < 1$ and thus where $l \in [log_2 n, log_2 n + 1)$.

We consider the ALA methods in which applicable regions are close to the range $[-1, 2)$. Let $n = \frac{1}{4}, \frac{1}{2}, 1, 2$, and 4; ALA methods can be obtained as:

$$2^l \cong \begin{cases} \frac{l+3}{4}, & -2 \leq l < -1 \\ \frac{l+2}{2}, & -1 \leq l < 0 \\ l+1, & 0 \leq l < 1 \\ 2l, & 1 \leq l < 2 \\ 4l-4, & 2 \leq l < 3 \end{cases} . \tag{15}$$

As shown in Fig. 4, these ALA methods always overestimate the results, whereas negative errors are introduced out of the applicable regions.

The ALA method is developed based on five candidate ALA methods (using $\frac{l+3}{4}$, $\frac{l+2}{2}$, $l + 1$, $2l$, and $4l - 4$) for $l \in [-1, 2)$ to introduce double-sided errors.

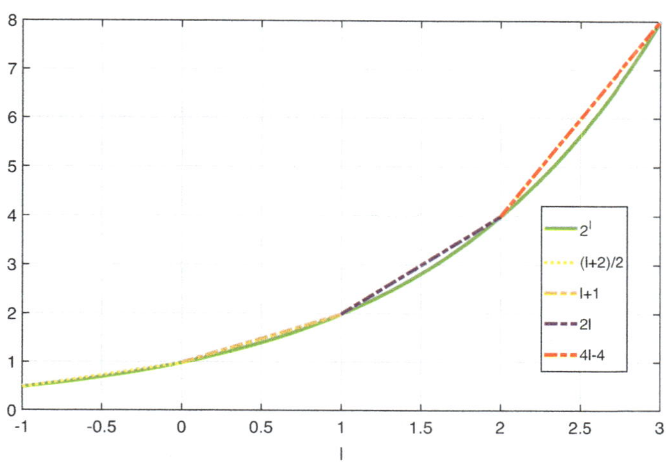

Fig. 4 ALA in (15)

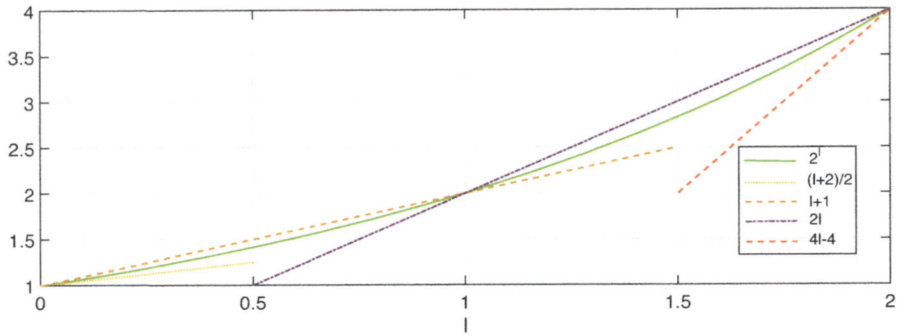

Fig. 5 Piecewise anti-logarithm approximation methods

Table 3 Piecewise anti-logarithm approximation

l		$[-1, -0.5)$	$[-0.5, 0)$	$[0, 0.5)$	$[0.5, 1)$	$[1, 1.5)$	$[1.5, 2)$
2^l	Positive	$\frac{l+2}{2}$	$\frac{l+2}{2}$	$l+1$	$l+1$	$2l$	$2l$
	Negative	$\frac{l+3}{4}$	$l+1$	$\frac{l+2}{2}$	$2l$	$l+1$	$4l-4$

Similar to the PWLA method, two approximation methods are considered in each sub-region, one with positive errors and another with negative errors. Consider that two logarithm results in the LA process are both obtained from the approximation based on FP754, I would be in the range of [0,2) Then, the ALA methods can be found in Fig. 5. The piecewise anti-logarithm approximation (PWALA) method takes 0.5 as the width of each sub-region, as per Table 3.

3.3 Logarithmic FP Multiplication

The logarithmic FP multiplication method based on piecewise approximation is developed from approximation in the logarithm and anti-logarithm methods, as shown in Algorithm 1.

Consider the LA process. F indicates the use of FP754 or NPFP2. The $\boldsymbol{C_{x_A}}$ (or $\boldsymbol{C_{x_B}}$) contains four elements, each taking either "p" or "n" to indicate using the approximation method with positive or negative errors in each interval. The use of NPFP2 converts x_A in the domain of $[0.5, 1)$ to $[-0.25, 0)$, resulting in the same configuration for $C_{x_A}(2)$ and $C_{x_A}(3)$ (or as $C_{x_B}(2)$ and $C_{x_B}(3)$). When x_A (or x_B) is located within $[0, 0.5)$, no matter which FP format is used, the LA candidate methods are the same. When x_A (or x_B) is in the interval $[0.5, 1)$, NPFP2 modifies E_A (or E_B) to $E_A + 1$ (or $E_B + 1$) and performs the LA using either $\frac{x_A - 1}{2}$ or $x_A - 1$. Consider the ALA process. The $\boldsymbol{C_{a_l}}$ has six elements, each with either "p" or "n." First, X_{AB} is determined by the configuration for PWALA methods. Then, E_P and X_P are obtained to satisfy the requirements of the FP754 format depending on X_{AB}.

A total of 5744 piecewise-based LFPM (PWLM) designs were generated using these approximation methods. Different configurations of PWLA methods for x_A and x_B and PWALA methods can lead to possible error compensation. Therefore, the respective accuracy of LA and ALA methods cannot predict the overall error characteristics of PWLMs. The overall error characteristics of these numbers of PWLMs are analyzed in Sect. 4.

4 Hardware Implementation

In this section, the generic circuit block diagram is introduced, and the simplifications for each circuit block are investigated. Since the computation process can be simplified in different manners, the hardware implementation can be specially designed for each PWLM.

4.1 The Generic Circuit

Figure 6 presents the generic circuit blocks for implementing these PWLMs. The sign S, the exponent E, and the mantissa M of the FP number can be obtained directly. The w-bit E is given by $E[w-1]E[w-2]\cdots E[0]$. $1.M$ is used to denote the actual mantissa. Here, M contains q bits, e.g., 23 bits for single-precision, after the binary point and the hidden "1." $1.M$ denotes the actual mantissa and is given as

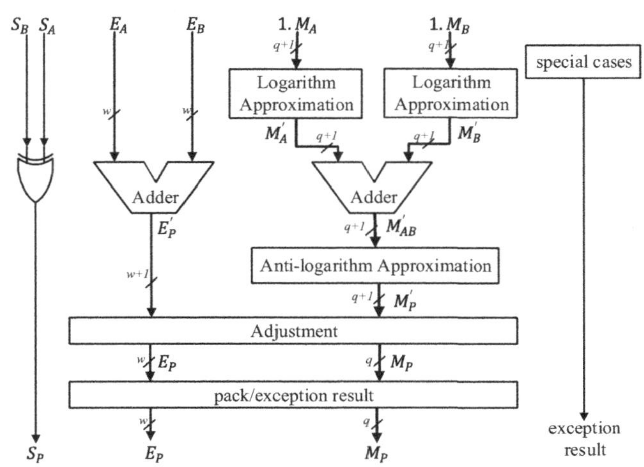

Fig. 6 The circuit block diagram for the generic implementation of PWLMs

Algorithm 1 Logarithmic FP multiplication based on piecewise approximation

Require: FP format: F; Sign bits: S_A and S_B; Exponents: E_A and E_B; Mantissas: x_A and x_B; Configuration for logarithm approximation: $\boldsymbol{C_{x_A}}$ and $\boldsymbol{C_{x_B}}$; Configuration for anti-logarithm approximation: $\boldsymbol{C_{al}}$

Ensure: An approximate FP multiplication result: S_p, E_p, and X_p.

```
S_p = S_A ⊕ S_B
//LA for x_A (and x_B)
ME_A = E_A
if x_A < 0.25 then
    if C_{x_A}(0) is p then
        log_{x_A} = 2x_A
    else
        log_{x_A} = x_A
    end if
else if x_A < 0.5 then
    if C_{x_A}(1) is p then
        log_{x_A} = (x_A+1)/2
    else
        log_{x_A} = x_A
    end if
else
    if F is NPFP2 then
        ME_A = E_A + 1
        if C_{x_A}(2) is p then
            log_{x_A} = (x_A-1)/2
        else
            log_{x_A} = x_A - 1
        end if
    else
        if x_A < 0.75 then
            if C_{x_A}(2) is p then
                log_{x_A} = (x_A+1)/2
            else
                log_{x_A} = x_A
            end if
        else
            if C_{x_A}(3) is p then
                log_{x_A} = (x_A+1)/2
            else
                log_{x_A} = x_A
            end if
        end if
    end if
end if
//ALA
log_s = log_{x_A} + log_{x_B}
ME_s = ME_A + ME_B
if log_s < -0.5 then
    if C_al(0) is p then
        X_AB = (log_s+2)/2
    else
        X_AB = (log_s+3)/4
    end if
else if log_s < 0 then
    if C_al(1) is p then
        X_AB = (log_s+2)/2
    else
        X_AB = l + 1
    end if
else if log_s < 0.5 then
    if C_al(2) is p then
        X_AB = log_s + 1
    else
        X_AB = (log_s+2)/2
    end if
else if log_s < 1 then
    if C_al(3) is p then
        X_AB = log_s + 1
    else
        X_AB = 2log_s
    end if
else if log_s < 1.5 then
    if C_al(4) is p then
        X_AB = 2log_s
    else
        X_AB = log_s + 1
    end if
else
    if C_al(5) is p then
        X_AB = 2log_s
    else
        X_AB = 4log_s - 4
    end if
end if
//compute E_P and X_P
if X_AB < 1 then
    E_P = ME_s - 1
    X_P = 2X_AB
else if X_AB < 2 then
    E_P = ME_s
    X_P = X_AB
else
    E_P = ME_s + 1
    X_P = X_AB/2
end if
```

Notes: ME_A and ME_B denote the modified exponents for A and B, respectively; log_{x_A} and log_{x_B} denote the logarithm mantissas for x_A and x_B, respectively; ME_s is the sum of ME_A and ME_B; log_s is the sum of log_{x_A} and log_{x_B}; X_{AB} is the anti-logarithm result of log_s.

$M[q].M[q-1]M[q-2]\cdots M[1]M[0]$, where $M[q]$ denotes the constant hidden one. Note that $1.M$ represents $1+x$ in (1).

The computation for the sign bit of the product, S_P, is implemented by an XOR gate with two input signs, S_A and S_B. To obtain the mantissa of the product, the approximate logarithmic values of $1.M_A$ and $1.M_B$, denoted as M'_A and M'_B, respectively, are computed in the logarithm approximation block. The two obtained values are then summed together since the multiplication is replaced by the addition in the logarithm domain. The sum of M'_A and M'_B, denoted by M'_{AB}, is propagated to the anti-logarithm approximation block to compute its approximate anti-logarithm, denoted by M'_P. The exponent of the product is firstly computed by the addition of E_A and E_B, denoted by E'_P. When the value of M'_P exceeds the range of the mantissa of the product, given by $[1, 2)$, M'_P and E'_P will be adjusted in the adjustment block. Finally, E_P and M_P are obtained. E_P is augmented with a constant bias to comply with the standard IEEE format. Special exception values (such as overflow, underflow, and "not a number") are reported depending on the input operand and on the result of a calculation. Note that the two operands are checked for exceptions at the beginning of the computation. Some circuit blocks in Fig. 6 can be combined into a single module, and certain computational components can be implemented with less hardware complexity.

4.1.1 Logarithm Approximation Block

According to Tables 1 and 2, the logarithm approximation block includes implementations for five different candidate LA methods and a multiplexer, as shown in Fig. 7, where M indicates M_A or M_B and M' indicates M'_A or M'_B.

A multiplexer is used to implement the PWLA as shown in Fig. 7a. In particular, when using a one-piece LA (i.e., $\log_2 1+k \cong k$), the multiplexer is not required. A 2-to-1 multiplexer is used for a two-piece LA. The select signals are $M[q-2]|M[q-1]$, $M[q-1]$, and $M[q-1]\&M[q-2]$, respectively, when 0.25, 0.5, and 0.75 are the boundary values of two sub-regions. Similarly, a 4-to-1 multiplexer is required for three-piece or four-piece LAs, and the boundary value is determined by $M[q-1]M[q-2]$. For example, $M[q-1]M[q-2] = 0.11$ (base-2) represents 0.75.

The five candidate LA methods using k, $2k$, $\frac{k+1}{2}$, $\frac{k-1}{2}$, and $k-1$ are implemented to serve as the inputs of the multiplexer as shown in Fig. 7b. Instead of directly implementing the equations, which would require adders and shifters, simple operations can be adopted to simplify the hardware. Since $1+k$ is implemented by $1.M[q-1]M[q-2]\cdots M[1]M[0]$, k is obtained as $0.M[q-1]M[q-2]\cdots M[1]M[0]$. The left/right shifter for implementing $\times 2$ and $/2$ operations is replaced by wire routing. Thus, $\frac{k+1}{2}$ and $2k$ are obtained as $0.M[q]M[q-1]M[q-2]\cdots M[1]$ and $M[q-1].M[q-2]\cdots M[1]0$, respectively. Note that $2k$ is used when $0 \leq k < 0.25$, and thus, $M[q]$ is 0 and can be ignored after being shifted left. $\frac{k-1}{2}$ and $k-1$ are obtained as a negative number in 2's complement, respectively, i.e., $1.1M[q-1]\cdots M[1]$ and $1.M[q-1]\cdots M[0]$. Note that when PWLA based on

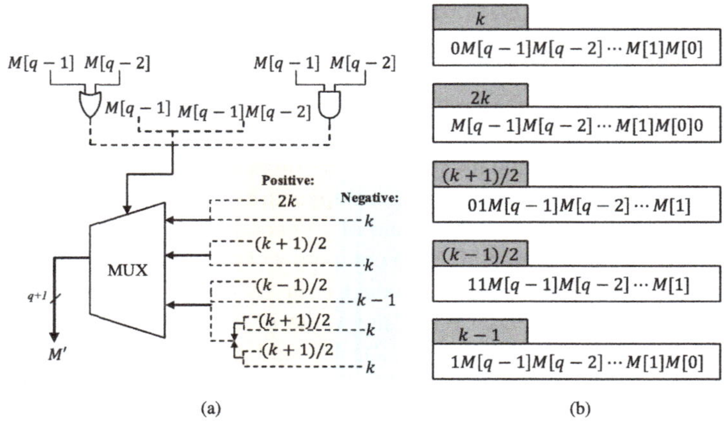

(a) (b)

Fig. 7 Hardware implementation for the logarithm approximation block: (**a**) circuit designs for PWLA methods and (**b**) inputs of the MUX in (**a**)

FP754 (in Table 1) and NPFP2 (in Table 2) are both used for the two input operands, M' is implemented in $q+2$ bits instead of $q+1$ bits, where the sign bit of candidate LA methods in Fig. 7b is extended by 1 bit.

4.1.2 Anti-logarithm Approximation Block

According to Table 3, a multiplexer and implementations for five candidate methods are required for computing the approximate anti-logarithm, M'_P, as shown in Fig. 8.

The PWALA is implemented by using a multiplexer, as shown in Fig. 8a. For the two-piece ALA, the select signals are $M'_{AB}[q-1]|M'_{AB}[q]$, $M'_{AB}[q]$, and $M'_{AB}[q]\&M'_{AB}[q-1]$, respectively, when 0.5, 1 or 0, and 1.5 or −0.5 are the boundary values. When using three-piece to five-piece ALAs, the boundary values are determined by $M'_{AB}[q]M'_{AB}[q-1]$, which is therefore the select signal for a 4-to-1 multiplexer. In particular, when PWLAs based on FP754 (in Table 1) and NPFP2 (in Table 2) are both used for the two input operands, the PWALA method is applied to $l \in [-0.5, 2)$. M'_{AB} is implemented in $q+2$ bits with an extended sign bit, denoted by $M'_{AB}[q+1]$, to be distinguishable for the ranges of $[-0.5, 0)$ and $[1.5, 2)$. In this case, for using three-piece, four-piece, and five-piece ALAs, $M'_{AB}[q+1]M'_{AB}[q]M'_{AB}[q-1]$ is used to determine the boundary values.

The four candidate ALA methods using $l+1$, $\frac{l+2}{2}$, $2l$, and $4l-4$ are implemented to obtain M'_P using wire routing. To simplify the circuit, the adjustment block is merged into the implementation of these ALA methods, where the adjusted value of M'_P, i.e., M_P, is used as the output for the multiplexer. The circuit for generating M_P is introduced in the next subsection.

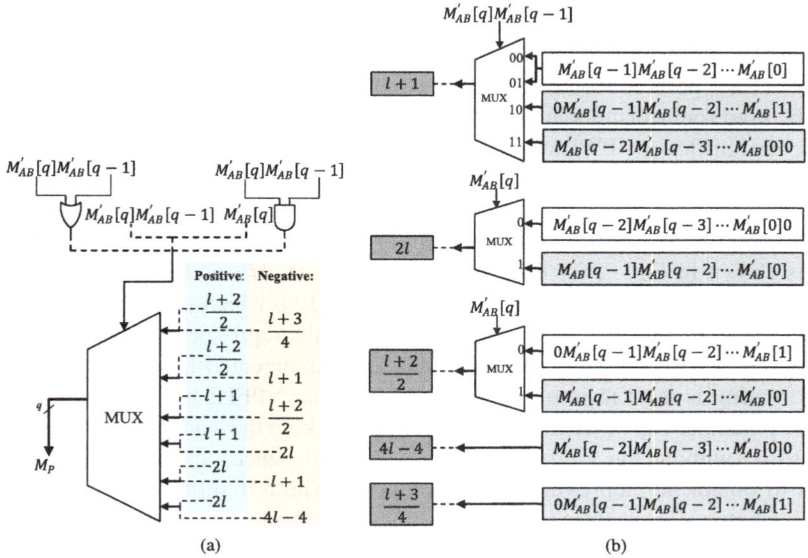

Fig. 8 Hardware implementation for the anti-logarithm approximation block and the adjustment block: (**a**) circuit designs for PWALA methods and (**b**) inputs of MUX in (**a**)

4.1.3 Adjustment Block

As shown in Fig. 8b, by merging the value adjustment into the ALA computation, the final mantissa, M_P, is directly implemented by M'_{AB} with a multiplexer in some cases.

According to Table 3, when $l \geq 1$, the values of $l+1$, $2l$, and $4l-4$ range from 2 to 4; when $l < 0$, the values of $l+1$, $\frac{l+2}{2}$, and $\frac{l+3}{4}$ range from 0.5 to 1. These values are divided by or multiplied by 2 to fit within the range $[1, 2)$. Therefore, the value adjustment is determined by checking $M'_{AB}[q]M'_{AB}[q-1]$ for $l+1$ and $M'_{AB}[q]$ for other candidate ALA methods. The implementations for the adjusted ALA methods are shown as gray boxes in Fig. 8b. Particularly, $l+1$ is implemented for the range $[-0.5, 0)$ when $M'_{AB}[q]M'_{AB}[q-1] = 11$ and for the range $[1, 1.5)$ when $M'_{AB}[q]M'_{AB}[q-1] = 0.10$ (base-2). $\frac{l+3}{4}$ and $4l-4$ are implemented without a multiplexer since they are used in one sub-region.

For the exponent, when ALA values are divided by or multiplied by 2, E'_P is added with or subtracted with one, respectively, to ensure correctness. When using PWLAs based on FP754 for the two inputs, to reduce the hardware complexity, a carry-in bit, implemented by $M'_{AB}[q]$, is used for the value adjustment of E'_P in the addition of E_A and E_B. When using PWLAs based on NPFP2, the final exponent is determined by both the NPFP2 conversion, as per (3), and the value adjustment for the ALA values.

5 Case Studies of PWLM Designs

The PWLMs provide various accuracies and hardware performance metrics, which can be used for error-tolerant applications with different requirements. To show the basic design flow of PWLMs and optimization techniques in detail, one-example designs (FPLM) are studied. In the FPLM, both the input operands are represented by NPFP2 format and use the same LA methods. The LA $log_2(1 + k) \approx k$ is considered for $k \in [-0.25, 0.5)$ and then for ALA, $2^l \approx l + 1$ when $l \in [-0.5, 1)$.

As shown in Fig. 9, for the two given mantissas, M_A and M_B are used as the inputs for the two FP logarithm estimators (FP-LEs), which compute the approximate logarithm of the mantissa. For FPLM, both the representation conversion and logarithm approximation for the mantissa are implemented by an FP-LE. Simple wire routing is used to implement the FP-LE. The NPFP2 format conversion for each FP number can be determined by simply checking the leading bit of the explicit mantissa (without the hidden 1), $M[q-1]$, and hence, a 2-to-1 multiplexer is used to obtain the approximate logarithm, M'. The left shifter for implementing the division by 2 is replaced by wire routing. Therefore, the format conversion, i.e., $(1+x)/2$, in (4) is implemented as $0.1M[q-1]\cdots M[1]$. When $M[q-1] = 1$, x' in (4), i.e., $(1+x)/2 - 1$, is obtained as a negative number in 2's complement, i.e., $M' = 1.1M[q-1]\cdots M[1]$; otherwise, M' is $0.M[q-1]\cdots M[0]$. In the case of $M[q-1] = 1$, the LSB of M', $M[0]$, is discarded to keep M' in $q+1$ bits. M'_A and M'_B are then summed using a $q+1$-bit adder.

The anti-logarithm approximation and value adjustment are implemented together using a multiplexer, as shown in Fig. 9. The $q+1$-bit M'_P obtained from the adder is the input for this block. Since the results of $M'_A + M'_B$ (i.e., l in (10)) can be negative in some cases, the MSB of M'_P, i.e., $M'_P[q]$, is the sign bit of M'_P in 2's complement. When $l < 0$, $M'_P[q] = 1$; otherwise, $M'_P[q] = 0$. Therefore, $M'_P[q]$ is used as the selection signal for the multiplexer. $l+1$ is implemented as $0.M'_P[q-1]\cdots M'_P[0]$ or $1.M'_P[q-1]\cdots M'_P[0]$ when $M'_P[q] = 1$ or $M'_P[q] = 0$, respectively. Moreover, $-0.5 \le l < 1$ means that $M'_P[q-1] = 1$ when $M'_P[q] = 1$. Therefore, in the case of $M'_P[q] = 1$, $1.M'_P[q-2]\cdots M'_P[0]0$ is obtained by performing the $\times 2$ operation on $0.M'_P[q-1]\cdots M'_P[0]$.

The circuits for the exponent conversion, addition, and value adjustment, shown as the green blocks in Fig. 6, are implemented together by integrating and simplifying multiple computations. For FPLM, according to (4), the exponents of the two operands are converted first, practically depending on $M_A[q-1]$ and $M_B[q-1]$. Then the converted exponents (denoted by E'_A and E'_B) are added subsequently, which is subtracted by 1 if $M'_P[q] = 1$, i.e., when $l < 0$. The required circuits to implement these operations are simplified to one adder with a carry-in bit, $Carry\_E$, that is determined by the modified value of E_A and E_B. When $M_A[q-1]M_B[q-1]$ are "00" or "11," $M'_P[q]$ can only be "0" or "1," respectively; otherwise, $M'_P[q]$ can be "0" or "1" in either case. When both mantissa values are smaller than 0.5 ($M_A[q-1]M_B[q-1] = 00$), $l \ge 0$ ($M'_P[q] = 0$), which means $E'_A + E'_B$ is not modified; hence, $Carry\_E = 0$. When both mantissas are greater

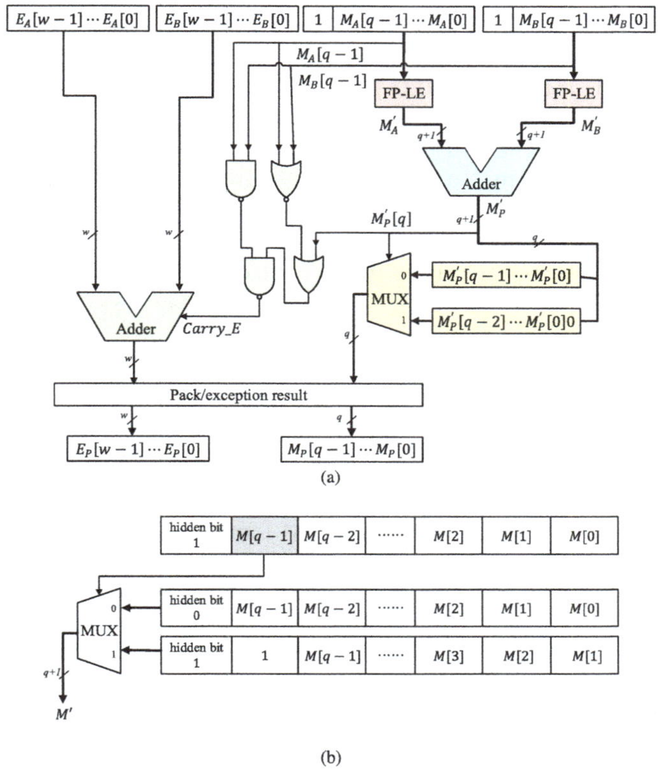

Fig. 9 The circuit design of FPLM: (**a**) the overall architecture and (**b**) the FP-LE circuit

than or equal to 0.5 ($M_A[q-1]M_B[q-1] = 11$), $l < 0$ ($M'_P[q] = 1$), which means that $E'_A + E'_B$ is subtracted by 1, i.e., $E_A + 1 + E_B + 1 - 1 = E_A + E_B + 1$; therefore, $Carry\_E = 1$.

Therefore, the $Carry\_E$ for FPLM is obtained by using (16). As shown in Fig. 9, four logic gates are used to generate $Carry\_E$.

$$Carry\_E = \overline{\overline{\left(M'_P[q] + \overline{(M_A[q-1] + M_B[q-1])}\right)} \cdot \overline{(M_A[q-1] \cdot M_B[q-1])}}\,. \tag{16}$$

6 Performance Evaluation and Neural Network Applications

6.1 Accuracy Evaluation

Two error metrics are considered to evaluate the error characteristics of the LFPMs.

Table 4 Error metrics for the multipliers

	Single-precision		Half-precision		Bfloat16		FP8	
	MRED	|AE|	MRED	|AE|	MRED	|AE|	MRED	|AE|
Normal distribution								
FPLM	0.0288	8.3×10^{-6}	0.0311	8.2×10^{-6}	0.0300	7.4×10^{-6}	0.2160	3.5×10^{-5}
LAM [25]	0.0382	1.2×10^{-5}	0.0410	1.3×10^{-5}	0.0434	4.3×10^{-5}	0.1891	4.8×10^{-5}
Uniform distribution								
FPLM	0.0289	3.2×10^{-5}	0.0289	0.0022	0.0302	0.0176	0.2311	0.5630
LAM [25]	0.0385	0.0833	0.0391	0.0848	0.0437	0.0950	0.1915	0.4380

- The mean relative error distance (MRED) is the average value of all possible relative absolute error distances.
- The average error (AE) is the average difference, which can be positive or negative between the exact and approximate products.

Four FP precision formats, 32-bit single-precision, 16-bit half-precision, brain floating-point (Bfloat16), and 8-bit FP (FP8) format in the form of (1, 5, 2) bits for the sign, exponent, and mantissa, are considered for the evaluation of each multiplier. The FP8 (1, 5, 2) format is chosen since it performs best with respect to the classification accuracy. It has been adopted to an application of NN training [24]. A sample of 10^7 random cases of uniform and standard normal distributions were generated, respectively, to obtain the results in Table 4. It is shown that, for a normal distribution, the FPLM performs up to 31% more accurately in 32- and 16-bit precisions with respect to MRED and has lower AE, compared to LAM. For a uniform distribution, the FPLM is more accurate in the single-precision, half-precision, and Bfloat16 formats with up to 45% smaller MRED and over $10^3\times$ smaller AE. For the FP8 format, both of the multipliers produce significantly larger errors; the FPLM is slightly less accurate than LAM.

6.2 *Hardware Evaluation*

The FPLM and the LAM in [25] were implemented in Verilog, and an EFM was obtained using the Synopsys DesignWare IP library (DW_fp_mult). All of the designs were synthesized using the Synopsys Design Compiler (DC) for STM's CMOS 28-nm process with a supply voltage of 1.0 V and a temperature of 25°C. The same process and the same optimization option were used to ensure a fair comparison. The evaluation results are shown in Table 5. All of the designs are evaluated with a 500-MHz clock frequency.

As shown in Table 5, the FPLM consumes 20.8× less PDP and 10.7× smaller area compared to the EFM in the single-precision format, while it consumes 23.5 × less energy and 6.5× smaller area in the FP8 format. Also, the FPLM achieves a shorter delay compared to the LAM [25] in the single-precision format. It is

Table 5 Circuit assessment of the FP multipliers

	FPLM	LAM [25]	EFM
	Single-precision		
Power (μW)	67.2	39.9	1366.9
Delay (ns)	1.66	1.69	1.70
Area (μm^2)	270.4	138.8	2910.3
PDP (fJ)	111.5	67.5	2323.7
	Half-precision		
Power (μW)	29.5	16.1	375.7
Delay (ns)	1.26	0.92	1.70
Area (μm^2)	116.8	76.5	1057.5
PDP (fJ)	37.2	14.8	638.6
	Bfloat16		
Power (μW)	30.4	18.0	252.2
Delay (ns)	1.22	0.92	1.70
Area (μm^2)	127.1	89.9	887.6
PDP (fJ)	37.1	16.5	428.7
	FP8		
Power (μW)	13.9	9.6	94.5
Delay (ns)	0.49	0.42	1.70
Area (μm^2)	58.5	48.1	385.6
PDP (fJ)	6.8	4.0	160.6

interesting to observe that the FPLM consumes less energy in Bfloat16 compared to the consumption in the half-precision format, while LAM is the opposite. Although LAM has the smallest PDP, its accuracy is lower as the trade-off. It is necessary to point out that, according to [26], the power consumption of the accurate FP multiplier is dominated by the mantissa multiplication for over 80% and the rounding unit for nearly 18%. Therefore, the power and area cost can be largely reduced due to the elimination of the mantissa multiplier and the rounding unit in the FPLM.

6.3 Neural Network Applications

6.3.1 Experimental Setup

The logarithmic FP multiplier is used in the arithmetic unit in a multilayer perceptron (MLP) to illustrate their performance in NNs with respect to the classification accuracy and hardware performance. It is evaluated against the EFM considered as the baseline arithmetic unit and the LAM in [25]. In the experiments, the exact multiplication is replaced with approximate designs in the training phase by using the PyTorch framework [27]. To fairly evaluate the effect of approximate multiplication on training, the multiplier used in the inference engine is an exact FP multiplier with the same precision as the approximate one. An NN employing approximate multipliers of four precisions was trained using the same number of epochs, and each training was repeated five times using random weight initializations. Since the code quality and GPU performance affect the acceleration of training, the training time is not eligible for comparison.

Three classification datasets, the fourclass [28], the HARS [29], and the MNIST [30], were used for the evaluation. A small MLP network was used for training the fourclass. The MLP networks used for the HARS and MNIST are (561, 40, 6) and (784, 128, 10) models, respectively.

6.3.2 Classification Accuracy Analysis

The comparison of the classification accuracy is shown in Fig. 10 to indicate the relative approximation error of the logarithmic FP multipliers. Since the benchmark NNs using EFM for training are considered as baseline models, the baseline accuracies for the four precisions are listed in Fig. 10 in the order of the single-precision, half-precision, Bfloat16, and FP8 formats.

The FPLM produces a higher accuracy than LAM for the three datasets in the single-precision, half-precision, and Bfloat16 format, while it degrades more in the FP8 format. For the MNIST, it is interesting to observe that the FPLM slightly

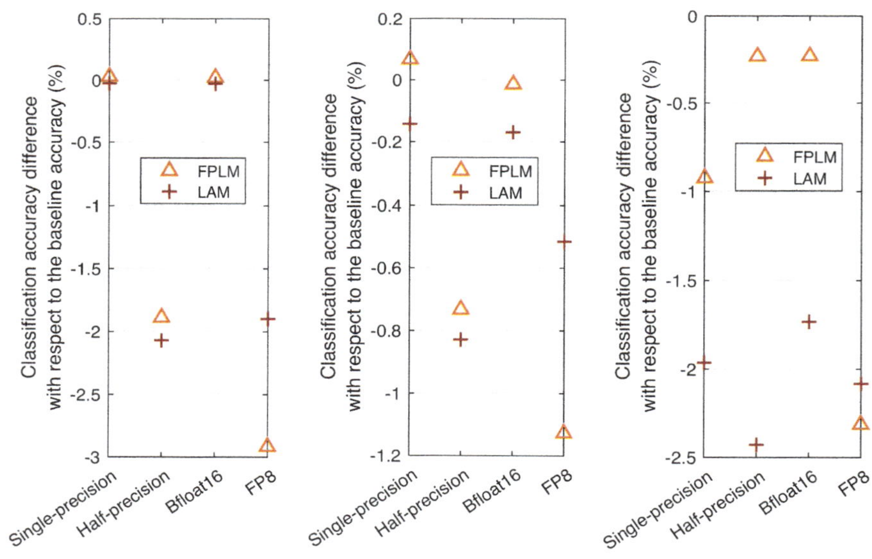

Fig. 10 Comparison of the average classification accuracy of three datasets with logarithmic multipliers for four precision levels: a negative percentage means a decrease, and a positive percentage means an increase in the accuracy with respect to using accurate multipliers. (**a**) For MNIST, (**b**) for HARS, and (**c**) for fourclass

Table 6 Circuit assessment of the artificial neuron

	FPLM	LAM [25]	EFM
Power (μW)	124.0	113.8	263.5
Delay (ns)	4.10	4.66	4.69
Area (μm^2)	674.8	606.9	1919.8
PDP (fJ)	508.4	416.5	1235.8

improves the classification accuracy in the single-precision and Bfloat16 formats. However, the classification accuracy for LAM-based training shows degradation in all the four precisions, with 2.068% in the half-precision. The FPLM shows up to 0.21% and 2.19% higher accuracy than LAM for the HARS and fourclass, respectively. It also indicates that the FPLM performs more accurately in the classification of larger dataset due to the offset effect of the double-sided errors.

6.3.3 Hardware Evaluation

The circuit of an artificial neuron with two inputs was measured to indicate the hardware cost of the implemented NNs. The FP adder used in the neuron was obtained using the Synopsys DesignWare IP library (DW_fp_add). The Bfloat16 was selected since LAM and FPLM performed very well for the classification (Fig. 10) with a low energy consumption (Table 5).

The simulation results in Table 6, obtained at a 200-MHz clock frequency, show that although the neuron using the LAM is more hardware efficient, the neuron using the FPLM consumes 2.4× less energy and is 2.8× smaller compared to the EFM-based neuron. In general, the hardware improvements are smaller than that in Table 5 since they are limited by the FP adder.

7 Conclusions

In this chapter, an approximation design framework using piecewise function approximations is presented for generating logarithmic FP multipliers with various accuracies and hardware complexities. The logarithmic FP multipliers are developed based on two FP number representation formats: the IEEE 754 Standard FP Format and the Nearest Power-of-Two FP Format. Consider both the logarithm and anti-logarithm functions; the applicable regions are divided into several intervals; then two approximation methods with positive or negative errors are considered for each interval. Both the LA and ALA can be simply implemented by shifting and multiplexing operations. Various configurations for approximation lead to approximate FP multiplier designs with different error characteristics, which support an assessment of error sensitivities of applications. The generic circuit design for those logarithmic FP multipliers is discussed with some simplification methods.

The library of logarithmic FP multipliers paves the way for efficient NN training by analyzing its inherent error tolerances for different error metrics. One example design with optimization methods is developed based on the approximation framework. Compared to using a conventional FP multiplier, the evaluation for several benchmark NNs reveals potential reductions in energy and area while achieving similar classification accuracies. Moreover, higher classification accuracies can be obtained by using the example design compared to using a conventional FP multiplier. We also found that no single FP logarithmic multiplier performed the best across all the datasets with different precisions. Therefore, the impact of FP logarithmic multipliers on the training process of NNs remains an important topic for future research.

Acknowledgments This work was supported by the Natural Sciences and Engineering Research Council (NSERC) of Canada (Project Number: RES0048688) and National Natural Science Foundation of China (NSFC) (Project Number: 62104127). T. Zhang is supported by a Ph.D. scholarship from the China Scholarship Council (CSC).

References

1. O. Gustafsson, N. Hellman, Approximate floating-point operations with integer units by processing in the logarithmic domain, in *ARITH* (2021), pp. 45–52
2. N. Wang, J. Choi, D. Brand, C.-Y. Chen, K. Gopalakrishnan, Training deep neural networks with 8-bit floating point numbers, *NIPS*, vol. 31 (2018)
3. J. Han, M. Orshansky, Approximate computing: An emerging paradigm for energy-efficient design, in *ETS* (IEEE, 2013), pp. 1–6
4. H. Jiang, F.J.H. Santiago, H. Mo, L. Liu, J. Han, Approximate arithmetic circuits: A survey, characterization, and recent applications, *Proc. IEEE*, vol. 108(12) (2020), pp. 2108–2135
5. D. Peroni, M. Imani, T.S. Rosing, Runtime efficiency-accuracy tradeoff using configurable floating point multiplier, *TCAD*, vol. 39(2) (2018), pp. 346–358
6. V. Camus, J. Schlachter, C. Enz, M. Gautschi, F.K. Gurkaynak, Approximate 32-bit floating-point unit design with 53% power-area product reduction, in *ESSCIRC* (IEEE, 2016), pp. 465–468
7. M. Franceschi, A. Nannarelli, M. Valle, Tunable floating-point for artificial neural networks, in *ICECS* (IEEE, 2018), pp. 289–292
8. P. Yin, C. Wang, W. Liu, E.E. Swartzlander, F. Lombardi, Designs of approximate floating-point multipliers with variable accuracy for error-tolerant applications. J. Signal Process. Syst. **90**(4), 641–654 (2018)
9. M. Imani, D. Peroni, T. Rosing, CFPU: Configurable floating point multiplier for energy-efficient computing, in *DAC* (IEEE, 2017), pp. 1–6
10. J.Y.F. Tong, D. Nagle, R.A. Rutenbar, Reducing power by optimizing the necessary precision/range of floating-point arithmetic, *TVLSI*, vol. 8(3) (2000), pp. 273–286
11. T. Zhang, Z. Niu, J. Han, A brief review of logarithmic multiplier designs, in *LATS* (IEEE, 2022), pp. 1–4
12. B. Xiong, S. Fan, X. He, T. Xu, Y. Chang, Small logarithmic floating-point multiplier based on FPGA and its application on MobileNet, *IEEE TCAS II* (2022)
13. C. Chen, W. Qian, M. Imani, X. Yin, C. Zhuo, PAM: A piecewise-linearly-approximated floating-point multiplier with unbiasedness and configurability, *IEEE TC* (2021)

14. T. Cheng, Y. Masuda, J. Chen, J. Yu, M. Hashimoto, Logarithm-approximate floating-point multiplier is applicable to power-efficient neural network training, *Integration*, vol. 74 (2020), pp. 19–31
15. Z. Niu, H. Jiang, M.S. Ansari, B.F. Cockburn, L. Liu, J. Han, A logarithmic floating-point multiplier for the efficient training of neural networks, in *GLSVLSI* (2021), pp. 65–70
16. X. Sun, N. Wang, C.-Y. Chen, J. Ni, A. Agrawal, X. Cui, S. Venkataramani, K. El Maghraoui, V.V. Srinivasan, K. Gopalakrishnan, Ultra-low precision 4-bit training of deep neural networks, *NeurIPS*, vol. 33 (2020), pp. 1796–1807
17. E.H. Lee, D. Miyashita, E. Chai, B. Murmann, S.S. Wong, Lognet: Energy-efficient neural networks using logarithmic computation, in *ICASSP* (2017), pp. 5900–5904
18. S. Ullah, S.S. Murthy, A. Kumar, SMApproxLib: Library of FPGA-based approximate multipliers, in *DAC* (IEEE, 2018), pp. 1–6
19. V. Mrazek, R. Hrbacek, Z. Vasicek, L. Sekanina, Evoapprox8b: Library of approximate adders and multipliers for circuit design and benchmarking of approximation methods, in *DATE* (IEEE, 2017), pp. 258–261
20. T. Zhang, Z. Niu, J. Han, A design framework for hardware-efficient logarithmic floating-point multipliers, in *TETC*. Under review (IEEE) (2024)
21. IEEE standard for floating-point arithmetic, *IEEE Std 754–2019 (Revision of IEEE 754-2008)* (2019), pp. 1–84
22. M.S. Ansari, B.F. Cockburn, J. Han, An improved logarithmic multiplier for energy-efficient neural computing, *IEEE TC*, vol. 70(4) (2020), pp. 614–625
23. J.N. Mitchell, Computer multiplication and division using binary logarithms. IRE Trans. Electron. Comput. **4**, 512–517 (1962)
24. N. Wang, J. Choi, D. Brand, C. Chen, K. Gopalakrishnan, Training deep neural networks with 8-bit floating point numbers, in *NeurIPS* (2018), pp. 7685–7694
25. T. Cheng, Y. Masuda, J. Chen, J. Yu, M. Hashimoto, Logarithm-approximate floating-point multiplier is applicable to power-efficient neural network training, in *Integration*, vol. 74 (2020), pp. 19–31
26. A. Gupta, S. Mandavalli, V.J. Mooney, K. Ling, A. Basu, H. Johan, , B. Tandianus, Low power probabilistic floating point multiplier design, in *ISVLSI* (2011), pp. 182–187
27. A. Paszke, S. Gross, F. Massa, A. Lerer, J. Bradbury, G. Chanan, T. Killeen, Z. Lin, N. Gimelshein, L. Antiga, A. Desmaison, A. Köpf, E. Yang, Z. DeVito, M. Raison, A. Tejani, S. Chilamkurthy, B. Steiner, L. Fang, J. Bai, S. Chintala, Pytorch: An imperative style, high-performance deep learning library, in *NeurIPS* (2019), pp. 8026–8037
28. C. Chang, C. Lin, Fourclass (1996). [Online]. Available: https://www.csie.ntu.edu.tw/cjlin/libs-vmtools/datasets/binary.html
29. U.M.L. Repository, Human activity recognition using smartphones data set (2012). [Online]. Available: https://archive.ics.uci.edu/ml/datasets/human+activity+recognition+using+smartphones
30. Y. LeCun, C. Cortes, C. Burges, MNIST handwritten digit database (2001). [Online]. Available: http://yann.lecun.com/exdb/mnist

Part IV
Quantum Computing and Other Emerging Computing

Cryogenic CMOS for Quantum Computing

Rubaya Absar, Hazem Elgabra, Dylan Ma, Yiju Zhao, and Lan Wei

1 Background

1.1 Brief Background on Quantum Computing

Classical processors have been revolutionized over the past decades, becoming a ubiquitous staple in modern technology [1]. Improvements in computational efficiency, in both power consumption and speed of computation, have been impressive. However, classical processors are still challenged by large combinatorial problems that require significant resource overhead, most commonly in terms of memory and computation time [2]. Compared to their classical counterparts, quantum processors have the potential to achieve significant computational speedups for specific problems, including prime number factorization (Shor's algorithm) [3], database search (Grover's algorithm) [4], certain classes of linear algebra problems [5], quantum cryptography protocols (BB84) [6], and more. A speedup in solving such problems is expected to impact a wide range of applications, including system security and encryption, drug discovery, financial portfolio optimization, and the analysis of rare failure events in advanced manufacturing processes.

Quantum processors use quantum bits (qubits) composed of two-level quantum systems exhibiting non-classical properties. A qubit can be prepared in a superposition of states and entangled with other qubits. Quantum algorithms take advantage of these non-classical features and provide exponential speedups for conventionally challenging applications. In the early 1980s, Richard Feynman and Yuri Manin first proposed the theory of quantum computers. They envisioned developing a quantum mechanical Turing machine to simulate objects of a quantum nature. David

R. Absar · H. Elgabra · D. Ma · Y. Zhao · L. Wei (✉)
University of Waterloo, Wateroo, ON, Canada
e-mail: lan.wei@uwaterloo.ca

W. Liu et al. (eds.), *Design and Applications of Emerging Computer Systems*,
https://doi.org/10.1007/978-3-031-42478-6_22

Deutsch later characterized the quantum Turing computer as an abstract computing system based on quantum mechanics [7]. In the following two decades, numerous theoretical and experimental advancements were made in the field of quantum computing. Several algorithms were developed for quantum computers to solve specific problems with less complexity and increased speed. One of the most notable examples was the prime number factorization algorithm by Peter Shor in 1994 [3]. He demonstrated that quantum computers could factor integers into prime roots in polynomial time exponentially faster than classical computers. In 1996, Lov Grover developed a quantum database search technique that provided a quadratic speedup for a wide range of tasks [4]. Any problem solved by a random search can now be solved at a four-times faster speed by employing this algorithm. Although Grover's algorithm delivered a less significant speedup than Shor's, it caught the attention of many people due to the extensive use of search-based techniques.

In 1998, the first 2-qubit quantum computer was developed by Mark Kubinec of the University of California, Isaac Chuang of the Los Alamos National Laboratory, and Neil Gershenfeld of the Massachusetts Institute of Technology (MIT) [8]. They implemented Grover's search algorithm for a four-state system using nuclear magnetic resonance (NMR) techniques. Since then, different physical systems have been explored to realize qubit creation and manipulation, including superconducting circuits [9], ion trap systems [10], electron spins in quantum dots (QDs) [11], topological nanowires [12], defects in solids [13], photonic systems [14], and others. Each physical implementation of a quantum processor requires the integration of classical electronics for communication, control, and/or readout. Some quantum technology (e.g., spin QD) can achieve all electrical operations, while others (e.g., trapped ion) may require non-electrical (e.g., optical) operations. One of the benefits of the all electrical operations is the potential of quantum/CMOS integration, which would take full advantages of the capability of technology scaling and large-scale integration achieved by the semiconductor industry.

In the past decade, significant advancements have been made in demonstrating quantum supremacy [15]. Quantum supremacy has arguably been claimed for limited examples, by Google using superconducting technology in 2019 and by the USTC, China, in 2020 using photonic technologies. In the near future, however, we will still be living in the noisy intermediate-scale quantum (NISQ) era, where we are limited in the number of physical qubits, gate depths, quantum error corrections (QEC), and so on. Undoubtedly, the frontier of quantum computing will continue to advance, with collaborative efforts to improve both hardware and software. For hardware, it is expected that the quantum systems will continue to improve on aspects such as scalability, controllability, and coherence.

1.2 Cryo-CMOS in Si QD Quantum Processor

The main components of a quantum processing unit (QPU) are illustrated in Fig. 1, regardless of the temperature. The quantum algorithm is executed on a

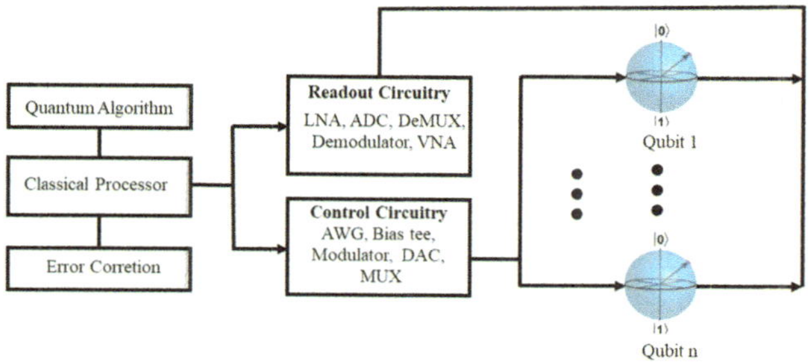

Fig. 1 The architecture of a QPU using spin QD. Though the manipulation and measurements are done differently for different qubit technology, in general, the quantum algorithm and the error-correcting scheme are first executed on a classical processor. In a spin QD system, the classical processor gives the instruction to the control and readout circuity based on the quantum algorithm and error correction methods. The control and readout circuitry (consisted of AWG, DAC, etc.) will then communicate with the qubits to perform necessary operations

classical processor that governs the control and readout operations of the qubits. The susceptibility of quantum systems to noise and other disturbances poses a significant barrier to developing a large-scale quantum processor. The information stored in a quantum system is readily lost to decoherence, i.e., by uncontrolled interactions between the system and its environment. Decoherence in solid-state devices can come from different sources, including thermal radiation, charge noise, control errors, and qubit crosstalk. Consequently, a fundamental prerequisite for a fault-tolerant quantum processor is quantum error correction (QEC). However, the error correction overhead for building a fully fault-tolerant device is exceedingly high, needing $\sim 10^4$ physical qubits for each logical qubit. As a result, millions of physical qubits will be required to construct fault-tolerant systems with enough logical qubits ($\sim$ hundreds) to handle issues that conventional processors cannot resolve. Integrating, connecting, and controlling such a large number of qubits is extremely challenging, regardless of the chosen platform.

One type of qubit technologies particularly gaining momentum lately is the silicon spin QD (Si QD), which has the advantages of high fidelity and long coherence time and most importantly is potentially scalable due to its manufacturing being compatible with existing CMOS fabrication process. In Si QD system, qubit control may be achieved through either electron spin resonance (ESR) or electron dipole spin resonance (EDSR) [16]. ESR operates by passing a large alternating current through a microstrip to generate a magnetic field to control the electron spin [16]. The alternative—EDSR—controls the spin state of the electron by applying microwave pulses of a specific frequency, f_{MW}, to the gate of the quantum well [16, 17]. The readout of each individual quantum bit (qubit) is achieved through a "frequency multiplexing technique" wherein a RF quantum point contact (QPC) and dispersive gate sensing (DGS) are used in conjunction with a multiplexer (MUX) [18]. The RF QPC can detect the presence of charge through the reflection/transmission characteristics of a carrier wave, while the DGS detects carriers tunneling in/out of the quantum well through changes in the quantum capacitance which changes the frequency response of a RF resonator [18, 19].

Figure 1 shows a generic architecture of a QPU using Si spin QD. The qubits are controlled using high-frequency ($\sim$ GHz) voltage pulses generated by an arbitrary waveform generator (AWG). A digital-to-analog converter (DAC) provides the necessary bias voltage required by the qubits. The output signal is amplified and digitalized using low-noise amplifiers (LNAs) and analog-to-digital converters (ADCs). The readout circuitry requires a vector network analyzer (VNA) to probe the reflected amplitude and phase of the qubits. Some multiplexing (MUX) and demultiplexing (DeMUX) circuits are used between the qubits and the electronics to reduce the wiring overhead.

Quantum processors require high-fidelity qubits with a control and readout circuit interface that maintain smooth communication between the qubits and high-precision electronics. The qubits operate at very low temperatures ($\sim$mK) as any thermal disturbance destroys information stored in quantum systems. In the current implementation, the qubits ($\sim$mK) are coupled to room temperature (RT) electronics for the control and readout operations (Fig. 2a), requiring a large number of wires to feed into the sub-kelvin temperature, causing a thermal load that severely limits the number of control lines. This approach is not scalable with one or more control lines per qubit.

The current research [20] adds an intermediate stage at 1–4 K temperature to place the control and readout electronics. The $\sim$1 W power budget at 1–4 K temperature is sufficient to implement circuitry for thousands of qubits. Some MUX and DeMUX circuits were also used to reduce the number of interconnects from the 4 K stage to RT. This architecture, shown in Fig. 2b, is a step toward scalability, but it still needs to address the issue of interconnecting the 4 K and lower temperature stages. Due to insufficient cooling power below 1 K, the electronics are limited to the 1–4 K range. Although MUX and DeMUX circuits can be placed at the 100 mK or lower temperature stages due to their ultra-low power consumption, as shown in Fig. 3c, the low 20 μW power budget will eventually limit the amount of multiplexing that is feasible. It is also not ideal to depend on heavy multiplexing and demultiplexing, as depending on whether time- or frequency-division multiplexing is employed, it either increases the delay between the electronics and qubits or imposes additional performance requirements on the electronics and qubits. To achieve such scalable system, many challenges should be addressed, from quantum technology, classical electronics, as well as quantum/CMOS integration. In the following sections, we will highlight the recent research on a few selected cryogenic CMOS topics, including the transport, high-frequency noise, and numerical simulations.

The ultimate scalable solution is to operate the control/readout electronics alongside the qubits at 1–4 K and considerably reduce the number of interconnects between the qubits and electronics. This arrangement is shown in Fig. 3d. The qubits and electronics both need to work at 1–4 K temperatures to implement such a system. Among the different qubit technologies mentioned in Sect. 1.1, silicon-based electron spin qubits can be considered as a candidate for developing such

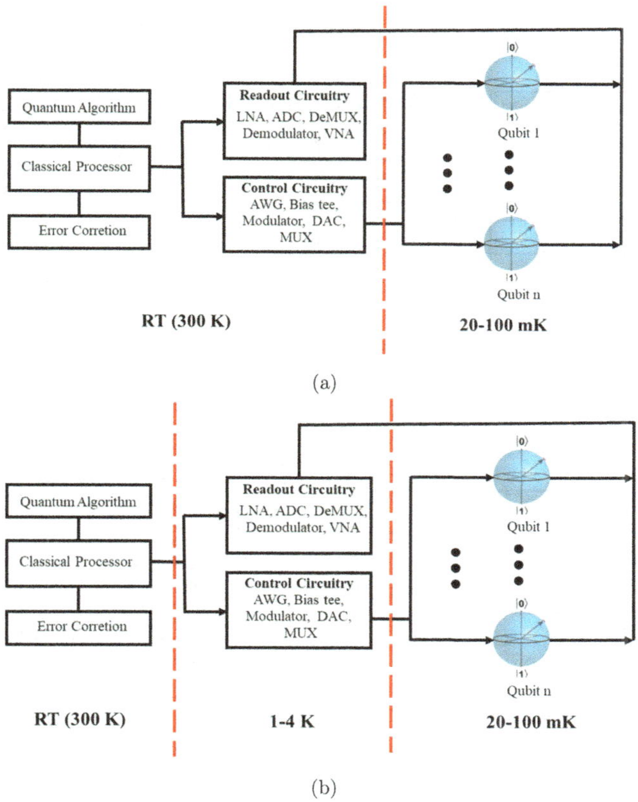

Fig. 2 (**a**) State-of-the-art quantum processors with qubits at 20–100 mK interfacing with RT electronics. (**b**) Proposed quantum processor with cryogenic electronics at 1–4K, which relaxes the cooling requirement at mK and reduces number of wires going from mK to RT

a scalable quantum processor. The silicon-based electron spin qubits are capable of operating at temperatures as high as ~2 K [21–23] with coherence periods reaching ~ 1 second. A long coherence time reflects a low probability of qubit errors, requiring less error correction overhead. The control reliabilities can reach up to 99.5% when isotopically enriched silicon (^{28}Si) is used for qubit realization [24–26]. The qubits can be closely packed as they have a small footprint on the order of 100×100 nm^2.

In terms of cryogenic electronics, several technologies have exhibited functionality at cryogenic temperatures, such as the junction field-effect transistors (JFET), high-electron-mobility transistors (HEMT), superconducting devices based on Josephson junctions, compound semiconductors (e.g., GaAs), and CMOS transistors [27, 28]. Table 1 shows some technologies and their lowest working temperatures. However, only complementary metal-oxide-semiconductor (CMOS) technology can provide low power consumption and functionality down to 30 mK

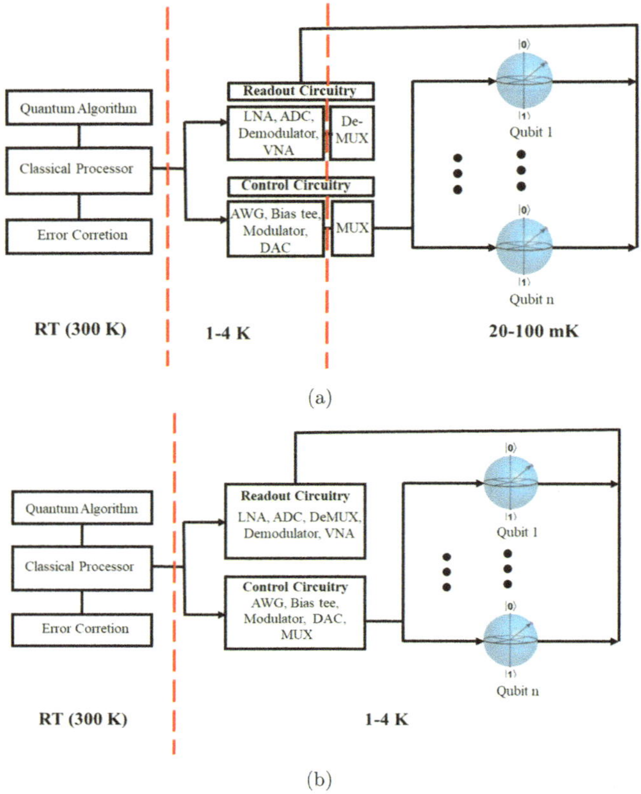

Fig. 3 (**a**) Quantum processor with some ultra-low-power multiplexing and demultiplexing circuits integrated with qubits at mK temperature to reduce wiring overhead from mK to 1–4 K stages. (**b**) Quantum processor with qubits and electronics integrated into the same die at 1–4 K

Table 1 Technologies and their lowest working temperatures [27, 28]

Semiconductors	Minimum temp.
Si BJT	100 K
Ge BJT	20 K
SiGe HBT	< 1 K
GaAs MESFET	< 4 K
CMOS	30 mK or below

[20] while integrating billions of transistors on a single chip. Recent experiments [29–31] show that CMOS at cryogenic temperatures (cryo-CMOS) can function with an increase in drain current of approximately 20–30% and an improvement in subthreshold slope (SS) from > 60 mV/dec at RT to as low as 10 mV/dec. Despite some second-order issues, cryo-CMOS can fulfill the jobs required for qubit control and readout operations [20, 32, 33].

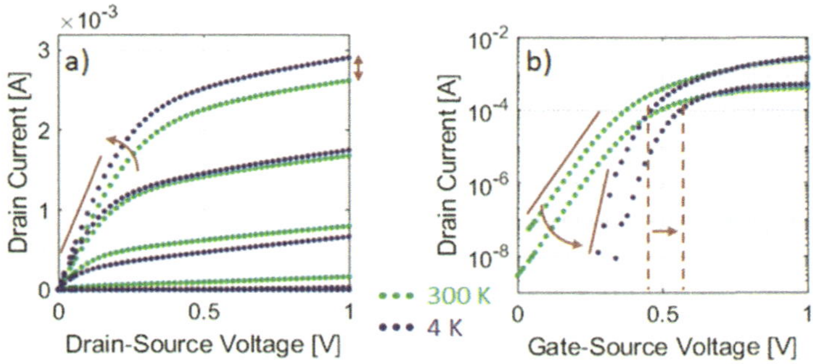

Fig. 4 Measured I-V characteristics of a 65-nm bulk NMOS device at RT and cryoT. (**a**) Output characteristics with $V_{GS} = [0, 0.2, 0.4, 0.6, 0.8, 1]$V. (**b**) Transfer characteristics with $V_{DS} = [0.05, 0.8]$V

The ultimate aim in establishing scalable, reliable, and high-performing quantum computing is undoubtedly a quantum integrated circuit (QIC), in which the array of qubits is integrated on the same chip as the CMOS electronics. The silicon-based qubits can be implemented in a manner similar to field-effect transistors (FETs) [34–36], presenting an opportunity to harness the semiconductor industry's integration capabilities to meet the upscaling difficulty. Silicon-based qubits integrated with CMOS control and readout circuits can increase overall system scalability while reducing operation latency.

Cryogenic CMOS refers to the design and operation of complementary metal-oxide-semiconductor (CMOS) integrated circuits at cryogenic temperatures, typically below 77K. This technology is critical for a range of applications in fields such as quantum computing, particle physics, radio astronomy, and deep space exploration, among others. In recent years, the need for more accurate numerical simulations of cryogenic CMOS circuits has become increasingly evident, due to progress in quantum computing research, in particular toward the integration with semiconducting Si QD-based quantum technology as explained above [16, 37]. Fortunately, the general behaviors of MOSFETs maintain at cryo-T—the device still behaves as a transistor whose gate voltage controls the formation of the channel through the field effect and drain voltage moves the carriers from source to drain to form channel current. Figure 4 shows the device I-V characteristics of a 65nm bulk MOSFET at RT and cryo-T showing qualitative similarity trends in the I-V dependence. Many aspects of the device physics at RT are also applicable at deep cryo-T. For example, the carriers would still follow the Fermi-Dirac distribution, although the distribution is sufficiently "sharp" at cryo-T as expected [37]. There are of course new physical effects and new challenges due to the extremely low operating temperate, some of which are highlighted in Sect. 2 (focusing on cryo-T transport behavior), Sect. 3 (focusing on cryo-T MOSFET high-frequency noise), and Sect. 4 (focusing on cryo-T numerical simulation). However, the fact that the MOSFET behaves similarly at RT and cryo-T (to the very first order) allows us to take advantage of the significant amount of knowledge, process, tools, and infrastructures developed and accumulated by the well-developed semiconductor industry in the past decades.

2 Transport in Cryogenic CMOS

2.1 *Cryogenic MOSFET Characteristics*

Although classical MOSFET models incorporate operating temperature as a parameter, the underlying theories were formulated with a focus on temperature ranges significantly above cryo-T. This generates two types of inherit shortcomings in such models: approximations based on RT assumptions without considering the implications of operating at low temperatures, and exclusion of cryo-T effects imperceptible at RT.

Consider the performance parameter change presented in Fig. 4 when the operating temperature of a 65-nm bulk NMOS device is reduced from 300 K to 4 K. At 4K, the device exhibits a higher threshold voltage (V_{th}), sharper subthreshold slope (SS), lower off-state current (IOFF) and higher on-state current (ION), which are generally observed in deep-submicron MOSFETs. The higher ION combined with lower IOFF are generally desired for IC, meaning that the circuit can enjoy better energy efficiency without sacrificing the speed. In fact, various proposals to use CMOS at cool temperature have been explored. Achieving high ION and low IOFF simultaneously by reducing the temperature is very attractive for circuit design. Focusing on device behavior, at first glance, some of these trends at cryo-T seem to corroborate classical theories predictions. However, closer inspection reveals the necessity to re-examine widely accepted models.

The subthreshold swing (S) is classically modeled as linearly related to temperature:

$$m = \frac{\partial V_{GS}}{\partial \phi_S} \tag{1a}$$

$$S = \frac{\partial V_G}{\partial log(I_D)} = mln(10)\phi_T \tag{1b}$$

where the subthreshold factor m is a dimensionless constant that describes the effect of V_{GS} on the surface potential ϕ_S. The thermal voltage $\phi_T = k_B T/q$ is related to the Boltzmann constant k_B and the elementary charge q. As its name suggests, ϕ_T is directly proportional to T, while the other values are constants and unaffected by T ($k_B = 1.38 \times 10^{-23}$J/K and $q = 1.6 \times 10^{-19}$C). Therefore, it is expected for S to improve by 75 times when transitioning from $T = 300$K to $T = 4$K, but the results shown in Fig. 4b indicate an improvement of only ~ 6 times. In fact, Fig. 5a reveals a subthreshold saturation effect in bulk CMOS at deep cryo-T. This effect is

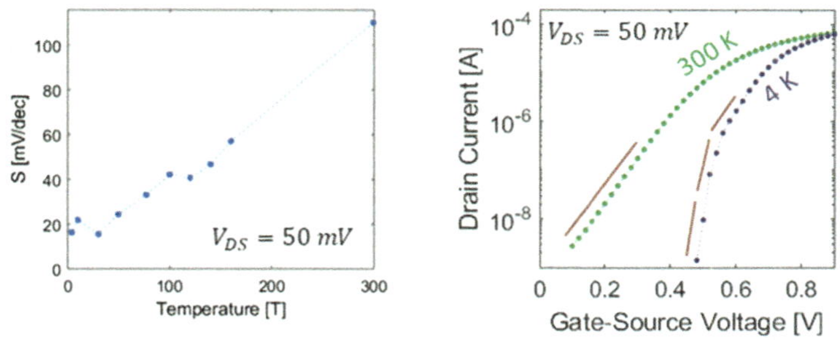

Fig. 5 (**a**) Measured subthreshold swing vs. temperature (colored solid line) compared against the classical model with $m = 1.48$ (solid black line). (**b**) Measured transfer characteristics at RT and cryoT, showing the variable m effect. Measurements were conducted on a 65-nm bulk NMOS sample with $V_{DS} = 50\text{mV}$

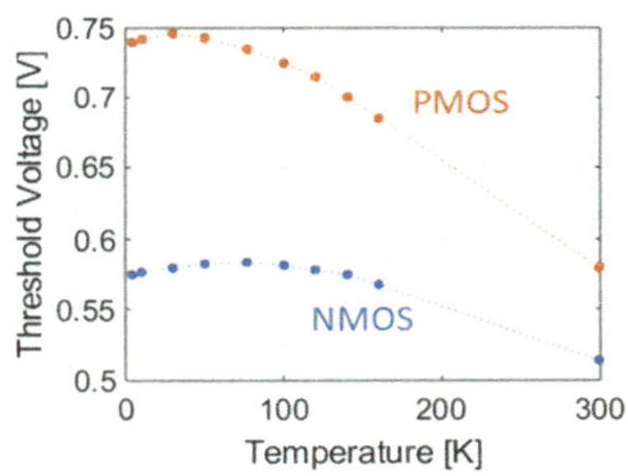

Fig. 6 Threshold voltage relation to temperature for 65-nm bulk devices NMOS

also observed in more advanced nodes such as FDSOI [38]. Moreover, evidence of bias-dependence at cryo-T have been observed in cryogenic I-V measurements [39]. A constant subthreshold factor leads to a constant S as seen in the RT transfer curve (Fig. 5b). However, variability in S can be observed well below V_{th} in the cryoT ID-VGS curve in the same figure, suggesting a gate-voltage dependence.

Another example of deviation from the expected behaviour is the threshold voltage response to decreasing temperature. V_{th} is commonly modeled to respond linearly with temperature [40]:

$$V_{th}(T) = V_{tho} + \alpha(T - T_o) \tag{2}$$

where, T_o is the reference temperature, V_{tho} is the reference threshold voltage, and, finally, α is a temperature coefficient, often around -2mV/K to 2mV/K for Si devices, primarily due to bandgap increase and mobility incease with decreasing temperature. Measurement results of V_{th} down to cryo-T are displayed in Fig. 6 where the linear relationship is apparent for temperatures above $\sim$ 150K, but the trend drastically changes at cryogenic temperatures. Indeed, the simple model Eq. 2 relies on multiple approximations that discount physical terms negligible at a higher temperature range, which can account for the observed discrepancy. Hence, some of the commonly adopted approximation at RT, such as complete ionization and Maxwell-Boltzmann distribution, should be re-visited when developing CryoMOSFET models. For an NMOS device with p-type bulk, the electrostatic expression of the threshold voltage is [41]:

$$V_{th} = 2\phi_F + \Phi_{MS} - \frac{Q(V_{th})}{C_{ox}} \tag{3}$$

where ϕ_F is the Fermi potential, Φ_{MS} is the metal-semiconductor work function, and $Q(V_{th})$ is the semiconductor charge developed due to the application of a gate-bulk voltage $V_{GB} = V_{th}$, and it includes mobile channel charges as well as localized depletion and interface charges. To evaluate $V_{th}(T)$, one should consider the temperature effect on of all terms in Eq. 3 as will be presented in Sect. 2.3.

Regarding transport parameters, carrier mobility, μ, is generally expected to improve with reduced temperature due to reduced phonon scattering [45] corroborating an empirical model provided for doped semiconductor in [46]. Measurement results (Fig. 7) reveal that while this seems to be generally the case, there are a few exceptions. The mobility starts to exhibit a saturation (Fig. 7a) or a non-monotonic behavior, as temperature approaches deep cryo-T (Fig. 7b) or even temperature independence for all T (Fig. 7c). This can be attributed to the existence of other scattering mechanisms that have different relations to temperature. For example, Coulomb scattering is inversely proportional to temperature, making its mobility component proportional to temperature [45], while neutral defects scattering is unrelated to temperature and heavily dominates mobility in small channel devices [44]. Moreover, the effective and ballistic mobility components are directly proportional to temperature through thermal velocity, v_t, [47]. Therefore, μ is expected to be a strong function of the manufacturing process. Similarly, carrier velocity displays similar inconsistencies across literature [48] which can be affected by the thermal velocity and those similar factors impacting mobility.

Finally, an unexpected effect that appears in cryoT known as the is the kink effect (Fig. 8a). As illustrated in Fig. 8b, the effect originates from increased bulk resistance at cryo-T due to the incomplete ionization of the substrate dopants,

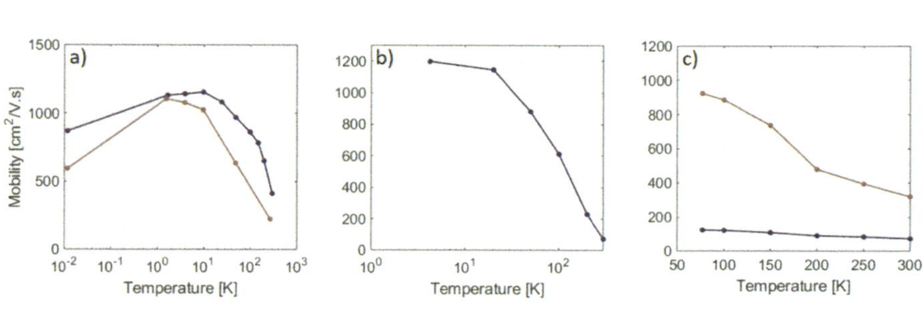

Fig. 7 Mobility vs. temperature. The data in (**a**), (**b**), (**c**) are reconstruction from [42], [43], and [44], respectively

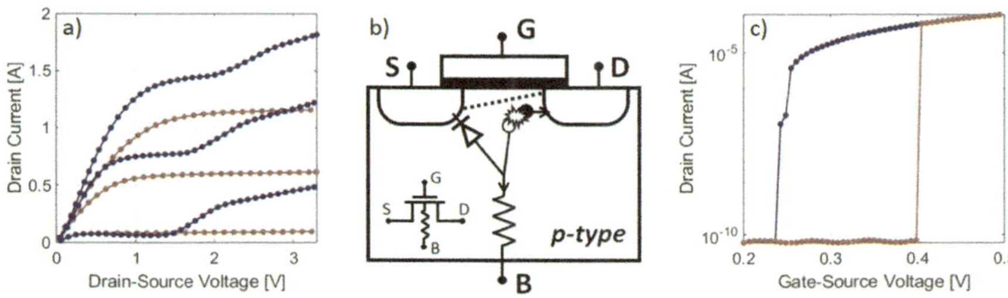

Fig. 8 (**a**) Output characteristics of a 160-nm bulk NMOS device at 300 K (dashed) and 4 K (solid). (**b**) NMOS cross section representation of the origin of the kink effect at cryo-T. (**c**) Transfer characteristics 4 K showing the hysteresis effect. These figures are reconstructed from [49]

which induces higher bulk potential, reducing V_{th} through the body effect, and increasing I_D. The hysteresis shown in Fig. 8c is a byproduct of this effect where the forward sweep captures the true V_{th} before the kink effect activates, while the backward sweep captures the reduced V_{th}. However, the kink effect requires high drain voltages to activate $V_{DS} \geq 1.5$V which are much higher than the operating voltages of modern process nodes ($\tilde{1}$ V) [49].

Hence, this effect is not included in the model as it does not occur within the possible operation range.

2.2 *Compact Models*

Compact models are a vital component in the integrated circuit (IC) design process and are an indispensable tool for circuit designers in today's fast-paced and ever-changing technology landscape. They serve as a bridge between the complex physical behavior of an individual device and circuit simulators that simulates the behavior of large numbers of devices integrated together into a complex circuit. This is accomplished by describing the physics behind a device performance and response to operating conditions into a set of simple but accurate mathematical expressions.

Compact models must meet several requirements to ensure that designers can confidently and efficiently simulate large circuits. They must be flexible enough to support the wide range of design parameters expected in large-scale ICs (e.g., device width and length) as well as process parameters' variation used by circuit designers to gauge circuit reliability against process uncertainties. In order for designers to have confidence in their simulations, compact models must also be accurate and validated over a wide range of operating conditions while still being computationally efficient to allow for a large number of devices in complex systems. Finally, compact model must have the computational efficiency and robustness to run and converge in different simulators. Thus, they must not rely on a specific

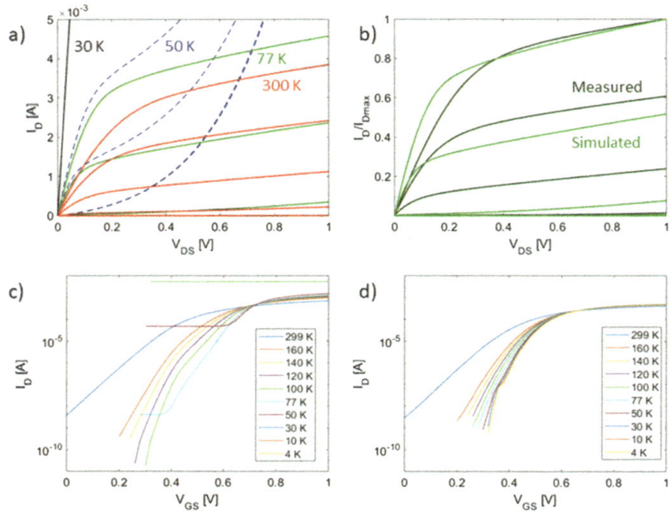

Fig. 9 I-V curves of a 65-nm NMOS device with W/L : 5μm/65nm, where (**a**, **b**) display the output characteristics and (**c**, **d**) display the transfer characteristics. Data for (**a**) and (**c**) are simulation results using the model provided in the PDK, while (**d**) is sourced from experimental measurements. (**b**) contains the simulated and measured data for the same devices at the same operating temperature ($T = 77$K) to highlight the discrepancy between them

commercial simulator's principles and be able to produce the same accurate results in different simulators, as well as ensure smoothness, efficiency, and backward compatibility.

Compact models are essential in process design kits (PDKs) provided by semiconductor foundries, which are carefully calibrated to match the measured behaviors of the physical components within the target range of operating conditions (e.g., voltage, temperature, etc.). The models are not meant to be used beyond the calibrated temperature range, which is well above 200 K. The cryogenic capability of a 65-nm bulk technology from a major commercial foundry is tested. The simulation results are shown in Fig. 9, where unrealistic results started appearing around below 77K (Fig. 9a). Experimental measurements and simulated results of the same device at 77K, which is beyond the calibrated temperature range, are compared together in Fig. 9. It is clear from Fig. 9b that the device parameters (such as mobility and threshold voltage) are different between simulated and actual values. Moreover, the exaggerated V_{th} shift and subthreshold performance inaccuracies are apparent from Fig. 9c and d.

2.3 Progress of Cryogenic MOSFET Models

As demonstrated in Sect. 2.2, commercial models are incapable of predicting device performance at cryogenic temperatures, rendering cryo-circuit simulations invalid. To overcome this limitation, low-temperature circuit development utilized other methods to produce results while cryoMOSFET model development initiated. Initially, several groups resorted to manually adjusting the parameters of the process-provided model (e.g., EKV, BSIM3 [50–55]) to represent the device performance at cryo-T. This approach requires measurement and parameter extraction for every cryo-T temperature point as a separate exercise resulting in different sets of parameter values at different temperatures. In other words, fitting the same device for n cryo-T points requires comparable labor to fit n different devices from n different process nodes. Simple effects exclusive to cryo-T operation, such as the kink effect, are either avoided [53] or added to the model (PSP and BSIM3) as a nonlinear resistor series with bulk [49]. There has been also notable effort to include temperature dependency to some parameters (e.g., μ) to avoid the need for specific parameter sets for each temperature. Other complex effects such as those discussed in Sect. 2.1 require more substantial modifications to existing models [56–59] or even developing new ones from scratch [38, 60, 61]. Overall, the cryoMOSFET model is an active and exciting research area, with dedicated efforts and progress seen on a wide range of commercial models, including BSIM, XMOS, EKV, and MVS [62–65]. In the following sections, progress on subthreshold swing and threshold voltages are particularly discussed.

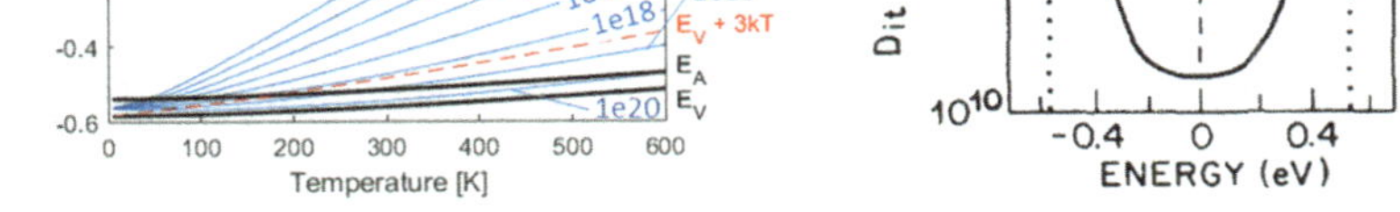

Fig. 10 (**a**) The Fermi level E_F positioning in Si as a function of temperature for select doping concentrations [46]. (**b**) Typical interface trap densities within the bandgap at the Si-SiO_2 interface [66]. E_C, E_V, E_i, E_A, and E_D are the bottom of the conduction band, top of the valance band, intrinsic Fermi level, and acceptor and donor energy level, respectively

2.3.1 Subthreshold Swing S

To this day, extensive effort is put forward to produce a cryogenic model suitable for circuit simulations. The performance parameter that received the most attention is the subthreshold swing. Since the Fermi level (E_F) significantly shifts toward the band edges (Fig. 10a), it was quickly determined that interface trap density D_{it} that increases exponentially toward the band edges (Fig. 10b) is responsible for the S effects observed in Sect. 2.1. However, a full understanding of cryogenic S phenomena and a complete model are yet to be achieved. These phenomena can be accounted for by turning the constant subthreshold factor in eq. 1 into a temperature-dependent property. The S discrepancy was initially compensated by a ratio between measured m at RT and cryo-T [49, 51] or an additional temperature-dependent m term [50, 61, 66, 67]. These models were used in literature until recently challenged by Bohuslavskyi et al. in 2019 [68].

Bohuslavskyi argues that this approach leads to unrealistic Dit values at deep cryoT (T =< 4K) that exceed the silicon density of states (DoS) of free carriers, and prompted the need for another theory. Band broadening-induced S saturation was first described in [68] where band tails are placed at the edges of the bandgap removing the bandgap in the ideal $DOS(E)$ and affecting the mobile carrier concentration. These band tails are characterized with a characteristic energy Wtail $= k_B T_S$, where T_S is the saturation temperature. The band tail is attributed to intrinsic mechanisms such as electron-phonon scattering, electron-electron and electron-hole interactions, and finite crystalline periodicity [56, 57]. The S saturation effect is modeled using the Gauss hypergeometric function, $F_1(a, b; c; z)$ [56, 57]:

$$S(T) = m_o \phi_T ln(10) \times \frac{\theta^{-1} F_1(1, \theta; \theta + 1; z)}{-z(1+\theta)^{-1} F_1(2, \theta + 1; \theta + 2; z)} \tag{4}$$

where $\theta =$ kBT/Wtail is unrelated to θ in eq. 4.11b, and $z = -$ exp[(EC-EF)/kBT] at the surface.

Moreover, the gate voltage dependence of m has been documented in literature [39, 58, 68] and attributed to localized disorder-induced interface traps. The most recent model to date is expressed using the Gaussian distribution, chosen to mathematically limit the distribution's growth at and beyond the band edges (EC and EV) [58]:

$$Q_o = q \int_{-\infty}^{\infty} G(N_o, W_o) f(E)\, dE \quad Q_o = q \frac{N_o}{2} \left[\text{erf} \left(\frac{E_F - E_C}{W_o \sqrt{2}} \right) + 1 \right] \tag{5}$$

where $G(N_o, W_o)$ is the Gaussian distribution with magnitude N_o and width W_o and $f(E)$ is the Fermi function. Equation 5 is only valid at deep cryo-T where $f(E)$ can be approximated by a Heaviside step function [69].

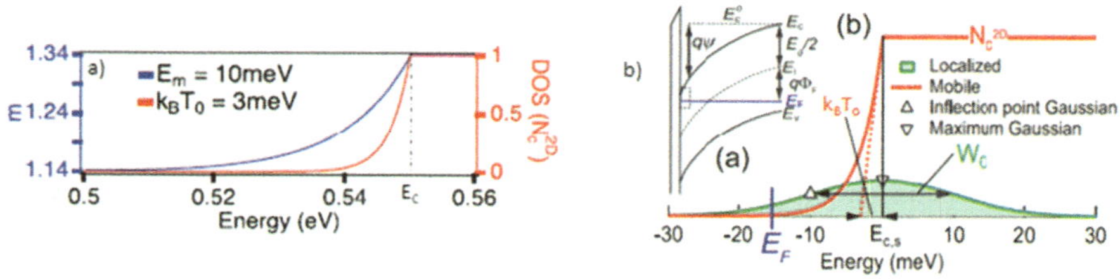

Fig. 11 Modified visualizations of the mobile and immobile band tails within the bandgap as theorized by (**a**) [68] and (**b**) [58]

2.3.2 Threshold Voltage V_{th}

Threshold voltage is another performance parameter that received a lot of attention for cryoMOSFET models. Electrostatic physics are employed to accurately model the V_{th} response to temperature variation producing similar models (Eqs. 6 and 7):

$$V_{th}(T) = \phi_F + \Phi_M - \frac{Q(V_{th})}{C_{ox}} + \gamma\sqrt{2\phi_F} + \Delta V_G \tag{6}$$

$$V_{th}(T) = \phi_F + \Phi_M - \Phi_{S_i} + \frac{Q_{D_i}}{C_{ox}}\sqrt{2\phi_F + \Delta\phi_F} + \frac{Q_{it}(\phi_F)}{C_{ox}} \tag{7}$$

Dao et al. [60] proposed Eq. 6 where γ is the body effect coefficient and ΔV_G represents the additional gate voltage required to completely ionize the dopants near the surface through field-assisted ionization. The second and third terms are assumed constant with temperature, while the third term models the body effect (dependent on ϕ_F). The field-assisted ionization term involves further expressions and fitting parameters. The majority of the temperature response is governed by ϕ_F. The detailed physical derivation including incomplete ionization is provided in [46, 61]. The energy levels term in the derived $\phi_F(T)$ is replaced in [60] with a fitting parameter to avoid exponential calculations.

Meanwhile, the model provided by Beckers et al. [59] in Eq. 7 preserved the full form of $\phi_F(T)$, included the temperature response of depletion and interface charge, and did not consider the body effect. The T in Eq. 7 denotes the temperature-independent portions of a parameter. The model also assumed constant metal work function and C_{ox} with temperature, similar to [60] due to lack of measurement data in literature. Finally, to account for field-assisted ionization, the model redefines the inversion threshold to be $2\phi_F + \Delta\phi_F$ where $\Delta\phi_F$ is the incomplete ionization term in $\phi_F(T)$.

3 High-Frequency Noise in Cryogenic CMOS

As described in Sect. 1.2, there is a great desire to integrate and operate CMOS at cryogenic temperatures in the push of scaling up the quantum computing system. This has highlighted a variety of effects exclusive to this temperature regime. In addition to various transport issues (such as change in threshold voltage, partial ionization of dopants, etc.), the noise performance of short-channel metal-oxide-semiconductor field-effect transistors (MOSFETs) is poorly understood. This noise generated by the MOSFETs may be transferred to the qubits which are extremely sensitive to noise. While the noise of a long-channel device is expected to scale down linearly with temperature, these long-channel devices have low device density and performance making them un-appealing for integration onto a quantum processor. The issue is that as the channel length shrank below 100 nm, the measured noise greatly exceeded thermal noise model predictions—a phenomena called "excess noise" [70]. This increased noise also did not follow a thermal dependence as the increase in noise seen at room temperature becomes more pronounced at 77 K [71]. As such, the use of short-channel devices—while desirable for their improved performance—may introduce intolerable amounts of excess non-thermal noise into the quantum system. Thus, understanding and modeling the potentially non-temperature-dependent noise is important to aid in the design of cryogenic qubit control and readout circuits.

3.1 High-Frequency Noise in MOSFET

Before discussing the challenges with modeling this "excess" noise, it's important to review how noise is quantified. First, the time-dependent current can be written as $I(t) = \langle I \rangle + \Delta I(t)$ where $\langle I \rangle$ is the average current and $\Delta I(t)$ is the fluctuation/noise in the current. To quantify how "much" noise there is at a specific frequency, the power spectral density (PSD) is defined:

$$S(\omega) = \lim_{\tau \to \infty} \frac{2|\Delta I(\omega)^2|}{\tau} \quad (8)$$

This has units of A^2/Hz and V^2/Hz for current and voltage PSDs, respectively. Here, $|\Delta I(\omega)|$ is the Fourier transform of the fluctuation, $\Delta I(t)$, over the period of measurement, $-\tau/2 \leq t \leq \tau/2$. Notable in Eq. 8 is a dependence on frequency—noise at different frequencies has different properties due to their different physical origins. For the RF regime of interest, the PSD is "white noise," which is "flat" in frequency—that is to say, the PSD is independent of frequency. Two sources of white noise are of interest here: thermal (Eq. 9) and shot (Eq. 10) noise:

$$S_I(\omega) = 4k_B T G \quad (9)$$

$$S_I(\omega) = 2q \langle I \rangle \cdot F \tag{10}$$

Here, k_B is the Boltzmann constant, T is the conductor's temperature, G is the conductance, q is the electron charge, $\langle I \rangle$ is the average current, and F is the "Fano factor" which determines how "suppressed" the shot noise is [72]. This suppression factor is used to model a variety of effects which may reduce the shot noise. Thermal noise for a conductor was first experimentally observed by John B. Johnson in 1928 and theoretically explained by Harry Nyquist later in the same here (and hence thermal noise is sometimes referred to as the Johnson-Nyquist noise) [73, 74]. This thermal noise is ascribed to the random Brownian motion of electrons in a conductor causing fluctuations in current/voltage. In contrast, Walter Schottky proposed shot noise in 1918 for carriers emitted by vacuum tubes [75]. The origin may be viewed as arising due to the discrete nature of carriers—electrons incident on a potential barrier may be randomly reflected or transmitted depending on the properties of the potential barrier.

With respect to noise in MOSFETs, the inversion layer may be viewed as (to the first order) a resistive layer. From this viewpoint, many long-channel noise models arose: like random thermal motion in a resistor governed by Eq. 9, the random thermal motion of carriers in the inversion layer gives rise to fluctuations in current and voltage [76, 77]. To this end, one of the first models for thermal noise was proposed by van der Ziel for a junction-gate field-effect transistor (JFET) and was later adapted by Sah, Jordan, and Jordan for MOSFETs [77, 78]:

$$S_I(\omega) = 4k_B T \gamma g_{d0} \qquad \gamma = \frac{1 - v - \frac{1}{3}v^2}{1 - \frac{1}{2}v} \qquad v = \frac{V_{DS}}{V_{GS} - V_t} \tag{11}$$

Here, k_B is the Boltzmann constant, V_t is the threshold voltage, T is the MOSFET temperature, and g_{d0} is the drain conductance at zero source-drain bias. γ is known as the "noise factor," a fitting parameter which transitions from $\gamma = 1$ in equilibrium to $\gamma = 2/3$ in saturation. An alternative model for thermal noise is the classic Klaassen-Prins equation [71]:

$$S_I(\omega) = \frac{4k_B T}{I_{DS} L^2} \int_0^{V_{DS}} g^2(V) dV \tag{12}$$

where L is the channel length, $g(v)$ is the local channel conductance, and I_{DS} is the drain current. This model for thermal noise is the one of the most commonly used thermal noise models in research and is widely implemented in many MOSFET compact models [70, 77]. Alternative models such as that of Tsividis' also exist [79]:

$$S_I(\omega) = \frac{4k_B T \mu_{eff} Q_{inv}}{L_{elec}^2} \tag{13}$$

where L_{elec} is the channel length, μ_{eff} is the effective mobility, and Q_{inv} is the total inversion layer charge in the channel. These thermal models have been extensively experimentally verified for long-channel devices—some examples are presented in [80, 81]. To model noise in subthreshold, a shot noise-based view is taken. For example, the noise in subthreshold may be modeled as [79]:

$$S_I(\omega) = 2qI_{DS}\coth\left(V_{DS}/2\phi_t\right) \tag{14}$$

Here, V_{DS} is the drain-to-source voltage, and ϕ_t is the thermal voltage k_BT/q. It is worth noting that in equilibrium as $V_{DS} \rightarrow 0$, the hyperbolic function can be approximated as $\coth(V_{DS}/2\phi_T) \rightarrow 2\phi_T/V_{DS}$. That is to say, Eq. 14 recovers the thermal noise equation, Eq. 11, under equilibrium.

This recovery of thermal noise under equilibrium highlights the key challenge to modeling noise in short-channel devices: the thermal noise equations (Eqs. 11, 12, and 13) are all derived with the assumption that the carriers exist at or very close to thermal equilibrium [70]. These thermal noise models have also never been validated for non-equilibrium transport [82]. However, as has been confirmed by various numerical studies, as channel lengths continue to shrink, transport in MOSFETs becomes increasingly non-equilibrium in nature [82–84]. This non-equilibrium nature of noise has been explored by some numerical studies utilizing non-equilibrium hydrodynamic transport models with the impedance field method which has been able to model this shot noise-like behavior [85–87]. Ultimately, the use of numerical models is unsuitable for circuit design due to their computational expense, and thus, a compact model for noise is desired. However, existing compact models for transport in short-channel devices are derived considering only the region "near equilibrium" (at the top of the source-channel potential barrier) as it is this region which bottlenecks the current and a non-equilibrium compact model would be necessary to model the noise behavior [47].

3.2 Noise in a Mesoscopic View

Much of the theoretical background for the non-equilibrium nature of noise arose in the 1980s and 1990s from the field of mesoscopic physics. The fields of transport and noise in mesoscopic devices are extremely extensive and cannot be covered in sufficient detail here, and excellent reviews of the mesoscopic origins of noise are provided in [88–91]. However, the key takeaway is that for a two-terminal nanoscale device where only elastic scattering occurs, the noise in the device is [92]:

$$S_I(\omega) = \frac{4q^2}{h}\int dE\, \mathcal{T}_n(E)\left[f_S(1-f_S) + f_D(1-f_D)\right] + \mathcal{T}_n(E)[1-\mathcal{T}_n(E)](f_S - f_D)^2 \tag{15}$$

where h is Planck's constant, $\mathcal{T}_n(E)$ is the energy-dependent transmission eigenvalues, and $f_{S/D}$ is the Fermi-Dirac distribution for the source/drain. For energy-independent transmission probabilities, $\mathcal{T}_n$, Eq. 15 may be evaluated to be:

$$S_I(\omega) = 4k_B T_e G_{eq} + 2q\langle I\rangle F\left[\coth\left(\frac{qV}{k_B T_e}\right) - \frac{2k_B T_e}{qV}\right] \qquad F = \frac{\sum_n \mathcal{T}_n(1-\mathcal{T}_n)}{\sum_n \mathcal{T}_n} \tag{16}$$

where T_n is the transmission probability for the mode n, T_e is the temperature, and $\langle I\rangle$ is the average current. G_{eq} is the conductance of the sample under equilibrium, and V is the bias voltage applied across the two-terminal device.

Equations 15 and 16 can be studied to highlight the origin of the equilibrium and non-equilibrium noise in a device. First, for the equilibrium case ($V \ll k_B T/q$), Eq. 16 reduces to the Johnson-Nyquist noise equation (Eq. 9). This can be understood by examining Eq. 15: the $4k_B T_e G_{eq}$ term arises from evaluating the integral of $\mathcal{T}_n(E)\,[f_S(1-f_S) + f_D(1-f_D)]$. To understand this term, we may view the injection of carriers by the source/drain contacts as a Bernoulli process where the probability of "success" (i.e., carrier injection) at energy "E" is simply the Fermi-Dirac distribution: $p = f(E)$. Then it can be seen that the variance of this Bernoulli process (i.e., variance in carrier injection) is simply $\sigma_p^2 = f(E)[1 - f(E)]$. As such, it can be seen that the thermal noise in Eq. 15 arises from the variance in random thermal injection of carriers into the device.

Conversely, for the "excess noise" or the case far out of equilibrium ($V \gg k_B T/q$), the shot noise term in Eqs. 15 and 16 dominates (second term—see Eq. 10). This term arises from the elastic scattering in the device which in turn causes the "partitioning" of carriers—a fraction of carriers injected by either the source or drain is transmitted, while the rest are reflected. This may be understood by treating the transport as a Bernoulli process with the added caveat that the number of trials, n, may vary (i.e., the number of carriers injected may vary). Here, we may define "success" as the event that the carrier transmits from source to drain in which case $p = \mathcal{T}(E)$. The variance in the number of carriers measured, "k," may be found by applying the Burgess variance theorem [93]:

$$\sigma_k^2 = \sigma_n^2 p^2 + \langle n\rangle\, p(1-p) \tag{17}$$

where k indicates the number of carriers measured (i.e., current), n is the number of carriers injected, and p is the probability of successful transmission. As before, the average number of carriers injected is determined by the Fermi-Dirac distribution, $\langle n\rangle = f(E)$, while the variance in the number of carriers injected is $\sigma_n^2 = f(E)[1 - f(E)]$. Then the variance in the number of carriers measured in one electrode due to electrons emitted from the other is:

$$\sigma^2 = \mathcal{T}^2(f_{1/2}(E)(1 - f_{1/2}(E))) + \mathcal{T}(1-\mathcal{T})(f_{1/2}(E)(1 - f_{2/1}(E))) \tag{18}$$

which can be expanded to recover Eq. 15. These equations have been experimentally verified for quantum point contacts and metallic diffusive wires [94–97].

These models describe the noise in a device with purely elastic scattering, a picture reminiscent of a short-channel MOSFET. In short-channel MOSFETs, the most prominent scattering mechanism is the elastic scattering of carriers off the channel potential near the source-channel junction [98]. Additionally, carriers injected near the source primarily interact with acoustic phonons which scatter the carriers elastically as they have insufficient energy to interact with phonons which would scatter the carriers inelastically [99]. Additional mechanisms near the source are also elastic; thus, the dominant scattering mechanisms in short-channel devices are elastic scatterers focused near the top of the barrier (ToB) or the "virtual source" [47, 100]. Based on this, it is evident that the elastic scattering in a short-channel device partitions carriers to produce shot noise which is the non-thermal "excess noise" measured as MOSFETs shrank below 100 nm [101].

The question now is why wasn't this excess noise measured in long-channel devices? Again, headway toward answers for this question was made by the field of mesoscopic physics in the 1990s. In essence, the combination of carriers losing energy as well as the Fermionic nature of electrons causes a change in the electron distribution and a reduction in the independence of scattering events, thereby reducing the independence of transport [82, 88, 93, 102]. But how does this lack of independence affect noise? First, it's important to note that "full" shot noise (i.e., Eq. 10 with $F = 1$) requires independent carrier emission [75]. For example, for carriers independently emitted from a source separated by time "τ," the distribution of carriers takes on a Poissonian distribution [89]:

$$P_N(t) = \frac{t^N e^{-t/\tau}}{\tau^N N!} \tag{19}$$

where N is the number carriers in timeframe t. However, scattering events—both elastic and inelastic—can affect the independence of transport, thereby reducing the measured shot noise. This can be seen in the case of elastic scattering where the suppression or "Fano factor" depends on the probability of transmission $\mathcal{T}$ (see Eq. 16).

Electrons in a MOSFET also experience a similar force causing carriers to lose energy and approach an equilibrium distribution: inelastic scattering. In particular, as carriers travel further beyond the virtual source, they continue to gain energy and may interact with g-longitudinal optical phonons with $\hbar\omega = 61\,\text{mV}$ which inelastically scatter these carriers causing them to lose energy in multiples of $k_B T$ [99, 100]. In short-channel devices, this inelastic scattering is rare as any carriers injected from the source are immediately shuttled out to the drain before it can gain sufficient energy to be inelastically scattered. However, as the channel length increases, the carrier must travel further to the drain and thus continues to gain more energy and experiences more inelastic scattering events. These inelastic scattering events are unlikely to cause the carriers to back-scatter to the drain as they lack the energy to overcome the multiple $k_B T$ of energy they have lost and thus are swept to the drain [103]. The increased optical phonon scattering also increases the charge density which causes increased inelastic carrier-carrier scattering [98, 99]. As such, as the channel length increases, the increased inelastic scattering causes increased shot noise suppression; thus, the only noise measured in the device is the thermal noise (see Eq. 16 as $F \rightarrow 0$) [101, 104]. In the other limit, as the channel length and number of inelastic scattering events have decreased, the shot noise becomes greater than the thermal noise and exhibits itself as the measured "excess noise."

Because fundamentally noise is a combination of both thermal and shot noise, this presents a potential problem for cryogenic MOSFETs. As mentioned, classically, the noise is expected to scale down linearly with temperature. However, the above suggests that as the temperature scales down, the "excess noise" or thermal noise would overtake the thermal noise. Looking at Eqs. 15 and 16, it would appear that the noise does not scale with temperature which would mean that MOSFETs operating at cryogenic temperatures may inject much more noise into the qubits than initially anticipated. Because this shot noise is also suppressed by inelastic scattering, the shot noise may increase at cryogenic temperatures further complicating cryogenic MOSFETs. Finally as a note, experiments studying shot noise in mesoscopic physics are often conducted at cryogenic temperatures to remove the effects of thermal noise [90].

3.3 Experimental Verification

While this theory for shot noise was developed prior to the 2000s, it wouldn't be until 2009 when the first measurement of shot noise in MOSFETs was demonstrated by Jeon et al. [105]. In their work, the noise in MOSFETs with channel lengths ranging from 10 nm to 180 nm was measured at 12 GHz [105]. Importantly, it was seen that for a 10 nm device operating at $V_G = 0.9\,\text{V}$, nearly full shot noise was measured [105]. Additionally, as the gate bias increased, the Fano factor decreases which may be due the increased lateral electric field inducing stronger surface scattering. Finally, it was demonstrated that as the channel length increases, the noise in the device appears to transition from shot noise to thermal noise [105].

Further work was done in 2018 by Wang et al. where they studied a 40 nm and 120 nm device at 10 GHz and 300 K/77 K [106]. This experiment corroborated the observation that a decrease in shot noise is observed with increasing V_G [106]. As the temperature was decreased from 300 K $\rightarrow$ 77 K, a slight decrease in shot noise was also observed [106]. A similar gate bias dependence was also recently demonstrated by Ohmori and Amakawa [107].

Das and Bardin have also recently extracted the Fano factor for a variety of channel lengths and device technologies (ranging from 12 nm FinFET to 180 nm bulk CMOS) [108]. Their results also indicate that as the channel length and bias increase, the Fano factor increases [108]. Das and Bardin also postulate that due to partial ionization at cryogenic temperatures, the Fano factor may decrease (though they note that further research is required) [108].

3.4 Progress and Challenges on Modeling High-Frequency Noise

Early models for this "excess noise" attempted to modify existing thermal noise models. One early model suggested this excess noise originated from "hot electrons" near the drain where velocity saturation occurs, though Chen and Deen point out that this region should not produce any noise [80]. Other effects such as channel length modulation (CLM) were proposed as increasing the drain conductance which would increase the thermal noise [77]. This view has been incorporated into the Klaassen-Prins model, though Smit et al. have suggested the CLM should have minimal effect on the thermal noise [70]. Other modifications to the thermal noise models are reviewed by Asgaran and Deen in [76], though as mentioned these thermal noise models fail to capture the non-equilibrium nature of noise.

A variety of models have attempted to capture the more non-equilibrium nature of noise. For example, Tsividis' model (Eq. 14) bears a striking resemblance to the second term of Eq. 16. Tsividis has also proposed a model based on the structure of a short-channel device [79]:

$$S_I(\omega) = 2q\frac{W\mu_{eff}C_{ox}(n-1)v_{th}^2}{L_{eff}}\exp\left(\frac{V_{GT}-V_{off}}{n\cdot\phi_t}\right)\left(1+e^{-V_{DS}/v_{th}}\right) \tag{20}$$

where $V_{off} = V_{T,sub} - V_T$ is the offset voltage as characterized by I-V in subthreshold, $V_{GT} = V_{GS} - V_T$, $V_T = V_{T0} - \text{DIBL}\cdot V_{DS}$ is the threshold voltage accounting for DIBL, and n is the subthreshold factor.

Spathis et al. have also proposed a model based on thermonic emission theory [109]. Their model is a combination of both shot and thermal noise, where the thermal noise is associated with the channel resistance, while the shot noise is associated with the thermonic emission of carriers over the barrier [109]. This model does require the height of the source-channel electrostatic barrier which would require experimental measurements or complex numerical simulations though Spathis et al. note that compact algorithms exist to calculate the barrier height [109]. The value predicted by this model, $F = 0.3$, has only been validated in strong inversion by Wang et al. (who measured a Fano factor of $F = 0.301$) [106]. Wang et al. have also proposed a similar expression which accounts for the effects of noise due to substrate capacitive coupling [106]:

$$S_I = 2qI_{DS}\cdot F + 4k_BTg_m^2R_g + 4k_BT(\omega C_{db})^2R_b \tag{21}$$

though no expression for the Fano factor is given [106]. Shen et al. have proposed a similar model though their model requires the use of fitting and measured parameters reducing its usefulness for design [110].

Chen et al. have also provided a model for the Fano factor which can model the effects of both thermal noise for $L > 180\,\text{nm}$ and the shot noise suppression for $L < 180\,\text{nm}$ [111]:

$$F = \frac{8\phi_t}{3}\left(\frac{\alpha}{V_{GS} - V_T} + \frac{1}{E_c L_{eff}}\right) \tag{22}$$

Here, $E_c = 2v_{sat}/\mu_{eff}$ is the critical electrical field, L_{eff} is the effective channel length, and α is the body effect coefficient. While this model is only valid in strong inversion and saturation, it has been verified for devices with channel lengths of 180 nm, 420 nm, and 970 nm [112].

While all these models treat the excess noise as shot noise, they fail to capture the increased suppression due to inelastic scattering. Furthermore, many of these models are only valid in specific bias regimes limiting their usefulness. Also, none of these models have been experimentally verified at cryogenic temperatures preventing them from being used to inform cryogenic CMOS design. Thus, it would be desirable to have a noise model valid for MOSEFTs operating at a range of temperatures, channel lengths, and biases.

4 Numerical Simulation for Cryogenic CMOS

Before concluding the chapter, we would like to briefly discuss on the needs and challenges posed to numerical simulation. With the excitement in exploring cryogenic CMOS in recent years, the need for more accurate numerical simulations of cryogenic CMOS has become increasingly evident. At these temperatures, the properties of transistors, as well as other semiconducting devices, perform differently from those at room temperature. Complementing experimental exploration, numerical simulation has always been an effective tool in studying material property and device physics. Such tool will arguably play a more important role for cryogenic CMOS given the significant amount of efforts and resources required for experiment and characterization at cryo-T.

Cryogenic CMOS numerical simulation, as expected, involves modeling the behavior of CMOS devices at extremely low temperatures, from a few kelvin down to below 100 kelvin. Besides the physical effects (e.g., incomplete ionization) significant at cryo-T to be included in cryo-TCAD tools, the very small temperature value poses numerical challenges for computation and convergence. For example, Newton's method is a numerical technique commonly used to solve the nonlinear system of equations arising from the discretization of the drift-diffusion equations. However, this technique leads to poor convergence at low temperatures because Newton's method is prone to overshooting as the length of the correction step for any given iteration is overestimated. This can lead to slow convergence and require numerous iterations to obtain a solution. To address this issue, a damping parameter can be used to restrict the size of each correction step and improve convergence. For each bias condition, an initial approximation is required for Newton's method to find the solution. However, if the initial guess does not adequately approximate the actual solution (i.e., if it is not within the solution's region of attraction), Newton's method will fail to converge.

Loss of numerical precision can also cause Newton's method to fail to converge, as subtracting nearly equal numbers using finite-precision arithmetic may result in corrupted update vectors that move the solution approximation outside of the solution's region of attraction. At low temperatures, double-precision arithmetic (8-byte) is not accurate enough to resolve solution updates for a given iteration. Moreover, the device equations involve both positive and negative exponential functions of temperature, making the system more ill-conditioned and sensitive to precision loss and other sources of numerical error. Consequently, convergence becomes increasingly challenging with decreasing temperature. To overcome this issue, critical calculations are performed with extended (16-byte) precision at temperatures below 150K. While extended-precision arithmetic may slow down the computation compared to normal arithmetic, it is necessary for obtaining accurate low-temperature solutions.

At present, commercial TCAD (technology computer-aided design) software, commonly used in device research at RT and above, fail to simulate at deep-cryogenic temperatures due to numerical challenges and the lack of important cryogenic effects. TCAD specifically designed for deep-cryogenic temperatures is still in its early stages. Companies, such as Nanoacademic Technologies, are actively pushing the frontier in developing commercial TCAD tools targeting cryogenic and quantum applications.

References

1. T. Haigh, P.E. Ceruzzi, *A New History of Modern Computing* (MIT Press, Cambridge, 2021)
2. F. Bova, A. Goldfarb, R.G. Melko, Commercial applications of quantum computing. EPJ Quantum Technol. **8**(1), 2 (2021)
3. P. Shor, Algorithms for quantum computation: Discrete logarithms and factoring, eng, in *Proceedings 35th Annual Symposium on Foundations of Computer Science* (IEEE Comput. Soc. Press, 1994), pp. 124–134. ISBN: 0818665807
4. L. Grover, A fast quantum mechanical algorithm for database search, eng, in *Proceedings of the Twenty-Eighth Annual ACM Symposium on Theory of Computing*, ser. STOC '96, vol. 129452 (ACM, 1996), pp. 212–219. ISBN: 0897917855
5. A.W. Harrow, A. Hassidim, S. Lloyd, Quantum algorithm for linear systems of equations. Phys. Rev. Lett. **103**(15), 150,502 (2009). [Online]. Available: https://link.aps.org/doi/10.1103/PhysRevLett.103.150502
6. C.H. Bennett, G. Brassard, Quantum cryptography: Public key distribution and coin tossing. Preprint (2020). arXiv:2003.06557
7. D. Deutsch, Quantum theory, the church–turing principle and the universal quantum computer. Proc. R. Soc. Lond. A. Math. Phys. Sci. **400**(1818), 97–117 (1985)
8. I.L. Chuang, N. Gershenfeld, M. Kubinec, Experimental implementation of fast quantum searching. Phys. Rev. Lett. **80**(15), 3408 (1998)
9. M. Kjaergaard, M.E. Schwartz, J. Braumuüller, et al., Superconducting qubits: Current state of play. Annu. Rev. Condens. Matter Phys. **11**(1), 369–395 (2020). ISSN: 1947–5454. [Online]. Available: https://doi.org/10.1146/annurev-conmatphys-031119-050605

10. D. Kielpinski, C. Monroe, D.J. Wineland, Architecture for a large-scale ion-trap quantum computer. Nature **417**(6890), 709–711 (2002). ISSN: 1476–4687. [Online]. Available: https://doi.org/10.1038/nature00784
11. C. Kloeffel, D. Loss, Prospects for spin-based quantum computing in quantum dots. Annu. Rev. Condens. Matter Phys. **4**(1), 51–81 (2013). ISSN: 1947–5454. https://doi.org/10.1146/annurev-conmatphys-030212-184248
12. T.W. Larsen, K.D. Petersson, F. Kuemmeth, et al., Semiconductor-nanowire-based superconducting qubit. Phys. Rev. Lett. **115**(12), 127,001–127,001 (2015). ISSN: 0031-9007. https://doi.org/10.1103/PhysRevLett.115.127001
13. H. Seo, H. Ma, M. Govoni, G. Galli, Designing defect-based qubit candidates in wide-gap binary semiconductors for solid-state quantum technologies. Phys. Rev. Mater. **1**(7), 075002 (2017). [Online]. Available: https://doi.org/10.1103/physrevmaterials.1.075002
14. P. Kok, W.J. Munro, K. Nemoto, T.C. Ralph, J.P. Dowling, G.J. Milburn, Linear optical quantum computing with photonic qubits. Rev. Modern Phys. **79**(1), 135–174 (2007). ISSN: 0034-6861. https://doi.org/10.1103/RevModPhys.79.135
15. F. Arute, K. Arya, R. Babbush, et al., Quantum supremacy using a programmable superconducting processor. Nature **574**(7779), 505–510 (2019)
16. L. Hutin, B. Bertrand, R. Maurand, et al., Si MOS technology for spin-based quantum computing, in *2018 48th European Solid-State Device Research Conference (ESSDERC)* (Sep. 2018), pp. 12–17. https://doi.org/10.1109/ESSDERC.2018.8486863
17. E. Kawakami, P. Scarlino, D.R. Ward, et al., Electrical control of a long-lived spin qubit in a Si/SiGe quantum dot, en. Nat. Nanotechnol. **9**(9), 666–670 (2014). ISSN: 1748–3395. https://doi.org/10.1038/nnano.2014.153. [Online]. Available: https://www.nature.com/articles/nnano.2014.153
18. J.M. Hornibrook, J.I. Colless, A.C. Mahoney, et al., Frequency multiplexing for readout of spin qubits. Appl. Phys. Lett. **104**(10), 103108 (2014). ISSN: 0003-6951. https://doi.org/10.1063/1.4868107. [Online]. Available: https://aipscitation-org.proxy.lib.uwaterloo.ca/doi/full/10.1063/1.4868107
19. R. Hanson, J. Elzerman, L. Willems van Beveren, L. Vandersypen, L. Kouwen-hoven, Electron spin qubits in quantum dots, in *IEDM Technical Digest. IEEE International Electron Devices Meeting, 2004* (Dec. 2004), pp. 533–536. https://doi.org/10.1109/IEDM.2004.1419211
20. E. Charbon, F. Sebastiano, A. Vladimirescu, et al., Cryo-CMOS for quantum computing, eng, in *2016 IEEE International Electron Devices Meeting (IEDM)* (IEEE, 2016), pp. 13.5.1–13.5.4. ISBN: 9781509039029
21. C.H. Yang, R.C.C. Leon, J.C.C. Hwang, et al., Operation of a silicon quantum processor unit cell above one kelvin. Nature (London) **580**(7803), 350–354 (2020). ISSN: 0028-0836. https://doi.org/10.1038/s41586-020-2171-6
22. L. Petit, H.G.J. Eenink, M. Russ, et al., Universal quantum logic in hot silicon qubits. Nature (London) **580**(7803), 355–359 (2020). ISSN: 0028-0836. https://doi.org/10.1038/s41586-020-2170-7
23. L.M.K. Vandersypen, H. Bluhm, J.S. Clarke, et al., Interfacing spin qubits in quantum dots and donors—hot, dense, and coherent. npj Quantum Inf. **3**(1), 34 (2017). ISSN: 2056-6387. [Online]. Available: https://doi.org/10.1038/s41534-017-0038-y
24. P. Cerfontaine, T. Botzem, J. Ritzmann, et al., Closed-loop control of a gaas-based singlet-triplet spin qubit with 99.5% gate fidelity and low leakage. Nat. Commun. **11**(1), 4144–4144 (2020). ISSN: 2041–1723. https://doi.org/10.1038/s41467-020-17865-3
25. A. Noiri, K. Takeda, T. Nakajima, et al., Fast universal quantum gate above the fault-tolerance threshold in silicon. Nature (London) **601**(7893), 338–342 (2022). ISSN: 1476–4687. https://doi.org/10.1038/s41586-021-04182-y
26. X. Xue, M. Russ, N. Samkharadze, et al., Quantum logic with spin qubits crossing the surface code threshold. Nature **601**(7893), 343–347 (2022). [Online]. Available: https://doi.org/10.1038/s41586-021-04273-w

27. H. Mantooth, J.D. Cressler, *Extreme Environment Electronics* (Taylor & Francis, London, 2013)
28. R. Kirschman, Survey of low-temperature electronics, in *Workshop on Low-Temp. Electronics WOLTE11* (2014)
29. A. Beckers, F. Jazaeri, C. Enz, Characterization and modeling of 28-nm bulk CMOS technology down to 4.2 k. IEEE J. Electron Dev. Soc. **6**, 1007–1018 (2018)
30. B. Xie, B. Li, J. Bi, et al., Effect of cryogenic temperature characteristics on 0.18-μm silicon-on-insulator devices. Chin. Phys. B **25**(7), 078,501 (2016)
31. H. Elgabra, B. Buonacorsi, C. Chen, J. Watt, J. Baugh, L. Wei, Virtual source based iv model for cryogenic CMOS devices, in *2019 International Symposium on VLSI Technology, Systems and Application (VLSI-TSA)* (IEEE, 2019), pp. 1–2
32. B. Patra, R.M. Incandela, J.P. Van Dijk, et al., Cryo-CMOS circuits and systems for quantum computing applications. IEEE J. Solid State Circuits **53**(1), 309–321, 2017.
33. F. Sebastiano, H. Homulle, B. Patra, et al., Cryo-CMOS electronic control for scalable quantum computing, in *Proceedings of the 54th Annual Design Automation Conference 2017* (2017), pp. 1–6
34. R. Maurand, X. Jehl, D. Kotekar-Patil, et al., A CMOS silicon spin qubit. Nat. Commun. **7**(1), 13,575–13,575 (2016). ISSN: 2041–1723. https://doi.org/10.1038/ncomms13575
35. A. Zwerver, T. Krähenmann, T. Watson, et al., Qubits made by advanced semiconductor manufacturing. Nat. Electron. **5**(3), 184–190 (2022)
36. L.C. Camenzind, S. Geyer, A. Fuhrer, R.J. Warburton, D.M. Zumbuühl, A.V. Kuhlmann, A hole spin qubit in a fin field-effect transistor above 4 kelvin. Nat. Electron. **5**(3), 178–183 (2022)
37. L.M.K. Vandersypen, M.A. Eriksson, Quantum computing with semiconductor spins. Phys. Today **72**(8), 38–45 (2019). ISSN: 0031-9228. https://doi.org/10.1063/PT.3.4270. [Online]. Available: https://physicstoday-scitation-org.proxy.lib.uwaterloo.ca/doi/full/10.1063/pt.3.4270
38. E. Arnold, Disorder-induced carrier localization in silicon surface inversion layers, en. Appl. Phys. Lett. **25**(12), 705–707 (1974). ISSN: 0003-6951, 1077-3118. https://doi.org/10.1063/1.1655369. [Online]. Available: http://aip.scitation.org/doi/10.1063/1.1655369
39. H. Elgabra, B. Buonacorsi, C. Chen, J. Watt, J. Baugh, L. Wei, Virtual source based I-V model for cryogenic CMOS devices, en, in *2019 International Symposium on VLSI Technology, Systems and Application (VLSI-TSA)* (IEEE, Hsinchu, Taiwan, Apr. 2019), pp. 1–2. ISBN: 978-1-72810-942-8. https://doi.org/10.1109/VLSI-TSA.2019.8804661. [Online]. Available: https://ieeexplore.ieee.org/document/8804661/
40. A.H. Talukder, B. Smith, M. Akbulut, F. Dirisaglik, H. Silva, A. Gokirmak, Temperature-dependent characteristics and electrostatic threshold voltage tuning of accumulated body mosfets. IEEE Trans. Electron Dev. **69**(8), 4138–4143 (2022)
41. Y. Taur, T.H. Ning, *Fundamentals of Modern VLSI Devices*, en, 2nd edn. (Cambridge University Press, Cambridge, UK; New York, 2009). ISBN: 978-0-521-83294-6
42. P. Galy, J. Camirand Lemyre, P. Lemieux, F. Arnaud, D. Drouin, M. Pioro-Ladriere, Cryogenic temperature characterization of a 28-nm FD-SOI dedicated structure for advanced CMOS and quantum technologies co-integration, en. IEEE J. Electron Dev. Soc. **6**, 594–600 (2018). ISSN: 2168-6734. https://doi.org/10.1109/JEDS.2018.2828465. [Online]. Available: https://ieeexplore.ieee.org/document/8370029/
43. K. Eng, G.A. Ten Eyck, L. Tracy, et al., Steps towards fabricating cryogenic CMOS compatible single electron devices, en, in *2008 8th IEEE Conference on Nanotechnology* (IEEE, Arlington, Texas, USA, Aug. 2008), pp. 496–499. ISBN: 978-1-4244-2103-9. https://doi.org/10.1109/NANO.2008.149. [Online]. Available: http://ieeexplore.ieee.org/document/4617131/
44. M. Shin, M. Shi, M. Mouis, et al., In depth characterization of electron transport in 14 nm FD-SOI CMOS devices, en. Solid State Electron. **112**, 13–18 (2015). ISSN: 00381101. https://doi.org/10.1016/j.sse.2015.02.012. [Online]. Available: https://linkinghub.elsevier.com/retrieve/pii/S003811011500057X

45. B. Cretu, D. Boudier, E. Simoen, A. Veloso, N. Collaert, Assessment of DC and low-frequency noise performances of triple-gate FinFETs at cryogenic temperatures, en. Semiconductor Sci. Technol. **31**(12), 124,006 (2016). ISSN: 0268-1242, 1361-6641. https://doi.org/10.1088/0268-1242/31/12/124006. [Online]. Available: https://iopscience.iop.org/article/10.1088/0268-1242/31/12/124006
46. R.F. Pierret, *Advanced Semiconductor Fundamentals* (Modular Series on Solid State Devices v. 6), 2nd edn. (Prentice Hall, Upper Saddle River, NJ, 2003). ISBN: 978-0-13-061792-7
47. M.S. Lundstrom, D.A. Antoniadis, Compact models and the physics of nanoscale FETs. IEEE Trans. Electron Dev. **61**(2), 225–233 (2014). ISSN: 1557–9646. https://doi.org/10.1109/TED.2013.2283253
48. W. Chakraborty, K. Ni, J. Smith, A. Raychowdhury, S. Datta, An empirically validated virtual source FET model for deeply scaled cool CMOS, in *2019 IEEE International Electron Devices Meeting (IEDM)* (Dec. 2019), pp. 39.4.1–39.4.4. https://doi.org/10.1109/IEDM19573.2019.8993666
49. R.M. Incandela, L. Song, H. Homulle, E. Charbon, A. Vladimirescu, F. Sebas-tiano, Characterization and compact modeling of nanometer CMOS transistors at deep-cryogenic temperatures, en. IEEE J. Electron Dev. Soc. **6**, 996–1006 (2018). ISSN: 2168–6734. https://doi.org/10.1109/JEDS.2018.2821763. [Online]. Available: https://ieeexplore.ieee.org/document/8329135/
50. A. Beckers, F. Jazaeri, C. Enz, Characterization and Modeling of 28-nm Bulk CMOS Technology Down to 4.2 K, en. IEEE J. Electron Dev. Soc. **6**, 1007–1018 (2018). ISSN: 2168-6734. https://doi.org/10.1109/JEDS.2018.2817458. [Online]. Available: https://ieeexplore.ieee.org/document/8326483/
51. A. Beckers, F. Jazaeri, A. Ruffino, C. Bruschini, A. Baschirotto, C. Enz, Cryogenic characterization of 28 nm bulk CMOS technology for quantum computing, en, in *2017 47th European Solid-State Device Research Conference (ESSDERC)* (IEEE, Leuven, Belgium, Sep. 2017), pp. 62–65. ISBN: 978-1-5090-5978-2. https://doi.org/10.1109/ESSDERC.2017.8066592. [Online]. Available: http://ieeexplore.ieee.org/document/8066592/
52. G.S. Fonseca, L.B. de Sá, A.C. Mesquita, Extraction of static parameters to extend the EKV model to cryogenic temperatures, en, in eds. by B.F. Andresen, G.F. Fulop, C.M. Hanson, J.L. Miller, P.R. Norton, Eds., Baltimore, Maryland, United States, May 2016, 98192B. https://doi.org/10.1117/12.2219734. [Online]. Available: http://proceedings.spiedigitallibrary.org/proceeding.aspx?doi=10.1117/12.2219734
53. Y. Creten, P. Merken, W. Sansen, R.P. Mertens, C. Van Hoof, An 8-bit flash analog-to-digital converter in standard CMOS technology functional from 4.2 K to 300 K, en. IEEE J. Solid State Circuits **44**(7), 2019–2025 (2009). ISSN: 0018-9200. [Online]. Available: http://ieeexplore.ieee.org/document/5109798/
54. C. Luo, Z. Li, T.-T. Lu, J. Xu, G.-P. Guo, MOSFET characterization and modeling at cryogenic temperatures, en. Cryogenics **98**, 12–17 (2019). ISSN: 00112275. https://doi.org/10.1016/j.cryogenics.2018.12.009. [Online]. Available: https://linkinghub.elsevier.com/retrieve/pii/S0011227518301413
55. L. Varizat, G. Sou, M. Mansour, D. Alison, A. Rhouni, A low temperature 0.35μm CMOS technology BSIM3.3 model for space instrumentation: Application to a voltage reference design, in *2017 IEEE International Workshop on Metrology for AeroSpace (MetroAeroSpace)* (IEEE, Padua, Italy, Jun. 2017), pp. 74–78. ISBN: 978-1-5090-4234-0. https://doi.org/10.1109/MetroAeroSpace.2017.7999541. [Online]. Available: http://ieeexplore.ieee.org/document/7999541/
56. A. Beckers, F. Jazaeri, C. Enz, Theoretical limit of low temperature sub-threshold swing in field-effect transistors, en. IEEE Electron Dev. Lett. **41**(2), 276–279 (2020). ISSN: 0741-3106, 1558-0563. https://doi.org/10.1109/LED.2019.2963379. [Online]. Available: https://ieeexplore.ieee.org/document/8946710/
57. A. Beckers, F. Jazaeri, C. Enz, *Revised Theoretical Limit of Subthreshold Swing in Field-Effect Transistors*, en, Dec. 2019. [Online]. Available: http://arxiv.org/abs/1811.09146

58. A. Beckers, F. Jazaeri, C. Enz, Inflection phenomenon in cryogenic MOSFET behavior, en. IEEE Trans. Electron Dev. **67**(3), 1357–1360 (2020). ISSN: 0018-9383, 1557-9646. https://doi.org/10.1109/TED.2020.2965475. [Online]. Available: https://ieeexplore.ieee.org/document/8974242/
59. A. Beckers, F. Jazaeri, A. Grill, S. Narasimhamoorthy, B. Parvais, C. Enz, Physical model of low-temperature to cryogenic threshold voltage in MOSFETs, en. IEEE J. Electron Dev. Soc. **8**, 780–788 (2020). ISSN: 2168-6734. https://doi.org/10.1109/JEDS.2020.2989629. [Online]. Available: https://ieeexplore.ieee.org/document/9076206/
60. N.C. Dao, A.E. Kass, M.R. Azghadi, C.T. Jin, J. Scott, P.H. Leong, An enhanced MOSFET threshold voltage model for the 6–300 K temperature range, en. Microelectron. Reliab. **69**, 36–39 (2017). ISSN: 00262714. https://doi.org/10.1016/j.microrel.2016.12.007. [Online]. Available: https://linkinghub.elsevier.com/retrieve/pii/S0026271416304437
61. A. Beckers, F. Jazaeri, C. Enz, Cryogenic MOS transistor model, en. IEEE Trans. Electron Dev. **65**(9), 3617–3625 (2018). ISSN: 0018-9383, 1557–9646. https://doi.org/10.1109/TED.2018.2854701. [Online]. Available: https://ieeexplore.ieee.org/document/8424046/
62. A. Akturk, M. Holloway, S. Potbhare, et al., Compact and distributed modeling of cryogenic bulk MOSFET operation. IEEE Trans. Electron Dev. **57**(6), 1334–1342 (2010). ISSN: 0018-9383. https://doi.org/10.1109/TED.2010.2046458. [Online]. Available: http://ieeexplore.ieee.org/document/5456141/
63. A. Akturk, M. Peckerar, K. Eng, et al., Compact modeling of 0.35μm SOI CMOS technology node for 4K DC operation using Verilog-A, en. Microelectron. Eng. **87**(12), 2518–2524 (2010). ISSN: 01679317. https://doi.org/10.1016/j.mee.2010.06.005. [Online]. Available: https://linkinghub.elsevier.com/retrieve/pii/S0167931710001917
64. S.K. Singh, S. Gupta, R.A. Vega, A. Dixit, Accurate modeling of cryogenic temperature effects in 10-nm bulk CMOS FinFETs using the BSIM-CMG model, en. IEEE Electron Dev. Lett. **43**(5), 689–692 (2022). https://doi.org/10.110/LED.2022.3158495. [Online]. Available: https://ieeexplore.ieee.org/document/9732438/
65. S. O'uchi, K. Endo, M. Maezawa, et al., Cryogenic operation of double-gate FinFET and demonstration of analog circuit at 4.2K, en, in *2012 IEEE International SOI Conference (SOI)* (IEEE, Napa, CA, USA, Oct. 2012), pp. 1–2. https://doi.org/10.1109/SOI.2012.6404376. [Online]. Available: http://ieeexplore.ieee.org/document/6404376/
66. S. Tewksbury, Attojoule MOSFET logic devices using low voltage swings and low temperature, en. Solid State Electron. **28**(3), 255–276 (1985). ISSN: 00381101. https://doi.org/10.1016/0038-1101(85)90006-1. [Online]. Available: https://linkinghub.elsevier.com/retrieve/pii/0038110185900061
67. I.M. Hafez, G. Ghibaudo, F. Balestra, Assessment of interface state density in silicon metal-oxide-semiconductor transistors at room, liquid-nitrogen, and liquid- helium temperatures, en. J. Appl. Phys. **67**(4), 1950–1952 (1990). ISSN: 0021-8979, 1089-7550. https://doi.org/10.1063/1.345572. [Online]. Available: http://aip.scitation.org/doi/10.1063/1.345572
68. H. Bohuslavskyi, A.G.M. Jansen, S. Barraud, et al., Cryogenic subthreshold swing saturation in FD-SOI MOSFETs described with band broadening, en. IEEE Electron Dev. Lett. **40**(5), 784–787 (2019). ISSN: 0741-3106, 1558-0563. https://doi.org/10.1109/LED.2019.2903111. [Online]. Available: https://ieeexplore.ieee.org/document/8660508/
69. G. Paasch, S. Scheinert, Charge carrier density of organics with Gaussian density of states: Analytical approximation for the Gauss–Fermi integral, en. J. Appl. Phys. **107**(10), 104,501 (2010). ISSN: 0021-8979, 1089-7550. https://doi.org/10.1063/1.3374475. [Online]. Available: http://aip.scitation.org/doi/10.1063/1.3374475
70. G.D.J. Smit, A.J. Scholten, R.M.T. Pijper, L.F. Tiemeijer, R. van der Toorn, D.B.M. Klaassen, RF-noise modeling in advanced CMOS technologies. IEEE Trans. Electron Dev. **61**(2), 245–254 (2014). ISSN: 1557-9646. https://doi.org/10.1109/TED.2013.2282960
71. F. Klaassen, J. Prins, Thermal noise of MOS transistors. Philips Res. Rep. **22**, 504–514 (1967)
72. U. Fano, Ionization yield of radiations. II. The fluctuations of the number of ions. Phys. Rev. **72**(1), 26–29 (1947). https://doi.org/10.1103/PhysRev.72.26. [Online]. Available: https://link.aps.org/doi/10.1103/PhysRev.72.26

73. J.B. Johnson, Thermal agitation of electricity in conductors. Phys. Rev. **32**(1), 97–109 (1928). https://doi.org/10.1103/PhysRev.32.97. [Online]. Available: https://link.aps.org/doi/10.1103/PhysRev.32.97
74. H. Nyquist, Thermal agitation of electric charge in conductors. Phys. Rev. **32**(1), 110–113 (1928). https://doi.org/10.1103/PhysRev.32.110. [Online]. Available: https://link.aps.org/doi/10.1103/PhysRev.32.110
75. W. Schottky, Über spontane Stromschwankungen in verschiedenen Elektrizitätsleitern, en. Annalen der Physik **362**(23), 541–567 (1918). ISSN: 1521–3889. https://doi.org/10.1002/andp.19183622304. [Online]. Available: https://onlinelibrary.wiley.com/doi/abs/10.1002/andp.19183622304
76. A. Asgaran, M.J. Deen, RF noise models of MOSFETs-A review, en-US. TechConnect Briefs **2**(2004), 96–101 (2004). [Online]. Available: https://briefs.techconnect.org/papers/rf-noise-models-of-mosfets-a-review/
77. R.P. Jindal, Compact noise models for MOSFETs. IEEE Trans. Electron Dev. **53**(9), 2051–2061 (2006). ISSN: 1557-9646. https://doi.org/10.1109/TED.2006.880368
78. A. Van der Ziel, *Noise in Solid State Devices and Circuits* (Wiley, New York, 1986). ISBN: 978-0-471-83234-8. [Online]. Available: https://catalog.hathitrust.org/Record/000667771
79. Y. Tsividis, C. McAndrew, *Operation and Modeling of the MOS Transistor* (The Oxford Series in Electrical and Computer Engineering), en, 3rd edn. (Oxford University Press, New York, NY, USA, 2011). ISBN: 978-0-19-517015-3
80. C.-H. Chen, M. Deen, Channel noise modeling of deep submicron MOSFETs. IEEE Trans. Electron Dev. **49**(8), 1484–1487 (2002). ISSN: 1557–9646. https://doi.org/10.1109/TED.2002.801229
81. K. Han, H. Shin, K. Lee, Analytical drain thermal noise current model valid for deep submicron MOSFETs. IEEE Trans. Electron Dev. **51**(2), 261–269 (2004). ISSN: 1557–9646. https://doi.org/10.1109/TED.2003.821708
82. R. Navid, R. Dutton, The physical phenomena responsible for excess noise in short-channel MOS devices, in *International Conference on Simulation of Semiconductor Processes and Devices* (Sep. 2002), pp. 75–78. https://doi.org/10.1109/SISPAD.2002.1034520
83. R. Venugopal, M. Paulsson, S. Goasguen, S. Datta, M.S. Lundstrom, A simple quantum mechanical treatment of scattering in nanoscale transistors. J. Appl. Phys. **93**(9), 5613–5625 (2003). ISSN: 0021-8979. https://doi.org/10.1063/1.1563298. [Online]. Available: http://aip.scitation.org/doi/10.1063/1.1563298
84. D. Ruić, C. Jungemann, Microscopic noise simulation of long- and short-channel nMOSFETs by a deterministic approach, en. J. Comput. Electron. **15**(3), 809–819 (2016). ISSN: 1572-8137. [Online]. Available: https://doi.org/10.1007/s10825-016-0840-3
85. J.-S. Goo, C.-H. Choi, A. Abramo, et al., Physical origin of the excess thermal noise in short channel MOSFETs. IEEE Electron Dev. Lett. **22**(2), 101–103 (2001). ISSN: 1558-0563. https://doi.org/10.1109/55.902845
86. M.S. Obrecht, T. Manku, M.I. Elmasry, Simulation of temperature dependence of microwave noise in metal-oxide-semiconductor field-effect transistors, en. Jpn. J. Appl. Phys. **39**(4R), 1690 (2000). ISSN: 1347-4065. https://doi.org/10.1143/JJAP.39.1690. [Online]. Available: http://iopscience.iop.org/article/10.1143/JJAP.39.1690/meta
87. M. Obrecht, E. Abou-Allam, T. Manku, Diffusion current and its effect on noise in submicron MOSFETs. IEEE Trans. Electron Dev. **49**(3), 524–526 (2002). ISSN: 1557-9646. https://doi.org/10.1109/16.987126
88. Y.M. Blanter, M. Büttiker, Shot noise in mesoscopic conductors, en. Phys. Rep. **336**(1), 1–166 (2000). ISSN: 0370-1573. https://doi.org/10.1016/S0370-1573(99)00123-4. [Online]. Available: https://www.sciencedirect.com/science/article/pii/S0370157399001234
89. T. Martin, Noise in mesoscopic physics (Jan. 2005). arXiv:cond-mat/0501208. [Online]. Available: http://arxiv.org/abs/cond-mat/0501208
90. K. Kobayashi, M. Hashisaka, Shot noise in mesoscopic systems: from single particles to quantum liquids. J. Phys. Soc. Jpn. **90**(10), 102,001 (2021). ISSN: 0031-9015. https://doi.org/10.7566/JPSJ.90.102001. [Online]. Available: https://journals.jps.jp/doi/10.7566/JPSJ.90.102001

91. T.T. Heikkilä, *The Physics of Nanoelectronics: Transport and Fluctuation Phenomena at Low Temperatures*, en. (OUP Oxford, Oxford, 2013). ISBN: 978-0-19-959244-9
92. T. Martin, R. Landauer, Wave-packet approach to noise in multichannel meso-scopic systems. Phys. Rev. B **45**(4), 1742–1755 (1992). https://doi.org/10.1103/PhysRevB.45.1742. [Online]. Available: https://link.aps.org/doi/10.1103/PhysRevB.45.1742
93. R.C. Liu, Quantum noise in mesoscopic electron transport, English, Ph.D. Stanford University, United States—California, Dec. 1997. [Online]. Available: https://www.proquest.com/docview/304455433/abstract/C36AEE51CC15484APQ/1
94. R.C. Liu, B. Odom, Y. Yamamoto, S. Tarucha, Quantum interference in electron collision, en. Nature **391**(6664), 263–265 (1998). ISSN: 1476-4687. https://doi.org/10.1038/34611. [Online]. Available: http://www.nature.com/articles/34611
95. F. Liefrink, J.I. Dijkhuis, M.J.M. de Jong, L.W. Molenkamp, H. van Houten, Experimental study of reduced shot noise in a diffusive mesoscopic conductor. Phys. Rev. B **49**(19), 14,066–14,069 (1994). https://doi.org/10.1103/PhysRevB.49.14066. [Online]. Available: https://link.aps.org/doi/10.1103/PhysRevB.49.14066
96. A.H. Steinbach, J.M. Martinis, M.H. Devoret, Observation of hot-electron shot noise in a metallic resistor. Phys. Rev. Lett. **76**(20), 3806–3809, May 1996. https://doi.org/10.1103/PhysRevLett.76.3806. [Online]. Available: https://link.aps.org/doi/10.1103/PhysRevLett.76.3806
97. R.J. Schoelkopf, P.J. Burke, A.A. Kozhevnikov, D.E. Prober, M.J. Rooks, Frequency dependence of shot noise in a diffusive mesoscopic conductor. Phys. Rev. Lett. **78**(17), 3370–3373 (1997). https://doi.org/10.1103/PhysRevLett.78.3370. [Online]. Available: https://link.aps.org/doi/10.1103/PhysRevLett.78.3370
98. M. Lundstrom, Z. Ren, Essential physics of carrier transport in nanoscale MOS-FETs. IEEE Trans. Electron Dev. **49**(1), 133–141 (2002). ISSN: 1557-9646. https://doi.org/10.1109/16.974760
99. H. Tsuchiya, S.-i. Takagi, Influence of elastic and inelastic phonon scattering on the drive current of quasi-ballistic MOSFETs. IEEE Trans. Electron Dev. **55**(9), 2397–2402 (2008). ISSN: 1557-9646. https://doi.org/10.1109/TED.2008.927384
100. K. Natori, Compact modeling of quasi-ballistic silicon nanowire MOSFETs. IEEE Trans. Electron Dev. **59**(1), 79–86 (2012). ISSN: 1557-9646. https://doi.org/10.1109/TED.2011.2172612
101. R. Navid, C. Jungemann, T.H. Lee, R.W. Dutton, High-frequency noise in nanoscale metal oxide semiconductor field effect transistors. J. Appl. Phys. **101**(12), 124,501 (2007). ISSN: 0021-8979. https://doi.org/10.1063/1.2740345. [Online]. Available: http://aip.scitation.org/doi/full/10.1063/1.2740345
102. A. Shimizu, M. Ueda, Effects of dephasing and dissipation on quantum noise in conductors. Phys. Rev. Lett. **69**(9), 1403–1406 (1992). https://doi.org/10.1103/PhysRevLett.69.1403. [Online]. Available: https://link.aps.org/doi/10.1103/PhysRevLett.69.1403
103. M. Lundstrom, *Nanohub-u: Fundamentals of Nanotransistors*, 2nd edn. (Dec. 2015). [Online]. Available: https://nanohub.org/courses/nt
104. R.C. Liu, Y. Yamamoto, Nyquist noise in the transition from mesoscopic to macroscopic transport. Phys. Rev. B **50**(23), 17,411–17,414 (1994). https://doi.org/10.1103/PhysRevB.50.17411. [Online]. Available: https://link.aps.org/doi/10.1103/PhysRevB.50.17411
105. J. Jeon, J. Lee, J. Kim, et al., The first observation of shot noise characteristics in 10-nm scale MOSFETs, in *2009 Symposium on VLSI Technology* (Jun. 2009), pp. 48–49
106. J. Wang, X.-M. Peng, Z.-J. Liu, L. Wang, Z. Luo, D.-D. Wang, Observation of nonconservation characteristics of radio frequency noise mechanism of 40-nm n- MOSFET, en. Chin. Phys. B **27**(2), 027,201 (2018). ISSN: 1674-1056. [Online]. Available: https://doi.org/10.1088/1674-1056/27/2/027201
107. K. Ohmori, S. Amakawa, Direct white noise characterization of short-channel MOSFETs. IEEE Trans. Electron Dev. **68**(4), 1478–1482 (2021). ISSN: 1557-9646. https://doi.org/10.1109/TED.2021.3059720

108. S. Das, J.C. Bardin, Characterization of shot noise suppression in nanometer MOSFETs, in *2021 IEEE MTT-S International Microwave Symposium (IMS)* (Jun. 2021), pp. 892–895. https://doi.org/10.1109/IMS19712.2021.9574931
109. C. Spathis, A. Birbas, K. Georgakopoulou, Semi-classical noise investigation for sub-40nm metal-oxide-semiconductor field-effect transistors. AIP Adv. **5**(8), 087,114 (2015). https://doi.org/10.1063/1.4928424. [Online]. Available: https://aip.scitation.org/doi/full/10.1063/1.4928424
110. Y. Shen, J. Cui, S. Mohammadi, An accurate model for predicting high frequency noise of nanoscale NMOS SOI transistors, en. Solid State Electron. **131**, 45–52 (2017). ISSN: 0038-1101. https://doi.org/10.1016/j.sse.2017.02.005. [Online]. Available: https://www.sciencedirect.com/science/article/pii/S0038110116303197
111. X. Chen, C.-H. Chen, M.J. Deen, Short noise suppression factor for nano-scale MOSFETs working in the saturation region, in *2017 International Conference on Noise and Fluctuations (ICNF)* (Jun. 2017), pp. 1–4. https://doi.org/10.1109/ICNF.2017.7986017
112. X. Chen, C.-H. Chen, R. Lee, Fast evaluation of the high-frequency channel noise in nanoscale MOSFETs. IEEE Trans. Electron Dev. **65**(4), 1502–1509 (2018). ISSN: 1557-9646. https://doi.org/10.1109/TED.2018.2808184

Quantum Computing on Memristor Crossbars

Iosif-Angelos Fyrigos, Panagiotis Dimitrakis, and Georgios Ch. Sirakoulis

1 Introduction

In the field of quantum computing, the utilization of quantum mechanics has enabled the development of advanced computational systems that surpass the capabilities of traditional high-performance computation systems. This advancement is known as the second quantum revolution [1]. The fundamental unit of information in quantum computing is the quantum bit (or qubit), which differs from classical bits by its ability to exist in a linear combination of basis states denoted by $|0\rangle$ and $|1\rangle$, respectively. This phenomenon is called superposition, which enables for massive quantum parallelism and dense coding. Entanglement is another important property of quantum systems that allows for the correlation of multiple qubits in a way that their quantum states cannot be described independent of one another, even when they are separated by great distances [2]. When a qubit is measured, it collapses back into a single basis state with the probability of collapsing into a particular basis state given by the absolute square of the coefficients multiplying its basis states. These two quantum mechanical resources, superposition and entanglement, have no classical physics counterpart and are the driving forces behind the potential of quantum computers to perform computations that are intractable for classical computers. It is expected that novel quantum algorithms will harness these resources to realize complex computational tasks [3].

I.-A. Fyrigos · G. Ch. Sirakoulis (✉)
Department of Electrical and Computer Engineering, Democritus University of Thrace, Xanthi, Greece
e-mail: ifyrigos@ee.duth.gr; gsirak@ee.duth.gr

P. Dimitrakis
Institute of Nanoscience and Nanotechnology, NCSR "Demokritos", Athens, Greece
e-mail: p.dimitrakis@inn.demokritos.gr

W. Liu et al. (eds.), *Design and Applications of Emerging Computer Systems*,
https://doi.org/10.1007/978-3-031-42478-6_23

1.1 Quantum Computers' Challenges and the Need of Simulators

Quantum computers can be considered still in their infancy, and as such, there are only a few actual quantum computers in existence. Furthermore, they consume high amounts of energy due to their significant cooling requirements. In the field of quantum computing, the current platforms are limited to a few dozen qubits, which are mainly implemented through superconducting circuits [4], ion traps [5], and photonic circuits [6]. However, quantum algorithms that are developed using quantum programming languages and decomposed into quantum circuits comprising qubits and quantum gates [7] require a large number of qubits. In addition, testing and development of these quantum algorithms on fabricated quantum computers are challenging tasks due to the decoherence of quantum systems. Decoherence occurs when the environment interacts with the qubits, uncontrollably changing their quantum states and causing the loss of information stored by the quantum computer. Even small interactions with the environment due to temperature or electromagnetic signals can distort the probabilities of different outcomes of the quantum computer. Moreover, decoherence can act on states like measurement, prematurely ending the quantum computation. Additionally, the scientific community has limited access to fabricated quantum systems. Therefore, simulation of quantum computers plays an essential role in the progress of quantum information science, both for numerical studies of quantum algorithms and for modeling noise and errors. Hence, there is a fast increasing need to develop quantum computer simulation due to the aforementioned challenges.

Various quantum computer simulators are currently available for researchers and developers to experiment with quantum algorithms and quantum gates. These simulators are mathematical frameworks that can emulate the behavior of a quantum computer. They are essential for quantum computing research because they enable scientists to explore the operation of quantum computers without having access to a physical one, which simplifies the task and reduces power consumption. Researchers can manipulate different parameters and observe how they affect the outcome of a quantum computation. Simulators can also facilitate the testing and debugging of algorithms and programs that are going to be executed on quantum computers, enabling faster adjustments to the algorithm. Additionally, simulators can be used to examine the impact of decoherence on quantum computers.

In the field of quantum computing, most simulators are software-based and run on central processing units (CPUs) of classical computers. Three well-known paradigms of these software-based simulators are Microsoft's Liquid [8], IBM's Qiskit [9], and ProjectQ [10] quantum simulators. Another software-based implementation of quantum computer simulator is the Quantum Exact Simulation Toolkit (QuEST) [11], which runs on a graphics processing unit (GPU). Additionally, there is an FPGA-based hardware implementation of a quantum computer simulator reported in literature [12]. Classical computers can simulate quantum algorithms, but they face an exponential slowdown as the number of qubits increases. To overcome this challenge, quantum simulators are being developed. These simulators not only facilitate the development of novel quantum algorithms but are also an essential part of the quantum computer stack, serving as hybrid solution for programming and executing algorithms on quantum computing platforms.

In quantum computer simulators, qubits are represented as vectors in Hilbert space, and quantum gates are represented by matrices that act on these qubits. To simulate quantum computations, mathematical operations such as matrix multiplications, matrix-vector multiplications, and tensor products are utilized. As the number of qubits (n) increases, the dimensions of the matrices representing quantum computations grow exponentially to $2^n \times 2^n$. Therefore, to enable the development of efficient quantum algorithms for practical applications, quantum simulators must perform these operations as efficiently as possible.

1.2 Memristor Crossbars as Hardware Accelerators

Memristor arrays arranged in a crossbar configuration are capable of performing matrix-vector multiplication (MVM) effectively along with low power consumption, small area footprint, and non-volatility as their main features [13–15]. These two-terminal nanoscale electronic elements can change their internal state or electrical properties, according to external signals, and their internal state history, giving them the ability of memory, such as having different resistances under the same applied voltage. In memristor modeling, the term "internal state" refers to a variable that provides the memristor with memory. A wide range of memristor devices exists, each characterized by unique switching mechanisms and material selections [16, 17]. Of these devices, RRAM memristors are the most commonly employed for the implementation of MVM circuits. In RRAM-based memristor models, the internal state variable is typically the size of the conductive filament formed between the two memristor contacts [18, 19].

While existing quantum simulators suffer from exponential execution times and memory requirements, as the number of qubits increases, the proposed approach of using memristor crossbar circuits enables efficient parallel quantum computation with a constant time complexity of $\mathcal{O}(1)$, regardless of system size, and reduces significantly the resulting hardware complexity [20–23]. The high density of memristor crossbars mitigates the memory problem by allowing more qubits to be simulated in the same die area as a conventional computing chip. In its quiver of assets, the memristor crossbar presents also inherent reconfigurality, so it can implement different quantum algorithms by using the same hardware, unlike, for example, a corresponding potential ASIC implementation of a quantum simulator.

This chapter proposes the use of memristor circuits as accelerators in the quantum computing stack and more specifically a memristive crossbar circuit capable of simulating quantum algorithms with high efficiency [24, 25]. Due to their ability to integrate the quantum properties of entanglement and superposition, the proposed circuit can implement any quantum algorithm based on dense coding and massive parallelism, including all the quantum gates that form the universal quantum gate set. The implementation of Deutsch and Grover quantum algorithms on the proposed memristor crossbar configuration demonstrates that the circuit supports essential quantum properties.

2 Basics of Quantum Computations

Quantum computing is a type of computation that utilizes qubits, unlike traditional digital computing which uses bits. Qubits are quantum mechanical two-state systems that can exist in a superposition of states, meaning they can take on any intermediate value between the states "0" and "1." The probability of being in either state can range from 0 to 1, but the probabilities always add up to 1 due to the qubit's two-state nature. The performance of a quantum computer increases exponentially as the number of qubits increases, unlike in a typical digital computer where it increases linearly. This increase in performance is due to the entanglement of numerous qubits. A quantum computer performs a quantum algorithm to modify qubit superpositions and entanglements to enhance some probabilities and decrease others. When a qubit is measured, its state and the states of entangled qubits collapse to one of the basis states of "0" or "1" with a probability dependent on the state of the qubit at the time of measurement. The objective is to maximize the probability of measuring the correct answer.

2.1 Quantum Computations in Simulators

Quantum simulators refer to software programs that function as the target machine for quantum algorithms and can be executed on classical computers to facilitate the testing and development of such algorithms. These simulators simulate the behavior of qubits under various operations and provide a reliable environment for running and evaluating quantum programs. Quantum simulators have numerous applications, including the development of new quantum algorithms and the assessment of proposed quantum computers. Various quantum computer simulators exist, each of which with its own strengths and limitations. Typically, these simulators are developed as software programs and are executed on classical computers via the CPU. However, recent efforts have been made to take advantage of the parallel capabilities of GPUs and FPGAs [11, 12]. An overview of available quantum computer simulators grouped by programming language can be found in the literature [26].

2.2 Gates and Qubit Representation

The quantum simulator employs matrices to represent quantum gates and vectors to represent qubits. A qubit is represented by a Hilbert space vector composed of probability amplitudes α and β, which are complex numbers as given by Eq. (1):

$$\alpha\,|0\rangle + \beta\,|1\rangle = \alpha\begin{bmatrix}1\\0\end{bmatrix} + \beta\begin{bmatrix}0\\1\end{bmatrix} = \begin{bmatrix}\alpha\\\beta\end{bmatrix} \tag{1}$$

The basis vectors $|0\rangle$ and $|1\rangle$ are also involved in this representation. To maintain a constant cumulative probability, it is necessary to satisfy the condition $|\alpha|^2 + |\beta|^2 = 1$. The application of quantum gates to qubits is characterized by a multiplication of the qubit vectors with a corresponding matrix. These gates are represented by Hilbert space operators, and as unitary matrices represent them, they are all unitary. This ensures that the probability of all outcomes adds up to 1 for each occurrence. The Hadamard gate is an example of a quantum gate, and its matrix representation is provided by Eq. (2):

$$H = \frac{1}{\sqrt{2}} \times \begin{bmatrix}1 & 1\\1 & -1\end{bmatrix} \tag{2}$$

The action of the Hadamard (H) gate on a qubit, which maps $|0\rangle \rightarrow \frac{|0\rangle - |1\rangle}{\sqrt{2}}$ and drives the qubit into superposition, is:

$$H \times |1\rangle = \frac{1}{\sqrt{2}}\begin{bmatrix}1 & 1\\1 & -1\end{bmatrix}\begin{bmatrix}0\\1\end{bmatrix} = \begin{bmatrix}\frac{1}{\sqrt{2}}\\-\frac{1}{\sqrt{2}}\end{bmatrix} \tag{3}$$

The matrix-vector multiplication is the method used for the quantum gate to act on the qubit's state. This operation can be carried out efficiently by the proposed crossbar circuit by converting the qubit's state to a voltage vector input and reprogramming the conductive-bridge random-access memory (CBRAM) crossbars to replicate the functionality of quantum gates as is briefly represented in Fig. 1. A detailed description of the product operation between qubits, as well as between quantum operators, is given in the following Appendix at the end of the chapter for readability reasons.

3 Performing Quantum Computations on Memristive Crossbars

The memristor crossbar setup is proficient at carrying out matrix-vector multiplication, which is the primary operation in quantum computing. By utilizing the

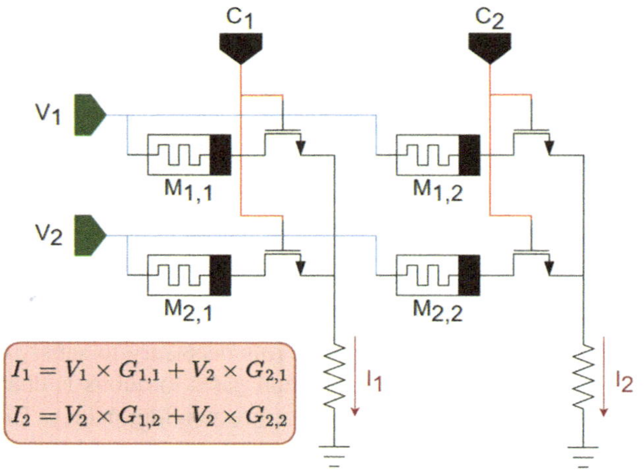

Fig. 1 Crossbar vector-matrix multiplication example

cross-point structure, row voltages are applied as the input vector $\mathbf{x}$, the memristors' conductance is used as values of the weight matrix $\mathbf{W}$, and the column current serves as the output vector $\mathbf{y} = \mathbf{W} \times \mathbf{x}$, allowing for efficient matrix-vector multiplication with $\mathcal{O}(1)$ time complexity. Kirchhoff's law is applied to obtain the desired output, as cross-point currents are accumulated instantly in each column. By reprogramming the memristors, a variety of quantum gates can be implemented using the same array.

While memristors are multi-state devices, programming them to multiple discrete resistance states can be challenging. However, binary programming with only two states, R_{on} and R_{off}, is easier and more accurate. In [27, 28], a universal set of quantum gates capable of being implemented through binary memristor crossbar arrays was discovered. This set includes the Hadamard and Controlled-Controlled-NOT (CCNOT) gates, which can create a superposition of basis states and evolve the superposition state to an entangled state, respectively. The only missing feature in this set is the ability to generate imaginary numbers. Bernstein and Vazirani [29] demonstrated that by adding an extra qubit to the circuit, only real-valued matrices are necessary to cover all quantum computing operations. The additional qubit indicates whether the system's state is in the real or imaginary part of the Hilbert space.

3.1 Simple Example of the Proposed Circuit Operation

To demonstrate the functionality of the memristor crossbar, the operation of the first two columns of a crossbar is shown in Fig. 2, where the circuit is capable of

Fig. 2 Basic crossbar circuit with differential amplifier

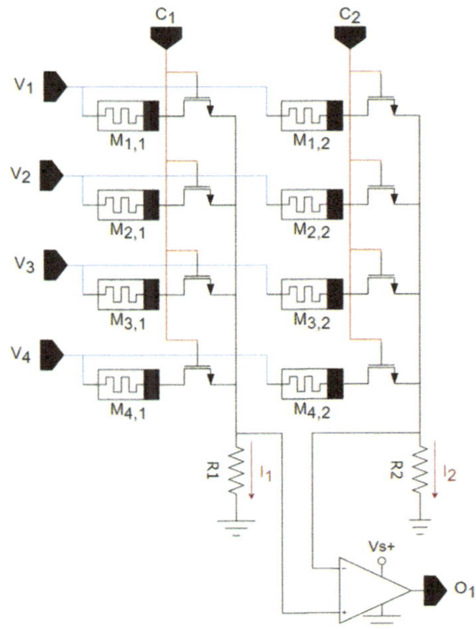

Table 1 Mapping column vector $[1, 0, -1, 0]$ to the memristor crossbar

Ternary value	$M_{i,0}$	$M_{i,1}$
1	R_{on}	R_{off}
0	R_{off}	R_{off}
−1	R_{off}	R_{on}
0	R_{off}	R_{off}

performing the multiplication between a row and column vector. The column vector is encoded in the memristors of the crossbar as shown in Table 1, while the row vector is represented as input voltages $[V_1, V_2, V_3, V_4]$.

Initially, the output of the differential amplifier ($Vout_1$) of Fig. 2 can be expressed as:

$$O_1 = G(V_{R1} - V_{R2}) \tag{4}$$

where G is the gain of the differential amplifier and V_{R1}, V_{R2} are the positive and negative terminal inputs of the amplifier, respectively. Because $R = R_1 = R_2$, then Eq. (4) can be written as:

$$O_1 = GR(I_{R1} - I_{R2}) \tag{5}$$

To avoid affecting the analog multiplication process, the resistance value, represented by R, should be significantly smaller than the column resistance of the crossbar. The column resistance reaches its worst-case scenario when all the memristors in the column are in the R_{on} state and when the same signal is applied to all crossbar inputs. For instance, in Fig. 2, if the voltage $V_i = 1\text{V}$ is assigned to the crossbar input for each i value in the set 1, 2, 3, 4, the minimum value for the column resistance is obtained. In this case, the memristors in the column are connected in parallel, and their equivalent resistance is R_{on}/N, where N represents the number of memristors in a column or the number of rows in the crossbar. Therefore, it is necessary that R, R_1, and R_2 are much smaller than R_{on}/N, and as a result, I_{R1} and I_{R2} can be calculated using the following equation:

$$I_{R1} = \frac{V_1}{M_{1,1}} + \frac{V_2}{M_{2,1}} + \frac{V_3}{M_{3,1}} + \frac{V_4}{M_{4,1}} \tag{6}$$

$$I_{R2} = \frac{V_1}{M_{1,2}} + \frac{V_2}{M_{2,2}} + \frac{V_3}{M_{3,2}} + \frac{V_4}{M_{4,2}} \tag{7}$$

By replacing $M_{i,j}$ with the corresponding values of Table 1, we can extract $I_{R1} - I_{R2}$:

$$\begin{aligned} I_{R1} - I_{R2} &= V_1\left(\frac{1}{R_{on}} - \frac{1}{R_{off}}\right) + V_2\left(\frac{1}{R_{off}} - \frac{1}{R_{off}}\right) \\ &+ V_3\left(\frac{1}{R_{off}} - \frac{1}{R_{on}}\right) + V_4\left(\frac{1}{R_{off}} - \frac{1}{R_{off}}\right) \\ &= \left(\frac{1}{R_{on}} - \frac{1}{R_{off}}\right)(V_1 - V_3) \end{aligned} \tag{8}$$

Combining Eqs. (5) with (8), we can derive:

$$O_1 = GR\left(\frac{1}{R_{on}} - \frac{1}{R_{off}}\right)(V_1 - V_3) \tag{9}$$

$$D = GR\left(\frac{1}{R_{on}} - \frac{1}{R_{off}}\right) \tag{10}$$

Because the first term of (9) are only constant variables, they can be grouped into one parameter D which has been set equal to 1 through proper gain selection. Finally, we extract the final output expression of the memristor crossbar which performs the multiplication between the input row vector $[V_0, V_1, V_2, V_3]$ and the column vector $[1, 0, -1, 0]$ that has been encoded in the crossbar:

$$O_1 = 1 \times V_1 + 0 \times V_2 - 1 \times V_3 + 0 \times V_4 \tag{11}$$

The aforementioned explanation is identical for every pair of columns of the crossbar leading to a memristor crossbar that is capable of performing vector-matrix multiplication.

In order to decide in which state the memristors have to be set to encode the appropriate value of the ternary system, it should be noticed that the differential amplifiers' inputs are mutually negated, in other words mapped to 0, when both memristors are found in the same state (R_{off}). To encode the values -1 (1), respectively, the memristor that is connected to the negative (positive) input of the differential amplifier has to be in R_{on} state, while the other one in R_{off} state, accordingly. Moreover, the necessary coefficients, which are multiplied with the whole matrix [i.e., in (2)], can be encoded in the gain of the output differential amplifier of the crossbar, mitigating the problem of the exact programming of memristors to the design of conventional circuit elements.

3.2 Multiple-Crossbar Configuration

The general representation of a crossbar capable of implementing an N qubit gate is depicted in Fig. 3a. Each cross-point consists of one transistor in series with one memristor (1T1R) cells. The crossbar dimension is $S \times 2S$, where S represents the number of possible basis states of the qubits.

More specifically, an $S \times 2S$ memristor array is employed to encode each quantum gate, where $S = 2^Q$ is the number of basis states that can be represented with Q qubits. To implement the quantum algorithms, multiple-crossbar arrays are utilized, which are connected in series. This connection is such that the S outputs of each crossbar, denoted $O_{1:S,\{i\}}$, are driving the inputs $in_{1:S,\{i+1\}}$ of the crossbars that are connected in series, where $i \in [1, T] \cap \mathbb{Z}$ and T is the total number of

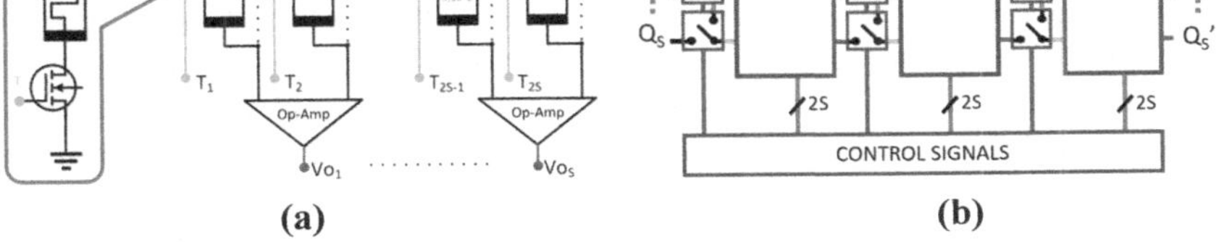

Fig. 3 (**a**) Proposed memristor crossbar circuit that can represent one quantum gate. (**b**) Proposed multiple-crossbar configuration that can compute quantum algorithms

crossbars. Figure 3b shows that the qubits' basis state is applied to $in_{1:S,\{1\}}$ and the final output is measured at $O_{1:S,\{T\}}$. At the intersection of the crossbar array, 1T1R cells are implemented to ensure that the sneak path problem is eliminated. This 1T1R configuration is preferred over other methods, such as the memristor-only crossbar [30], because it enables better control of the current flowing through the memristor, resulting in a more accurate programming of the memristor to a specific resistance.

To program the memristors and encode the matrix values of the quantum gate, a combination of `WRITE` and `CONTROL` signals is used. During the programming phase of each crossbar, the write voltage pulses ($P_{1:S,\{i\}}$) are used instead of the input signals ($in_{1:S,\{i\}}$) to manipulate the memristors' resistance and program the memristor crossbar. The `CONTROL` signals play two roles. The first one is to schedule the column-by-column programming of the memristor crossbar, which is handled by the signals $T_{1:2S}$ and $P_{1:S,\{i\}}$. The second one is to switch the operation of the circuit between programming and computation phases, which is controlled by the signal `State` that operates the switches of Fig. 3b. When `State` is set to 1, the inputs of the crossbars are connected to the programming signals, while when `State` is set to 0, the crossbars are connected in series and ready to compute the quantum algorithm.

4 Simulation Results of the Universal Set of Quantum Gates

Two 1T1R crossbars have been simulated that implement the universal set of Hadamard and CCNOT gates, accordingly. The programming process of the crossbar and the matrix-vector multiplication (computation) process are demonstrated here. All simulations shown in this section were held in Cadence's Virtuoso Suite.

4.1 The Hadamard Gate

The crossbar equivalent of the 1-qubit Hadamard gate is presented in Fig. 4a and consists of a 2×4 memristor crossbar configuration and 2 differential amplifiers [31].

At the beginning of the process, the crossbar needs to be set up to function as a Hadamard gate through programming. Figure 4a illustrates the state of each memristor in the crossbar. Figure 4b shows the results of simulating the programming process. Initially, all memristors are initialized in the R_{off} state by applying a negative voltage (V_{RST}) to every row (I_1, I_2). At the same time, the selector-transistors are activated by applying a voltage of 5V (V_{SEL}) to their gates ($T_1 - T_4$). The crossbar is then programmed column by column. For the first column, the memristors are switched to R_{on} by applying a voltage (V_{SET}) to I_1 and I_2 while selecting T_1 through V_{SEL}. For the second column, both memristors remain in the

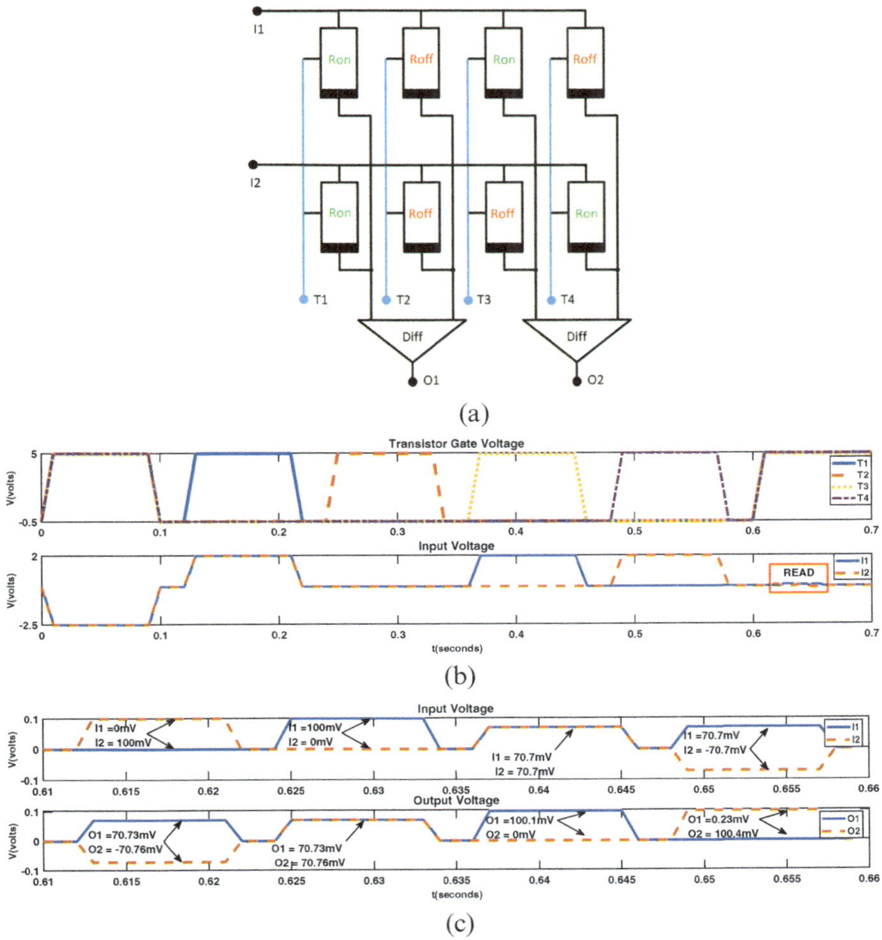

Fig. 4 (**a**) Memristive crossbar representing a Hadamard gate. (**b**) Programming process of the crossbar column by column to map the Hadamard gate. (**c**) Applying the Hadamard gate to different qubit states to evaluate its proper behavior [Computation (READ) Process] (Adapted from [32])

R_{off} state, so no input voltage is applied. In the third column, V_{SEL} is applied to T_3, and V_{SET} is applied to I_1, causing the upper memristor to switch. Finally, for the fourth column, V_{SEL} is applied to T_4, and V_{SET} is applied to I_2, causing the lower memristor to switch. This programming process only needs to occur once and is necessary again only if the same crossbar needs to be reprogrammed to a different quantum gate.

In the process of calculating the vector-matrix multiplication, all memristors in the crossbar are involved in the calculation, with each column selected by applying V_{SEL} to $(T_1 - T_4)$. The qubit state is represented as a voltage signal in

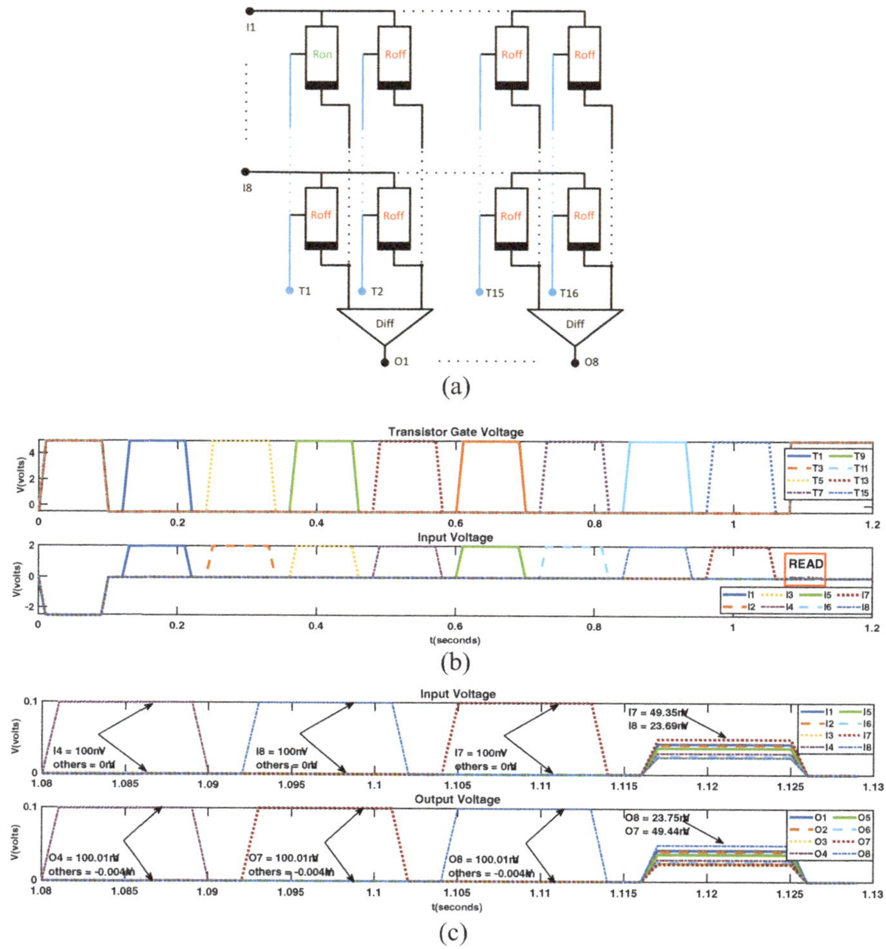

Fig. 5 (**a**) Memristive crossbar representing a CCNOT gate. (**b**) Programming process of the crossbar column by column to map the CCNOT gate. (**c**) Applying the CCNOT gate to different qubit states to evaluate its proper behavior [Computation (READ) Process] (Adapted from [32])

I_1 and I_2, while the final result of the calculation is represented by the output voltages of the differential amplifiers (O_1 and O_2), indicating the new state of the qubit. To accurately map the qubit basis state coefficients, input and output voltages range between -100mV and 100 mV. Higher input voltages are avoided to prevent memristors from switching during computation and to maintain low power consumption. The simulation of the computation process can be found at the end of Fig. 4b, marked with a red square and labeled READ. A more detailed depiction of the computation process is shown in Fig. 4c. Four different input combinations were tested, and the output voltages confirm that the memristor crossbar operates as a Hadamard gate.

4.2 *The CCNOT Gate*

The second quantum gate of the universal set that has been implemented in the crossbar grid is the CCNOT gate, the action of which is represented by the following matrix:

$$CCNOT = \begin{bmatrix} 1&0&0&0&0&0&0&0 \\ 0&1&0&0&0&0&0&0 \\ 0&0&1&0&0&0&0&0 \\ 0&0&0&1&0&0&0&0 \\ 0&0&0&0&1&0&0&0 \\ 0&0&0&0&0&1&0&0 \\ 0&0&0&0&0&0&0&1 \\ 0&0&0&0&0&0&1&0 \end{bmatrix} \tag{12}$$

Using the same programming-computing method, the CCNOT gate was implemented and tested on an 8×16 memristor crossbar as depicted in Fig. 5a. Since the gate only has values of 0 and 1, there was no need to map negative values onto the memristor crossbar, leaving half of the crossbar's columns unchanged during the programming process. The entire programming process is shown in Fig. 5b, and the computation process is demonstrated in Fig. 5c. Four different input combinations were tested, with the outputs O_7 and O_8 being reversed only when at least one of the inputs I_7 or I_8 was present, indicating proper operation of the CCNOT gate.

5 Simulation Result of Quantum Algorithms' Implementation

In this section, the programming of three crossbars is demonstrated to represent two well-known quantum algorithms, namely, Deutsch and Grover. All the simulations, as well as the circuit design, were held in Cadence Virtuoso Platform. Spectre was selected to perform the circuit simulation. The simulation results were extracted in CSV format, while the plotting of these data has been processed in MATLAB.

5.1 Utilized Memristor and Transistor Models

The memristor model utilized in the crossbar array is the ASU RRAM Compact Model [33]. It is a physics-based RRAM device model with compact equation set, developed in Verilog-A that enables the large-scale circuit simulations. The model properly captures the device's change in resistance due to the filamentary switching found in fabricated RRAM devices, and it is fitted to the experimental data of IMEC HfO_x-based RRAM devices [34, 35], which are capable for fast transitions among their resistance states (10kΩ–3MΩ) in the range of nanoseconds.

An in-series transistor (selector) is connected with each memristor at the cross-points of the array (Fig. 3a), constituting the 1T1R configuration shown in Fig. 3b. In particular, the bottom electrode of the MIM memristor is connected to the drain contact a MOSFET, while the gate voltage is used to adjust the resistance of the memristor, i.e., allow the programming of the memristor. Obviously, the gate controls the maximum current that flows from memristor's top electrode to transistor's source. The switching behavior (resistance) of one 1T1R cell is showcased in Fig. 3a when applying both −2.5V (`reset`) and 2.5V (`set`) voltage signals. The same behavior is observed for exactly every memristor of the crossbar. Apart from the aforementioned advantages, 1T1R configuration harms the array size scaling and increases the design complexity; thus, new integration approaches such as vertical stacking, where the transistors are fabricated in a different layer than memristors, are currently under extensive research. The selection of the enabled cross-points is performed column-wise, while the behavior of each transistor is simulated by the BSIM3V3 transistor model [36].

5.2 Deutsch Algorithm

To implement Deutsch's algorithm (Fig. 6a), the C_1 crossbar in Fig. 3c is programmed to represent $H \otimes H$, while C_2 is programmed to F and must be reconfigured each time a different function $F_a - F_d$ is evaluated. The four different cases of the oracle F that has to be encoded in the C_2 crossbar can be observed in (13)–(16).

$$F_a = \begin{bmatrix} 0\,1\,0\,0 \\ 1\,0\,0\,0 \\ 0\,0\,0\,1 \\ 0\,0\,1\,0 \end{bmatrix} \quad (13) \qquad F_b = \begin{bmatrix} 1\,0\,0\,0 \\ 0\,1\,0\,0 \\ 0\,0\,1\,0 \\ 0\,0\,0\,1 \end{bmatrix} \quad (14)$$

$$F_c = \begin{bmatrix} 0\,1\,0\,0 \\ 1\,0\,0\,0 \\ 0\,0\,1\,0 \\ 0\,0\,0\,1 \end{bmatrix} \quad (15) \qquad F_d = \begin{bmatrix} 1\,0\,0\,0 \\ 0\,1\,0\,0 \\ 0\,0\,0\,1 \\ 0\,0\,1\,0 \end{bmatrix} \quad (16)$$

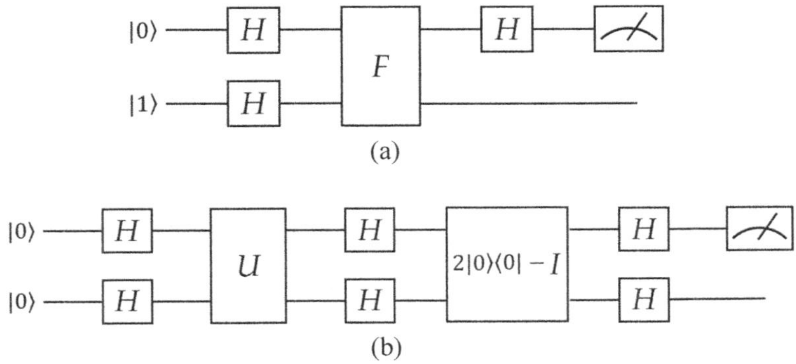

Fig. 6 (**a**) Block diagram of the Deutsch algorithm quantum circuit. (**b**) Block diagram of the Grover algorithm quantum circuit

C_3 is programmed to $H \otimes I$. The signals applied (input voltages $in_{1:4,1}$) are selected to represent the basis state $|01\rangle$ (where $a = 0, b = 1, c = 0$, and $d = 0$).

The correct functioning of the crossbar circuit for executing the Deutsch algorithm is demonstrated in Fig.7a, where the input and output of each crossbar are shown, with the output of C_3 ($O_{1:4,3}$) proving the correct operation of the circuit. To aid comparison between the computations of the crossbar and the mathematical calculation of the quantum algorithm, a detailed representation of the amplitudes of the computing pulses is provided in Fig. 7b. This figure also contains four different tables, each corresponding to a different function under investigation. The simulation results agree with the mathematical solution of the Deutsch algorithm presented in Eqs. (17)–(20):

$$|\psi 1\rangle = |01\rangle \tag{17}$$

$$\begin{aligned} |\psi 2\rangle &= (H \otimes H) \times |\psi 1\rangle \\ &= 0.5\,|00\rangle - 0.5\,|01\rangle + 0.5\,|10\rangle - 0.5\,|11\rangle \end{aligned} \tag{18}$$

$$|\psi 3\rangle = F_{a:d} \times |\psi 2\rangle = \begin{cases} \pm 0.5\,|00\rangle \mp 0.5\,|01\rangle \pm 0.5\,|10\rangle \mp 0.5\,|11\rangle \\ \pm 0.5\,|00\rangle \mp 0.5\,|01\rangle \mp 0.5\,|10\rangle \pm 0.5\,|11\rangle \end{cases} \tag{19}$$

$$\begin{aligned} |\psi 4\rangle &= (H \otimes I) \times |\psi 3\rangle \\ &= \begin{cases} \pm 0.707\,|00\rangle \mp 0.707\,|01\rangle \\ \pm 0.707\,|10\rangle \mp 0.707\,|11\rangle \end{cases} \end{aligned} \tag{20}$$

The basis state coefficients of $|\psi 1\rangle$, $|\psi 2\rangle$, $|\psi 3\rangle$, $|\psi 4\rangle$ correspond to the scaled $Input$, $C1\ Output$, $C2\ Output$, and $C3\ Output$ voltages of the circuit displayed

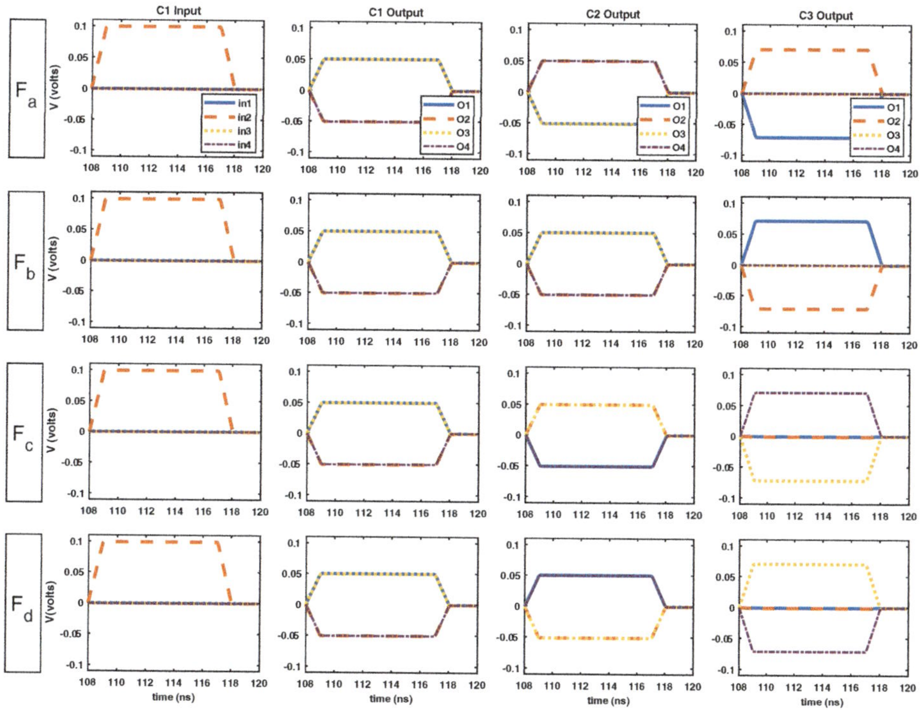

(a)

F_a

	Input	C1 Out	C2 Out	C3 Out
$\|00\rangle$	0.0000	0.0501	-0.0505	-0.0712
$\|01\rangle$	0.1000	-0.0501	0.0503	0.0707
$\|10\rangle$	0.0000	0.0501	-0.0505	0.0000
$\|11\rangle$	0.0000	-0.0501	0.0503	0.0000

F_b

	Input	C1 Out	C2 Out	C3 Out
$\|00\rangle$	0.0000	0.0501	0.0505	0.0712
$\|01\rangle$	0.1000	-0.0501	-0.0503	-0.0707
$\|10\rangle$	0.0000	0.0501	0.0505	0.0000
$\|11\rangle$	0.0000	-0.0501	-0.0503	0.0000

F_c

	Input	C1 Out	C2 Out	C3 Out
$\|00\rangle$	0.0000	0.0501	-0.0505	-0.0002
$\|01\rangle$	0.1000	-0.0501	0.0503	-0.0003
$\|10\rangle$	0.0000	0.0501	0.0503	-0.0710
$\|11\rangle$	0.0000	-0.0501	-0.0505	0.0710

F_d

	Input	C1 Out	C2 Out	C3 Out
$\|00\rangle$	0.0000	0.0501	0.0503	-0.0003
$\|01\rangle$	0.1000	-0.0501	-0.0505	-0.0002
$\|10\rangle$	0.0000	0.0501	-0.0505	0.0710
$\|11\rangle$	0.0000	-0.0501	0.0503	-0.0710

(b)

Fig. 7 (**a**) Crossbars' output voltages during the computation phase of the Deutsch algorithm for the different cases of evaluated functions ($F_a - F_d$). (**b**) Table representation of the output voltages during the computation phase (Adapted from [37]) in Fig. 7b. To determine whether the functions are balanced or constant, the first qubit must be measured. Measuring the first qubit results in a state of $|0\rangle$ or $|1\rangle$, indicating whether the function is constant or balanced, respectively. In the case of F_a and F_b, the C_3 $Output$ indicates that only the basis states $|00\rangle$ and $|01\rangle$ have a high probability, adding up to one $((\pm 0.7)^2 + (\pm 0.7)^2 = 1)$. This means that the first qubit will be in the $|0\rangle$ state during measurement, resulting in F_a and F_b being classified as constant. For F_c and F_d, only the basis states $|10\rangle$ and $|11\rangle$ have a high probability, adding up to one $((\pm 0.7)^2 + (\pm 0.7)^2 = 1)$, indicating that the first qubit will be in the $|1\rangle$ state during measurement, and thus, F_c and F_d are classified as balanced. This is illustrated in Fig. 6a.

5.3 Grover Algorithm

By properly programming C_1–C_3 crossbar arrays, the Grover algorithm of (21) (Fig. 6b) can be implemented using the same setup as the Deutsch algorithm.

$$|\psi\rangle = (H \otimes H)(2\,|0\rangle\,\langle 0| - I)(H \otimes H)U(H \otimes H)\,|00\rangle \tag{21}$$

The required quantum gates are represented by programming the memristor crossbars, accordingly. The operators U_a–U_d encode the different lists, and each time a different list is being investigated, C_2 crossbar must be reprogrammed to represent the corresponding operator. The programming procedure for the Grover algorithm follows the same approach as for the Deutsch algorithm. More specifically, to implement Grover's algorithm of Fig. 6b, the circuit components are programmed as follows: C_1 represents $H \otimes H$, and C_2 represents U and is reconfigured each time a different function Ua-Ud is evaluated, with their matrix equivalents presented in the following Eqs. (22)–(25).

$$U_a = \begin{bmatrix} -1 & 0 & 0 & 0 \\ 0 & 1 & 0 & 0 \\ 0 & 0 & 1 & 0 \\ 0 & 0 & 0 & 1 \end{bmatrix} \tag{22}$$

$$U_c = \begin{bmatrix} 1 & 0 & 0 & 0 \\ 0 & 1 & 0 & 0 \\ 0 & 0 & -1 & 0 \\ 0 & 0 & 0 & 1 \end{bmatrix} \tag{24}$$

$$U_b = \begin{bmatrix} 1 & 0 & 0 & 0 \\ 0 & -1 & 0 & 0 \\ 0 & 0 & 1 & 0 \\ 0 & 0 & 0 & 1 \end{bmatrix} \tag{23}$$

$$U_d = \begin{bmatrix} 1 & 0 & 0 & 0 \\ 0 & 1 & 0 & 0 \\ 0 & 0 & 1 & 0 \\ 0 & 0 & 0 & -1 \end{bmatrix} \tag{25}$$

C_3 represents M as defined in Eq. (26). The qubits are initialized to the basis state $|00\rangle$, which corresponds to $(a = 1,\ b = 0,\ c = 0,\ d = 0)$.

$$M = (H \otimes H)(2\,|0\rangle\,\langle 0| - I)(H \otimes H) \tag{26}$$

Once the crossbar circuit is programmed to represent the quantum gates of the Grover algorithm, the computation of the algorithm can be performed. The correct operation of the crossbar circuit to execute the Grover algorithm is demonstrated in Fig. 8a through the input and output pulses of each crossbar, with the output of C_3 ($O_{1:4,\{3\}}$) showcasing the result of the computation. Additionally, Fig. 8b provides a clear visualization of the amplitudes of the computing pulses, which makes it easier to compare the calculation of the crossbar to the mathematical calculation of the quantum algorithm. The four different tables in Fig. 8b correspond to the different lists that contain the in-search item. To illustrate the matching of the crossbar output voltages to the solution of the Grover algorithm, the equations that correspond to the Grover algorithm when the in-search item is $|11\rangle$ (list U_d) are presented below as an example:

$$|\psi 1\rangle = |00\rangle \tag{27}$$

$$\begin{aligned} |\psi 2\rangle &= (H \otimes H) \times |\psi 1\rangle \\ &= 0.5\,|00\rangle + 0.5\,|01\rangle + 0.5\,|10\rangle + 0.5\,|11\rangle \end{aligned} \tag{28}$$

$$|\psi 3\rangle = U_d \times |\psi 2\rangle = 0.5\,|00\rangle + 0.5\,|01\rangle + 0.5\,|10\rangle - 0.5\,|11\rangle \tag{29}$$

$$\begin{aligned} |\psi 4\rangle = M \times |\psi 3\rangle &= (H \otimes H)(2\,|0\rangle\,\langle 0| - I)(H \otimes H) \times |\psi 3\rangle \\ &= 0.0\,|00\rangle + 0.0\,|01\rangle + 0.0\,|10\rangle + 1.0\,|11\rangle \end{aligned} \tag{30}$$

The basis state coefficients of $|\psi 1\rangle$, $|\psi 2\rangle$, $|\psi 3\rangle$, $|\psi 4\rangle$ correspond to the $Input$, $C1\ Output$, $C2\ Output$, and $C3\ Output$ signals of the circuit, shown in the U_d Table of Fig. 8b, with a scaling factor of 10. Each of the four tables in Fig. 8b represents the algorithm's execution for a different list, enabling the identification of the in-search item within the list. The output vector of C_3 ($O_{1:4,\{3\}}$) contains a basis state coefficient equal to 1 ($0.1V$) that indicates the location of the searched item in the list.

6 Framework

In order to accelerate the development and analysis of memristive crossbar configurations toward quantum computing, a graphical user interface (GUI) has been designed that would make it easy for users to design, implement, and analyze universal quantum computations on memristor crossbars [38]. The GUI's startup screen (shown in Fig. 9) displays all the available options for quantum computations, such as the number of required qubits, the number of computational steps, and the gates applied at each step. The leftmost column of the display represents the row number of each qubit, while the other columns represent the quantum gates applied at each step. The user has the flexibility to input a range of quantum gates. More

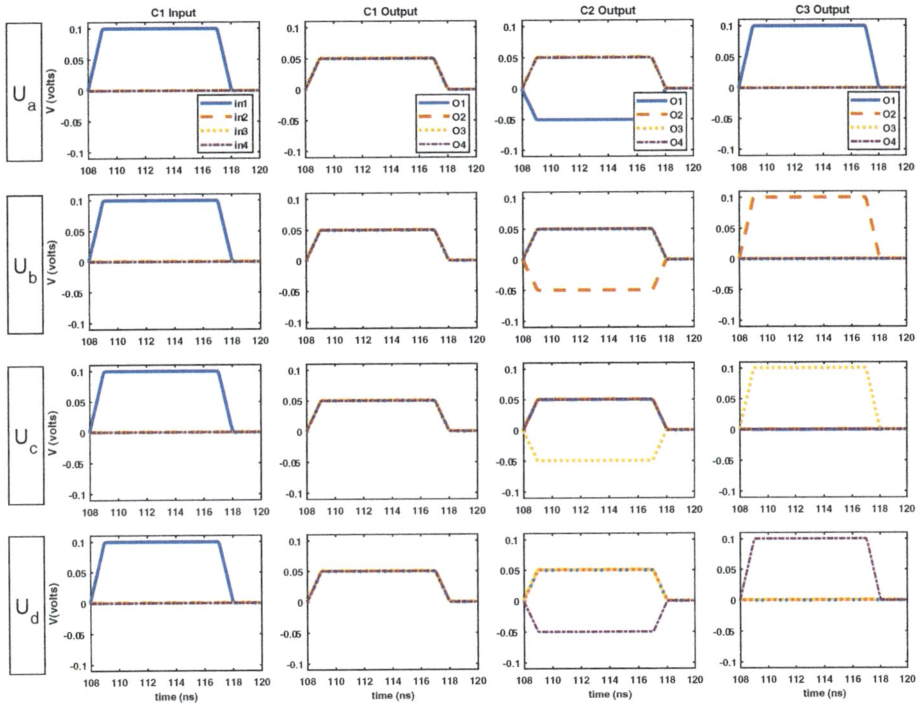

(a)

U_a

	Input	C1 Out	C2 Out	C3 Out
|00>	0.1000	0.0501	-0.0501	0.1000
|01>	0.0000	0.0501	0.0501	0.0000
|10>	0.0000	0.0501	0.0501	0.0000
|11>	0.0000	0.0501	0.0501	0.0000

U_b

	Input	C1 Out	C2 Out	C3 Out
|00>	0.1000	0.0501	0.0501	0.0000
|01>	0.0000	0.0501	-0.0501	0.1000
|10>	0.0000	0.0501	0.0501	0.0001
|11>	0.0000	0.0501	0.0501	0.0001

U_c

	Input	C1 Out	C2 Out	C3 Out
|00>	0.1000	0.0501	0.0501	0.0000
|01>	0.0000	0.0501	0.0501	0.0001
|10>	0.0000	0.0501	-0.0501	0.1000
|11>	0.0000	0.0501	0.0501	0.0001

U_d

	Input	C1 Out	C2 Out	C3 Out
|00>	0.1000	0.0501	0.0501	0.0000
|01>	0.0000	0.0501	0.0501	0.0001
|10>	0.0000	0.0501	0.0501	0.0001
|11>	0.0000	0.0501	-0.0501	0.1000

(b)

Fig. 8 (**a**) Crossbars' output voltages during the computation phase of the Grover algorithm for the different cases of the in-search item ($U_a - U_d$). (**b**) Table representation of the output voltages during the computation phase (Adapted from [37]) specifically, the user can select one qubit gates like Hadamard (H) and Pauli (X, Z). Hadamard drives the qubit state to an equal superposition, while Pauli gates (X, Z) perform a rotation around the x and z axes of the Bloch sphere by π radians accordingly. Moreover, the user can select two qubit gates, like the controlled gates (CZ, CNOT) and SWAP. In controlled gates, one qubit serves as the control for a specific operation that will be applied to a second qubit, but only if certain conditions are met. Additionally, the SWAP gate can exchange the states of two qubits. Finally, the user can input three qubit gates like Toffoli (CCNOT) and Fredkin (CSWAP) which have similar functionality with their two qubit counterparts. By clicking the "GENERATE" button, the designed memristive circuit is extracted.

Once the quantum circuit is generated, the user can access more detailed information by navigating through the top tabs, namely, "Memristor Crossbar," "Write Voltage Figures," and "Quantum Computations." In the "Memristor Crossbar" tab, the simulated memristor crossbar is displayed along with all the memristances, following the crossbar programming. In the next tab, "Write Voltage Figures," the input voltage and transistor gate voltage are illustrated for each computational step during the "WRITE" process. In the last tab, "Quantum Computations," the user can select the initial state of the qubit in each row and perform all the available computations that are illustrated as a figure of the output coefficients in each qubit basis state.

The GUI has been designed to be easily adaptable to different selections of the quantum circuit, as well as to different memristor devices. Thus, a final tab, "Memristor Settings," is available for the user to select all the memristor model parameters, as well as the SET/RESET parameters and the time step needed for successful simulations.

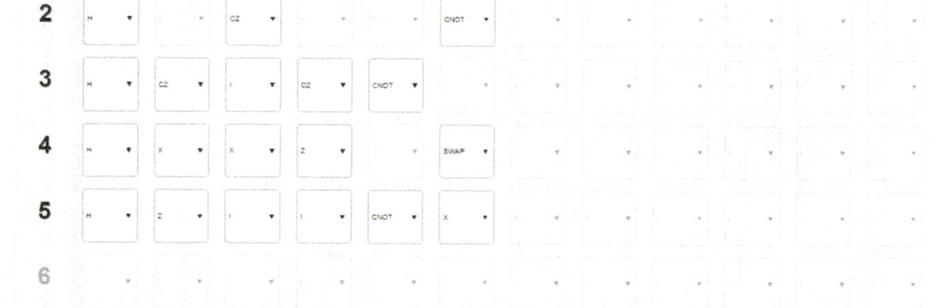

Fig. 9 Circuit setup screen of the graphical user interface (GUI)

The GUI was originally created in MATLAB app designer, but a standalone installation is available with the supplementary files [39]. The simplicity and versatility of the GUI, along with the adaptability and reprogrammability of the memristor crossbar, make it suitable for a wide range of quantum applications.

7 Discussion on the Variability and Stochastic Behavior of the Memristor

The quantum computer produces a probability distribution that favors obtaining the correct result over an incorrect one when running the algorithm. Measuring the qubit state determines the outcome, usually resulting in the highest probability basis state indicating the solution. However, since there is a possibility of incorrect results, additional runs or error correction may be necessary in a real quantum computer. In the proposed circuit, the probability amplitude values of each basis state of the qubits are accessible. This allows the basis state with the highest probability to be chosen as the solution without concerning the precise value of the probability. Therefore, even if the output is somewhat imprecise due to circuit variability, as long as the solution has a high enough voltage amplitude, it can be distinguished, providing fault tolerance to the circuit.

Furthermore, the reality of measurement errors is often disregarded in simulating quantum circuits, with many approaches still trying to fit within the deterministic mode of computation. In deterministic circuits, the results of the computation can be obtained without any measurement errors. The inherent variability of the memristors can be exploited to simulate the decoherence found on the fabricated quantum computers. By tuning the writing voltage of the memristors during programming, the probability of their switching could be manipulated until it matches the error rate of the quantum computer.

Finally, regarding the measurement procedure that takes place in quantum computers, it can be emulated by controllable stochastic devices, such as memristors, driven by the output voltages of the crossbar configuration that constitute the final state of the qubits. More specifically, in quantum simulators, we can observe the evolution of the qubits' state when applying different quantum gates without disturbing or affecting their state. However, in a real quantum computer, a measurement must be performed to extract the calculation of the quantum system, which leads to the collapse of the qubits' quantum state. The basis state that the qubits collapse into is a stochastic process, and the probability is dependent on the coefficients of the basis states just before the measurement. To simulate this behavior in our quantum simulator, we plan to use a stochastic device at the output of crossbar C3 that will switch stochastically to either 1 or 0 depending on the amplitude of the output voltages, which encode the final state of the qubits. We intend to exploit the stochastic switching of memristors for this operation and create a connection between the results of the quantum simulator and the actual behavior of a quantum computer after measurement.

8 Conclusions

This chapter explores the possibility of using reprogrammable memristor crossbars to emulate quantum gates. Since matrix-vector multiplication is the most significant algebraic operation in quantum computing, the unique properties of memristive grids are utilized to perform circuit-level quantum computations. The Hadamard and CCNOT gates, which together form a universal quantum set, are effectively mapped onto memristor crossbars, and their accurate operation is demonstrated. Also, the implementation of two well-known quantum algorithms has been demonstrated. The simulation results presented in this study provide a basis for further investigation into the use of memristor devices for hardware acceleration of quantum computations. Finally, a framework is presented toward accelerating the development and analysis of memristive crossbar configurations for quantum computing.

Appendix

The *bra-ket* notation is used to represent qubits, where a qubit is expressed as a *ket* $|a\rangle$, while a *bra* $\langle b|$ is used to represent linear functions that convert qubits to complex numbers in the $\mathbb{C}$ plane. In vector form, the *ket* is a column vector represented as $|a\rangle = [a_1, a_2]^\mathsf{T}$, and the *bra* is a row vector represented as $\langle b| = [b_1^*, b_2^*]$. This work utilizes the outer and Kronecker products of the *bra* and *ket* mathematical operations in quantum algorithms. More specifically, the outer product $|a\rangle\langle b|$ is used to create a linear operator:

$$|a\rangle\langle b| = \begin{bmatrix} a_1 b_1^* & a_1 b_2^* \\ a_2 b_1^* & a_2 b_2^* \end{bmatrix} \tag{31}$$

Moreover, the Kronecker product ($\otimes$) between two matrices **A** and **B** is defined as:

$$\mathbf{A} \otimes \mathbf{B} = \begin{bmatrix} A_{1,1}\mathbf{B} & A_{1,2}\mathbf{B} \\ A_{2,1}\mathbf{B} & A_{2,2}\mathbf{B} \end{bmatrix} = \tag{32}$$

$$\begin{bmatrix} A_{1,1} \begin{bmatrix} B_{1,1} & B_{1,2} \\ B_{2,1} & B_{2,2} \end{bmatrix} & A_{1,2} \begin{bmatrix} B_{1,1} & B_{1,2} \\ B_{2,1} & B_{2,2} \end{bmatrix} \\ A_{2,1} \begin{bmatrix} B_{1,1} & B_{1,2} \\ B_{2,1} & B_{2,2} \end{bmatrix} & A_{2,2} \begin{bmatrix} B_{1,1} & B_{1,2} \\ B_{2,1} & B_{2,2} \end{bmatrix} \end{bmatrix} \tag{33}$$

and constitutes the main product operation between qubits. In particular, the combination of two qubits can be performed through the Kronecker product such as:

$$|ab\rangle = |a\rangle \otimes |b\rangle = \begin{bmatrix} a_1 \\ a_2 \end{bmatrix} \otimes \begin{bmatrix} b_1 \\ b_2 \end{bmatrix} = \tag{34}$$

$$\begin{bmatrix} a_1 \begin{bmatrix} b_1 \\ b_2 \end{bmatrix} \\ a_2 \begin{bmatrix} b_1 \\ b_2 \end{bmatrix} \end{bmatrix} = \begin{bmatrix} a_1 b_1 \\ a_1 b_2 \\ a_2 b_1 \\ a_2 b_2 \end{bmatrix} \tag{35}$$

Additionally, when expressed in vector form, quantum operators are represented by matrices, such as the Hadamard gate ($H = [1, 1; 1, -1]/\sqrt{2}$) and the *NOT* gate ($X = [0, 1; 1, 0]$). They are subject to multiplication operations, specifically matrix-matrix multiplication and the Kronecker product. Matrix-matrix multiplication is used to apply consecutive quantum gates on qubits, while the Kronecker product is a crucial operation utilized to combine quantum gates that are applied simultaneously on multiple qubits. For instance, the subsequent application of a Hadamard and a *NOT* gate on a single qubit $|a\rangle$ is accomplished using the operator:

$$X \times H = \frac{1}{\sqrt{2}} \begin{bmatrix} 1 & 1 \\ 1 & -1 \end{bmatrix} \times \begin{bmatrix} 0 & 1 \\ 1 & 0 \end{bmatrix} = \frac{1}{\sqrt{2}} \begin{bmatrix} 1 & 1 \\ -1 & 1 \end{bmatrix} \tag{36}$$

On the other hand, when two Hadamard quantum gates are applied concurrently to two qubits, the operator is constructed through the Kronecker product:

$$H \otimes H = \frac{1}{2} \begin{bmatrix} 1 & 1 & 1 & 1 \\ 1 & -1 & 1 & -1 \\ 1 & 1 & -1 & -1 \\ 1 & -1 & -1 & 1 \end{bmatrix} \tag{37}$$

References

1. T. Ladd, F. Jelezko, R. Laflamme, Y. Nakamura, C. Monroe, J. O'Brien, Quantum computers. Nature **464**, 45–53 (2010)
2. M.A. Nielsen, I. Chuang, *Quantum computation and quantum information* (Cambridge University Press, Cambridge, 2010)
3. A. Montanaro, Quantum algorithms: an overview. npj Quantum Inf. **2**, 15023 (2016)
4. J. Clarke, F.K. Wilhelm, Superconducting quantum bits. Nature **453**, 1031–1042 (2008)
5. H. Häffner, C.F. Roos, B. Blatt, Quantum computing with trapped ions. Phys. Rep. **469**, 155–203 (2008)
6. L.M. Duan, H.J. Kimble, Scalable photonic quantum computation through cavity-assisted interactions. Phys. Rev. Lett. **92**, 127902 (2004)
7. V. Silva, *Practical Quantum Computing for Developers* (Apress, New York, 2018)
8. D. Wecker, K.M. Svore, LIQUiD: A software design architecture and domain-specific language for quantum computing. Preprint (2014). arXiv:1402.4467
9. Qiskit python api. https://qiskit.org/, accessed: 2021-07-01
10. Qproject python api. https://projectq.ch/, accessed: 2021-07-01
11. T. Jones, A. Brown, I. Bush, S.C. Benjamin, Quest and high performance simulation of quantum computers. Sci. Rep. **9**(1), 1–11 (2019)
12. J. Pilch, J. Długopolski, An FPGA-based real quantum computer emulator. J. Comput. Electron. **18**(1), 329–342 (2019)

13. L. Chua, Memristor-The missing circuit element. IEEE Trans. Circuit Theory **18**(5), 507–519 (1971)
14. D.B. Strukov, G.S. Snider, D.R. Stewart, R.S. Williams, The missing memristor found. Nature **453**(7191), 80 (2008)
15. L. Chua, G. Sirakoulis, A. Adamatzky, *Handbook of Memristor Networks* (Springer International Publishing, New York, 2019)
16. E. Tsipas, T.P. Chatzinikolaou, K.-A. Tsakalos, K. Rallis, R.-E. Karamani, I.-A. Fyrigos, S. Kitsios, P. Bousoulas, D. Tsoukalas, G.C. Sirakoulis, Unconventional memristive nanodevices. IEEE Nanotechnol. Mag. **16**(6), 34–45 (2022)
17. Y. Li, Z. Wang, R. Midya, Q. Xia, J.J. Yang, Review of memristor devices in neuromorphic computing: materials sciences and device challenges. J. Phys. D Appl. Phys. **51**(50), 503002 (2018)
18. I.-A. Fyrigos, V. Ntinas, G.C. Sirakoulis, P. Dimitrakis, I.G. Karafyllidis, Quantum mechanical model for filament formation in metal-insulator-metal memristors. IEEE Trans. Nanotechnol. **20**, 113–122 (2021)
19. I.-A. Fyrigos, T.P. Chatzinikolaou, V. Ntinas, S. Kitsios, P. Bousoulas, M.-A. Tsompanas, D. Tsoukalas, A. Adamatzky, A. Rubio, G.C. Sirakoulis, Compact thermo-diffusion based physical memristor model, in *2022 IEEE International Symposium on Circuits and Systems (ISCAS)* (IEEE, 2022), pp. 2237–2241
20. P. Bousoulas, S. Kitsios, T.P. Chatzinikolaou, I.-A. Fyrigos, V. Ntinas, M.-A. Tsompanas, G.C. Sirakoulis, D. Tsoukalas, Material design strategies for emulating neuromorphic functionalities with resistive switching memories. Jpn. J. Appl. Phys. **61**, SM0806 (2022)
21. N. Vasileiadis, V. Ntinas, I.-A. Fyrigos, R.-E. Karamani, V. Ioannou-Sougleridis, P. Normand, I. Karafyllidis, G. C. Sirakoulis, P. Dimitrakis, A new 1p1r image sensor with in-memory computing properties based on silicon nitride devices, in *2021 IEEE International Symposium on Circuits and Systems (ISCAS)* (IEEE, 2021), pp. 1–5
22. T.P. Chatzinikolaou, I.-A. Fyrigos, G.C. Sirakoulis, Image shifting tracking leveraging memristive devices, in *2022 11th International Conference on Modern Circuits and Systems Technologies (MOCAST)* (IEEE, 2022), pp. 1–4
23. C. Tsioustas, P. Bousoulas, J. Hadfield, T.P. Chatzinikolaou, I.-A. Fyrigos, V. Ntinas, M.-A. Tsompanas, G.C. Sirakoulis, D. Tsoukalas, Simulation of low power self-selective memristive neural networks for in situ digital and analogue artificial neural network applications. IEEE Trans. Nanotechnol. **21**, 505–513 (2022)
24. D. Deutsch, R. Jozsa, Rapid solution of problems by quantum computation. Proc. R. Soc. Lond. A Math. Phys. Sci. **439**(1907), 553–558 (1992)
25. L.K. Grover, A fast quantum mechanical algorithm for database search, in *Proceedings of the Twenty-Eighth Annual ACM Symposium on Theory of Computing* (1996), pp. 212–219
26. List of QC simulators, https://www.quantiki.org/wiki/list-qc-simulators, accessed: 2022-13-10
27. D. Aharonov, A simple proof that toffoli and hadamard are quantum universal. Preprint (2003). arXiv:quant-ph/0301040
28. Y. Shi, Both Toffoli and controlled-NOT need little help to do universal quantum computation. Preprint (2002). arXiv:quant-ph/0205115
29. E. Bernstein, U. Vazirani, Quantum complexity theory. SIAM J. Comput. **26**(5), 1411–1473 (1997)
30. M.A. Zidan, H.A.H. Fahmy, M.M. Hussain, K.N. Salama, Memristor-based memory: The sneak paths problem and solutions. Microelectron. J. **44**(2), 176–183 (2013)
31. T.M. Taha, R. Hasan, C. Yakopcic, Memristor crossbar based multicore neuromorphic processors, in *2014 27th IEEE International System-on-Chip Conference (SOCC)* (IEEE, 2014), pp. 383–389
32. I.-A. Fyrigos, V. Ntinas, G.C. Sirakoulis, P. Dimitrakis, I. Karafyllidis, Memristor hardware accelerator of quantum computations, in *2019 26th IEEE International Conference on Electronics, Circuits and Systems (ICECS)* (IEEE, 2019), pp. 799–802
33. P.-Y. Chen, S. Yu, Compact modeling of RRAM devices and its applications in 1T1R and 1S1R array design. IEEE Trans. Electron Dev. **62**(12), 4022–4028 (2015)

34. Y.Y. Chen, B. Govoreanu, L. Goux, R. Degraeve, A. Fantini, G.S. Kar, D.J. Wouters, G. Groeseneken, J.A. Kittl, M. Jurczak, et al., Balancing SET/RESET pulse for $> 10^{10}$ endurance in HfO_2/Hf 1T1R bipolar RRAM. IEEE Trans. Electron Dev. **59**(12), 3243–3249 (2012)
35. Y.Y. Chen, M. Komura, R. Degraeve, B. Govoreanu, L. Goux, A. Fantini, N. Raghavan, S. Clima, L. Zhang, A. Belmonte, et al., Improvement of data retention in HfO_2/Hf 1T1R RRAM cell under low operating current, in *2013 IEEE International Electron Devices Meeting* (IEEE, 2013), pp. 10–1
36. Y. Cheng, C. Hu, *MOSFET Modeling & BSIM3 User's Guide* (Springer Science & Business Media, New York, 1999)
37. I.-A. Fyrigos, V. Ntinas, N. Vasileiadis, G.C. Sirakoulis, P. Dimitrakis, Y. Zhang, I.G. Karafyllidis, Memristor crossbar arrays performing quantum algorithms. IEEE Trans. Circuits Syst. I Reg. Pap. **69**(2), 552–563 (2021)
38. I.-A. Fyrigos, T.P. Chatzinikolaou, V. Ntinas, N. Vasileiadis, P. Dimitrakis, I. Karafyllidis, G.C. Sirakoulis, Memristor crossbar design framework for quantum computing, in *2021 IEEE International Symposium on Circuits and Systems (ISCAS)* (IEEE, 2021), pp. 1–5
39. Quantum computing on memristor crossbars framework, https://github.com/DUTh-FET/Team-Research/tree/master/4.%20Technical%20Material, accessed: 2023-09-03

A Review of Posit Arithmetic for Energy-Efficient Computation: Methodologies, Applications, and Challenges

Hao Zhang, Zhiqiang Wei, Bo Yin, and Seok-Bum Ko

1 Introduction

The IEEE floating-point format, defined in the IEEE-754 standard [1], has been long used as the standard numeric format for many computer systems and applications. It has a large bit-width exponent component to provide large dynamic range and a large bit-width mantissa component to provide high precision. Almost all applications, including image processing, digital signal processing, and scientific computing, can get benefit from IEEE floating-point format.

However, due to the complexity of the floating-point format, the arithmetic operations based on floating-point formats have many problems that are difficult to be solved. One of them is the rounding error which can lead to different computation results for a series of floating-point additions when the order of the additions change. In addition, the exception handling of the floating-point format is complex which makes the floating-point arithmetic units difficult to be fully compliant to the IEEE standard and difficult to be fully verified.

In recent years, with the emerging of machine learning and edge computing, energy-efficient computation becomes vital for digital applications. Choosing an appropriate numeric format for their computation is important to achieve the performance requirements while maintaining a high energy efficiency for computation. The data of many applications, such as machine learning, have tapered distribution where most of the data are distributed around zero, while few values are large in

H. Zhang · Z. Wei · B. Yin
Faculty of Information Science and Engineering, Ocean University of China, Qingdao, China
e-mail: hao.zhang@ouc.edu.cn; weizhiqiang@ouc.edu.cn; ybfirst@ouc.edu.cn

S.-B. Ko (✉)
Department of Electrical and Computer Engineering, University of Saskatchewan, Saskatoon, SK, Canada
e-mail: seokbum.ko@usask.ca

W. Liu et al. (eds.), *Design and Applications of Emerging Computer Systems*,
https://doi.org/10.1007/978-3-031-42478-6_24

magnitude. For this kind of data distribution, although both high numeric precision and large representation range are needed, they are not required at the same time. For small values, as their magnitude is small, the exponent field does not need to be long in bit-width. However, as the density of small values is large, we still need a high precision to distinguish them which leads to a mantissa field with large bit-width. Similarly, for large values, we need a large bit-width exponent, but the bit-width of mantissa can be reduced. When using floating-point format in these computations, as the bit-width of each component is pre-determined by the IEEE-754 standard, to meet the requirements of both numeric precision and representation range, a high-precision floating-point format, such as single-precision or double-precision, is needed. However, as the hardware cost of large floating-point format is quite high, the efficiency of the computation will be quite limited. Therefore, the floating-point format may not be the optimal choice for these emerging applications.

To deal with the problems of the floating-point format, the universal number system is proposed by Dr. Gustafson [2]. It uses an extra bit to accommodate all cases that the results are not accurate. The new format effectively deals with the error problems of the floating-point format, but it is hardware expensive. It is later revised to the second generation with more hardware-friendly features. In 2017, the posit format which is the third generation of the universal number format is proposed [3]. Compared to the floating-point format, the posit format solved the error issue, and it has only one exception value which makes the design and verification process much easier. In terms of data distribution, the posit format can provide a much larger dynamic range than the floating-point format with the same total bit-width. Moreover, the posit format has a tapered number distribution which makes it especially suitable in machine learning, edge computing, and many other applications. Due to these benefits, the posit format is quickly adopted in many fields of applications, and many researchers tend to design hardware components based on the posit format.

This chapter provides a comprehensive review of posit arithmetic and their applications. We will begin with a detailed description of the posit numeric format in Sect. 2. Then the deployment of the posit format in modern computing system is going to be presented. In Sect. 3, applications that are benefited from the posit format are presented. The developing tools of posit arithmetic is presented in Sect. 4. Section 5 presents several designs of posit-based arithmetic units. The hardware processors designed based on the posit format are presented in Sect. 6. Design challenges and possible future directions of posit format are presented in Sect. 7. Finally, Sect. 8 concludes the whole chapter.

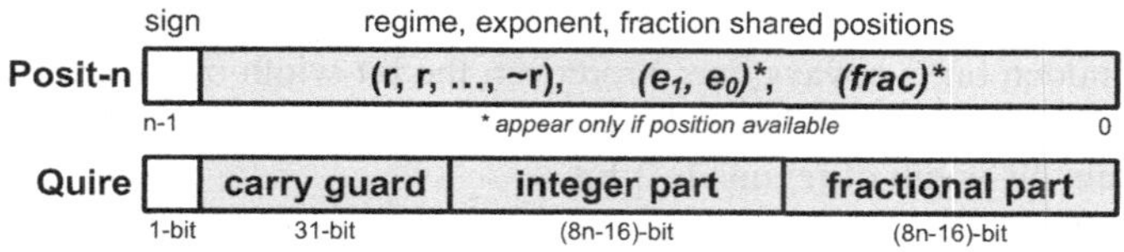

Fig. 1 Posit and quire format

2 Posit Numeric Format

2.1 General Format

The general numeric format of a posit number is shown in Fig. 1. A posit format can be defined with the total bit-width n, which is denoted as Posit-n. It is composed of four components: sign, regime, exponent, and fraction. Note that in the original format definition in [3], the bit-width of the exponent is not a constant as well, and the format is defined with both total bit-width and exponent bit-width as Posit(n, es). However, with application development experience, 2-bit exponent can provide enough dynamic representation range for most of the applications. As a result, in the recent revised posit standard document [4], the exponent is fixed to 2-bit, and thus, the format is denoted with the total bit-width only. As most of the designs in the literature are developed based on the original format, in the description below, we will still use es to represent the bit-width of the exponent. But it is easy to adapt to the newly revised format by replacing es with 2.

In a posit format, the sign bit s always occupies the most significant bit. The remaining $n - 1$-bit are shared by regime, exponent, and fraction. These three components have variable bit-width. The regime part is considered first. When there are still bit positions available after the regime part, the exponent and then the fraction can appear in the format. The regime part is a series of zeros or ones terminated by an opposite bit. As a result, the minimum bit-width of regime part is 2-bit. Sometimes, the series of ones or zeros can occupy the whole $(n - 1)$-bit positions, and thus, the opposite bit will not appear in these cases. The value of the regime can be considered as the magnitude expansion of the exponent. Different decoding and encoding methods of the posit format can give different regime values. We will discuss regime value later in this section with specific decoding and encoding method.

When the regime part does not occupy all the bit positions, the exponent part appears in the format after the regime vector. The exponent here is an unsigned value of es-bit. Different from the IEEE floating-point format, the exponent part in posit format is not biased. In the newly revised posit format standard [4], es is fixed to 2-bit which can reduce the complexity of posit arithmetic unit while still providing good enough dynamic range for a variety of applications.

The definition of the fraction part in posit format is similar to that in IEEE floating-point format. There is also a hidden bit in addition to the fraction bits which explicitly appeared in the format. However, as there are no subnormal numbers in posit, the hidden bit is always one. Moreover, the bit-width of the fraction is not a constant, and the maximum bit-width is $(n - es - 3)$-bit without the hidden bit as the minimum bit-width of regime is 2-bit.

In posit format, there are only two exception values, zero and not-a-real (NaR). Zero is represented as all zero bits in the format. NaR is represented with single one bit followed by all zero bits.

In the posit standard [4], in addition to the normal posit format, a quire format is also defined. The quire format can be used as an accurate accumulator which is long enough to support the exact accumulation of a certain amount of posit numbers. The purpose of this quire format is similar to the Kulish accumulator [5] in floating-point format. With $es = 2$, the quire format is shown in Fig. 1 as well. The quire is a signed number. When using the quire accumulation, the posit number is first converted to a long fixed-point number which will occupy the sign bit, the integer part, and the fractional part of the quire format. The 31-bit carry guard is to guarantee an exact accumulation of up to 2^{31} posit numbers without overflow which is large enough for most of the applications.

The posit format is treated as in sign-magnitude form, where the sign is separately encoded and the fraction is treated as an unsigned value, which is similar to IEEE floating-point format. The decoding and encoding of posit format under this form are discussed in [3]. But if only one form is kept, then two's will be better. In the following subsections, we are going to discuss both of them.

2.2 *Sign-Magnitude Form Decoding Method*

Posit format is designed as an alternative to the IEEE floating-point format. Therefore, the posit format is intuitively treated as in sign-magnitude format. In sign-magnitude format, the fraction is treated as an unsigned number. Before decoding the posit components, the sign bit needs to be considered first. If the sign bit is one, then two's complement needs to be performed on the posit vector where each bit of the vector is inverted and then a single one bit is added to the least significant bit (LSB) of the vector.

The regime part is then evaluated. The length of the series of identical bits is calculated first, which is denoted as r. Then, if the identical bits are zeros, then regime value $rg = -r$. On the other hand, if the identical bits are ones, then $rg = r - 1$. After removing the regime part, the exponent e and the fraction f can then be evaluated. The value of the posit number can be calculated with Eq. (1):

$$value = (-1)^s \times useed^{rg} \times 2^e \times (1 + f) \tag{1}$$

where $useed = 2^{2^{es}}$ which equals to 16 with the new standard [4].

After decoding, the following computation can use similar hardware as the floating-point units where an unsigned arithmetic unit can be applied to the fraction component.

2.3 Two's Complement Form Decoding Method

In the sign-magnitude form, complementing is needed when the posit number is a negative number. This adds extra cost in the hardware's perspective. To eliminate this extra cost, Yonemoto, who is the co-author of the original posit paper [3], proposed to decode the posit number in a different way. For negative numbers, the most significant digit, which is the hidden bit in sign-magnitude form, is treated as −2 instead of 1. With this method, there is no need to perform two's complement first for the negative number.

The decoding of regime and exponent also need slight changes. For regime $\widetilde{rg}$, if the identical bits are the same as the sign bit s, then $\widetilde{rg} = -r$. Otherwise, $\widetilde{rg}$ equals to $r - 1$. For exponent $\widetilde{e}$, the original exponent e from the posit format needs to be XORed with the sign s and is then used to calculated the value of the posit number. Then the value of the posit number can be calculated with Eq. (2):

$$value = useed^{\widetilde{rg}} \times 2^{\widetilde{e}} \times (1 - 3s + f) \tag{2}$$

In two's complement decoding, as the fraction is eventually a signed number, a signed arithmetic unit is needed for the computation of the fraction part in the following computation phases.

2.4 Posit to Quire Conversion

When an accurate accumulation is required in posit computation, the quire format can be used, and the conversion from posit format to quire format is needed. As the quire format is encoded in two's complement fixed-point format, the conversion from posit to quire is not difficult. Specifically, a shifting of the fraction according to the regime and exponent values can be used for the conversion. As the bit-width of the quire is large enough to support many accumulations without overflow, usually sign extension of the shifted fraction and zero padding to the right of the shifted fraction are needed to fill the whole bit positions.

2.5 Quire to Posit Conversion

After performing exact accumulation, before sending the computation result back to memory, the quire to posit conversion is needed to keep the computation result in posit format. Overflow detection is needed to check whether the accumulated result can be represented by the target posit format. If no overflow happens, the scaling factor of the fraction can be generated by counting the leading zeros (or leading ones when quire is negative) in the integer part of the quire format. The scaling factor is the combination of regime value and the exponent value, and after obtaining the scaling factor, the regime and the exponent can be extracted from the scaling factor. After removing the leading bits, the fraction part can be obtained. The sign, regime, exponent, and fraction are then packed together. Note that if the total bit-width is larger than the target posit bit-width, a rounding operation on the fraction or truncation of the fraction and exponent components may be required.

3 Posit Applications

The posit format is designed as an alternative to the IEEE floating-point format. Therefore, theoretically, any applications that use floating-point format in their computation can be potential applications of the posit format. However, the tapered number distribution makes posit especially suitable for some applications, such as machine learning [3]. In addition, within the same bit-width, the posit format can provide much larger dynamic range than the floating-point format. As a result, a small bit-width posit format is expected to be used to replace large bit-width floating-point format for some applications to achieve better energy efficiency. Therefore, posit format is also suitable for low-power computation scenarios, such as edge computing [6].

Machine learning is one of the most popular applications of posit format. Two main reasons make the posit successful in machine learning, especially deep learning and computation. The first reason is the large dynamic range provided by the posit format. Many research works reveal that the dynamic range of a numeric format is more important than the representation precision for deep learning model accuracy. As large dynamic range is one of the advantages of posit format, it is widely adopted in deep learning computation. The second reason is the tapered number distribution of posit format. For many deep learning models, their feature maps and weight parameters follow normal distribution which is quite similar to the tapered precision distribution of posit format. When the value is small, as the scale of the value is small, the regime can leave more bit positions for the fraction which leads to a higher representation precision. On the other hand, when the value is large, the regime will occupy more positions. Therefore, within a small bit-width, the posit format can use the regime part to trade off the precision for larger dynamic range, while when a more precise representation is needed to distinguish many different numbers, the regime part can occupy fewer bit positions. As a result, the posit format is more efficient in deep learning computation than IEEE floating-point format. Usually, an 8-bit posit format is good enough for deep learning computation to take the place of 32-bit IEEE floating-point format [3, 7–12]. In addition, many non-linear activation functions, such the sigmoid function, can be efficiently approximated using posit format operations [3]. Due to these reasons, posit formats are widely used in deep learning training [13–15] and inference [16–19]. In both training and inference computations, small bit-width posit numbers are used to achieve better energy efficiency than floating-point numbers, and even compared with fixed-point implementation, due to the larger dynamic range, the posit implementation can achieve better accuracy.

In addition to the machine learning, many edge computing [6, 14], digital signal processing applications [20, 21], optical flow estimation [22], autonomous driving [23], weather and climate modeling [24, 25], scientific data analysis [26], and model predictive control [27, 28] also used the posit format for better efficiency.

4 Posit Developing Tools

To use posit format in an application, researchers and engineers need to write the program with posit format arithmetic operations. The application tools discussed in this section are used to ease the posit programming process.

As a newly proposed numeric format, posit still does not have any native support by high-level programming languages, such as C/C++. One can use fixed-point operations and integer operations to emulate the computation process of posit arithmetic operators. However, the emulation-based method is complex and easy to make mistakes during development. Moreover, as it is a series of consecutive operations, the speed performance is relatively low compared to other natively supported operators.

To help improve the efficiency of posit development, researchers have built many useful libraries or frameworks that can be integrated with mainstream programming languages and development tools. The SoftPosit library [29] proposed by Leong is a posit arithmetic library developed for C and C++ languages. It is mainly developed for Posit(8,0), Posit(16,1), and Posit(32,2) formats and supports arithmetic operations, format conversion, and many other operations. The library is later extended support Posit(n, es) format with arbitrary total bit-width (n) and exponent bit-width (es). In order to be compatible with the current posit standard [4], one can set the es to 2 and generate the corresponding operators. In addition to the support for C/C++, SoftPosit also has a Python version, named SoftPosit-Python [30], to support program with posit format in Python environments. However, at the time of writing this chapter, SoftPosit-Python only provides support for Posit(8,0), Posit(16,1), and Posit(32,2) formats which may not be completely compatible with the current Posit standard [4].

In addition to the libraries, such as SoftPosit, for general-purpose development, as machine learning is one of the most popular applications for posit format, researchers also developed posit-based deep learning frameworks to help in deep learning model development [31, 32]. The [31] framework is developed for the Keras deep learning framework. It uses emulation-based method to simulate the calculation process of posit operators. By using quantization method for each layer's computation, the PositNN framework supports deep learning inference using 5-bit to 8-bit posit numbers. However, deep learning training is not yet supported by this framework. Another deep learning framework, Deep PeNSieve [32], is developed based on the TensorFlow framework. The computation of posit numbers is still achieved through software emulation since there is no native posit arithmetic support in CPUs and GPUs. Deep PeNSieve supports both deep learning training and inference. For training, it supports Posit(16,1) and Posit(32,2) format. After training, a post-training quantization process is performed so that the inference operations can be done with Posit(8,0) with quire accumulation.

The posit-based development libraries and frameworks discussed in this section enable the easy adoption of posit formats and operations in many different applications. However, as there is no native support of posit arithmetic operations in CPUs and GPUs, these libraries and frameworks all use emulation-based method to accomplish posit computation. The emulation-based methods significantly affect the speed performance of the computation. Therefore, to speed up the posit computation and to fully utilize the benefits of posit formats, hardware processors that have native supports for posit operations are required. In the next two sections, the design of posit hardware arithmetic units and hardware processors will be discussed.

5 Posit-Based Arithmetic Units

Arithmetic unit is one of the most important components in a hardware processor. To design an efficient posit processor, high-performance and energy-efficient posit arithmetic units are necessary. In the past few years, many research efforts have been put on the design of efficient posit arithmetic units.

Posit format is similar to the floating-point format. Therefore, the technologies developed for the floating-point arithmetic can be used in posit arithmetic as well. The big difference is the decoding (unpacking) process, where each component is extracted from the format, and the encoding (packing) process, where the computing result is packed into the specific format. Each component in floating-point format has fixed component bit-width, and therefore, the unpacking and packing are easy to be performed. However, for posit format, as the bit-width of the regime and the fraction are not constants, complex unpacking and packing operations are required before and after doing the core computation. This makes the posit arithmetic unit more hardware expensive than the floating-point counterpart [33–35]. To reduce the hardware overhead, many posit arithmetic unit architectures are proposed in the literature. For the rest of this section, we are going to discuss the hardware design of the posit decoding and encoding modules first and then review several posit arithmetic unit designs available in the literature.

5.1 Posit Decoding and Encoding Module

Posit decoder is designed following the rules of posit format, discussed in Sect. 2. The posit decoder module under the sign-magnitude format is shown in Fig. 2a. In Fig. 2, we assume the bit-width of the operand posit number is nb-bit; $pvec$ is the vector after removing the sign bit from the operand. When doing the decoding, a two's complement operation is performed first. The complemented vector is then split into two paths: one assumes the identical bits in regime are zeros and the other one assumes the identical bits in regime are ones. In each path, by using leading bit detection, the bit-width of the regime vector and thus the regime value can be obtained. The correct path is then selected by using the identical bit in regime. Then, a left shifting operation is performed to remove the regime vector from the operand vector, and then the exponent and the fraction can be extracted from the operand vector. The exponent and the regime are combined as the scale of the fraction. The hidden bit of the fraction is prefixed to form the complete fraction part. Note that for simplicity, the processing for sign bit and exceptional handling (detecting whether operand is zero or NaR) are not shown in this figure but their operations are straightforward.

As shown in Fig. 2a, the decoder uses both leading zero detector (LZD) and leading one detector (LOD) and two sets of incrementer or decrementer to calculate for the regime bit-width and regime value. This duplicated design consumes a lot of logic resource. An improved decoder design is shown in Fig. 2b. In this design, when the identical bits in regime are ones, the operand vector is first inverted, and thus, the identical bits are all inverted to zeros. Then, the leading bit counting can be realized by using a LZD only, which reduces the resource consumption.

The encoder logic under the sign-magnitude format is shown in Fig. 3, which is used to pack the computation result into the posit format. The basic idea is to first create a vector that follows the regime format (a sequence of identical bits terminated by the opposite valued bit). It is then combined with the exponent and the fraction from the direct computation result. A right shifting is then performed according to the resulting regime value. Rounding of the result is performed if there are bits shifted out of the LSB position. Finally, two's complement is performed if the sign of the result is negative. This gives the final result of the computation in posit format.

With two's complement format, as there is no need to perform complement operation for operand, the critical path delay is expected to be reduced. The decoder module and the encoder module under two's complement format are shown in Fig. 4. The general datapath is similar to that under sign-magnitude format. The main difference is that there is no complement circuit at the beginning of the decoder and

Fig. 2 Posit decoder module under sign-magnitude format. (**a**) Sign-magnitude decoder. (**b**) Optimized sign-magnitude decoder

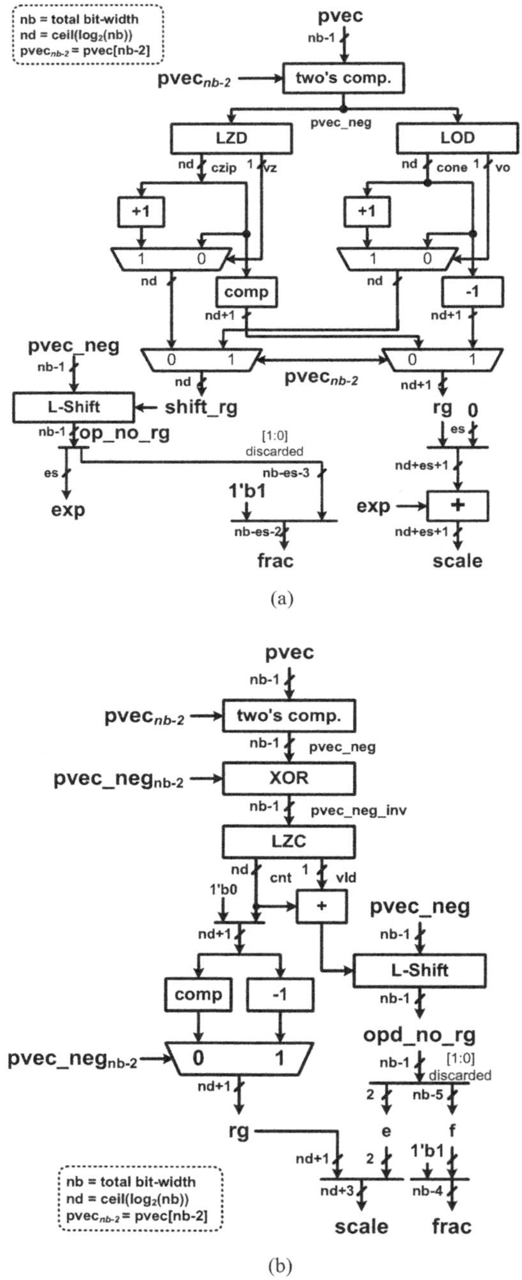

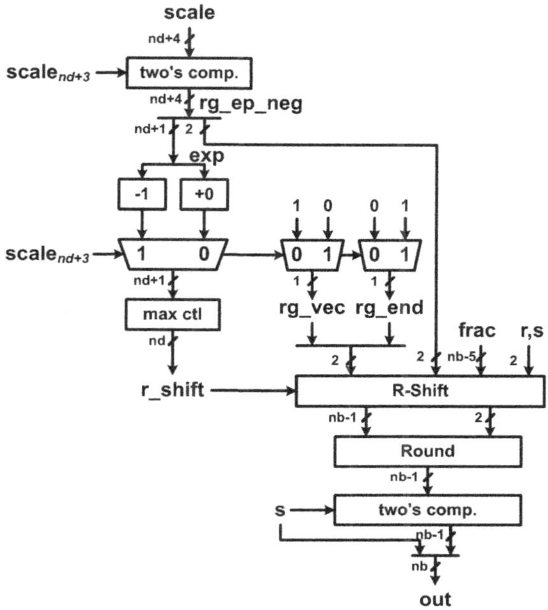

Fig. 3 Posit encoder under sign-magnitude format

at the end of the encoder. In addition, two bits are prefixed in front of the fraction as the implicit digit is treated as -2 for negative posit numbers.

Generally speaking, using two's complement decoding and encoding method can effectively reduce the delay and the hardware cost of the decoder and encoder units. However, as the fraction part is treated as a signed number, the complexity of the fraction multiplier is increased which leads to slightly more area and power, but the overall delay is still reduced [36]. For posit adder, the overall delay and resource consumption still get improved [36].

5.2 *Posit Arithmetic Units*

In the literature, researchers have investigated the design of posit adder, posit multiplier, posit divider, and posit fused unit. In the remaining part of this section, we are going to give an overview of each of them. As the posit format is similar to the IEEE floating-point format, many designs use the floating-point computation cores after the decoding module, as shown in Fig. 5 which presents a datapath of a standard posit multiplier.

Unlike IEEE floating-point numbers, posit format does not have pre-defined bit-widths. Therefore, any bit-width that is suitable for the application can be treated as a legal format. According to this characteristic, parameterized posit arithmetic units are usually proposed by the researcher, so that users can choose a suitable bit-width

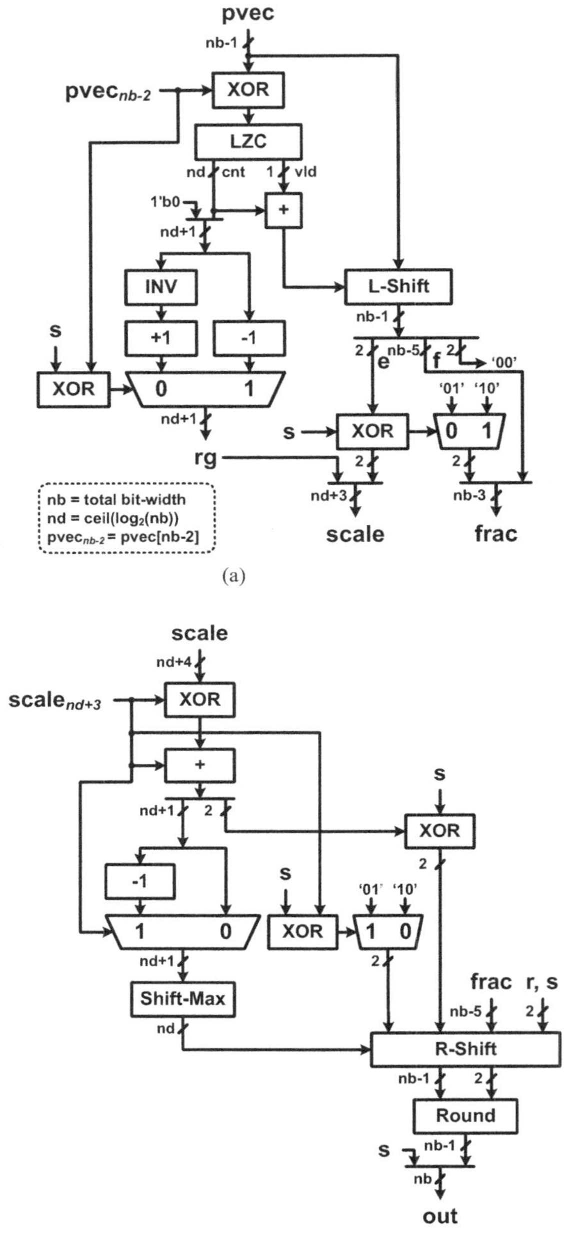

Fig. 4 Posit decoder and encoder module under two's complement format. (**a**) Posit decoder. (**b**) Posit encoder

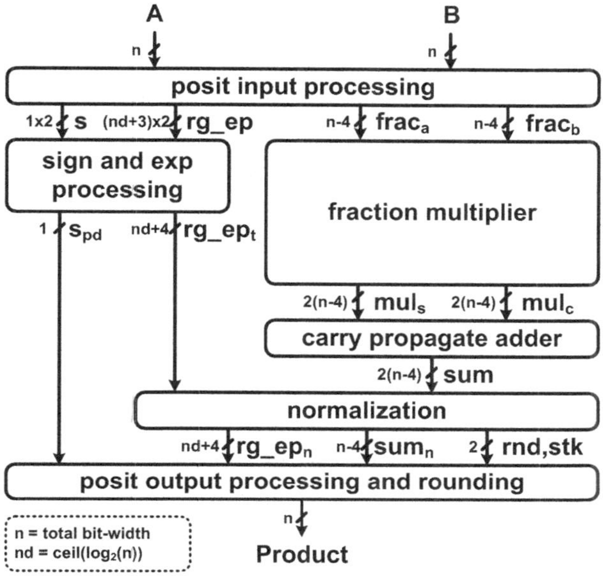

Fig. 5 Datapath of a normal posit multiplier

for their applications. In [37], architecture generator for posit adder is proposed. In [38, 39], and [40], parameterized posit adder and posit multiplier architectures are proposed for FPGA devices. In the design of PACoGen [41], in addition to the parameterized combinational adder and multiplier designs, the pipeline strategies for the posit adder and posit multiplier are also proposed. The proposed adder and multiplier architectures are evaluated on both the FPGA platform and the ASIC platform. In [42], posit multiplier and adder are implemented and integrated with the FloPoCo core generator [43], which is a famous floating-point arithmetic unit generator for FPGA devices. A multiple-precision posit multiplier architecture, which supports multiplications of posit formats with different bit-width, is proposed in [44]. In addition to the adder and multiplier, posit-based division and square root units are also proposed in the literature [45, 46].

Besides the basic arithmetic units mentioned above, fused arithmetic units, such as fused multiply-add (FMA) unit and multiply-accumulate (MAC) unit, are also proposed for posit format. With fused arithmetic units, multiple arithmetic operations can share hardware resources which leads to a reduced area consumption. In addition, rounding operation is only performed once at the end of the fused operation which brings better computation precision. For posit format, as the encoder and decoder are expensive, by using fused operations, multiple arithmetic operations can be performed after decoding operation and before encoding operation which further reduces the resource consumption. Moreover, many applications, such as image processing and deep learning computation, use fused arithmetic

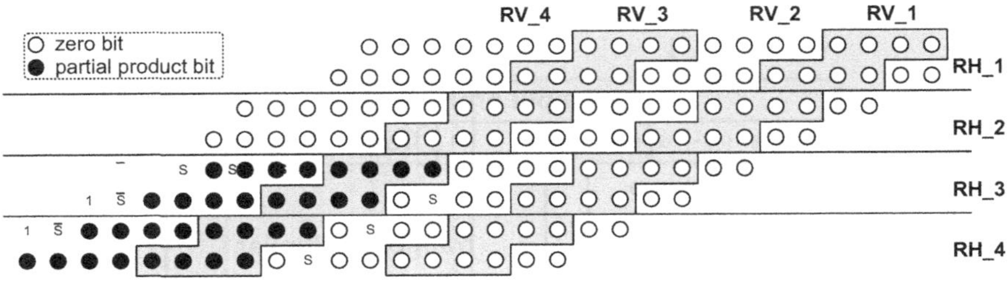

Fig. 6 Partial product array of the decomposed fraction multiplier in [50]

operations, and thus, designing a fused arithmetic unit can better fit the computation requirements of applications.

In [47], architecture generator for posit MAC unit is proposed for deep learning computation. In this design, the product of two posit numbers are accumulated to another posit number to complete the fused arithmetic operations. Another MAC unit is proposed in [48] where the accumulation is performed with quire accumulator. The product of two posit numbers are converted to the quire format to perform accumulation, and at the very last accumulation step, the accumulated result is converted back to the posit format. A multiple-precision posit MAC unit is proposed in [49]. This design also uses quire to perform accumulation. IEEE floating-point MAC operations are also supported in the same architecture.

Posit arithmetic units are generally more hardware expensive than the floating-point counterparts [33–35]. On the one hand, as the bit-widths of components in posit format are not constants, expensive decoder and encoder are needed to unpacking posit number and packing computation results to posit format. On the other hand, hardware design needs to consider the maximum possible bit-width of a signal. As the regime can potentially occupy almost all the bit positions, a large bit-width leading bit counter and a large bit-width shifter are needed to unpack the regime component. Similarly, as the fraction bit-width can also be large, a large bit-width multiplier is needed for the fraction part in a posit multiplier. Both reasons make the posit computation slow and power-consuming.

In order to improve the efficiency of posit computation, researchers have proposed many optimization techniques to either improve the computation speed [50] or reduce the energy consumption [51–55]. In [50], an associative memory-based posit decoding architecture is proposed. The proposed decoding architecture uses a ternary content addressable memory (TCAM) to speed up the decoding process.

In [51], a power-efficient posit multiplier architecture is proposed. The proposed architecture is designed based on the fact that the bit-width of the fraction of a posit operand cannot always be the maximum value. According to this phenomenon, the authors proposed to decompose the fraction multiplier into multiple small portions, as shown in Fig. 6. At runtime, only those which contain effective fraction bits are activated. Other parts of the multiplier are disabled to reduce the power consumption. The control signal to activate or disable a multiplier region can be easily generated during posit decoding with simple combinational logic.

The dynamic bit-width of posit components brings extra cost for posit arithmetic units. To reduce this cost, the authors in [52] proposed to use a fixed bit-width for regime and exponent and thus the fraction. After this modification, significant improvement in energy efficiency is achieved. However, the fixed bit-width limits the dynamic representation range and may lead to precision degradation in some applications.

Approximate computing is another effective method to reduce the cost of an arithmetic unit. In [53], the authors propose an approximate posit multiplier architecture. The basic idea of this design is to compute for a fixed bit-width fraction. The bit-width of fraction is empirically chosen as 12-bit. When the actual fraction is longer than 12-bit, its LSBs will be truncated. On the other hand, if the fraction is shorter than 12-bit, then zero padding is performed to keep a 12-bit fraction. The computation is performed iteratively to compensate for the approximation error.

The design proposed in [54] is another approximate posit multiplier available in the literature. In this design, the approximation is performed in the logarithm domain with the Mitchell logarithm multiplier algorithm [56]. According to the logarithm computation rule, a multiplication in the linear domain can be performed with simple addition in the logarithm domain, which significantly reduces the cost of the multiplication. However, the conversion between the linear domain and the logarithm domain is expensive. To reduce the cost of this conversion process, Mitchell proposed to use approximation in the conversion process. Specifically, assume $x \in [0, 1)$ and then $log_2(1 + x) \approx x$. As a result, the fraction, without the implicit bit, can be considered as the logarithm domain value of the complete fraction. Then, the multiplication of two posit numbers can be converted to a simple addition of the two fractions. By using this method, the cost of the approximated posit multiplier can be comparable with the floating-point counterpart. However, the drawback of this method is the error of logarithm approximate multiplier is relatively large which leads to a large degradation in accuracy in neural network computation.

In [55], a variable precision approximate multiplier is proposed. For the multiplication of the fraction part, the bit-width of the product is the sum of the bit-width of both fractions which is large. However, as the final product is also packed into the posit format, the bit-width of the fraction in the final product could be limited, and those extra bits in the fraction product do not have much contribution for the final result. According to this analysis, the core idea of this proposed design is to truncate the fraction multiplier according to the scale of the product. A mask signal is generated based on the product scale, and it is applied to the partial product array to truncate unnecessary partial product columns, as shown in Fig. 7. By using this method, significant improvement in energy consumption is achieved, while the error generated by this approximate computation is small. A logarithm domain approximate multiplier is also proposed in this chapter with the same truncation method.

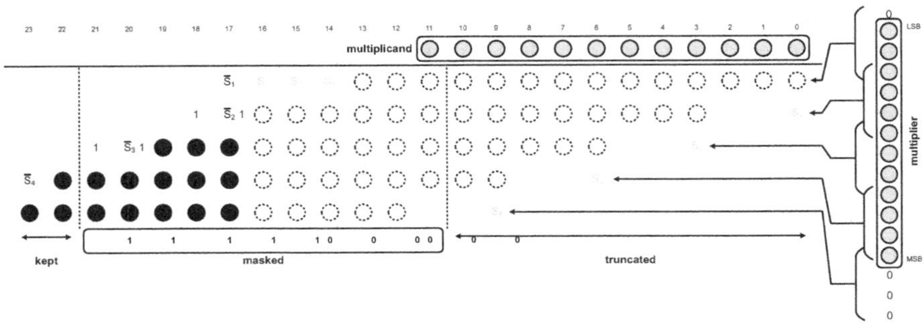

Fig. 7 Partial product array of the variable precision approximate posit multiplier in [55]

6 Posit-Based Hardware Processors

With the posit arithmetic units discussed in the last section, many posit-based hardware processors are available in the literature. Some of them are proposed for general-purpose computation, while others are designed for specific applications.

For the general-purpose processor, RISC-V instruction sets are usually used. The RISC-V extension for posit numbers [57] is proposed by Dr. Gustafson in 2018. After that, many posit-enabled processors are designed by using the RISC-V instruction set. CLARINET [58] is the first RISC-V-based posit processor available in the literature. It supports fused multiply-accumulate or divide-accumulate with quire and conversion between posit and floating-point. However, most of the operations are still performed with floating-point cores, and data is converted to posit when needed. PERC [38] is another posit-enabled processor designed based on the open-source Rocket Chip. It uses a posit arithmetic unit to replace floating-point units available in the Rocket Chip. However, quire is not supported in this design. A configurable posit-enabled RISC-V core, named PERI, which support multiple-precision computation at runtime is proposed in [59]. It supports dynamic switching between posit formats; however, quire is also not supported in this design. A more recent work, PERCIVAL [60], is the first posit-enabled core that integrates the complete posit instruction set. It supports all posit-based operations and quire accumulation. In addition, floating-point operations are also supported in the same core.

In addition to the abovementioned general-purpose processors, as posit is widely adopted in deep learning computation, there are also many posit-based deep learning processors available in the literature. Deep Positron [61] is a posit-enabled deep learning processing engine which support low-precision posit multiplication with exact accumulation. A RISC-V core with vectorized posit operations to speed up the deep neural network is proposed in [62]. In [63], the authors use posit representation to store data with reduced bit-width. During computation, the posit format data is converted to floating-point format and is then computed with floating-point unit. In [64], a posit-based processing element supporting approximate computing is proposed for deep learning accelerators. It uses a posit format that has fixed bit-width for regime, exponent, and fraction similar as the idea proposed in [52]. In [65], a deep learning training processor is proposed for edge devices. It contains posit-based logarithm domain processing elements. A reconfigurable dataflow and adaptive compression method are proposed to further reduce the energy consumption for edge devices. Besides the complete deep learning cores, variable precision posit arithmetic is also applied in designing tensor computation unit [66, 67] and directed acyclic graphs processing [68].

7 Discussion and Perspectives

We have so far discussed the posit applications, development tools, arithmetic unit, and hardware processors in the literature. Due to the advantages of posit format, many research efforts have been put in this field to improve the performance and efficiency of posit computation. However, there is still a big space to further improve the current designs. In this section, we are going to discuss the main challenges in posit arithmetic and to propose some potential future works.

7.1 Improving the Latency of Posit Arithmetic Operations

So far, many research efforts have been put on improving the energy efficiency of the posit arithmetic unit. By using approximate computing or decomposing the computation resources, significant improvement in energy efficiency is achieved. However, improving the latency of posit arithmetic operations is still challenging. Take posit multiplier as an example; the critical path contains two slow operations: shifter used to decode regime and fraction multiplier. Finding an effective way to overlap the latency of these two operations is expected to be a solution to reduce the total latency of the computation. In addition, proposing a new algorithm to decode the posit format may also be helpful.

7.2 Developing a Practical Tool for Posit Verification

Verification is necessary to ensure the functions of the proposed design all work correctly. For posit development, although there exist some libraries or frameworks to emulate the posit computation process, there are many limitations in these libraries. One problem is the limited format support since most of the tools only support, for example, Posit(8,0), Posit(16,1), and Posit(32,2), without effective methods to extend to other customized format. Another problem is the emulation speed is slow. Therefore, it takes a long time for a large set of simulation. As a result, it is expected to have an efficient tool, which is able to generate a large amount of testing vectors conveniently and perform emulation with high speed, to help in the verification process.

7.3 Designing a Flexible Posit Processor for Applications

Posit format is popularly used in deep learning computation, and as discussed in Sect. 6, many posit-enabled processors are proposed for deep learning computation. However, with the widely adoption of deep learning applications in more fields, the diversity of deep learning models becomes more severe, such as convolutional neural network, recurrent neural networks, transformers, graph neural networks, and neural architecture search (NAS)-generated neural networks. In addition, many application scenarios require the adoption of multiple deep learning models simultaneously. For different deep learning models, the optimal computation flow or datapath may not be the same. Therefore, to efficiently process all the deep learning workloads, a flexible posit-enabled hardware processor is required.

7.4 Exploring the Use of Posit in More Fields of Applications

Currently, posit format is widely used in machine learning applications because the adoption of posit format in machine learning is straightforward. However, for other applications, they may have different computation requirements and thus may need some specific techniques, for example, multiple-precision computation and mixed-precision computation, to use posit format. Due to the advantages of the posit format, it is worth for researchers to investigate the use of posit format in more fields of applications.

8 Conclusion

In this chapter, posit arithmetic for energy efficiency computation is reviewed. We start with the introduction of posit format. Then, we go through all the technology layers of posit computing system, including the application, the development tool, and the hardware processor. Finally, we discuss the current design challenges in the field of posit computation and propose several potential future works accordingly. This chapter is expected to help readers to understand basics of posit format and posit arithmetic, to get a general understanding of the current development status in each layer, and to motivate further research projects to solve the design challenges.

Acknowledgments The authors would like to thank the Ocean University of China, the University of Saskatchewan, and the Natural Sciences and Engineering Research Council of Canada (NSERC) for their financial support for the related projects and the writing of this chapter.

References

1. IEEE Standard for Floating-Point Arithmetic, *IEEE Std 754-2019 (Revision of IEEE 754-2008)*, pp. 1–84 (Jul 2019)
2. J.L. Gustafson, *The End of Error: Unum Computing* (Chapman & Hall/CRC Computational Science, London, 2015)
3. J.L. Gustafson, I.T. Yonemoto, Beating floating point at its own game: posit arithmetic. Supercomput. Front. Innov. Int. J. **4**(2), 71–86 (2017)
4. Posit Working Group, *Standard for Posit Arithmetic (2022)*, pp. 1–12 (Mar 2022)
5. U. Kulisch, *Computer Arithmetic and Validity: Theory, Implementation, and Applications* (De Gruyter, Berlin, 2008)
6. A. Guntoro, C. De La Parra, F. Merchant, F. De Dinechin, J.L. Gustafson, M. Langhammer, R. Leupers, S. Nambiar, Next generation arithmetic for edge computing, in *2020 Design, Automation and Test in Europe Conference and Exhibition (DATE)* (Mar 2020), pp. 1357–1365
7. S.H. Fatemi Langroudi, T. Pandit, D. Kudithipudi, Deep learning inference on embedded devices: fixed-point vs posit, in *2018 1st Workshop on Energy Efficient Machine Learning and Cognitive Computing for Embedded Applications (EMC2)* (Mar 2018), pp. 19–23
8. J. Johnson, Rethinking floating point for deep learning, in *Workshop on Systems for ML and Open Source Software at NeurIPS 2018* (Dec 2018), pp. 1–8
9. A.Y. Romanov, A.L. Stempkovsky, I.V. Lariushkin, G.E. Novoselov, R.A. Solovyev, V.A. Starykh, I.I. Romanova, D.V. Telpukhov, I.A. Mkrtchan, Analysis of posit and bfloat arithmetic of real numbers for machine learning. IEEE Access **9**, 82,318–82,324 (2021)
10. S.M. Mishra, A. Tiwari, H.S. Shekhawat, P. Guha, G. Trivedi, P. Jan, Z. Nemec, Comparison of floating-point representations for the efficient implementation of machine learning algorithms, in *2022 32nd International Conference Radioelektronika (RADIOELEKTRONIKA)* (Apr 2022), pp. 1–6
11. L. Sommer, L. Weber, M. Kumm, A. Koch, Comparison of arithmetic number formats for inference in sum-product networks on FPGAs, in *2020 IEEE 28th Annual International Symposium on Field-Programmable Custom Computing Machines (FCCM)* (May 2020), pp. 75–83
12. H.F. Langroudi, V. Karia, J.L. Gustafson, D. Kudithipudi, Adaptive Posit: parameter aware numerical format for deep learning inference on the edge, in *2020 IEEE/CVF Conference on Computer Vision and Pattern Recognition Workshops (CVPRW)* (Jun 2020), pp. 3123–3131
13. J. Lu, C. Fang, M. Xu, J. Lin, Z. Wang, Evaluations on deep neural networks training using posit number system. IEEE Trans. Comput. **70**(2), 174–187 (2021)
14. Y. Wang, D. Deng, L. Liu, S. Wei, S. Yin, LPE: Logarithm posit processing element for energy-efficient edge-device training, in *2021 IEEE 3rd International Conference on Artificial Intelligence Circuits and Systems (AICAS)* (Jun 2021), pp. 1–4
15. G. Raposo, P. Tomás, N. Roma, Positnn: Training deep neural networks with mixed low-precision posit, in *ICASSP 2021 - 2021 IEEE International Conference on Acoustics, Speech and Signal Processing (ICASSP)* (Jun 2021), pp. 7908–7912
16. J. Lu, S. Lu, Z. Wang, C. Fang, J. Lin, Z. Wang, L. Du, Training deep neural networks using posit number system, in *2019 32nd IEEE International System-on-Chip Conference (SOCC)* (Sep 2019), pp. 62–67
17. M. Cococcioni, F. Rossi, E. Ruffaldi, S. Saponara, Fast deep neural networks for image processing using posits and ARM scalable vector extension. J. Real Time Image Process. **17**, 759–771 (2020)

18. S. Nambi, S. Ullah, S.S. Sahoo, A. Lohana, F. Merchant, A. Kumar, ExPAN(N)D: Exploring posits for efficient artificial neural network design in FPGA-based systems. IEEE Access **9**, 103,691–103,708 (2021)
19. H.F. Langroudi, V. Karia, Z. Carmichael, A. Zyarah, T. Pandit, J.L. Gustafson, D. Kudithipudi, Alps: Adaptive quantization of deep neural networks with generalized positS, in *2021 IEEE/CVF Conference on Computer Vision and Pattern Recognition Workshops (CVPRW)* (Jun 2021), pp. 3094–3103
20. N. Neves, P. Tomás, N. Roma, Dynamic fused multiply-accumulate posit unit with variable exponent size for low-precision DSP applications, in *2020 IEEE Workshop on Signal Processing Systems (SiPS)* (Oct 2020), pp. 1–6
21. M. Kant, R. Thakur, Implementation and performance improvement of POSIT multiplier for advance DSP applications, in *2021 Fifth International Conference on I-SMAC (IoT in Social, Mobile, Analytics and Cloud) (I-SMAC)* (Nov 2021), pp. 1730–1736
22. V. Saxena, A. Reddy, J. Neudorfer, J. Gustafson, S. Nambiar, R. Leupers, F. Merchant, Brightening the optical flow through posit arithmetic, in *2021 22nd International Symposium on Quality Electronic Design (ISQED)* (Apr 2021), pp. 463–468
23. M. Cococcioni, F. Rossi, E. Ruffaldi, S. Saponara, B. Dupont de Dinechin, Novel arithmetics in deep neural networks signal processing for autonomous driving: challenges and opportunities. IEEE Signal Process. Mag. **38**(1), 97–110 (2021)
24. M. Klower, P.D. Duben, T.N. Palmer, Posits as an alternative to floats for weather and climate models, in *Proceedings of the Conference for Next Generation Arithmetic 2019* (2019)
25. M. Klower, P.D. Duben, T.N. Palmer, Number formats, error mitigation, and scope for 16-bit arithmetics in weather and climate modeling analyzed with a shallow water model. J. Adv. Model. Earth Syst. **12**(10), 1–17 (2020)
26. J. Hou, Y. Zhu, S. Du, S. Song, Enhancing accuracy and dynamic range of scientific data analytics by implementing posit arithmetic on FPGA. J. Signal Process. Syst. **91**, 1137–1148 (2019)
27. C. Jugade, D. Ingole, D. Sonawane, M. Kvasnica, J. Gustafson, A memory-efficient explicit model predictive control using posits, in *2019 Sixth Indian Control Conference (ICC)* (Dec 2019), pp. 188–193
28. C. Jugade, D. Ingole, D. Sonawane, M. Kvasnica, J. Gustafson, A framework for embedded model predictive control using posits, in *2020 59th IEEE Conference on Decision and Control (CDC)* (Dec 2020), pp. 2509–2514
29. C. Leong, SoftPosit, https://gitlab.com/cerlane/SoftPosit, accessed: Oct 2022
30. C. Leong, SoftPosit-Python, https://gitlab.com/cerlane/SoftPosit-Python, accessed: Oct 2022
31. H.F. Langroudi, Z. Carmichael, J.L. Gustafson, D. Kudithipudi, PositNN Framework: tapered precision deep learning inference for the edge, in *2019 IEEE Space Computing Conference (SCC)* (Jul 2019), pp. 53–59
32. R. Murillo, A.A. Del Barrio, G. Botella, Deep PeNSieve: A deep learning framework based on the posit number system. Digit. Signal Process. **102**, 102762 (2020)
33. F. de Dinechin, L. Forget, J.-M. Muller, Y. Uguen, Posits: The good, the bad and the ugly, in *Proceedings of the Conference for Next Generation Arithmetic 2019* (Mar 2019), pp. 1–10
34. S.D. Ciocirlan, D. Loghin, L. Ramapantulu, N. Ţăpuş, Y.M. Teo, The accuracy and efficiency of posit arithmetic, in *2021 IEEE 39th International Conference on Computer Design (ICCD)* (Oct 2021), pp. 83–87
35. Y. Uguen, L. Forget, F. de Dinechin, Evaluating the hardware cost of the posit number system, in *2019 29th International Conference on Field Programmable Logic and Applications (FPL)* (Sep 2019), pp. 106–113
36. R. Murillo, D. Mallasen, A.A.D. Barrio, G. Botella, Comparing different decodings for posit arithmetic, in *Proceedings of the Conference for Next Generation Arithmetic 2022* (Jul 2022), pp. 84–99
37. M.K. Jaiswal, H.K.-H. So, Architecture generator for type-3 unum posit adder/subtractor, in *2018 IEEE International Symposium on Circuits and Systems (ISCAS)* (May 2018), pp. 1–5

38. R. Chaurasiya, J. Gustafson, R. Shrestha, J. Neudorfer, S. Nambiar, K. Niyogi, F. Merchant, R. Leupers, Parameterized posit arithmetic hardware generator, in *2018 IEEE 36th International Conference on Computer Design (ICCD)* (Oct 2018), pp. 334–341
39. A. Podobas, S. Matsuoka, Hardware implementation of POSITs and their application in FPGAs, in *2018 IEEE International Parallel and Distributed Processing Symposium Workshops (IPDPSW)* (May 2018), pp. 138–145
40. M.K. Jaiswal, H.K.-H. So, Universal number posit arithmetic generator on FPGA, in *2018 Design, Automation and Test in Europe Conference and Exhibition (DATE)* (Mar 2018), pp. 1159–1162
41. M.K. Jaiswal, H.K.-H. So, PACoGen: A hardware posit arithmetic core generator. IEEE Access **7**, 74,586–74,601 (2019)
42. R. Murillo, A.A. Del Barrio, G. Botella, Customized posit adders and multipliers using the FloPoCo core generator, in *2020 IEEE International Symposium on Circuits and Systems (ISCAS)* (Oct 2020), pp. 1–5
43. F. de Dinechin, B. Pasca, Designing custom arithmetic data paths with FloPoCo. IEEE Des. Test Comput. **28**(4), 18–27 (2011)
44. H. Zhang, S.-B. Ko, Efficient multiple-precision posit multiplier, in *2021 IEEE International Symposium on Circuits and Systems (ISCAS)* (May 2021), pp. 1–5
45. F. Xiao, F. Liang, B. Wu, J. Liang, S. Cheng, G. Zhang, Posit arithmetic hardware implementations with the minimum cost divider and SquareRoot. Electronics **9**(10), 1622 (2020)
46. A. Raveendran, S. Jean, J. Mervin, D. Vivian, D. Selvakumar, A novel parametrized fused division and square-root POSIT arithmetic architecture, in *2020 33rd International Conference on VLSI Design and 2020 19th International Conference on Embedded Systems (VLSID)* (Jan 2020), pp. 207–212
47. H. Zhang, J. He, S.-B. Ko, Efficient posit multiply-accumulate unit generator for deep learning applications, in *2019 IEEE International Symposium on Circuits and Systems (ISCAS)* (May 2019), pp. 1–5
48. R. Murillo, D. Mallasén, A.A. Del Barrio, G. Botella, Energy-efficient MAC units for fused posit arithmetic, in *2021 IEEE 39th International Conference on Computer Design (ICCD)* (Oct 2021), pp. 138–145
49. L. Crespo, P. Tomás, N. Roma, N. Neves, Unified Posit/IEEE-754 vector MAC unit for transprecision computing. IEEE Trans. Circuits Syst. II Exp. Briefs **69**(5), 2478–2482 (2022)
50. S. Sarkar, P.M. Velayuthan, M.D. Gomony, A reconfigurable architecture for posit arithmetic, in *2019 22nd Euromicro Conference on Digital System Design (DSD)* (Aug 2019), pp. 82–87
51. H. Zhang, S.-B. Ko, Design of power efficient posit multiplier. IEEE Trans. Circuits Syst. II: Exp. Briefs **67**(5), 861–865 (2020)
52. V. Gohil, S. Walia, J. Mekie, M. Awasthi, Fixed-posit: A floating-point representation for error-resilient applications. IEEE Trans. Circuits Syst. II: Exp. Briefs **68**(10), 3341–3345 (2021)
53. C.J. Norris, S. Kim, An approximate and iterative posit multiplier architecture for FPGAs, in *2021 IEEE International Symposium on Circuits and Systems (ISCAS)* (May 2021), pp. 1–5
54. R. Murillo, A.A. Del Barrio, G. Botella, M.S. Kim, H. Kim, N. Bagherzadeh, PLAM: A posit logarithm-approximate multiplier. IEEE Trans. Emerg. Top. Comput. **10**(4), 2079–2085 (2022)
55. H. Zhang, S.-B. Ko, Efficient approximate posit multipliers for deep learning computation. IEEE J. Emerg. Sel. Top. Circuits Syst., 1–1 (2022)
56. J.N. Mitchell, Computer multiplication and division using binary logarithms. IRE Trans. Electron. Comput. **EC-11**(4), 512–517 (1962)
57. J.L. Gustafson, RISC-V proposed extension for 32-bit posits, https://posithub.org/docs/RISC-V/RISC-V.htm, June 2018
58. R. Jain, N. Sharma, F. Merchant, S. Patkar, R. Leupers, CLARINET: A RISC-V based framework for posit arithmetic empiricism. CoRR, abs/2006.00364 (2020)
59. S. Tiwari, N. Gala, C. Rebeiro, V. Kamakoti, PERI: A configurable posit enabled RISC-V core. ACM Trans. Archit. Code Optim. **18**(3), Article 25 (2021)

60. D. Mallasén, R. Murillo, A.A.D. Barrio, G. Botella, L. Piñuel, M. Prieto-Matias, PERCIVAL: Open-source posit RISC-V core with quire capability. IEEE Trans. Emerg. Top. Comput. **10**(3), 1241–1252 (2022)
61. Z. Carmichael, H.F. Langroudi, C. Khazanov, J. Lillie, J.L. Gustafson, D. Kudithipudi, Deep positron: A deep neural network using the posit number system, in *2019 Design, Automation and Test in Europe Conference and Exhibition (DATE)* (Mar 2019), pp. 1421–1426
62. M. Cococcioni, F. Rossi, E. Ruffaldi, S. Saponara, Vectorizing posit operations on RISC-V for faster deep neural networks: experiments and comparison with ARM SVE. Neural Comput. Appl. **33**, 10575–10585 (2021)
63. M. Cococcioni, F. Rossi, E. Ruffaldi, S. Saponara, A lightweight posit processing unit for RISC-V processors in deep neural network applications. IEEE Trans. Emerg. Top. Comput. **10**(4), 1898–1908 (2022)
64. M. Zolfagharinejad, M. Kamal, A. Afzali-Khusha, M. Pedram, Posit process element for using in energy-efficient DNN accelerators. IEEE Trans. Very Large Scale Integr. (VLSI) Syst. **30**(6), 844–848 (2022)
65. Y. Wang, D. Deng, L. Liu, S. Wei, S. Yin, PL-NPU: An energy-efficient edge-device DNN training processor with posit-based logarithm-domain computing. IEEE Trans. Circuits Syst. I: Reg. Papers **69**(10), 4042–4055 (2022)
66. N. Neves, P. Tomás, N. Roma, Reconfigurable stream-based tensor unit with variable-precision posit arithmetic, in *2020 IEEE 31st International Conference on Application-specific Systems, Architectures and Processors (ASAP)* (Jul 2020), pp. 149–156
67. N. Neves, P. Tomas, N. Roma, A reconfigurable posit tensor unit with variable-precision arithmetic and automatic data streaming. J. Signal Process. Syst. **93**, 1365–1385 (2021)
68. N. Shah, L.I.G. Olascoaga, S. Zhao, W. Meert, M. Verhelst, DPU: DAG processing unit for irregular graphs with precision-scalable posit arithmetic in 28 nm. IEEE J. Solid State Circuits **57**(8), 2586–2596 (2022)

Designing Fault-Tolerant Digital Circuits in Quantum-Dot Cellular Automata

R. Marshal, K. Raja Sekar, Lakshminarayanan Gopalakrishnan, Anantharaj Thalaimalai Vanaraj, and Seok-Bum Ko

1 Introduction

Fabrication defects are probable to arise in the synthesis and deposition stages of QCA [1–3, 5–7]. Based on the type of defect, the fault tolerance of QCA circuits is analyzed in the literature. In this chapter, a detailed analysis of the existing fault tolerance measurement techniques are analyzed. The parameters that contribute to develop fault-resistant circuits are analyzed and the design considerations that can be followed to improve the fault tolerance are presented.

The remainder of the chapter is partitioned as follows. Section 2 gives an overview of QCA working and its performance metrics. Section 3 provides an insight into the fabrication defects and fault tolerance analysis in QCA. Section 4 presents the parameters and design considerations for designing fault-tolerant circuits. Section 5 provides the conclusion.

R. Marshal
Indian Computer Emergency Response Team, New Delhi, India

K. Raja Sekar
Centre for Development of Advanced Computing, Bengaluru, India

L. Gopalakrishnan
National Institute of Technology, Tiruchirappalli, Tamil Nadu, India
e-mail: laksh@nitt.edu

A. T. Vanaraj
R&D, Western Digital, San Jose, CA, USA

S.-B. Ko (✉)
Department of Electrical and Computer Engineering, University of Saskatchewan, Saskatoon, Canada
e-mail: seokbum.ko@usask.ca

W. Liu et al. (eds.), *Design and Applications of Emerging Computer Systems*,
https://doi.org/10.1007/978-3-031-42478-6_25

2 QCA Operation

2.1 *Cell Components and Logic*

Every QCA cell incorporates two electrons placed inside four quantum dots that are transportable within the dots. The electronic repulsion forces the electrons to get fixed in any one of the diagonal positions. A normal cell quantum dot position has 45° orientation difference with a rotated cell. The dot orientation difference along with the electron occupation is presented in Fig. 1. The binary encoding of electron occupation also called as polarization of the cell is calculated by (1).

$$P = \frac{(\rho_1 + \rho_3) - (\rho_2 + \rho_4)}{(\rho_1 + \rho_2 + \rho_3 + \rho_4)} \tag{1}$$

Where, $\rho_i = 0$, if the quantum dot does not hold an electron, and 1, if the quantum dot holds an electron (i = 1, 2, 3, 4). $P = -1$ and $+1$ are set as logic '0' and '1,' respectively.

2.2 *Radius of Effect*

The influence of the columbic repulsion is dependent on the radius of effect. The electrons inside a cell will be able to influence other cells based on the radius of effect value. The radius of effect is the radius of the circular region computed from the center of the cell as presented in Fig. 2 and any cell placed in that region can have its logic and electron position influenced by the electrons present in the cell. It is a critical value as it can determine the location of cells and clocking used in circuits [8].

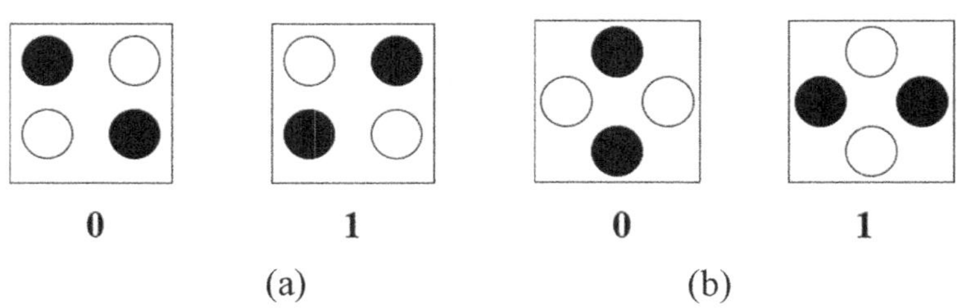

Fig. 1 Quantum dot orientation in cells: (**a**) normal and (**b**) rotated

Fig. 2 Radius of effect

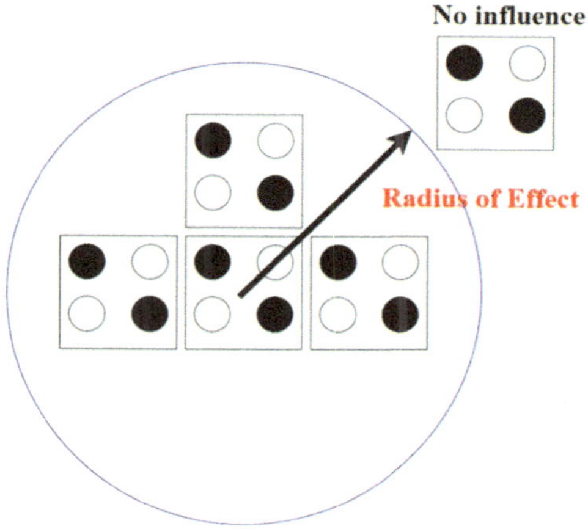

2.3 Kink Energy

Kink energy ($E^k_{i,j}$) reflects the repulsion and influence between the cells and its value drops inversely with the cell-cell distance [4]. Dots in the cells i and j have an electrostatic interaction represented by ($E_{i,j}$) as given in (2):

$$E_{i,j} = \frac{1}{4\pi \varepsilon_0 \varepsilon_{\mathrm{r}}} \frac{q_i q_j}{\left|r_i - r_j\right|} \tag{2}$$

Where, ε_0 and ε_{r} denote the free space and relative permittivity of the material, respectively. $|r_i - r_j|$ denotes the distance between dots in the cells. Computation of $E^k_{i,j}$ is done by calculating the electrostatic energy difference between the cells when they have same and opposite polarization. The Kink energy also drops similar to radius of effect as the cell-cell distance increases.

2.4 Wires

To perform information transmission, wires are formed using cells. Placement of one normal cell with another will result in identical replication of logic both cells as presented in Fig. 3a. Hence, a group of normal cells when joined in an array form a normal wire that will carry an unchanged logic from one end to another. Placement of rotated cell with another rotated cell will result in each cell having opposite logic due to their dot orientation and electron repulsion. Hence, a cell in a rotated wire will have opposite logic in either of its side as presented in Fig. 3b.

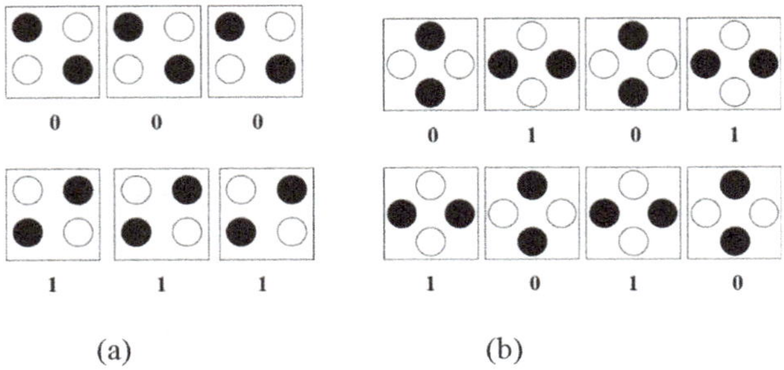

Fig. 3 Wire: (**a**) normal and (**b**) rotated

Fig. 4 3-input MV

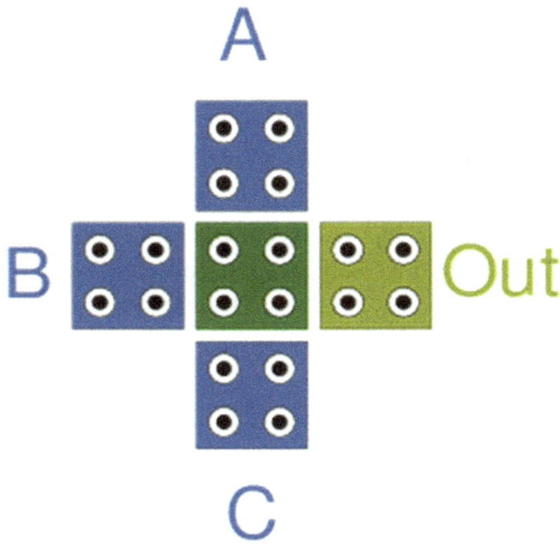

2.5 *Logic Gates*

Logical circuits are deployed using majority gates (also called majority voter) and inverter. The columbic repulsion between cells is utilized to form majority gates and inverter. Majority voters (MVs) have odd number of inputs and single output. The output will be dependent on the logic of the majority of the inputs. The MVs may have 3 or 5 inputs, and they are used to realize the conventional Boolean operations. A few arrangements of MVs used in the literature are presented in Figs. 4 and 5. To perform logic inversion, inverters are deployed using different cell arrangements as presented in Fig. 6.

Fig. 5 5-input MV as shown (**a**) in Refs. [9] and (**b**) [10]

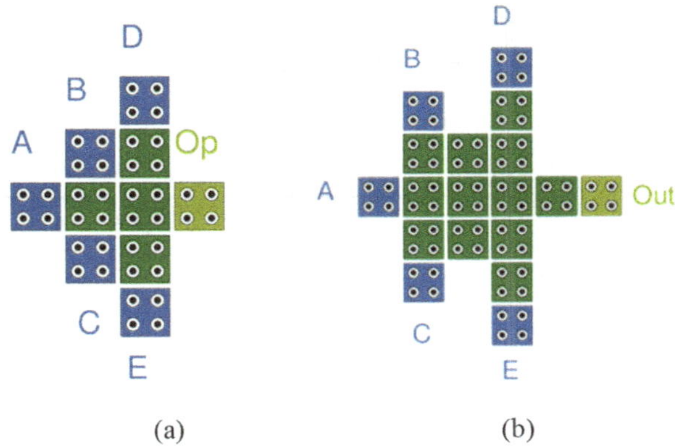

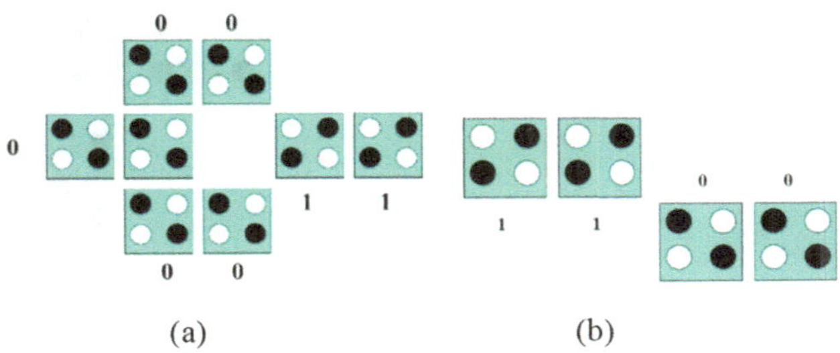

Fig. 6 Inverters: (**a**) double path and (**b**) simple

2.6 Clocking

To ensure directional and controlled information flow, clocking is used in QCA. The clock signal is split into four clocks with a ninety-degree difference in phase with the adjacent clocks. The clocks are connected to zones in the circuits to ensure directionality and pipelined synchronization of information flow. The state of the cells in the zones depends upon the state of the clock signal. The states of the cell are categorized as switch, release, relax, and hold. During these states, the inter dot barriers changes to allow switching, holding and relaxing the electrons inside the cells. QCADesigner and QCADesignerE are the tools used in the literature to design QCA circuits [11]. To differentiate and identify the clocks connected to the cells, the cells are represented in different colors, as presented in Fig. 7. A clock cycle is considered to have four clock phases, and the delay of the circuit is expressed in terms of clock phases and clock cycles.

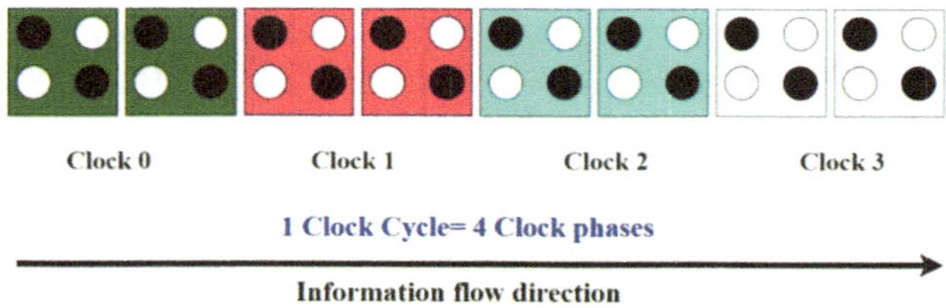

Fig. 7 Cell zones

Fig. 8 Crossover using normal and rotated cell non-interaction

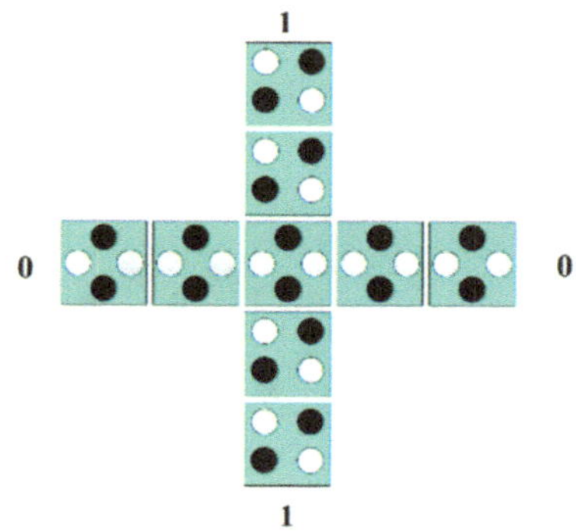

2.7 Crossover

To facilitate crossing of wires, single and multilayer crossovers are used. Non-interaction between rotated and normal cells (rotated crossover), non-adjacent clock zone cells (clock zone crossover) are deployed in single layer crossovers as presented in Figs. 8 and 9, respectively. In multilayer, cells are placed in layers above the main layer, and interaction between the cells in three-dimensional format is used to carry the information across the layers. In most cases, three layers are used for crossings as presented in Fig. 10.

2.8 Performance Metrics

The performance of QCA circuits is generally compared using cell count, area, and delay of the circuit. However, these metrics do not reflect all the design parameters involved in QCA. To effectively compare the performance of QCA circuits, a cost function was proposed in [12] as given in (3),

$$\mathrm{Cost}_{\mathrm{QCA}} = \left(M^k + I + C^l\right) T^p, 1 \leq k, l, p \tag{3}$$

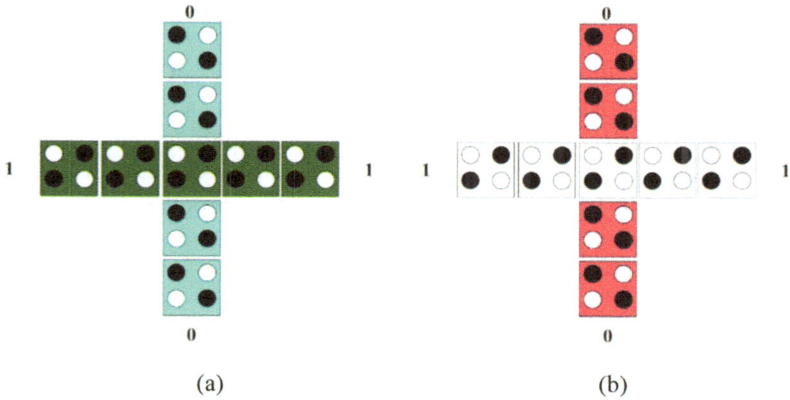

Fig. 9 Crossover using non-interaction between non-adjacent clock zones

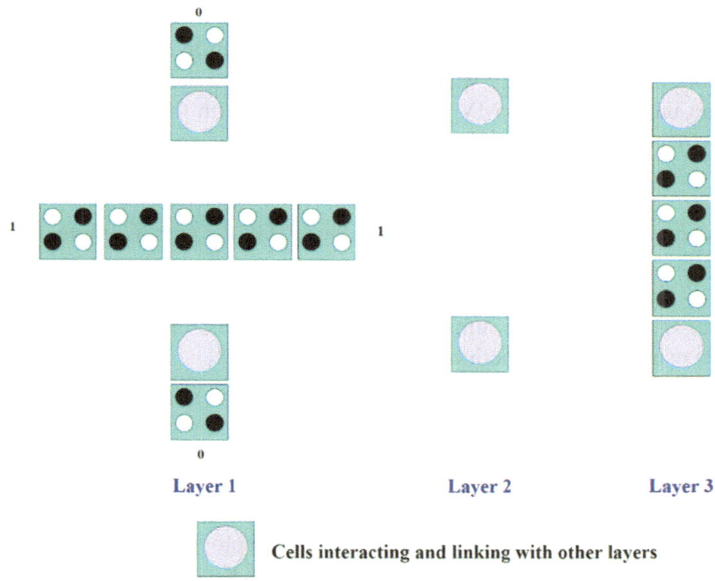

Fig. 10 Crossover using multiple layers

Where, *M, I, C* denote the count of MVs, inverters, and crossovers, respectively, *T* denotes the delay, and *k, l, p* are the exponential weightings which are generally set as 1. The Eq. (3) comprises almost all the parameters that influence the design of QCA circuits.

However, in some cases, the cost of the circuit is calculated by using the area-delay product (ADP), and in some cases, the cost of the circuit is computed by multiplying the ADP with the number of the layers as given in (4).

$$\text{Cost}_{\text{QCA}} = L \times A \times T \quad (4)$$

Where, L denotes the count of layers, A denotes the area, and T denotes the delay.

Apart from MVs, dedicated structures are proposed in QCA to perform operations such as exclusive-or, and-or-inverter, and other multifunctional gates. These gates are developed by using cell-cell interaction and displaced cells. In [8], to compute the cost function of dedicated gates, and gates with cell displacement from ideal positions was proposed as given in (5).

$$\text{Cost}_{\text{gate}} = \left(M^k + I^n + C^l + D^m + R^q\right) T^p, 1 \leq k, l, m, n, p, q \quad (5)$$

Where, D and R denote the count of displaced and rotated cells, respectively. The exponential weightings n, m, q were considered as 2 and parameters k, l, p were considered as 1. A higher weighting value was applied to displaced and rotated cells to highlight their fabrication complexity.

3-input MVs occupy lesser area and cells compared to 5-input MVs. As crossovers have an impact on delay as well as fabrication constraints, a parameter called fabrication complexity was introduced in [13] to incorporate the fabrication issues and the influence of MV gate type and count on the design complexity as given in (6).

$$\text{Fabrication Complexity (FC)} = \text{Gate Count (GC)} \times \text{Cost Complexity (CC)} \quad (6)$$

Where, GC denotes the sum of gates and inverters, and CC denotes the cost computed using the cost function in [12].

2.9 Simulation Settings

As mentioned earlier, QCADesigner is the widely deployed tool for designing QCA designs. The default settings used in QCADesigner/QCADesignerE are provided in Table 1.

The same is considered in this chapter for analyzing the fault tolerance characteristics of circuits. These parameters are used as default parameters in most cases. However, few works have analyzed the performance of designs by changing the radius of effect, temperature, and cell dimensions.

Table 1 Simulation settings

Parameter	Value
Radius of effect	8×10^{-8} m
Clock high	0.098×10^{-24} J
Clock low	0.038×10^{-24} J
Temperature	1 K
Width of the cell	1.8×10^{-8} m
Height of the cell	1.8×10^{-8} m
ε_r	12.9
Dot radius	2.5×10^{-9} m
Cell-cell spacing	2.5×10^{-9} m
Dot-dot spacing	0.0040×10^{-6} m
Conduction band effective mass	$0.067m_e$

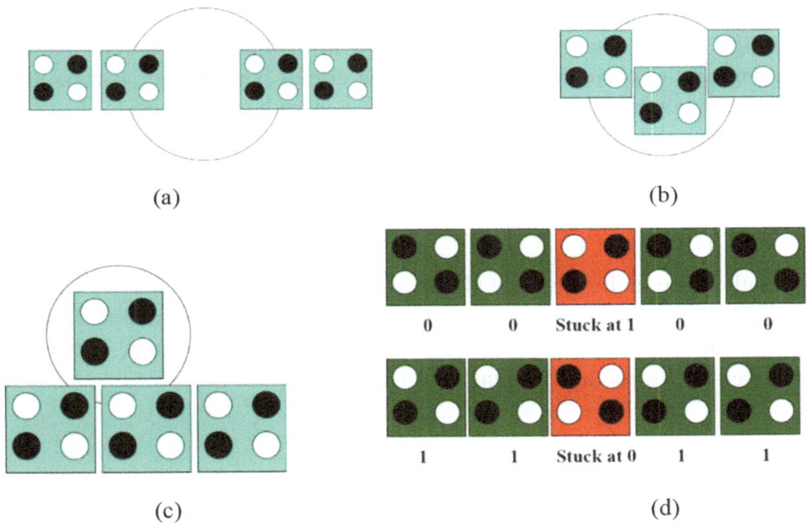

Fig. 11 Fabrication defects in a wire: (**a**) missing/omission, (**b**) misalignment/displacement, (**c**) additional/extra, and (**d**) stuck at fault

3 Fabrication Defects

The stages of fabrication in QCA comprises synthesis and deposition periods. In the period of synthesis, the cells are formed by placing the quantum dots and fixing the electrons inside the cells with tunneling features. In the period of deposition, the cells are positioned in the circuit according to the functionality desired. However, defects are likely to occur during this period due to the small size of the circuit and the difficulty in placing the cells at ideal positions. The different defects are missing/omission cell, additional/extra cell, misalignment/displacement, and stuck at fault defects. The different possible defects are presented in Fig. 11.

Missing/omission defect arises when a cell placement in its position is failed. Depending upon the position of the cell, the circuit performance can get affected. These kinds of defects may be crucial at the corners as missing cell can lead to logic inversion at the corners. Additional/extra cell defect arises when an additional/extra cell gets placed to a cell. Depending upon the cell zone, it can impact the normal operation of the circuit.

Misalignment/displacement defect arises when the cell gets positioned but not in the ideal position. If the cell gets placed within the admissible range, it may not affect the performance. However, when it gets placed above a certain admissible range, it might lead to undesirable performance of the circuit. The small size of the cells and the extremely small spacing between the cells are major challenges in the placement of cells. These defects have greater probability as it is not possible to have accurate placement at such small-scale levels.

Stuck at fault is a defect arising out of cells getting fixed to a particular polarization and not changing in accordance with its neighboring cells and inputs. Depending on the position of the defect, the circuit performance can be affected drastically. Depending upon the value, the stuck at fault can be stuck at 1 or stuck at 0 defect. In general, the fault percentage of a circuit is computed by using (7).

$$\text{Fault percentage} = \frac{\text{No.of faulty outputs}}{\text{Total numbe of possible outputs}} \times 100 \tag{7}$$

To analyze the fault resistance ability of circuits, a detailed analysis was carried out in [14] by briefing on the methodologies on how the defect and fault tolerance can be calculated for different defects. In addition, it proposed factors such as critical factor, tolerance factor, and immunity factor.

3.1 Critical Factor

A cell is a critical cell if a fabrication defect that occurs within a cell can impact the performance. For missing/omission defect critical cell, the outputs are 25% or more faulty due to the missing/omission of the cell from the circuit. For misalignment/displacement defect critical cell, the displacement of the cells is more than 25% of the minimum cell-cell distance between the cells. For additional/extra defect critical cell, any additional/extra cell when placed near it may lead to more than 25% of faulty outputs. For stuck at 0 or stuck at 1, a cell is defined as critical if 25% or more outputs become faulty due to the presence of stuck at defect at the particular cell. Based on these factors, the critical factor of the circuit for a particular defect is given by (8). The critical factor for a good fault-tolerant circuit must be as small as possible, and ideally it should be 0 for a completely fault-tolerant circuit.

$$\text{Critical Factor (CF)} = \frac{\text{No.of critical cells}}{\text{Total no.of cells}} \tag{8}$$

3.2 Tolerance Factor

A cell is a tolerant cell if the cell is more resistant to defects and the defect at that cell will not have an impact on the output. For missing/omission defect tolerant cell, the output is 100% fault free even after the defect. For additional/extra defect tolerant cell, the additional/extra cell placed near the cell will not alter any of the output. For misalignment defect tolerant cell, the misaligned cell will not affect any of the output even if the cell gets placed beyond the minimum cell-cell distance between cells. For stuck at 0 and stuck at 1 fault, a tolerant cell does not impact any of the output. The tolerance factor of the circuit for a particular defect is computed using (9). For a good fault-tolerant circuit, the tolerant factor must be high, and for an ideal fault resistant circuit, the tolerance factor must be 1.

$$\text{Tolerance factor (TF)} = \frac{\text{No. of tolerant cells}}{\text{Total no. of cells}} \tag{9}$$

3.3 Immunity Percentage

To estimate the fault-tolerant capability of the circuits, immunity percentage is computed by using the critical factor and tolerant factor of the circuits by using (10):

$$\text{Immunity percentage (IP)} = (1 - \text{CF}) \times \text{TF} \times 100 \tag{10}$$

For a good fault resistant circuit, the immunity percentage must be high, and for an ideal fault resistant circuit, the immunity percentage must be 100%.

4 Design Considerations for Fault Tolerance

Based on the analysis carried out in [14] and other works [6, 15–31] which focused on developing fault-tolerant circuits and design techniques to achieve more reliability, the following factors were found to play a key role in realizing reliable circuits.

4.1 Wires

The wire length affects the fault tolerance ability of the circuit. According to the design rules in [23], each zone must have two cells in it for ensuring effective design of QCA circuits. However, considering the missing/omission defects, the cells in a wire must have more than two cells to increase the fault-tolerant capability. Use

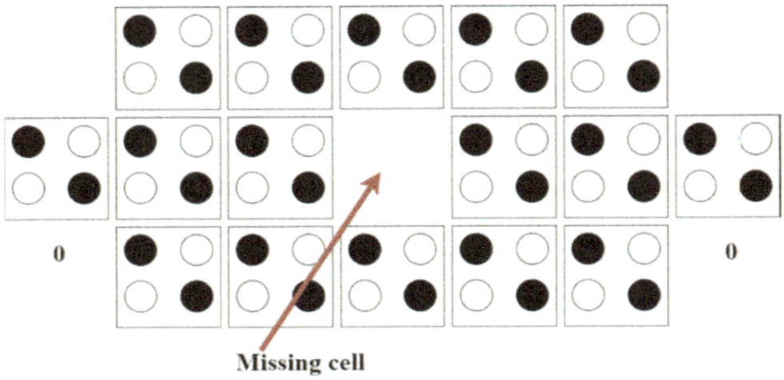

Fig. 12 Improving fault tolerance by thickening wires

of three cells will reduce the fault probability as the cells in the same clock zone will respond better to the neighboring cell which will be within the radius of effect. Fault-tolerant ability of the circuits can also be increased by increasing the thickness of the wires [24, 25] as presented in Fig. 12.

Adding additional lines of wires along with the wires will increase the fault-tolerant capability of the circuits. Due to the increased thickness of the wire, missing cell defect will not have an impact as additional wires will ensure efficient transmission of information without error. However, they have a drawback of increased area and cell count. It also increased the connection complexity with the gates as bending, wire crossovers will become highly complex.

4.2 Gates

Another major consideration that designers need to consider is the gate chosen for realizing the circuits. In [17], an extensive analysis is made on the MVs and inverters and their fault-tolerant capability of gates. It was observed that the 5-input MVs had better fault tolerance compared to 3-input MVs. The 5 MV with greater fault tolerance as observed by [17] is available in [26] and is presented in Fig. 13. It was also observed that the double path design among the inverters had better fault tolerance due to the additional inverter travel path in it. Deployment of all the logic functionalities using these gates can lead to increased area and cell count. However, a trade-off or compromise can be made by designing with other MVs or dedicated gates that have better fault tolerance with fewer cells, area, and better simplicity for interconnection with other gates.

4.3 Crossovers

Crossovers are critical in determining the cost function and fabrication complexity parameters. As it is evident from the cost function and fabrication complexity

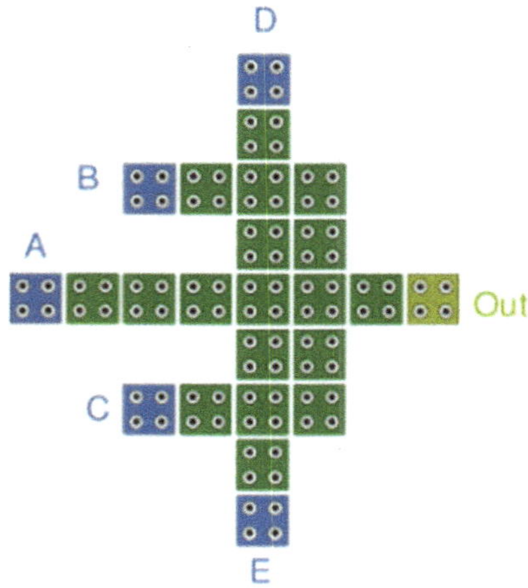

Fig. 13 Fault-tolerant 5 MV in [26]

expressions, it can be seen that the addition of crossovers increases the fabrication defect probability. Rotated crossovers also increase the fault probability as it requires a mix of normal and rotated cells to be placed in ideal positions to ensure perfect transfer of information. Among the techniques, clock zone-based methodology to cross wires proves to be more effective. However, it also comes with a compromise on the delay of the circuit as crossing cannot be done between cells operating in the same zone.

4.4 Clocking

Clocking crossings are also a serious issue in fabrication. To overcome this challenge, many clocking schemes are proposed in the literature [27–31]. These schemes reduce the crossing complexity. However, they increase the layout complexity and increase the difficulty in designing sequential circuits where providing feedback paths get more complex and increase the delay of circuits. The widely used universal, scalable, and efficient clocking scheme (USE) clocking methodology in [29] is presented in Fig. 14. The structure is repeated again and again for deploying larger circuits. It provides better feedback path for designing circuits.

In addition to these challenges, the larger clock zone-based arrangement of clocking zones increases the area and cell count. It also increases the wire crossing complexities of the circuits. However, these schemes try to enhance the realizability chances of QCA with reduced fabrication challenges. Another major challenge with these clocking schemes is the deployment of crossovers using clock zones as it is not possible to have two clock zone cells in a same zone. These schemes support only rotated cell crossings and multiple layer crossings. However, they need to have a trade-off with the fabrication complexity and cost of the circuit.

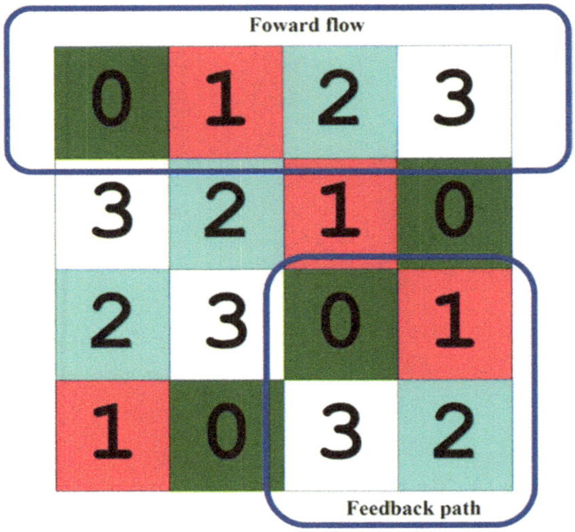

Fig. 14 USE clocking zone placement

4.5 Layout Challenges

The lack of automated layout tools and the lack of availability of libraries that consider the fault tolerance, clocking schemes increase the design complexity of fault tolerant circuits [20]. With the limited capabilities of QCADesigner/QCADesignerE, it finally comes to the designer to design effective QCA circuits manually. During the layout, the designer must consider all these factors into consideration. However, shifting more toward fault tolerance will increase the delay, area, and cell count. More techniques need to be evolved to design fault-tolerant techniques without increasing delay, area, and cell count. A trade-off/compromise should be made to design fault-tolerant circuits.

5 Conclusion

QCA paradigm focuses on designing high complex circuits. Existing QCA circuits focus more on reducing the design complexity by reducing the logic complexity, cells, delay, and area. However, fabrication defect is another major area to be considered during the design of QCA circuits. In this chapter, the fabrication defects that can deter the performance of QCA circuits and metrics to measure fault tolerance are discussed in detail. The parameters that influence the fault tolerance and the design considerations that need to be followed for designing effective fault tolerant circuits are also presented. In general, fault tolerance is achieved by having a trade-off with area and cell count. It is observed that the lack of automated tools and fault-tolerant libraries poses a big challenge in fault-tolerant circuit design. Modern techniques need to be developed without affecting the delay, area, and cell count. The design considerations may be integrated in the layout constraints to automate fault-tolerant circuits in QCA.

References

1. A. Malinowski, J. Chen, S.K. Mishra, S. Samavedam, D. Sohn, What is killing Moore's law? Challenges in advanced FinFET technology integration, in 2019 MIXDES - 26th International Conference "Mixed Design of Integrated Circuits and Systems", Rzeszow, 2019, pp. 46–51. https://doi.org/10.23919/MIXDES.2019.8787084
2. M. Orlowski, CMOS challenges of keeping up with Moore's Law, in 2005 13th International Conference on Advanced Thermal Processing of Semiconductors, Santa Barbara, 2005, pp. 19. https://doi.org/10.1109/RTP.2005.1613679
3. B. Sheu, K. Wilcox, A.M. Keshavarzi, D. Antoniadis, EP1: Moore's law challenges below 10nm: Technology, design and economic implications, in 2015 IEEE International Solid-State Circuits Conference - (ISSCC) Digest of Technical Papers, San Francisco, 2015, pp. 1–1. https://doi.org/10.1109/ISSCC.2015.7063150
4. C.S. Lent, P.D. Tougaw, A device architecture for computing with quantum dots. Proc. IEEE **85**(4), 541–557 (1997). https://doi.org/10.1109/5.573740
5. M. Momenzadeh, M. Ottavi, F. Lombardi, Modeling QCA defects at molecular-level in combinational circuits, in 20th IEEE International Symposium on Defect and Fault Tolerance in VLSI Systems (DFT'05), Monterey, 2005, pp. 208–216. https://doi.org/10.1109/DFTVS.2005.46
6. M. Momenzadeh, J. Huang, M.B. Tahoori, F. Lombardi, On the evaluation of scaling of QCA devices in the presence of defects at manufacturing. IEEE Trans. Nanotechnol. **4**(6), 740–743 (2005). https://doi.org/10.1109/TNANO.2005.858611
7. V. Dhare, U. Mehta, Defect characterization and testing of QCA devices and circuits: A survey, in 2015 19th International Symposium on VLSI Design and Test, Ahmedabad, 2015, pp. 1–2. https://doi.org/10.1109/ISVDAT.2015.7208060
8. M. Raj, L. Gopalakrishnan, S.B. Ko, Reliable SRAM using NAND-NOR gate in beyond-CMOS QCA technology. IET Comput. Digit. Tech. **15**, 202–213. https://doi.org/10.1049/cdt2.12012
9. K. Navi, R. Farazkish, S. Sayedsalehi, M.R. Azghadi, A new quantum-dot cellular automata full-adder. Microelectron. J. **41**(12), 820–826 (2010). https://doi.org/10.1016/j.mejo.2010.07.003
10. R. Akeela, M.D. Wagh, A five-input majority gate in quantum-dot cellular automata. NSTI Nanotech **2**, 978–981 (2011)
11. K. Walus, T.J. Dysart, G.A. Jullien, R.A. Budiman, QCADesigner: a rapid design and simulation tool for quantum-dot cellular automata. IEEE Trans. Nanotechnol. **3**(1), 26–31 (2004). https://doi.org/10.1109/TNANO.2003.820815
12. W. Liu, L. Lu, M. O'Neill, E.E. Swartzlander, A first step toward cost functions for quantum-dot cellular automata designs. IEEE Trans. Nanotechnol. **13**(3), 476–487 (2014). https://doi.org/10.1109/TNANO.2014.2306754
13. K.R. Sekar, R. Marshal, G. Lakshminarayanan, Reliable adder and multipliers in QCA technology. Semicond. Sci. Technol. **37**(9), 095006 (2022). https://doi.org/10.1088/1361-6641/ac796a
14. M. Raj, L. Gopalakrishnan, S.B. Ko, Design and analysis of novel QCA full adder-subtractor. Int. J. Electron. Lett. **9**(3), 287–300 (2021). https://doi.org/10.1080/21681724.2020.1726479
15. H. Du, H. Lv, Y. Zhang, F. Peng, G. Xie, Design and analysis of new fault tolerant majority gate for quantum dot cellular automata. J. Comput. Electron. **15**(4), 1484–1497 (2016). https://doi.org/10.1007/s10825-016-0918-y

16. M. Goswami, B. Sen, B.K. Sikdar, Design of low power 5-input majority voter in quantum-dot cellular automata with effective error resilience, in 2016 Sixth International Symposium on Embedded Computing and System Design (ISED), Patna, 2016, pp. 101–105. https://doi.org/10.1109/ISED.2016.7977063
17. A.T. Vanaraj, M. Raj, L. Gopalakrishnan, Reliable coplanar full adder in quantum-dot cellular automata using five-input majority logic. J. Nanophoton. **14**(2), 026017 (2020). https://doi.org/10.1117/1.JNP.14.026017
18. M. Sun, H. Lv, Y. Zhang, G. Xie, The fundamental primitives with fault tolerance in quantum dot cellular automata. J. Electron. Test. **34**(2), 109–122 (2018). https://doi.org/10.1007/s10836-018-5723-z
19. S.S. Ahmadpour, M. Mosleh, S.R. Heikalabad, Robust QCA full-adders using an efficient fault-tolerant five-input majority gate. Int. J. Circuit Theory Appl. **47**, 1037–1056 (2019)
20. K.R. Sekar, R. Marshal, G. Lakshminarayanan, Framework for QCA layout generation and rules for rotated cell design. J. Circuits Syst. Comput. (2022). https://doi.org/10.1142/S0218126623501141
21. M. Momenzadeh, J. Huang, M.B. Tahoori, F. Lombardi, Characterization, test, and logic synthesis of and-or-inverter (AOI) gate design for QCA implementation. IEEE Trans. Comput.-Aided Des. Integr. Circuits Syst. **24**(12), 1881–1893 (2005). https://doi.org/10.1109/TCAD.2005.852667
22. R. Marshal, G. Lakshminarayanan, Fault resistant coplanar QCA full adder-subtractor using clock zone-based crossover. IETE J. Res. **69**(1), 584–591 (2020). https://doi.org/10.1080/03772063.2020.1838340
23. W. Liu, L. Lu, M. O'Neill, E.E. Swartzlander, Design rules for Quantum-dot Cellular Automata, in 2011 IEEE International Symposium of Circuits and Systems (ISCAS), Rio de Janeiro, 2011, pp. 2361–2364. https://doi.org/10.1109/ISCAS.2011.5938077
24. G. Singh, B. Raj, R.K. Sarin, Fault-tolerant design and analysis of QCA-based circuits. IET Circuits Devices Syst. **12**, 638–644 (2018). https://doi.org/10.1049/iet-cds.2017.0505
25. Y. Mahmoodi, M.A. Tehrani, Novel fault tolerant QCA circuits, in 2014 22nd Iranian Conference on Electrical Engineering (ICEE), Tehran, 2014, pp. 959–964. https://doi.org/10.1109/IranianCEE.2014.6999674
26. K. Navi, S. Sayedsalehi, R. Farazkish, M.R. Azghadi, Five-input majority gate, a new device for quantum-dot cellular automata. J. Computat. Theory Nanosci. **7**, 1546–1553 (2010)
27. S. Rani, T.N. Sasamal, Design of QCA circuits using new 1D clocking scheme, in 2017 2nd International Conference on Telecommunication and Networks (TEL-NET), Noida, 2017, pp. 1–6. https://doi.org/10.1109/TEL-NET.2017.8343540
28. V. Vankamamidi, M. Ottavi, F. Lombardi, Two-dimensional schemes for clocking/timing of QCA circuits. IEEE Trans. Comput.-Aided Des. Integr. Circuits Syst. **27**(1), 34–44 (2008). https://doi.org/10.1109/TCAD.2007.907020
29. C.A.T. Campos, A.L. Marciano, O.P. Vilela Neto, F.S. Torres, USE: A universal, scalable, and efficient clocking scheme for QCA. IEEE Trans. Comput.-Aided Des. Integr. Circuits Syst. **35**(3), 513–517 (2016). https://doi.org/10.1109/TCAD.2015.2471996
30. M. Goswami, A. Mondal, M.H. Mahalat, B. Sen, B.K. Sikdar, An efficient clocking scheme for quantum-dot cellular automata. Int. J. Electron. Lett. **8**(1), 1–14 (2019). https://doi.org/10.1080/21681724.2019.1570551
31. J. Pal, A.K. Pramanik, J.S. Sharma, A.K. Saha, B. Sen, An efficient, scalable, regular clocking scheme based on quantum dot cellular automata. Analog Integr. Circuits Signal Process. **107**, 659–670 (2021). https://doi.org/10.1007/s10470-020-01760-4

Ising Machines Using Parallel Spin Updating Algorithms for Solving Traveling Salesman Problems

Tingting Zhang, Qichao Tao, Bailiang Liu, and Jie Han

1 Introduction

Combinatorial optimization (CO) is an important task in various social and industrial applications, such as machine learning, chip design, and data mining [1]. However, CO problems are non-deterministic polynomial time (or NP)-hard, characterized by the exponentially increasing number of candidate solutions as the problem size increases. It is very challenging to solve such problems by using enumeration. For example, an enumeration method needs to traverse all $(M-1)!$ possible routes to solve a traveling salesman problem (TSP) of M cities, which is prohibitive when M is large.

The Ising model has recently emerged as an efficient method to solve a CO problem by maximizing (or minimizing) an evaluation function under a given set of constraints. It describes the ferromagnetism of magnetic spins in statistical mechanics [1]. An Ising machine aims to find the ground state (i.e., the lowest-energy state) of an Ising model. Various Ising machines have been designed, including those implemented in superconducting circuits based on quantum annealing [2] and the coherent Ising machines (CIMs) implemented using optical parametric oscillators [3]. However, it is challenging to build those systems due to the requirements of cryogenic environments or a long optical fiber. Therefore, classical Ising machines [4–6] have been developed to offer inexpensive implementations and an easier integration with complementary metal-oxide-semiconductor (CMOS) circuits. At the core of an Ising machine, the algorithm plays an important role in the solution search process.

T. Zhang · Q. Tao · B. Liu · J. Han (✉)
University of Alberta, Edmonton, AB, Canada
e-mail: ttzhang@ualberta.ca; qichao@ualberta.ca; bailiang@ualberta.ca; jhan8@ualberta.ca

W. Liu et al. (eds.), *Design and Applications of Emerging Computer Systems*,
https://doi.org/10.1007/978-3-031-42478-6_26

Simulation algorithms, which emulate certain physical phenomena on classical computers, have been developed for solving CO problems by decreasing the energy of the Ising model. Simulated annealing (SA) [7] serves as the basis of various annealing algorithms. Like the thermal annealing in metallurgy, it is aimed to converge the energy to a minimum value. A random flip, or a stochastic state transition, is applied to prevent the Ising model from being stuck in a local minimum. However, the states of neighbor spins cannot be simultaneously updated; therefore, many designs of Ising machines implement a sparse spin-to-spin structure, such as the 2-D lattice topology, the 3-D lattice topology, and King's graph topology. As shown in Fig. 1a, each spin in a 2-D lattice topology has four connections with its neighbor spins (i.e., the connections with the north, east, south, and west spins). The dotted lines indicate the connections between the spins at the edge and those not shown in this figure. A 3-D lattice topology consists of two (or more) layers of 2-D lattices. As shown in Fig. 1b, each spin in a 3-D lattice topology has five connections with its neighbor spins (i.e., the connections with the north, east, south, west, and front/back spins). King's graph topology is also a planar structure, but it is more complex than the 2-D lattice topology. Figure 1c shows the structure of King's graph topology. Each spin in a King's graph topology has eight connections with its neighbor spins (i.e., the connections with the north, east, south, west, northeast, northwest, southeast, and southwest spins). However, spins for solving Ising problems can have more interactions than those in sparse spin-to-spin structures. Therefore, Ising machines with sparse spin-to-spin structures need an embedding process to convert general Ising problems to the available physical Ising models, in which redundant spins in the physical Ising model are used to represent one spin in the logical Ising model.

Ising machines with a complete graph topology adapt to solving any kind of Ising problems, as shown in Fig. 1d. The number of a spin interactions depends on the scale of an Ising machine. To mitigate the issue of increased search time due to sequential update of connected spins, there are various algorithms with parallel spin update developed for building fully connected Ising machines. Momentum annealing (MA) [8] and stochastic cellular automata annealing (SCA) [5] leverage a two-layer spin structure to achieve parallel spin update. They belong to the class of parallel annealing (PA) algorithms. A quantum mechanics-inspired

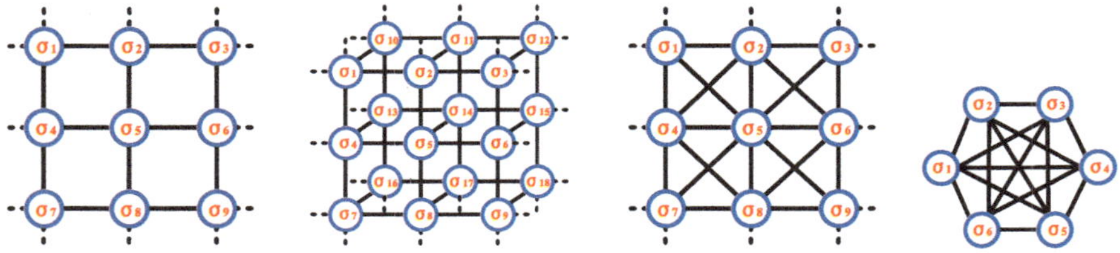

(a) (b) (c) (d)

Fig. 1 Topologies of Ising machines: (**a**) a 2-D lattice topology, (**b**) a 3-D lattice topology, (**c**) a King's graph topology, and (**d**) a complete graph topology algorithm referred to as simulated bifurcation (SB) simulates the quantum adiabatic optimization of Kerr-nonlinear parametric oscillator networks to realize massive parallelism in computation [9]. These parallel algorithms can achieve a fast energy convergence speed for solving unconstrained problems. However, they face with the problems of escaping from local minimum states when solving constrained COPs due to the limited fluctuations in energy.

This chapter presents studies on algorithms, called improved PA and improved SB, to solve constrained combinatorial optimization problems such as the TSP using parallel fully connected Ising machines. The effectiveness is specifically discussed and evaluated by comparing the solution quality to recent annealing algorithms. Experiments on benchmark datasets show that these improved parallel algorithms-based TSP solvers offer superior performance in solution quality and achieve a significant speed-up in runtime than recent annealing algorithms-based ones.

The remainder of this chapter is organized as follows. Section 2 presents the basics. Sections 3 and 4 discuss an improved parallel annealing algorithm and an improved simulated bifurcation algorithm for efficiently solving TSPs. Section 5 concludes this chapter.

2 Background

2.1 *Problem-Solving via Ising Machines*

In the Ising model, each spin can be in either an upward (+1) or downward (−1) state. The interactions among the spins and external magnetic fields affect the states of spins. In an N-spin system, the Hamiltonian in an Ising model is defined as [10]:

$$H(\sigma_1, \ldots, \sigma_N) = -\sum_{i,j} J_{ij}\sigma_i\sigma_j - \sum_i h_i\sigma_i, \tag{1}$$

where σ_i ($\in \{-1, +1\}$, $i \in \{1, 2, \ldots, N\}$) denotes the state of the ith spin, J_{ij} indicates the interaction between the ith spin and the jth spin, and h_i is the external magnetic field for the ith spin.

CO searches for the solution that indicates the maxima (or minima) of an objective function and that satisfies the given constraints. It can be formulated as a quadratic unconstrained binary optimization (QUBO) problem. Let $\mathbb{B} = \{0, +1\}$ be a binary set, $\mathbb{N}$ be a set of integers, and $\mathbb{R}$ be a set of real numbers. Given $\mathbb{B}$, $\mathbb{N}$, and $\mathbb{R}$, a QUBO problem can be described as [11]:

$$\min_{\boldsymbol{x}\in\mathbb{B}^N} F(\boldsymbol{x}) = \boldsymbol{x}^T\boldsymbol{A}\boldsymbol{x} + \boldsymbol{x}^T\boldsymbol{b} + c, \tag{2}$$

where $\boldsymbol{x}$ ($\boldsymbol{x} \in \mathbb{B}^N$, $N \in \mathbb{N}$) is a vector of binary variables and $\boldsymbol{A}$ ($\in \mathbb{R}^{N\times N}$), $\boldsymbol{b}$ ($\in \mathbb{R}^N$), and c ($\in \mathbb{R}$) denote the weight matrix, the weight vector, and a constant scalar, respectively.

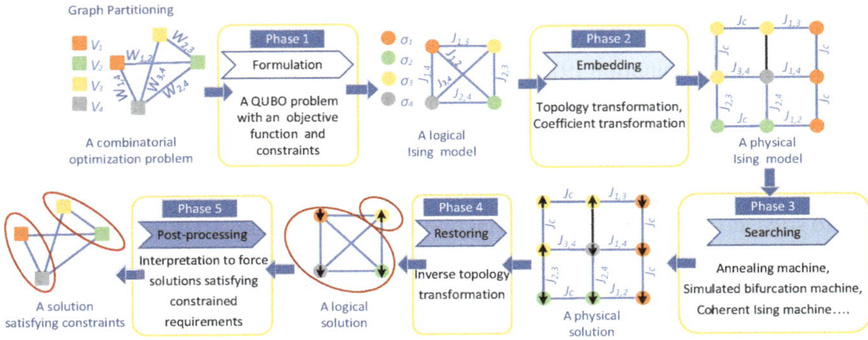

Fig. 2 Solving a combinatorial optimization problem using an Ising machine [12]. An example of graph partitioning is given at the top. Given an undirected graph with an even number of vertices (V) and edges with weights (W), graph partitioning divides the vertices into two subsets of an equal size (as the constraint) with the sum of the weights of edges between the vertices belonging to different subsets minimized (as the objective). To this end, the problem is first mapped to a logical Ising model, where the weights of edges (W) are converted to the interactions between spins (J) by considering the objective and the constraint. Then, the logical Ising model is embedded into the topology of an Ising machine, where J_C is the coupling strength between duplicated spins to ensure that they are in the same states [13]. The solution found in Phase 3 is interpolated back to the logical Ising model, which may provide a solution that violates the constraint. Finally, the solution is modified to satisfy the constraint in Phase 5

By using $\boldsymbol{x} = \frac{1+\boldsymbol{\sigma}}{2}$, where $\boldsymbol{\sigma} \in \{-1, +1\}^N$, (2) can be converted to the expression in (1). Then, a QUBO problem can easily be mapped to an Ising model. The configuration of spins at the ground state provides the optimal solution.

As shown in Fig. 2, solving a CO problem using an Ising machine consists of the following five phases [12]:

Phase 1: A CO problem (COP) formulated as a quadratic unconstrained binary optimization (QUBO) problem is mapped to a logical Ising model that describes the problem without any restriction in the topology and the precision of coefficients [10].

Phase 2: This logical Ising model is converted to a physical Ising model that can readily be embedded into the topology of an Ising machine and that meets the requirement for the bit width of coefficients [10, 13–15].

Phase 3: A specific algorithm searches for the ground state of the Hamiltonian of the physical Ising model [16–20].

Phase 4: The result from Phase 3 is restored back to the logical Ising model using the inverse steps that are specified in the embedding method in Phase 2 [14].

Phase 5: The solution found may not meet the constraints due to the stochastic behavior of the Ising model. Thus, an interpretation method is applied to force the solution to satisfy the constraints by identifying and modifying the states of spins that cause the violation of constraints [21].

2.2 Mapping the Traveling Salesman Problem

A traveling salesman problem (TSP) is to find the shortest route to visiting all cities. Each city can only be visited once, and the salesman should return to the starting point [22]. An n-city TSP can be formulated as a COP using n^2 variables in a lattice, as [23]:

$$H_{TSP} = A\sum_{k \neq l}\sum_{i} W_{kl}a_{ik}a_{(i+1)l} + B\sum_{i}(\sum_{k} a_{ik} - 1)^2 + C\sum_{k}(\sum_{i} a_{ik} - 1)^2, \quad (3)$$

where a_{ik} ($\in \{0, +1\}$) indicates whether the kth city is visited (+1) or not (0) at the ith step and W_{kl} denotes the distance between the kth city and the lth city. The first term in (3) is the objective function of the TSP, which computes the total distance of the route. The second and the third terms in (3) are constraints that prevent visiting multiple cities in one step and visiting a city more than once, respectively. These two terms take the minimum value 0 when $\sum_k a_{ik} = \sum_i a_{ik} = 1$. A, B, and C are the parameters (with positive values) that balance the weights between the objective function and the constraints.

Then the TSP can be mapped to the Ising model by converting $a_{ik}(\in \{0, +1\})$ to $\sigma_{ik}(\in \{-1, +1\})$ as follows [23]:

$$\begin{aligned} H_{TSP} = &\frac{A}{4}\sum_{k \neq l}\sum_{i} W_{kl}\sigma_{ik}\sigma_{(i+1)l} + \frac{A}{2}\sum_{k \neq l}\sum_{i} W_{kl}\sigma_{ik} \\ &+\frac{B}{4}\sum_{i}\sum_{k}\sum_{l}\sigma_{ik}\sigma_{il} + \frac{(n-2)B}{2}\sum_{i}\sum_{k}\sigma_{ik} + \frac{C}{4}\sum_{i}\sum_{k}\sum_{j}\sigma_{ik}\sigma_{jk} \\ &+\frac{(n-2)C}{2}\sum_{i}\sum_{k}\sigma_{ik} + \frac{A}{4}\sum_{k \neq l}\sum_{i} W_{kl} + (\frac{n^3}{4} - n^2 + n)(B + C). \end{aligned} \quad (4)$$

The last two constant terms in (4) unrelated to the states of spins are ignored when minimizing H_{TSP}. The first, third, and fifth terms correspond to the interaction term in (1), and the other terms correspond to the external field term in (1).

Equation (4) can be rewritten to adapt to the Ising formulation as in (1), given by:

$$H_{tsp} = -\sum_{i=1}^{n}\sum_{k=1}^{n}\sum_{j=1}^{n}\sum_{l=1}^{n} J_{ikjl}\sigma_{ik}\sigma_{jl} - \sum_{i=1}^{n}\sum_{k=1}^{n} h_{ik}\sigma_{ik}. \quad (5)$$

3 Improved Parallel Annealing for TSPs

This section presents an improved parallel annealing (IPA) algorithm to solve the TSP [24]: (1) using an exponential temperature function with a dynamic offset

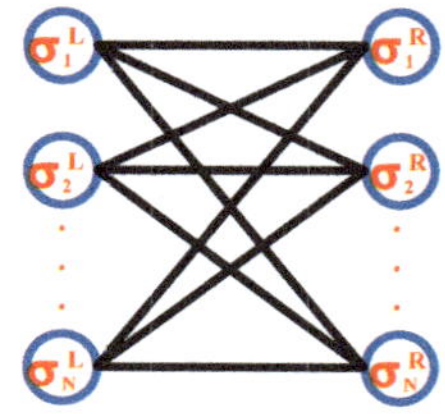

Fig. 3 A two-layer spin structure for PA machines

and (2) using k-medoids in Ising model-based TSP solvers to preprocess data for improving the quality of solutions.

3.1 Parallel Annealing

In the PA with the two-layer spin structure, as shown in Fig. 3, the couplings between σ_i^L and σ_j^R are denoted as $J_{ij}(i \neq j)$, whereas the couplings between σ_i^L and σ_i^R are called self-interactions (denoted as ω_i). Thus, the Hamiltonian based on PA, H_P, is given by [5, 8]:

$$H_P = -\sum_{i,j} J_{ij}\sigma_i^L\sigma_j^R - \frac{1}{2}\sum_i h_i(\sigma_i^L + \sigma_i^R) + \omega_i \sum_i (1 - \sigma_i^L\sigma_i^R). \tag{6}$$

Only when the self-interactions ω_i are sufficiently large are the spin configurations in both layers the same, i.e., $\sigma_i^R = \sigma_i^L$. Thus, the third term in (6) can be eliminated, so (6) becomes the same as (1) [5, 8]. ω_i is given by [8]:

$$\omega_i = \begin{cases} \sum_{s_j \in S} |J_{ij}| - \frac{1}{2}\sum_{s_j \in C} |J_{ij}| & (s_i \in C) \\ \frac{\lambda}{2} & (s_i \notin C) \end{cases}, \tag{7}$$

where λ is the largest eigenvalue of $-\boldsymbol{J}$ ($\boldsymbol{J}$ is a matrix of J_{ij}), s_i is the ith spin, C is a subset of the set of all spins S, and C satisfies $C = \{s_i | \lambda \geq \sum_{s_j \in S} |J_{ij}|\}$.

The spin-flip probability is calculated using the Metropolis algorithm [25]. If σ_i^L is flipped, the total energy will be increased by:

$$\Delta_i = 2\sigma_i^L\left(\frac{h_i}{2} + \sum_j J_{ij}\sigma_j^R + \omega_i\sigma_i^R\right). \tag{8}$$

Then, the new spin-flip probability is $\min\{1, exp(-\Delta_i/T)\}$, where T is the temperature.

To improve the efficiency of annealing, dropout and momentum scaling are introduced in [8]. The dropout sets each ω_i as "0" with a decreasing probability.

The momentum scaling multiplies every interaction ω_i by an increasing factor from 0 to 1. Thus, at the end of momentum annealing, every interaction ω_i will return to the value computed in (7) to ensure $\sigma_i^R = \sigma_i^L$ in (6).

3.2 Improved Parallel Annealing

The IPA for solving TSPs is shown in Algorithm 1. An exponential temperature function is used in the IPA, and a dynamic offset is applied to the temperature function. Firstly, the state of a spin is randomly initialized to "-1" or "$+1$," and the temperature increment (ΔT) due to the dynamic offset is initialized to "0." In Fig. 1c, spins in the left layer are updated when the current step s is odd; otherwise, the spins in the right layer are updated. In each iteration ($s \in [1, iternum]$), the dropout rate (p_s) and the momentum scaling factor (c_s) are updated, where the $iternum$ is the total number of iterations. The temperature (T_s) is then recalculated, where r in Algorithm 1 is the cooling coefficient. During the annealing, the self-interaction (ω_{ik}) is set to "0" with the probability p_s or decreased to $c_s \cdot \omega_{ik}$. Then, the energy variation (Δ_{ik}) when σ_{ik} is flipped is evaluated using the spin interaction (J_{ikjl}) and the updated ω_{ik}. Subsequently, the spin-flip probability (P_{ik}) is calculated using the Metropolis algorithm. If P_{ik} is larger than a randomly generated number within (0, 1), the spin will be flipped. Otherwise, the spin will remain unchanged. After each iteration, if no spin is flipped, ΔT will increase. Otherwise, ΔT will be reset to "0." Finally, the spin configuration (σ) at the end of iterations is output as the solution to the COP found by the IPA.

3.3 A Temperature Function

The classical annealing with parallel spin update uses a logarithmic function in (9) as the temperature function to solve the max-cut problem [8]:

$$T_s = \frac{1}{\beta_0 \ln(1+s)}, \tag{9}$$

where β_0 is a scaling factor for the inverted value of temperature and T_s denotes the temperature in the sth iteration. When the Ising model reaches a local minimum or ground state and T_s is sufficiently small, the flip probability for each spin, P_{ik}, is considered to be close to 0 (see lines 15 and 16 in Algorithm 1). With a proper β_0, however, the temperature only decreases to a value that results in low flip probabilities for all spins. Hence, it is possible for the Ising model to escape from local minima when solving max-cut problems.

To solve a TSP, the temperature required for maintaining low spin-flip probabilities is larger because Δ_{ik} is larger due to the constraints. However, an Ising model cannot reach a local minimum or ground state or meet the constraints at

Algorithm 1 Improved parallel annealing for TSPs

Require: spin interaction: $\boldsymbol{J}$; external magnetic field: $\boldsymbol{h}$;
the number of cities: M; self-interaction: $\boldsymbol{\omega}$;
hyperparameters: $iternum$, T_{init}, T_{inc}, r
Ensure: spin configuration (σ)
Initialize spin configurations
$T_s \Leftarrow T_{init}$
$\Delta T \Leftarrow 0$
for $s = 1$ to $iternum$ **do**
  if s is odd **then**
    $A \Leftarrow L, B \Leftarrow R$
  else
    $A \Leftarrow R, B \Leftarrow L$
  end if
  Update p_s and c_s
  $T_s \Leftarrow (T_s + \Delta T) \cdot r^{s-1}$
  for $i = 1$ to M **do**
    for $k = 1$ to M **do**
      Temporarily set $\omega_{ik} \Leftarrow 0$ with the probability p_s, and temporarily decrease $\omega_{ik} \Leftarrow c_s \cdot \omega_{ik}$
      $\Delta_{ik} \Leftarrow 2\sigma_{ik}^A(\frac{h_{ik}}{2} + \sum_{j,l} J_{ikjl}\sigma_{jl}^B + \omega_{ik}\sigma_{ik}^B)$
      $P_{ik} \Leftarrow \min\{1, exp(-\Delta_{ik}/T_s)\}$
      if $P_{ik} > rand$ **then**
        $\sigma_{ik}^A \Leftarrow -\sigma_{ik}^A$
      end if
    end for
  end for
  if no spin is flipped **then**
    $\Delta T \Leftarrow \Delta T + T_{inc}$
  else
    $\Delta T \Leftarrow 0$
  end if
end for

such a temperature. Thus, the temperature needs to be sufficiently low at the end of annealing when solving TSPs. It will, therefore, be difficult for the Ising model to escape from a local minimum. Moreover, the temperature using a logarithmic function rapidly decreases, so it will prevent the Ising model from traversing additional local minima, thereby reducing the quality of solutions. Hence, an exponential function is used as the temperature function to solve the TSP, as:

$$T_s = T_{init} \cdot r^{s-1}, \tag{10}$$

where T_{init} is the initial temperature and r is the cooling rate. The slower decreasing rate of the exponential function makes the Ising model stay longer at a high temperature, therefore improving the quality of solutions.

Considering that the number of local minima increases with the TSP scale, the Ising model is prone to be stuck in a local minimum during annealing. Reducing the time spent in a local minimum can improve efficiency. Thus, we consider introducing a dynamic offset, as in [26–28], into the temperature function. To increase the probability of escaping from a local minimum, the temperature T_s needs to be very large. Therefore, ΔT is added to T_s, where ΔT is increased by T_{inc} if the spin configuration is unchanged. Lastly, ΔT is reset to zero after a change of the spin state has occurred.

The improvement of the solution quality after using an exponential function with a dynamic offset and the details for setting a proper T_{inc} are discussed in Sect. 4.

3.4 A Clustering Approach

The solution quality drastically deteriorates when the number of cities in the TSP is large. We further consider a clustering approach [23] to improve the quality of the solution. The basic idea is to group the nearby cities into one cluster and use the centric point to represent each cluster. Then the TSP consisting of those central points can be solved by using the IPA. After learning the visiting order of each cluster, the original TSP can be more efficiently solved. This Ising machine can avoid producing solutions that conform to the constraints but with very long travel distances. For example, if cluster A contains three cities and is first visited among the clusters, then the visiting order of these three cities will be confined to the first three steps.

The k-medoids and k-means are two typical clustering approaches. To divide M vertices into k clusters, the first step of the k-means is to randomly generate k new vertices as the central points of the k clusters, while the k-medoids method chooses k vertices from the original set as the central points. In the second step, after the k central points are obtained, the other $(M - k)$ vertices in the set are assigned to the closest central point and form a cluster. In the third step, the k-means method generates a new central point for each cluster according to the mean value of the coordinates of the vertices in the cluster, while the k-medoids method chooses the vertex with the smallest sum of distances from the other vertices in the same cluster as the new central point. Then, the second and the third steps are repeated until there is no change in any cluster.

Compared with the k-means, the disadvantage of k-medoids is the computation time for each cluster in the third step, $O(m^2)$, where m is the number of vertices in one cluster. However, there are $M \times M$ accumulators in the circuit of an Ising model for $M \times M$ spins that solves an M-city TSP, where $M = \sum_{i=1}^{k} m_i$ and m_i is the number of vertices in the ith cluster. Thus, this computation time can be reduced to $O(m)$ as it can be calculated in parallel with m accumulators. Furthermore, calculation of the distances between the vertices is not required in an Ising machine as the distance values are included in the system's input, i.e., in the spin interaction matrix $\boldsymbol{J}$. In contrast, the k-means method needs extra arithmetic units to compute the distances between the vertices and the central points. Therefore, using k-medoids for clustering can achieve a higher hardware efficiency than using k-means with no performance trade-off in an Ising machine.

Algorithm 2 The k-medoids clustering

Require: distance matrix: $\boldsymbol{W}(M \times M)$; the number of clusters: k
Ensure: vertex indexes (with cluster labels)

```
Step 1
for i = 1 to M do
    D_i = \sum_{j=1}^{M} W_{ij}
end for
Choose k vertices (v) with the first k smallest D as the central points
Step 2
for i = 1 to M do
    Assign v_i to the closest central point and mark the label of v_i with the index of the
    corresponding central point
end for
Step 3
for each cluster do
    for each v_i in the cluster do
        d_i = \sum_j W_{ij}
    end for
    Choose v with the smallest d to be the new central point of the current cluster
end for
Repeat Step 2 and Step 3 until there is no change of elements in each cluster
```

The k-medoids clustering was proposed in [29], but it is first applied in an Ising model-based TSP solver in this work, as shown in Algorithm 2. A strategy for choosing the k vertices at the center of the map as initial central points is applied to improve the efficiency of k-medoids. The k vertices with the k smallest sums of distances from all the other vertices are selected as initial central points.

To implement the visiting restrictions, an M-by-M matrix $\boldsymbol{h_p}$ is added to the external magnetic field matrix, $\boldsymbol{h}$. For example, if the first three cities are confined to be visited at the first three steps, $\boldsymbol{h_p} = \begin{pmatrix} \boldsymbol{a} & \boldsymbol{b} \\ \boldsymbol{c} & \boldsymbol{d} \end{pmatrix}$, where $\boldsymbol{a}$ and $\boldsymbol{d}$ are 3-by-3 and $(M-3)$-by-$(M-3)$ zero matrices, respectively, and $\boldsymbol{b}$ and $\boldsymbol{c}$ are 3-by-$(M-3)$ and $(M-3)$-by-3 matrices of $M \cdot \max\{abs(\boldsymbol{J})\}$ (as a large value), respectively. The $\max\{abs(\boldsymbol{J})\}$ returns the entry in matrix $\boldsymbol{J}$ with the largest absolute value.

4 Experimental Results

4.1 Experiment Setup

Three benchmark datasets from the TSPLIB benchmark are used in the experiments, including $burma14$, $ulysses16$, and $ulysses22$ [30]. The results are obtained with parameters $A = 1$, $B = C = \max\{\boldsymbol{W}\}$, $r = 0.97$, and $T_{init} = 1 \times 10^7$. The average (Ave), maximum (Max), minimum (Min), and standard deviation (Std) of the travel distances are obtained after performing annealing by 100 times with an iteration number of $10k$.

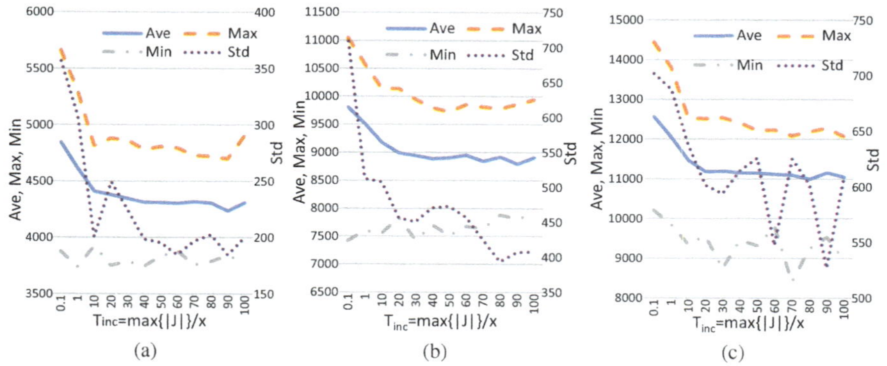

Fig. 4 The effect of T_{inc} on the quality of solutions: (**a**) for the benchmark *burma*14, (**b**) for the benchmark *ulysses*16, and (**c**) for the benchmark *ulysses*22

4.2 Using Different Incremental Temperatures

As a key to increasing the probability to escape from a local minimum, an appropriate setting of T_{inc} can optimize the efficiency of an Ising model. Therefore, we investigate the effect of different T_{inc} values on the Ave and Std of solutions found by the IPA. As shown in Fig. 4, the Ave, Max, and Std of the travel distances found by the Ising models decrease when T_{inc} decreases from $10 \times \max\{abs(\boldsymbol{J})\}$ to $\max\{abs(\boldsymbol{J})\}/10$. Furthermore, the Ave and Max tend to be stable when T_{inc} is between $\max\{abs(\boldsymbol{J})\}/10$ and $\frac{\max\{abs(\boldsymbol{J})\}}{90}$. The further decrease of T_{inc} results in an increase of Ave, so it degrades the quality of solutions. A small T_{inc} reduces the chance of the Ising models escaping from local minima, and thus, a larger iteration number is required for finding a suboptimal solution. However, no significant increase of Ave is observed for *ulysses*22, due to the limited performance of IPA so far in solving large-scale TSPs.

4.3 Comparison

To evaluate the performance of the proposed methods, MA [8] and DA [28] are considered for comparison. The MA implements parallel spin update but using a logarithmic temperature function, while the DA employs a dynamic offset but without parallel spin update. The results are obtained with $T_{inc} = \frac{max\{|\boldsymbol{J}|\}}{90}$ for the IPA and DA. For the MA, $\beta_0 = 9 \times 10^{-4}$ for *burma*14, $\beta_0 = 8 \times 10^{-4}$ for *ulysses*16, and $\beta_0 = 5 \times 10^{-4}$ for *ulysses*22. These β_0 values are chosen to produce the best solution quality in the experiment. Table 1 shows the performance of the IPA, MA, and DA for solving the TSP. The algorithms with parallel spin update (IPA and MA) obtain lower Ave than DA for all three benchmarks. However, the Ave

Table 1 (Unitless) travel distances by using the IPA, MA, and DA for solving the TSP

Metrics	$iternum = 10k$			$iternum = 50k$			Using clustering in the IPA
	IPA	MA	DA	IPA	MA	DA	
				burma14			
Ave	4241.6	5322.4	8832.9	4018.5	5133.3	6451.8	3813.8
Max	4703.0	7524.0	9655.0	4423.0	7443.0	8009.0	4334.0
Min	3839.0	4178.0	7507.0	3580.0	4099.0	4945.0	3345.0
Std	185.1	683.7	379.8	159.9	547.7	696.4	268.9
				ulysses16			
Ave	8804.2	11,513.0	12,722.0	8387.6	11,451.0	12,040.0	7705.0
Max	9869.0	13,992.0	15,454.0	9218.0	14,366.0	14,669.0	8923.0
Min	7816.0	8859.0	9827.0	7554.0	9242.0	8815.0	6686.0
Std	407.9	1113.6	1141.5	303.3	1057.4	1240.4	440.9
				ulysses22			
Ave	11,170.0	13,811.0	16,619.0	10,389.0	13,367.0	16,435.0	8011.4
Max	12,301.0	18,914.0	20,316.0	11,167.0	16,799.0	18,862.0	9371.0
Min	9527.0	11,154.0	13,224.0	9163.0	9363.0	13,000.0	7219.0
Std	527.3	1622.3	1425.6	433.7	1284.8	1156.4	433.2

Using clustering in the IPA: $k_1 = 7$, $k_2 = 4$ for $burma14$; $k_1 = 8$, $k_2 = 4$ for $ulysses16$; and $k_1 = 10$, $k_2 = 6$ for $ulysses22$

obtained by MA can hardly be improved by increasing the number of iterations. This occurs because the annealing algorithm using a logarithmic temperature function is easily stuck in a local minimum when solving TSPs. On the contrary, the algorithms that employ a dynamic offset, such as the IPA and DA, can find shorter distances when the iteration number increases. For solving $burma14$, when the iteration number is $10k$, the IPA decreases the Ave by 52.0% compared with DA and by 20.3% compared with MA. The required iterations for the DA, MA, and IPA to produce an Ave around 4920 are $250k$, $20k$, and $1k$, respectively, while the runtimes are 4.44 seconds, 1.99 seconds, and 0.10 seconds, respectively. The IPA achieves a 44.4× speed-up in runtime compared with DA and 19.9× compared with MA.

Further decrease in Ave is obtained by using the clustering approach. We applied k-medoids twice for each benchmark. The original TSP is clustered into a second-level TSP with k_1 centric points, and the second-level TSP is clustered into a third-level TSP with k_2 centric points. The iteration number for solving the third-level TSP is 1000; it is 2500 for the second-level TSP and 3000 for the original TSP, so the total iteration number is 6500. As shown in Table 1, the reduction in Ave is 51.8% or 42.0% compared to DA or MA for $ulysses22$ ($iternum = 10k$), 39.4% or 33.1% for $ulysses16$, and 56.8% or 28.3% for $burma14$, respectively.

5 Improved Simulated Bifurcation for TSPs

This section presents an efficient ballistic SB (bSB)-based TSP solver with several improvement strategies by taking advantage of the adiabatic evolution in bSB [31, 32]: (1) dynamically configuring time steps used in the integrator and (2) evolving the redundant position during the search process.

5.1 Simulated Bifurcation

A quantum mechanics-inspired algorithm referred to as SB can realize massive parallelism in computation [20]. By simulating the quantum adiabatic optimization of Kerr-nonlinear parametric oscillator networks, SB searches for an approximate solution by solving a pair of differential equations [19]. Two branches of the bifurcation (indicated by the sign of an oscillator's position) are considered as the two states of a spin. To restrain the errors introduced due to the use of continuous variables (for the positions), the bSB introduces hard thresholds to limit the evolution of the oscillators' positions to quickly find suboptimal solutions [9]. To solve an Ising problem in (1) but without the external field (h_i), the classical Hamiltonian for the Ising model with bSB and the Hamiltonian equations of motion are given by [9]:

$$H_{bSB} = \sum_i \frac{a_0}{2} {y_i}^2 + \frac{a_0 - a(t)}{2} \sum_i {x_i}^2 - c_0 \sum_{i,k} J_{ik} x_i x_j, \tag{11}$$

$$\dot{x_i} = \frac{\partial H_{bSB}}{\partial y_i} = a_0 y_i, \tag{12}$$

$$\dot{y_i} = -\frac{\partial H_{bSB}}{\partial x_i} = -\{a_0 - a(t)\} x_i + 2c_0 \sum_{j=1}^{N} J_{ij} x_j, \tag{13}$$

where x_i and y_i are the position and the momentum of the ith oscillator, $\dot{x_i}$ and $\dot{y_i}$ denote the derivatives with respect to time, and a_0 and c_0 are manually tuned constants. $a(t)$ is a time-dependent variable to guarantee the adiabatic evolution. In bSB, x_i is replaced by its sign and $y_i = 0$ when $|x_i| > 1$.

An Ising model-based solver with bSB utilizes the semi-implicit Euler method as an integrator to solve the pair of differential equations, (12) and (13). At the end of the search, the sign of x_i indicates the state of the ith spin.

5.2 TSP Solvers Using the Ising Model Without External Fields

5.2.1 Reformulation of the TSP

To formulate the TSP for bSB, a redundant spin with the state $\sigma_{(n+1)(n+1)}$ fixed to "+1" is first introduced to (5) as:

$$H_{tsp} = -\sum_{i=1}^{n}\sum_{k=1}^{n}\sum_{j=1}^{n}\sum_{l=1}^{n} J_{ikjl}\sigma_{ik}\sigma_{jl} - \sum_{i=1}^{n}\sum_{k=1}^{n} h_{ik}\sigma_{ik}\sigma_{(n+1)(n+1)} \tag{14}$$

Then, each h_{ik} is divided by 2 to convert the external magnetic fields to the coupling coefficients between n^2 spins and the redundant one. Different from the mapping in (5), therefore, an n-city TSP is reformulated as an Ising problem without external magnetic fields by expanding n^2 spins to $(n+1)^2$ spins in a lattice, as:

$$\begin{aligned} H_{tsp} = & -\sum_{i=1}^{n}\sum_{k=1}^{n}\sum_{j=1}^{n}\sum_{l=1}^{n} J_{ikjl}\sigma_{ik}\sigma_{jl} - \sum_{i=1}^{n}\sum_{k=1}^{n} \frac{h_{ik}}{2}\sigma_{ik}\sigma_{(n+1)(n+1)} \\ & -\sum_{j=1}^{n}\sum_{l=1}^{n} \frac{h_{jl}}{2}\sigma_{(n+1)(n+1)}\sigma_{jl} = -\sum_{i=1}^{n+1}\sum_{k=1}^{n+1}\sum_{j=1}^{n+1}\sum_{l=1}^{n+1} J'_{ikjl}\sigma_{ik}\sigma_{jl}, \end{aligned} \tag{15}$$

where:

$$J'_{ikjl} = \begin{cases} J_{ikjl} & i,k,j,l \in \{1,2,\ldots,n\} \\ \frac{h_{ik}}{2} & i,k \in \{1,2,\ldots,n\}\ and\ j = l = n+1 \\ \frac{h_{jl}}{2} & j,l \in \{1,2,\ldots,n\}\ and\ i = k = n+1 \\ 0 & otherwise \end{cases}. \tag{16}$$

To satisfy the constraint that there is only one spin with an up state ("+1") in the same row and the same column, $\sigma_{(n+1)(n+1)}$ is fixed to "+1," and the states of the other spins in the $(n+1)$th dimension are fixed to "−1."

5.2.2 Solving the TSP with bSB

Following (11)–(13), the classical Hamiltonian for the Ising model in (15) using bSB to solve TSPs ($H_{tsp_{bSB}}$) and the corresponding pair of differential equations are given by:

$$H_{tsp_{bSB}} = \sum_{i,k} \frac{a_0}{2} {y_{ik}}^2 + \frac{a_0 - a(t)}{2} \sum_i {x_{ik}}^2 - c_0 \sum_{i,j,k,l} J'_{ikjl} x_{ik} x_{jl}, \tag{17}$$

$$\dot{x_{ik}} = a_0 y_{ik}, \tag{18}$$

$$\begin{aligned} \dot{y_{ik}} &= -\{a_0 - a(t)\}x_{ik} + 2c_0 \sum_{j=1}^{n+1}\sum_{l=1}^{n+1} J'_{ikjl}x_{jl} \\ &= -\{a_0 - a(t)\}x_{ik} + 2c_0 \sum_{j=1}^{n}\sum_{l=1}^{n} J_{ikjl}x_{jl} + c_0 \sum_{j=1}^{n}\sum_{l=1}^{n} h_{ik}x_{(n+1)(n+1)}, \end{aligned} \tag{19}$$

where x_{ik} and y_{ik} ($i, k \in \{1, 2, \ldots, n\}$) are the position and the momentum of the oscillator in the ith row and kth column in a lattice. $x_{(n+1)(n+1)}$ is expected to be 1 at the end of the search to ensure the spin state $\sigma_{(n+1)(n+1)}$ to be "+1."

With (18) and (19), a TSP can efficiently be solved by using bSB.

5.3 *Improvement Strategies*

5.3.1 Dynamic Time Steps

To accelerate the convergence of Hamiltonian, the dynamic configuration of the time step (DTS) is considered to solve the pair of differential equations (18) and (19). In hardware, the multiplication with 0.5 can be implemented by using a shift operation, and the multiplication with 1 does not need any specific processing. Therefore, for an efficient hardware implementation, the time step (denoted by Δ_t) is selected to be 0.5 or 1 by using a piecewise function during a given iteration (denoted by $iter$) in the update of the spin states. Four different dynamic configurations of the time step are considered, as shown in Table 2.

Since it is more challenging to skip the local minimum as time increases, a small time step $\Delta_t = 0.5$ is used at the beginning of a search to ensure the solution quality, and a large time step $\Delta_t = 1$ is used near the end of the search to increase the probability of changing the state of a spin. Three configurations are developed by using different proportions of small and large time steps during the update of the spin states. As a basic configuration, DTS1 employs equally distributed time steps by taking the value of either 0.5 or 1. The large time step is preferred in the last two-thirds of iterations by using the configuration referred to as DTS2, whereas the configuration referred to as DTS3 uses small time steps during the first two-thirds of iterations.

The state of a spin (σ_{ik}) is determined by the sign of the related position (x_{ik}), which is difficult to change at the beginning before the bifurcation occurs. Therefore, the large time step is used at both the beginning and the end of search in the configuration referred to as DTS4.

Table 2 Different dynamic configurations of the time step (DTS) in the Ising model-based solver with bSB

Dynamic configurations		Formulation
Small-large	DTS1: equally distributed	$\Delta_t = \begin{cases} 0.5 & r < \frac{iter}{2} \\ 1 & r \geq \frac{iter}{2} \end{cases}$
	DTS2: large Δ_t preferred	$\Delta_t = \begin{cases} 0.5 & r < \frac{iter}{3} \\ 1 & r \geq \frac{iter}{3} \end{cases}$
	DTS3: small Δ_t preferred	$\Delta_t = \begin{cases} 0.5 & r < \frac{2\times iter}{3} \\ 1 & r \geq \frac{2\times iter}{3} \end{cases}$
Large-small-large	DTS4	$\Delta_t = \begin{cases} 0.5 & \frac{iter}{3} < r < \frac{2\times iter}{3} \\ 1 & otherwise \end{cases}$

Table 3 Different evolution approaches (EAs) for $x_{(n+1)(n+1)}$ in the Ising model-based solver with bSB

Evolution		Formulation
Linear	EA1	$x_{(n+1)(n+1)} = \frac{r}{2iter} + 0.5$
Constant-linear	EA2	$x_{(n+1)(n+1)} = \begin{cases} 0.5 & r < \frac{iter}{2} \\ \frac{r}{iter} & r \geq \frac{iter}{2} \end{cases}$
Linear-constant	EA3	$x_{(n+1)(n+1)} = \begin{cases} \frac{r}{iter} & r < \frac{iter}{2} \\ 1 & r \geq \frac{iter}{2} \end{cases}$
Constant-constant	EA4	$x_{(n+1)(n+1)} = \begin{cases} 0.5 & r < \frac{iter}{2} \\ 1 & otherwise \end{cases}$
Exponential	EA5	$x_{(n+1)(n+1)} = 0.5 + \frac{r^2}{2iter^2}$

5.3.2 Evolution of $x_{(n+1)(n+1)}$

If $x_{(n+1)(n+1)}$ is fixed at 1 during the entire simulation, the effect of the external field on the spin is more significant than the interaction between spins. Moreover, $\sigma_{(n+1)(n+1)}$ is "+1" as long as the sign of $x_{(n+1)(n+1)}$ is positive. Therefore, different evolution approaches (EAs) are considered for $x_{(n+1)(n+1)}$ to change gradually from a positive value to 1, as presented in Table 3.

In EA1 and EA5, $x_{(n+1)(n+1)}$ increases linearly and exponentially, respectively, from 0.5 to 1 with time. Using EA2, $x_{(n+1)(n+1)}$ remains at 0.5 in the first half of the iteration and then linearly increases to 1 in the second half, whereas in EA3, $x_{(n+1)(n+1)}$ linearly increases in the first half of iterations and stays unchanged as 1 in the second half. $x_{(n+1)(n+1)}$ takes 0.5 at first and then takes 1 using EA4.

Table 4 (Unitless) travel distances by using the bSB with different DTS and EAs for solving the TSP

	Using different DTS						Using different EAs				
Metrics	$\Delta_t = 0.5$	$\Delta_t = 1$	DTS1	DTS2	DTS3	DTS4	EA1	EA2	EA3	EA4	EA5
burma14											
Ave	3707	4091	4005	4011	3775	3679	3780	3836	3939	3992	3792
Max	4217	4647	4936	4776	4331	4150	4525	4751	4757	4841	4903
Min	3371	3672	3536	3511	3413	3417	3323	3346	3511	3454	3323
Std	211	207	263	297	255	230	269	464	274	260	400
ulysses16											
Ave	7678	8619	8393	8389	7865	7479	7999	8015	8448	8529	7997
Max	8637	9853	9785	9853	9668	8496	9224	10,027	9914	9763	10,718
Min	6803	7180	7172	7439	7079	6863	7062	7027	7318	7332	6857
Std	433	643	633	559	572	459	539	702	608	575	674
ulysses22											
Ave	8441	9577	9292	9547	8258	8267	8646	8810	9410	9485	8499
Max	10,353	12,109	11,301	11,701	9575	9273	10,012	10,397	11,029	11,191	10,296
Min	7203	8208	7779	8210	7249	7419	7330	7545	7993	8000	7451
Std	584	945	830	803	597	489	608	642	642	721	672

5.4 Experimental Results

5.4.1 Experiment Setup

Consider (18) and (19), where we set $a_0 = 1$ and $a(t)$ increases from 0 to 2 to ensure the adiabatic evolution. $\boldsymbol{x}$ and $\boldsymbol{y}$ are initialized to a zero matrix and a random matrix with entries within $[-0.1, +0.1]$, respectively. The parameters A, B, and C are, respectively, set to 1, $\max\{W\}$, and $\max\{W\}$. Three datasets from the TSPLIB benchmark [30] are considered. The solution quality is evaluated by Ave, Max, Min, and Std of the obtained route distances from 100 trials.

5.5 Using Different Dynamic Configurations of the Time Step

The quality of the TSP solvers based on the Ising model with bSB using different dynamic configurations of the time step is presented in Table 4 with the default value 1 for $x_{(n+1)(n+1)}$. Using $\Delta_t = 0.5$ leads to a higher solution quality than using $\Delta_t = 1$. An improvement of 9.4%, 10.9%, and 11.8% in Ave has been achieved, respectively, when solving the problems in $burma14$, $ulysses16$, and $ulysses22$.

Compared with using a constant time step, the use of dynamic configurations of the time step leads to a higher solution quality (with smaller Ave, Max, Min). As seen from the results obtained by using DTS1, DTS2, and DTS3 in Table 4, the solution quality is improved with the increased use of the small time step ($\Delta_t =$

0.5). Compared with DTS1 and DTS2, DTS3 prefers small Δ_t in most iterations during the search, so it shows a higher solution quality. Moreover, DTS4 improves the efficiency and stability in the solutions by exploiting the inherent characteristics of the bSB algorithm. Although DTS4 employs the small time step ($\Delta_t = 0.5$) in only one-third of all updates, it achieves a reduction up to 2.5% in Ave and 10.4% in Max, but with a slight increase up to 2.9% in Min than using the constant $\Delta_t = 0.5$; it also achieves a reduction up to 13.6% in Ave, 23.4% in Max, and 9.6% in Min than using the constant time step $\Delta_t = 1$.

5.5.1 Using Different Evolution Approaches for $x_{(n+1)(n+1)}$

Table 4 presents the solution quality of using different evolution approaches. Δ_t is fixed to 1 to simply the hardware implementation. No matter which evolution approach is employed, the bSB-based TSP solver achieves an improvement in the solution quality. The EA1 and EA5, where $x_{(n+1)(n+1)}$ increases linearly and exponentially, respectively, lead to a significant improvement, while a slight improvement is obtained by using EA4. Using EA1 and EA5 obtains a similar improvement in Ave, at least by 7.3%, 7.1%, and 9.7% for solving $burma14$, $ulysses16$, and $ulysses22$, respectively. Moreover, using the bSB-based TSP solver with EA1 can find a solution with higher stability (with a smaller Std).

5.5.2 Comparison

Figure 5 shows a comparison of the performance of using the improved SA [33] and the proposed bSB. The bSB-based TSP solvers with the DTS4 and the EA1 are referred to as bSB-DTS4 and bSB-EA1, respectively. The number of iterations is 50k in the improved SA and 2k in the bSB.

The Ising model-based solver with bSB-DTS4 offers solutions of higher quality (with smaller Ave and Max) than that with bSB-EA1. Compared with the recent SA [33], due to the parallel spin update and faster convergence of Hamiltonian, bSB-DTS4 can significantly improve the Ave and the Std by 42% and 66%, respectively, with about 7.34× shorter runtime in solving $burma14$. It further reduces the Ave and the Std by 37% and 62%, respectively, with about 13.44× shorter runtime in solving $ulysses16$. Finally, it reduces the Ave and the Std by up to 47% and 67%, respectively, with about 7.69× shorter runtime in solving $ulysses22$. A similar solution quality can be obtained by using the SA [33] if the iteration increases, but at the cost of more than 100× runtime.

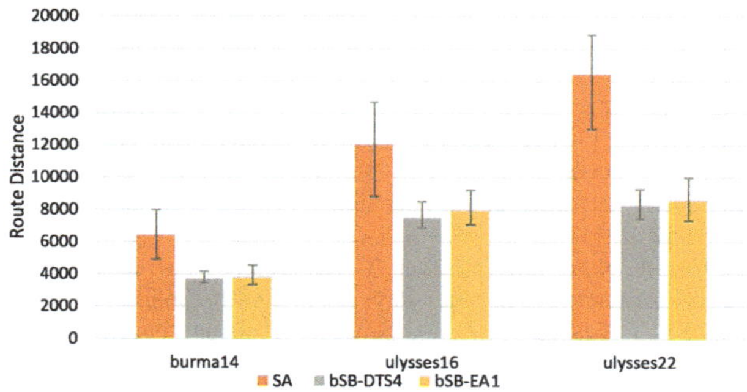

Fig. 5 Comparison of the TSP solvers using the improved SA [33] and the bSB

6 Conclusions

This chapter presents efficient solutions for constrained combinatorial optimization problems such as the TSP using Ising machines with parallel update. Specifically, an improved parallel annealing algorithm and an improved simulated bifurcation algorithm are discussed. The improved parallel annealing leverages an exponential temperature function with a dynamic offset and a k-medoids clustering approach. The exponential temperature function with a dynamic offset can alleviate the problem of being stuck in local minima, while the k-medoids clustering significantly reduces the average travel distance. In the improved simulated bifurcation, the external magnetic fields are first considered as the coefficients between spins by introducing a redundant spin as the position of an oscillator in simulated bifurcation. In this way, the TSP is mapped to an Ising problem without external magnetic fields. Then, the improved simulated bifurcation utilizes dynamic configurations of time steps and evolutions of a redundant oscillator position for efficiently updating the spins' states by leveraging the features of simulated bifurcation. Due to the massive parallel processing capacity, the improved algorithms are at least an order of magnitude faster than the conventional annealing algorithms to find a similar average travel distance for solving the TSP. These results pave the way for the development of energy-efficient circuit architectures for solving combinatorial optimization problems using the Ising model. The hardware design of a parallel fully connected Ising machine can be carried out in future work.

Acknowledgments The work was supported by the University of Alberta (Project Number: RES0049590), the Natural Sciences and Engineering Research Council (NSERC) of Canada (Project Numbers: RES0048688, RES0051374 and RES0054326) and Alberta Innovates (Project Number: RES0053965). T. Zhang is supported by a Ph.D. scholarship from the China Scholarship Council (CSC).

References

1. A.S. Andrae, T. Edler, On global electricity usage of communication technology: trends to 2030. Challenges **6**(1), 117–157 (2015)
2. M.W. Johnson, M.H.S. Amin, S. Gildert, T. Lanting, F. Hamze, N. Dickson, R. Harris, A.J. Berkley, J. Johansson, P. Bunyk et al., Quantum annealing with manufactured spins. Nature **473**(7346), 194–198 (2011)
3. Z. Wang, A. Marandi, K. Wen, R. L. Byer, Y. Yamamoto, Coherent Ising machine based on degenerate optical parametric oscillators. Phys. Rev. A **88**(6), 063853 (2013)
4. K. Tatsumura, M. Yamasaki, H. Goto, Scaling out Ising machines using a multi-chip architecture for simulated bifurcation. Nat. Electron. **4**(3), 208–217 (2021)
5. K. Yamamoto, K. Kawamura, K. Ando, N. Mertig, T. Takemoto, M. Yamaoka, H. Teramoto, A. Sakai, S. Takamaeda-Yamazaki, M. Motomura, STATICA: A 512-spin 0.25 M-weight annealing processor with an all-spin-updates-at-once architecture for combinatorial optimization with complete spin–spin interactions. IEEE J. Solid State Circuits **56**(1), 165–178 (2020)
6. N. Hiroshi, K. Junpei, Y. Noboru, M. Toshiyuki, *Description: Third Generation Digital Annealer Technology* (Fujitsu Limited, Tokyo, 2021)
7. R.A. Rutenbar, Simulated annealing algorithms: An overview. IEEE Circuits Dev. Mag. **5**(1), 19–26 (1989)
8. T. Okuyama, T. Sonobe, K.-i. Kawarabayashi, M. Yamaoka, Binary optimization by momentum annealing. Phys. Rev. E **100**(1), 012111 (2019)
9. H. Goto, K. Endo, M. Suzuki, Y. Sakai, T. Kanao, Y. Hamakawa, R. Hidaka, M. Yamasaki, K. Tatsumura, High-performance combinatorial optimization based on classical mechanics. Sci. Adv. **7**(6), eabe7953 (2021)
10. A. Lucas, Ising formulations of many NP problems. Front. Phys. **2**, 5 (2014)
11. F. Glover, G. Kochenberger, Y. Du, A tutorial on formulating and using qubo models. Preprint (2018). arXiv:1811.11538
12. T. Zhang, Q. Tao, B. Liu, J. Han, A review of simulation algorithms of classical Ising machines for combinatorial optimization, in *2022 IEEE International Symposium on Circuits and Systems (ISCAS)* (IEEE, 2022), pp. 1877–1881
13. T. Shirai, S. Tanaka, N. Togawa, Guiding principle for minor-embedding in simulated-annealing-based Ising machines. IEEE Access **8**, 210,490–210,502 (2020)
14. D. Oku, K. Terada, M. Hayashi, M. Yamaoka, S. Tanaka, N. Togawa, A fully-connected Ising model embedding method and its evaluation for CMOS annealing machines. IEICE Trans. Inf. Syst. **102**(9), 1696–1706 (2019)
15. K. Tamura, T. Shirai, H. Katsura, S. Tanaka, N. Togawa, Performance comparison of typical binary-integer encodings in an Ising machine. IEEE Access **9**, 81,032–81,039 (2021)
16. S. Kirkpatrick, C.D. Gelatt Jr, M.P. Vecchi, Optimization by simulated annealing. Science **220**(4598), 671–680 (1983)
17. T. Okuyama, M. Hayashi, M. Yamaoka, An Ising computer based on simulated quantum annealing by path integral monte carlo method, in *2017 IEEE International Conference on Rebooting Computing (ICRC)* (IEEE, 2017), pp. 1–6
18. E.S. Tiunov, A.E. Ulanov, A. Lvovsky, Annealing by simulating the coherent Ising machine. Opt. Exp. **27**(7), 10,288–10,295 (2019)
19. H. Goto, Bifurcation-based adiabatic quantum computation with a nonlinear oscillator network. Sci. Rep. **6**(1), 1–8 (2016)
20. H. Goto, K. Tatsumura, A.R. Dixon, Combinatorial optimization by simulating adiabatic bifurcations in nonlinear hamiltonian systems. Sci. Adv. **5**(4), eaav2372 (2019)
21. S. Kanamaru, K. Kawamura, S. Tanaka, Y. Tomita, H. Matsuoka, K. Kawamura, N. Togawai, Mapping constrained slot-placement problems to Ising models and its evaluations by an Ising machine, in *2019 IEEE 9th International Conference on Consumer Electronics (ICCE-Berlin)* (IEEE, 2019), pp. 221–226

22. K.L. Hoffman, M. Padberg, G. Rinaldi et al., Traveling salesman problem. Encyclopedia Oper. Res. Manag. Sci. **1**, 1573–1578 (2013)
23. A. Dan, R. Shimizu, T. Nishikawa, S. Bian, T. Sato, Clustering approach for solving traveling salesman problems via Ising model based solver, in *2020 57th ACM/IEEE Design Automation Conference (DAC)* (2020)
24. Q. Tao, J. Han, Solving traveling salesman problems via a parallel fully connected Ising machine, in *Proceedings of the 59th ACM/IEEE Design Automation Conference* (2022), pp. 1123–1128
25. T. Takemoto, K. Yamamoto, C. Yoshimura, M. Hayashi, M. Tada, H. Saito, M. Mashimo, M. Yamaoka, 4.6 a 144kb annealing system composed of 9 × 16kb annealing processor chips with scalable chip-to-chip connections for large-scale combinatorial optimization problems, in *2021 IEEE International Solid- State Circuits Conference (ISSCC)* (2021)
26. M. Aramon, G. Rosenberg, E. Valiante, T. Miyazawa, H. Tamura, H.G. Katzgraber, Physics-inspired optimization for quadratic unconstrained problems using a digital annealer. Front. Phys. **7**, 48 (2019)
27. S. Matsubara, M. Takatsu, T. Miyazawa, T. Shibasaki, Y. Watanabe, K. Takemoto, H. Tamura, Digital annealer for high-speed solving of combinatorial optimization problems and its applications, in *2020 25th Asia and South Pacific Design Automation Conference (ASP-DAC)* (2020)
28. S. Tsukamoto, M. Takatsu, S. Matsubara, H. Tamura, An accelerator architecture for combinatorial optimization problems. Fujitsu Sci. Tech. J. **53**(5), 8–13 (2017)
29. H.-S. Park, C.-H. Jun, A simple and fast algorithm for k-medoids clustering. Expert Syst. Appl. **36**(2), 3336–3341 (2009)
30. http://comopt.ifi.uni-heidelberg.de/software/TSPLIB9
31. T. Zhang, Q. Tao, J. Han, Solving traveling salesman problems using ising models with simulated bifurcation, in *2021 18th International SoC Design Conference (ISOCC)* (IEEE, 2021), pp. 288–289
32. T. Zhang, J. Han, Efficient traveling salesman problem solvers using the Ising model with simulated bifurcation, in *2022 Design, Automation & Test in Europe Conference & Exhibition (DATE)* (IEEE, 2022), pp. 548–551
33. A. Minamisawa, R. Iimura, T. Kawahara, High-speed sparse Ising model on FPGA, in *2019 IEEE 62nd International Midwest Symposium on Circuits and Systems (MWSCAS)* (IEEE, 2019), pp. 670–673

Approximate Communication in Network-on-Chips for Training and Inference of Image Classification Models

Yuechen Chen, Ahmed Louri, Shanshan Liu, and Fabrizio Lombardi

1 Introduction

Image classification applications widely use deep convolutional neural networks (CNNs) and are deployed from cloud to edge computational frameworks for a variety of scenarios, such as virtual reality and self-driving cars [1, 2]. As the complexity of these applications and the resolution of images continue to increase, conventional homogeneous architectures (such as multicore CPU/GPU) are constrained due to excessive communication latencies and significant power dissipation [3–5]. To efficiently process these applications, heterogeneous architectures have been proposed with preprocessing and accelerator cores [3–8].

- Preprocessing cores are designed to prepare data by resizing the raw image and then normalizing the value for each pixel into a specific range.
- Accelerator cores are designed to fetch the processed data and parameters of the CNN model to perform inference or training.

Network-on-chips (NoCs) have been widely used to efficiently connect cores, memory interfaces, and caches in these architectures [9]. Recent research [3,

Y. Chen (✉) · A. Louri
The George Washington University, Department of Electrical and Computer Engineering, Washington, DC, USA
e-mail: yuechen@gwu.edu; louri@gwu.edu

S. Liu
New Mexico State University, Klipsch School of Electrical and Computer Engineering, Las Cruces, NM, USA
e-mail: ssliu@nmsu.edu

F. Lombardi
Northeastern University, Department of Electrical and Computer Engineering, Boston, MA, USA
e-mail: lombardi@ece.neu.edu

W. Liu et al. (eds.), *Design and Applications of Emerging Computer Systems*,
https://doi.org/10.1007/978-3-031-42478-6_27

10] has shown that with a heterogeneous architecture, data transfer can account for up to 34% of the execution time and up to 40% of the overall chip power consumption. Since image classification applications can tolerate errors in the parameters and the inputs, approximation communication is an emerging design methodology for reducing data transfer, thus reducing network latency and power consumption. Existing approximate communication techniques [11–16] reduce communication latency and power consumption by utilizing packet approximation in NoCs. However, existing techniques only rely on relative error for data approximation. Since relative error tolerance is limited for image classification applications, only a few packets can be approximated using existing approximate communication techniques.

To address the above issues, an approximate communication technique (ACT) that enhances communication efficiency for image classification model training and inference is proposed for heterogeneous systems; it leverages the error-tolerance of the image classification models to reduce the transmitted packet size, thus reducing power consumption and network latency. ACT utilizes two approximate communication schemes: an approximate communication for the preprocessing cores (ACT-P) and an approximate communication for the accelerator cores during inference (ACT-I). The ACT-P and ACT-I utilize the contrast reduction method and quantization method to reduce on-chip communication during model inference. Each scheme includes quality control and data approximation mechanisms to leverage the error tolerance in multiple steps of the image classification process. The proposed ACT technique is implemented with a dual-matrix compression method to take advantage of the sparse matrix during training and inference. Specifically, the contributions of this chapter are as follows.

- The proposed approximate communication technique (ACT) is utilized in the preprocessing cores (ACT-P) and the accelerator cores (ACT-I) to reduce network latency and dynamic power consumption for image classification applications by leveraging the error tolerance of the application.
- The ACT is implemented with software-hardware co-design with a dual-matrix compression method for lower on-chip communication during training and inference.
- Performance evaluation results show that compared to the existing approximate communication techniques, ACT reduces dynamic power consumption and network latency by 35% and 31%, respectively, for training. In terms of inference, the proposed method achieves 27% and 26% reduction in dynamic power consumption and network latency, respectively, with less than 0.99% accuracy loss.

This paper is organized as follows. Section 2 presents a background for the proposed technique; Sect. 3 outlines the basic operational principles of the ACT. The implementation is presented in detail in Sect. 4, while Sect. 5 deals with its extensive evaluation. Section 6 concludes this manuscript.

2 Backgrounds

2.1 Existing Approximate Communication Techniques

Approximate communication is considered to be an effective approach to improve network performance when an application can tolerate errors [11–16]. With a reduced accuracy during communication, approximation techniques significantly reduce network latency and power consumption for on-chip communication.

Figure 1 shows an approximate communication NoC [11–16] implemented in a heterogeneous multicore system [3–8] with an L2 shared cache for inference or training of image classification models. The data approximation module in the network interface reduces the packet size by truncation or lossy compression according to the approximation information, which includes variable error tolerance and type (e.g., integer or floating-point). Consider a cache miss during a memory load or store operation by an X86 CPU for image preprocessing.

- *Miss on load operation*

When a cache miss occurs during a memory load operation, a read request packet is sent to the memory or the shared cache through the NoC. The memory or shared cache uses a read reply packet to send the required data back to the core. If the memory load can be approximated, then the read request also carries approximation information. Subsequently, the data approximation module at the memory/shared cache node reduces the size of the read reply packet accordingly. The approximated read reply packet carries the approximated data and the approximate information to the core. When the approximated read reply reaches the core, the data recovery module recovers the approximated data to the original length in accordance with the approximation information.

- *Miss on store operation*

When a cache miss occurs during a memory store operation, the data is incorporated into a write request packet and sent to the memory or shared cache through the NoC. The data approximation module reduces the size of the write request packet according to the approximation information if the memory store can be approximated. The approximated write request packet carries the approximated data and approximation information to the memory or shared cache node. When the approximated write request reaches the memory or shared cache node, the data recovery module recovers the approximated data to the original length in accordance with the approximation information. After the memory or shared cache has received the data, a write reply is sent back to the core to confirm a successful memory write.

Various data approximation methods [12, 13, 15–17] have been proposed to reduce the packet size according to the approximation information. However, existing techniques achieve limited improvement when models are utilized for image classification because the parameters of the model and the inputs cannot be approximated using methods based on relative error. Figure 2 shows the network

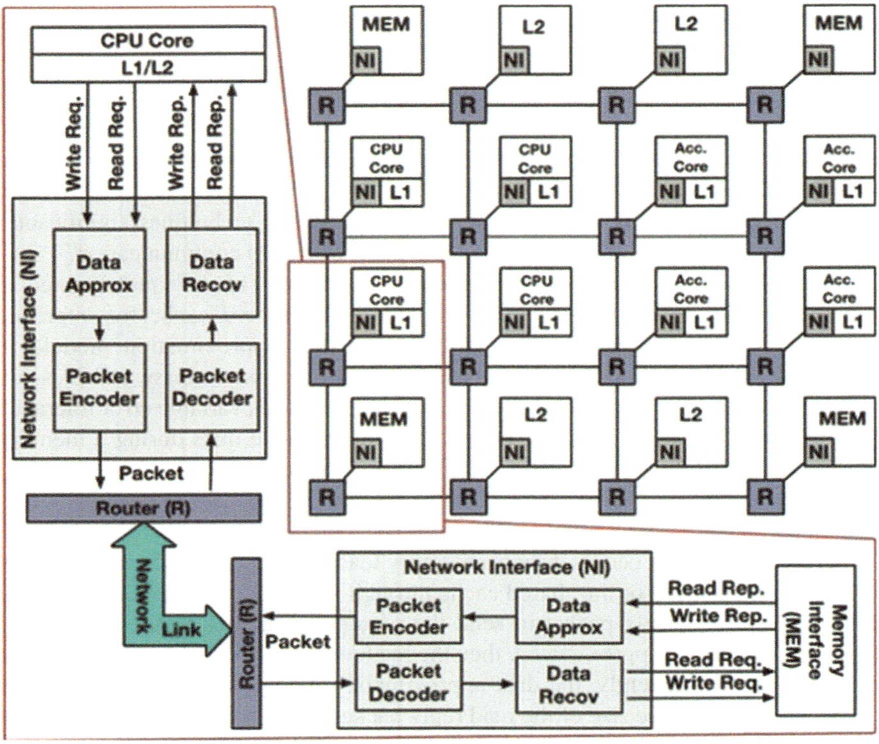

Fig. 1 Heterogeneous multicore architecture with CPU cores, accelerator (Acc.) cores, and approximate communication NoC

latency reduction normalized to the network with no approximation (baseline) for the existing approximate communication framework (ACF) [16]. This figure indicates that ACF only reduces network latency by less than 5% when applied to various state-of-the-art image classification applications [18–29] as executed on heterogeneous multicore architectures with an NoC.

2.2 Existing Sparse Matrix Compression Techniques

The matrix sparsification is a promising approximation scheme for improving training and inference efficiency without substantively affecting accuracy [30–33]. Model and ephemeral sparsification are the two categories of sparsification processes for training.

- The model sparsification process sets selected weights to zeros [31, 32]. The iterative model pruning is widely used as a model sparsification method due to

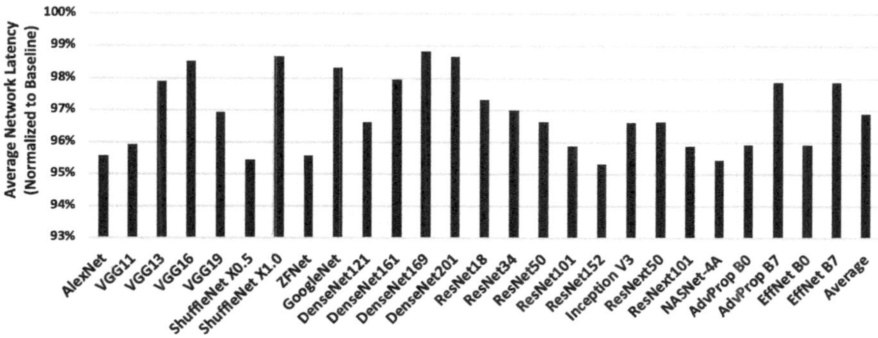

Fig. 2 Network latency reduction by employing the existing approximate communication technique [16] to different image classification applications during inference

its high accuracy and low training cost. First, the model is trained for several epochs to reach a certain accuracy. Then the pruned weights are selected by comparing each weight to a threshold, which is determined based on a user-defined sparsity for this epoch. After pruning, the values of pruned weights are kept at zero in future training epochs. This process continues until reaching the desired sparsity and accuracy.

- The ephemeral sparsification process sets selected activations to zero during each training epoch [33–35]. In terms of ephemeral sparsification, this chapter mainly focuses on the dropout and ReLU functions [36, 37]. The dropout function randomly nullifies activations during the training process, whereas ReLU changes the negative activations to zero. Considering a CNN model with multiple hidden layers, the dropout function randomly sets the activations for a layer to zero for each training sample. The ReLU function sets the negative activation for a layer to zero during FP.

The matrix sparsification provides the potential for highly efficient training and inference by skipping zeros during memory access for efficient on-chip communication [36–38]. Figure 3 shows two widely used compressed sparse matrix formats, namely, compressed sparse row (CSR) [38, 39] and bitmap format (BF) [38, 40]. Both compression formats contain two parts, which are nonzero elements and indices. The indices of CSR feature two integer arrays for column and row indices to indicate the location of the nonzero elements. The indices of BF are represented with a binary bitmap. Using these matrix formats in training and inference eliminates the memory access of the zeros from off-chip DRAM and reduces on-chip communication.

However, relying on either CSR or BF limits the potential savings on the memory traffic during sparse training and inference. The problem is mainly due to the relationship between the size of the indices and matrix sparsity. With a known matrix dimension and sparsity, the BF has a fixed-size bitmap, which is the same dimension as the original matrix. On the other hand, the CSR format depends on the

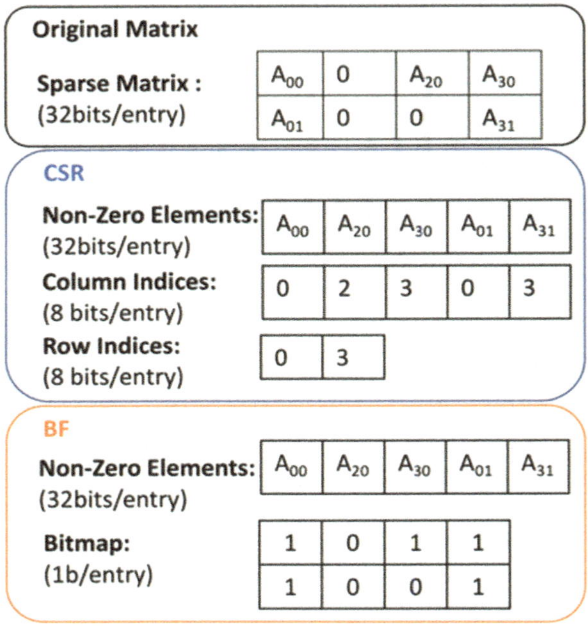

Fig. 3 CSR and BF are used by existing sparse-training accelerators

number of nonzero elements in the matrices. More nonzero elements result in more entries in the column index array. As the matrix sparsity changes during the training process, two different indices formats have different impacts on the memory traffic. For example, when matrix sparsity is low, the CSR format requires more memory access to store indices compared to BF due to a longer column array. When matrix sparsity is high, the CSR format achieves a smaller data size compared to BF if the size of two integer arrays is smaller than the bitmap.

Moreover, the existing sparsification processes result in random zeros with dynamic sparsity in the matrices. Figure 4 shows the distribution of zeros in these matrices when iterative model pruning is applied to Resnet18 with a target pruning rate of 92%. It can be seen that all four matrices are at different sparsity levels during different training epochs. Considering the memory traffic via NoC dominates the computation time and on-chip power consumption, it is essential to develop a compression method with a better compression rate under different matrix sparsity.

3 Proposed Approximate Communication Technique

The proposed approximate communication technique (ACT) reduces network latency and power consumption of on-chip communication in NoCs. This is mainly accomplished by reducing the size of each packet and exploiting the error-tolerant features of image classification applications. The image classification applications

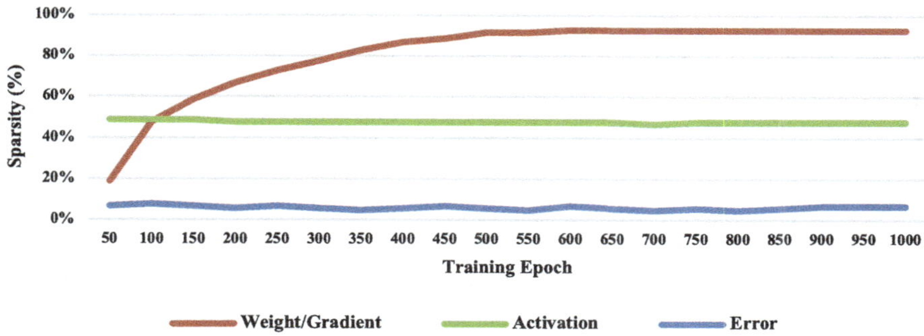

Fig. 4 Matrix sparsity for Resnet18 under iterative pruning

tolerate two types of errors during the inference process [41–43]: The first type is image contrast reduction during image pre-processing; the second type is quantization errors in the fully connected layer during model inference. Thus, ACT includes two sets of approximate communication techniques to leverage the error tolerance of image classification applications.

1. The *approximate communication for image pre-processing* (ACT-P) includes quality control and data approximation mechanisms.
    - The *quality control mechanism* dynamically adjusts the image contrast and monitors the accuracy of the application to balance it with communication efficiency.
    - The *data approximation mechanism* for image preprocessing reduces the data size by reducing the image contrast.
2. The *approximate communication for model inference* (ACT-I) includes quality control and data approximation mechanisms.
    - The *quality control mechanism* monitors the values of the variables when a fully connected layer is processed.
    - After recording the maximum/minimum values of the variables by the quality control mechanism, the *data approximation mechanism* utilizes data quantization to reduce data size.

3.1 Approximate Communication for Image Preprocessing (ACT-P)

Recent research has shown that image classification applications are resilient to contrast reduction on the raw image prior to inference [41, 42]. In this chapter, it is assumed that the level of contrast C ranges from -255 to infinity. When $C = 0$, there is no adjustment to the image, but when $C \in (-255,0)$, the image contrast is

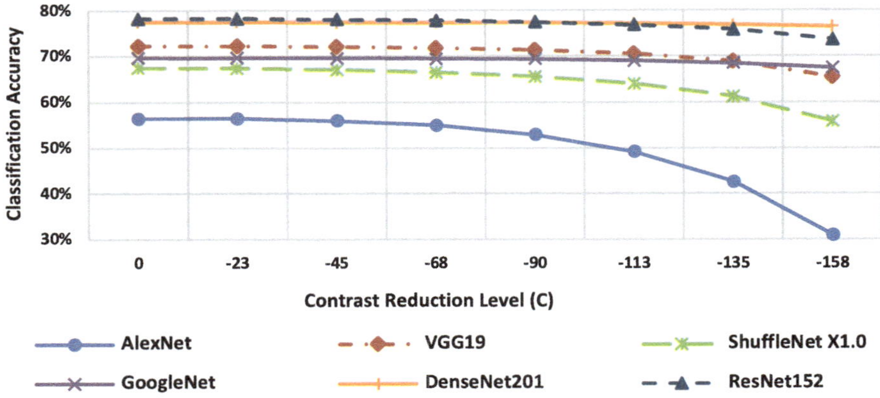

Fig. 5 Classification accuracy versus contrast reduction level (C)

reduced. When $C = -255$, all values of the pixels (R, G, B) in an image are 128, making the image of a solid grey color. Hence, Eq. (1) describes the relationship between the contrast correction factor F and the level of contrast C.

$$F = \frac{259\,(C + 255)}{255\,(259 - C)} \tag{1}$$

As per F above, the contrast reduction for each pixel is performed by Eq. (2), in which the variable P is the value of a color of a pixel (in a range from 0 to 255) and $P^{'}$ represents the corresponding value with contrast reduction.

$$P' = \text{round}\,(F\,(P - 128) + 128) \tag{2}$$

Figure 5 shows the classification accuracy for a few widely-used image classification applications [2] versus the level of contrast reduction; image classification applications can tolerate 23 levels of contrast reduction (i.e., $C = -23$) with negligible accuracy reduction (0.07% accuracy reduction on average). Figure 5 also shows that different image classification applications have different accuracy tolerance for image contrast reduction; for example, for a classification accuracy loss of up to 1%, AlexNet [19] can tolerate 23 levels of contrast reduction, while VGG19 can tolerate −90 levels. Thus, a quality control mechanism is needed to select the appropriate contrast reduction level for the different image classification applications to avoid a significant loss in classification accuracy.

3.1.1 Quality Control

A quality control mechanism for image pre-processing is utilized to maintain the accuracy of image classification. Figure 6 shows the proposed design of the quality

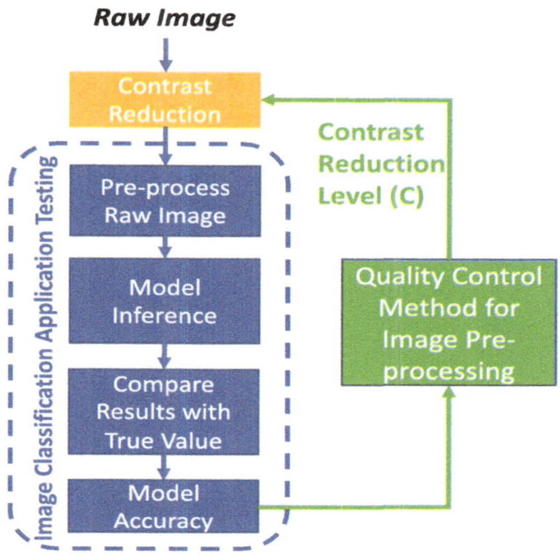

Fig. 6 Proposed quality control mechanism for image pre-processing

control for image classification. This mechanism adjusts the contrast level during the testing process. This is the last step prior to the classification of the images by the application. Testing includes three phases.

1. The raw images in the test data set are processed by the core. Different from the images that the application processes, the data set contains the query data (raw image) and the true value for each image (label).
2. The model inference is then accomplished by fetching the processed data and the classification model.
3. Finally, the generated result is processed by the core to compare it with the true value. The model accuracy is calculated by comparing the predictions generated by the model with the true value.

The quality control mechanism utilizes the accuracy calculated by the core to adjust the image contrast. When considering the potential accuracy reduction caused by applying approximate communication for model inference, the accuracy reduction due to the image contrast reduction is limited to less than 1%. The proposed quality control mechanism supports eight contrast reduction levels that are shown in the left column of Table 1. Thus, the following novel procedure is proposed to determine the image contrast reduction level:

1. During the first phase of the test process, the classification accuracy of the image application is calculated with no image contrast reduction. The classification accuracy is calculated as in Eq. (3).

$$\text{Classification accuracy} = \frac{\text{Number of correct classifications}}{\text{Total number of classificaitons}} \tag{3}$$

Table 1 Relationships between quality control supported contrast reduction level (C) and contrast correction factor (F)

Supported contrast reduction level (C)	Contrast correction factor (F)
0	1
−23	0.835
−45	0.701
−68	0.581
−90	0.480
−113	0.388
−135	0.309
−158	0.236

The correct classification is defined as the image category (e.g., cat, dog, and car) with the highest probability (as predicted by the model); this must be exactly the same as the expected answer (label).

2. The quality control mechanism gradually reduces the image contrast levels by choosing a contrast reduction level according to the left column of Table 1 until there is more than 1% loss (as threshold) in the classification accuracy compared to the estimated base accuracy in the next testing process.
3. The value prior to the last contrast reduction level is chosen for image classification. For example, if the classification error exceeds 1% at level − 68, the lower consecutive level (−45 according to Table 1) is selected for image classification.

Section 5.5 discusses the impact on classification accuracy if different accuracy loss thresholds are chosen. During image classification (i.e., model inference), the true value for each query image is not available, so the contrast reduction level is fixed and registered in the network interface prior to the classification.

3.1.2 Data Approximation

The data approximation mechanism reduces the amount of transmitted data for image contrast reduction. The so-called base-delta approximation mechanism is proposed to take advantage of the reduced image contrast for data reduction. Since the difference in values between pixels in an image is small, the base-delta compression mechanism can significantly reduce the number of bits needed to represent each pixel. Moreover, the proposed image contrast reduction process further reduces the difference between pixels, so the data size can be substantially reduced by only transmitting the difference between pixels. Figure 5 shows the design of the proposed base-delta approximation mechanism for image pre-processing. The data approximation process consists of two steps.

Step 1: The image contrast reduction operation is activated with a contrast reduction level.

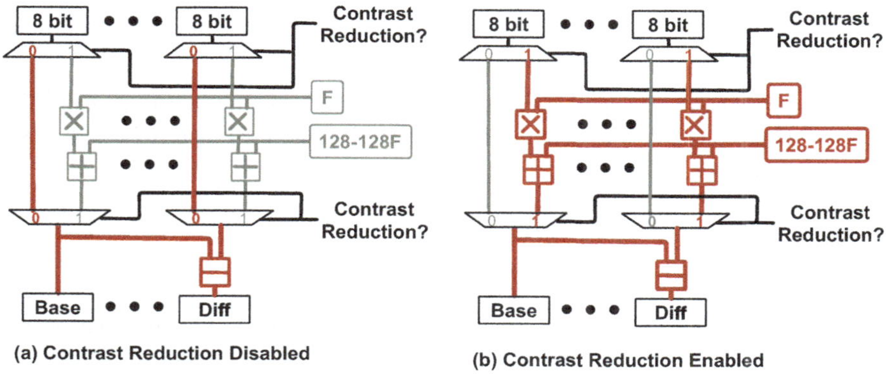

Fig. 7 Design of data approximation for image pre-processing

Step 2: The multipliers and adders then adjust the value for each pixel based on Eqs. (1) and (2).

To reduce its complexity, ACT-P supports eight levels of image contrast reduction. Table 1 shows the mapping of the conversion of the supported contrast reduction level (C) into the contrast correction factor (F). The first 8-bit data is chosen as the base; the remaining data are represented as the distance to the base. Figure 7a shows the approximation process when the image contrast reduction is deactivated (Contrast Reduction Level = 0). The data bypass the contrast reduction operation and is compressed with full accuracy. Figure 7b shows the approximation process when the image contrast reduction is activated.

For example, suppose a packet contains three pixels with values 128 (10000000), 192 (11000000), and 100 (1100100). Also, these values can tolerate −68 levels of contrast reduction. After approximation, the packet only contains 128 (10000000), −38 (1100110), and 6 (110), respectively. After the compressed data arrives at the destination, the data are recovered by adding the delta and the base.

3.2 *Approximate Communication for Model Inference (ACT-I)*

[44] has shown that the classification accuracy reduction in image classification applications is negligible after applying quantization to the parameters and activation of the fully connected layers (inputs). As floating-point data type is widely used in image classification applications [18–29] to represent parameters and activation, the quantization process consists of mapping a floating-point value $x \in [\alpha, \beta]$ to a b-bit integer $x_q \in [\alpha_q, \beta_q]$; this is computed as per Eq. (4) (where c and d are variables).

$$x_q = \text{floor}\left(\frac{1}{c}x - d\right) \tag{4}$$

Note that when performing quantization, the floating-point 0 must be mapped to a b-bit integer 0. Thus, the relationship between c, d and the ranges of x and x_q are given as follows.

$$\begin{cases} \beta = c\left(\beta_q + d\right) \\ \alpha = c\left(\alpha_q + d\right) \end{cases} \tag{5}$$

In Eq. (5), α and β are the minimum and maximum of the floating-point value, respectively. α_q and β_q are the minimum and maximum of the integer value (i.e., quantized floating-point value), respectively. c and d are the two variables that must be solved for the quantization process. Equation (6) illustrates the solution for Eq. (5).

$$\begin{cases} c = \frac{\beta-\alpha}{\beta_q-\alpha_q} \\ d = \frac{\alpha\beta_q-\beta\alpha_q}{\beta-\alpha} \end{cases} \tag{6}$$

[44] has shown that the accuracy reduction caused by quantizing the IEEE standard 32-bit floating-point data into 8-bit integers in fully connected layers is negligible. This is shown in Fig. 8 for different image classification applications in which the parameters and inputs of the fully connected layers are quantized. The quantization process only reduces the classification accuracy by 0.017% on average. This observation suggests that data quantization is an attractive solution to significantly reduce on-chip communication when the model has fully connected layers.

However, as per Eq. (6), the range of x (i.e., α and β) must be considered when performing data quantization. Since data (i.e., x) exceeding the range is

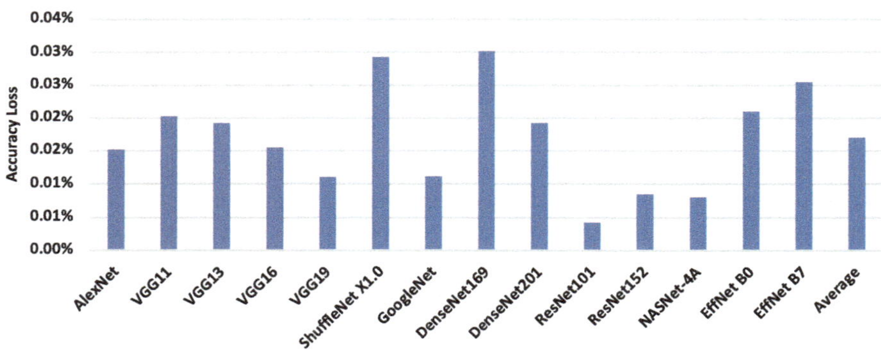

Fig. 8 Quantization of fully connected layers affects classification accuracy

basically clipped (by truncation) during the quantization process, the range must be dynamically determined for different data items. Otherwise, many data items could be clipped, thus negatively impacting the model accuracy. Therefore, a novel quality control mechanism is developed to estimate the range of inputs and parameters in the proposed scheme.

Also, once the data range is determined, floating-point operations (such as multiplication, division, and subtraction) must be performed to map the data (as per Eqs. (4) and (6)). These operations are not always acceptable because they could incur significant communication overheads for latency and power consumption. Therefore, the proposed data approximation mechanism uses a variable i to quantize data, as described next.

3.2.1 Quality Control

Figure 9 shows the proposed process of quality control for model inference. The proposed quality control mechanism constantly monitors the parameters and inputs of the fully connected layer. To reduce the complexity of data quantization, a new variable i is introduced based on the following observations.

Observation 1: Quantization maps data from the original range to another range with different granularity, thus causing quantization errors. For example, when quantizing 32-bit floating-point data into 8-bit integers, the granularity of the data range increases from 1/16777216 to 1/255; in this case, the error originates from the decimal part.

Observation 2: For an integer, a deviation within (0, 1) (i.e., adding the integer with a decimal value) is only reflected on a few lower mantissa bits in its floating-point representation, and it has an almost negligible impact on all upper bits. Moreover, the changed mantissa bits are separated from the upper bits related to the sign and the integer part of the data.

Observation 3: The expansion of a floating-point value by 2^i times only changes its exponent bits (where i is a positive integer). Consider Eqs. (7) and (8) that

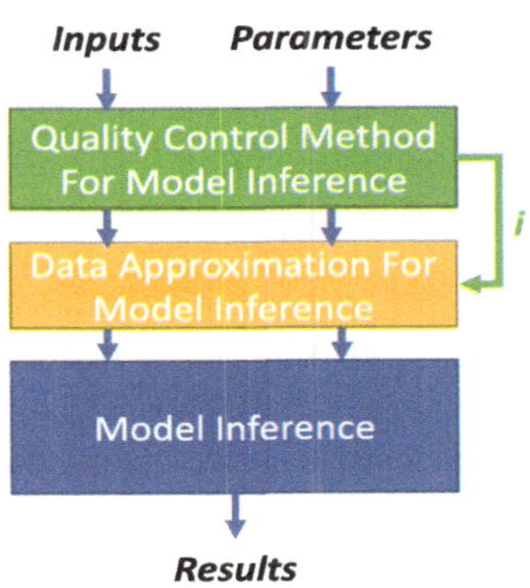

Fig. 9 Quality control mechanism for image classification model inference

calculate the value of a 32-bit floating-point data D [45] and the corresponding enlarged value with 2^i times, respectively; only the exponent value increases by i, while the sign and mantissa remain the same.

$$D = (-1)^{\text{sign}} \cdot 2^{\text{exponent}-127} \cdot (1 + \text{mantissa}) \quad (7)$$

$$D \cdot 2^i = (-1)^{\text{sign}} \cdot 2^{(\text{exponent}+i)-127} \cdot (1 + \text{mantissa}) \quad (8)$$

Therefore, as per the above observations, a conventional data quantization approach can be replaced by expanding the original data by 2^i times and then rounding it down (i.e., mapping the floating-point data to an integer).

Thus, when the quality control mechanism receives the minimum (α) and the maximum (β) values of the weights and biases, i is calculated based on α and the β using Eq. (9).

$$i = \log 2 \left(\max \left(|\alpha|, |\beta|\right)\right) \quad (9)$$

However, the dynamic range of the inputs is not fixed because the query image is changed after each image classification. Thus, the inputs of the fully connected layer are constantly monitored during the image classification to establish the dynamic range of the input to calculate i.

To reduce the hardware overhead, i is limited to 8 bits, and the initial i values for the inputs and parameters are calculated during the image classification application testing and registered in the network interface before processing images. Since the increase of the value range leads to a decrease of i, the quality control mechanism automatically reduces i by 1 when an input value exceeds the dynamic range during the classification. Then, the data approximation mechanism reduces the data size based on i.

3.2.2 Data Approximation

The proposed data approximation mechanism quantizes data by enlarging the original data by 2^i times and then rounding it down. Thus, the quantization error is bounded within a few lower mantissa bits. Only the sign bit, the exponent bits, and a few upper mantissa bits need to be transmitted because they are separated from the lower bits. Moreover, to perform the multiplication with 2^i, a binary sequence of i needs to be added to the exponent part, that is, only a binary addition operation is required, rather than floating-point arithmetic operations. As the ranges of inputs and parameters of the fully connected layers can be determined by utilizing the quality control mechanism proposed in the previous section, then the value of i is adjusted to guarantee that the quantized data belong to an integer range with an acceptable granularity (so to provide a good classification accuracy, such as the 8-bit integer range [43]).

Table 2 Mapping between 8-bit exponent patterns and 3-bit symbols

Integer	Exponent patten	Symbol
0	00000000	000
2^0	01111111	001
$[2^1,2^2)$	10,000,000	010
$[2^2,2^3)$	10,000,001	011
$[2^3,2^4)$	10,000,010	100
$[2^4,2^5)$	10,000,011	101
$[2^5,2^6)$	10,000,100	110
$[2^6,2^7)$	10,000,101	111

To further reduce the size of the transmitted data, the exponent part is compressed by mapping the data patterns into symbols with a shorter length. Since all integers within the range of $[2^j, 2^{j+1})$ share the same exponent pattern as per Eq. (7) (where $j = 1, 2, ..., 127$), then only a few patterns are used for representing the quantized data that belong to a range significantly smaller than the entire floating-point field. For example, when quantizing data into the 8-bit integer range, only eight exponent patterns may appear (i.e., 00000000 for value 0, 01111111 for value 2^0, 10,000,000 for values within $[2^1, 2^2)$, 10000001 for values within $[2^2, 2^3)$, etc.). In this case, 3-bit symbols that provide eight different combinations can be used for mapping all possible exponent patterns (and so transmitted), thus reducing the size for each exponent from 8 to 3 bits.

Next, an example of quantizing 32-bit floating-point values within [−0.0903250, 0.0882086] (i.e., the range of parameters for one fully connected layer of the trained VGG11 [20]) into 8-bit integers is shown as an application of the proposed mechanism. To perform quantization, the original data are enlarged by 2^{10} (i.e., $i = 10$) times because $i = 11$ makes the quantized data exceeding the range of $[-127, 127]$. Therefore, only 15 bits, including 1 sign bit, 8 exponent bits, and 6 upper mantissa bits, are sufficient for all quantized data. Then, the exponent part is further compressed by utilizing 3-bit symbols as per the mapping given in Table 2 because only eight exponent patterns are used to represent the quantized data. Overall, the data size is reduced from 32 bits to 10 bits by employing the proposed mechanism, achieving a reduction of 68.75%.

The hardware design for the proposed data approximation mechanism for data quantization is illustrated in Fig. 10. Once data approximation is enabled, an 8-bit adder is utilized to perform the binary addition between the original exponent and the binary sequence of i obtained by the quality control logic for quantization (i.e., enlarge the data by 2^i times); then the mapping hardware compresses the quantized exponent (Table 2) to further reduce the data size. Finally, the approximated data are sent to the packet encoder for transmission. Note that once the data arrives for neuron computation, the compressed exponent is decompressed according to Table 2, and a few bits with 0 are padded to the mantissa to recover the format of the quantized data back into the standard 32-bit floating-point format for subsequent computation.

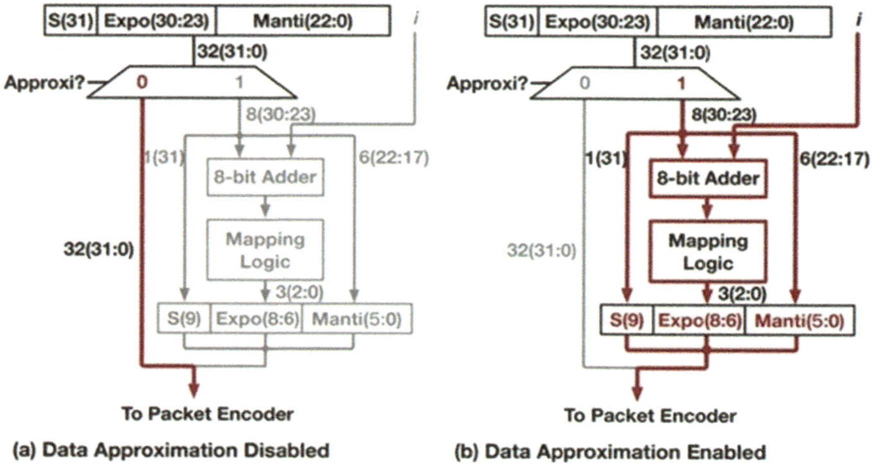

Fig. 10 Hardware design of data approximation for model inference

4 Implementation of the Approximate Communication Technique (ACT)

An architecture based on hardware-software co-design is proposed in this section to implement ACT for image classification applications. The proposed implementation includes a software interface and an architectural design. The software interface is designed to identify the variables that need to be monitored or approximated during image classification. The network interfaces in the heterogeneous architecture are augmented with data approximation, quality control, and sparse matrix compression.

4.1 Dual-Matrix Compression Method for Sparse Matrix

The proposed compression method is targeted to reduce on-chip traffic for efficient training and inference. The design of proposed method is based on the following observations.

Observation 1: Matrix sparsification processes are widely used in both training and inference.

Observation 2: Existing CSR or BF compression methods has limited compression rate under certain matrix sparsity.

For lower on-chip communication, the proposed compression method is implemented with two features. (1) Row-partitioned CSR (RP-CSR) and row-partitioned

BF (RP-BF) formats are developed for a unified data format, which allows for seamless switching between the two formats. (2) A selection algorithm is developed to choose the compression format with a smaller size during training or inference for lower off-chip memory traffic.

4.1.1 Selection Algorithm Design

Equation (10) illustrates the total number of bits for a CSR format with a given sparsity $s \in [0, 1]$ and height (Y) / width (X) of a matrix. Assuming each non-zero element is 32-bits and 8-bits for each entry in column index and row index arrays.

$$\text{Size}_{\text{CSR}} = \text{Size}_{\text{non}} - \text{zero} + \text{Size}_{\text{col}}\text{index} + \text{Size}_{\text{row}}\text{index} \tag{10}$$

$$\text{Size}_{\text{non-zero}} = 32 \times XY \times (1 - s) \tag{11}$$

$$\text{Size}_{\text{col}}\text{index} + \text{Size}_{\text{row}}\text{index} = 8XY \times (1 - s) + 8Y \tag{12}$$

Equation (13) illustrates the total number of bits for a BF.

$$\text{Size}_{\text{BF}} = \text{Size}_{\text{non-zero}} + \text{Size}_{\text{bitmap}} \tag{13}$$

$$\text{Size}_{\text{bitmap}} = XY \tag{14}$$

From Eqs. (11), (12), and (14), the ratios of nonzero elements size (Eq. (11)) to index size (Eqs. (12) and (14)) for both formats are based on the matrix sparsity. Therefore, a unified on-chip SRAM design, which stores both nonzero elements and indices, is preferred, as the matrices are not limited by either the size of indices or the size of nonzero elements. With unified SRAM, the accelerator is able to achieve a better data reuse rate and lower off-chip memory access.

From Eqs. (10) and (13), the size of a compressed sparse matrix depends on the dimension of the matrix and sparsity. To find out which compression format achieves a smaller size under a given sparsity, the equation ($\text{Size}_{\text{CSR}} = \text{Size}_{\text{BF}}$) is solved to find the threshold sparsity. Equation (15) shows the solution for the equation.

$$S_{\text{threshold}} = 0.875 + \frac{1}{X} \tag{15}$$

Considering the width of a matrix (X) is very large during training, the threshold for sparsity is approximated to 88%. Figure 11 shows the matrix size reduction for CSR and BF of a matrix with a dimension of 500 width and 500 height under

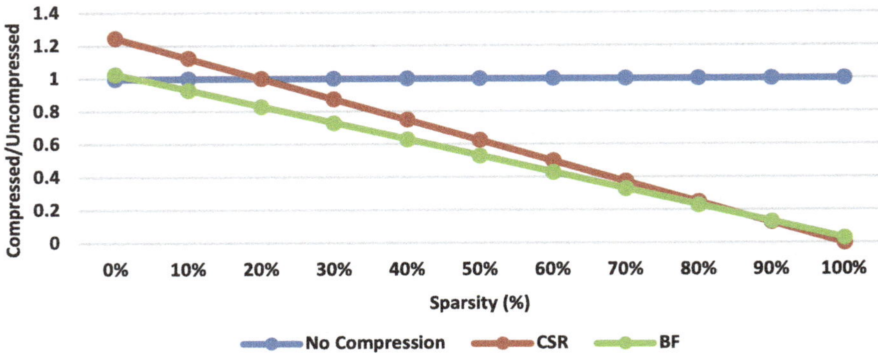

Fig. 11 Matrix size reductions under different sparsity for CSR and BF

different sparsity. It can be seen that when matrix sparsity is less than 88%, BF is smaller in matrix size.

Equation (15) also implies that the threshold is not affected by the height (Y) of the matrix. This indicates: If the matrix is row-partitioned, each row can determine its own compression format by comparing the sparsity of the row with the threshold.

This further increases the compression rate, as the location of zeros in all the matrices is dynamically changing during sparse training. Therefore, the selection algorithm is described as follows. When the sparsity of a row is lower than 88% ($S_{\text{row}} \leq S_{\text{threshold}}$), the RP-BF format is selected for better compression rate, otherwise ($S_{\text{row}} > S_{\text{threshold}}$) the RP-CSR format is selected.

4.1.2 Row-Partitioned Compression Format

Although supporting dual matrix compression achieves a higher compression rate across all sparsity levels, seamlessly switching between two data formats is a major challenge for the proposed implementation. By analyzing the CSR, it can be seen that it is naturally partitioned by row according to the row index array. On the other hand, the indices of BF are not partitioned. Thus, to seamlessly switch between two formats, BF also needs to be row partitioned to match the data format of CSR.

Figure 12 shows the proposed row-partitioned CSR (RP-CSR) and row-partitioned BF (RP-BF) with an example, which is based on the same sparse matrix shown in Fig. 3. Compared to the existing BF (Fig. 3), the RP-BF puts the bitmap at the beginning of each row with non-zero elements. Compared to existing CSR (Fig. 3), the RP-CSR places the column index at the beginning of each row, then the row indices, followed by the nonzero elements. The proposed row-partitioned compression format provides the following benefits compared to conventional compression formats (i.e., CSR and BF).

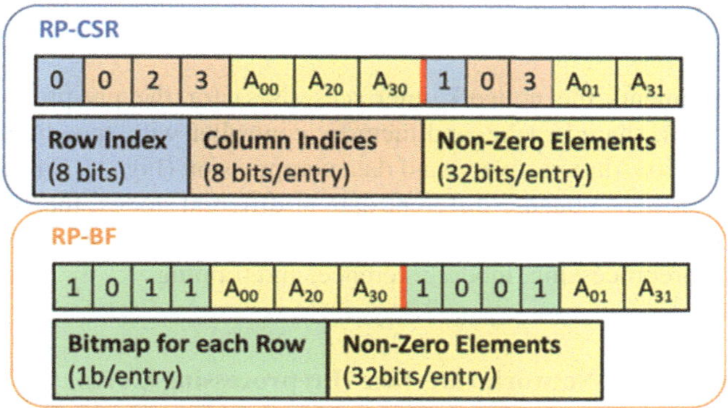

Fig. 12 Proposed RP-CSR and RP-BF formats with examples

1. The row-partitioned compression formats only change the sequence of the data, and it does not add extra bits to the packet.
2. The proposed formats allow mixed RP-CSR and RP-BF for the same matrix, as every row is independent of the other. This feature not only allows seamless format switching, but also further increases the compression rate with less off-chip memory access.

4.2 Software Interface for Approximate Communication

ACT approximates pixels in the images when the pre-processing cores convert the raw image. Also, ACT quantizes the inputs and parameters when the inference cores process the fully connected layers. Hence, ACT monitors and approximates the pixels, inputs, and parameters when the image classification application is executed on the heterogeneous architecture. Two specialized instructions are developed to identify these variables in the source code and the on-chip communication. When the application designer programs the preprocessing cores, the variables (which store the images) are separately annotated in the application. For the preprocessing cores that are X86 CPUs, once the program is compiled into X86 instructions, then the load-and-store of an image pixel (*mov dist, src*) is replaced with (*amov dist, src*) for the network interface to identify the image pixels that can be approximated. Similarly, the loading process of the parameters and the inputs for the fully connected layer (*ld dist, src*) are replaced using specialized instructions (*ald dist, src*). During the execution of an application, these new instructions allow the network interface to identify these variables in the requests or replies.

4.3 Architecture Design of ACT

The ACT arguments the network interfaces (NIs) for the pre-processing cores, accelerator cores, shared cache, and memory controller with specific hardware for data approximation, data recovery, and data compression (Fig. 1). Since the approximation logic needs to handle different data at different nodes, the approximation and recovery logics are specifically designed according to the functionality of the node, such as preprocessing, model inference, and training.

4.3.1 Approximate Network Interface (Preprocessing Cores)

To support the ACT-P, the data approximation logic approximates image pixels according to the contrast reduction level. Since images must be processed by the preprocessing core, the write requests and read replies carry image pixels and data in these packets can be approximated.

Figure 13 shows the proposed approximation logic for the pre-processing core. The approximation logic includes the data approximation logic and the quality control logic to adjust the image contrast. The design of the data approximation logic for a pre-processing core is described in Sect. 3.1.2. For clarity, only the control signal for the quality control logic is shown in Fig. 13. The quality control logic monitors the write requests. If the write requests contain raw images, then the quality control logic instructs the data approximation logic to approximate the requests according to the current contrast reduction level. 3 bits are used to represent the contrast reduction level to support 8 contrast reduction levels (0 to -158). If the write request cannot be approximated, the data approximation logic applies base-delta compression without contrast reduction (level 0). Then the quality control logic checks the length of the write requests. If the length is larger than the original write request (Approx. Size > Org. Size), the original request is sent to the packet encoder. Once the memory or shared cache has received the data, a write reply is sent back to the core to confirm a successful memory write.

During the image load, the quality control logic attaches the information of contrast reduction mode (3 bits) to the read requests. Once the read reply packet arrives at the core, the data recovery logic recovers the data into its original form if the packet is compressed. Otherwise, the data recovery logic directly sends the read reply to the core. The data recovery logic for the preprocessing cores decompresses the data by adding the delta back to the base.

Since the cores neither apply matrix sparsification during image preprocessing nor read sparse matrices from memory, the data in memory write and read packets have a limited number of zeros. Thus, the dual-matrix compression method is not implemented with the approximation logic and recovery logic for the preprocessing cores.

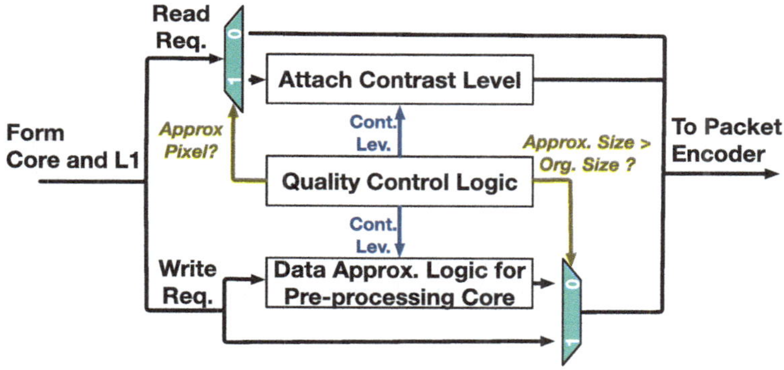

Fig. 13 Approximation logic for preprocessing cores

4.3.2 Approximate Network Interface (Accelerator Cores)

Since the core directly loads and stores data from/to memory or shared cache, the read and write requests are generated by the node and sent to the memory controller or shared cache. To support ACT-I, the data approximation logic monitors the write requests and read replies to update the dynamic range of the parameters and the inputs for the fully connected layer.

Figure 14 shows the proposed approximation logic for the model inference and training. The quality control logic monitors all requests and replies to update i for the inputs; it also controls two demultiplexers and the data approximation logic. Since the destination of the write request could be another node for model inference or a memory controller or a shared cache, i (monitored at a specific node) can be the dynamic range of a section of the inputs for the fully connected layer. To find the dynamic range of the inputs for the entire layer, the following procedure is proposed. (1) The quality control logic attaches i of the inputs to the read request packet if the destination of the packet is the memory controller or shared cache. (2) The quality control logic constantly monitors the i of the write reply packets from the memory controller or shared cache. If the received i is smaller than the current i, the value of i for the inputs in the current node is updated. Therefore, the i in the memory controller and shared cache node has the dynamic range of all the inputs when the core loads the data. When the core stores the result, the i in each node is updated with the i in the memory controller and shared cache.

As an accelerator core needs to fetch images, parameters, and inputs, the data recovery logic contains two decompression functions. The decompression function for images is the same function used in the preprocessing core. The decompression function for the parameters and inputs recovers the data based on Table 2, and a few bits (of values 0's) are padded to the mantissa to recover the format of the quantized data back into the standard 32-bit floating-point format for subsequent computation.

Since accelerator cores contain matrix sparsification units, the proposed dual-matrix compression logic is implemented in the network interface to further

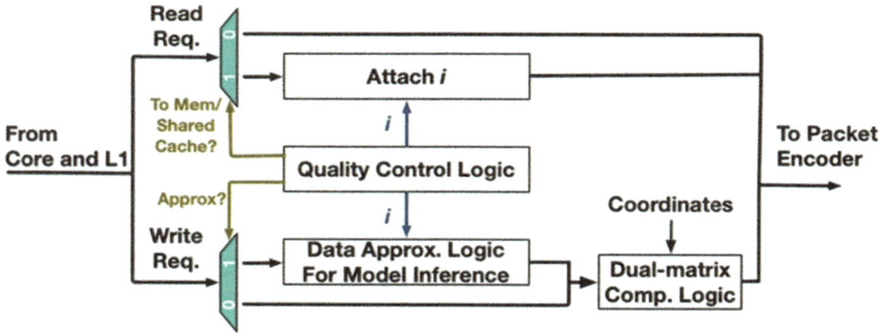

Fig. 14 Approximation logic for accelerator cores

reduce the on-chip communication. Since the proposed compression method needs information on the coordinates of the values, the core provides the coordinates of the values for each element in the packets. Then the matrix compression unit decided which compression format to use based on the sparsity of the matrix. For the read reply packet, the decompression unit is added before the data are sent to the data recovery unit. Since the proposed sparse matrix compression unit only eliminates the zeros in the data packets, the training and inference accuracy is not affected.

4.3.3 Approximate Network Interface (Memory Controller and Shared Cache)

Since the memory controller and shared cache handles requests from both preprocessing and accelerator cores, this interface performs data approximation and recovery functions for both tasks. Also, the network interface carries the quality control logic for both preprocessing, inference, and training.

Figure 15 shows the approximation logic for the memory controller and shared cache. The approximation logic consists of data approximation and quality control logic. The quality control logic monitors the read request packets for the i value from the node for inputs. If the i value is smaller than the value stored in the quality control logic, the stored i is updated. The updated i is attached to the write replies to update i stored in the network interface at the node for model inference. The quality control logic also monitors the read request packets for receiving the contrast level for the read reply packet approximation. When the read reply has the data for image preprocessing or model inference, the corresponding data approximation logic is activated to approximate the data based on the contrast level or i. Similar to the quality control logic in the preprocessing core, the quality control logic checks the length of the read reply to the preprocessing core; if the length is greater than the original read reply after base-delta compression, the original reply is sent to the packet encoder. Since the traffic contains the pixels, model parameters, and inputs,

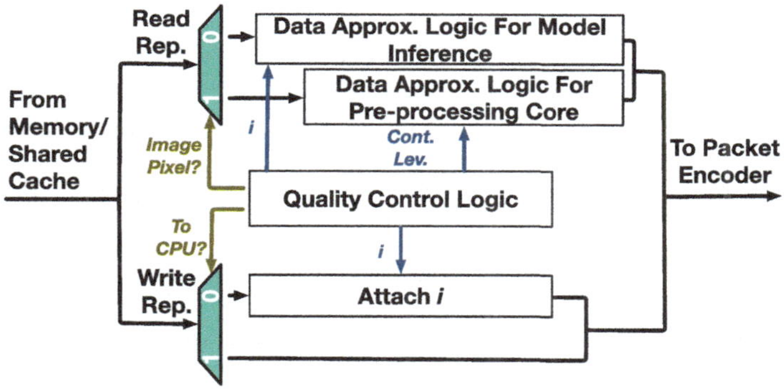

Fig. 15 Approximation logic for memory controllers and shared caches

the data recovery logic has recovery functions for both model inference and pre-processing.

Considering that the memory controller and shared cache store the data, the compressed sparse matrices can be stored in the memory for lower memory consumption. Thus, the decompression process is not necessary for the network interface. Compared to the approximation logic in other nodes, the approximation logic in the current node needs to identify and approximate nonzero values in the packet.

5 Evaluation

In this section, the performance of the approximate communication technique (ACT) is evaluated by using the SMAUG [3] simulator. The SMAUG simulation model is modified to support the ACT and heterogeneous architectures for image classification. Table 3 shows the settings for the SMAUG simulator. The hardware for data approximation, data recovery, and quality control is implemented in the network interface. The heterogeneous architecture [3, 4] is modified for the training and inference of image classification models. The heterogeneous architecture is based on Simba [4] with matrix sparsification unit. All the cores are connected using 6×6 2D mesh NoC. Table 4 shows the executed image classification models with their classification accuracy [18–29] and the corresponding contrast reduction levels (C).

For model inference, we evaluate the proposed technique by comparing it with approximate communication framework (ACF) [16], Approx-NoC [12], AxBA [15], and the baseline (i.e., NoC with no approximation and sparse matrix compression) from the communication efficiency perspective, which includes network latency and dynamic power consumption. For model training, we evaluate the proposed technique by comparing it with CSR [39], BF [40], and ACF [16].

Table 3 Simulation environment

Architectures	CPU/NDLA
Preprocessing cores	X86 CPU * 8
Model-inference cores	NVIDIA Deep Learning Accelerator(NDLA) * 28 [46]
NoC parameter	Network type: Garnet; topology: 6 × 6 2D mesh; data packet size: 5 flits; link width: 128 bits; routing algorithm: X-Y routing; flow control: wormhole switching; number of router pipeline stage: 6
System parameter	32 kB L1 instruction cache; 32 kB L1 data cache; 8-bank fully shared 16 MB L2 cache
Data set	ImageNet Large Scale Visual Recognition Challenge [2]
Approximation techniques for inference	Approximate communication Framework(ACF) [16]; Approx-NoC [12]; AxBA [15]; proposed technique (ACT)
Sparse matrix compression technique	SCR [39]; BF [40]; proposed dual-matrix compression (ACT)

Table 4 Image classification models

Name	Accuracy	C	Name	Accuracy	C
AlexNet [19]	56.55%	−45	DensNet169 [18]	77.20%	−158
VGG11 [20]	69.02%	−68	DensNet201 [18]	77.65%	−135
VGG13 [20]	69.93%	−68	ResNet101 [23]	77.37%	−45
VGG16 [20]	71.59%	−68	ResNet152 [23]	78.31%	−45
VGG19 [20]	72.38%	−90	NASNet-4A [27]	74.00%	−135
ShuffleNet X1.0 [21]	67.60%	−45	EfficientNet B0 [29]	76.30%	−68
GoogleNet [22]	69.78%	−113	EfficientNet B7 [29]	84.40%	−23

5.1 Network Latency

Network latency is defined as the number of clock cycles elapsed between sending a packet at the source node and the successful delivery of the packet to the destination. Thus, the network latency includes the time of three procedures: packet generation at the source node, packet transmission in the network, and data extraction at the destination node. Next, ACT is compared with the baseline, ACF, Approx-NoC, and AxBA.

- *The Heterogeneous system for model inference*

Figure 16 shows the results for the network latency normalized with respect to the baseline. ACT achieves an average network latency reduction of 29% and 26% compared to the baseline and ACF, respectively. This occurs because image classification applications have limited tolerance to the relative error for a smaller reduction in data size compared to ACT. Moreover, the duel-matrix compression method

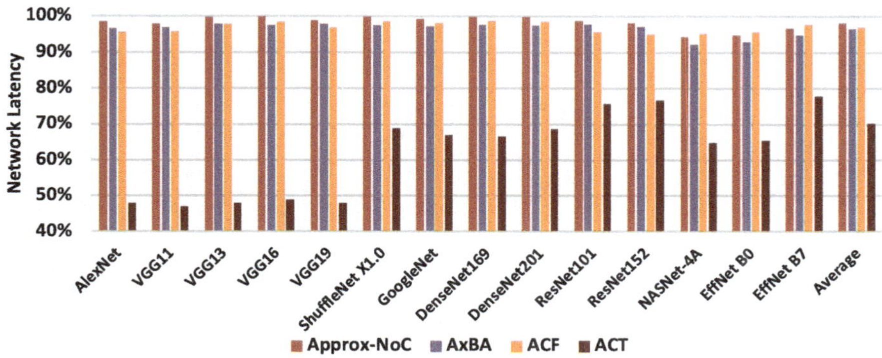

Fig. 16 Network latency for model inference (normalized to the baseline)

further reduces the on-chip communication for sparse activation matrices after ReLU. The largest network latency reduction achieved by ACT in the experiment is VGG11 (53% reduction compared to the baseline), while the smallest network latency improvement is obtained for EffcientNet B7 (22% reduction compared to baseline).

- *The Heterogeneous system for model training*

Figure 17 shows the results for the average network latency normalized with respect to the baseline for the heterogeneous system. The pruning approximation method is applied and executed by the matrix sparsification units during the training with a target pruning rate of 95%. The ACT achieves an average network latency reduction of 31% compared to the ACF. The significant improvement in network latency is mainly due to the fact that ACF lacks the capability of compressing sparse matrices. Compared to the system using CSR and BF for sparse matrix compression, the ACT achieves 14% and 16% reduction in network latency, respectively. This occurs due to the better sparse matrix compression method, which achieves a high compression rate under any matrix sparsity.

Compared to the baseline, existing approximate communication techniques (e.g., Approx-NoC, AxBA, and ACF) achieve marginal improvement in network latency (less than 5% on average), as these techniques only rely on the relative error to approximate data. As a result, existing techniques miss the opportunity of data approximation for image classification applications; however, ACT can achieve a significant latency reduction due to the proposed approximate communication scheme. Moreover, the proposed technique significantly reduces the network latency when the model frequently uses the fully connected layer and can tolerate a significant image contrast loss. For example, Fig. 18 shows the size of the fully connected layer in the image classification models. VGG11 uses 86% of the data, which includes inputs and parameters for the fully connected layers. As Table 4 shows that VGGs can tolerate −68 levels of contrast reduction ($C = -68$) with minimal accuracy loss, then the combined effect of two packet approximation

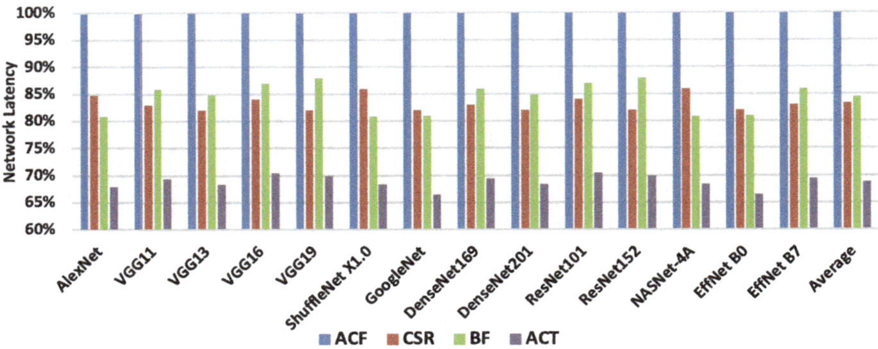

Fig. 17 Network latency for model training (normalized to the baseline)

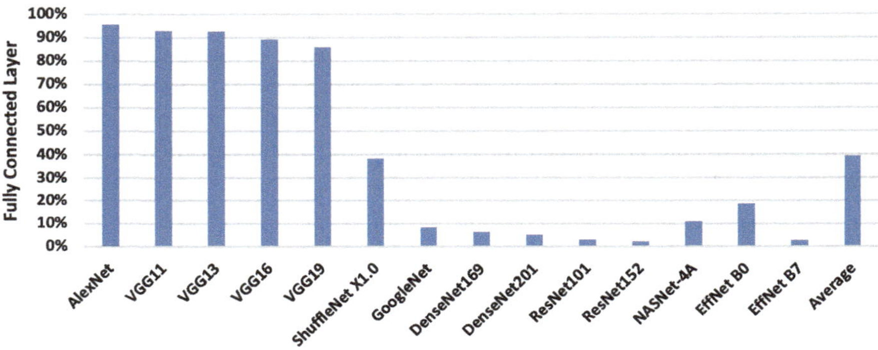

Fig. 18 The size of a fully connected layer in the image classification models

mechanisms leads to a high reduction in packet size when VGG11 is executed on the heterogeneous system with ACT. Therefore, the proposed technique achieves the smallest network latency improvement when EfficientNet B7 is executed on the heterogeneous system.

5.2 *Dynamic Power Consumption*

Dynamic power includes the power consumed by the switching activity for all transistors in the NIs and routers. For all on-chip communication, the results are normalized with respect to the baseline. Figure 19 shows the dynamic power consumption for the heterogeneous system during inference. ACT achieves an average dynamic power reduction of 32% and 27% compared with the baseline and ACF, respectively. The power reduction for the rest of the applications is between 58% and 29% compared to the baseline. Figure 20 compares the dynamic

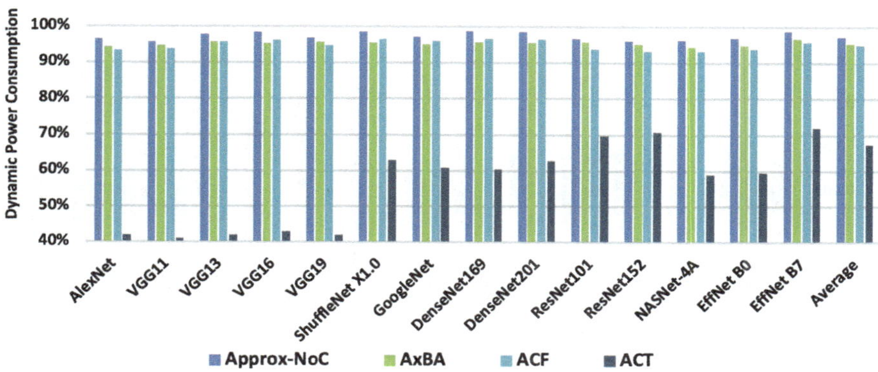

Fig. 19 Dynamic power consumption for the heterogeneous system during inference (normalized to the baseline)

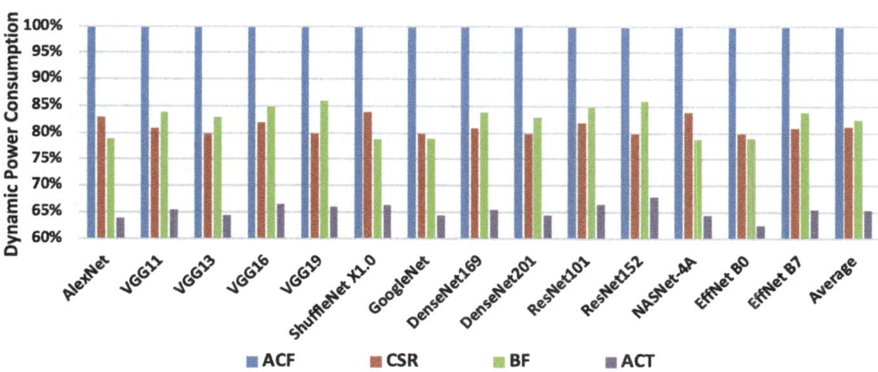

Fig. 20 Dynamic power consumption during training (normalized to the baseline)

power consumption during training. As the proposed technique effectively compresses sparse matrices during training, the proposed technique achieves an average dynamic power reduction of 35%, 16%, and 17% compared with the ACF, CSR, and BF, respectively.

5.3 Accuracy Loss

Figure 21 shows the accuracy loss (i.e., loss of classification accuracy) for image classification model inference when ACT and ACF are applied to the heterogeneous systems. The classification accuracy is measured using the testing data set of ImageNet [2]. 512 randomly selected images from the testing data set are used for testing and setting the contrast reduction level. The rest of the images are used to

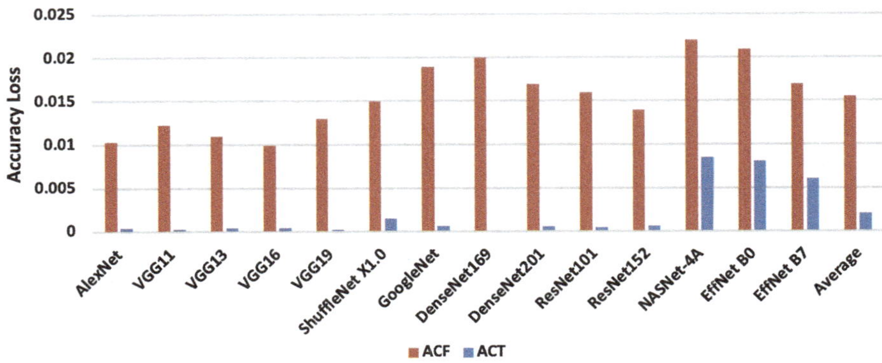

Fig. 21 Accuracy loss for image classification applications

measure the accuracy loss of the application. The accuracy loss for all applications is less than 0.99% across all considered heterogeneous systems for the ACT. However, ACF has a significantly higher quality loss compared to ACT. The highest accuracy loss (2.2%) is observed when NASNet-4A is executed on heterogeneous systems with ACF. This is mainly due to the low relative error tolerance of the image classification application. The highest accuracy loss (0.85%) is observed when NASNet-4A is executed with ACT. Moreover, the incurred accuracy loss is consistent across all systems, thus indicating that the proposed quality control mechanisms are effective in maintaining a low accuracy loss during approximate communication.

5.4 Overall System Performance Evaluation

The ACT is implemented using Verilog to evaluate the area, static power, and latency. The entire system is synthesized with 32 nm technology using Synopsys Design Vision software. The synthesis results show that for each NI, the proposed hardware implementation incurs in an area of 4.80 μm^2. When the supply voltage is 1.0 Volt, the proposed technique incurs a static power overhead of 1.7 mW for each NI. For a 6 $\times$ 6 2D mesh NoC, the ACT modules occupy 1.7% of the total NoC area and consume 4.7% of the total static power consumption. As for the latency, the approximation process and data recovery for preprocessing cores require one cycle each. Also, the approximation process, sparse matrix compression, and data recovery for the mode-inference cores require one cycle each. As for the overhead of this process, five iterations of testing are needed on average for the quality control mechanism to choose the appropriate contrast reduction level. Testing overhead can be further reduced for the proposed technique by using a small test data set or a predetermined contrast reduction level.

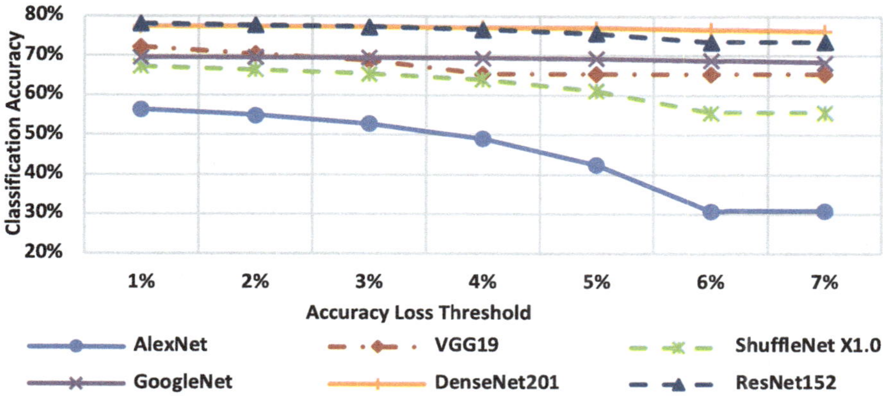

Fig. 22 Image classification accuracy versus accuracy loss threshold

5.5 Sensitivity Study

Figure 22 shows the accuracy loss of the image classification applications when the threshold for accuracy loss changes from 1% to 7% during inference. If the threshold accuracy loss is more than 1%, then the approximation mechanism incurs more than a 1% accuracy loss for the applications that are sensitive to image contrast reduction. Thus, 1% is chosen to be the threshold for the quality control mechanism for image preprocessing.

6 Conclusion

In this work, we have proposed an approximate communication technique (ACT) to enhance on-chip communication efficiency for the inference and training of image classification models. The proposed technique leverages the error tolerance of image classification applications to enhance communication efficiency during the execution of an application. ACT-P and ACT-I are developed for preprocessing and inference, respectively, thus reducing the transmitted data while maintaining the image classification accuracy during inference. Novel approximate network interfaces for the preprocessing core, inference core, memory controller, and shared cache have been proposed to implement ACT in NoCs. Dual-matrix compression method is implemented to further reduce the transmitted data for sparse matrices during training and inference. Compared to existing approximate communication techniques, ACT significantly reduces the transmitted data by efficiently approximating image classification applications. The detailed evaluation shows that compared to the state-of-the-art approximate communication technique (ACF) [16], ACT reduces dynamic power consumption and network latency by 35% and 31%, respectively, for training. In terms of inference, the proposed method achieves 27% and 26% reduction in dynamic power consumption and network latency, respectively, with less than 0.99% accuracy loss.

References

1. Y. LeCun, Y. Bengio, G. Hinton, Deep learning. Nature **521**(7553), 7553 (2015)
2. J. Deng, W. Dong, R. Socher, L.-J. Li, K. Li, L. Fei-Fei, ImageNet: a large-scale hierarchical image database, in 2009 IEEE Conference on Computer Vision and Pattern Recognition, 2009, pp. 248–255
3. S.(L.). Xi, Y. Yao, K. Bhardwaj, P. Whatmough, G.-Y. Wei, D. Brooks, SMAUG: end-to-end full-stack simulation infrastructure for deep learning workloads. ACM Trans. Archit. Code Optim. **17**(4), 39:1–39:26 (2020)
4. Y.S. Shao et al., Simba: scaling deep-learning inference with multi-chip-module-based architecture, in Proceedings of the 52nd Annual IEEE/ACM International Symposium on Microarchitecture, New York, 2019, pp. 14–27
5. D. Shin, J. Lee, J. Lee, J. Lee, H.-J. Yoo, DNPU: an energy-efficient deep-learning processor with heterogeneous multi-Core architecture. IEEE Micro **38**(5), 85–93 (2018)
6. N. Chandramoorthy et al., Exploring architectural heterogeneity in intelligent vision systems, in 2015 IEEE 21st International Symposium on High Performance Computer Architecture (HPCA), 2015, pp. 1–12
7. N. Bohm Agostini et al., Design space exploration of accelerators and end-to-end DNN evaluation with TFLITE-SOC, in 2020 IEEE 32nd International Symposium on Computer Architecture and High Performance Computing (SBAC-PAD), 2020, pp. 10–19
8. S. Venkataramani et al., ScaleDeep: a scalable compute architecture for learning and evaluating deep networks, in Proceedings of the 44th Annual International Symposium on Computer Architecture, New York, 2017, pp. 13–26
9. H. Zheng, A. Louri, Agile: a learning-enabled power and performance-efficient network-on-chip design. IEEE Trans. Emerg. Top. Comput., 1–1 (2020)
10. H. Zheng, K. Wang, A. Louri, Adapt-NoC: a flexible network-on-chip design for heterogeneous manycore architectures, in 2021 IEEE International Symposium on High-Performance Computer Architecture (HPCA), 2021, pp. 723–735
11. F. Betzel, K. Khatamifard, H. Suresh, D.J. Lilja, J. Sartori, U. Karpuzcu, Approximate communication: techniques for reducing communication bottlenecks in large-scale parallel systems. ACM Comput. Surv. **51**(1), 1:1–1:32 (2018)
12. R. Boyapati, J. Huang, P. Majumder, K.H. Yum, E.J. Kim, APPROX-NoC: a data approximation framework for network-on-chip architectures, in Proceedings of the 44th Annual International Symposium on Computer Architecture, Toronto, 2017, pp. 666–677
13. L. Wang, X. Wang, Y. Wang, ABDTR: approximation-based dynamic traffic regulation for networks-on-chip systems, in 2017 IEEE International Conference on Computer Design (ICCD), 2017, pp. 153–160
14. V. Fernando, A. Franques, S. Abadal, S. Misailovic, J. Torrellas, Replica: a wireless manycore for communication-intensive and approximate data, in Proceedings of the Twenty-Fourth International Conference on Architectural Support for Programming Languages and Operating Systems, New York, 2019, pp. 849–863
15. J.R. Stevens, A. Ranjan, A. Raghunathan, AxBA: an approximate bus architecture framework, in Proceedings of the International Conference on Computer-Aided Design, San Diego California, 2018, pp. 1–8
16. Y. Chen, A. Louri, An approximate communication framework for network-on-chips. IEEE Trans. Parallel Distrib. Syst. **31**(6), 1434–1446 (2020)

17. S. Xiao, X. Wang, M. Palesi, A. K. Singh, T. Mak, ACDC: an accuracy- and congestion-aware dynamic traffic control method for networks-on-chip, in 2019 Design, Automation Test in Europe Conference Exhibition (DATE), 2019, pp. 630–633
18. G. Huang, Z. Liu, L. van der Maaten, K. Q. Weinberger, Densely connected convolutional networks (2018), ArXiv160806993 Cs
19. A. Krizhevsky, One weird trick for parallelizing convolutional neural networks (2014), *ArXiv14045997 Cs*
20. K. Simonyan, A. Zisserman, Very deep convolutional networks for large-scale image recognition 2015, ArXiv14091556 Cs
21. X. Zhang, X. Zhou, M. Lin, J. Sun, ShuffleNet: an extremely efficient convolutional neural network for mobile devices (2017), ArXiv170701083 Cs,
22. C. Szegedy et al., Going deeper with convolutions, in 2015 IEEE Conference on Computer Vision and Pattern Recognition (CVPR), 2015, pp. 1–9
23. K. He, X. Zhang, S. Ren, J. Sun, Deep residual learning for image recognition. Presented at the Proceedings of the IEEE Conference on Computer Vision and Pattern Recognition, 2016, pp. 770–778
24. S. Xie, R. Girshick, P. Dollar, Z. Tu, K. He, Aggregated residual transformations for deep neural networks, in 2017 IEEE Conference on Computer Vision and Pattern Recognition (CVPR), Honolulu, 2017, pp. 5987–5995
25. C. Szegedy, V. Vanhoucke, S. Ioffe, J. Shlens, Z. Wojna, Rethinking the inception architecture for computer vision. Presented at the Proceedings of the IEEE Conference on Computer Vision and Pattern Recognition, 2016, pp. 2818–2826
26. M.D. Zeiler, R. Fergus, Visualizing and understanding convolutional networks, in *Computer Vision – ECCV 2014*, ed. by D. Fleet, T. Pajdla, B. Schiele, T. Tuytelaars, vol. 8689, (Springer International Publishing, Cham, 2014), pp. 818–833
27. B. Zoph, V. Vasudevan, J. Shlens, Q. V. Le, Learning transferable architectures for scalable image recognition, in 2018 IEEE/CVF Conference on Computer Vision and Pattern Recognition, Salt Lake City, 2018, pp. 8697–8710
28. C. Xie, M. Tan, B. Gong, J. Wang, A. Yuille, Q. V. Le, Adversarial examples improve image recognition (2020), ArXiv191109665 Cs
29. M. Tan, Q. Le, EfficientNet: rethinking model scaling for convolutional neural networks, in International Conference on Machine Learning, 2019, pp. 6105–6114
30. Y. LeCun, J.S. Denker, S.A. Solla, Optimal brain damage, p. 8
31. S. Han, H. Mao, W.J. Dally, Deep compression: compressing deep neural networks with pruning, trained quantization and Huffman coding (2016), ArXiv151000149 Cs
32. S. Han, J. Pool, J. Tran, W. Dally, Learning both weights and connections for efficient neural network, in Proceedings of the 28th International Conference on Neural Information Processing Systems - Volume 1, Cambridge, 2015, pp. 1135–1143
33. Z. Song et al., Approximate Random Dropout for DNN training acceleration in GPGPU, in 2019 Design, Automation & Test in Europe Conference & Exhibition (DATE), 2019, pp. 108–113
34. N. Srivastava, G. Hinton, A. Krizhevsky, I. Sutskever, R. Salakhutdinov, Dropout: a simple way to prevent neural networks from overfitting. J. Mach. Learn. Res. **15**(56), 1929–1958 (2014)
35. L. Wan, M. Zeiler, S. Zhang, Y.L. Cun, R. Fergus, Regularization of neural networks using DropConnect, in Proceedings of the 30th International Conference on Machine Learning, 2013, pp. 1058–1066
36. M.A. Raihan, T. Aamodt, Sparse weight activation training, in *Advances in Neural Information Processing Systems*, vol. 33, (2020), pp. 15625–15638
37. S. Dave, R. Baghdadi, T. Nowatzki, S. Avancha, A. Shrivastava, B. Li, Hardware acceleration of sparse and irregular tensor computations of ML models: a survey and insights. Proc. IEEE **109**(10), 1706–1752 (2021)
38. P. Dai et al., SparseTrain: exploiting dataflow sparsity for efficient convolutional neural networks training, in 2020 57th ACM/IEEE Design Automation Conference (DAC), 2020, pp. 1–6

39. J.S. Lew, Y. Liu, W. Gong, N. Goli, R.D. Evans, T. M. Aamodt, Anticipating and eliminating redundant computations in accelerated sparse training, in Proceedings of the 49th Annual International Symposium on Computer Architecture, New York, 2022, pp. 536–551
40. E. Qin et al., SIGMA: a sparse and irregular GEMM accelerator with flexible interconnects for DNN training, in 2020 IEEE International Symposium on High Performance Computer Architecture (HPCA), 2020, pp. 58–70
41. S. Dodge, L. Karam, Understanding how image quality affects deep neural networks, in 2016 Eighth International Conference on Quality of Multimedia Experience (QoMEX), 2016, pp. 1–6
42. S.F. Dodge, L.J. Karam, Quality robust mixtures of deep neural networks. IEEE Trans. Image Process. **27**(11), 5553–5562 (2018)
43. J. Qiu et al., Going deeper with embedded FPGA platform for convolutional neural network, in Proceedings of the 2016 ACM/SIGDA International Symposium on Field-Programmable Gate Arrays, New York, 2016, pp. 26–35
44. T.-J. Yang, Y.-H. Chen, V. Sze, Designing energy-efficient convolutional neural networks using energy-aware pruning. Presented at the Proceedings of the IEEE Conference on Computer Vision and Pattern Recognition, 2017, pp. 5687–5695
45. "IEEE Standard for Floating-Point Arithmetic," IEEE Std 754-2008, pp. 1–70, Aug. 2008
46. "NVIDIA Deep Learning Accelerator.". http://nvdla.org/

Index

A
Accelerators, 3, 32, 79, 98, 146, 256, 288, 315, 355, 384, 421, 473, 502, 625, 665, 709
Accuracy improvement, 256, 315, 373, 457
Active storage, 143–178
Adders, 39, 122, 238, 269, 283, 323, 353, 386, 421, 453, 474, 504, 532, 659, 719
Annealing, 265, 268–272, 278, 389, 687–696, 698, 705
Application-oriented arithmetic circuit design, 353–377
Approximate accelerators, 372, 444–449, 501–526
Approximate arithmetic circuits, vii, 286, 421–449
Approximate circuits, 286, 373, 385, 386, 408, 414, 422, 426, 446–448, 492, 512, 554, 562
Approximate computing (AC), vii, 281–300, 304, 353, 355, 366, 383–386, 442, 447, 453, 463, 473–498, 501–503, 505, 531–562, 567, 663, 665
Approximate decision trees (DTs), 413
Approximate multiplier (AM), 282, 285, 286, 294, 296, 299, 353, 356, 357, 360–362, 368–370, 373, 387, 405, 407, 424–430, 433, 447, 448, 454–460, 463, 465, 480, 503–505, 515, 516, 520, 522, 543, 663
Approximate neural networks (NN), 362, 367, 373, 477–481
Approximation, 104, 266, 353, 383, 423, 453, 473, 502, 531, 567, 598, 663, 710
Automated generation, 353–377

C
CBRAM circuit, 627
Channel estimation, 532, 542, 548–549
CIM architecture, 24, 31–61, 67
CIM classification, 32–38
CIM design flow, 59–61
Classifiers, 60, 185–205, 370, 384, 408–415, 491
Clocking, 672, 675–676, 683–684
Clustering, 317, 436, 509, 695–696, 698, 705
CMOS invertible logic (CIL), 265, 266, 272–278
Combinatorial optimization (CO), 687–690, 705
Compact model, 601–602, 607, 608
Computation-in-Memory (CIM), 4, 31–61
Convolutional neural networks (CNNs), 23, 39, 41, 50, 52–53, 78–81, 107, 109, 117, 129, 134, 146, 149, 152, 287, 291, 292, 295, 299, 300, 305, 306, 309, 310, 315–319, 321, 324, 325, 404–407, 414, 436–438, 469, 470, 709, 713
Convolutional operations, 81, 287, 296
Correlation, 211–213, 221, 225, 231, 242–243, 283, 284, 290–292, 304, 305, 307, 310, 314, 315, 317–319, 333, 335–337, 339, 344–346, 348, 369, 370, 447, 511, 523
Cryogenic CMOS, 591–614
Cryogenic TCAD, 613, 614

D
Data management, 143–178

W. Liu et al. (eds.), *Design and Applications of Emerging Computer Systems*, https://doi.org/10.1007/978-3-031-42478-6

Data retrieval, 144–146, 148–150, 153, 155, 157, 159, 166, 173
Decision tree (DT), 195, 196, 205, 384, 387, 389, 403, 404, 408–415, 506–509, 511, 514, 516, 519–523, 525
Deep learning applications, 281, 282, 286, 288–289, 292, 295–299, 666
Deep neural networks (DNNs), 50, 92, 93, 106, 108, 109, 111, 146, 148–150, 152, 153, 156, 157, 159, 160, 169, 171, 175, 305, 353, 355, 361, 365–370, 372–375, 384, 387, 389, 403–408, 410, 414, 435–449, 664
Defect, 67, 133, 422, 592, 600, 671, 679–684
Design space exploration (DSE), 359, 360, 363, 383, 385, 386, 388–393, 397–400, 402, 404, 406–408, 411–414
Deterministic approach, 253–254, 258–260
D flip-flop (DFF), 19, 20, 26, 211–234, 243, 277, 334, 336
Digital circuit, 61, 94, 111, 265, 266, 268, 272, 323, 391, 449, 671–685
Digital filters, 430–435, 449
Discrete cosine transform (DCT), 395, 397–403, 414, 504

E
Edge computing, 325, 473, 649, 650, 654, 655
Emerging technologies, 3–26
Energy efficiency (EE), 3, 14, 32, 53, 60, 79, 92, 104–111, 151, 152, 174–176, 244, 248, 249, 278, 304, 309, 315, 316, 319, 320, 324, 325, 353, 357, 376, 410, 445, 453, 468, 470, 481, 488, 491, 493, 515, 531, 533, 540, 598, 650, 654, 655, 663, 665, 666
Energy efficient computation, 649–666
Ensemble classifiers, 194–198

F
Fabrication, 593, 671, 678–680, 682–684
Fast Fourier transform (FFT), 453–471, 532, 542–549, 562
Fault, 110, 116, 126–138, 186, 205, 211, 237, 249, 253, 262, 281, 284–286, 331, 533, 591, 643, 671–685
Fault tolerance, 130–132, 211, 262, 331, 643, 671, 678, 680–684
Ferroelectric Field Effect Transistor (FeFET), 5, 17–26
FFT processors, 532, 542–549, 562
Field-programmable gate array (FPGA), 45, 79, 125, 151, 168–170, 173–176, 187, 266, 277, 278, 285, 289, 293, 295, 296, 298, 310, 315, 325, 363, 364, 367, 400, 402, 403, 406–408, 410, 413, 414, 507, 509, 517–520, 522–526, 624, 626, 661
Floating-point (FP) arithmetic, 309, 396, 531–542, 546, 567, 650, 656, 661, 722
Floating-point multiplier, 455–457, 460, 534, 537–542, 567–586

G
Genetic algorithm, 357, 363, 389, 558, 559

H
Hardware accelerators, 3, 98, 103–109, 146, 159, 288, 289, 315, 384, 386, 393, 397, 403, 405, 409–410, 414, 503, 504, 625–626
Hardware neural networks (HNN), 303, 309
Hard-wired connection, 249, 263

I
Image classification, 23, 78, 102, 157, 435, 709–738
Image processing, 39, 50, 56, 211, 250–253, 256, 259, 260, 281, 295, 305, 320, 321, 332, 384, 389, 391, 393–403, 407, 414, 453, 501, 504, 509–511, 517, 523, 524, 537, 650, 661
In-memory computing (IMC), v, vi, 4, 7–14, 16–17, 22–25, 84, 85
Input-aware approximation, 507, 517, 525
Ising model, vii, 265–279, 687–697, 699, 700, 702–705

K
K nearest neighbors (KNNs), 119, 163, 186–200, 205

L
Linear feedback shift register (LFSR), 212, 244, 283, 307, 333–337
Logarithmic multiplier (LM), 424, 428–430, 433, 434, 441–443, 584, 586
Low discrepancy sequences, 212, 231, 244, 333

M

Machine learning (ML), vi, vii, 3, 104, 111–139, 147, 151, 165, 185–205, 289, 304, 321, 353, 361, 363, 365–368, 370, 371, 376, 377, 384, 385, 391, 409, 414, 453, 501–526, 649, 650, 654–656, 666, 687
Machines, 42, 85, 105, 115, 143, 198, 244, 266, 289, 304, 342, 353, 384, 435, 453, 501, 591, 626, 650, 687
Magnetic random access memory (MRAM), 5, 11–14, 25, 32, 35, 36, 50, 55, 60, 68–74, 76–88
Memory architectures, 7, 12, 16, 18–22, 53–55, 67–88
Memristor, 5, 11, 35, 38, 41–46, 48, 52, 61, 104, 623–645
Memristor crossbar, 42, 52, 623–645
Morphological neural networks (MNNs), 304–306, 320–325
Multi-branch neural network, 116–138
Multi-layer perceptrons (MLPs), 117–119, 121–125, 129, 132, 134, 137, 293, 305, 308, 310–313, 315, 583, 584
Multi-objective optimization, 383–415
Multipliers, 39, 80, 122, 214, 237, 272, 282, 311, 333, 353, 386, 423, 453, 480, 503, 532, 567, 659, 719
MVM accelerator, 625

N

NAND flash, 7, 68, 147, 155–163, 166–168
Network-on-chips (NoCs), viii, 709–738
Neural networks (NNs), 10, 50, 67, 92, 115, 144, 186, 211, 237, 266, 282, 303, 332, 356, 404, 422, 469, 474, 504, 538, 567, 663
Neuromorphic computing, vi, 32, 91–112
NN acceleration, 475

O

Optimization, 8, 18, 102, 103, 105, 106, 108, 119, 120, 124, 167, 172–174, 265–267, 309, 317, 359–364, 371, 376, 383–415, 453–455, 458, 460, 463–468, 470, 473, 476, 477, 481, 482, 492, 498, 504, 524, 545, 568, 580, 582, 586, 591, 662, 689, 690, 699, 705

P

Parallel computing, 67, 245, 444
Phase-change material, 14–16
Piecewise function, 568, 572, 585, 701
Pipeline, 48, 52, 105, 106, 125, 162, 410, 445, 446, 454, 455, 458, 459, 462, 464–466, 470, 661, 732
Polar decoder, 531, 532, 549–561
Posit applications, 654–655, 665
Posit arithmetic unit, 651, 656, 657, 659–665
Posit decoding, 657–659, 662, 663
Posit developing tools, 655–656
Posit encoding, 657–659
Posit format, 650–657, 659, 661–666
Posit MAC unit, 662
Posit multiplier, 659, 661–665
Posit processor, 656, 664, 666

Q

Quality assurance, 502, 507, 520, 526
Quality of result (QoR) evaluation, 353, 385, 390, 392
Quantum computer simulator, 624–626
Quantum computing, vii, 591–614, 623–645
Quantum computing accelerator, 625–626
Quantum-dot cellular automata (QCA), 671–685

R

Radial-basis neural networks, 312–315
Random number source (RNS), 211–234, 254
Reliability, 31, 126, 186, 542, 595, 601, 681
Resistive random access memory (ReRAM/RRAM), 5–11, 13, 16, 25, 32, 35, 36, 38, 39, 42–44, 46–48, 50, 53–58, 60, 68, 84, 85, 625, 636

S

Siamese network (SNs), 115–139
Similarity measuring, 115, 139, 260
Simulated annealing (SA), 265–272, 278, 389, 688
Simulated bifurcation (SB), 689, 690, 699–705
Sobol sequences, 212, 228, 231, 232, 256, 339–344, 346, 347
Soft errors, 116, 126, 129, 188, 501
Spiking neural networks (SNNs), 53, 92, 95, 97–112, 293
Spin qubit, 594, 595
Spin-transfer torque MRAM (STT-MRAM), 5, 11–14, 25, 32, 55–56, 60, 68, 69, 73, 74, 83, 84
Spintronics, v, 67–88, 309

Stochastic computing (SC), 7, 211–234, 265–278, 281–300, 303–326, 331–348
Stochastic number generator (SNG), 212, 214, 243–246, 256–260, 283, 298, 331–348
Support vector, 119, 187, 198–205, 313

T
Tolerance, 7, 116, 126–139, 186–188, 191, 198, 202, 205, 211, 262, 304, 306, 331, 361, 369, 372, 374, 375, 421, 422, 454, 455, 458, 459, 462, 464, 465, 470, 475, 586, 643, 671, 678, 680–682, 684, 710, 711, 715, 716, 732, 736, 737
Top-down, 104, 453–471, 562
Traveling salesman problem (TSP), 687–705
Triplet networks (TNs), 115, 117–122, 139

U
Unstructured data, 143–150, 152, 153, 155, 159, 178

V
Virtual-source model, 610
VLSI design, 133, 265, 319

W
Winograd algorithm, 287, 291, 295, 296
Wireless communication, vii, 531–562

The manufacturer's authorised representative in the EU is Springer Nature Customer Service Centre GmbH, Europaplatz 3, 69115 Heidelberg, Germany. If you have any concerns regarding our products, please contact ProductSafety@springernature.com

Printed and bound by CPI Group (UK) Ltd, Croydon, CR0 4YY
28/09/2026
02224056-0019